ADVANCES IN ANALYTICAL CHEMISTRY AND INSTRUMENTATION
Volume 10

Electroanalytical Chemistry

Advances in
Analytical Chemistry and Instrumentation

EDITED BY CHARLES N. REILLEY,
Department of Chemistry, University of North Carolina, Chapel Hill, North Carolina

AND ROYCE W. MURRAY,
Department of Chemistry, University of North Carolina, Chapel Hill, North Carolina

ADVISORY BOARD

Electroanalytical Chemistry

Edited by

H. W. Nürnberg

Central Institute of Analytical Chemistry
Nuclear Research Establishment Jülich (KFA)
Federal Republic Germany

An Interscience ® Publication

JOHN WILEY AND SONS
London · New York · Sydney · Toronto

An Interscience ® Publication

Library of Congress Cataloging in Publication Data:

Main entry under title:

Electroanalytical chemistry.

(Advances in analytical chemistry and instrumentation, v. 10)

1. Electrochemical analysis. I. Nürnberg, H. W., ed. II. Series.

QD71.A2 vol. 10 [QD115] 543'.008s [543'.087] 73-15061

ISBN 0 471 65234 2

Printed in Great Britain by
J. W. Arrowsmith Ltd., Winterstoke Road, Bristol

INTRODUCTION TO THE SERIES

The scope and even the purpose of analytical chemistry is growing so amazingly that even the dedicated scientist with time on his hands cannot follow the significant developments which appear now in ever increasing numbers. Analytical chemistry has its new and wider roles and these new developments must become everyday working knowledge and be translated into practice. At present a serious time lag still exists between evolution and practice. This new venture aims to bridge the hiatus by presenting a continuing series of volumes whose chapters deal not only with significant new developments in ideas and techniques, but also with critical evaluations and the present status of important, but more classical, methods and approaches. The chapters will be contributed by outstanding workers having intimate knowledge and experience with their subject.

It is the hope and belief that *Advances in Analytical Chemistry and Instrumentation* will offer a new medium for the exchange of ideas and will help assist effective, fruitful communication between the various disciplines of analytical chemistry.

Additionally, articles of this series will discuss developments in other fields, such as biology, physics, electronics, and mathematics, which are within the scope of analytical chemistry in this broader context. These would include, for example, applications of kinetics, isotopic tracers, computers, and modern physical tools to studies of complex molecules, short-lived species, ultraclean systems, continuous plant streams, and living organisms.

These volumes contain articles covering a variety of topics presented from the standpoint of the nonspecialist but retaining a scholarly level of treatment. Although a reasonably complete review of recent developments is given, a dry and terse cataloguing of the literature without description or evaluation is avoided. The scope of the *Advances* is flexible and broad, hoping to be of service to the modern analytical chemist whose profession each day demands broader perspectives and solution of problems with increased complexity. The periodical literature is inherently specialized and the appearance of suitable monographs takes place only after many years. Reviews are frequently directed to the specialist and often lack adequate description or evaluation. *Advances* hope to fill in the resulting need for critical comprehensive articles surveying various topics on a high level satisfying the specialist and nonspecialist alike. Comments and suggestions from readers are heartily welcome.

THE EDITORS

PREFACE

The present volume deals with three branches of electroanalytical chemistry, i.e. voltammetry and polarography, coulometry and pH-measurements. All three areas are characterized by the fact that on one side they look back to a long history of development and numerous and widespread applications rendering them into the group of well established classical branches of electroanalytical chemistry and on the other hand typical properties are their still tremendous potential for further development, their continuous expansion to new fields of application and thus their growing importance for science and technology in the future.

In chemistry, pharmacy, biology and medicine as well as in development of materials and new technologies for energy presentation, especially with respect to nuclear and electrochemical approaches, and in the wide range of environmental pollution and related problems of geology and oceanography a steadily growing number of key problems depends on the answers analytical chemistry is able to give. In fact analytical chemistry with its spectrum of instrumental techniques has become one of the most important and stimulating applications of physical chemistry. From the viewpoint of analytical chemistry to a large extent the problems to be tackled fall into the range of micro- or trace analysis.

Well established methods and techniques are demanded which offer to a sufficient extent: selectivity, sensitivity and reliability due to their accuracy and precision. Their application range should be rather wide, their mode of operation convenient and as far as possible simple and rapid and especially for routine and control applications a reasonable ratio between costs and output of analytical informations is desired. Growing significance gains further the adaptability of certain analytical techniques to automatic handling and data processing.

Many of the techniques of modern instrumental analysis correspond to a remarkable extent to some of the mentioned demands but only few, among them the electroanalytical branches treated in this volume, are able to meet to a satisfactory and still further improvable degree all the demands enumerated in the previous section. In addition the voltammetric, polarographic and coulometric techniques have compared with many other powerful analytical methods the inherent advantage that they are free of basic standardization problems because the quantity of substance to be determined is strictly related according to Faraday's laws to the amount of electrical charge which is transferred through the interface electrode/solution in the course of the electrode reaction and which is also always the measured entity in this group of faradaic electroanalytical techniques. Also the electroanalytical techniques treated in this volume are especially distinguished by the fact that they can rely on a remarkably well developed and exact theoretical basis with respect to the state reached by electrochemical thermodynamics and kinetics.

Of course even with respect to the areas selected from the large field of electroanalytical chemistry the chapters of this volume cannot cover comprehensively all important aspects of development and applications. This was also not the intention for a volume in a series of "Advances". Rather the editor had in mind to select some more general and some more special topics in fields of current interest with respect to the achieved state of development, importance, potentialities and expanding trends of applications in various areas of science and technology. Thus, a number of international experts contributing actively with their collaborators to the treated fields and being well known for their hitherto achieved results have been persuaded to devote time to the task to write a critical, matured and authoritative essay on their fields of research rather than a mere progress report. It remains to be hoped that the reader will find valuable informations, thoughts and conclusions in the following chapters stimulating him to further research with respect to new developments and applications in the treated branches of electroanalytical chemistry.

Jülich
August 1973   H. W. NÜRNBERG

CONTRIBUTORS

JANINE BADOZ-LAMBLING, *Laboratory of General Analytical Chemistry, C.N.R.S. E.P.C.I., Paris*

G. CAUQUIS, *Department of Fundamental Research, Nuclear Research Centre, Grenoble, France*

PHILIP J. ELVING, *The University of Michigan, Ann Arbor, Michigan*

DALE J. FISHER, *Nuclear Medicine, Veterans Administration Hospital, Gainsville, Florida*

H. HOFFMANN, *Institute of Pharmacy, University of Frankfurt/M., Federal Republic Germany*

B. KASTENING, *Research Group Applied Electrochemistry, Central Institute of Analytical Chemistry, Nuclear Research Establishment Jülich (KFA), Federal Republic Germany*

G. W. C. MILNER, *Actinide Analysis Group, Atomic Energy Research Establishment, Harwell, U.K.*

G. PHILLIPS, *Actinide Analysis Group, Atomic Energy Research Establishment, Harwell, U.K.*

K. SCHWABE, *Institute for Electrochemistry and Physical Chemistry, Technical University, Dresden, G.D.R.*

J. VOLKE, *'J. Heyrovský'-Institute of Physical Chemistry and Electrochemistry, Czechoslovakian Academy of Sciences, Prague, C.S.S.R.*

CONTENTS

CHAPTER I

Advances in Instrumentation for DC Polarography and Coulometry

DALE J. FISHER, *Nuclear Medicine, Veterans Administration Hospital, Gainesville, Florida*

1

I. OBJECTIVES, SCOPE AND TERMINOLOGY

The principal objectives of this chapter are the following. The first is to present the material in this chapter from a viewpoint that stresses analytical chemistry considerations rather than an electrical engineering orientation. A second objective is to describe system concepts. On the basis of experience, it is believed that the best and most efficient means of pursuing the goals of achieving excellence and advances in the design of instrumentation for measurements and in the use of instrumentation is through the application of system concepts. This philosophy of design arises from and takes advantage of the fact that all elements of a measurement system are important and are interrelated. These concepts organize research to create better systems and better measurements. Examples are given of the application of system concepts to the design and use of instrumentation designed for dc polarography and coulometry. Although not every pertinent topic is included among the examples, there are enough examples included to illustrate the systems *modus operandi*. Because of its fundamental importance, substantial space has been devoted to the subject of the definition and the application of practical electrical and electrochemical performance tests of polarographic and coulometric measurement systems. The application of operational amplifier sub-systems is included. Third, a number of specific suggestions are given for projections of research in instrumentation systems that may result in future improvements in polarographic and coulometric measurements and in advancement of the state of the instrumentation art. It is hoped that the material in this chapter will also provide a basis for the generation of ideas for future work. Fourth, the cataloguing of all pertinent recent work in instrumentation for dc polarography and coulometry and the critical reviewing of all important work on this subject has not been done for this chapter. Exhaustive reviews of developments in these subjects are published at frequent intervals and are available.

There are four principal subject divisions in this chapter: system concepts, measurement systems for dc polarography, instrumentation for coulometry and digital systems for electrochemistry. The major topics that are included are tabulated in the chapter list of contents.

The results of his discovery and first investigations of polarography were published by Jaroslav Heyrovský in 1922. Since that time, the field of work has grown so very large that it has become necessary to define the scope included and certain of the terminology in any specific discussion of polarography. In this chapter, dc polarography is voltammetry (measurements based on current–voltage relations) with a dropping mercury electrode (DME), where the DME voltage is scanned linearly with respect to time but at such a rate that the change is no more than a few millivolts during the life of each drop (i.e. DME voltage is quasi-constant) and the current is measured as a function of the DME voltage. The measurement of current versus time at a constant voltage is also included. In controlled-potential dc polarography, a potentiostat establishes the DME potential relative to the potential of a reference electrode, and the tip of the reference electrode determines the site in the solution at which the voltage exists. We have called those dc polarograms in which the undamped or the average or the maximum current is recorded versus voltage 'regular' dc polarograms. This meaning of these three varieties of a regular dc polarogram is used in this chapter. Nth derivative dc polarograms consist of the Nth (first, second, third, . . . , N) derivative of the average value of the diffusion current with respect to the voltage (also called potential) of the DME, recorded as a function of that voltage. Current-averaging filter circuits are used to obtain the average value of the diffusion current as a function of voltage. Active analog computer time derivative circuits are used to obtain these derivative polarograms under conditions such that the scan rate is a known and constant value. The average-current regular dc polarogram is the input to the first derivative circuit, and the output of the first derivative circuit is the input to the second derivative circuit. In this chapter, differential polarography refers to polarographic methods in which two cells are used and the difference in the currents in the two cells is measured. Modes of differential polarography include subtractive, comparative and ΔE-differential. Comparative polarography refers to a high precision ($\sim 0 \cdot 1$ per cent) method applied at about millimolar concentrations in which variables that affect the current at the DME are closely controlled and the difference between the current for a sample and a standard (whose concentration is about 5 to 10 per cent different) is measured on an expanded scale. Both one and two cell methods of comparative dc polarography exist. As illustrated in this chapter, modern dc polarographic instrumentation has enabled sensitivity, precision and resolution to be obtained that are substantially better than those obtained with classical or conventional dc polarography.

In coulometry, analyses are based upon measuring the quantity of electricity (coulombs) that is associated with a reaction for the species being measured. Section IV of this chapter is concerned specifically with controlled-potential coulometry which is also called coulometric analysis at controlled-potential, or potentiostatic coulometry, or controlled-potential coulometric titration or coulometry at controlled electrode potential. The two principal components of the circuitry portion of the measurement system (which is called the controlled-potential coulometric titrator) are the coulometer and the potentiostat. The coulometer is a sub-system that produces an output signal that is directly proportional to coulombs. The potentiostat maintains the voltage of the working electrode (relative to the potential of the reference electrode) at (or below) the constant voltage that has been selected to drive a reaction in the cell. Particular attention is directed toward the design, the testing and the operation of the measurement system so that the selectivity that is ideally inherent in controlled-potential coulometry can be realized, as well as low error and high precision at relatively low concentrations.

II. SYSTEM CONCEPTS

With appropriate, excellent instrumentation systems, one or more of the following objectives can be reached. New or better information can be obtained. New or better analytical measurement methods arise. New, simpler, faster or automated applications are possible. Moreover, the progress of science and of scientific knowledge, including the evolution and the testing of theory, depends upon the quality and kinds of information that can be obtained through analytical measurements. Systems are appropriately designed and used if they can reach the above objectives and when the output information is of the highest quality that is justified by real needs and by economic and technical compromises. Objectives can be attained in some instances because barriers between compartments of technology or science have been bridged within the system. Since these objectives are desirable, the design of systems to reach them is critically important.

Experience has demonstrated that system concepts are a sound and excellent basis for the design of instrumentation, including that for dc polarography and for coulometry and for its use. At the 1964 ACS Symposium on Chemical Measurements through System Control, an invited paper was presented that summarized system concepts, as developed up to that time (1). For the Award address, 1969 American Chemical Society Award in Chemical Instrumentation, system concepts, as evolved to that time, were described at the Award Symposium, with illustrations of their application (2). Section II of this chapter is a presentation of system concepts, with special attention to systems for dc polarography and for coulometry. The principles of applying these concepts to the design of instrument systems for measurements of highest sensitivity have been described (3).

A. Premises

Whenever measurements are made, information is obtained by means of a measurement system. The system performs as a whole. All elements of the system are important and are interrelated. Hence, all parts of the system must be recognized and rationally considered, both individually and as a part of the whole, during the design and during applications of the system.

A system is significant only when it is used. Therefore, since the instrument user is an element of the system, means for effective communication are essential.

Problems may arise because expressions that formulate the predicted most probable behaviour of physical or chemical entities cannot be conveniently changed by mandate when they cause measurement problems or limitations. However, these chemical measurement difficulties and limitations of the performance of specific components can be circumvented. Such problems are circumvented as follows: (*a*) through the appropriate selection and application of mathematical, physical, chemical and engineering laws and principles; (*b*) by either selecting, designing, modifying or controlling elements; (*c*) by coordinating elements.

The system is satisfactory if needed information is obtained that is of high quality because the system is appropriately designed and used. There is concern that presumed needs be shown to be real needs and that credible trade-off options be identified so that their comparative consequences— both bad and good—can be considered.

B. System Elements

A measurement system consists of a specific combination of the following five interrelated elements: (1) instrument user, (2) measurement method, (3) transducer and contents (e.g. cell, electrodes, reactions), (4) circuit elements and circuit sub-systems for signal processing and transformation, and (5) information readout device. The instrument portion of the system can be thought of as the combination of the latter three of these five elements. In a too narrow sense, the fourth element is the instrument. The essential fact is that every one of the above five factors are elements; not any of them is merely an accessory, a thing of secondary importance. An instrument can be designed in a rational way only in context of its use. To create a satisfactory system, the designer must recognize and respond to the facts that the instrument user and the measurement method are elements of the system. Moreover, specific or typical applications must be considered and provided for. Also, the designer must know what information and accuracy and precision are needed.

Briefly, the elements are designed as follows. The fact that the instrument user is a functional element of the measurement system whenever it is used to obtain information cannot be overlooked. The improvement of communication is the design objective that is applicable to this first element. The design of the second element (measurement method) includes the appropriate selection and application of physical and chemical principles and of measurement techniques that can most effectively yield the desired information (i.e. the selection of the method to be used). It also includes the design of the conditions of the measurement, including the minimization of and correction for background or other undesired responses. The third element, the coupled transducer, is particularly important in electrochemical measurement systems since the cell and electrode design and electrode placement and chemical reactions in the cell have very significant effects upon the electrical operation of the circuitry and upon the results obtained with the system, including the attained signal-to-noise ratio. A good example of a design aspect of the fourth element (circuit elements and sub-systems) is the present widespread use of operational amplifiers and linear and digital integrated circuits in analytical instrumentation. An example of a design area that is of current interest for the fifth element (information readout device) is the increasing use of digital systems, including hardware and software programmed digital computers and including computer-assisted and directed automated analyses.

C. Instrument User

New or better information is obtained most efficiently when a well-designed instrument is operating within its design specifications and is used properly and to best advantage. Therefore, the measurement method and the instrument user are integral parts of the measurement system and means are essential for effective two-way communication between the user and the other elements. An inherently valid instrumentation idea can be useless or only partially realized if it is applied in the absence of good communication. The user should be able to communicate with the remainder of the measurement system without unreasonable effort and the remainder must be capable of communicating with the user and with maintenance personnel. (The number of people involved in the separate functions of design, use and maintenance is immaterial and can vary from one to many.) The improvement of communication is accomplished by instrument design steps that take advantage of rather than being haphazardly limited by the characteristics of the human (i.e. by the application of human engineering principles), by the design of both electrical and functional tests that are soundly based on theory and that are sensible and practical, by built-in calibration and electrical performance tests (or built-in provision for the use of external calibration or test instrumentation yielding higher precision or accuracy than can be available at reasonable cost with built-in circuitry), by providing means such as functional rather than or in addition to electrical block diagrams, circuit diagrams, written procedures for the electrical and the functional (e.g. electrochemical) testing and for the maintenance of performance (with specifications, i.e. a definition of normal performance), and for the use of the instrument, and by providing, whenever feasible, by design, outputs that are direct reading and unambiguous in terms of the specific information that is to be obtained in the measurement and/or in terms of applicable electrochemical or other equations. Good, two way information flow between the human and the other portions of the system substantially assists in achieving correlation between designed-for results (system design objectives) and obtained results (system performance). Feedback from applications and from maintenance experience (good records should be kept) closes the information loop for the designer. Ideally, this feedback results in future system design improvements.

D. Importance of Circuit Diagram

A circuit diagram has important values other than serving as an aid for the repair or the duplication of an instrument. A detailed circuit diagram (including directly or indirectly the identification of parts used) is an indispensable part of a description of an instrument. This is so because electronic component and circuit performance characteristics cannot be prevented from affecting the results obtained with the system even though design steps are taken to minimize the extent of limitations caused by this dependence and to allow interchangeability of components, at least for the specific applications of the system that are anticipated. On occasion, the interpretation of data obtained with the instrument requires a knowledge of the specific circuit and its performance characteristics in order to decide whether an evident unusual feature in the data is legitimate for chemical reasons (is a departure from ideal behaviour or an unexpected chemical detail that is in fact present) or is illegitimate (an artifact caused by some circuit or component performance characteristic or limitation or misapplication). This need for a detailed knowledge of the circuit may arise in future work with the instrument, or to reinterpret data obtained with the instrument in the past. A block diagram is a generalized description of functional relationships within a system or sub-system. A circuit diagram is a specific but concise description of a particular sub-system or instrument.

E. Importance of Construction Practices

The inclusion of a compendium of recommended construction practices is not an objective of this chapter but the Figures at the end of Section II.E of this chapter provide some illustrations of good construction practices. Information on these subjects is available in textbook sources. The need for the application of human engineering principles as a means for providing good communication has been pointed out in Section II.C of this chapter. Other aspects of the importance of the subject of construction practices are as follows.

Excellent design concepts have no doubt been rejected due to poorly constructed prototypes, with the rejection being caused by erroneously attributing observed performance deficiencies to inadequacies of the concept itself. In other less drastic but also unfortunate cases, the performance of the realized system has not been in accordance with the inherent value of the concept. In other cases, due to poor construction practices that result in unrecognized circuit couplings or due to the use of circuits that depend upon a maximum performance of components or due to unsuspected insufficient information about the essential construction details, a prototype of a system may perform to the satisfaction of the designer but a presumed replicate of the prototype may perform unsatisfactorily. When feasible, it is well to regard the status of the design of a system as not fully established until it is known that the quality of construction drawings is such that replicates of the prototype instrument also perform within design specifications.

Major manufacturers of kits for the at-home construction of electronic apparatus have learned that the majority of malfunctions in newly constructed kits are due to poor soldering techniques. Excellent illustrated brochures that fully describe the proper soldering techniques for discrete component wiring and for printed circuit wiring are available from these manufacturers. Inadequate or dirty contacts in switches and in various kinds of plug-in connectors cause similar problems in instrument systems. Related but more insidious problems are prevalent in so-called 'dry circuits' (e.g. in electromechanical chopper circuits) where switching is done at such low currents that resistive films build up on the contacts and cause serious malfunctions. The resistance of wires connecting low-valued precision resistors or the thermal generation of potentials at junctions of unlike metals are sometimes significant problems. In recent times, better components such as reliable switches and connectors and wires and cables having high quality insulation have become available. The use of such components, even if their immediate cost is high, may save time and money.

In the beginning stages of the design of a new instrument, it is customary to survey the literature to learn the status of published designs of pertinent elements and of instruments for similar purposes. The question can arise as to the extent to which a presumably satisfactory published circuit, or a portion of it, should be duplicated. Due to the rapid proliferation of new components, amplifiers, power supplies and other modules of improved performance, better performance and economies can often be achieved by the use of new devices. One is generally well advised to survey the models and parts available at the time of design and construction before duplicating a published circuit to the extent of using the same components. On the other hand, particularly if a published circuit gives satisfactory performance but its performance is significantly affected by constructional details, an overall advantage (e.g. saving of time and development costs) can be achieved by duplication of portions or all of it.

The quality of the design and of the construction of an instrument system (i.e. the performance of the system) is more significant than whether the latest modules and components are used. Performance is valued more than novelty. Broad statements such as the following are open to question: transistors are superior to vacuum tubes; integrated modules are superior to discrete modules; printed circuits are superior to wired circuits; digital techniques are superior to analog methods. These statements are not meaningful except in the context of specific embodiments of these competing technologies at a point in time and in terms of specific criteria for 'superior'. The specific criteria should be related to realistic needs and values in projected applications. Components, modules or techniques of improved performance or reliability that enable the design and construction of better measurement systems obviously should be used where justified by both technical and economic considerations. Recent performance and economic advances, available because of integrated circuits, are particularly valuable. The availability of modules (operational amplifiers and operational sub-systems, digital logic modules, and power supplies and regulators, are outstanding examples) and digital computers make the design of measurement systems of advanced performance not only easier but economically feasible. If facilities are available, the use of techniques such as printed circuit board or etched wiring board circuits and photo panels should be considered, particularly if duplication of a prototype is expected. The benefits of both analog and digital functions are now readily available.

The construction of the instrument portion of a measurement system on a modular basis can provide worthwhile advantages, particularly if the modules are organized on a functional basis and if no substantial compromises of system performance result from the compartmentalization and from the requirements for interconnecting cables and terminals. (Multipurpose instruments, consisting of a main frame with plug-in modules, with each module converting the instrument for a specific kind of measurement, are a special case.) For example if functional sub-system portions of an instrument such as power supplies, voltage and scan sources, polarographic current-averaging filters, coulometric current integrators or potentiostats were built as individual modules within the instrument, it might be more practical to upgrade the performance of the system on an individual module basis, should it become possible to design or purchase a module of improved performance to replace the original module, rather than on the basis of having to redesign an entire instrument. Also, if extra modules that are known to be operating correctly are available, these can greatly assist, by substitution, the trouble-shooting (isolation and identification or verification of location of fault) and repair of a malfunctioning instrument. In addition, it might also facilitate the adaptation of the system to special future problems. Alternative modules might be selected to form an assembly that is optimized or specialized for specific applications or objectives. Some individual modules might be used in more than one kind of instrument. At the time of design of an instrument, modular construction should be considered for the above reasons.

Illustrations are provided in Figures 1, 2, 3, 4, 5 and 6 of construction practices used by the Product Design and Fabrication Group, Electronics Section, Instrument Department, Instrumentation and Controls Division, Oak Ridge National Laboratory. The instruments seen in these Figures were designed by members of the Instrumentation Group, Analytical Chemistry Division, ORNL.

Fig. 1. Front panel view, ORNL model Q-2792 polarograph.

Fig. 2. Top view, ORNL model Q-2792 polarograph.

Fig. 3. Rear chassis view, ORNL Q-2792 polarograph.

Fig. 4. Bottom view, ORNL model Q-2792 polarograph.

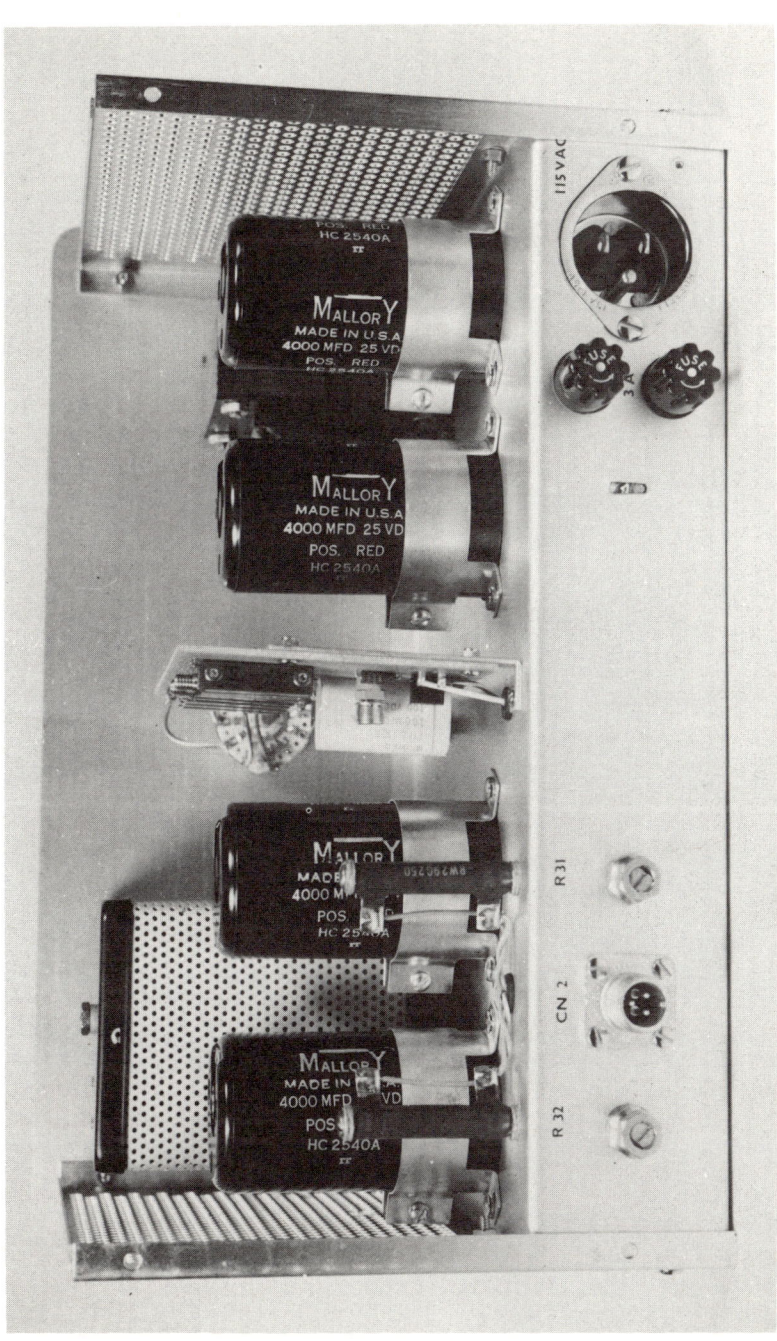

Fig. 5.   Rear view, ORNL model Q-2942 DME drop time controller.

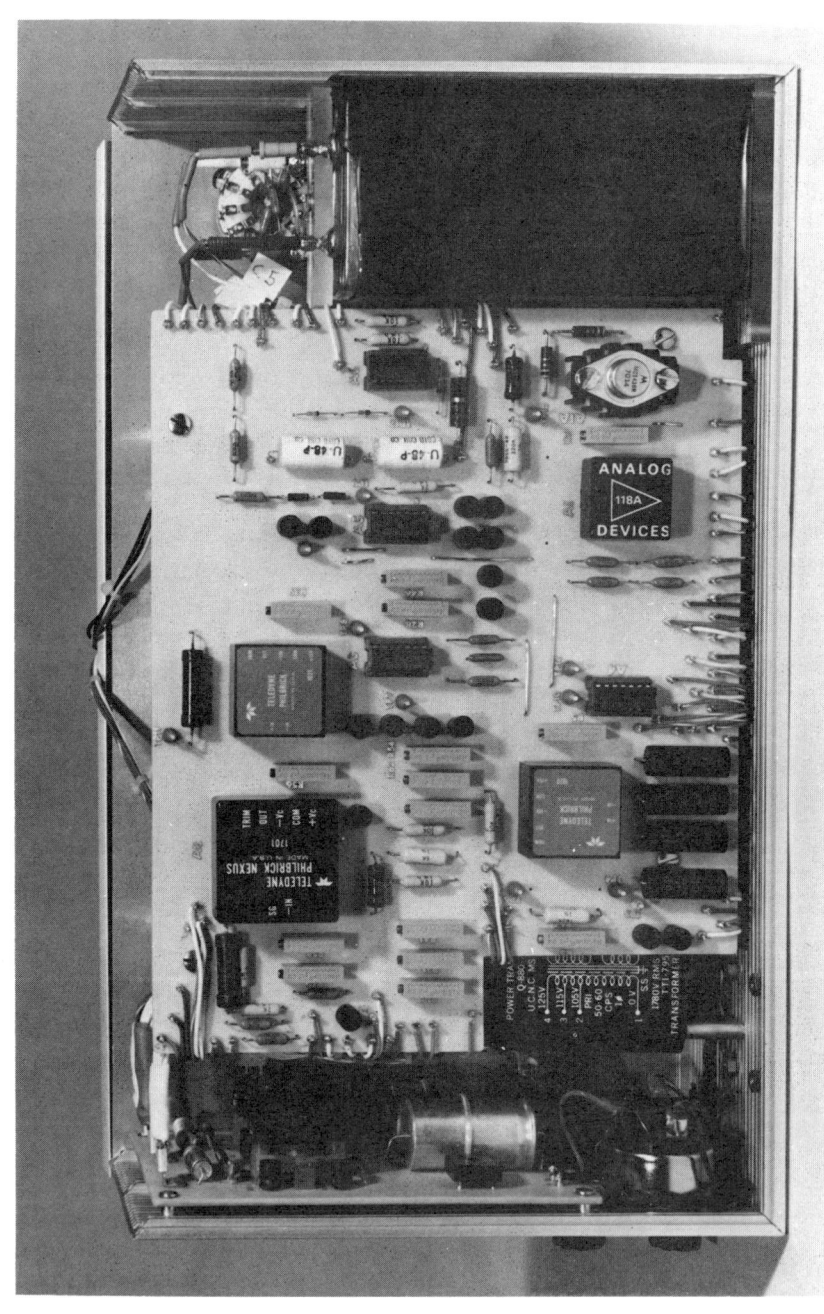

Fig. 6. Interior view showing etched wiring board construction.

Components and wiring and construction details of sufficiently high quality are specified so that the performance and reliability of the measurement system are accordingly enhanced and ensured. Mechanical drawings are prepared detailing the fabrication of all chassis, brackets, panels, etc., and Mylar film negatives are made for producing etched wiring boards and photo panels and tags for labels. Controls are named, labelled and arranged to facilitate the ease with which the instrument may be used. Labels are applied to identify components and service adjustments. Panels and some tags are made by the Metalphoto process. Mounting boards and terminals are used for parts such as resistors, capacitors, diodes and transistors. Teflon insulated wire is used. Leads are cabled. The dual tuned parallel-T filter and the fourth-order filter for the model Q-2792 polarograph are constructed as individual plug-in modules. These are placed in the empty sockets seen in Figures 2 and 3. The solid state operational amplifiers used in the model Q-2792 polarograph are commercial plug-in modules and may be seen in Figures 2 and 3. Commercial modular power supplies are used when appropriate. These have been used in the model Q-2792 polarograph (see Figures 2 and 3) and in the model Q-2942 DME drop-time controller (see Figure 5). In Figure 5, a commercial modular power supply is located at the distant corner of the left side. Components of another power supply are mounted on the chassis. At the centre of the figure, the other circuitry module is seen, built with an etched wiring board. Identifying symbols are formed on the etched board, adjacent to the parts.

Figure 6 is a photograph of the interior of an automatic coulometric titrator whose fabrication was completed early in 1972. At the left of Figure 6, a modular power supply may be seen. The central part of the figure is the etched wiring board. A number of integrated circuit operational amplifiers are used. Exclusive of the power supply, there are fourteen operational amplifier sub-systems in this titrator by means of which unique advantages are realized. Many circuit test points are provided, with identification printed on the board, and many of the components are identified in the same manner. This is the prototype instrument, built directly from the drawing of the circuit which was designed.

F. Sensitivity

Signals are inevitably accompanied by noise. The signal, S, is the wanted information-bearing component of the observable total, signal-with-noise. Noise, N, consists of the undesired portions of observed measurements of S-with-N that adversely affect the measurement and use of S. At the output of a measurement system, N is the unwanted fallacious perturbation of the output that is present along with S. Noise may originate at the point of generation of S and at various other points within the measurement system.

Noise masks or obscures S when $S/N \leq 1$ and its presence is a fundamental limitation to the transmission, measurement and use of S. The sensitivity with which S can be measured depends upon the characteristics of the S time series, the statistical characteristics of N and upon S/N; it cannot be increased indefinitely in proportion to the amplification provided in the measurement system because of the presence of N along with S. The sensitivity of electrical measurements made under the most ideal conditions is significantly, adversely and inescapably affected by the presence of thermal N, and, particularly when the dc level of S is of interest, by finite dc drift and gain variation N during the measurement interval. If several signals are present, some of which are not mutually resolved, this is a resolution problem rather than a sensitivity problem. When ultimate resolution is needed in polarography, the choice of first derivative dc polarography is one of the most effective solutions. In the case of dc polarography, we have found that, although the DME is a relatively ideal type of polarographic working electrode, with well designed polarographic circuitry, sensitivity (and precision as well) is limited more by the S/N at the DME (where N includes all contributions to a lack of perfect reproducibility of the current at the DME rather than being only the magnitude of background current) than by the kind of polarography chosen for the measurement of concentration. In the case of coulometry, since only total coulombs can be measured and faradaic coulombs for the reaction of interest cannot be directly measured, we have found that sensitivity is limited primarily by the degrees to which the ratio of faradaic to total coulombs can be maximized by system design and to which the optimized observed total coulombs can be corrected accurately or precisely for the remaining background coulombs.

Solutions to the problem of retrieving S from S-with-N involve taking advantage of the comparative characters of the mathematical functions respectively describing S and N. From a frequency viewpoint, in addition to the relative amplitudes of S and N, it is a matter of how much the frequency spectra of S and N overlap. Unfortunately, N spectra usually extend from very low to very high frequencies. Even if S is dc the use of a low pass filter only partially separates S from N since N has very low frequency components. Similarly, even if S is a single sinusoidal frequency (or is a dc signal chopped at constant frequency), the use of a narrow, tuned bandpass filter does not eliminate the portion of the N spectrum that passes through the finite width bandpass of the filter and through the filter skirts. These problems are particularly serious when $S/N \leq 1$. From a time domain viewpoint, if N is described mathematically by a normal random distribution function, its average value is zero and it does not correlate with itself or with any periodic S. S is coherent and does correlate with itself or with other periodic signals of the same frequency.

The characters of S, N and S/N and the applicability and the degree of physical realizability of ideal circuit elements and sub-systems affects the actual improvements in sensitivity that can be achieved by various design steps. The character of S (whether ergodic, periodic, repeatable, simple or complex and slow or fast wave form), the required S information (average amplitude only or undistorted wave form), the type of N (whether ergodic, random white, random but amplitude increasing toward lower frequencies, sporadic impulse or interference) and the magnitude of S/N in a single observation (whether $\gg 1$ or ≤ 1) all markedly affect whether each element or sub-system can be used and if so, how well it will perform. Where both S and N are ergodic time series, at least over the measurement times of interest, continuous S can in principle be measured discretely (with due regard to the requirements of the Sampling Theorem) by analog-to-digital converters (ADC's) without loss of information, ensembles can be obtained for repeatable S, and ensemble averages are equal to time averages. The technique of enhancing S/N by summing a series of measurements of S in register via a trigger that is time locked to S in the presence of N that is not time locked to S (ensemble averaging with digital systems) should be applicable to any measurement where one variable can be repeatedly measured against another.

From a measurement system viewpoint, when sensitivity is of interest, the system elements are designed individually and as a part of the whole to optimize and to enhance S/N wherever technically and economically feasible. The appropriate selection of methods (including choices of S) and conditions of measurement and the design of methods and the selection and design of transducers optimizes S/N in the method and transducer system elements. For the circuit elements and circuit sub-systems portion of the measurement system, design steps are used that increase S (without increasing N in equal proportion) or that decrease or discriminate against N (without decreasing S in the same degree). These steps include choices of circuit elements, circuit sub-systems and components and choices of transformations of S (e.g. current-averaging and derivative polarography) and the use of correlation techniques in the frequency domain (e.g. filtering) and in the time domain (e.g. ensemble averaging). Correlation techniques, including ensemble averaging with digital systems, can be applied in the readout portion of the system to enhance the optimized S/N. To obtain measurements of highest sensitivity, design steps to optimize S/N throughout the system are required; the further enhancement of S/N at the readout element is generally advantageous and should also be done whenever applicable and economically justified. The limit of measurement sensitivity is set by the statistical character of the N inevitably present with S but communication theory and statistical techniques (including correlation in the frequency and time domains) can be applied to increase sensitivity. New or better quality measurements become possible when advances can be made in optimizing S/N or in processing an optimized S/N to enhance it.

A paper has been prepared that describes a method for the design of systems for measurements of highest sensitivity (3). In this paper, the above points are discussed in detail from a systems viewpoint. The characteristics of various kinds of N are described. Exemplary applications of and advantages obtained by ensemble averaging and other correlation methods are cited for biological, physical, chemical and technical measurements. Also, the application is described of these system concepts to our design of flame photometric measurement systems. Another paper reports a demonstration of the use of ensemble averaging to substantially increase the sensitivity of measurements with flame spectrophotometers and points out other advantages of ensemble averaging in flame photometry (4).

In subsequent portions of this chapter, illustrations are given of the application of these system concepts to dc polarography and to coulometry, including applications of correlation techniques.

G. Needs for Both Electrical and Specific Performance Tests

Experience has shown that both electrical and specific performance tests of measurement systems are essential. The definition of 'performance' is involved and cannot be entirely divorced from the anticipated and typical uses of the system. It is important to note that electrical tests with the transducer absent or simulated are essential but are not sufficient tests. The method and the coupled transducer (e.g. cell, electrodes, reactions) are elements of the system whenever it is used and so the system including these elements must also be tested. In practical work, tests including a typical instrument user (the first element of the system) are not trivial. These may reveal inadequacies in or places to improve the design of two way system communication. Theory is used as a guide for the design and the analysis of measurement systems and for the design of electrical and specific performance test experiments and for the interpretation of experimental evidence of performance in order to judge the suitability of the measurement systems. Neither theory alone or experiment alone is satisfactory. Reasoned judgment is the basic or fundamental means of reaching decisions about system performance; it is based upon theory, test results and experience.

In the case of systems for electrochemical measurements, the approach is used that those portions of the system whose performance can be clearly established by means of electrical tests are tested first with electrical tests designed for the purpose. Then, after it is known that these test criteria are met, electrochemical tests are used to test the entire system. Unfortunately, there are substantial difficulties inherent in the design of electrochemical tests and in the interpretation of the results of these tests. If significant discrepancies are found between expected and observed results, the reasons for these anomalies must be learned so that corrective action can be taken; the possibility that the tests are not valid (or are not designed so that unequivocal evaluation of test results can be made) must not be overlooked.

Usually, the circuit and readout elements of a measurement system are linear elements (i.e. the principle of superposition is valid) and so the important and necessary tests of their performance can be designed directly from a knowledge of electrical engineering principles. There are well-developed engineering procedures for the analysis and testing of these elements which can be readily applied to specific cases. If a potentiostat is connected to a dummy cell built from linear electrical components, electrical tests of its performance (e.g. of stability and of potential control) are not difficult since an active but linear system is involved. However, when a real and coupled transducer is attached to the potentiostat, several complications arise so that both electrical and electrochemical tests of the performance of the system including the coupled transducer are necessary.

In the case of systems for polarography and for coulometry, performance tests are electrochemical tests; the definition, design and correct interpretation of these tests is as important as is the design of the system itself. Electrical and electrochemical tests that are useful for evaluating the performance of dc polarographs and of controlled-potential coulometric titrators and of their method and transducer elements are described in corresponding subsequent sections of this chapter. In the remainder of this section of this chapter, the significance of and reasons for the needs for both electrical tests and electrochemical performance tests of systems for measurements in polarography and coulometry are given as illustrations of the general need for both types of tests.

The design of practical electrochemical tests that can be clearly interpreted in terms of system performance is rather complicated but possible. Various well-known equations exist for current–time and for current– or coulomb–voltage relations in polarography and in coulometry that are fairly exact only as restricted by the assumptions made for various limiting cases in the derivation of the equations. The problem is to design the tests so that experimental electrochemical conditions fulfill fairly exactly the applicable set of assumptions so that the performance of the instrument portion of the system can be discerned. It should, however, be appreciated that the design of experimental conditions to obtain conformity with assumptions involved in a selected equation can consist in arranging conditions such that errors are negligible or so that errors tend to cancel. (A well-known example of the latter case is the fact that a DME can be operated so that, for non-first drops, fairly good agreement is obtained with the Ilkovic current–time equation. Conversely, however, if the Ilkovic equation were selected for the test and test results with a DME did not agree with the equation, it cannot be assumed from this alone that the DME needs further design.) If one selects an electro-chemical equation and designs the test so that the electrochemical assumptions inherently required for valid interpretation of results on the basis of the equation are presumably met, small differences may be observed between theory and experimental results. It can be difficult to prove through further tests whether these discrepancies are due to imperfections in the total system, not revealed through the electrical tests of the sub-systems, or are due to the finite degree of correspondence to be expected between theory and experiment when the system performance is good enough so that it does not limit the extent of the correspondence. Electroanalytical applications seem to be on firmer ground than the degree of exactness of the theory. Fortunately, when practical applications of these measurement systems are guided by theory, the utilities of the many applications are not seriously limited by this dilemma. Correspondingly, the degree of validity of the tests of the system or of a system element only has to be compatible with the required degrees of accuracy and precision of the measurements to be made with the system. In the design of electrochemical tests, insofar as is possible, test reactions and conditions must be selected that are believed to conform closely to the selected equation.

The difficulties of judging the performance of a system or of an element of the system on the basis of electrochemical tests designed for a comparison of experimental results with those predicted by one of the usual electro-chemical equations may be brought out more clearly by the following illustrative examples. Test conditions for a DME can be designed as well as can be arranged to conform with the inherent assumptions (e.g. constant potential and constant mercury flow rate and diffusion-controlled current) underlying a chosen current–time equation. It would be anticipated in the case of the use of a current–time equation that the latter portion of the observed current–time curve would conform more closely to the equation than the very first part. It may be uncertain, however, as to whether a portion of the observed difference in the entire curve is caused by imperfections in the circuitry part of the system (such as excessive damping or overshoot due to the potentiostat or to the current amplifier) rather than being entirely due to inexactness of the equation itself. A similar situation exists when the test is based on polarographic diffusion coefficients. Moreover, a question that is basic to the definition of 'performance' is the significance of such differences in terms of the use, for example, of the system to measure concen-tration either with high precision or with high sensitivity. In the case of coulometry, a system might be tested by comparing the coulombs calculated from the amount present from a standard in the cell (assuming that with careful work the amount is known with greater precision and accuracy than the needed precision and accuracy for the test results) with the observed number of total coulombs, minus the calculated corrections for background coulombs, if this correction is judged to be significant for the conditions of the test. If the difference between the coulombs calculated to be required from the amount present and the observed (with calculated corrections) coulombs exceeds the precision or error requirements of the test, it is necessary to establish whether the difference is due to circuitry inadequacies or to an inadequate background correction or to other electrochemical problems. While the performance of the coulometer portion of the system can be tested electrically rather well, there could be performance limitations of the potentiostat portion of the system that were not revealed by electrical tests and which are significant when the real transducer element is in use under the conditions of the electrochemical test.

Potentiostats are used in controlled-potential dc polarography and in controlled-potential coulometry. Potentiostats, in common with other servo systems, must be stable. Both electrical and electrochemical tests of stability are needed because of the presence of the transducer system element, including the electrode reactions, in the feedback loop whenever the system is used for analytical experiments or measurements. Also, the existence of iR voltage drops in solution and thus the design and arrangement of the cell electrodes not only has effects on stability but also affects the degree to which the reaction driving voltage is equal to the command voltage set into the potentiostat. A potentiostat can only balance the potentials as presented to its circuitry.

In addition to testing a potentiostat for stability, it must also be tested for its ability to control the reaction driving voltage. In polarography, it is desirable to measure a function of the current (instantaneous, maximum or average value, corrected for residual current, or the Nth derivative of the average value with respect to voltage) against the corresponding voltage that is driving the reaction at the DME rather than against the voltage that exists at some point in the circuit. The use of a potentiostat to provide and control the reaction voltage at the DME to a known value is not particularly difficult in the usual low specific-resistance solutions where there is a large excess of supporting electrolyte present in order to minimize the migration current of the species of interest to a negligible value and at the usual levels of cell current. On the other hand, in high specific-resistance solutions or at high cell currents, this desired relationship is closely approached with a potentiostat only in conjunction with cell and DME design. In analysis applications of controlled-potential coulometry, the intended advantage of electrochemical selectivity is retained as long as the reaction driving voltage at the working electrode never (except for fast brief transients) exceeds the command value set into the potentiostat. However, due to the fact that iR drops in solution in coulometry, particularly during the first portion of an electrolysis, may be quite substantial, these iR drops in solution have several undesirable effects. In addition to the possible introduction of stability problems, these include the extension of the time required for the electrolysis which is one of the factors that adversely affects the ratio of faradaic to background coulombs. Here also, the design of the transducer is important so that the potentiostat can efficiently perform its intended function. If an electrochemical test for controlled-potential coulometry showed a positive error due to iR effects other than loss of stability, there might be a question as to whether it were due to loss of potential control or to the prolongation of the titration time with a resulting inaccuracy of the background correction. Particularly in the presence of cell current, even a Luggin probe cannot be used for an exact electrical test of potential control at the site of the reaction; electrochemical tests have to be devised in addition.

The amount of material that has reacted in a coulometric cell can indeed be calculated by means of Faraday's laws of electrolysis to a sufficient degree of accuracy and precision in well-controlled situations. However, the coulometer portion of the system can only observe and measure the total coulombs used (or a number proportional to the total); it cannot directly measure only those coulombs used in the reaction of interest. In practice, by system design, especially of the method and transducer elements of the system, either the portion of the total coulombs that is due to other reactions is caused to be negligible in terms of the amount of material of interest reacted or else the total coulombs are corrected for background coulombs with accuracy and precision sufficient for the analyses in question. Even when the adequacy of whichever approach is used is judged through electrochemical tests with standardized test solutions, there is always the danger that samples will contain impurities that cause errors not present in the tests with standards. For this reason, an investigation has been planned at Oak Ridge National Laboratory of the utility of standard addition as a solution to this problem. (See Section IV-A of this chapter.) The greatest danger present in this problem is that the existence of this kind of error for some samples may be neither apparent or known.

III. MEASUREMENT SYSTEMS FOR DC POLAROGRAPHY

A. Design of Polarographic Systems to Provide Good Two-way Communication between Instrument User and other System Elements

The Instrument User is an integral element of a measurement system when it is obtaining information. The effectiveness of the combination of the Instrument User and the other elements, working together as a measurement system, depends upon the quality of the two-way communication between the User and the remainder of the system. Communication of good quality exists when needed and correctly understood information can be acquired. Good communication expedites the efficient acquisition of credible information. Good communication results in a close rapport between the User and the other elements. Also, good communication is needed for fabrication, for testing and for trouble-shooting and maintenance. The following three factors are needed and must be provided through system design in order to ensure good two-way communication : (a) system outputs that can be directly interpreted in terms of needed polarographic information, (b) performance tests, (c) descriptive information. In Sections III-A-1, III-A-2 and III-A-3 of this chapter, which follow, examples are given of the provision of these communication factors for polarographic measurement systems. To facilitate communication, the Function switch in recent ORNL polarographs may be thrown from the Operate position to either the Cell Open or the Test position. The Cell Open position identifies on the recorder the scale position corresponding to zero cell current. The Test position expedites the testing of the performance of the circuit sub-systems, apart from the transducer, by methods discussed in Section III-A-2 of this chapter.

1. Directly Interpretable System Outputs

It is essential that the design of the polarographic system be such that the outputs can be directly and specifically interpreted in terms of information that is needed. Directly interpretable system outputs are a means of or factor for good communication between the Instrument User and the other system elements. This factor is obtained by instrument and method design.

For all kinds of polarography, the attained sensitivity of concentration measurement, though predicted from the calculated ultimate signal levels, depends also upon the attained signal-to-noise ratio with the actual system. Equally good performance is not obtained with all instrument portions that are designed for a given kind of polarography. Examples of some comparisons of ultimate signal levels and ultimate resolutions have been presented (5). With good instruments, the signal-to-noise ratio tends to be set by the performance of the DME. When concentration is the information that is required in cases where the ultimate degree of resolution is not necessary, then a better signal-to-noise ratio in derivative dc polarography is achieved in recent ORNL polarographs with faster scan rates, though the wave form is then not in exact agreement with the applicable equations. However, the peak heights are highly reproducible and directly proportional to concentration at each given fast scan rate so that the output is directly interpretable in terms of the desired information.

The usual equations for regular and for derivative dc polarographic wave forms and the equations for the wave forms for other kinds of polarography and voltammetry are distribution functions. The relative ultimate resolving powers of different kinds of polarography can be meaningfully defined and compared on the basis of these equations (5). The ultimate resolution for each kind of polarography is achieved to the extent that a system is available so that polarograms for that kind of polarography can be recorded that are in close agreement with the applicable equation. Also, with such systems, the recorded polarograms can most easily and directly be interpreted for information such as polarographic reversibility, n values and $E_{1/2}$ values.

With recent ORNL polarographs, operating conditions may be chosen to optimize either wave form or sensitivity, as described in Sections III-E and III-F of this chapter. The choice is made in accordance with the nature of the information that is wanted. Regular and derivative polarograms can be recorded whose forms agree very closely with the corresponding equations.

The three elements constituting the instrument portion must be designed for good communication in order to achieve such results. Examples follow of design steps that have accomplished this objective for dc polarographic instruments that have been designed at ORNL.

Recent ORNL dc polarographs have been designed so that, for the ordinary conditions with low specific resistance supporting electrolyte and where uncompensated iR_{inner} is negligible (13, 14), electrolysis proceeds at controlled potential and the output is related to that potential instead of to an applied voltage at some point in the circuit. The potentiostat consists primarily of two operational amplifier sub-systems: a current amplifier and a potential control amplifier. The current amplifier holds the DME at ground potential and provides stable and sensitive current measurement without allowing the DME potential to be affected by the voltage across a series current-measuring resistor. Through negative feedback, the potential control amplifier supplies current and voltage to a counter electrode (sometimes called the third electrode), as demanded, so that the potential difference between the terminals of the DME and the reference electrodes is kept equal to the desired control potential. The control potential is the sum of the initial and the scan potentials. Detailed descriptions of ORNL polarographic potentiostats are available elsewhere: model Q-1988 (6); model Q-1988-ES, Q-1988A and Q-1988-FES (7–9); controlled-potential differential dc polarograph (10); Q-2792 (11). The DME potential that is seen and controlled by the potentiostat is that relative to the equipotential surface in solution defined by the location of the tip of the reference electrode. The resistance of the mercury thread within the DME capillary, about 50 ohms, does not introduce a significant iR error at the usual, low polarographic currents. The DME can be built so that less resistance is included in the circuit. In high specific-resistance solutions and/or at unusually high cell currents, a significant potential error may result from the inner resistance at the DME. However, the error can be reduced to an insignificant value, even in rather high specific resistance solutions, by means of design and arrangement of the transducer element. A Smoler 90°, vertical orifice DME that is tapered by grinding its tip is used so that the small tip of a reference electrode or probe can be placed very close above the upper surface of the mercury drops, within 0·1 of the maximum drop radius (13, 14). See Section III-B-1 of this chapter. The performance advantages that have been achieved by means of controlled-potential (also called potentiostatic) electrolysis are summarized in Section III-D of this chapter. Other operational-amplifier sub-systems that have been provided in potentiostat sections include an electronic scan generator (7–9) (see Section III-C-2 of this chapter), a compensator for the linear-with-potential component of the residual current (linear compensator) (7, 8), a single or cyclic scan source with adjustable scan limit (11), and a filter lag compensation circuit (11, 12). An initial potential source and an initial current compensation source are also included.

Sub-systems have been designed to obtain the maximum value of the current at the DME as a function of the DME potential. The more recent sub-system is a peak follower (6, 8).

Sub-systems have been designed to obtain the average value of the current at the DME as a function of E_{DME}. These sub-systems are current-averaging low-pass filters. Average current regular dc polarograms of good form are of advantage as such and, in addition, are necessary as inputs to derivative computers. Also, these sub-systems eliminate the requirement to synchronize the drop fall from the two DME's in differential two cell methods (10, 15). The performance advantages that have been obtained through a current-averaging system are summarized in Section III-E of this chapter. Several current-averaging systems have been devised, including the quadruple parallel-T filter which was designed for use with 2 to 5 second drop time DME's (6, 8, 16, 17), the tuned and untuned dual parallel-T filters which were designed for use with 0·5 second and 1 second DME's (17), and an active filter consisting of two tuned parallel-T sections and a fourth-order low-pass RC filter which was designed for use with constant, controlled drop time DME's (11, 12).

Operational amplifier sub-systems have been designed to obtain first and second derivative dc polarograms. A first derivative polarogram is obtained by means of a practical active analog derivative computer and an inverter stage. The average-current dc polarographic wave is the input to the computers. Another active analog derivative computer, with the first derivative polarogram as its input, is used to obtain a second derivative polarogram. The design of polarographic derivative computers is discussed in Section III-C-3 of this chapter. The relative advantages of regular and of derivative dc polarograms are described in Section III-F of this chapter.

The two most recent ORNL dc polarographs, model Q-1988-FES (17) and model Q-2972 (11, 12), are designed so that the Instrument User may select conditions that optimize either resolution or sensitivity. The circuit features through which these objectives are reached are described, respectively, in the above-cited papers. Highly reproducible polarograms are obtained with these polarographs. In the cases of regular and derivative dc polarography, to obtain an optimized wave form and resolution, scan rates must be slow enough so that the voltage change per drop life and the time lags in the current-averaging and derivative computing sections have a negligible effect upon the recorded wave forms. Diagnostic criteria have been described for testing and establishing that the forms of regular and derivative dc polarograms are not adversely affected by limitations of instrument performance (18). See Section III-A-2 of this chapter. At slower scan rates (~ 0.1 volt per minute) for which derivative polarograms of wave form in good agreement with the usual equations and of calibrated magnitude are recorded, the high resolution and the direct interpretability of the polarograms from the usual equations are distinct practical advantages. Resolution equal to the ultimate resolution of first derivative dc polarography (identical to that of, e.g. sinusoidal ac, square wave and pulse polarography) is available. Information such as n values and conformity to polarographic reversibility is available. For reaction and concentration ranges for which the heights of regular dc polarograms are directly proportional to concentration, the derivative polarogram peak heights are also directly proportional to concentration. At faster scan rates, excellent, optimized, high sensitivity is available with the first and second derivative modes.

The current-averaging and the active analog derivative circuits substantially contribute to the improved measurement and interpretation of recorded dc polarograms. That is these circuits improve the quality of communication. This is illustrated more specifically in Section III-E-1 of this chapter.

Sub-systems have also been designed for improved communication for two cell differential controlled-potential dc polarography (10); modes that have been evaluated include ΔE-differential (19), comparative (20, 21) and subtractive (22). With a well-designed system and with careful control of experimental conditions, a relative standard deviation of 0·1 per cent can be obtained at the millimolar concentration level with one cell comparative polarography; the two cell differential mode is not required for obtaining high precision information (20). Of the two cell modes, the subtractive two cell differential mode (22) is of particular interest from the standpoint of providing outputs that can be directly interpreted in terms of needed information. This mode allows the measurement of wanted signals in the presence of unwanted signals, and with improved sensitivity and selectivity over those of regular dc polarography. A subtractive regular polarogram of $10^{-7}M$ cadmium is shown in Figure 2 of reference 22.

A properly designed transducer, particularly the cell and the electrodes, also contributes to the clear interpretability of the results that are obtained. This is discussed in Section III-B of this chapter.

The design of the polarographic method or procedure that is used, in addition to circuitry sub-system and transducer design, is important. A properly designed and applied method certainly is intrinsically required in order to obtain credible results in specific applications in sample analysis or in measurements for research work. This is a most significant reason for including the measurement method as an element of the system. Method design steps to enhance interpretability (i.e. communication) include: choices of media, supporting electrolyte, maximum suppressor, complexing agents, etc.; uses of preliminary separations such as solvent extraction, chromatography, etc.; techniques of measuring recorded polarograms.

2. Performance Tests

It is necessary to have means for establishing that the information obtained as the output of the system is credible, i.e. trustworthy. The fulfilling of this need is an important part of the work that needs to be done in order to develop superior polarographic instrumentation systems. To provide these means, performance criteria must be defined and performance tests must be designed for elements and for the system. Such tests, with tabulations defining normal performance limits for a specific system, provide means for showing that a measurement system is performing in a normal way, without malfunctions that might cause output errors or artifacts, or for locating faults so that repairs may be made, as the case may be. Moreover, criteria and tests are needed to show that electrochemical conditions are as assumed in the equations.

In Section II-G of this chapter, the point was made that both electrical and specific performance tests are needed, the general approach used for testing systems for electrochemical measurements was described, and examples were listed of specific needs for both electrical and electrochemical performance tests of polarographic and coulometric systems.

In this portion, Section III-A-2, polarographic performance tests are presented and discussed in some detail. The results of the application of some of these tests to the ORNL model Q-1988-FES polarograph have been described (17).

(a) Purposes of electrical and electrochemical tests.   Performance criteria and performance tests are essential guides during research aimed at advancing the quality of polarographic systems or system elements. The performance of a polarographic system or element can only be defined, described and specified in terms of specific test results. These results also provide a basis for comparing performances achieved with various kinds of polarographs. When a polarograph, a cell or an electrode is newly constructed, when it is placed in service, its performance should be tested and found to be satisfactory before work is begun. When work of special importance is to be performed, it is prudent to test the measurement system before these new measurements are begun. If the quality of results that are obtained with a system that has been in use become in doubt or poor, performance tests must be made. In this last case, the first information that needs to be obtained by means of the tests is the location of the trouble. Malfunctioning may originate in the polaro-graph itself or may be due to electrochemical problems specific to the sample just introduced and thus be an information problem or a measure-ment method problem. The readout device and the circuit sub-system elements of the instrument can be checked out by means of electrical tests. When it is known that these elements are performing within specifications, electrochemical tests made with test reactions of known characteristics can establish whether the problem originates in the cell, electrodes or sample. Such criteria and tests are valuable tools for performing routine maintenance, including preventative maintenance, on existing equipment.

(b) *Electrical tests of polarographs.* It is worthwhile to take into account the possibility that the operation of the information readout device may have become faulty. The readout device should be tested in accordance with information and specifications furnished by the manufacturer before it is used to obtain results of tests of other elements of the system. Similarly, it is prudent to test the power supply before testing other circuits within the instrument.

The design of electrical tests of linear circuit sub-systems within the polarograph is relatively simple, since it is based upon a direct application of electrical engineering principles. The specification of realistic, i.e. meaning-ful and needed, electrical performance limits, however, requires some common sense judgment based upon a knowledge of electrochemistry and of the projected applications of the system.

It is intended in the following paragraphs to point out examples of needed electrical tests rather than to give stepwise test procedures. Detailed test procedures have been prepared for ORNL polarographs (23, 24).

Němec has clearly described the importance of ensuring that only linear damping takes place in the processes of measuring and recording dc polaro-grams (25). Non-linear damping will occur, for example, if the readout recorder servo is unable to keep up with the varying input signal presented to it. This must be prevented through the selection and appropriate use of the readout device. For dc polarography, a current-averaging sub-system can produce true average current regular polarograms; moreover, the response rate demands upon the recorder are minimized. Step and ramp tests have been made which show that the ORNL current-averaging sub-systems are linear devices. Němec points out that if a damping system is linear, when a step input is applied, the output at steady state equals the magnitude of the input step, and, also, the time required to reach any specific fraction of the input magnitude is independent of the magnitude of the input step. Němec has shown that the displacement of the recorded half-wave potential of a regular dc polarographic wave from the true half-wave potential that is caused by the time-lag of a linear damping system can be predicted suficiently well from the calculated response of the system to a ramp input and that the calculated displacement is independent of the slope of the ramp. Ramp tests have shown that the steady-state outputs of the parallel-T current-averaging sub-systems have the same slopes as those of the input ramps and that the voltage axis shifts correspond to the filter time-lags.

The following are examples of other needed electrical tests. Power supply tests, circuit voltage and/or resistance checks, calibration and output stability of sources (e.g. of initial potential, initial current, derivative offset and voltage scan sources), current range accuracy, tests for internal ac pickup, adjustment of bias and tests of offset of amplifiers, response of current amplifier and potential control amplifier to a step input signal, tests of operation of other sub-systems (e.g. linear compensator, peak-follower, scan limit), tuning of filters and the use of a ramp input to calibrate and to specify the noise level of derivative computers and to specify the noise level of filters. The potentiostat performance can be partially evaluated by an electrical test, by inserting resistors in series with the counter electrode lead (26). The form of the recorded polarographic wave should not be affected by the presence of the added resistance. However, this is only a test of the ability of the potentiostat to prevent simulated bulk resistance from affecting the wave forms. With a DME, a substantial fraction of the cell resistance, the inner resistance, is located within a few drop radii of the surface of the DME drops. Hence, electrochemical tests of potentiostat performance are needed.

Electrical tests of the performance of all recent ORNL polarographs are facilitated by the fact that an electrical test function is built in. In this mode the cell is switched out of the circuit and a precision 1-megohm resistor is switched in. A test step current of $1 \mu A V^{-1}$ can be generated at the input of the polarograph by means of the initial potential source. A test ramp signal can be generated by means of the scan voltage source. Separate, prior electrical tests are provided of the performance of these sources (23, 24).

(c) *DME test criteria.* The quality of performance of the DME that is used is important. With well-designed circuitry, the performance of the DME may be the principal determinant of overall system performance. One of the primary objectives of polarographic circuit design is, in fact, to cause this to be the case. It is necessary to relate the choice and definition of DME test criteria to specific performance needs. In some polarographic work, for the output to be directly interpretable, it is necessary to know whether the current at the DME is diffusion controlled. When the measurement of concentration with high sensitivity or high precision is needed, the signal-to-noise ratio and the reproducibility over a time span of the current at the DME enter importantly into the results that can be achieved. Since all

DME's are not of equal quality, DME test criteria are needed. Examples of difficulties in clearly establishing such criteria and in arranging operating conditions such that the assumptions involved in the derivation of the various polarographic equations are fulfilled were mentioned in Section II-G of this chapter.

Much time is saved by examining new DME's with a microscope, prior to installation. DME's which have cracks or spurs or other orifice defects must be rejected and bores that contain glass or other particles must be cleaned prior to installation. Also, it is as well to verify, at several voltages within the working range, that the drop time of the installed DME is not sporadically irregular, due to dirt, tilting, vibration, ac pickup, etc. In this connection, the recording and examination of epochs at each of several voltages of undamped current–time curves is useful. Sporadic irregularities should not be present.

Operating conditions such as the mercury head (h), the mercury flow rate (m) and sporadic mechanical shocks and vibration also affect the performance of a given DME. In the first two examples, the experimental optimization of these parameters and the specification thereof are appropriate. It appears to be advisable to have a high mercury head to minimize back pressure effects and to arrange for an average mercury flow rate no greater than a few milligrams per second to reduce the likelihood of maxima of the second kind. In the latter example, other than the control of environmental conditions, precise DME drop time control is useful since the controlled and early termination of drop life minimizes deleterious effects of vibration which are greater as the drop size increases.

The composition of the standard test and sample solutions also affect the performance of a DME. Although trouble can be caused through the improper use of maximum suppressors, it appears to be best to use a maximum suppressor. Schmid and Reilley have discussed the effect of surface-active substances on polarographic currents, the question of the optimum concentration of maximum suppressors and the use of current–time wave forms for diagnosing electrode processes (27).

The convective increase of wave height due to the existence of small maxima of the second kind may not be apparent from the slightly altered shape of the wave. Precautions to minimize the probability of occurrence of such maxima, including the use of maximum suppressors and low average mercury flow rates, seem warranted.

We have observed that for $\sim 10^{-3}$ M Cd^{2+} test solutions, in addition to adding a maximum suppressor (e.g. 0·001 per cent Triton X-100) and adding HCl to a concentration of 0·001 M to suppress hydrolysis, that 1 M KCl is superior to 0·1 M KCl as supporting electrolyte. Better wave forms are obtained with the thousand-fold excess of supporting electrolyte. This improvement may be due in part to the lower specific resistance and concomitant lowering of uncompensated iR_{inner} rather than only to the further reduction of the transference number for Cd^{2+} or to changes in the complexes that are present.

Another problem that can be significant when half-wave potentials are of interest are the junction potential(s) within reference electrode–salt bridge assemblies and at the tip–solution interface. Belew and Raaen have pointed out and have presented data showing that the magnitudes of liquid junction potentials can be quite large, particularly when the assemblies are not arranged to avoid junctions between several different electrolytes (28).

The agreement of measured polarographic D values or of current–time curves with generally accepted values or equations would be an intellectually satisfying criteria of DME performance. Yet, in practice, even with careful arrangement of the operating conditions, there are likely to be small differences that are not necessarily a valid indication that the DME should be rejected. This fact greatly complicates the selection of test criteria of DME performance that are valid, useful and unambiguous. Substantial experimental difficulties seem inevitable in obtaining polarographic D or current–time data, including the exact correction of observed diffusion current for residual current, the elimination of all current perturbation by vibration, and the effects of DME design and operation upon the magnitudes of corrections for, for example, spherical diffusion, shielding, depletion with non-first drops, mercury flow rate variation during each drop life and delays at the beginning of a new drop. Very careful work is required in order to obtain data that can be used with these criteria. Presumably, over a range of flow rates and drop times where diffusion control exists, a constant value of D would be observed. Jones and Fritsche describe a procedure for obtaining D values (29).

Apparatus has been designed, ORNL model Q-2942, to electromagnetically control the drop time of a DME to a precise, constant value (30). A drop time of 0·5 second is advantageous for derivative dc polarography because current-averaging filters and derivative computers having small time-lags (relative to those designed for a 5 second DME) can be used so that first derivative polarograms having excellent wave form and ultimate high resolution can be recorded at scan rates of about 0·1 volt per minute and so that excellent high sensitivity can be obtained at scan rates of about 1 volt per minute. The latter takes net advantage of the dependence of Nth derivative magnitude upon the Nth power of the scan rate. No disadvantage need result from operating with a drop time of 0·5 second unless kinetics of processes (e.g. adsorption) occurring at the DME are at this time scale. The current–time curve at a fixed potential on the wave for the Q-2942 controlled drop is in close agreement with the first 0·5 second of the uncontrolled DME current–time curve. The short drop time is obtained at a low average mercury flow rate. The performance criterion for the controller is that it should not cause any significant change to the net current–time relationship other than the desired early, constant termination of the drop life. The current at the controlled DME is not significantly increased due to convective disturbances from the operation of the controller.

A definition of DME test criteria can be reached through another viewpoint. Performance characteristics that may be needed in polarographic work include high precision and high sensitivity. The following two DME test criteria have been found to be useful for evaluating a DME and the conditions of its operation in terms of the ability to obtain data meeting these application needs. As a specific test for reproducibility (precision), triplicate regular and first and second derivative dc polarograms are recorded with the millimolar Cd^{2+} test solution at a scan rate of 0·1 volt per minute. The solution must be very well sparged and at constant temperature. The test is sensitive at this scan rate; high reproducibility is much easier to obtain at a scan rate of 1 volt per minute. With the ORNL model Q-2792 polarograph (11) and Q-2942 drop time controller (30) operating at 0·5 second, the polarograms respectively reproduce with a good DME to within a recorder pen width. As a specific relative measure of sensitivity, replicate first and second derivative dc polarograms are recorded at a scan rate of 1 volt per minute using 10^{-6} and $10^{-7} M$ Cd^{2+} test solutions. See, for example, Figure 28, Section III-E-3.

(d) *Needs for and methods of electrochemical testing of circuit sub-systems.* Electrochemical tests have been devised and used to test the performance of circuit sub-systems and of various combinations of them. Examples of needs for such tests are given first followed by discussions of tests that have been used. The composition of standard test solutions and the operation for tests of DME's are discussed in Section III-A-2-(c) of this chapter.

Electrochemical tests are needed in addition to electrical tests. Examples of needs for such tests follow. The potentiostat transient response must be examined and found to be satisfactory and the potentiostat must be stable. Since the transducer and its contents are in the feedback loop when polarograms are being obtained, transient response and stability tests must include electrochemical tests with the coupled transducer in operation. See Section IV-B of this chapter. A substantial fraction of the iR drop in the solution occurs within a few drop radii of the surface of the DME drops. These iR_{inner} drops can be of significant magnitude and cause errors in potential and in scan rate in high specific-resistance, i.e. above 5000 ohm-cm solutions (31), or in low specific resistance solutions at unusually high cell currents (32). The design and arrangement of the electrodes (see Section III-B-1 of this chapter) as well as the electrical performance of the potentiostat, enters into the degree to which this iR problem can be circumvented. Thus, electrochemical tests are required in order to establish the conditions under which polarograms can be recorded that are substantially free of iR loss distortion. If concentration is to be measured, electrochemical tests must be used to establish that each of the various available modes has an output that is directly proportional to concentration. Other tests must show that the sub-systems for each mode do in fact produce the intended response (e.g. maximum or average current; first or second derivative). Moreover, a criterion of ideal instrument performance is that it should be possible to select instrument operating modes such that the recorded polarograms are not only highly reproducible but that the forms are dictated by the electrochemical processes and are not adversely affected by the performance of the circuits used to obtain the polarograms. Indeed, these operating modes are required when ultimate resolution is needed. (On the other hand, when the main purpose is the measurement of concentration without the need for ultimate resolution, with recent ORNL polarographs, better derivative signal-to-noise ratio can be attained at higher scan rates, in spite of the fact that the forms of the waves are then affected by circuit time-lags.) Therefore, electrochemical wave form tests are needed to establish that these operating modes are available.

The transient response is examined and the stability of the potentiostat is verified through oscilloscopic observations. Typically, a potentiostat consists of a potential control amplifier (together with a voltage follower in some cases) and a current amplifier, together with initial and scan potential sources. The cell is in the circuit with a test solution present and observations are made at voltages corresponding to the foot of the wave and to points on the wave. If a test or standby mode is provided wherein the cell can be switched out of the circuit, the stability of the potentiostat under these conditions must also be verified. Summing points should be held at reference potential (e.g. at ground potential), steady-state oscillation of outputs should not occur, and the output of the current amplifier should agree with relevant current–time equations to within design specifications (23, 33). The transient response of ORNL dc polarographic potentiostats has been tailored by means of capacitive feedback. This prevents the current amplifier from following a sharp charging surge just after DME drop detachment which would drive the current amplifier to its limit on sensitive current ranges (34).

The placement of large valued resistors in series with the counter electrode lead will serve to test the electrical performance of the potentiostat in correcting for bulk iR loss, as discussed in Section III-A-2-(b) of this chapter. A properly functioning potentiostat can only establish the DME voltage as seen by the tip of the reference electrode. Hence electrochemical wave form tests of the performance of the potentiostat are needed. An important advantage of the Smoler 90°, vertical orifice DME over the usual horizontal orifice DME is that a potentiostat reference electrode probe tip can be placed very close to the upper surface of the drops, an arrangement which minimizes the amount of iR error not compensated for by the potentiostat (14, 35, 36). The upper surface is nearly stationary. Uncompensated iR error delays but does not prevent the attainment of the correct wave height in a regular dc polarogram (37). The usual Nernst-type semi-log plot of data from a regular dc polarogram will show whether conditions are such that the slopes of the rising part of the wave were insignificantly affected by uncompensated iR loss. The results of such a test have been published (38, 39). With negligible uncompensated iR, a straight line having the correct slope and intersection is obtained. The test has to be made under conditions (sufficiently slow scan rate) such that the time-lag of the current-averaging filter has a negligible effect upon wave form. Schaap and McKinney have reported that the deviation from theory (as calculated from i_d and scan rate) of the peak height of a derivative polarogram is a convenient and sensitive measure of uncompensated iR. The uncompensated iR decreases the effective scan rate and thus decreases the peak height. They have published an equation by means of which the uncompensated, average resistance can be calculated (32). It is possible to record first derivative polarograms with usual aqueous media whose peak heights are in good agreement with those calculated from the wave form equation. Conditions can be arranged so that both uncompensated iR and circuit time-lags have little effect upon the recorded wave form. Examples of such data are shown in Table I. Thomas and Schaap have evaluated another method for testing for and measuring uncompensated iR (40). Their method is based upon the comparison of the observed potential difference between two points (e.g. the quarter and three-quarter height points) on the rising part of a regular dc polarogram with that calculated from the Heyrovský-Ilkovič wave form equation. Uncompensated iR decreases the slopes of the rising part of the wave and increases the potential difference above the calculated amount.

TABLE I

Examples of Accuracy of Observed First Derivative Peak Height

Polarograph ORNL model	Scan rate V min^{-1}	First derivative peak height error, %
Q-1988-FES[a,b]	0·3	−2
Q-2792[c,d]	0·1	−0·9
	0·2	−1·9
	0·3	−4·2

[a] D. J. Fisher, W. L. Belew and M. T. Kelley, *Chem. Instrum.*, **1**, 181 (1968); p. 208.

[b] Dual parallel-T current-averaging filter (designed for and used with 0·5 second DME).

[c] H. C. Jones, W. L. Belew, R. W. Stelzner, T. R. Mueller and D. J. Fisher, *Anal. Chem.*, **41**, 772 (1969); Table I.

[d] Active current-averaging filter selected to optimize S/N for high sensitivity measurements (filter with constants calculated for 1 second DME used with 0·5 second DME).

Both the reproducibility of recorded regular and first and second derivative dc polarograms and the linearity of recorded wave and peak heights with concentration of reducible species are of practical interest. Hence electrochemical tests of reproducibility and of linearity are required. The degree of reproducibility is conveniently observed by superimposing replicate polarograms on an XY recorder. At millimolar concentrations, polarograms should reproduce within a recorder-pen line width (17). The peak height of a recorded first derivative dc polarogram at a given scan rate is directly proportional to the average-current wave height, i_d. At slow scan rates, where circuit time-lag is small, and under conditions such that there is a negligible amount of uncompensated iR loss, the proportionality factor should agree with that calculated from the wave form equation (17). At each of the faster scan rates, the proportionality factor is less than the calculated value but the relationship should remain linear (17). Second derivative polarograms are usually recorded at faster scan rates to obtain polarograms having a more nearly optimum signal-to-noise ratio. These polarograms are most useful for measuring very low concentrations at high sensitivity, whereas first derivative polarograms are also useful when recorded at slower scan rates for obtaining $E_{1/2}$, n values or ultimate resolution (17). The heights (either peak or peak to peak) of second derivative polarograms recorded at a given scan rate should be directly proportional to concentration (17). Regular, first derivative, and second derivative polarogram heights are measured with appropriate corrections for residual values (41). The first derivative mode is of particular utility when ultimate resolution is needed. Electrochemical tests of wave form, given in Section III-A-2-(e) of this chapter, are used to establish that ultimate resolution can be obtained with a polarograph, i.e. that single, resolved first derivative waves can be recorded having wave forms in good agreement with the accepted equations.

Electrochemical tests have been made that show that a first derivative polarogram that consists of several partially overlapped waves (no interreactions) is equal to the sum of the constituent waves (42). See Figure 21, Section III-E-1 of this chapter.

The peak follower sub-system can provide regular dc polarograms that consist of the maximum current as a function of potential (43). Through oscilloscopic observation, it was shown that the output of this system is equal to the maximum current value reached at the end of a drop life (44). The peak follower is useful in situations where the peak values envelope is constant or increasing with potential.

The highly efficient filtering that is obtained by means of the ORNL current-averaging sub-systems has been documented (11, 45). Electrochemical testing has been done to show that the output is the average-current value (46–48). The test consists of measuring the ratio of the wave height obtained with a current-averaging filter to that obtained with the maximum currents sub-system, the peak follower. Oscilloscopic testing has shown that the latter sub-system can deliver true maximum current values. The measured ratio is in good agreement with that obtained by others by independent means. Data obtained for the quadruple parallel-T current averaging sub-system are shown in Table II (48).

TABLE II

Ratio of Average to Maximum Value of Polarographic Diffusion Current
Measured by Means of Active Analog Circuits Using Operational Amplifiers

Solution	i_d Quad. 11-T
	i_d Peak follower
10^{-2} M Cd^{2+} in 1 N HCl	0·813 and 0·803
10^{-3} M Cd^{2+} in 1 N HCl	0·81
$1·5 \times 10^{-3}$ M Tl^+ in 0·1 M KCl	0·805

Theoretical value, Ikovič equation 0·859
Literature values, from oscillographic measurement, Millimolar Cd^{2+} by J. K. Taylor, R. E. Smith and I. L. Cooter, *J. Res. N.B.S.*, **42**, 387 (1949). 0·810 and 0·805
Conditions
    Model Q-1988 controlled-potential and derivative polarograph 24°C

The performances of the first and second derivative computer sub-systems have been extensively evaluated as to output wave form fidelity by means of electrochemical testing. Descriptions of tests for this purpose are included in Section III-A-2-(e) of this chapter.

(e) *Diagnostic criteria and electrochemical tests of fidelity of wave forms of regular and derivative dc polarograms.* Interpretations of recorded polarograms are simplified when mass transfer is diffusion controlled, with contributions due to migration or convection being negligible, and when the reaction is simple and polarographically reversible. The Nernst equation remains applicable if convection contributions are appreciable, but it is generally believed that diffusion controlled current is more easily reproduced from measurement to measurement. When the primary purpose of interpretations of recorded polarograms is to evaluate the performance of the measurement system or a portion of it, it has been our practice to use simple test reactions under experimental conditions that are arranged to be such that recorded forms should agree well with the usual wave form equations, if the forms of the recorded polarograms are not adversely affected by circuitry performance. Pertinent considerations involved in the arrangement of DME operating and test solution composition conditions are discussed in Section III-A-2-(c) of this chapter. Wave form tests are made at relatively high concentrations, $\geq$ millimolar, in order to minimize the possibility that observed disagreements are caused by inexact correction for residual currents rather than by circuit performance factors. Practices used for measuring polarograms have been documented (41). With substantially lower concentrations, at the higher measuring sensitivities necessarily used, the non-linearity and polarity reversal with potential of the residual current has a substantial effect on the form of the net (i.e. the as observed whole, free of spurious effects) polarogram. Current-averaging and derivative sub-systems certainly simplify the measurement of such concentrations. Even with well-sparged millimolar solutions in well-sealed cells, residual current contributions to the net wave may appreciably affect the wave form and/or height unless properly accounted for in the measurement technique. Without proper measurement, variations thereof over a time span are quite likely to affect the precision of replicate heights, when precise differential or peak follower height measurement methods are used. The net polarogram consists of the residual current versus voltage function(s) plus the faradaic current versus voltage function(s) (or the Nth derivatives thereof) in the absence of chemical inter reactions (49). The wave or the peak height (and magnitudes at other potentials) of polarograms can be measured by superimposing the foot of the net polarogram upon the corresponding voltage portion of a separately obtained regular or derivative polarogram of the supporting electrolyte alone. (Or by extrapolating the foot with a line passing through it and parallel to the blank polarogram.) While this method of measurement may not exactly compensate for residual current, it certainly is to be preferred to the linear extrapolation of the foot of the wave (50). (See Figure 20, Section III-E-1.) It is particularly convenient to use an XY recorder for the wave form tests, since then the X-axis scale (in millivolts per inch) is independent of the scan rate in use, and since polarograms made at various scan rates and in either scan direction can easily be superimposed. With the diagnostic criteria that follow, it is readily possible to ensure by means of electrochemical wave form tests that instrument operating modes and conditions have been selected that provide for the recording of polarograms of excellent wave form. When the recorded form of the polarogram is of itself of interest in applications of the instrument, it is recommended that such modes and conditions be used.

Wave form equations for regular and for first, second and third derivative dc polarograms (simple, polarographically reversible, diffusion controlled reductions) have been published (51). Equations that are useful for electrochemical tests of wave form include the following (constants are evaluated at 25°C):

Regular wave

$$\bar{i} = \frac{\bar{i}_d}{1 + e^{-[(nF/RT)(E_{1/2} - E)]}} \tag{1}$$

$$E = E_{1/2} - \frac{RT}{nF} \ln \frac{i}{i_d - i} \tag{2}$$

$$|E_{3/4} - E_{1/4}| = \frac{0.0564}{n} \tag{3}$$

First derivative wave

$$\frac{d\bar{i}}{dt} = \frac{-i_d(nF/RT)(dE/dt)}{e^{-(E_{1/2}-E)(nF/RT)} + 2 + e^{(E_{1/2}-E)(nF/RT)}} \tag{4}$$

maximum value (peak height) at $E_{1/2}$

$$= \left(\frac{d\bar{i}}{dt}\right)_{max} = \frac{-i_d}{4} \frac{nF}{RT} \frac{dE}{dt} \tag{5}$$

full peak width at half peak height, millivolts,

$$W_{1/2} = \frac{2000RT}{nF} \cosh^{-1} 3 \tag{6}$$

$$= \frac{90.5}{n}$$

ratio of peak height in $\mu A \, min^{-1}$ to $\bar{i}_d$, diffusion current in μA of corresponding regular polarogram; (dE/dt) in $V \, min^{-1}$

$$= \frac{nF}{4RT} \frac{dE}{dt} = 9.73 \frac{dE}{dt} n \tag{7}$$

Second derivative wave

$$\frac{d^2\bar{i}}{dt^2} = \frac{i_d[(nF/RT)dE/dt]^2[e^{-(E_{1/2}-E)(nF/RT)} - e^{(E_{1/2}-E)(nF/RT)}]}{[e^{-(E_{1/2}-E)(nF/RT)} + 2 + e^{(E_{1/2}-E)(nF/RT)}]^2} \tag{8}$$

ratio of height of either peak in $\mu A \, min^{-2}$ to diffusion current in μA of corresponding regular polarogram, $\bar{i}_d$

$$= \frac{n^2}{10.39}\left(\frac{F}{RT}\right)^2\left(\frac{dE}{dt}\right)^2 \tag{9}$$

At $25°C$, $[F^2/10.39(RT)^2]$ is approximately 145.8.

Third derivative wave

$$\frac{d^3\bar{i}}{dt^3} = \frac{i_d[(nF/RT)(dE/dt)]^3[e^{-(E_{1/2}-E)(nF/RT)} + e^{(E_{1/2}-E)(nF/RT)} - 4]}{[e^{-(E_{1/2}-E)(nF/RT)} + 2 + e^{(E_{1/2}-E)(nF/RT)}]^2} \tag{10}$$

ratio of middle peak height in $\mu A \, min^{-3}$ to diffusion current, i_d, in μA

$$= \frac{n^3}{8}\left(\frac{F}{RT}\right)^3\left(\frac{dE}{dt}\right)^3 \tag{11}$$

A plot (solid line) of the wave form equation of a first derivative dc polarogram is shown in Figure 7. A plot of the wave form equation of a second derivative dc polarogram is shown in Figure 8.

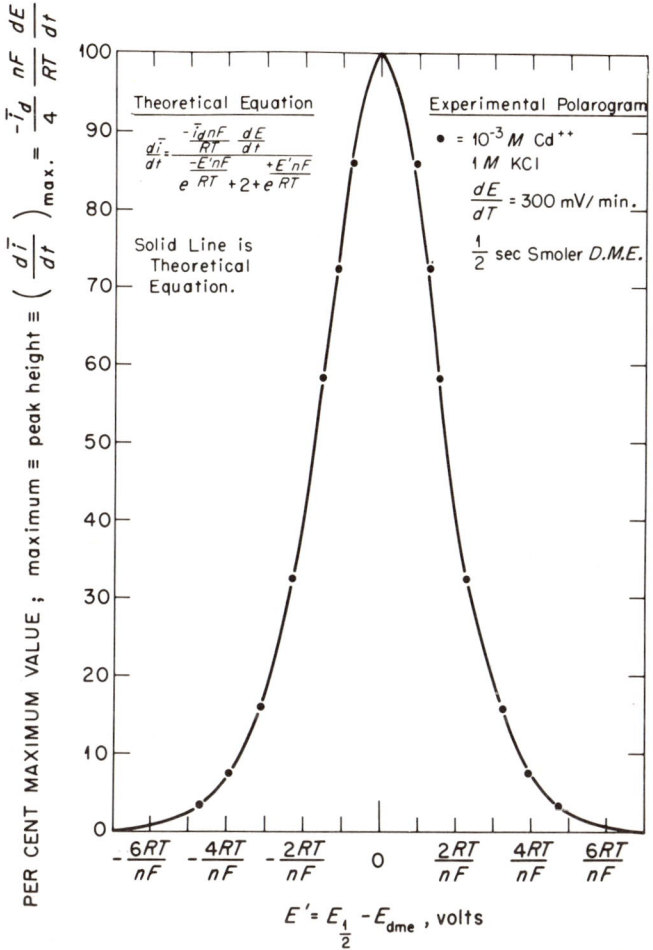

Fig. 7. Plot (solid line) of wave form equation of first derivative dc polarogram and test (dots) of fidelity of experimental polarogram. Reprinted from D. J. Fisher, W. L. Belew and M. T. Kelley, *Chem. Instrum.*, **1**, 181 (1968), Figure 4, by courtesy of Marcell Dekker, Inc.

The time-lags in the linear current-averaging and derivative computing sub-systems cause recorded wave forms of regular and first derivative dc polarograms for simple diffusion controlled, polarographically reversible reactions to deviate from the forms according to the usual equations in the following ways. The regular wave is affected in two ways. The recorded half-wave potential is displaced by a constant time interval in the direction

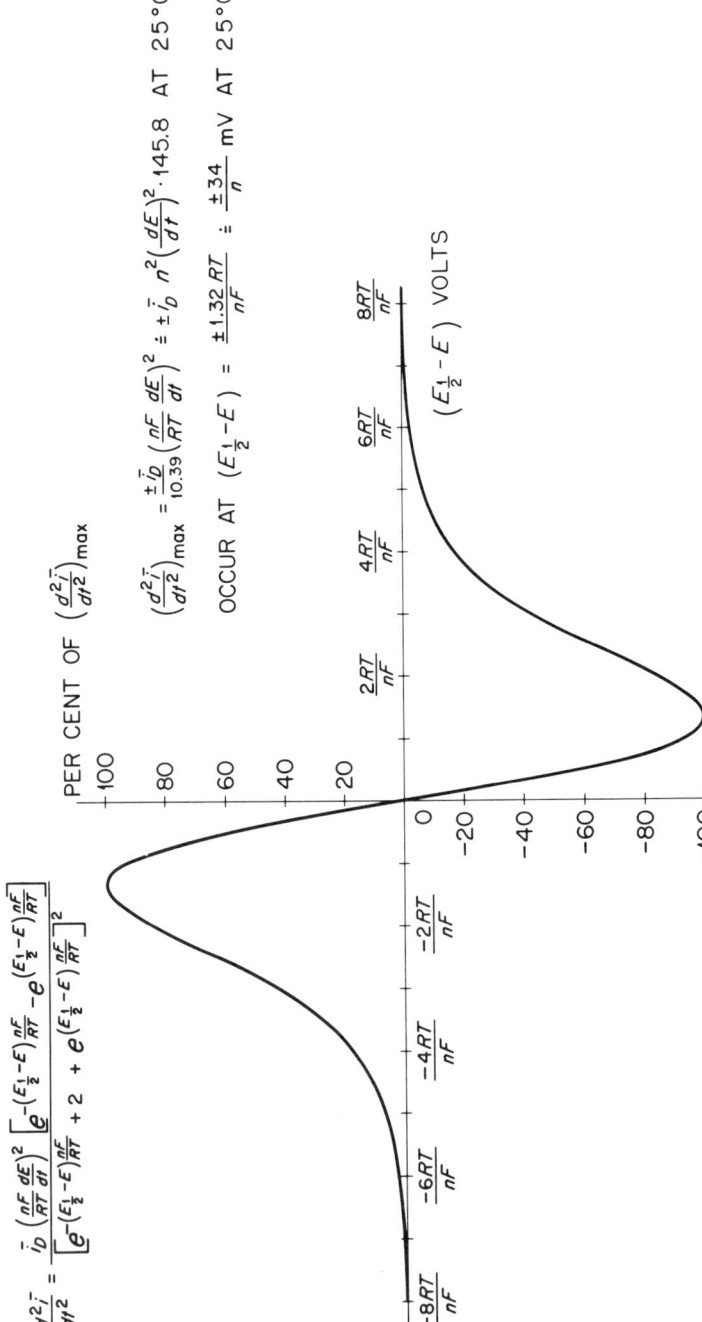

Fig. 8. Theoretical form of second derivative dc polarographic wave (simple, reversible, diffusion-controlled process).

of scan from the actual half-wave potential (52). The significance of the lag in terms of millivolts is directly dependent on the scan rate and is negligible at sufficiently slow rates. Secondly, the recorded wave height is the average diffusion current value, i_d, but the rising portion of the wave is drawn out with its slopes being less than the correct values. Time-lag affects the form of the first derivative wave in four ways. The voltage corresponding to the maximum value (peak height) is displaced in the direction of scan from the half-wave potential. The total amount of displacement at the peak is that due to the time-lag in the current-averaging filter plus that due to the lag in the derivative computer (52). Secondly, the peak height is less than the correct value which can be computed from i_d by means of equation (5) or (7). Thirdly, the derivative polarogram is broader than it should be so that a value of n calculated (equation (6)) from the width at half-height, $W_{1/2}$, is too small. The degree of broadening depends both upon the scan rate and upon n. Fourthly, the recorded derivative peak is not symmetrical about the peak potential.

On the other hand, when first or second derivative polarograms are recorded with a DME, an advantage of the time-lag in the current-averaging filter is that the sufficiently slow averaging process prevents the status (size) of a DME drop at the moment of traversing through the potentials that respectively correspond to derivative peak locations from affecting the recorded magnitude (and reproducibility for replicates) of the peak heights. Moreover, best signal-to-noise ratio is obtained at the high sensitivities needed to record derivative polarograms of low ($< 10^{-5}$ M) concentrations at the faster scan rates where lag has greater effects on wave form.

As has been pointed out in Section III-A-2-(d) of this chapter, uncompensated iR losses in solution have similar effects upon the recorded forms of polarograms.

In the following paragraphs, explicit diagnostic criteria and electrochemical tests of wave form fidelity are given.

In the case of regular average-current dc polarograms, time-lags introduced by the linear current-averaging circuits delay the attainment of the average-current wave height, i_d, but do not cause an error in it. Polarograms made at various scan rates will show the scan rate below which the time-lag shift in the direction of scan of the recorded $E_{1/2}$ is negligible. At scan rates for which superimposed polarograms made in each scan direction coincide, the recorded polarograms are insignificantly affected by time-lag (16). Tomes' method, equation (3), of evaluating n from the $E_{1/4}$ and $E_{3/4}$ points of recorded polarograms and n and $E_{1/2}$ determinations from the usual Nernst semilog plot, equation (2), can be used as diagnostic criteria of circuit performance. Test solutions are used in which uncompensated iR loss does not cause a significant delay through the rising part of the wave. Testing to show conditions where the potentiostat is accomplishing this objective is discussed in Section III-A-2-(d) of this chapter. If the semilog plot is linear and has the correct slope (n value) and intersection ($E_{1/2}$ value), then the instrumentation is producing regular polarograms in which the combined effects of iR in solution and time-lags in circuitry sub-systems cause negligible distortion to the wave form. For example with the ORNL Q-1988-FES polarograph (0·5 second DME and dual parallel-T current-averaging system), values of n obtained from such plots (millimolar Cd^{2+}, 1 M KCl) are 1·98 and 1·96 at scan rates of, respectively, 0·1 and 0·3 volts per minute.

In the case of derivative dc polarograms, it has been found that a particularly sensitive criterion of circuit performance is a comparison of the recorded ratios of the first or second derivative peak heights to i_d with the respective calculated values. Expressions for calculating ultimate derivative ratios have been given, equations (7), (9) and (11). The derivation of the calibration of the derivative scales of the ORNL Model Q-1988-FES polarograph, examples of the calculation of signal ratios and data showing that for scan rates at which circuit time-lags are insignificant the recorded ratios are in good agreement with the calculated ratios have been presented (53). Values obtained with ORNL polarographs models Q-1988-FES and Q-2792 are quoted in Table I of this chapter.

TABLE III

Example of n Values Obtained from $W_{1/2}$

Polarograph ORNL model	DME drop time, sec	$\dfrac{dE}{dt}$ V min^{-1}	n True	n Observed
Q-1988-ES or A[a,b]	~4·5	0·02	1	1·00
	~4·5	0·05	1	0·964
	~4·5	0·01	2	2·00
	~4·5	0·02	2	1·92
	~4·5	0·01	3	2·94
Q-1988-FES[a,c]	~0·5	0·3	1	1·00
	~0·5	0·3	2	1·96
	~0·5	0·3	2	1·97
	~0·5	0·3	3	2·84
Q-2792[d,e]	0·5[f]	0·1	2	1·99
	0·5[f]	0·2	2	1·95
	0·5[f]	0·3	2	1·95

[a] D. J. Fisher, W. L. Belew and M. T. Kelley, *Chem. Instrum.*, **1**, 181 (1968); pp. 199–201.

[b] Quadruple parallel-T, RC current-averaging filter, slow derivative computer.

[c] Dual parallel-T current-averaging filter, fast derivative computer.

[d] H. C. Jones, W. L. Belew, R. W. Stelzner, T. R. Mueller and D. J. Fisher, *Anal. Chem.*, **41**, 772 (1969); Table I.

[e] Active current-averaging filter selected to optimize S/N for high sensitivity measurements (filter with constants calculated for 1 second DME used with 0·5 second DME).

[f] Controlled with ORNL model Q-2942 electromechanical apparatus. (30).

As a second test of first derivative dc polarograms, the calculation of n from $W_{1/2}$ by means of equation (6) is a sensitive criterion of circuit performance. The method of evaluating the fidelity of a first derivative dc polarogram by determining n for a test reaction from the measurement of $W_{1/2}$ is shown in Figure 9. An experimental value, 2·96, with an indium test polarogram made with the ORNL model Q-1988-ES polarograph is shown on that figure.

Examples of n values obtained from $W_{1/2}$ with models Q-1988A, Q-1988-FES and Q-2792 are shown in Table III.

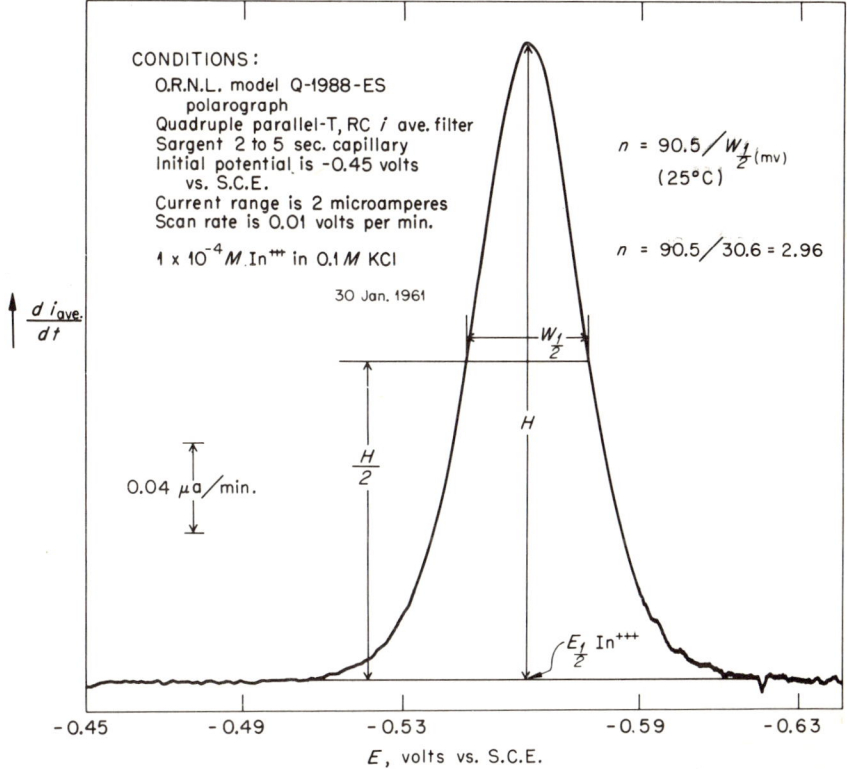

CONDITIONS:
O.R.N.L. model Q-1988-ES polarograph
Quadruple parallel-T, RC i ave. filter
Sargent 2 to 5 sec. capillary
Initial potential is -0.45 volts vs. S.C.E.
Current range is 2 microamperes
Scan rate is 0.01 volts per min.

$1 \times 10^{-4} M . In^{+++}$ in $0.1 M$ KCl

30 Jan. 1961

$n = 90.5 / W_{\frac{1}{2}} (mv)$
(25°C)

$n = 90.5 / 30.6 = 2.96$

$\frac{d\,i_{ave.}}{dt}$

$0.04\ \mu a / min.$

$\frac{H}{2}$

$W_{\frac{1}{2}}$

H

$E_{\frac{1}{2}} In^{+++}$

-0.45   -0.49   -0.53   -0.59   -0.63

E, volts vs. S.C.E.

Fig. 9. Evaluation of wave form of first derivative dc polarogram by determining n from $W_{1/2}$. Reprinted from: D. J. Fisher, W. L. Belew and M. T. Kelley, pp. 89–134 in G. J. Hills, Ed., *Polarography 1964*, Vol. 1, Macmillan, London, 1966, Figure 7 by permission of The Polarographic Society and Macmillan & Co. Ltd.

As a third test of first derivative wave form, if waves are recorded at each of several scan rates, a criterion of negligible time-lag is that the recorded peak heights are directly proportional to the scan rates; for example if waves are recorded superimposed on current ranges having the same ratios as the scan rates, the waves should coincide.

A fourth test of the ability of the circuitry to produce first derivative polarograms of good wave form is a comparison of a plot of the percent of the maximum value of the recorded first derivative wave against E with the plot of the corresponding wave form equation. This test is less sensitive than the first three since it is normalized; a good fit will be seen for first derivative waves even if the peak height is lower than the value calculated from i_d by ~ 1 per cent. The result of such a test is shown in Figure 7. The results of the application of this test to the quadruple parallel-T current-averaging and slow derivative computer sub-systems combination, designed for use with a 2 to 5 second DME, have been published (54). A figure showing results of this test of the performance of the ORNL model Q-2792 polarograph has been published (55).

Other sensitive tests for time-lag effects in first derivative polarograms are to note whether or not the recorded voltage at which a first derivative peak occurs is independent of scan direction and to make comparisons of recorded E_{peak} values obtained at various scan rates with $E_{1/2}$. If time-lag and iR-loss effects are negligible, E_{peak} coincides with $E_{1/2}$. A line dropped from the peak of a first derivative polarogram (at E_{peak}) perpendicular to the voltage axis should divide the peak into two symmetrical parts and $W_{1/2}$ into two equal parts; this is a relatively insensitive test.

In the case of second derivative dc polarograms, with excellent circuit performance, the polarogram passes through zero at $E_{1/2}$, the two peak heights are equal and occur at $\pm 1.32RT/nF$ (i.e. at $\pm 34/n$ mV at 25°C) from $E_{1/2}$, and the peak heights are directly proportional to the second power of the scan rate. These relationships are shown in equations (8) and (9) and are illustrated in Figure 8. Also, a normalized plot of points from these recorded polarograms (per cent. of maximum value against E) can be compared with the calculated wave form equation. The ultimate resolution of second derivative dc polarography is only slightly greater than that of first derivative dc polarography (56). At scan rates that are slow enough so that second derivative polarograms of good form are obtained, the signal-to-noise ratio is not as good as can be obtained with first derivative polarograms that have good form. With the ORNL model Q-1988-FES polarograph (millimolar Cd^{2+} in 1 M KCl, dual parallel-T current-averaging filter, 0.5 second DME, fast derivative computers, 0.3 V min^{-1}), a value of n calculated from the relationship $n = 68/\Delta$ peaks, mV is 1.94 (56). Under the conditions listed in the preceding sentence, the calculated ratio of either peak of a second derivative polarogram to i_d, equation (9), is 52.5 and the recorded ratio for the first peak is 52.5 (56). At faster scan rates, the second derivative mode is of particular value for the measurement of concentrations down to low values (57).

Tests of the wave form fidelity of third derivative dc polarograms, analogous to those just described for first and for second derivative polarograms, can be based on the criteria implicit in equations (10) and (11). According to equation (10), the ratio of the height of the middle peak (which occurs at the half-wave potential, $E_{1/2}$) to that of the two equal peaks (of opposite polarity to the middle peak) that occur at $\pm 2 \cdot 29(RT/nF)$ volts from $E_{1/2}$ is 3. The polarogram crosses through zero magnitude at $\pm 1 \cdot 32(RT/nF)$ volts from $E_{1/2}$. Experience has not been gained in the application of such tests, although the addition of a third derivative mode (with a readout device fast enough to not introduce non-linear distortion) to the ORNL model Q-2792 polarograph was planned. This would enable learning whether with current-averaging and derivative computing sub-systems that are optimized for a 0·5 second drop time DME, third derivative dc polarograms could be obtained that are useful in terms of resolution or sensitivity.

If the necessary digital equipment is available, another approach to testing the fidelity of output wave forms of some polarographic sub-systems becomes possible: a direct comparison of output *versus* input. For example the output of a derivative computer sub-system can be digitally sampled and stored, digitally integrated, and then compared, point by point, against the actual input wave form. (See Section V-D of this chapter.) The closeness of fit of the stored data to an appropriate equation can also be examined, with a digital computer that is programmed to perform the work.

3. Descriptive Information

The descriptive information that is provided for the ORNL polarographs includes: functional block diagram; circuit and mechanical construction drawings; construction specifications and parts list; check out and main-tenance procedures; operating and trouble-shooting procedures; papers describing the principles of design, the performance features and the per-formance capabilities of the system; maintenance records. The need for and value of such descriptive information as a factor of communication have been discussed in Sections II-C, II-D, II-E and II-G of this chapter.

B. Design of Transducer

In Section III-A-2-(c) of this chapter, chosen DME test criteria and selected proper conditions of operation of the DME were discussed. In the present Section III-B, the arrangement of the electrodes within the trans-ducer and the design of the DME and of the cell to obtain better polaro-graphic results are considered.

1. Electrode Arrangement

With potentiostatic (also named controlled-potential) electrolysis, three electrodes are used: a working (also named controlled) electrode which is a DME in the case of dc polarography, a reference electrode and a counter (also named auxiliary or third) electrode. Except in high specific resistance solutions where the combination of a Luggin probe or a QRE and a tapered Smoler 90°, vertical orifice DME is useful to minimize uncompensated iR_{inner}, the reference electrode that is usually used is a commercial SCE and salt bridge assembly, of the type commonly used with pH meters. A silver–silver chloride reference electrode and the quasi-reference electrode, QRE (35), have also been used. For the counter electrode, platinum or gold wires or graphite rods have been used, dipping directly into the solution, without barrier isolation. Since traces of platinum in mercury substantially lower the overvoltage of hydrogen discharge on mercury in highly acid aqueous solutions, for such work, the use of platinum for the counter electrode or for the contact to the mercury reservoir may have to be avoided.

Under conditions commonly encountered in polarography, i.e. where excess ionized supporting electrolyte is present at a concentration as high as 0·1 or 1 M and the specific resistance of the solution is low, the relative arrangement of the three electrodes in the cell is not critical. It has been our practice to position the counter and reference electrodes on opposite sides of the DME (58). The conditions under which iR_{inner} could represent a significant source of potential error can be predicted from the Ilkovič inner resistance equation (13). With the usual DME having its orifice in a horizontal plane, it is impractical to locate and maintain the tip of the reference electrode within a small fraction of the drop radius from the drop surface. This is a disadvantage: If the tip is several radii away from the surface, most of the iR_{inner} cannot be compensated for by the potentiostat, although it does compensate for iR_{bulk}. This disadvantage can be circumvented: As has been pointed out in Section III-A-1 of this chapter, the use of a Smoler 90°, vertical orifice DME whose tip is tapered by grinding permits the placement of the tip of a Luggin–Haber probe or a miniature QRE at a very close, within 0·1 of the maximum drop radius, nearly constant, distance from the upper surface of the drops (14). This electrode arrangement substantially minimizes the uncompensated iR_{inner} loss. In very high specific resistance solutions or at unusually high currents, the possibility that uncompensated iR_{inner} is not negligible should be investigated. Electrochemical tests for iR loss are given in Sections III-A-2-(d) and III-A-2-(e) of this chapter.

The design, arrangement and positioning of the sharpened Smoler 90° vertical orifice DME and the reference electrode tip so that the potentiostat can minimize the uncompensated iR_{inner} are shown in Figure 10 (14). A photograph of this apparatus is shown in Figure 11 (14).

Advantages accrue in dc polarography from the use of potentiostatic electrolysis via controlled-potential polarographic circuitry. These advantages are summarized in Section III-D of this chapter. Descriptions of ORNL polarographs and references to instrumental details are found in Sections III-A-1 and III-H of this chapter.

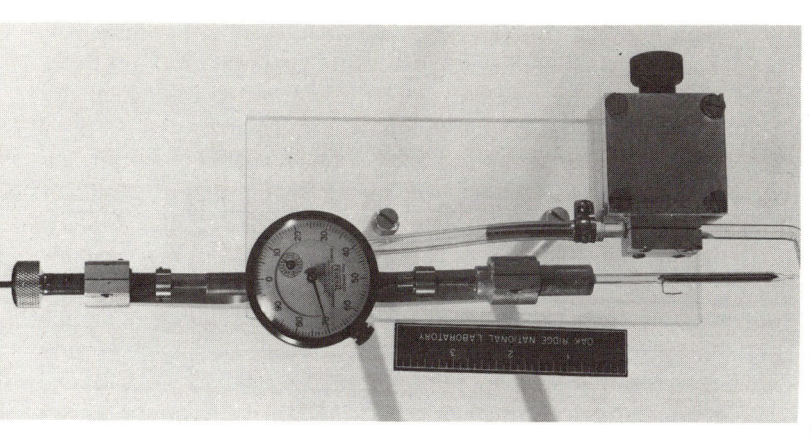

Fig. 11. Photograph of apparatus to position electrodes. Reprinted from W. L. Belew, D. J. Fisher, M. T. Kelley and J. A. Dean, *Chem. Instrum.*, **2**, 297 (1970), Figure 2, by courtesy of Marcel Dekker, Inc.

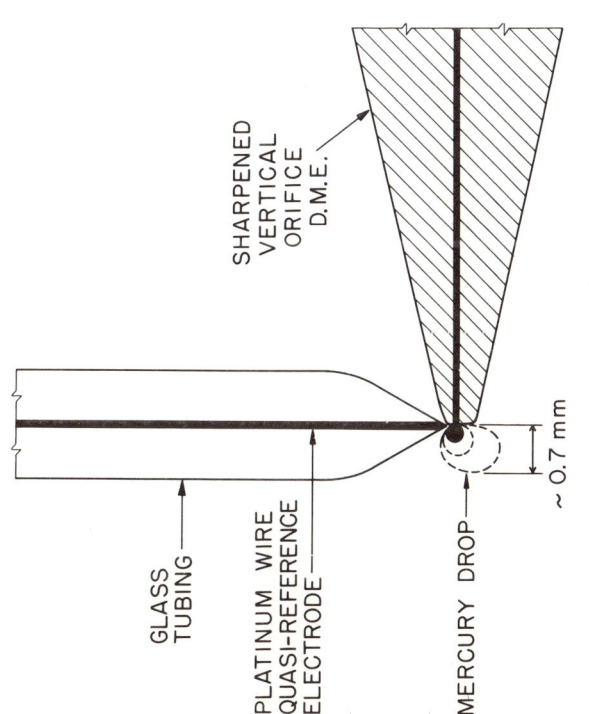

Fig. 10. Design, arrangement and positioning of DME and reference electrode to minimize uncompensated iR_{inner}. Reprinted from W. L. Belew, D. J. Fisher, M. T. Kelley and J. A. Dean, *Chem. Instrum.*, **2**, 297 (1970), Figure 1, by courtesy of Marcel Dekker, Inc.

2. Dropping Mercury Electrode, DME

Since the performance with well-designed circuitry may be limited by that of the DME itself, efforts have been made and are in progress at ORNL to optimize the performance of DME's. The minimization of current disturbances from noise sources, i.e. the optimization of signal-to-noise ratio, is one objective of such work. Another area of DME design at ORNL has been the development by H. P. Raaen of Teflon DME's that are useful in hydrofluoric acid and other media that corrode glass (59–67). A slight design modification of a Raaen Teflon DME and its use in concentrated HF solutions have been reported (68). In order to approach more closely the attainment of the ultimate superior sensitivity and resolution of derivative dc polarography, it is necessary to use current-averaging filter subsystems having a smaller time-lag than those designed for ~ 5 second DME's; this can only be done through the use of DME's having a shorter drop time than 5 seconds. In retrospect, it is of interest to remember that with ~ 5 second DME's and corresponding filtering, we once concluded that the actually attained signal-to-noise ratio of second derivative was inferior to that of first derivative dc polarography. With 0.5 second DME's and corresponding fast circuits and scan rates, the inherent sensitivity advantage of the second derivative mode is realized to a useful degree, and, the excellent ultimate resolution of the first derivative mode is attained at convenient (~ 0.1 volt per minute) scan rates. In order to utilize the advantages of a constant drop time as short as 0.5 second at low average mercury flow rates and without net convective disturbances, we have developed a precise electromagnetic drop time controller, ORNL model Q-2942 (30). The advantages of a constant drop time include the independence of drop time from changes in surface tension associated with scanning E or with the presence of surfactants, a relative freedom from noise caused by sporadic mechanical shocks and vibrations, and independent adjustment of drop time and average mercury flow rate. Three devices have been designed and built to control the average mercury flow rate, $\bar{m}$, to a constant value throughout the drop life. This work is discussed in Section III-G of this chapter. Ensemble averaging and timed current sampling are also under investigation at ORNL; see Section V-D of this chapter. Digital polarography affords an opportunity to investigate and utilize the possibility that improved sensitivity or precision can be obtained through timed sample and hold sub-systems. Perhaps, during each current-time drop life, the signal-to-noise ratio is poorest just as the old drop is about to fall and just as the new drop is extruded into the solution. If this were the case, a sampling measurement method could be used to prevent current during these portions of the drop life from being measured. DME operating conditions that tend to optimize DME performance are considered in Section III-A-2-(c) of this chapter.

In the following paragraphs of Section III-B-2, the meaning and the definition of polarographic signal-to-noise ratio, sources of and ways to minimize imperfect reproducibility of current at a DME, the Smoler 90°, vertical orifice, 0·5 second DME and advantages of precise control of the DME drop time are discussed.

The ultimate resolving powers and the ultimate signal levels of various kinds of polarography can be calculated from the respective wave form equations. The relative sensitivities are determined by the relative signal-to-noise ratios that can be attained with specific implementations of each method, not just by the ultimate signal levels.

The ratio of the faradaic response to the measured residual response is often given in the literature as a definition of polarographic signal-to-noise ratio. The signal-to-noise ratio, as the determinant of attained sensitivity, is more meaningfully defined as follows (69). It is the ratio of the faradaic response to the *range* of measured values of the total response plus the *range* of measured values of the residual response. The faradaic response in regular dc polarography is the wave height, i_d, and is the peak height in derivative dc polarography. The total response in regular dc polarography is the cell current on the limiting plateau and the residual value is the measured, not the extrapolated, value. The significance of our definition of polarographic signal-to-noise ratio is that the denominator includes all forms of noise in and lack of perfect reproducibility of the measured responses rather than being the average value of the residual response. If the average value of the residual response were exactly reproducible and the same in a sample and in a blank, it would merely be a zero offset, non-linear with respect to potential. The residual response is smaller in first derivative than in regular dc polarography since the linear-with-E component is eliminated as a zero offset and is smaller yet in second derivative dc polarography. The slope of the residual current versus potential relation (residual response for first derivative dc polarogram) is more reproducible than its magnitude (residual response for regular dc polarogram). This definition of polarographic signal-to-noise ratio includes both of these factors, namely, magnitude and reproducibility.

The reproducibility of the current is relatively favourable at a DME, as compared to that of other kinds of mercury working electrodes (70). Nevertheless, the design of a polarographic measurement system for attaining high sensitivity should include steps to minimize noise sources (imperfect current reproducibility) at the DME.

Sources of imperfect reproducibility of the current at the DME include some that can be minimized by control of environmental factors (e.g. sporadic mechanical shocks and vibration, temperature dependence of current, surfactant contaminants, dissolved oxygen concentration after sparging, changes in cell solution volume caused by addition or loss of solvent during sparging, motion of the solution caused by excessive sparge gas overflow after sparging), some that can be minimized by circuit and instrument design (ac voltage pickup of tens of millivolts on DME dc voltage, instability of E sources and of current measurement or transformation circuits), and some that can be minimized by DME selection, design and operation (imperfections of orifice including cracks or spurs, glass or dirt particles in the capillary bore, solute in bore due to interruptions of mercury flow in presence of solution, application of excessive voltage, convective disturbances at the interface due to poorly designed or operated devices for the electromechanical control of drop time, variations in average mercury flow rate, $\bar{m}$, which are not necessarily controlled by the use of a high, constant head, h, convective or turbulent disturbances at the interface caused by excessive values of $\bar{m}$, variations in back pressure, depletion effect, periodic ejection of solution that has crept upwards into the capillary from the orifice). The magnitude and frequency of occurrence of the latter can be minimized by the use of a DME that has been drawn out so that the diameter of the orifice is reduced by a factor of about three to a value of about 0·025 millimetres (71). With the aid of a microscope, it can be seen that a sporadic noise pulse occurs when solution that has crept up into the capillary forms a pocket and is ejected from the capillary. This phenomenon has been discussed by Meites (72) and by De Levie (73). With the drawn-out DME, very reproducible first derivative dc polarograms were obtained at the 10^{-6} M level (71). Koryta and coworkers have reported that the use of this drawn-out DME was advantageous in their chronocoulometric investigation of double-layer structure because it considerably decreased the residual current, a main obstacle to the method (74).

The significant advantage of a DME in providing a highly reproducible and a renewed working electrode with the formation of each successive mercury drop seems to be realized most effectively when each new electrode surface forms smoothly, with minimum swirling of the mercury itself and with minimum convective disturbance of the solution layer at the interface. The Smoler 90°, horizontal (i.e. vertical orifice) DME has been described (75–78). It can provide a drop time as short as 0·5 second but with a low mercury flow rate and at a reasonably high mercury head (70). The current appears to be diffusion controlled; it is highly reproducible. The drop size is relatively small. Obtaining a short drop time with the usual horizontal orifice DME by increasing the mercury head is not satisfactory because the flow rate becomes so high that maxima of the second kind are encouraged and convective contributions to the current are likely to be significant. Smoler has shown that the small amount of laminar solution flow at his 90° vertical orifice DME is not harmful and actually has advantages in eliminating the depletion effect and in removing electrolysis products from the interface. Smoler 90° DME's having a drop time of 0·5 second can be fabricated by bending Sargent 2 to 5 second DME's near the centre to a right angle and breaking the capillary about $\frac{1}{4}$ to $\frac{1}{2}$ inch away from the bend. With some selection of capillaries, it is possible to obtain a drop time of 0·5 second at a mercury flow rate of no more than 5 milligrams per second for mercury heads from 0·6 to 1·2 metre (70). Smaller flow rates are preferred. In conjunction with the ORNL model Q-2942 drop time controller (30), since drop time and flow rate are independently selectable, it is easily possible to operate a Smoler 90° DME at a drop time of 0·5 second and at a flow rate of, for example, 1 milligram per second. Three advantages result from the use of vertical (90° Smoler) instead of horizontal orifice DME's in this apparatus. First, sparge gas bubbles are not trapped after sparging against the bottom of the capillary. Second, it is much easier to obtain coincident corresponding portions of uncontrolled and controlled drop time current–time curves at low average mercury flow rate values. Third, when also tapered by grinding the tip, it facilitates the minimization of iR_{inner} error in high specific resistance solutions (14). See Figures 10 and 11. No disadvantage from the use of Smoler 90° vertical orifice DME's has been observed.

Extremely reproducible polarograms are obtained with the combination of the ORNL model Q-2792 polarograph (11), the model Q-2942 drop time controller (30) and a Smoler 90° DME controlled at 0·5 second. The model Q-2942 device precisely controls the drop time, without introducing disturbances from convection. Since the current is sampled and averaged throughout the drop life in the Q-2792 polarograph, the requirements for the performance of the drop time control device are much more exacting than is the case with controllers used in conjunction with polarographs that utilize current sampling near the end of drop life, after disturbances introduced by such devices have damped out. An earlier model electromagnetic drop time controller did not improve the signal-to-noise ratio (with current-averaging throughout the drop lives) (70). Proper, precise electromagnetic control of DME drop time is an excellent method of retaining conditions necessary for superior results with dc polarography (e.g. negligible change of E per drop life, diffusion-controlled mass transport, high signal-to-noise ratio, low average mercury flow rate with sufficiently high mercury head) while attaining the extra advantages of a short and a constant drop time. Criteria for testing the performance of DME's controlled with the Q-2942 apparatus are given in Section III-A-2-(c) of this chapter. Drop time is controlled at values from $\frac{1}{4}$ to 10 seconds, each with a long term relative standard deviation of 0·05 per cent. An appropriately damped movement of the DME is provided. The capillary is displaced horizontally in one direction to shear the drop, held at that position for a fixed period of time, and then displaced in the opposite direction to shear the next drop. Several duplicate Q-2942 devices have been fabricated at ORNL and elsewhere from the ORNL drawings and all operate very well. We have been informed

that others are extremely pleased with the combined performance of a
Q-2942 controller and an ORNL model Q-1988-FES polarograph that they
built from our drawings (79). The mechanical DME drop time controller
marketed by Princeton Applied Research is patterned after the ORNL
model Q-2942 (80). The piezoelectric properties of barium titanate have been
utilized in a DME time drop controller. The design of this controller has
been briefly described (81).

Several workers have reported finding that maxima are less likely to
occur with properly obtained short drop times. This is ascribed to the fact
that time is required after drop termination for streaming to get underway.

A large part of the time expenditure in polarographic analysis is consumed
by sample and standard preparation and in sparging prepared sample,
standard (or spiked) and blank (supporting electrolyte) solutions. In rapid
dc polarography, with a 0·5 second DME, at a scan rate of 3 volts per minute,
a regular or a first or a second derivative average-current dc polarogram can
be recorded with a fast response recorder in about 10 seconds. This is comparable to the time required to obtain a polarogram in oscilloscopic polarographic methods but provides a large-scale recording. More important
advantages, however, are the relative insignificance of the increased time
required per sample analysis in order to record replicate polarograms, the
advantages of measuring superimposed replicate polarograms recorded on
an XY recorder, and the greater signal-to-noise ratios obtained at faster
scan rates in derivative dc polarography (82).

3. Polarographic Cell

A third aspect of transducer design to obtain better polarographic results
is the design of the cell itself. Pertinent considerations include the following.

It has been observed that best reproducibility and highest signal-to-noise
ratio at high measuring sensitivity are obtained with well-sparged solutions
(83, 84). The inert, pure sparge gas (e.g. argon) must be led to the cell through
tubing that is impervious to atmospheric oxygen, e.g. through glass tubing.
After sparging, it is necessary but not sufficient to maintain the pressure of
the sparge gas overflow over the solution above atmospheric pressure. The
driving force for leakage of atmospheric oxygen back into the cell solution
is related to the great difference in oxygen partial pressure in these two
regions. To ensure good sparging to a very low dissolved oxygen concentration, the cell can be built so that it is well-sealed. The cell top can be machined
from Teflon and held tightly, with a gasket, against the glass cell bottom. As
a gasket, a ring of Apiezon sealing compound Q can be used. For the cells
used with the ORNL model Q-2942 drop time controller, a rubber boot
sealed to the cell top allows the DME to be freely moved in the sealed cell

(30). In one such cell, one of the glass sparge gas inlet tubes is terminated near the bottom of the cell for sparging through a 10 millimetre diameter medium porosity Pyrex stick filter. While the use of this frit reduces sparge time substantially, in some cases, it might complicate the cleaning of the cell between samples. It is possible, instead, to draw out the glass sparging tube to a fine point. The inlet sparge gas can be passed through a bubbler to saturate the sparge gas with the solvent prior to its introduction into the cell. This is particularly useful if the solvent is a volatile non-aqueous liquid. The sparge gas should exit from the cell through a water (or other solvent in use) trap to seal the cell against the entry of atmospheric oxygen. Gas flow meters can be placed in inlet and exit lines; equal readings indicate that the cell is well sealed. However, equal rates of bubbling in the inlet bubbler and the exit trap are usually a sufficient indication that the cell is well sealed. Precautions such as excellent sparging and precise solution temperature control are not necessary for less exacting work where highest sensitivity or precision is not needed. In some practical applications, sparging prior to the admission of mercury to an aerated solution could be appropriate.

When results having high reproducibility are needed, it can be of advantage to use a rather large cell. A large cell, used with the ORNL model Q-2942 drop time controller, has been described (30). The glass cell bottom is a high form weighing bottle, with a 40/12 taper top, and is 40 and 80 millimetres in, respectively, inside diameter and height. It is held up against the cell top by tension applied with an adjustable clamp underneath the cell bottom. The cell solution volume is ~50 millilitres and its upper surface is ~2·5 centimetres below the cell top. Under these conditions, DME movement, sparge gas overflow and mercury drops collecting on the cell bottom do not cause convective disturbance at the DME. Also, with such a large solution volume, concentration changes due to evaporation or addition of solvent during sparging are minimized. For less demanding analyses, a smaller cell that contains ~10 millilitres is used.

A small cell can be constructed very easily for use with the quasi-reference electrode, QRE. A cell that contained 0·3 millilitres was assembled, using a funnel with a conical shape as the cell bottom. The electrodes were a Sargent 2 to 5 second DME and two bright platinum wires, one as a counter electrode and one as the QRE. The wave shape obtained under these conditions is entirely conventional (85).

A cell has been described that contains a solution volume of 0·5 millilitres (86). This cell was used with an electromechanical drop time controller and the QRE. Uranium was extracted into an organic solvent and increased sensitivity (relative to the concentration in the sample prior to extraction) was obtained through the use of a high aqueous to organic volume ratio in the extraction step. Polarograms were obtained directly in the organic extract (after the addition of supporting electrolyte). Advantages of the QRE in this application include its small size and the absence of leakage of electrolyte from a true reference electrode and the absence of precipitation of KCl (or other salt used in the bridge) in the junction.

The cells that were designed for use with the ORNL controlled-potential differential dc polarograph have been described (87). These cells were designed so as to expedite the obtaining of highly reproducible currents in the two cells for differential polarographic modes. For example the counter and reference electrodes are each in a separate compartment, the cells are tightly sealed, equal sparge gas (high purity argon) flow rates are provided to the two cells, equal sparge times are used, the sparge gas is water saturated ahead of the cells, Smoler $90°$ DME's with an uncontrolled drop time of ~ 1 second ($\overline{m} \sim 2$ mg sec^{-1}, $h \sim 60$ cm) are used that are fabricated from a single piece of capillary so that mating orifices are employed, the cells are thermostatted by passing constant temperature ($25.00 \pm 0.05°C$) water through cell water jackets, and a substantial (~ 35 millilitres) solution volume is used.

Details of the electrode design and arrangement in a cell designed so that the potentiostat can minimize the uncompensated iR_{inner} were summarized in Section III-B-1 of this chapter (14). See Figures 10 and 11.

C. Circuit Sub-systems for Signal Processing and Transformations

As implied by their name, operational amplifiers were once used primarily to perform mathematical operations in systems for analog simulation or computation. Each amplifier, with associated input, feedback and bias networks, performed a mathematical operation such as addition or integration, primarily at dc and low frequencies. Since about 1955, the application has also become very widespread of operational amplifier sub-systems in the signal processing and data transformation element of measurement systems for chemical, biological, and physical research, chemical analysis and process monitoring. The use of operational amplifiers has the advantage that the design and construction of instrument circuitry sub-systems remains fairly simple whereas it would be relatively complex and expensive if the design and wiring of each amplifier were also required. These important practical advantages are extended greatly by recent technology, including printed circuit board construction and integrated circuit modules. Furthermore, if the operational amplifier sub-systems are designed with sufficient attention to detail, to biasing and to other application information provided by the more reliable manufacturer's data sheets for the particular amplifier, advanced performance and advantageous data transformations are easily obtained, and at a reasonable cost. Numerous choices now exist among amplifiers having excellent performance characteristics. Also, the commercial availability is of interest of unit operational sub-system modules such as active filters, exponential and logarithmic, comparator, and sample and hold. The present goal of design is to use optimum mixes of analog *and* digital functions and modules. For these reasons, a brief treatment of the topic of operational amplifier sub-systems was selected for Section III-C of this chapter. Representative early uses that have been made at ORNL of these sub-systems in instrumentation for chemical research and analyses have been described (9, 88, 89). A very substantial literature attests to the many uses and developments of operational amplifier sub-systems.

1. Operational Amplifier Sub-systems for Polarographs

(a) *Ideal performance and practical considerations.*   Operational amplifiers are designed to function over a frequency range that includes dc. The input may be single-ended or differential. The unit may be operated with negative or positive feedback or open loop.

Ideally, the open loop gain is high enough so that the summing point is held at an error voltage, e_x, that is very close to reference (e.g. ground) potential and is rolled off with frequency in a simple manner that results in a stable system. The effects of finite input current and voltage offsets, leakage and noise and of dc drift are negligible. Amplifier input and output impedances are respectively so high and so low and signal source impedance is so low that their magnitudes do not have to be considered in circuit analysis. The amplifier is not driven to its current or to its voltage limit or caused to latchup. Common mode rejection is very high for different input units. Stray reactances are considered to be of negligible importance in the circuit. The sub-system is assumed to operate in a linear fashion. The degree to which the above listed idealized characteristics are valid in a specific application of a specific amplifier depends strongly upon the specific performance characteristics of the amplifier that is selected or under consideration and upon the performance demands of the application. A functional block diagram of an idealized single-ended input negative feedback operational amplifier sub-system is shown in Figure 12.

In Figure 12 and in others that follow in Section III-C of this chapter, the reference is shown as ground. A ground reference is frequently used in instrumentation circuitry, but is entirely arbitrary. The reference bus can be floated (i.e. isolated) from chassis and/or earth ground or connected at only one point to prevent ground loops. In the idealized case, which with good design can be very closely approached in many practical applications, through node or loop equations, it is seen that the steady-state output, e_{out}, is equal to the negative of the input signal, e_{in}, times a transfer function. The transfer function is the ratio of the operational impedances, Z_{out}/Z_{in}. Tabulations of transfer functions and of corresponding input and output

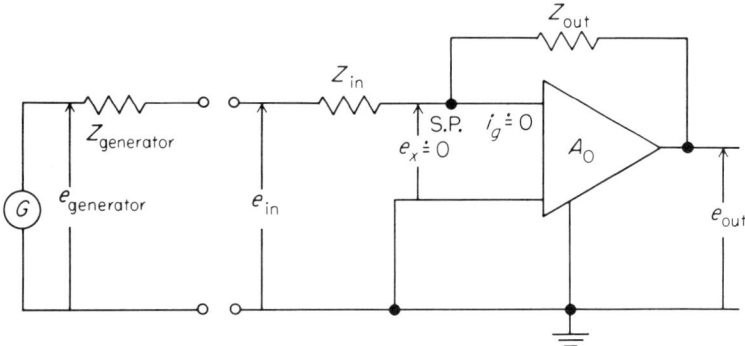

Fig. 12. Idealized single-ended input negative feedback operational
amplifier sub-system.

networks that are made up of lumped linear circuit elements are available in
handbooks. The operational impedance of a circuit element is equal to the
ratio of the Laplace transform of the voltage across the element to the
Laplace transform of the current through the element. Non-zero initial
conditions are written in as such in loop or nodal equations since they are
not included in the transfer function.

It is always a risk during design to consider circuit elements only in terms
of their ideal behaviour. Ideal performance or a sufficient approximation to
it is assured in practical circuits only when specific design steps and checks
are taken to establish that no more than insignificant departures occur from
the intended operating characteristics. Operational amplifier sub-systems
are no exception. In analysing these circuits, those facets of non-ideal
behaviour that are not negligible in the specific application must be taken
into account. The open loop gain is shown as A_0, the value at dc, in Figure 12.
Possibly the most significant thing that distinguishes even high quality real
from ideal operational amplifiers is the fact that the open loop gain is
vectorial rather than scalar. The magnitude not only varies with signal frequency but also according to signal level and output loading and current
and voltage demands; the phase shift is frequency dependent rather than
being always $-180°$. Also, the transient performance of the sub-system may
be significant, in addition to the steady-state performance. Circuit node and
loop equations, with operational calculus, predict both steady-state and
transient response. The effects upon the output voltage of normal variations
in open loop gain and in input offset voltage and input current magnitudes,
drifts and variations should be calculated, particularly for the lower limit
of the input voltage signal. In certain cases, the input and output networks
that are associated with the operational amplifier are modified from those
for the simple, ideal transformation in order that the output has more nearly
the intended properties. The addition of bounding and current and voltage

bias circuits are simple examples. For another example see Section III-C-3 of this chapter, active analog derivative computation. Other practical considerations include taking into account the effects upon performance of specific roll-off characteristics, finite source and amplifier input and output impedances, common mode rejection (especially in applications such as voltage followers), and latchup and limiting characteristics. As discussed in Section II-E of this chapter, construction practices and network component and power supply qualities are also important.

(b) *Specific uses.*   The performance limits of classical dc polarography (concentrations down to about 10^{-5} M, relative standard deviation, S, of 1–2 per cent, for replicate measurements and a resolution of $\Delta E_{1/2} = 9.19$ RT/nF for two successive regular waves of equal heights and n values overlapping by 1 per cent) have been substantially improved by modern instrumentation based upon operational amplifier sub-systems. At Oak Ridge National Laboratory (ORNL) a series of controlled-potential and derivative dc polarographs have been designed with operational amplifier sub-systems. (See Section III-H of this chapter.) By means of these simple circuits, information is fed out of the polarographs in forms that are more useful than undamped or RC damped regular polarograms. Through these sub-systems, better voltage sources, better current measuring, potentiostatic electrolysis, average or maximum current regular polarograms and first and second derivative polarograms are available.

Two sub-systems, the analog integrator and the active derivative computer, are described in greater detail, as representative examples, in Sections III-C-2 and III-C-3 of this chapter. Papers that describe the ORNL polarographs, referenced in Section III-H of this chapter, provide details about these and other polarographic sub-systems. Advantages that have been achieved through the use of these sub-systems are discussed in Sections III-D, III-E, III-F and III-G of this chapter. Also, a review of dc polarographic methods and instruments by Nürnberg and Wolff (90) includes discussions of advantages that accrue from design features such as electronic scan, potentiostatic electrolysis, parallel-T and peak follower damping, the use of a short-drop-time DME, and first- and higher-order derivative polarography (particularly when obtained by means of sharp cutoff, low-pass current-averaging filters followed by active analog derivative computers).

In Figure 13, a block diagram presents principal features of the ORNL model Q-1988-FES polarograph (91). This diagram is drawn for the Function switch in the Operate position. For clarity, the initial current compensation and the linear compensator circuits and the Test and the Cell Open functions are omitted from the block diagram. Although this polarograph could be described as a modification of the ORNL model Q-1988A polarograph, it is important to note that the additional circuitry is designed to take

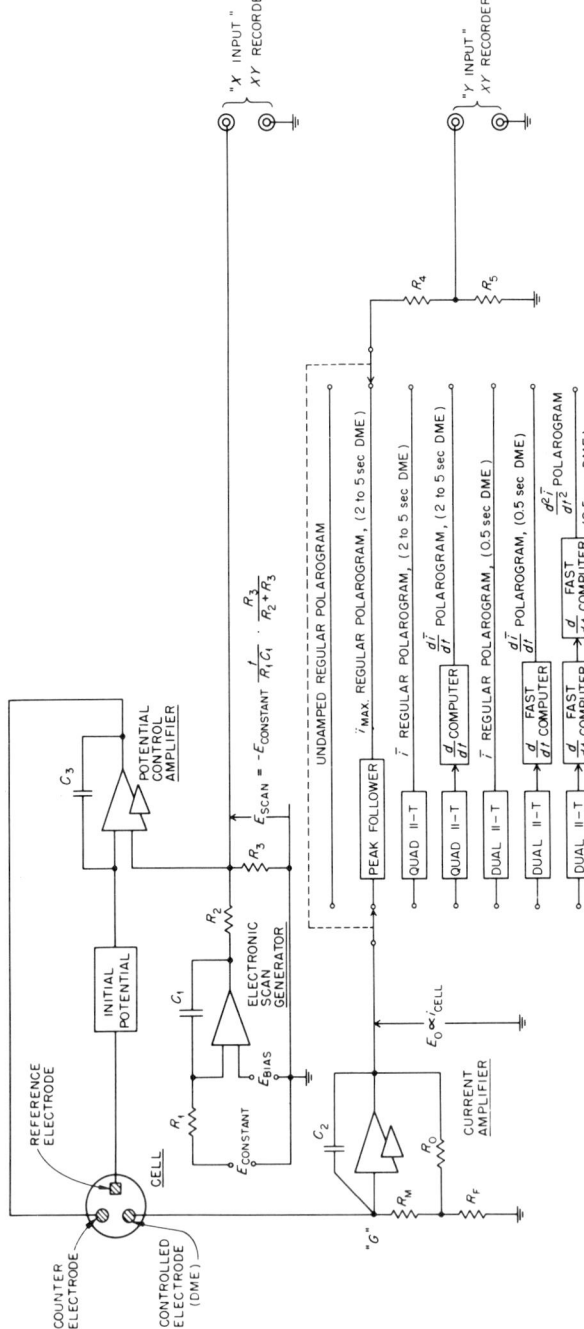

Fig. 13. Block diagram showing features of ORNL model Q-1988-FES polarograph. Reprinted from D. J. Fisher, W. L. Belew and M. T. Kelley, *Chem. Instrum.*, **1**, 181 (1968), Figure 1, by courtesy of Marcel Dekker, Inc.

increased advantage of the inherently superior sensitivity and resolution of derivative dc polarography. The types of dc polarograms that can be recorded, and the sub-systems that are respectively used to obtain them, are shown in the figure.

A block diagram of the ORNL controlled-potential differential dc polarograph is shown in Figure 14 (92). The filters (parallel-T current-averaging filters designed for use with 1 second DME's) eliminate the need to synchronize the dropping of the two electrodes and also provide average-current regular polarograms. The filters are terminated by voltage follower stages.

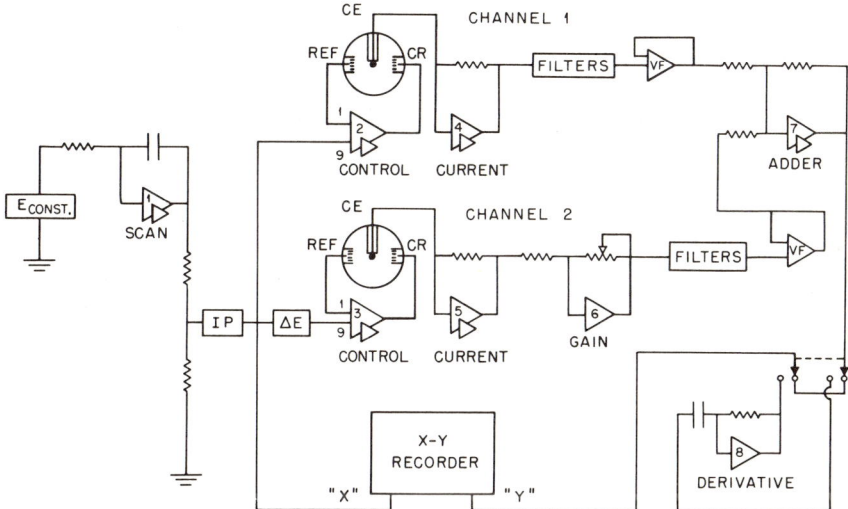

Fig. 14. Block diagram of ORNL controlled-potential differential dc polarograph. Reprinted from W. D. Shults, D. J. Fisher, H. C. Jones, M. T. Kelley and W. B. Schaap, *Z. Anal. Chem.*, **224**, 1 (1967), Figure 1, by permission from Springer-Verlag.

The gain sub-system in channel 2 (amplifier no. 6) provides for electrical matching of the characteristics of the two DME's and also inverts the polarity of channel 2. The ΔE circuit allows the potentials of the two DME's to be exactly matched or to be offset for the ΔE-differential mode. In this figure CE stands for 'controlled electrode', i.e. the working electrode, the DME and CR stands for 'counter electrode'. The output of either channel, (through switching, not shown) or its derivative, the differential output or its derivative may be recorded. Three differential polarography modes (ΔE—differential, comparative and subtractive) are provided.

A block diagram of the ORNL model Q-2792 polarograph is shown in Figure 15 (11). An adding type potentiostat is used. Through a voltage crossing detector (Scan Limit Detector) and associated relays, the scan may

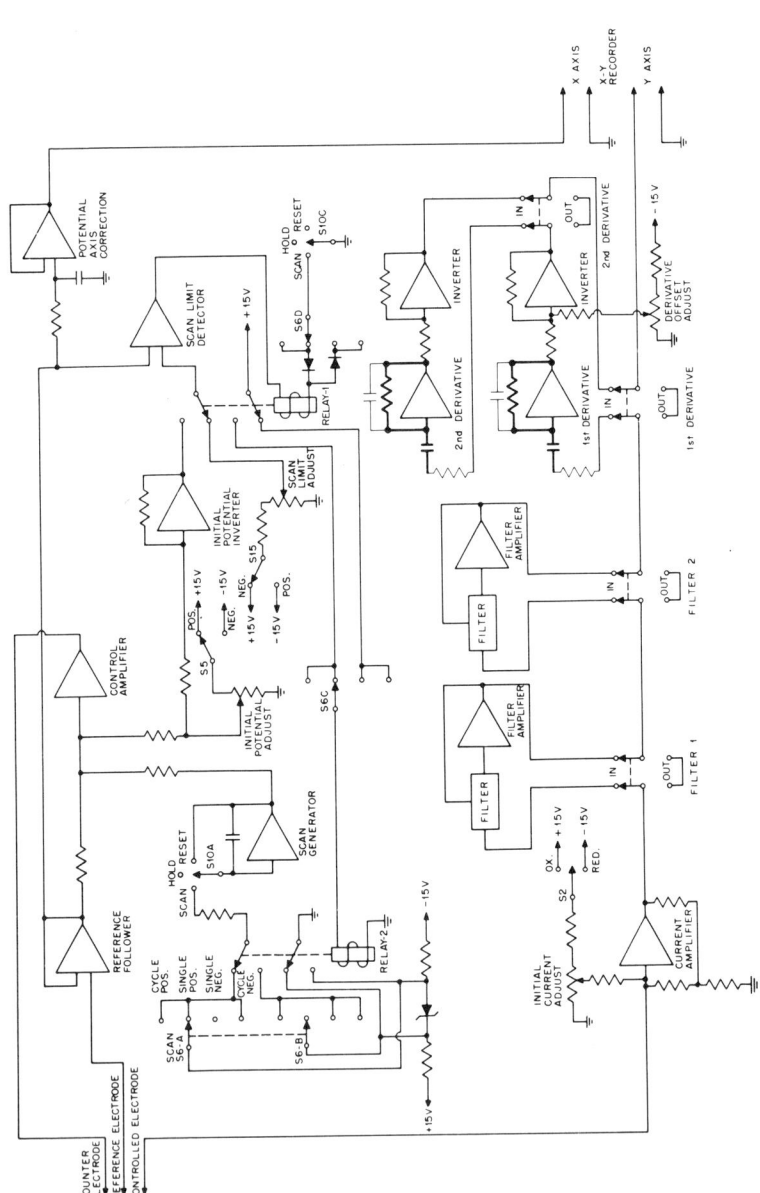

Fig. 15. Block diagram of ORNL model Q-2792 polarograph. Reprinted from H. C. Jones, W. L. Belew, R. W. Stelzner, T. R. Mueller and D. J. Fisher, *Anal. Chem.* **41**, 772 (1969); Figure 1. Copyright 1969 by the American Chemical Society. Reprinted by permission of the copyright owner.

be terminated automatically at a preselected voltage limit or the scan may be cycled continuously and linearly between this limit and the initial potential. Scan rates are stable to 0·1 per cent relative standard deviation. The optional potential axis correction sub-system introduces a time-lag in the drive to the X input of the XY recorder that is adjusted so that the peak potentials for first derivative dc polarograms appear at $E_{1/2}$ at all scan rates. Peak height and form are not changed. Cell Open and Test $-1\,\mu A/V$ functions are provided. The performances of the sub-systems and other design features of this polarograph have been described (11, 12).

2. Analog Integrator

Analog integrators, built with operational amplifiers, are used as sub-systems for polarography and for coulometry. The following discussion, and that in Section IV-C of this chapter, of the design of the analog integrator sub-system have been included, moreover, as a vehicle to point out the general requirement and importance of in-depth and specific design that includes practical considerations in operational amplifier sub-systems. It is not enough to consider only idealized properties and idealized performance in order to design satisfactory operational amplifier sub-systems. Moreover, the requirements of the specific applications must always be kept in mind because some trade-offs are required in obtaining a feasible optimization.

The block diagram of an analog integrator is shown in Figure 16. A capacitor, C, is used as the feedback element in a single ended input inverting

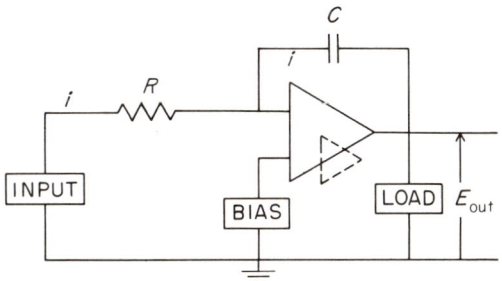

INPUT

Electrolysis Cell : $E_{out} = \dfrac{Q}{C} = \dfrac{i}{pC} = \dfrac{1}{C}\displaystyle\int^{t} i\,dt$

Constant Current : $E_{out} = \dfrac{i}{C}\cdot(\text{time}) = k\cdot(\text{time})$

Current integrator or linear-voltage-scan generator.

Fig. 16. Block diagram of analog integrator.

amplifier. Also shown on the figure are the equations relating output to input for two typical applications of analog integrators: Coulometer and scan generator. In normal operation, the output voltage, E_{out}, equals the voltage across the capacitor terminals because the operational amplifier holds its summing point very close to ground potential by means of negative feedback. The voltage at the summing point equals the output voltage divided by the open loop gain and so is of negligible magnitude relative to those of the input and output voltages. It is the input voltage at the amplifier, the error voltage. The voltage across a capacitor equals the charge stored, Q, divided by the capacitance, C. Since current cannot flow into the operational amplifier at the summing point, the amplifier must develop an output voltage such that the input current, i, flows through the feedback impedance. Thus, in the analog integrator, the input current is stored in C. A coulometer can be obtained by using an analog integrator to integrate all or a known fraction of the current at a working electrode in an electrolysis cell. This application is discussed in more detail in Section IV-C of this chapter. If a second resistor is added to the circuit, across the points labelled 'Input' in Figure 16, then it is effectively in parallel with R. A fraction of the total current is then integrated, in accordance with the relative conductances of the two resistors. Examples of block diagrams of controlled-potential coulometric titrators that include analog integrator coulometers are shown in Figures 29, 30 and 32. If a constant input voltage source is applied to R so that i is constant, then the output voltage is directly proportional to time. Because the summing point is very close to ground potential, i equals the constant input voltage divided by R. In this case, the analog integrator is a linear voltage scan generator (also called a ramp or a sweep generator). This application is frequently made in polarographic instruments. Examples of block diagrams of polarographs that include analog integrator scan generators are shown in Figures 13, 14 and 15. In Figure 13, the equation describing the generation of the scan voltage is also shown. As in the case with many operational amplifier sub-systems, not only does the operational amplifier activate passive input and feedback networks having the desired transfer function so that the desired mathematical operation upon the input signal is obtained, but also it delivers current to a load, without disturbing the transformation, so that the output voltage can be observed and used in subsequent portions of the system.

A well-designed analog integrator is an excellent voltage scan generator for polarography. Single or cyclic scans of either polarity or with cyclic polarity reversal and having a wide range of slopes are readily obtained. Cyclic operation can be obtained through the use of a voltage-crossing detector operational amplifier and relay sub-system as a periodic reset device. See Figure 15. Alternatively, by combining the analog integrator and digital logic sequence control (programming) circuitry, linear sweep (ramp) or triangular (cyclic scan) voltage sources can be obtained. Moreover, the digital logic circuitry can program the time of application of the voltage source to the cell relative to some feature of the experimental variable and it can program the timing of the measurement of the signal as well. In the digital logic sequence circuitry portion, integrated circuit digital logic modules are used rather than analog modules such as, for example, voltage-crossing detectors. Digital logic programming has the advantage that various steps including the initiation and the programming of the scan voltage and the measurement of the signal can be directly commanded by output from systems that include a digital computer. Such sub-systems are inherently appropriate for digital systems for electrochemistry, particularly those involving computer aided and directed automated analyses. See Section V of this chapter. The earlier use of multiturn potentiometers driven by synchronous motors to obtain a voltage scan generator has been supplanted by analog integrators: Scan sources of higher quality are obtained. Data showing that the output of an electronic scan generator is less noisy than that of a motor-driven multiturn potentiometer have been published (93). With analog integrators that utilize chopper stabilized high gain operational amplifiers, scan rates can be obtained that each vary 0·1 per cent or less. Since the magnitude including the peak height of a Nth derivative dc polarogram is directly proportional to the Nth power of the scan rate, the generation of low-noise, reproducible, scan rates is one of the requirements for quantitative derivative dc polarography.

The dashed triangle in Figure 16 implies that a chopper stabilized operational amplifier may be selected. These are more suitable for critical integrator applications due to having lower input offset voltage and input current magnitudes and drifts. A model having high open loop dc gain must be selected, particularly if fast rise time signals have to be accurately integrated. Non-chopper stabilized amplifiers can be selected that are satisfactory for some scan generator applications. Sufficient output current must be available for very fast scan rates. It is possible to select a solid-state operational amplifier whose characteristics are such that when used with attention to practical details, essentially ideal behaviour can be obtained in many applications of integrators.

The reduction of output drift to a negligible amount is a principal integrator design problem. Output drifts are variations with time and temperature from the output value that corresponds to the ideally transformed input signal. Drift can be caused by non-ideal behaviour (especially temperature-dependent input voltage and current offsets and drifts) of the operational amplifier itself, by spurious inputs from external sources, and by capacitor and other component imperfections and temperature coefficients. An amplifier must be selected that has gains that are very high at the frequencies of the components of the input signal so that the magnitude of the error signal to which the amplifier can hold its input summing point is very small. In practice, at least at a particular integrator operating temperature, drifts caused by small but finite amplifier input current and input offset voltage can be cancelled out by the introduction of bias compensation as indicated in Figure 16. The amplifier manufacturer's data sheets should include recommended voltage and current bias circuits. The usual method of adjusting the voltage offset bias is, while not applying an input signal, to adjust the external 'trim' potentiometer which is connected to well-regulated power supply voltages to a value such that the output subsequently remains sufficiently close to zero for the longest time of interest in the applications. Then, if required, the input current offset is biased out by adjusting the compensating current fed to the input. It is good design practice to arrange, when possible, that the input signal is large relative to the amplifier offset voltage and to stray input leakages. Another way to make the portion of the output caused by spurious inputs to be insignificant as compared to that corresponding to the input signal is through the application of good construction practices. Such practices provide needed insulation, guarding and shielding to all input connections and eliminate ground loops that could result in spurious inputs. Ground loop errors are usually caused because current from power supply and load connections or from other sub-systems is allowed to flow through the same ground wire that carries signal current. This results in applying an iR_{wire} ground loop voltage in series with the signal. Separate wires all the way to power supply common should be used for signal return, amplifier common pin and load return connections. It is normal practice to select a capacitor of very high quality. The absence of significant leakage at and within the capacitor can be established through applying a signal until a charge of about full scale value has been accumulated, removing the input, and then monitoring the output over the longest time of interest. If leakage is not significant, the output variation with time will not differ from the drift rate with no signal having been applied. The output must be as constant as is necessary in the particular application. Another imperfection of capacitors that can cause trouble in critical applications is dielectric absorption. After the output of an integrator has been read, or at the end of a voltage scan, the integrator is reset to zero by connecting a small resistance across the capacitor with a reset switch or relay until the capacitor has been discharged. In the absence of dielectric absorption, after the switch is reopened, the output of the integrator will remain at zero. Tests should be made to show that the length of time during which the reset switch is closed is long enough so that the capacitor is entirely discharged. As long as three minutes may be needed. Wire wound resistors of very high quality are selected for R. When a constant voltage source is applied to R to generate a linear voltage scan, a ramp, the source voltage must be sufficiently constant; sources of very high quality are available. Since any of the critical components can change or fail, it is prudent to periodically test the performance of the integrator when it is used in critical applications.

Practical considerations that are particularly pertinent to the especially critical integration or coulometer application are discussed in further detail in Section IV-C of this chapter. Also, detailed circuit diagrams and parts lists for instruments designed at ORNL are referenced in Sections III-H, ORNL DC Polarographs, and IV-D, ORNL Controlled-Potential Coulometric Titrators.

3. Derivative Transformation Methods

Certain polarographic methods that impose an ac signal on the dc scan voltage (e.g. square wave, pulse, radio frequency, faradaic rectification) produce a wave that closely matches the form of the corresponding calculated wave only when the amplitude of the ac is very small. To obtain better sensitivity instead of highest resolution, larger amplitudes are used. This is analogous to the practice at ORNL with polarographs models Q-1988-FES and Q-2792 and 0·5 second DME's of using scan rates of about 0·1 volt per minute when derivative wave form or highest resolution is of interest and of about 1 volt per minute when sensitivity is of particular interest. With these instruments, true first derivative dc wave forms can be recorded which is an important advantage in providing better communication (Section III-A-1 of this chapter) and ultimate resolution as well. On the other hand, the use of methods that can produce only approximate derivative wave forms is disadvantageous because of the difficulty of interpreting the pseudo derivative polarograms and the inability to obtain either true first derivative wave forms for ultimate high resolution or to obtain high sensitivity relative to that of regular dc polarography (see Section III-C-3-(a)).

The attainment of derivative dc polarograms of good form at scan rates up to about 0·1 volt per minute (first derivative has greatest utility when ultimate resolution is wanted) and a net advantage at scan rates of about 1 volt per minute of the dependence of the peak height of the Nth derivative upon the Nth power of the scan rate (and upon n^{N+1} times concentration) and of the inherent minimization of the effects of residual current is due to several factors. These include the use of 0·5 second drop times so that faster current-averaging sub-systems can be used, a potentiostat to facilitate a constant scan rate, a sensitive and stable current amplifier and practical operational amplifier time derivative sub-systems.

(a) Pseudo methods.   The label 'pseudo methods' is strong language. It is chosen because the results obtained by such methods have in the past, incorrectly, labelled derivative dc polarography itself rather than the method being used as insensitive or as producing recorded wave forms that are empirical. Results obtained by modern active analog methods with an operational amplifier sub-system (Section III-C-3-(b)) are quite different and are highly useful.

On two occasions and with two models of differential dc polarographs, the ΔE-differential method has been investigated at ORNL (15, 19). Two unsynchronized and uncontrolled drop time DME's are used for simplicity instead of synchronized drop time DME's but with current-averaging filters so that the beats between the DME's do not cause a problem. Two cells each containing portions of the same sample solution are used, one of the DME's is maintained at a potential that is a constant few millivolts, ΔE, different from the other, and the difference between the two average current regular dc polarograms is recorded. A block diagram of the latter of these two polarographs is shown in Figure 14. Shults and Schaap showed that for simple, polarographically reversible reactions, the maximum value of a ΔE-differential polarogram occurs at $\Delta E/2$ away from rather than at $E_{1/2}$ and that the maximum value approaches zero with ΔE approaching zero and approaches (but never attains) $\bar{i}_d$ for very large values of ΔE and is independent of scan rate (19). They found that ΔE values from 10 to 50 millivolts give optimum results but that still better sensitivity and resolution can be obtained with an operational amplifier practical active analog derivative computer sub-system. They also noted that the ΔE circuit is useful for balancing the DME potentials in the comparative and the subtractive modes of differential polarography.

Another pseudo method is the first of the analog methods described in the next section (Section III-C-3-(b)), the passive RC network method.

(b) *Analog methods.* The basis of analog methods of computing the time derivative of a voltage signal is the fundamental property of a capacitor that:

$$Q = CE \qquad (12)$$

where Q is the charge stored in a capacitor of capacitance C and E is the voltage developed across the capacitor (94). By differentiating equation (12) with respect to time, one obtains:

$$dQ/dt = i_c = C\,dE/dt \qquad (13)$$

Real capacitors are readily available whose properties, at frequencies of interest in derivative polarography, very closely approximate those of an ideal capacitor. Consequently, if the signal voltage, E, is impressed across a capacitor, C, the current passing through the capacitor, i_c, is directly proportional to dE/dt, as shown in equation (13).

The underlying problems of analog methods of obtaining derivative transformations of signals include the following. First, a means must be provided to impress all of the signal across C. Second, a means of measuring i_c is required. Third, signals of interest are inevitably accompanied by noise.

The principal constituents of the noise portion are commonly of higher frequency than those of the signal. Hence the signal-to-noise ratio (S/N) of a true time derivative of signal plus noise is typically worse (lower) than that of the original signal plus noise, and the higher-frequency components of the noise are particularly emphasized. Thus it is necessary to design a practical active analog derivative computer to have only sufficiently fast response to obtain the true derivative of the signal, which is the regular average current polarographic wave.

In polarographic applications, the derivative of the DME current with respect to DME potential is wanted. The derivative of the current with respect to time is directly proportional to the required transformation because the scan rate is constant.

Passive RC network. The simplest but least satisfactory method of implementing the principle of equation (13) is to use a passive RC network consisting of a resistor in series with a capacitor (94). The output signal, E_{out}, is the voltage across R. For a simple example, where $E = kt$,

$$E_{out} = RC(1 - e^{-t/RC})\frac{d}{dt}E \tag{14}$$

From equation (14) it is seen that there are three conflicting criteria for the choice of RC. First, to obtain a large output, RC should be large. Second, to obtain a true output, i.e. to have the transient term, $e^{-t/RC}$, small, RC should be small. Because of the transient term, E_{out} will lag behind and be unequal to the true derivative signal which is $RC(d/dt)E$. Also, the presence of R makes it impossible to impress all of E across C. Moreover, a finite current is needed by whatever device is used to measure the voltage developed across R. Furthermore, there is always noise present together with real E signals, and the derivatives of the noise components are greater than those of the input signal, i.e. the traditional disadvantage of the derivative transformation of a signal is the degradation of the signal-to-noise ratio, S/N. Hence, thirdly, to minimize the degradation of S/N, RC must be large. In the case of derivative dc polarography, it is therefore not possible to successfully compromise among these conflicting criteria with only a passive RC network and the inherent advantages of sensitivity and resolution will not be achieved. Hence, the passive RC analog method is a pseudo method. In their investigation of the use of a passive RC network, Lingane and Williams found that high quality derivative dc polarograms could not be obtained because it was necessary to use a large time constant RC (95). A typical first derivative dc polarogram obtained with an optimized compromise for RC but without selective current-averaging filtering, is shown in Figure 17 (15). It is apparent that the form is far from that of a true first derivative dc polarogram. A current-averaging filter system (instead of heavy

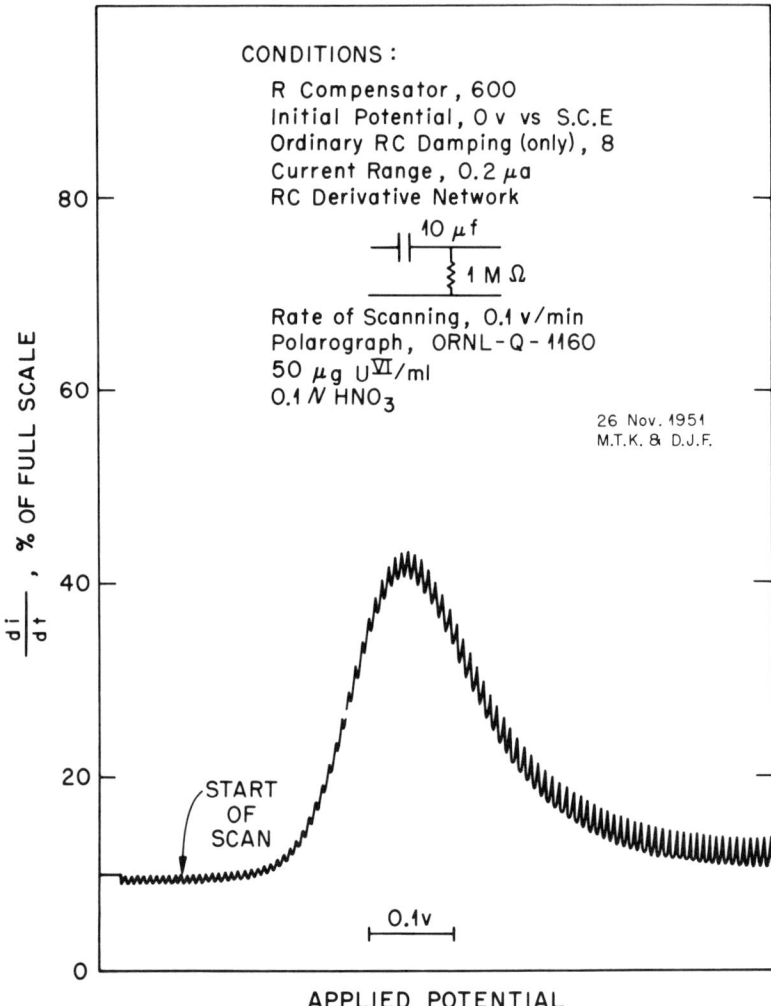

Fig. 17. Pseudo first derivative polarogram obtained with passive *RC* network. Reprinted from M. T. Kelley and D. J. Fisher, *Anal. Chem.*, **30**, 929 (1958), Figure 4. Copyright 1958 by the American Chemical Society. Reprinted by permission of the copyright owner.

RC damping that is otherwise required) results in some but not enough improvement (15).

Simple (ideal) operational amplifier derivative computer. Some improvement over the performance attainable with only a passive *RC* network is possible by use of a simple (ideal) operational amplifier derivative computer, consisting of a capacitor and a resistor in series and an operational amplifier

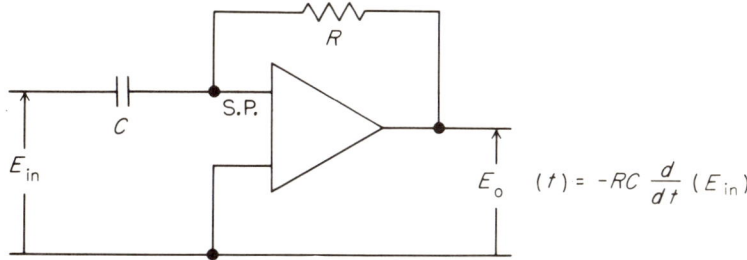

Fig. 18. Idealized active derivative computer.

(94). This sub-system is shown in Figure 18. The common connection of C and R is connected to the summing point (S.P.) of the amplifier and the other terminal of R is connected to the output terminal of the amplifier. The input signal, E_{in}, is connected from ground to the free terminal of C and the output, E_0, is the voltage from ground to the output terminal of the amplifier. By negative feedback, the amplifier holds its summing point very close to ground potential whereby all of the input signal, E_{in}, is impressed across the capacitor; also, by holding the summing point very close to ground potential, the amplifier controls its output voltage to a value such that all of the current passing through the capacitor as a result of the impressed signal, i_c, flows through R. The value of R can be high in order to obtain high sensitivity; it must be high enough to prevent loading of the output stage of the amplifier, but since all of the amplifier input noise current (offset and leakage currents) will also be driven through R, the value must not be too high to prevent degradation of S/N. The output voltage of the amplifier, E_0, is:

$$E_0 = -RC\frac{d}{dt}E_{in} \qquad (15)$$

which shows that E_0 is a true derivative of the input signal, E_{in}. The advantages of the simple, ideal, operational amplifier derivative computer are as follows. There is no time-lag because all of the input signal is impressed across C. The value of RC may be high in order to obtain high sensitivity without the introduction of time-lag. Another advantage is that a useful load current may be drawn from the amplifier without affecting the accuracy of the derivative computation. However, the disadvantage of this system is that it does not prevent the degradation of S/N because noise originates ahead of the computer and is fed in along with E_{in}, the input signal of interest.

Practical active derivative computer. The practical active derivative computer differs from the simple, ideal operational amplifier derivative computer in that the input network consists of a resistor, R_1, and a capacitor, C_1, in series and the feedback network consists of a resistor, R_2, and a capacitor, C_2, in parallel (94). Usually, but not necessarily, $R_1C_1 = R_2C_2$, and the cut-off frequency is:

$$f_{co} = 1/2\pi R_1 C_1 = 1/2\pi R_2 C_2 \tag{16}$$

The output voltage, $E_{out}(p)$, is:

$$E_{out}(p) = -\frac{R_2 C_1 p E_{in}(p)}{(1 + pR_1 C_1)(1 + pR_2 C_2)} \tag{17}$$

where p is the Laplace transform variable equivalent here to the time derivative operator. For input signals, $E_{in}(p)$, having frequencies $>f_{co}$, the output of the computer decreases from the true derivative of this input signal at 12 db per octave. For input signals having frequencies $\ll f_{co}$, the output of the computer, $E_{out}(t)$, is:

$$E_{out}(t) = -R_2 C_1 \frac{d}{dt} E_{in}(t) \tag{18}$$

The computer output is the true (within ~ 1 per cent.) derivative of the input signal from dc to $0.1\, f_{co}$. A block diagram of this sub-system is shown in Figure 19.

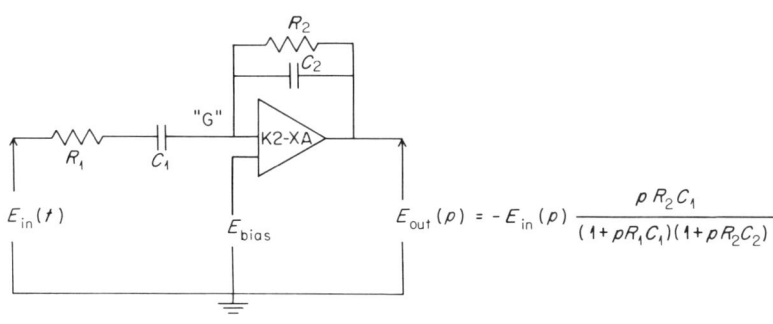

Fig. 19. Practical active derivative computer.

With this sub-system, derivative dc polarographic waves of excellent wave form (and thus of ultimate, high resolution) or of excellent, high signal-to-noise ratio (and thus of high sensitivity) can be recorded, as discussed in Sections III-E and III-F of this chapter. The basis of analog methods has been used in a way that solves the underlying problems. In a computer designed for derivative dc polarography, the value of f_{co} is chosen such that an accurate time derivative of the average current polarographic wave is recorded at lower scan rates and such that S/N is excellent at faster scan rates. For the slow computer designed for use with 2 to 5 second DME's and the quadruple parallel-T current-averaging assembly, f_{co} is equal to $1/4\pi$ or 0.08 Hz. In the fast first and second derivative computers in the Q-1988-FES polarograph for use with faster scan rates, 0.5 second DME's and the dual parallel-T current-averaging assembly, f_{co} is increased ten-fold (96). The numerical value of the transfer-function proportionality (gain) factor, R_2C_1, is the same in the slow and fast computers. The derivation of the calibration of first and second derivative scales has been described (97). More details concerning the choice of f_{co} are available (98). Stephens and Harrar have described the criteria that they used to choose f_{co} for derivative computers for fast-sweep polarography (99).

Criteria for and methods of testing the form of derivative dc polarograms are given in Sections III-A-2-(d) and III-A-2-(e) of this chapter. The fact that excellent first derivative wave forms can be recorded through the use of active practical operational amplifier analog sub-systems is shown by data in Table I, Table III, and Figure 7. The close agreement with the calculated form, seen in Figure 7, is in marked contrast to the wave form obtained by a passive RC network which is shown in Figure 17.

D. Advantages of Potentiostatic (Controlled-potential) Electrolysis

The advantages of potentiostatic (also called controlled-potential) electrolysis in polarography may be summarized as follows. A potentiostat eliminates the need to pass the cell current through a reference electrode and its salt bridge and junction so that a reference electrode may be used without the introduction of iR loss. Also, the current amplifier portion of the potentiostat prevents the iR drop across a current-measuring resistor from being seen by the DME as well as providing very sensitive and stable current measurement. The resistance of the mercury thread in the DME does not ordinarily result in an uncompensated iR loss of significant magnitude. In cases where this loss across the entire length is significant, the DME is designed so that the connection can be made near the orifice. It is important to note that a properly functioning potentiostat can only control the potential of the DME relative to that which is established (or seen) by the tip of the reference electrode (or Luggin probe) in solution. Even with the usual horizontal orifice DME, the tip can be located in a manner so that iR_{bulk} is compensated for by the potentiostat. The conditions, especially the specific resistance of the solution and the distance of the tip from the mercury surface in drop radii, under which iR_{inner} is of significant magnitude can be estimated from the Ilkovič equation for the inner resistance of a DME. Moreover, in conjunction with electrode design, e.g. the use of a Smoler 90° vertical orifice DME having a tapered tip with a Luggin probe or a QRE as described in Section III-B-1 and illustrated in Figures 10 and 11 of this chapter, iR_{inner} loss can be reduced to negligible magnitudes except in very high specific resistance solutions or at unusually high DME currents (14). Criteria and tests have been given in Sections III-A-2-(d) and III-A-2-(e) of this chapter for establishing or verifying the conditions under which iR losses and circuit time-lags have a negligible effect upon the forms of the recorded regular and derivative dc polarograms. In such situations, the minimization of iR loss by the potentiostat has contributed to the attainment of a constant effective scan rate which is one of the necessary conditions for obtaining derivative dc polarograms of theoretical form. Schaap and McKinney have demonstrated another polarographic advantage of a potentiostat : for quantitative purposes, polarographic analyses can in some cases be extended upward to at least $0 \cdot 1$ M concentrations of electroactive species (32). Values of i_d/C remained constant from 10^{-3} to 10^{-1} M. Of great practical importance is the fact that a potentiostat enables the use of small reference electrodes or small high resistance reference electrodes (6), or poorly poised reference electrodes (35, 100), or the use of a quasi-reference electrode, QRE (35). With a potentiostat, a reference electrode only needs to have an exchange current that is sufficient to prevent significant polarization by the potentiostatic transient and steady state input current rather than by the very much larger cell current in order to function as a true reference electrode. At one extreme is the reversible reference electrode that does not polarize significantly when even the full cell current is passed through it and at the other extreme is the QRE which has an undefined but stable potential when only very low currents are passed through it. The intermediate case is where the potential determining reaction at the electrode has a fairly low exchange current so that the electrode would be severely polarized by the passage of all of the cell current through it but so that it can sustain the very small current demanded by a potentiostat without significant polarization. This can be reason enough for the use of a potentiostat. Reference electrodes that can withstand the full cell current are not available for some practical applications and in other cases, it is difficult or impossible to use large electrodes. The use of a Luggin probe is possible with a potentiostat even though it is small and also may have very high resistance. The practical advantages of the use of a QRE have been illustrated elsewhere (35, 101–107). Relatively dilute supporting electrolyte can be used for trace-level analyses even though the resistance of the solutions is high; the impurities in the supporting electrolyte have thereby a lower concentration in the prepared solutions (108). In addition, potentiostatic polarographs facilitate polarography in conjunction with non-aqueous solvents, chromatography and solvent extraction. If the solutions have a very high specific resistance, it is advantageous to include a voltage follower in the potentiostat so that transient charging currents are not drawn through the reference electrode (109). If the application involves the measurement of concentration, it may not be necessary to utilize electrode design to minimize uncompensated iR_{DME} and iR_{inner} loss since useful concentration measurements can be made even if the forms of the polarograms are affected by iR loss. The fact that true reference electrodes that can withstand the full cell current are not needed can be a particularly advantageous feature of the potentiostat in non-aqueous solutions. It can be very useful to utilize solvent extraction to separate from interferences and/or to concentrate a reducible substance, particularly if the polarographic determination can be done directly in the extract, after addition of supporting electrolyte without substantial dilution of the extract. Some early polarographic work with solvent extracts, done without the benefits of potentiostatic electrolysis, resulted in polarographic methods for the determination of uranium after its separation by solvent extraction from ions that interfere and/or from hazardous radioactivity but the best results were not obtained by measurements directly in the extract; an ~ 20-fold dilution step was used (110). With potentiostatic polarographs, a substantial concentration advantage can be realized in addition to the selectivity advantage of extraction (111). A method for the determination of trace quantities of uranium by controlled-potential dc polarography in a tri-n-octylphosphine oxide extract has been described (102). Through the use of a cell that contained 0·5 milliliters and the QRE, it was possible to obtain up to a 250-fold concentration advantage so that the detection limit relative to the original aqueous sample is increased accordingly (103). Also, potentiostatic polarography facilitates investigations of substances that are insoluble in water or which cannot be determined polarographically in aqueous solution because of interferences or interactions due to the presence of water but which undergo useful electrochemical reactions in non-aqueous solvents.

E. Advantages of Use of Current-averaging and Active Analog Derivative Sub-systems

Although advantages of the use of current-averaging and active analog (practical operational amplifier) derivative sub-systems have been briefly mentioned at appropriate places in earlier parts of this chapter, it is of value to summarize these advantages herewith. The principal advantages are improvements in measurement and interpretation, in resolution, and in sensitivity.

1. Measurement and Interpretation

The improved measurement and the clearer and more straightforward interpretation of regular and derivative dc polarograms that are made

possible by the use of current-averaging and active analog derivative sub-
systems substantially contribute to the provision of good communication.
Another aspect is the relative ease of measurement of regular and of derivative
dc polarograms.

Less time and effort is required to measure and to interpret average-current
regular dc polarograms than non-average-current dc polarograms. Even at
only $2 \times 10^{-5}\, M$, the advantages of a current-averaging filter can already
be seen, as illustrated in Figure 20. In the unfiltered regular dc polarogram,
the fact that the charging current changes sign at the electrocapillary
maximum is evident. At potentials that are less negative (to the left), the
average values of the net (residual plus faradaic) current–time curves lie
near the tops. At more negative potentials (to the right), the average values
lie near the bottoms. In the latter region, the polarities of charging and
faradaic currents are the same. The current-averaging filter produces an
average-current regular dc polarogram which enables the improved measurement of $\bar{i}_d$ and which constitutes a satisfactory input for the active practical
analog derivative computer. The average-current polarograms for the
supporting electrolyte (labelied residual current in the figure) and for
cadmium are superimposed. It is seen that the extrapolated residual current,
found by linearly extrapolating the foot of the wave, is not a proper correction.
Moreover, the limiting part of the average current regular dc polarogram is
parallel to the true average residual current. The current-averaging filter
sub-system knows where to draw the line, as it were, very reproducibly, and
at an interpretable position. The use of this sub-system greatly simplifies
the measurement of dc polarograms and it is particularly needed at and
below $10^{-5}\, M$ (112). Although this method of correcting for residual current
may not be exact in every situation, it is certainly better than drawing lines
through some arbitrary per cent of recorded current–time curves and through
the foot of the wave. In Figure 20, the first and last parts of the first derivative
dc polarogram are of unequal magnitude because the foot of the regular
polarogram is not parallel to the limiting part. The error caused by measuring
peak height, $(di/dt)_{\max}$, relative to the extrapolated (dashed) line rather than
relative to the actual derivative of the residual current (solid line) is less than
that introduced by extrapolation on the regular dc polarogram. However,
at substantially higher measuring sensitivities, it is essential to measure first
derivative peak height relative to the superimposed derivative for the
supporting electrolyte alone (113). (Also shown later, in Figures 22 and 28
of this chapter.) This is made possible by these two sub-systems. The fact that
the derivative residual curves are more reproducible than the regular residual
curves is an important advantage of the use of these sub-systems since this
constitutes a sensitivity (signal-to-noise ratio) advantage.

With scan rates for which the forms of the recorded polarograms obey the
respective wave form equations, the measurement and interpretation of

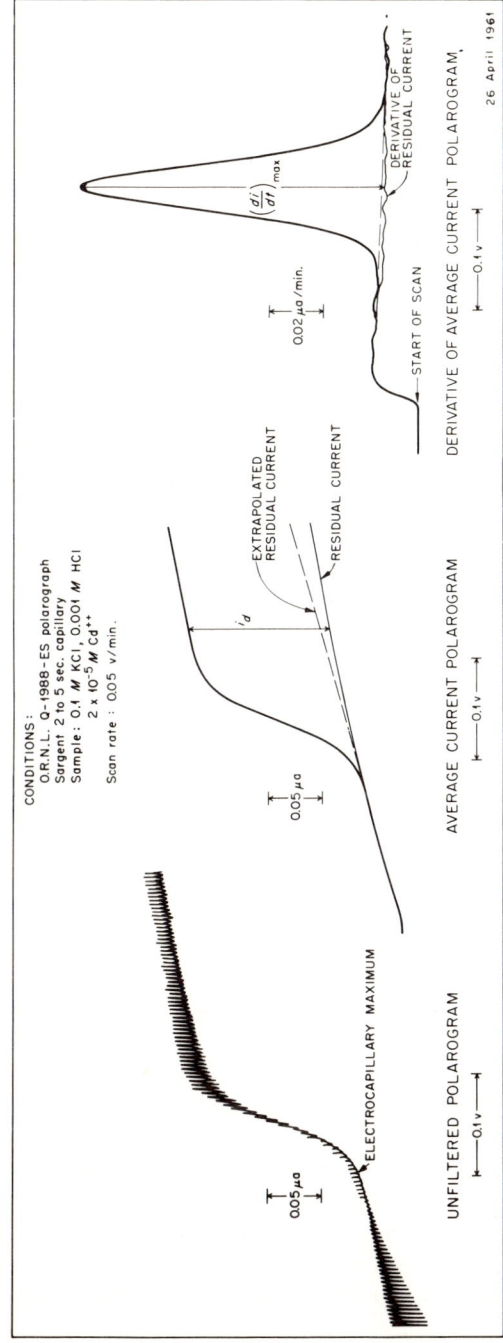

Fig. 20. Advantages of current-averaging filters and of active practical derivative computers in measurement of low concentraion polarograms. Reprinted from: D. J. Fisher, W. L. Belew and M. T. Kelley, pp. 89–134 *in* G. J. Hills, Ed., *Polarography 1964*, Vol. 1, Macmillan, London, 1966, Figure 1 by permission of The Polarographic Society and Macmillan & Co. Ltd.

recorded average-current regular and first and second derivative dc polarograms is much simpler than in the case of regular polarograms that are
distorted by non-linear damping and in the case of pseudo derivative
polarograms obtained by passive RC networks or by ΔE-differential methods.
Under these conditions, high ultimate resolution is obtained by first derivative
dc polarography, as shown in Figure 7. In all cases, interpretations are direct
implications of the respective wave form equations. Data in Table III show
that n values can be obtained from the width at half peak height, $W_{1/2}$, of
first derivative dc polarograms. Data in Table I show that the ratio of first
derivative peak heights to i_d is as calculated. Figure 21 further illustrates the
direct interpretability of polarograms obtained with these sub-systems. The

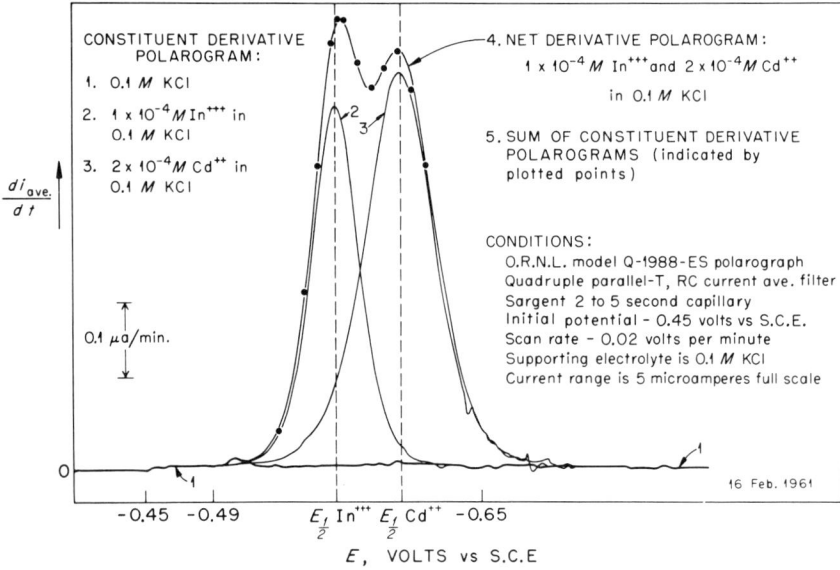

Fig. 21. Comparison of net derivative polarogram with sum of constituent derivative polarograms. Reprinted from: D. J. Fisher, W. L. Belew and M. T. Kelley, pp. 89–134 in G. J. Hills, Ed.,
Polarography 1964, Vol. 1, Macmillan, London, 1966, Figure 9, by permission of The Polarographic Society and Macmillan & Co. Ltd.

observed constituent polarograms (supporting electrolyte, number 1; for
indium, number 2; for cadmium, number 3) and the net (i.e. the as observed
whole, free of spurious effects) polarogram (both indium and cadmium in
the supporting electrolyte, number 4) are all shown. First, it is seen that the
algebraic sum of the constituent polarograms (plotted points, number 5)
fit right on the observed net polarogram. Secondly, it is seen that the shift to
the right of the voltage at which the first peak occurs in the net polarogram
from the $E_{1/2}$ for the more easily reduced constituent is a geometric effect

(114). This shift is not caused by an imperfection of the circuitry. Polarograms that overlap should therefore be measured at chosen potentials rather than at net peak or summit potentials. Thirdly, the enhancement of the two net peak heights is caused by the overlap of the constituent peaks, to the degree predicted and described by the wave form equations, and is not due to circuitry imperfections. It is possible to measure the concentration of each constituent in a sample in such cases by the use of simultaneous equations whose proportionality constants are evaluated through successive standard additions to the sample (115). For this purpose, first derivatives are better than regular dc polarograms (115). (See Section V-D of this chapter.)

The relative ease of measurement of regular and first and second derivative dc polarograms recorded at high measuring sensitivity is of practical interest. The signal-to-noise ratio (defined in Section III-B-2 of this chapter) rather than the ratio of the signal to the residual component is the significant determinant of the sensitivity that can be obtained in all types of polarographic measurements. However, the latter ratio does have an effect upon the relative ease of measurement of regular and derivative dc polarograms recorded at high sensitivity. Data have been reported that were obtained with the ORNL model Q-1988-FES polarograph that are measured values of the ratio of i_d to the average value of the residual current at $E_{1/2}$ (regular dc polarograms) and of the ratios of the first and second derivative peak values to the corresponding residual values for concentrations of indium ranging from 1×10^{-6} to 5×10^{-4} M (116). At each concentration of indium, the ratios are five-fold higher for first derivative than for regular dc polarography, twenty-fold higher for second derivative than for first derivative dc polarography, and one hundred-fold higher for second derivative than for regular dc polarography. The portion of the residual current that is linear with E becomes a constant offset in first derivative dc polarography and has a value of zero in second derivative dc polarography as has been mentioned by Nürnberg and Wolff (90). For the reason given in the second sentence of this paragraph, this does *not* mean that, for example, second derivative dc polarography is one hundred times more sensitive than regular dc polarography. The relatively greater ease of measurement of derivative dc polarograms (especially, of second derivative dc polarograms measured peak to peak) than of regular dc polarograms recorded at high sensitivity is shown in Figure 22. Regular wave heights and derivative peak heights for sample polarograms are most accurately measured relative to the corresponding blank polarograms for the supporting electrolyte alone, as illustrated in Figure 20. With regular and with first derivative dc polarograms, the sample and blank polarograms have to be superimposed for measurement. Regular, first derivative, and second derivative dc polarograms of 5×10^{-6} M indium (with replicate sample and blank polarograms superimposed) are shown in Figure 22. It was necessary (by trial, with the recorder pen lifted) to

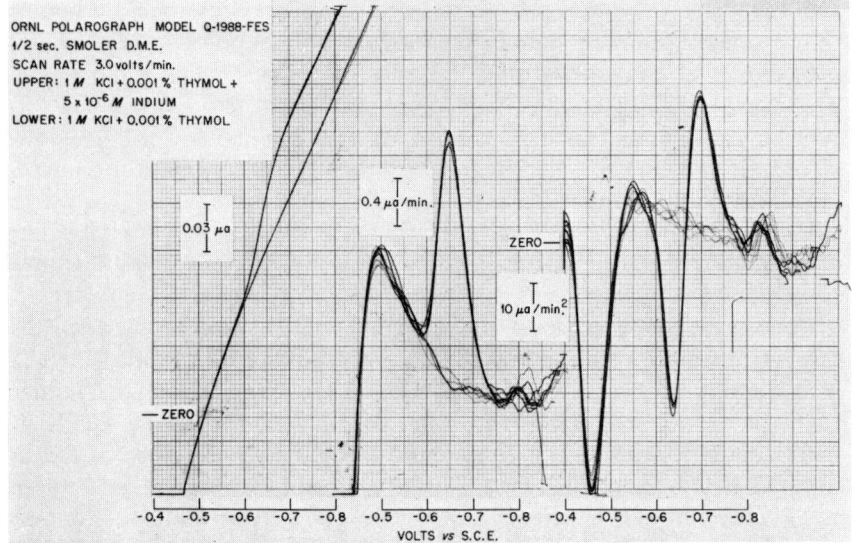

Fig. 22. Regular, first derivative and second derivative polarograms of indium, $5 \times 10^{-6} M$. Reprinted from D. J. Fisher, W. L. Belew and M. T. Kelley, *Chem. Instrum.*, **1**, 181 (1968), Figure 6, by courtesy of Marcel Dekker, Inc.

set the zero value for the first derivative polarograms off of scale so that the sensitive scale could be used to record the derivative waves. On the other hand, the zero position for second derivative polarograms of samples and blanks can be at mid-scale even on very sensitive ranges, as shown in Figure 22. It can readily be seen that the reproducibilities of replicate polarograms are very good from the fact that three regular and five first and five second derivative sample and blank polarograms are superimposed in Figure 22.

In subtractive mode differential dc (22) and in first and second derivative dc polarography, measurement problems caused by the presence of residual current along with faradaic current are minimized further than through the use of only a current-averaging system in regular dc polarography.

When resolution rather than sensitivity is of particular interest, not only is the resolution greater but overlapped first derivative dc polarograms are easier to measure than overlapped regular dc polarograms. The attainment of these features is another advantage of the use of current-averaging and active analog derivative sub-systems. Frisque, Meloche and Shain have shown that overlapped regular dc polarograms can be resolved mathematically and have given a criterion for measuring with minimum error (117). The mathematical resolution of overlapped first derivative dc polarograms has also been demonstrated (115). Before and after standard additions, net wave magnitudes are measured at a chosen pair of potentials, one or more of which is at or near a $E_{1/2}$ value. The error caused by a finite uncertainty in selecting the positions on the potential axis of the recorded polarogram is less in the case of first derivative polarography, as shown in Table IV. The rate of change of the magnitude of a first derivative wave with potential is relatively low near $E_{1/2}$ and is zero at $E_{1/2}$ but the rate of change of the magnitude of a regular dc wave with potential is relatively large near $E_{1/2}$ and is maximum at $E_{1/2}$. This is another example of a practical interpretation that can be made directly from the respective wave form equations, due to the use of these sub-systems.

TABLE IV

Relative Error of Mathematical Resolution of Overlapped Regular and of Overlapped First Derivative dc Polarograms

Shift from $E_{1/2}$ $\pm$ mV	Change of magnitude, $\pm$ %					
	Regular wave n			First derivative wave n		
	1	2	3	1	2	3
2	3·88	7·76	11·6	0·15	0·60	1·35
5	9·70	19·2	28·4	0·94	3·69	8·06

Reprinted with corrections from: D. J. Fisher, W. L. Belew and M. T. Kelley, pp. 89–134 in G. J. Hills, Ed., *Polarography 1964*, Vol. 1, Macmillan, London, 1966, Table 3, by permission of The Polarographic Society and Macmillan & Co. Ltd.

It is important to point out that, when highest resolution is not concurrently needed, it is not necessary or particularly desirable to choose a scan rate low enough so that the recorded forms agree closely with the calculated ideal forms. With rapid scan rates, e.g. 1 or 3 volts per minute, derivative polarograms recorded with a 0·5 second DME, the dual or the active current-averaging filter, and the fast derivative computers are more useful than those recorded at slow scan rates for concentration measurement; the more easily measured second derivative polarograms are particularly useful for measurements at very high sensitivity. The active practical analog derivative circuits prevent the loss of sensitivity, relative to that of regular dc polarography, associated with the use of passive RC networks or ΔE-differential methods and enable the realization of the inherent resolution advantage of first derivative dc polarography.

2. Resolution

The advantages of the use of current-averaging and active analog derivative sub-systems, from the standpoint of resolution, may be summarized as follows. In applications in which high resolution of adjacent polarographic waves is needed, first derivative dc polarography is of particular advantage (118). If instrumental conditions are selected that result in the recording of first derivative waves of good form, the attained resolution is the same as that obtained under optimum conditions in square-wave, pulse and sinusoidal ac methods of polarography (119). This high ultimate resolution is described by the fact that (simple, polarographically reversible process) the full peak width at half peak height, $W_{1/2}$, of a first derivative dc polarogram (and these other types of polarograms as well) in millivolts is:

$$W_{1/2} = \frac{2000RT}{nF} \cosh^{-1} 3$$

which at 25°C

$$= \frac{90 \cdot 5}{n}$$

The relative resolutions of regular, first derivative, and second derivative dc polarography are shown in Figure 23.

The required separations in the adjacent $E_{1/2}$ values are illustrated for the case where the i_d values are equal and the n values are each equal to 2 and the second polarogram overlaps (adds to the measured height or peak of) the first by 1 per cent. The resolution of first derivative dc polarography is substantially better than that of regular dc polarography. The same data are shown in tabular form in Table V (119). Additional examples of the relative resolutions of regular and first derivative dc polarograms have been given

TABLE V
Relative Resolution of Regular and of First and Second Derivative dc Polarography (Two Successive Waves of Equal n, Equal Height, 1 per cent Overlap)

Type of dc polarogram	$\Delta E_{1/2}$	
	Volts	Millivolts at 25°C
Regular	$\dfrac{9 \cdot 19RT}{nF}$	$\dfrac{236}{n}$
First derivative	$\dfrac{5 \cdot 99RT}{nF}$	$\dfrac{154}{n}$
Second derivative	$\dfrac{5 \cdot 58RT}{nF}$	$\dfrac{143}{n}$

Reprinted from: D. J. Fisher, W. L. Belew and M. T. Kelley, pp. 89–134, *in* G. J. Hills, Ed., *Polarography 1964*, Vol. 1, Macmillan, London, 1966, Table 1, by permission of The Polarographic Society and Macmillan & Co. Ltd.

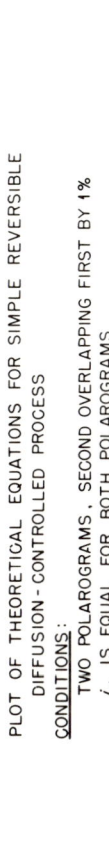

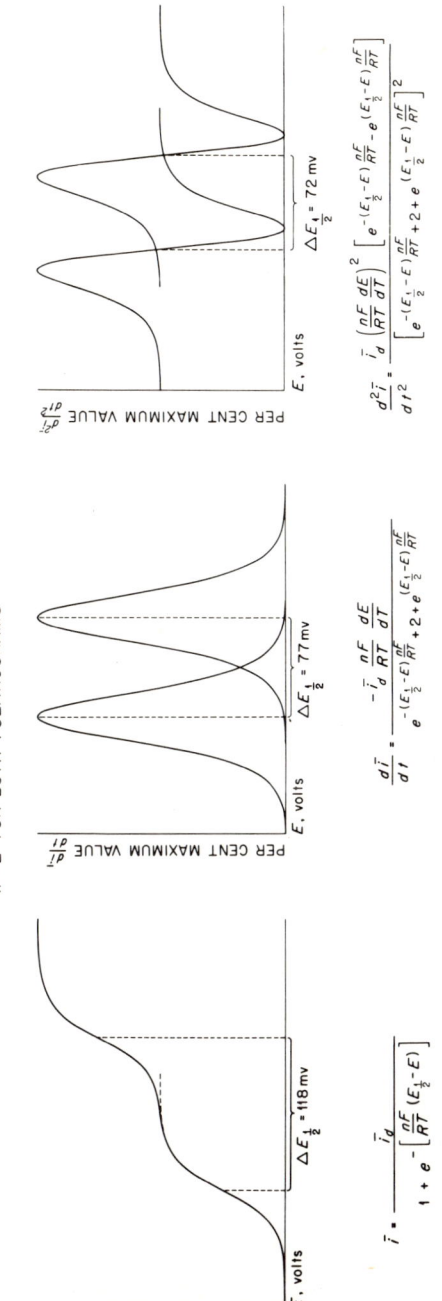

PLOT OF THEORETICAL EQUATIONS FOR SIMPLE REVERSIBLE
DIFFUSION-CONTROLLED PROCESS

CONDITIONS:

TWO POLAROGRAMS, SECOND OVERLAPPING FIRST BY 1%

i_d IS EQUAL FOR BOTH POLAROGRAMS

$n = 2$ FOR BOTH POLAROGRAMS

$$\bar{i} = \frac{i_d}{1 + e^{-\left[\frac{nF}{RT}(E_\frac{1}{2} - E)\right]}}$$

$$\frac{d\bar{i}}{dI} = \frac{-i_d \frac{nF}{RT} \frac{dE}{dT}}{e^{-(E_\frac{1}{2} - E)\frac{nF}{RT}} + 2 + e^{(E_\frac{1}{2} - E)\frac{nF}{RT}}}$$

$$\frac{d^2\bar{i}}{dt^2} = \frac{\bar{i}_d \left(\frac{nF}{RT} \frac{dE}{dT}\right)^2 \left[e^{-(E_1 - E)\frac{nF}{RT}} - e^{(E_\frac{1}{2} - E)\frac{nF}{RT}}\right]}{\left[e^{-(E_1 - E)\frac{nF}{RT}} + 2 + e^{(E_\frac{1}{2} - E)\frac{nF}{RT}}\right]^2}$$

Fig. 23. Relative resolution of regular, first derivative and second derivative polarograms. Reprinted from: D. J. Fisher, W. L. Belew and M. T. Kelley, pp. 89–134 in G. J. Hills, Ed., *Polarography 1964*, Vol. 1, Macmillan, London, 1966, Figure 4, by permission of The Polarographic Society and Macmillan & Co. Ltd.

by Müller (120). Even in the case where adjacent constituent waves (no interactions between primary and/or produced species) are so severely overlapped that the net first derivative dc polarogram appears to be a single broad
peak, the net wave can be resolved mathematically by simultaneous equations and the proportionality constants can be evaluated by standard additions. This has been demonstrated with active analog derivative computers
(see Section III-E-1 of this chapter). Digital dc polarography can provide
further refinements (see Section V-D of this chapter). At sufficiently but
inconveniently slow scan rates, it is possible to record first derivative dc
polarograms of good wave form with the quadruple parallel-T current-
averaging filter sub-assembly, slow derivative computer, 2 to 5 second DME
combination. Such polarograms have been published. For example a first
derivative polarogram of In^{3+} made at a scan rate of 0·01 volts per minute
has an n value calculated from $W_{1/2}$ of 2·96 (121). It is shown that points
from recorded first derivative polarograms of Cd^{2+} and of In^{3+} that were
made at scan rates of 20 and 10 millivolts per minute, respectively, fit very well
onto a plot of the theoretical equation for a first derivative polarogram (54).
Another example is shown as Figure 21 of this chapter. With such polarograms, the ultimate high resolution of first derivative dc polarography is
realized in practice. On the other hand, at such slow scan rates, the signal-
to-noise ratio is not as favourable as it is at higher scan rates, and it takes
25–50 minutes (during which sporadic DME noise pulses sometimes occur)
to traverse a 0·5 volt span. Considerable emphasis has been placed on the
fact that the time-lag in the quadruple parallel-T current-averaging filter
sub-system has a significant effect upon recorded wave form unless the scan
rate is sufficiently slow. However, the magnitude of this effect is very much
less than that associated with the usual RC damping used in many polarographs. In order to illustrate that quadruple parallel-T damping allows good
(though not optimum) resolution even at scan rates at which the time-lag
is significant, polarograms were recorded under the conditions that are
tabulated in Figure 24 (119). Polarograms recorded with solutions of about
the same composition by means of the Tast polarograph and by means of
a polarograph with ordinary RC damping have been published (122–124).
The constituent polarograms are not resolved with ordinary RC damping,
but they are resolved with the Tast polarograph and with the ORNL polarograph. With the ORNL polarograph, the undamped, average current, and
maximum current regular dc polarograms and the first derivative average
current dc polarogram are all resolved. It should be pointed out that,
although the ratio of Tl^+ to Cd^{2+} is twenty-five-fold, the half-wave potentials
are about 140 millivolts apart. Another example, with the pertinent conditions, is shown in Figure 25 (125). The quadrupole parallel-T current-
averaging filter is used with a 2 to 5 second DME and a scan rate of 0·1 volt
per minute. Two pairs of regular and first derivative dc polarograms are

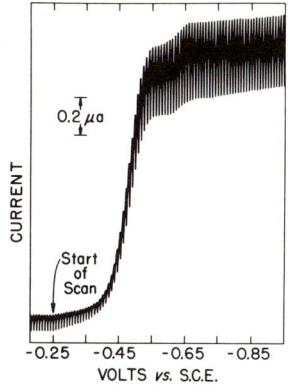

Undamped, Regular Polarogram

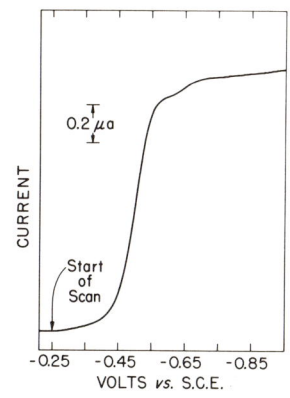

Average Current, Regular Polarogram
(quadruple parallel-T, RC filter)

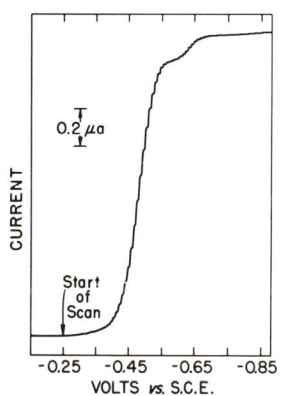

Maximum Current, Regular Polarogram
(peak follower)

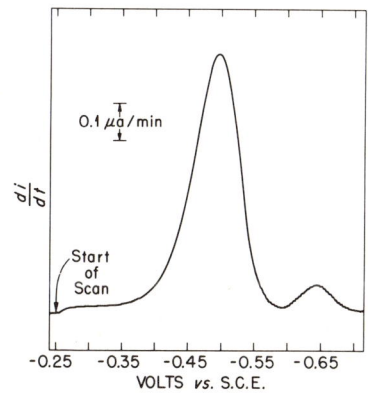

First Derivative Polarogram
(quadruple parallel-T, RC filter
and derivative computer)

SOLUTION : $2.5 \times 10^{-4} M$ Tl$^+$ and $1 \times 10^{-5} M$ Cd^{++}
$0.1 M$ KCl and $0.001 M$ HCl

CELL : Sargent 2 to 5 second DME

BECKMAN SCE REFERENCE ELECTRODE

Pt COUNTER ELECTRODE

ORNL model Q-1988-ES Controlled
Potential and Derivative Polarograph

SCAN RATES : Regular 0.100 volts/min
             Derivative 0.050 volts/min

8 May 1961   WL B

Fig. 24. Example showing that quadruple parallel-T, *RC* damping provides good resolution. Reprinted from: D. J. Fisher, W. L. Belew and M. T. Kelley, pp. 89–134 *in* G. J. Hills, Ed., *Polarography 1964*, Vol. 1, Macmillan, London, 1966, Figure 6, by permission of The Polarographic Society and Macmillan & Co. Ltd.

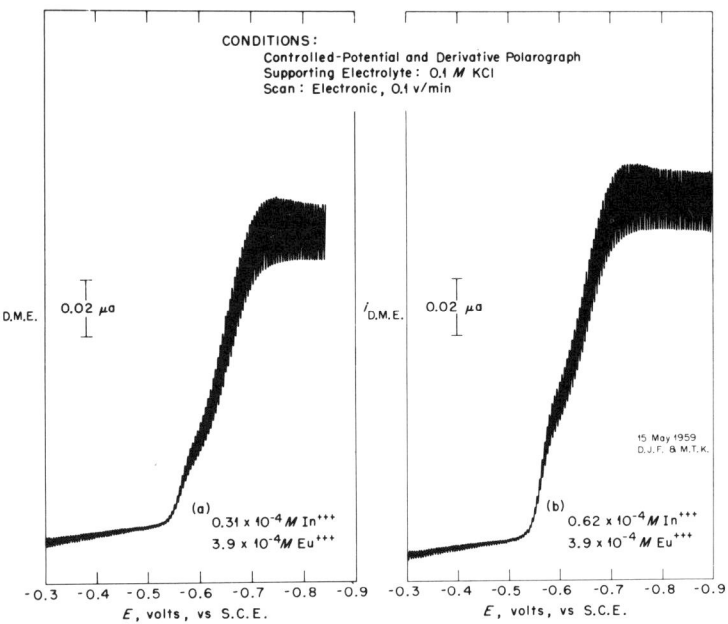

UNDAMPED REGULAR POLAROGRAMS OF In^{+++}AND Eu^{+++}

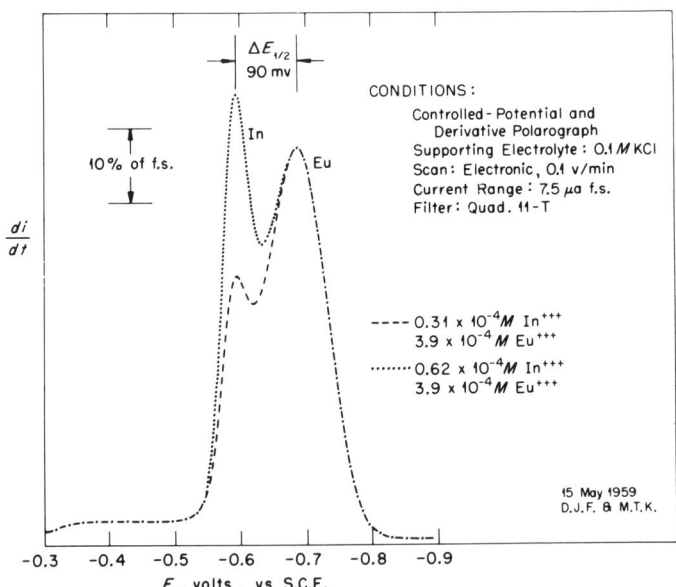

DERIVATIVE POLAROGRAMS, In^{+++}AND Eu^{+++}

Fig. 25. Undamped regular and first derivative dc polarograms of indium and europium. Reprinted with permission from M. T. Kelley, D. J. Fisher, W. D. Cooke and H. C. Jones, pp. 158–182 in I. S. Longmuir, Ed., *Advances in Polarography*, 1960, Figures 7 and 8, Vol. 1, Pergamon Press Ltd.

shown, with a fixed concentration of europium and two concentrations of indium. The narrow ($n = 3$) first derivative peak for indium is resolved from the broader ($n = 1$) peak for europium but the corresponding regular dc polarograms are overlapped too severely for direct measurement. The half wave potentials are about 90 millivolts apart. Another illustration of the relative resolution of regular and first derivative dc polarography is given in Figure 26 (119). In this case, the quadruple parallel-T current-averaging filter is used with a 2 to 5 second DME at a scan rate of 20 millivolts per minute. The separation of half-wave potentials is about 38 millivolts. In this example, the constituent regular dc polarograms are so severely overlapped that it is not evident from the net polarogram that two waves are present but two peaks are seen in the net first derivative dc polarogram. The recorded

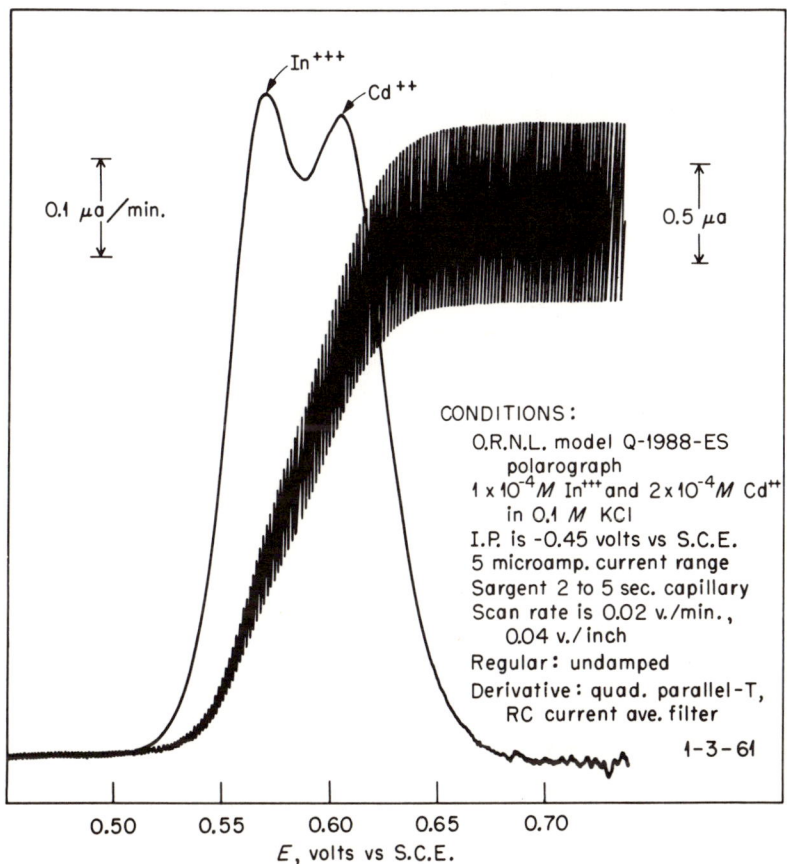

Fig. 26. Qualitative comparison of resolution by regular and derivative polarography. Reprinted from: D. J. Fisher, W. L. Belew and M. T. Kelley, pp. 89–134 *in* G. J. Hills, Ed., *Polarography 1964*, Vol. 1, Macmillan, London, 1966, Figure 5, by permission of The Polarographic Society and Macmillan & Co. Ltd.

wave forms of the constituent first derivative polarograms agree closely with calculated forms. Figures 24 and 25 are illustrations of cases where the concentration ratios, n values and $E_{1/2}$ separations are such that the broadening of $W_{1/2}$ by current-averaging filter lag is not enough to prevent the resolution of the first derivative dc polarograms. Figures 21 and 26 are illustrations of a case where two adjacent first derivative dc waves are overlapped but the enhancement of the two net peak heights is that dictated by the electrochemical equations and is not increased by circuitry imperfections. The substantially greater resolving power of first derivative dc polarography has been found advantageous in routine analyses. Even if adjacent peaks are not fully resolved, the Instrument User is much more likely to be able to tell that a net polarogram is for more than one constituent. When an unsuspected impurity is present in a sample and its wave severely overlaps that of the constituent being measured, the use of first derivative dc polarography helps to prevent the reporting of an unrecognized positive analysis error. Very significant practical advantage, particularly when resolution is of interest, is provided through the use of satisfactory 0·5 second DME's and fast current-averaging and derivative taking sub-systems. These features are provided with ORNL polarographs models Q-1988-FES (17) and Q-2792 (11, 12). The use of the ORNL model Q-2942 DME drop time controller (30) is also advantageous. First derivative dc polarograms can be recorded at

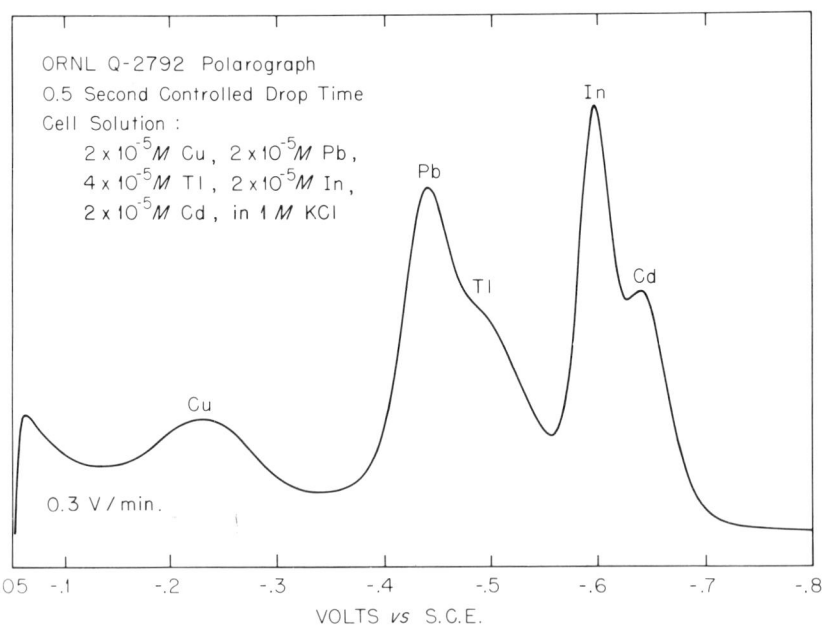

Fig. 27. Resolution of first derivative dc polarograms.

ten-fold faster scan rates whose forms agree very well with calculated wave forms. An example of a first derivative dc polarogram of good form, recorded at a scan rate of 0·3 volts per minute with the model Q-1988-FES polarograph is shown in Figure 7 (126). Another example, obtained with the Q-2792 polarograph and the Q-2942 controller operating at 0·5 second but with current-averaging filters as calculated for a 1 second drop time in order to optimize signal-to-noise ratio rather than resolution, and with scan rates from 0·3 to 1 volt per minute, has been published (127). The solution contains $2 \times 10^{-5} M \, Cu^{2+}$, $2 \times 10^{-5} M \, Pb^{2+}$, $4 \times 10^{-5} M \, Tl^+$, $2 \times 10^{-5} M \, In^{3+}$ and $2 \times 10^{-5} M \, Cd^{2+}$ in $1 M$ KCl. Polarograms obtained with similar solutions with other kinds of polarographs have been published elsewhere. There is very little distortion of the constituent first derivative dc waves even at 0·3 volt per minute. This net first derivative dc polarogram is shown in Figure 27. If ultimate resolution were desired, a scan rate of 0·1 volt per minute with a 0·5 second DME and fast sub-systems provides it, as illustrated in Tables I and III.

3. Sensitivity

Advantages of the use of current-averaging and active practical analog derivative sub-systems, from the standpoint of sensitivity, may be summarized as follows (128). The ultimate signal levels of the various kinds of polarography, voltammetry and other electrochemical methods can be calculated from the wave form equations. For example the ultimate signal level of the Nth derivative dc polarogram is directly proportional to concentration, the Nth power of the scan rate and n^{N+1}. For equations, see Section III-A-2-(e) of this chapter. Comparisons of these ultimate signal levels are useful in estimating the ultimate relative sensitivities of these methods in favourable cases and as restricted by the implied and assumed conditions, limitations and applicability of the equations used for the calculations. However, as with all kinds of measurements, polarographic sensitivity depends upon the signal-to-noise ratio that is achieved in practice with a specific implementation of each method. The amount of noise that is present depends upon the characteristics and the conditions of use of the DME (see Sections III-A-2-(c) and III-B-2 of this chapter) and of the cell (see Section III-B-3 of this chapter). Also, the qualities and features of the circuit sub-systems (see Section III-C of this chapter) and readout device (see Section V-D of this chapter) have an effect in determining the smallest signal magnitude that can be usefully measured in the presence of the noise. The meaning and the definition of polarographic signal-to-noise ratio have been given in Section III-B-2 of this chapter. It is the ratio of the average of the faradaic response to the *range* of the measured values of the total response plus the *range* of measured values of the residual response rather than being the ratio of the faradaic response to the residual response (129). The denominator includes all forms of noise in and lack of perfect reproducibility of the measured responses. Examples have been published of signal-to-noise ratios attained in regular and in first derivative dc polarography with the quadruple parallel-T current-averaging filter assembly, a 2 to 5 second DME, and the slow derivative computer at a scan rate of 0·1 volt per minute (130, 131). Examples have been published of signal-to-noise ratios attained in first derivative dc polarography with the quadruple and the quintuple parallel-T current-averaging filter assemblies, a drawnout 2 to 5 second DME, and the slow derivative computer (132). The peak heights on the recorder paper of the first derivative polarograms (132). are as follows: $5·7 \times 10^{-7} M$ $(0·037 \mu g \, ml^{-1})$ Zn^{2+}, 78 mm; $2 \times 10^{-6} M$ (0·41 p.p.m.) Pb^{2+}, 215 mm; average for first and second peaks, respectively, $8·5 \times 10^{-7} M$ $(0·050 \mu g/ml^{-1})$ Ni^{2+} (irreversible reaction), 165 and 80 mm. Examples have been published of signal-to-noise ratios attained in first and in second derivative dc polarography with the ORNL model Q-1988-FES polarograph (17) the dual parallel-T current-averaging filter assembly, a 0·5 second DME, and the fast derivative computers at a scan rate of 3 volts per minute (134, 135); see Figure 22 of this chapter. Data have been published that show that the peak-to-peak heights of second derivative dc polarograms are directly proportional to concentration over the range tested, 5×10^{-7} to $1 \times 10^{-3} M$ Pb^{2+} (133). By use of the ORNL controlled-potential differential dc polarograph (10) which uses current-averaging filters and includes an active analog derivative sub-system, characteristics of the subtractive mode with uncontrolled drop time DME's were demonstrated. For example whereas a single-cell first derivative dc polarogram of $10^{-6} M$ Cd^{2+} is superimposed on the non-linear with potential portion of the residual current, the background is rendered nearly constant with potential over a useful range through the use of the subtractive mode (136). However, better signal-to-noise ratios have been obtained through the use of 0·5 second controlled drop time DME's with the ORNL model Q-2792 polarograph (see below). A subtractive mode dc polarogram of $10^{-7} M$ Cd^{2+} has been published (137). Examples have been published of first derivative dc polarograms made with the ORNL model Q-2942 drop time controller (30) operating a Smoler 90° vertical orifice DME at 0·5 second and the ORNL model Q-2792 polarograph (11, 12) with an active current-averaging filter sub-system (as calculated for a 1 second drop time to optimize signal-to-noise ratio), a fast active analog derivative computer sub-system and a scan rate of 1 volt per minute (138) and 3 volts per minute (139). The signal-to-noise ratio that is obtained at the 10^{-6} $M Cd^{2+}$ level by first and by second derivative dc polarography is illustrated in Figure 28. Five superimposed replicate polarograms are shown in each case for the Cd^{2+} solution and for the supporting electrolyte alone.

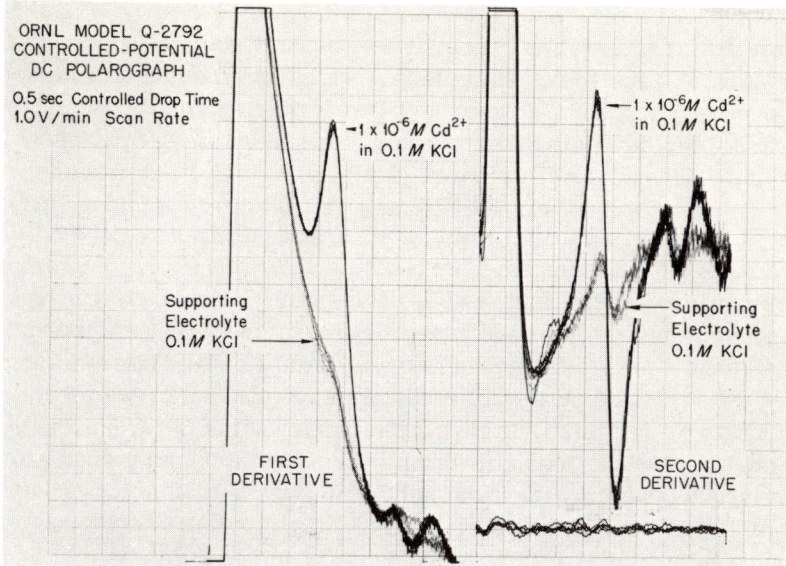

Fig. 28. First and second derivative dc polarograms of $1 \times 10^{-6} M$ cadmium.

It is clear that an increase in the sensitivity of dc polarography has been achieved by means of rapid (i.e. use of uncontrolled or, preferably, precisely electromagnetically controlled 0·5 second DME with low $\bar{m}$, current-averaging and fast active practical analog derivative computing sub-systems, fast scan rates) first and second derivative dc polarography. The ultimate sensitivity of first derivative dc polarography is more nearly attained than that of second derivative dc polarography, but the higher signal-to-noise ratios realized in the second derivative mode and the substantially greater ease of measurement at high sensitivity of second derivative dc polarograms (particularly from peak to peak) make rapid second derivative polarography the mode of choice for high-sensitivity dc polarography.

F. Relative Advantages of Regular and First and Second Derivative DC Polarography

The relative advantages of regular dc polarography and first and second derivative dc polarography are summarized in the following two sections of this chapter.

1. Practical Advantages of First Derivative over Regular DC Polarography

The practical advantages of first derivative over regular dc polarography may be summarized as follows (140). With a 0·5 second DME (such as the

Smoler 90° vertical orifice DME operated at low average mercury flow rates) and corresponding current-averaging and active practical analog derivative sub-systems, at slower (about 0·1 volt per minute) scan rates, the ultimate high resolution of first derivative dc polarography is fully realized while at faster scan rates (about 1 volt per minute) improved sensitivity is obtained. The ultimate resolution of first derivative dc polarography is substantially greater than that of regular dc polarography and is equal to that of other methods such as square wave and pulse polarography. When successive half wave potentials are close together, for various concentration ratios and n values, first derivative dc polarograms are less likely than regular dc polarograms to be so severely overlapped that the presence of the second constituent is not recognized and causes a positive analytical error. Moreover, especially at higher measuring sensitivities, the first derivative polarograms are easier to measure. The linear component of the residual current becomes a constant offset. The derivative blank is more reproducible than the regular blank. With polarograms of low concentration solutions ($\leq 10^{-5}$ M), as a rule of thumb, whenever the two plateaus of the regular polarogram or of the first derivative polarogram are not parallel or equal, respectively, measurements of wave or peak height by graphical extrapolation are not trustworthy; however, the error caused from measuring peaks by extrapolation is less than that from measuring wave heights by extrapolation (see Figure 20). Regular and derivative polarograms can be most accurately corrected for residual current by measuring the sample polarograms relative to the residual current or its derivative (regular or derivative blank polarogram of supporting electrolyte alone). At still lower concentrations ($\leq 10^{-6}$ M), it is essential to measure regular and first derivative polarograms by superimposing sample and blank polarograms (see Figures 22 and 28), because the actual non-linear character of residual current versus voltage is quite evident at the current ranges needed to record such polarograms. The inherently higher resolution and sensitivity of first derivative dc polarography, relative to those of regular dc polarography, are made available by means of the system elements that are provided in ORNL instruments models Q-1988-FES (17) and Q-2792 (11, 12) plus Q-2942 (30). Samples are routinely analysed by first derivative dc polarography at ORNL.

2. Relative Advantages of First and Second Derivative DC Polarography

The relative advantages of first and second derivative dc polarography may be summarized as follows (141). The ultimate resolution of second derivative dc polarography is somewhat greater than that of first derivative dc polarography but at slower scan rates, first derivative dc polarograms can be recorded whose forms conform more closely to the wave form equation and which also have higher signal-to-noise ratios. When the resolution of adjacent waves or interpretations based on wave form are needed, first derivative dc polarography is more useful than second derivative dc polarography. The greatest advantage of second derivative dc polarography is with faster scan rates when sensitivity rather than resolution is of primary interest. The non-linear character of the i_c versus E relation poses less of a problem over a potential range sufficient to record a polarogram at high measuring sensitivity with second derivative dc polarograms of acceptable form than with first derivative or regular polarograms. It has been found that the signal-to-noise ratios of second derivative dc polarograms recorded at high sensitivity with a 0·5 second DME (such as the Smoler 90° vertical orifice DME operated with low average mercury flow rates) and a current-averaging filter and the fast derivative computers at a scan rate of about 1 volt per minute are substantially superior to those obtained at slower scan rates. It has also been found that the measurement of these polarograms is indeed much easier than the measurement of the corresponding regular and first derivative dc polarograms. Although these polarograms are not of theoretical form because of the effects of circuit time-lags at high scan rates, they are very reproducible and very satisfactory for concentration measurement. See Figures 22 and 28. The problem of correcting second derivative dc polarograms for the residual component at high sensitivity is much less than that of similarly correcting first derivative dc polarograms, for the following reasons. To properly measure the first derivative polarograms at high sensitivity, the sample and blank polarograms have to be superimposed, during or after recording them. The second derivative polarograms can be measured from the value of the first peak to the value of the second peak and it is not necessary to superimpose the sample and blank second derivative polarograms to correct the sample values for the blank values. Even at high sensitivity, the zero value for second derivative polarograms can be placed at the centre of the recorder scale, but to superimpose the sample and blank first derivative polarograms during recording, the zero value will have to be shifted off the recorder scale. Also, the contribution of the second derivative residual component to the sample wave is insignificant at concentrations of reducible ion greater than approximately $5 \times 10^{-5} M$, and it is relatively small at all concentrations. When resolution or wave form is of particular interest, first derivative dc polarography is more useful than second derivative dc polarography, but when sensitivity is of particular interest, rapid second derivative dc polarography is advantageous with systems such as recent ORNL polarographic instrumentation models Q-1988-FES (17), Q-2792 (11, 12) and Q-2942 (30).

G. Polarographic Precision

We have found that polarographic precision (i.e. the reproducibility of replicate polarograms, the relative standard deviation, S, of wave heights or peak heights) can be meaningfully stipulated only in conjunction with the context, the time duration over which the measurements are gathered. Short-term precision, i.e. S for replicate polarograms recorded over a time duration long enough to record 3 to 10 polarograms of a sparged solution in succession, is quite a different matter from long-term precision, i.e. S for replicate sample analyses that require enough time for the several additional sparging, cleaning and calibrating or comparison steps required for recording replicate polarograms for a sample and for a spiked or standard solution. With careful work, the short-term precision as obtained by replicate measurements with a single cell filling and the long-term precision as obtained by replicate measurements with several repeated cell fillings seem to differ because of the different respective total durations of measurement: the time over which the DME calibration factor (characteristics) must remain unchanged. It is very much more difficult to reliably obtain an S value ≤ 0.1 per cent over the long time durations sufficient to perform replicate analyses than over the short time durations sufficient to obtain replicate polarograms of a given sparged solution.

Heyrovský, in a 1946 lecture titled 'The Fundamental Laws of Polarography', stated that the precision with which a component may be determined by regular dc polarography reaches in favourable cases ± 1 per cent of its absolute amount (142). He also said that the precision can be increased to 0.1 per cent by means of polarometric (also called amperometric) titrations in which i_d of the component is followed at a constant voltage as it is diminished during precipitation with a suitable reagent. Taylor has also assessed the precision for analyses by regular dc polarography as about 1 relative per cent in the most favourable cases (143). He lists practical tolerances that insure results reliable to 2 relative per cent (e.g. mercury head controlled to ± 0.3 per cent, mercury flow rate controlled to ± 1 per cent, temperature of cell controlled to $\pm 0.5°C$, drop time controlled to ± 2 per cent). Taylor points out that it would appear that these tolerances could be tightened so that an overall precision of 0.1 per cent would be obtained but that in practical analytical work other problems enter in such as maxima, incomplete resolution of consecutive waves, inexactness of residual current compensation and irregularities in electrode performance. Similarly, Davis and Shalgosky reported that with one cell operation, using a differential cathode-ray polarograph, the short-term reproducibility is excellent (successive traces exactly superimpose) but the long-term reproducibility is dependent on that of the DME (which is generally about 3 per cent (144)). On the other hand, Shalgosky and Watling demonstrated that with two cell differential operation, the comparative mode, using rapid scan oscilloscopic polarography, at the millimolar concentration level, the coefficient of variation of a polarographic determination can be as low as 0.04 per cent (145). It is important to note, however, that the various factors that enter into affecting current magnitude at and current balance of the two DME's were controlled very carefully in addition to employing the scale expansion feature of the comparative mode. A number of investigators have subsequently confirmed that this technique, with careful work and the expenditure of enough time, can yield analyses having a precision of about 0·1 per cent. We have found that at about millimolar concentrations, the short-term precision provided by the combination of the ORNL model Q-2942 electromechanical DME drop time controller (30) and the ORNL model Q-2792 polarograph (11, 12) is excellent. Superimposed triplicate regular and first and second derivative dc polarograms (run in succession on a sparged solution) have been published that each coincide throughout within the pen width of the XY recorder (146). With a recorder, the short-term precision is $S \sim 0·2$ per cent, that of the recorder (11). As pointed out in Section III-A-2-(c) of this chapter, this is a criterion of DME performance with a test solution that is met whenever the total system is operating in a normal way. Even at micromolar concentrations, derivative dc polarograms (at fast scan rates) can be reproduced rather well (see Figure 28 of this chapter).

In papers describing the excellent results that can be obtained with two cell oscilloscopic comparative polarography (at millimolar concentrations, precision of 0·1 per cent and accuracy dependent upon that of the standard solution), the ten-fold improvement in precision is usually attributed to two effects of using a second cell as a standard or reference. First, the difference between the cell currents can be measured to the usual ~ 1 per cent on a more sensitive current range and is added to the exactly known concentration of the standard (which is 5 to 10 per cent different from the sample) so that the precision relative to the total concentration of the sample becomes 0·1 per cent. Secondly, with the two capillaries closely matched for the duration of the experiment, there is said to be a cancellation of common mode errors. It appears that in practice the second factor is really the careful control of a number of conditions so that the two DME calibration factors are sufficiently constant for a long enough time. We have postulated that if this is the case, and one DME is used, the first factor, scale expansion, should be obtainable with the use of only one cell (147). It is apparent that with all methodology, the total time required for the sets of measurements required for an analysis should be as short as possible.

Measurements with a single cell for which the characteristics of only one DME need to be maintained sufficiently constant for a time duration long enough for analyses (including measurement of i_d's or of peak heights for samples and for standards or after standard additions) are inherently simpler than measurements with two cells and DME's. Investigations have been made and others are in progress at ORNL of the long-term precision attainable by one cell dc polarographic methods. A simpler high precision method would be attractive for analyses of samples.

The ORNL differential controlled-potential dc polarograph (10) has been used to obtain both one and two cell comparative polarographic results having a precision of 0·1 per cent at ∼millimolar concentrations (20, 21, 148). The ratio for the concentration of sample and standard was about 0·95 (20). In the two cell comparative method, in essence, regular dc polarograms are measured differentially at the same time in two cells with previously electrically balanced DME calibration factors and twenty-fold scale expansion. In the one cell comparative method, measurements with the sample and the standard solutions are made in succession, one at a time, in one cell, at fixed voltages. The value corresponding to the first i_d is nulled and then is subsequently subtracted by a stable current offset circuit while the i_d for the second is measured so that the difference is recorded on a sensitive (i.e. expanded) scale. It is necessary, as with oscilloscopic comparative polarography, to closely control experimental variables. Moreover, even at the millimolar level and with careful control of sparging, it was necessary to correct i_d values recorded at a voltage on the diffusion plateau of the regular dc polarograms for residual current by subtracting average current values recorded at a voltage on the foot of the wave. This is a 'zero shift' correction, not a 'gain' correction.

Further studies have been made at ORNL of factors influencing the precision attainable in one cell measurements. First derivative dc polarography was chosen because it was thought that the inherent rejection of the linear component of the residual current and the observed fact, for supporting electrolyte alone, that derivative polarograms are more reproducible long term than regular dc polarograms might minimize the 'zero shift' due to variations, for example, in dissolved oxygen content after sparging. The ORNL model Q-2942 DME drop time controller (30) and the ORNL model Q-2792 polarograph (11, 12) were used. At a drop time of 0·5 second and a scan rate of 1 volt per minute, each polarogram can be obtained in about 30 seconds (after sparging). A precision peak follower was built so that the output of the polarograph can be read with a five digit digital voltmeter instead of on the XY recorder (149). The cell assembly was placed in a thermostatted glove box; the cell solution and DME must be at a temperature constant to better than $\pm 0·05°C$ at the times of measurement. The DME drop time was controlled at 0·5 second ($S = 0·03$ per cent) and the mercury head, h, was maintained at a high constant value by manual adjustment. With this degree of control of the DME calibration factor, over a short term (seven first derivative peak heights for ∼millimolar Cd^{2+} obtained and measured in succession), $S = 0·03$ per cent. The peak height depends on the value of the scan rate, but a given scan rate in the Q-2792 polarograph can be repeated with $S < 0·1$ per cent (11). Also, the current amplifier, active current-averaging, and derivative computer gains are believed to be sufficiently constant. However, measurements (first derivative peak heights) of replicates of a standard placed in the cell in succession over a period of several hours had a precision of $S \sim 1$ per cent. Also, the precisions of analyses of synthetic samples made by alternating standards and samples were about 1 per cent. In other words, the long-term stability of the DME calibration factor was not good enough. These deviations seemed to result from sporadic changes in the average mercury flow rate, $\bar{m}$. The value of $\bar{m}$ depends in part on the back pressure, which is a function of the interfacial tension at the surface of the mercury drops.

Three devices have been designed and built that are intended to directly control $\bar{m}$ to a constant value of ~ 1 milligram per second (150). Each device replaces the usual gravity stand-pipe arrangement and is intended to force mercury at a constant rate through the DME capillary which is mounted in the Q-2942 drop time controller. The first device which utilized a commercial piston burette was found to be unsatisfactory. The second device was also found to be unsatisfactory. It utilized the thermal expansion of a secondary displacement fluid. The third device is a carefully constructed direct displacement syringe driven by a high precision leadscrew and a stepping motor. The plunger diameter is about 1 mm. The effect of directly controlling this additional variable, $\bar{m}$, upon the long-term stability of the DME calibration factor (and thus upon polarographic precision for actual sample analysis) and the form of current–time curves at constant voltage when $\bar{m}$ is controlled to a constant value are of interest.

Lingane has discussed the problems inherent in achieving high precision with polarography and the results that he achieved by careful experimental technique for the control of DME characteristics and with time-integration of the diffusion current for 50 or more successive DME drops at constant voltage (151). He verified that the most critical factor in precision polarography is the reproducibility of the current at the DME itself and that it is possible to obtain a precision of 0·1 per cent by a one-cell method.

With the above-mentioned (third ORNL model) device for the mechanical positive displacement control of the average flow rate of mercury through a DME, with the entire cell assembly housed in a controlled temperature glove box, and by recording polarograms only when the temperature of the solution near the DME orifice was constant, as verified with a thermistor, to $\pm 0.02°C$, the value of $\bar{m}$ was constant to 0·1 per cent. Over a period of several days, the polarographic calibration factor (measured for $5 \times 10^{-3} M$ Cd^{2+} either by first derivative peak height or averaged diffusion current on the plateau of a wave) is relatively controlled and constant: $S = 0.15$ per cent for 28 trials (152). These results have been reported orally in an invited paper (153).

Although these experimental results are a useful confirmation that the long-term reproducibility of the current at the DME is the principal factor involved in attaining high precision analyses, the fabrication of this apparatus for directly controlling $\overline{m}$ and the concomitant temperature control that is needed are fairly severe experimental requirements for the routine measurement of sample concentrations. It is probable that 0·1 per cent polarographic precision can be attained in a more practical manner with digital dc polarography: see Section V-D of this chapter.

H. ORNL DC Polarographs

A number of models of dc polarographs have been designed at ORNL. Publications are available that describe the features, operation and performance of each of these. A purpose of Section III-H of this chapter is to tabulate certain of these references so that these details can be found. In addition to published papers, engineering drawings are prepared and are available for construction and methods have been prepared for the ORNL Master Analytical Manual that contain details concerning the operation, use and maintenance of these instruments. Single copies of these methods are available from M. T. Kelley, Director, Analytical Chemistry Division, Oak Ridge National Laboratory, Oak Ridge, Tennessee 37830. Identifying numbers for such material are shown in Table VI. Block diagrams of recent ORNL dc polarographs model Q-1988-FES, the controlled-potential differential dc polarograph and model Q-2792 are shown, respectively, as Figures 13, 14 and 15 of this chapter.

TABLE VI

Polarograph model no.	Construction drawings	Master analytical manual method no.
Q-1160	Q-1160 series	1 003040 and 9 003040
Q-1338	Q-1338 series	1 003041 and 9 003041
Q-1673	Q-1673 series	1 003042 and 9 003042
Q-1988A	Q-1988A series	none
controlled-potential differential dc polarograph	circuits: ORNL-Dwg. 65-12178A ORNL-Dwg. 65-12442A	none
Q-1988-FES	Q-1988A series and, circuit, ORNL-Dwg. 65-11949B	none
Q-2792	Q-2792 series	1 003043 and 9 003043
Q-2942	Q-2942 series	1 003044 and 9 003044

One of the first dc polarographs designed at ORNL is model Q-1160 (154). It is an example of a chart recording polarograph for conventional dc polarography. The sensitivity of this instrument is high relative to that of others available at the time of its design. Later on, the quadruple parallel-T current averaging filter was added to the several Q-1160 instruments then in use at ORNL (16). The work at ORNL on the subject of first derivative dc polarography originated because of a need for better polarographic resolution. It soon led to polarograph models Q-1338 (155) and Q-1673 (156). In retrospect, it is interesting that operational amplifier type current amplifiers were designed and used in models Q-1338 and Q-1673. These were not fabricated by adding external networks to commercially available operational amplifiers, as can so easily be done today. Instead, Brown servo amplifiers were modified to function as operational amplifiers. Although the advantages that might result from the use of active analog derivative computing sub-systems were appreciated at the time of design, passive RC networks were used instead, for economy. Some performance refinements were made, however, through the use of the quadruple parallel-T filter and of amplification so that a smaller RC time constant could be used in the analog section (157). Similarly, the performance of the diode filter (for maximum current polarograms) in model Q-1673 is not as good as that achieved later with the addition of a voltage follower sub-system in model Q-1988. By the time that model Q-1988 was designed, it was highly feasible to use operational amplifiers.

Model Q-1988 is the first dc polarograph designed at ORNL that includes sub-systems built with commercially available operational amplifiers for functions such as improved voltage and current sources, potentiostatic electrolysis, current amplification, current-averaging (in conjunction with voltage followers), maximum current (peak-follower) and active analog derivative computing (6). A test function and a cell-open function are also provided in this and in later instruments. The polarizing circuitry of model Q-1988 includes an electromechanical voltage scan generator. The second model of this series, Q-1988-ES (ES means 'electronic scan'), has an all-electronic scan generator (analog integrator fed by constant current) and polarizing section of improved performance and retains the current amplifier and signal transformation sub-systems of the model Q-1988 (7, 8). In the model Q-1988A polarograph, an additional circuit for the compensation of initial current due to prior waves is added to the features of model Q-1988-ES. A complete set of electrical and mechanical drawings, photographs and a checkout and test procedure (ST-175A) were prepared for the model Q-1988A (158). Model Q-1988A is available internationally from commercial sources (159, 160). The controlled potential differential dc polarograph has been described (10) and its performance characteristics in the following modes of differential polarography have been evaluated: ΔE − differential (19), comparative (20, 21) and subtractive (22). Through the use of properly obtained 0·5 second DME drop times, in polarograph models Q-1988-FES (FES means 'fast circuits, electronic scan') (17) and Q-2792 (11, 12), faster current-averaging and derivative-taking circuits and the addition of the second derivative dc polarographic mode enable better advantages to be taken of the excellent ultimate resolution of first derivative dc polarography and the dependence of the Nth derivative peak height upon the Nth power of the scan rate. Advantages obtained with first and second derivative dc polarography are summarized in Sections III-E and III-F of this chapter. The model Q-2792 is built with solid-state components. New operational features of the model Q-2792 instrument include single or cyclic scanning, adjustable scan voltage limit control so that excessive voltage cannot be applied to the DME, optional time-lag correction for X axis (voltage axis) of XY recorder and plug-in active current-averaging filters so that a set can be plugged in to optimize either resolution or signal-to-noise ratio for a selected DME drop time or for use with stationary working electrodes. The ORNL model Q-2942 DME drop time controller precisely electromagnetically controls the drop time at constant values from 0·25 to 10 seconds so that drop time is independent of variables such as DME voltage (30). References to some work that has been done with models Q-1988, Q-1988-ES, Q-1988A and Q-1988-FES and references to some reported modifications of these instruments have been tabulated elsewhere (161). Some other recent applications have been noted of models Q-1988-ES (162), a modified Q-1988-ES (163), Q-1988A (164–167), Q-1988-FES (66, 168) and a high-current modification (17) of model Q-1988-FES (169). Papers have been published that describe some work that has been done with model Q-2792 (102, 103 and 170–176).

Work that has been done at ORNL with digital polarography is summarized in Section V-D of this chapter.

IV. INSTRUMENTATION FOR COULOMETRY

A. The Design of Measurement Systems to Solve Problems in Controlled-potential Coulometry

Controlled-potential coulometric titration (coulometry at controlled electrode potential) is of interest for analytical applications because, when used properly, this technique can offer advantages including selectivity and, also, high sensitivity or high precision and accuracy. The IUPAC definition of titration (177) can clearly encompass the electrochemical process involved (to completion) in controlled-potential coulometry.

However, the description of controlled-potential coulometry as having the advantage of being in principle an absolute analytical method, based upon the value of the Faraday, with electricity or electrons substituted for reagents, is true but an oversimplification of the facts. It is true that in many practical applications of coulometry measurement systems, analyses having the required precision and accuracy are obtained through calculations based upon the value of the Faraday. In other applications, particularly in the measurement of small quantities, very careful work with a well-designed measurement system (including well-designed method, transducer and instrument elements) and the use of empirically tested procedures for the correction of measured coulombs for background coulombs is required to obtain high precision and, especially, low error (each ≤ 0.1 per cent). Certain problem areas cause a need for both electrical and performance (electrochemical) tests of the system, as was discussed in Section II-G of this chapter.

Measurements systems for controlled-potential coulometry include two major circuit sub-systems: A potentiostat to control the working electrode potential and a coulometer to measure the coulombs consumed in the transducer. These sub-systems are discussed in Sections IV-B and IV-C of this chapter. T. W. Richards named the sub-system for accurately measuring coulombs (178). He perfected the electrochemical coulometer. At present, the use of an electronic coulometer to integrate current with time is much more convenient and can also produce accurate coulomb measurement, both with analog and with digital devices.

The fact that the coulometer can only measure all or a known fraction of the coulombs consumed in the transducer causes the designs of the measurement method and the transducer elements of the coulometry system, the method and the cell, to be extremely important. The design of the transducer element also must be such that the potentiostat can achieve a sufficiently good control of the voltage driving the reaction at the working, i.e. controlled, electrode in the transducer.

In applications where the reaction is carried, essentially, to completion, the signal (i.e. information) part of the observable quantity, total transducer coulombs, Q_T, is the faradaic coulombs corresponding to the completion of the reaction of interest, Q_F. The unwanted part of Q_T is the difference between Q_T and Q_F, background coulombs, Q_B. It is good practice to design the measurement system to maximize the ratio of Q_F to Q_B. In cases where the amount being titrated is large enough and the system design is optimized well enough so that Q_B is negligible as compared to Q_F (i.e. negligible in terms of the precision and the accuracy needed in the analyses), no corrections are needed for background. In some routine methods for the analysis of large numbers of similar samples, the analysis conditions are closely controlled (i.e. reproduced) so that it is possible to obtain the answer with good precision and accuracy without background corrections even though a small negative error due to the reaction not being quite completed at the selected cut-off current is cancelled by a positive error due to the small value of Q_B. An example of this case is the well-established controlled-potential coulometric titration at ORNL of about 5 milligrams of uranium until the cell current decreases from the initially limited value of about 65 mA to a cut-off value of 50 μA. The answer is obtained with a relative standard deviation of 0.1 per cent (see Figure 31) and with negligible bias from the electrical calibration of the coulometer element. The continuous background current is much less than 50 μA when air does not leak into the well-sparged solution and in the absence of other interferences that contribute to the background current. However, particularly when small amounts of material are reacted, it is essential to correct Q_T for Q_B in order to obtain Q_F with good precision and accuracy. In cases where background corrections are needed, controlled-potential coulometry is in reality a relative rather than an absolute method. The problem of correctly correcting Q_T for the minimized Q_B may be more difficult than the design of the coulometer or of the potentiostat parts of the system. Design steps that result in a short titration time (not over 10 to 15 minutes but less time if possible) tend to optimize Q_F/Q_B because the value of Q_F depends upon the amount titrated in the transducer but the value of Q_B depends upon the sum of the time integrals of each of the various kinds of background currents. Some of the background currents flow as long as the transducer is in operation and others also depend either upon the amount of the species of interest reacted or not yet reacted. If an interfering background reaction that is irreversible will eventually proceed to completion at any potential that is sufficient to cause a finite reaction rate, there is another incentive for minimizing titration time in order to maximize the ratio of Q_F to Q_B. In the case of uranium, in order to minimize Q_B, it is essential to have a cell that is tightly sealed from oxygen in the atmosphere or that is provided with secondary containment with an inert gas blanket. In general, the ratio of the working electrode area to the solution volume should be as high as possible to decrease the time needed for the reaction. The importance of good and efficient stirring, which also minimizes the time required for the reaction, and which delivers material uniformly and steadily to the working electrode, cannot be overemphasized. In the case of mercury pool working electrodes, both the mercury interface and the solution itself need to be very well stirred without disruption of either liquid. To the extent that it is practicable, the counter and working electrodes in coulometric cells should be arranged with symmetry so that the potential distribution at the working electrode tends to be uniform and the reference electrode tip should be close to the working electrode. If the tip of the reference electrode is placed in a position such that substantial iR voltage is picked up it tends to reduce the reaction-driving EMF and so to prolong the titration. If the first part of a titration proceeds at limited current, the reaction-driving voltage provided by the potentiostat will be less than the value set into the potentiostat and electrolysis will proceed at limited rather than at controlled-potential until the transducer current demand falls below the limited current value. However, the total titration time may be increased only insignificantly since from this point on the faradaic current decreases exponentially, in uncomplicated cases asymptotically approaching the constant background current component of the transducer current. As an example, in titrations of

~ 5 milligrams of uranium, beginning with a current limited to 65 mA and terminating at an arbitrary cut-off current of 50 μA, the titration time is from 12 to 15 minutes whereas in titrations of ~ 8 milligrams of uranium with the ORNL model Q-2564 titrator whose potentiostat cannot deliver more than about 8 mA to the transducer, even though the current is limited for the first 15 minutes of the titration, an arbitrary cut-off current of 10 μA is reached in only 20 minutes from the start of the titration. (The model Q-2564 instrument is intended to be used for the titration of lesser equivalent amounts than this.)

It can be difficult to ensure that the substantial contributions to Q_B are of the same types in titrations of standards and blanks and of every sample in a series, and to ensure that the desired reaction is as complete as required by the accuracy needed. Nevertheless, the validity of the method used for the correction for background is customarily tested by means of titrating standard solutions or synthetic sample solutions. It is certainly desirable as a test of system operation that there be agreement (within the desired precision and accuracy limits of the analyses) of an answer calculated with the electrical calibration factor of the coulometer portion of the system with the known amount of a standard whose concentration is known with sufficient accuracy by some independent means. Agreement ideally means either that corrections for background if not made were not needed or if made were sufficiently accurate for the specific titration conditions. Unfortunately, but realistically, sometimes the agreement is fortuitous because of a near cancellation of errors or meaningless because not all of the sources of titration error in the samples are present in the standard solution. Several types of background currents are dependent upon trace constituents in the matrix or upon the concentration of the constituent being titrated. For example, impurities that lower the overvoltage of hydrogen on a mercury working electrode may be present but unsuspected in a sample but not in a standard or blank and thus cause a positive error in the analysis of the sample. It is suggested that accuracy should also be tested, when possible, by means of comparisons with results obtained with typical samples by other independent analytical methods.

In applications where the required precision is high and the maximum allowable error is small and especially if only relatively small amounts of material are available for each analysis, the correction of the total measured coulomb value must be accurate. Presparging of solutions before admission of the mercury working electrode to the cell and, generally, pretitration at a potential as close as feasible to the chosen titration potential followed by discharging the integrator and then a change, without opening the cell circuit, to the latter potential and the use of smaller cells and solution volumes have been used to minimize the relative magnitude of the correction for background. Differential, i.e. two cell, methods have been used in order to obtain an accurate empirical correction by means of the second, reference cell. In the presence of some samples, even though a high concentration of electrolyte is present, the background contributions may be significantly different from those obtained with the reference cell containing either a blank or a known amount of standard. When this problem exists, the use of two cells does not necessarily result in valid tests of the background corrections just as serial titrations in a single cell of sample and standard may not. The above-mentioned example where such errors may arise, the case where an unanticipated impurity in the sample, not present in the standard or synthetic sample, causes a decrease in the overvoltage of hydrogen on mercury so that a significant positive analysis error is caused, has been a real problem in more than one laboratory. It is herewith proposed that standard addition, directly to the sample in the titration cell, and thus in the presence of the matrix introduced by the sample, may provide a way to reveal the existence of the problem or to establish its absence and to obtain more accurate background corrections as well.

An investigation at ORNL has been planned of the utility of standard addition in controlled-potential coulometry. After the titration of the amount first present in the cell, one or more standard additions, each followed by additional titration, should provide an in-situ calibration factor. In the absence of background complications, the in-situ calibration factor should be the same whether the first substance present is a pure standard or a sample and should be independent of the amount present and should agree with the electrical calibration factor of the coulometer. If background complications of certain types are present, standard addition should be a means of discovering the problem and of correcting for it.

Before the design of a system can be meaningfully organized or an evolving design can be evaluated, it is necessary to define and detail a number of specific performance objectives. These include items such as the needed values of potentiostat output voltage and current, the allowed amounts of drifts and the required accuracies of control voltages and of coulometer element readout, the needed speed of response of the potentiostat and coulometer, and the needed accuracy and precision for typical amounts to be titrated. Through performance tests designed to test for the achievement of these specific objectives, the prototype system is evaluated, and, as a minimum requirement, it must meet them. From an engineering viewpoint, it is more satisfying to design a system to have a performance that is limited only by the state of the art since the exact requirements for all future applications of the system cannot be foreseen at the time of design. The field of usefulness and the technological life of the system is thus optimized. Yet, some design trade off is forced by compromises between conflicting practical possibilities.

It is coulombs that count. This is the basis of the ultimate and final criterion for the judgment of the performance of a measurement system for coulometry. The overall objective is to design the entire system such that Q_F and the answer can be calculated from Q_T. The design is good if the problems are solved to a degree such that the problems have a negligible effect on the answers.

B. Potentiostat

1. Required Potentiostat Performance

The required potentiostat performance, on the basis of its use for chemical analysis by controlled-potential coulometry, is as follows. In the I–Q method of controlled-potential coulometry, during an early (but not necessarily the very first) portion of a reaction not to be carried to completion, the reaction must be driven by the potentiostat at controlled potential. Also, the rate of transport of material to the working electrode must be constant, temperature must be constant, and a compensation current, equal but opposite in sign to the background current, must be fed into the coulometer element. Thereby, I_F versus Q_F can be recorded and is linear so that the answer can be obtained sufficiently well by extrapolation. In the analysis by means of controlled-potential coulometry of the amount of the material present and to be entirely reacted in the transducer, the reaction may proceed to completion (within the error limit set for the analyses), driven at potentials equal to (within a few millivolts) or less than that selected and set into the potentiostat as a command signal to it. If the reaction is to be driven to completion in a short time, the final portion of the titration must proceed at controlled potential. While the reaction usually does not proceed at controlled potential during the early portion of the titration, the need for the reaction to be driven at '100 per cent current efficiency at all times' is real (although the specific need involved here is the design of the method and transducer elements so that Q_F can be calculated well enough from Q_T). In analytical applications, the early part of the reaction, just as in a manual procedure, can be driven at limited rather than at controlled potential. The fact that this is permissible is fortunate, since to avoid it, the transducer and the potentiostat and the electrodes would have to be designed so that the potentiostat could deliver a huge, fast initial charging transient and a large current at a substantial voltage and with the reaction driven at controlled potential at the instant that the titration is started. It is evident that, as with a manual method, which is carried on automatically through the use of a potentiostat, an important requirement is that the possibility of an interfering reaction being driven at a potential greater than the intended value must be avoided, so that the inherent selectivity virtue of the method (described by the Nernst equation, or by potentiocoulograms) is thereby ensured.

2. Control of Reaction-driving Potential

To obtain a system that provides good analytical answers, the careful design of the method and transducer elements of the system, as well as those of the potentiostat and coulometer, are required. Problems of controlling the reaction-driving potential cannot be solved through the design of only the potentiostat. From an electrical viewpoint, it is possible to design a potentiostat so that it can balance the voltages that are presented to its input. Unfortunately, the direct and specific observation of the voltage that is driving the reaction at the working electrode is not simple. The dimensions of the tip of a reference electrode are large as compared to the distance from the working electrode to the site of the reaction so that it is hard to place it close enough; yet if it is too close, the tip excessively shields the electrode surface. The use to combat this problem of an auxiliary probe having a small tip in conjunction with a reference electrode was first described by F. Haber in 1900 and Haber credited the idea to his friend, H. Luggin (179). A normally operating, balanced potentiostat can control the potential at the equi-potential surface in solution corresponding to the location of the tip of the reference electrode probe to the command value set into the potentiostat. Whether the potential at every reaction site, driving the reaction, is equal (within the number of millivolts necessary in the specific application) to or less than that value is less certain in cases where the working and counter electrode geometry is not symmetrical, in cases where there is non-uniform or fluctuating mass transfer to the working electrode and in cases where iR drops in solution are substantial. The probability of this difference being significant is minimized if the geometry of the working and counter electrode is symmetrical or effectively so, with good uniform stirring, and in low specific-resistance solutions. Potentiostats are usually not in balance throughout controlled-potential coulometric titrations, particularly during the first part of the titration, but this does not necessarily result in substantial prolongation of the titration time or in significant titration error due to loss of potential control. While an out of balance potentiostat applies less than the control voltage, the electrolysis proceeds at limited potential. Although the application because of a potentiostat malfunction of a value in excess of the control voltage might in some cases only drive the reaction of interest faster, if not otherwise rate limited, and not cause a titration error, this is an undesirable risk. If hangup of the potentiostat occurs, the probability for error is greater, as discussed in Section IV-B-4 of this chapter. Over the first part of a titration, cell current demand in excess of potentiostat output capacity does not cause hangup with well-designed systems and iR pickup at the reference electrode probe tip only results in electrolysis at limited potential. The tolerance of the potential variation at the working electrode from the control voltage set into the potentiostat as a command signal is much greater in analysis applications than in investigations of kinetic or other electrode reaction parameters or in electrochemical preparations of a species. However, for analysis, the selectivity advantage could be lost (or errors could be caused due to other irreversible changes in Q_B) if a significantly large excess potential could exist, unless, for example, an undesired reaction exactly reverses when potential control is gained during the final part of the titration. It is prudent, insofar as is possible, to design the system so that the possibility that the potentiostat can provide a potential for a reaction in excess of the command voltage is minimized. Also, signal lights can be provided to warn the Instrument User that an abnormally large potentiostat input unbalance exists or that the potentiostat has been driven to its output current or voltage limit.

In cases where the iR drop along the wire from the controlled electrode to the coulometer circuits within the instrument is substantial during the first part of a titration, one wire can be used to carry electrode current and a second wire, which carries essentially no current, can be added from the electrode to the potential sensing point of the potentiostat.

In an investigation of electrode potential gradients and current density distributions at platinum and at mercury working electrodes in coulometric cells (including studies with a sub-divided working electrode), Harrar and Shain found that it is possible for the electrode potential to exceed the control potential with certain solution conditions and geometrical arrangements of the three electrodes (180). They conclude that the safest location for the tip of the reference electrode is on the line of minimum separation between the counter and working electrodes. With coulometric cells having a deliberately highly unsymmetrical arrangement of the mercury pool working electrode and the counter electrode and with the tip of the reference electrode at various locations including far away from the line of minimum separation but with typical low specific-resistance solutions, we were unable to obtain any electrochemical evidence of loss of selectivity due to the electrode potential exceeding the control potential as a result of the improper relative arrangement of the electrodes (181, 182).

Recommendations for the arrangement of the three electrodes in cells for controlled-potential coulometry have been given by Lingane (183) and by Delahay (184). Their suggestions seem to be reasonable for the usual analytical conditions. Some of the cells used at ORNL for controlled-potential coulometry have been described by Shults (185).

Mumby and Perone have reviewed the problems involved and describe a new potentiostat-cell system design for application to the study of very rapid electrochemical processes (186). In this application, state-of-the-art potentiostatic performance and use is called for; their design provides sub-microsecond transient response with very high current capability. Very small diameter reference electrode probes are used. The performance of their system was evaluated by applying it to the study of several electrode processes whose behaviour could be predicted. An earlier review by Schroeder and Shain also contains much useful information (187).

Note added in proof. Harrar has prepared an excellent and detailed presentation of cell design for controlled-potential coulometry. He emphasizes the fact that a cell having appropriate design must be used in order to achieve good potential control and advantages such as selectivity and high accuracy. He discusses the effects of various cell design choices and practical compromises upon the control of potential at the working electrode. He describes specific cell design features to minimize potential control problems. Harrar recommends symmetrical geometry if possible, and, in any case, placing the reference electrode tip on the line of minimum separation between working and counter electrodes and close to the working electrode. The importance of good stirring is also brought out. See: Jackson E. Harrar, 'Techniques, Apparatus, and Analytical Applications of Controlled-Potential Coulometry,' a chapter in Vol. 8 of *Electroanalytical Chemistry*, A. J. Bard, Ed., Marcel Dekker, New York, 1974.

3. Needs for and Methods of Current Limiting

Electrochemical and circuit needs for current limiting. The maximum magnitude of the titration current may need to be limited to a value less than that which would otherwise be demanded according to the amount being titrated and the stirring efficiency, electrode area to solution volume ratio and other titration conditions. Current limiting may be needed in a given system for electrochemical reasons, circuit reasons or both. For electrochemical reasons, to avoid a positive bias, it may be necessary to limit the initial current that can be delivered to the electrochemical cell to a value smaller than would exist if the reaction were begun at the control potential (188). It has been repeatedly observed that current limiting prevents a positive bias in some applications in which no electrical (circuit) need for limiting, no malfunctioning of the integrator and no loss of potential control can be identified at higher currents. It is presumed that the need must be due to electrochemical reasons. Various mechanisms have been postulated to explain the emergence of this need at higher cell currents. These include: (a) an increased leakage of material from the counter electrode compartment and (b) the lowering of the overvoltage of hydrogen on the working electrode at sites of highest current density. For circuit reasons, current limiting may be necessary to prevent voltage or current overload or hangup. Also for circuit reasons, when two control amplifiers are used, as in the

ORNL model Q-2564 titrator, it may be necessary to ensure that one amplifier cannot deliver more current than can be handled accurately by the other amplifier (189).

Methods of obtaining current limiting.   One method of obtaining current limiting is to add a resistance in series with the counter electrode. If a high-valued current limiting resistance is added in series with the counter electrode, at higher cell currents, a control amplifier may be driven to its voltage limit and create a hangup problem. (See Section IV-B-4 of this chapter.) Also, instability problems may be encouraged (typically, at low cell currents) due to phase shift caused by the large valued resistor. The current that can be delivered to the cell can be limited by, instead, adding a small fixed or adjustable resistance in series with the working electrode (190). The electrolysis then begins at limited potential and finishes at controlled-potential. Thirdly, it is possible with adder-type potentiostats to use an active circuit to provide automatic current limiting action. Fourthly, current limiting can be achieved manually by starting a titration at the pretitration potential, and advancing the control potential slowly or in steps to the titration potential in a manner such that the cell current does not exceed the desired limit value. It is the latter (fourth) method that is in use at ORNL with the model Q-2005-X50 titrator for the determination of ~ 5 milligrams of uranium. The latter method was used also by Propst to avoid amplifier overload (191). A fifth method of obtaining current limiting (which may unavoidably be used to a degree in many practical cases) is the deliberate location of the tip of the reference electrode in a position where iR drop is sensed that is substantial only during the first portion of a titration (192). However, the tip should be small and should be placed below the counter electrode and rather close to the surface of the working electrode so that potential control is achieved during the last part of the reaction.

4. Hangup: Definition, Cause, Dangers from, Diagnosis for and Prevention

Definition and cause of hangup.   A potentiostat is an example of an operational amplifier servo system. In normal operation, through feedback, such systems maintain a balance that is intended to create the designed-for relationship between input and output quantities. In a balancing potentiostat, the potential difference between the terminals of the working and reference electrodes is maintained by a control amplifier sub-system at the selected input (control) voltage. If a control amplifier has been driven to its maximum output voltage limit and is saturated, or if an element within, e.g. the input stage, is saturated, balance will subsequently be lost if a need arises for a new output to balance a new input. At and after the time at which the servo recovers from saturation, normal operation should resume and continue. In practice, a finite time is required for the system to regain normal operation; this time is the hangup time. There is not an instant recovery from saturation during overload. Hangup is the delay during recovery. When recovery from saturation does not gradually occur and the system does not spontaneously regain its ability to balance, the condition is called latch up. After latch up, the amplifier may also oscillate or may motorboat. During design, an amplifier should be selected that has fast recovery from overload and saturation (negligible hangup) and that does not latch up.

An example of a cause of hangup is as follows. When the system is unable to maintain balance, the input(s) may depart from the null voltage(s) in which case amplifier filter networks become charged to values substantially different from normal values. The system cannot return instantaneously to balance because time is required for the amplifier to drive the networks back to normal values. Hangup time is typically longer (up to minutes instead of $\sim$ a millisecond) in the case of 60 Hz chopper stabilized operational amplifiers. When the summing point voltage of a chopper stabilized amplifier rises from the near zero (error) value, the capacitors in the input and output filters in the chopper stabilizing portion of the amplifier become charged accordingly. The lower the chopper frequency, the longer must be the time constants of these filters in order to remove the chopper frequency (and harmonics) from the error signal. After the input signal has changed to an in-range value, the chopper stabilizing portion of the amplifier requires time to discharge the long time constant filters to their normal voltage values. During recovery from hangup, the amplifier output voltage changes from the limit value to the lower value that establishes a balance.

Dangers from hangup. Titration error is not caused in every case where hangup occurs. In other cases, error is caused and so the system should be designed to prevent or, at least, to minimize the duration of hangup. The hangup duration of interest is from the time at which the cell current demand drops to the limit value to the time at which the potentiostat regains balance. During hangup, a control amplifier applies excess voltage to the cell and the cell current may temporarily continue at the limited value. At full amplifier recovery, the cell current reaches the proper value. During hangup, reactions other than the desired reaction may take place to supply that current in excess of the current demanded by the reaction of interest. Unless this extraneous reaction reverses after recovery so that the error in the accumulating Q_T is cancelled, a significant error in the calculated Q_F may remain, depending upon the titration conditions. Also, in some controlled-potential coulometric titrator circuits, the correct operation of the coulometer element of the titrator depends upon a control amplifier holding its summing point at reference (e.g. ground) potential.

Diagnosis for hangup. With a given potentiostat and transducer, hangup is most likely to occur when larger amounts are titrated. When larger amounts are titrated and when the greatest part of Q_B is due to the continuous background current, it is possible to arrange conditions such that the I–Q method can be used (conditions are summarized in Section IV-B-1 of this chapter). During the development of a titrator or of a titration method or of a transducer, a useful diagnostic criterion is provided by simultaneously recording (or, observing with a suitable dc oscilloscope) faradaic cell current (i.e. total current minus the compensating current) against Q_F and also the working-to-reference electrode voltage against Q_F (193). The compensating current is set equal to the experimentally observed continuous, constant background current value. In a typical, normal, controlled-potential coulometric titration, after the initial transient response, the current remains at first at a constant limited value. Beginning at the time at which the cell current demand drops to a value that can be supplied to the cell by the potentiostat, the I_F versus Q_F recording should change to a constant negative slope. The voltage is at first at a value less than the control value. Ideally, at the time when the faradaic current begins its exponential decrease with time, the voltage rises rapidly to and thereafter remains at the control value. If a significant hangup problem exists, during the hangup interval, the voltage rises to a value substantially above the control value and the current remains at the limited value. During the latter part of this interval, the voltage drops to the control value and the current drops to the correct value. After recovery, the voltage and current plots are normal. Typically, during the latter part of the hangup interval, the current drops rapidly from an excessive value to the correct value. However, the magnitude of the titration error that results from the hangup behaviour that is thus revealed depends upon the titration conditions, as just discussed.

Prevention of hangup.   A titrator circuit and the cells and methods used with it should be designed if possible such that hangup of a control amplifier does not occur. If impedances in series with the output of a control amplifier are low (values of series current limiting resistors; impedance of counter electrode assembly), it may be possible to select an amplifier having satisfactory characteristics and which will never be driven to its limit voltage, even at the maximum titration current demanded by the cell (amount titrated, stirring rate, etc.) or which does not hangup or latch up.

Methods of minimizing hangup duration: Selection of amplifiers, use of error and output voltage limiting circuits, and selection of method of current limiting.   Since all applications of a titrator cannot be foreseen, design steps that minimize hangup duration to a negligible value may be worthwhile since any titration error that can occur during hangup will thus also be reduced accordingly. Such design steps may also be useful in upgrading the performance of an existing titrator. Amplifiers require a finite time (hangup time) to recover from being driven to their voltage limit. In the past, 60 Hz chopper stabilized amplifiers were usually used for a control amplifier in order to obtain sufficient gain and low drift. Such amplifiers had a long hangup duration. Chopper stabilized amplifiers are now available that use electronic rather than electromechanical choppers; with the decreased filtering required at higher chopping frequencies, recovery time is faster. Moreover, at the present time, high gain, low drift, low input noise and offset non-chopper stabilized operational amplifiers exist that have a very fast recovery from limit voltage. In general, overload recovery can be speeded up by shunting the input of the amplifier with high quality silicon diodes which limit the excursion of the voltage at the summing point during hangup. Also, the amplifier can be prevented from being driven to its limit voltage by the addition of circuits using diodes and zener diodes in a feedback loop for the purpose of bounding the output voltage (both + and −) to a magnitude less than the limit value. This minimizes the problem of hangup by preventing the amplifier from being driven to its limit voltage. Ideally, the bounding circuits are dormant until the bound voltage value is attained; in practice, leakage currents before bounding and load currents during bounding may need to be considered in the design of the bounding circuit. Although the recovery from bounding can never be instantaneous, it can be sufficiently fast. The use of a limiter (bounding) circuit to minimize hangup time has been discussed by Harrar and Behrin (194). The data sheets supplied by the amplifier manufacturer should be consulted for specific recommendations. A third way to minimize the possibility that hangup may occur is to select the use of a small current limiting resistor in series with the working electrode instead of the use of a large current limiting resistor in series with the counter electrode, should it be desired to use one of these methods of current limiting. In this method, it may be necessary to adjust the value of the resistor for specific applications or to change the value during the titration because a single value large enough to prevent hangup may prolong the titration unnecessarily. A prior knowledge of the maximum current expected for particular applications can be used to select the smallest value of resistance required for current limiting; also, it may be small enough so that potential control is automatically obtained during the final part of the titration.

Prevention of hangup in the ORNL models Q-2005 and Q-2005-X50 electronic controlled-potential coulometric titrators. The ORNL models Q-2005 (188, 195–197) and Q-2005-X50 (188, 195–199) electronic controlled-potential coulometric titrators use a 60 Hz chopper stabilized operational amplifier (driving a transistor current booster) for the voltage control amplifier of the potentiostat. The circuit includes a 160 ohm current limiting resistor in series with the counter electrode. In the applications of these titrators at ORNL, this chopper stabilized amplifier (which has an output voltage capability of ∼ 100 volts) does not hangup because it is not driven to its voltage limit. It is possible, however, to create a hangup problem with these titrators by adding a large valued current limiting resistor in series with the counter electrode. For the past decade, thousands of 5 to 7 milligram uranium samples have, for example, been titrated at ORNL with the manual method of current limiting (to ~ 65 mA, for electrochemical reasons) with a relative standard deviation, S, of 0·1 per cent. (See Figure 31.)

Prevention of hangup in the ORNL model Q-2564 high-sensitivity coulometric titrator. The ORNL model Q-2564 high-sensitivity coulometric titrator also uses a 60 Hz chopper stabilized operational amplifier (amplifier no. 1) for a control amplifier (189, 200). This titrator was designed primarily for use at high sensitivity for the controlled-potential coulometric titration of small amounts of sample (< 40 μeq). In the unmodified circuit, a 13,000 ohm resistor, R_1, is in series with the counter electrode lead (189). The purpose of this resistor is to ensure that no more than ~ 8 mA will flow, which is less than the rated output current of amplifiers 1 and 2, so that the range of the control and current amplifier (amplifier no. 2) cannot be exceeded. There is a greater possibility of titration conditions being such as to cause hangup of the voltage control amplifier with this titrator than with models Q-2005 and Q-2005-X50 because of the high value of the 13,000 ohm current limiting resistor. In applications testing the performance of this instrument (189) and in other uses of it (103), hangup was not observed. However, later, it was discovered that when this titrator is used to analyse concentrated solutions, due to the large value of R_1, amplifier 1 can be driven to its output voltage limit so that hangup occurs. As a precautionary measure, this titrator has been modified in order to eliminate the risk of hangup causing a significant titration error (201). The absence of hangup and greater operating stability can be attained in the model Q-2564 titrator by means of a very simple modification of the published Q-2564 circuit. R_1 is removed from the counter electrode lead (to eliminate the likelihood of potentiostat hangup due to output voltage limit saturation; greater control loop stability also results) and a 27 ohm $\frac{1}{2}$ watt resistor is inserted at the controlled electrode lead (for circuit current limiting). Modified instruments were tested with the several designs of cells in use in the ORNL analytical laboratories during titrations of $\sim 0·5$ and ~ 5 milligrams of uranium (stirred mercury pool controlled electrode) and of $\sim 0·5$ and ~ 5 milligrams of iron (platinum gauze controlled electrode). Titration times are not unduly lengthened, the potentiostat is stable, and titration errors traceable to hangup have not recurred. The prototype and five other model Q-2564 titrators in use at ORNL have been field modified. Since current limiting is not desirable when I–Q readout is used, the current limiting resistor is bypassed by the Current Limit switch in this mode of operation and the concentration of the species to be determined is limited to a value that will not cause hangup of the control amplifier in the early stages of titration. In general, when initial currents greater than 8 mA are expected, the Current Limit switch is thrown to In so that the cell current is limited to 8 mA or less (202). See Figure 32.

5. Transient Overshoot of Working Electrode Potential above Control Value

Potentiostats for controlled-potential coulometric titration are designed to maintain the potential of the working (controlled) electrode at or below the value selected to drive the reaction at the electrode. If the command (control) voltage is stepped from the pretitration to the titration voltage (or from a lower to a higher titration voltage), during the establishment of the new steady state operating balance, there may be a short transient overshoot of the selected potential. During a titration, when the summing point is perturbed by variations in current demand due to fluctuations in stirring, there may be momentary transient overshoot. If an amplifier in the potentiostat is driven to its voltage limit, or is otherwise saturated, during recovery (hangup), there may be a transient overshoot of the selected potential of relatively long duration. Transient overshoot is a characteristic of servo systems. Yet, if the potentiostat applies a higher potential to the working (controlled) electrode than the value selected for the titration, the selectivity advantage of the titration may be but is not necessarily lost. To the degree that this advantage is lost, an error will accumulate in the total coulomb readout. If the potentiostat did not allow the potential to overshoot, a possible source of titration error would be eliminated. However, the requirement of zero transient overshoot in the response to a sudden change in command voltage would in general result in an unnecessarily slow potentiostat response to the change. (See Sections IV-B-9 and IV-B-10 of this chapter.) The objective, then, is to achieve negligible transient overshoot. The following criterion for negligible transient overshoot of the working electrode potential applies. It is clear that if the titration results (accuracy and precision) are not degraded below the performance specifications because of transient overshoot of the working electrode potential, the existing degree of overshoot is negligible.

6. Servo Basis of Description and Tests of Dynamic Characteristics of Potentiostats and other Operational Amplifier Systems

The prediction for designing and testing and the description of the behaviour of operational amplifier systems (or sub-systems) may be based upon node (current) or loop (voltage) linear network analysis. This analysis yields steady-state dc input–output transfer functions, and, with operational calculus, transient relations also. It is assumed that the gain of the amplifier is sufficiently high at dc and at all other frequencies that must be dealt with that the summing point is held within the negligibly small magnitude error signal value of reference potential (or, in the case of active differential-input amplifier circuits, that the inputs are held at equal potential).

It is assumed that the non-ideal details of actual amplifiers such as input and output impedances and input leakage currents and offset voltages have a negligible effect upon their use in the application in question or else that these details are accounted for in the analysis. These systems in many applications must also be active at frequencies above dc; their response to perturbation by fast signal or noise input changes is not instantaneous, from the former to the new steady state, so that time-lags and phase shifts occur. Most of these systems are closed loop, negative feedback systems; exceptions exist, e.g. voltage crossing detectors and potentiostats having a positive feedback loop to compensate for iR.

Potentiostats are an important example of such systems and operate a coupled transducer, in the feedback loop, having complex properties. In order to mathematically examine aspects of their dynamic performance such as stability and transient response characteristics, it is necessary to describe such systems as servo systems, even if the analysis is for expediency restricted to linear regions of operation and to piecewise-linear models of some of the system components that are present in the feedback loop. A linear system is defined as one in which the principle of superposition applies. Experimental tests of dynamic system behaviour also must be based on and interpreted by means of servo theory but need not involve linear approximations or restrictions—yet, the results of these tests are also subject to restrictions, since they are proven only for the specific system test conditions. Though the scope of applicability of the results of both mathematical and experimental tests thus is restricted, such tests are nevertheless a necessary part of system design.

When the input signal is perturbed to a new value, a servo system responds by attempting to re-establish the null balance at its summing point(s). The application of a step or a pulse change at the input and the detailed observation of the servo response is a good experimental method of testing a servo, if it is tested under real load conditions. The coupled transducer is an integral part of the potentiostat servo system whenever the system is used to obtain information. A question that must be asked is how quickly must the potentiostat be able to achieve a balance at its summing point(s) in order to enforce and maintain electrolysis at controlled potential or in order to control the potential at the working electrode in a manner including the achievement of negligible transient overshoot such that a good answer can be obtained by calculation from the coulometer element output which is proportional to Q_T. Capacitive elements are included in the feedback path. A capacitor cannot be instantly charged to a new voltage without supplying it with an infinitely large current, and real capacitances, especially electrolytic double layer capacitances, have to be charged at finite gain through cell resistances of significant value as well as through the dynamic output impedance of the potentiostat. The most stringent requirements are for potentiostats that are to be used for studies of reaction kinetics and mechanisms or for preparations. In such studies, minimum time must be used to change the potential of a double layer to a new value. In any potentiostat having an electrolytic cell as an intrinsic part of the system, the potentiostat is presented with a cacophonous mixture of voltage input signals, having various magnitudes, frequencies and phases, that it must sum to a null by means of current feedback through the transducer. At higher frequencies, the phase of the output from an operational amplifier increases from the ideal dc value of $-180°$. Phase shift through the reference electrode and stray impedances of it also cause problems in fast circuits. In addition, the relative magnitudes, frequencies and phases of the signals from the cell change and also the interface characteristics change (particularly, if adsorption processes are not negligible) at the working and at the counter electrodes as the electrolysis proceeds. Moreover, negative resistance (dynatron) characteristics can exist in the transducer in some applications. In general, the servo, in response to a sudden command change, may undershoot, be critically damped, overshoot or break into oscillation. It is only too possible for the potentiostat to become unstable, particularly if the potentiostat is too fast or too slow, or, with some transducers or transducer conditions. The best and usual approach is to use a potentiostat that is designed to be stable for as broad a spectrum of uses as possible and that is adjusted, if necessary, for any specific problem applications. The requirements for achieving stability and for maintaining a satisfactory control of the working electrode potential are much less severe for most analytical applications such as controlled-potential polarography and coulometry than for precise investigations of electrode reaction parameters.

Well developed stability criteria and electrical engineering test procedures based upon them exist for systems consisting of minimum phase shift networks of lumped linear elements. On this basis, since the potentiostat is a feedback servo system, it is stable only when the feedback existing in the path is of the designed polarity at all frequencies for which the loop gain is equal to or greater than unity. A stability margin is desirable. For these reasons, roll-off networks are built into most operational amplifiers; however, additional potentiostat response tailoring networks may be needed for some applications. With, for example, time variant systems, the procedures become more involved. The classical transfer function is defined only for linear systems. The transfer function of an electrolytic cell is not independent of the voltages that are applied to the cell and of the contents of the cell. A set of models may be required for an exact mathematical description of the transducer, each valid over a limited potential range. A very basic question is how closely must the model(s) that is (are) used approximate all details of the real situation in order to correctly predict stability and other dynamic characteristics. If gain-phase margin criteria (after the use of compensating networks, if needed) predict stability, will the real system be unstable in some applications because of non-trivial details in these situations not described by the model(s)? In some cases, it appears that a simple model accounting for the characteristics of the transducer with electrolyte present but not accounting for complications such as a faradaic reaction is useful as the worst case for predictions of dynamic characteristics. See Section IV-B-12. Several investigators, recognizing these difficulties, and appreciating the values of tests based upon linear servo control theory, have pointed out that tests consisting of oscilloscopic observations, throughout all typical applications with a given system, are also essential, and are, moreover, the minimum necessary verification by electrical tests of satisfactory dynamic performance. The solution to the dynamic characteristics prediction and test dilemma does not consist in deciding whether only one of these approaches is valid. The certain requirement for oscilloscopic verification of satisfactory dynamic performance can be justified either on the basis of establishing that the anticipated and tested real cases do not deviate significantly from the predictions and tests based upon a model(s) or on the basis that these tests eliminate the need to assume that the stability-margin, model approach is valid for these cases. Tests with all anticipated real cases (loads) cannot be avoided, and all valid tests are based upon and interpreted with servo theory.

7. Dynamic Characteristics of Servo Systems: Speed of Response, Relative Stability and Accuracy

The dynamic characteristics of a feedback system include speed of response, relative stability and accuracy (203). Speed of response and relative stability can be defined in both the time and in the frequency domain. The time domain characteristics are usually specified in terms of the ratio of the output (controlled variable) to a step function input versus time (closed-loop, with load connected). Similar specifications are described by means of impulse and ramp function responses and there are relationships between the responses to the three types of input perturbations. Speed of response is specified in the time domain by the time constant (time to reach 0·632 of the final value), the rise time (time to go from 0·1 to 0·9 of the final value) or by the settling time (time to reach a specified per cent of the final value). The time to reach the first peak of overshoot is related to the frequency of damped oscillation. Speed of response is specified in the frequency domain by means of bandwidth, cutoff frequency or frequency of peak overshoot. The relative stability can be specified in the time domain by means of the per cent of first overshoot and the number of overshoots and in the frequency domain by the peak value of the closed-loop output–input amplitude ratio versus frequency or by the gain and phase margins of the open-loop log gain and phase versus log frequency characteristics. Analytical and graphical techniques are used with calculated and with measured terms (204, 205); for example in the frequency domain, Bode, Nyquist and attenuation diagrams and Nichols' charts can be used for stability characterization. The accuracy of the system is the degree to which the controlled variable is equal to the input command signal. In the case of a potentiostat, the controlled variable is the voltage between the working and reference electrode terminals (from an electronic viewpoint; from an electrochemical viewpoint, the potential that drives the reaction of interest is of ultimate concern) and the input command signal is the sum of control voltages that are applied to the input of the potentiostat.

8. Qualitative Description of Stability

A qualitative, simple description of stability is as follows. At dc, in negative feedback applications, the output of an operational amplifier sub-system is exactly out of phase ($-180°$) with the input signal and some chosen fraction of the output is externally compared with and nulled against the input signal. For operational amplifiers that have a roll-off network, the magnitude of the open-loop gain decreases beyond the first break point with frequency. The phase shift of the output signal increases at higher frequencies of the input signal. With a potentiostat having the load in the feedback loop, additional phase shifts occur as a result of the complex load. If an additional $-180°$ of shift occurs in the fedback part of the output signal, it is in phase with the input signal and feedback is positive rather than negative. The magnitude of the output signal equals that of the error signal times the open-loop gain. If the wave form of the input signal is complex, the signal is mathematically equivalent to a sum of sinusoidal signals of various relative amplitudes and phases. If the system is linear, the response to the signal is the same as the sum of the responses to the component signals. If the open-loop gain has decreased below unity at a lower frequency than that at which the additional phase shift equals $-180°$, then it is to be expected that positive feedback cannot cause oscillation in the presence of these higher frequency components and that transient responses to such frequencies will not be reinforced or prolonged. For this reason, most operational amplifiers include internal networks providing roll-off (decrease beyond the first break point of the open-loop gain). If this is not sufficient in a specific application, additional external tailoring networks can be required. For potentiostats used in controlled-potential coulometry, plug-in networks, each adjusted for a particular cell and application, are useful. Frequency dependent time-lags (phase shifts) of signals arising from or passing through components of the total system are the underlying cause of an unsatisfactory degree of stability. However, the degree and the effect of phase and frequency cacophony of the error signal at the summing point can be controlled to obtain stable operation.

9. Significance of Relative Stability

In an absolute sense, a system is stable if steady-state oscillation is absent and if the transient responses to a perturbation die out eventually. In a practical sense, from an engineering and chemical viewpoint, the time required in attaining and the transient behaviour during the attainment of a satisfactory approach to the steady state are of prime importance. Hence, the relative stability is more significant than the necessary existence of absolute stability. The attainment of both high accuracy (high open-loop gain) and high relative stability requires some compromise and a stability margin should be provided. Response to changes in the input command signal and recovery from noise disturbances must be rapid but sufficiently smooth. If the system is under damped, there will be a temporary oscillatory transient response; if it is critically damped, there will be the most rapid transient response possible without overshoot. A degree of overshoot that does not degrade the results obtained with the system beyond specifications may be tolerated, or even desirable, in order to obtain a sufficient speed of response and sufficient accuracy.

10. Criterion for Satisfactory Stability of a Potentiostat for Controlled-potential Coulometry

It is generally agreed that a potentiostat is not stable unless steady-state oscillation is absent even though the presence of low magnitude steady-state oscillation does not in every case degrade the precision or accuracy of the analytical results. It is neither conventional nor realistic to restrict the definition of a stable potentiostat to those having zero transient overshoot. When a potentiostat is perturbed by a change in its input voltage (an intentional signal from operator changes of command titration voltage, signals resulting from electrochemical changes and mass transfer rate changes at the controlled and reference electrodes, and various unintentional or spurious transient pickup signals), its output response in general includes both transient and steady-state components and some portion of these will (and should) also be imposed on the potential of the controlled electrode. From the electrical viewpoint, the potentiostat is stable if the transient components die (damp) out; from the chemical viewpoint, the potentiostat stability is satisfactory if no significant error in the accumulating chemical information, total Q in the case of analysis, results from transients. If transient responses are small enough and die out fast enough so that the analytical results (accuracy and precision) are not degraded below specifications, it is reasonable to consider the potentiostat to have satisfactory stability. It is coulombs that count.

11. The Attainment of Satisfactory Relative Stability Must be Proved by Test Observations During the Entire Course of Typical Uses of the System

Test observations must be made during the entire course (including pretitrations) of typical examples of all controlled-potential coulometric titrations that are to be done with the system (and after design changes to portions of the system) in order to prove that satisfactory relative stability is attained by the system. Specific titration cells and their contents, including typical samples (the coupled transducer), are an integral part of a controlled-potential coulometric titration system and are not precisely simulated by a single dummy circuit made up of lumped constant circuit elements such as resistors and capacitors. The voltage between the terminals of the reference and working electrodes plays a dual role as the directly controlled variable and as a source of input and noise signals to the summing point(s) of the potentiostat. Stirring action can result in the application of transient noise perturbations to the summing point and demands caused by switching the command signal from one to another voltage are step perturbations. The relative stability of the potentiostat must be satisfactory in the presence of such disturbances. The specific design of the cell, including that of the stirrer and electrodes and the relative arrangement of the electrodes, not only affects the current demand and is a factor that determines whether the reaction driving voltage is equal to or less than the voltage between the terminals of the reference and working electrodes but also, since the cell is in the feedback loop, contributes to the problem of attaining stability. Stability and other dynamic response problems are particularly likely to be substantial at the initiation of the titration (at changes in command signals to the potentiostat), at the end of the first portion of the titration if the cell current demand exceeds the current or voltage output capability of the potentiostat, and during the last part of the titration as the transducer current approaches the background current value. Methods of making these test observations are considered in the next section of this chapter, Section IV-B-12.

12. Methods of Testing Dynamic Characteristics, Including Speed of Response, Relative Stability and Accuracy

A minimum and required degree of testing of the dynamic characteristics of the potentiostatic coulometry system includes oscilloscopic observations of speed of response and relative stability and of balance and electrical accuracy together with electrochemical tests of system precision and accuracy. The oscilloscopic tests, designed and interpreted on the basis of servo theory, and the electrochemical tests establish that the design of the system is satisfactory. They must be repeated whenever the design of a portion of the total system is modified. The necessity of oscilloscopic test observations has been pointed out (188, 194, 206, 207). With an oscilloscope, the dynamic performance of the system is monitored through observations of the transient and the steady-state balance at the summing point(s) and of the voltages at the cell working and reference electrode terminals with regard to transient overshoot and hangup, and the steady-state magnitude relative to the control potential and through looking for the absence of steady-state oscillation. These observations need to be made at the initiation of command signal changes and during the entire course of the test titrations and in the absence of deliberate faradaic reaction. A decision that the relative stability is satisfactory means that the potentiostat does not oscillate (steady state) and that it can be judged to establish and maintain appropriate control of the reaction-driving voltage throughout the typical examples of all titrations that are to be done with the system and that transient overshoot is negligible. The concurrent electrochemical tests of the precision and accuracy of the titration results are also necessary because accuracy is one of the dynamic characteristics but it is not sufficient to only compare the voltage between the electrode terminals with the control voltage, i.e. to observe electrical accuracy: The precision and accuracy of the answers that are calculated from Q_T are the ultimate criterion of system performance. It is coulombs that count. Complementary methods of evaluating and of arriving at a satisfactory dynamic performance have been used for such systems. The Bode attenuation diagram, transfer function approach has been applied to this problem (186, 187, 194, 208–227). Applications include the analysis of operational amplifier sub-systems (216) and the tailoring of their frequency response (215) and the analysis of the dynamic characteristics of potentiostats with a linear dummy load; Booman and coworkers and others have also used these methods of prediction and tailoring for the case where the cell is connected to the potentiostat. Potentiostatic instrumentation and the use of these techniques have been reviewed by Schroeder and Shain (187). Harrar recommends that the minimum electrical tests should, in addition, include the monitoring of the dynamic response of a given system to an impressed square wave (194). The use of square wave, step or unit impulse test signals is more than an empirical method of observing dynamic response. For example the frequency response function and the unit impulse response of a linear system form a Fourier transform pair. Both functions contain exactly the same information. Automated instrumentation exists for making such tests and digital computers have been used for such purposes. Tests with square waves or pulses or step functions have been described (225). The theory and procedure of these tests and also of sine-wave and noise tests and signal analysis by digital techniques have been described (227). Theory for the interpretation of stability from Bode plots is based upon the properties of linear fixed lumped parameter systems having minimum-phase transfer functions. According to Chestnut and Mayer, the Bode analysis may, with caution, be applied on a broader basis, and, it is possible for a servo system to have a minimum-phase overall transfer function even though a system element does not have a minimum-phase transfer function (224). At a specified working electrode voltage, for a particular cell and electrode and reaction combination, the validity of using a simple linear equivalent circuit to represent the coupled transducer sub-system and of analysing and tailoring for stability via Bode plots would seem to be on a sounder theoretical basis in the absence of a faradaic reaction than with a faradaic reaction and in the absence of complications such as adsorption. Also, faradaic reaction and double layer capacitance are not necessarily independent. It is usually assumed that the effect of the counter electrode, if it has a very low impedance, can be neglected in the analysis. The equivalent circuit used by Booman and coworkers is for the case where no faradaic reaction, adsorption or other complication is taking place. It is important to note that it is reported that a faradaic reaction reduces the phase shift in the transducer and has a stabilizing influence on the system (223). Thus, such calculations have been regarded as considering the worst case as far as stability is concerned, with the possible exception of cases where negative impedance electrode effects occur (223). We conclude that both electrical tests, guided by servo theory, and electrochemical tests are required.

C. Coulometer

A coulometer is a circuit sub-system that is designed to measure coulombs to the degree of accuracy and precision needed in its applications (178). The coulometer portion of coulometry measurement systems provides an output that is directly proportional to the total number of coulombs that were consumed in the reaction at the working electrode in the transducer. The output can be direct reading in terms of equivalents or in grams of a specified reactant by providing scaling factors with voltage dividers.

The coulometer can be an analog current–time integrator, built with a chopper stabilized operational amplifier into or as a module within the instrument and voltage and current biased to compensate for the particular integrator drift rates at the time of adjustment. See Section III-C-2, Analog Integrator, of this chapter. A notable performance limitation of integrators is their finite output drift rates, caused by input voltage and current offsets, which change from time to time and which are temperature dependent. These offsets are compensated at the time of bias adjustment, leaving only residual offsets due to drifts from the input values at the time of adjustment. The effect of the residual input voltage offset is minimized by choosing large values of RC and the effect of the residual input current offset is minimized by using a large value of C. Also, the source impedance, which includes R, must not be excessive since amplifier input current will be drawn through it and create an iR drop; consult the amplifier manufacturer's data sheets for the recommended size limit of R; however, R must be high enough to not overload the input signal source. The uncompensated part of the integrator output drift, if significant in terms of coulombs for the amount reacted in the transducer, degrades the precision and the accuracy of the analytical results. In many cases, the uncertainty in the value of background coulombs is a

greater source of error, however. It is emphasized that design steps that decrease titration time, in addition to enhancing the ratio of faradaic coulombs
to background coulombs, also minimize the small errors caused by the
changes (subsequent to the last adjustment of the drift correction circuitry) in
the integrator drift rate if the integrator output is read shortly after the termination of the reaction. For the resistors, we have used Daven type 1252
wire wound resistors which have a temperature coefficient of 0·001 per cent
$°C^{-1}$ and a stability of 0·01 per cent ΔR per year. The quality of the capacitor
may be the factor that limits the performance of an analog integrator, if an
operational amplifier of highest performance is selected—an example of a
suitable amplifier is Analog Devices model 232 K (228)—and if good design
and construction practices are used. Rather large, 10 μf, capacitors of very
high quality are used in the analog integrator coulometers in the ORNL controlled-potential coulometric titrators. The capacitor in model Q-2005-X50
is the Stabelex-D manufactured by Industrial Condenser Corporation,
Chicago, Illinois. In the model Q-3094 coulometer of the model Q-4010
coulometric titrator, it is a polystyrene dielectric capacitor, part no. 521-
1388 X, developed by Electronic Associates, Inc., West Long Branch, New
Jersey 07764 for their PACE analog computers (228). For the 521-1388 X
capacitor, from 21·1 to 29·4°C, the temperature coefficient is 0·01 per cent ΔC,
the stability is 0·01 per cent ΔC over 1 year, the dielectric absorption is 0·02
per cent and the insulation resistance in our experience appears to be as
high as 10^{12} ohms. Both internal and external leakage paths can exist across
the terminals of the capacitor. The latter are minimized through the use of
good insulation during the construction of the coulometer. High quality
insulation is particularly important in humid environments. Dielectric
absorption can cause errors that can be significant even with high quality
capacitors, especially at lower faradaic coulomb levels, unless enough time
is allowed, as much as 3 minutes, to discharge the capacitor so that dielectric
equilibrium is reached. Also the electrical calibration must be independent
of the voltage across the capacitor. All points that are directly connected to
the input terminal of the analog coulometer are locations where spurious
charges can be fed into the integrator. Such leakage is kept to a minimum
through good construction practices that produce excellent insulation and
negligible ground loops. Also, construction must avoid 'thermocouple'
voltage generation at the junctions of dissimilar metals. Low thermal EMF
solder should be used. The pickup of transient interference noise pulses can
cause the storage of a non-trivial charge. In environments in which pickup is
a problem, pickup must be minimized by shielding the source of interference or the input to the coulometer. Particularly if low-valued integrator
range switching resistors are used, low resistance, reliable switches are
essential. Gold-plated contacts are recommended. Characteristics of the
operational amplifier that affect the performance of the integrator include

input impedance, input current, input offset voltage, the sensitivity of input current and offset voltage to temperature and to power supply variations, and the gain at operating frequencies. Moreover, the amplifier must be able to deliver the required current. The use of presently available high quality operational amplifiers and resistors and capacitors and good design and construction practice can readily result in analog integrator coulometers that can be depended upon when the needed relative standard deviation or the error of the integration is as low as 0·05 per cent. With the ORNL model Q-2005-X50 titrator, the charge remaining in the charged coulometer 5 minutes after charging is 99·98 per cent. With the ORNL model Q-3094 coulometer portion of the model Q-4010 coulometric titrator, 99·998 per cent of the charge remains after 5 minutes. The electrical calibration of the model Q-3094 coulometer in each of the ORNL model Q-4010 coulometric titrators is routinely scheduled to be checked every 6 months when it is reset to $\pm 0·01$ per cent. The short-term stability is about $\pm 0·02$ per cent. Over 2 years of operation, the calibration has not changed more than $\pm 0·05$ per cent over 6 month intervals. The M-T model 3 controlled-potential system (229) of Harrar and Behrin (194) is also very stable; the integrator drift is $\pm 0·5$ mV/15 minutes and the calibration stability is $\pm 0·1$ per cent per year (229).

If high precision or accuracy is needed, digital methods of integration should also be considered. Devices having high immunity to noise interference must be selected. Several kinds of digital methods exist. Digital devices have been used to read the output voltage of an analog integrator; the performance achievable should be limited by that of the analog integrator itself. The necessity of using a voltage measuring device of sufficient accuracy and precision for this purpose, however, is sometimes overlooked. A digital method of interest involves the use of precision current-measuring resistors, voltage-to-frequency converters and accumulating digital counters. There are several embodiments of voltage-to-frequency converters. One type incorporates an analog integrator, a discriminator (voltage-level detector), and pulse circuitry. The pulse circuitry produces an output pulse for every unit increment of charge put into the integrator and simultaneously, by one of several methods, pumps a precise reset unit charge of opposite polarity into the input of the analog integrator. Some undesirable reset methods cause significant dead times. The analog integrator output voltage cycles over a small range. It is usually asserted that if the integrating capacitor is cycled over a very small voltage range, error due to capacitor hysteresis is minimized; also, a smaller, i.e. cheaper, capacitor can be used. It is apparent that a wide range of performance qualities are obtained with specific digital integrators, depending upon the characteristics of the sub-systems. Some digital integrators are not as good as the best analog integrators. A comparison of the relative merits of the analog and the digital kind of current–time integration

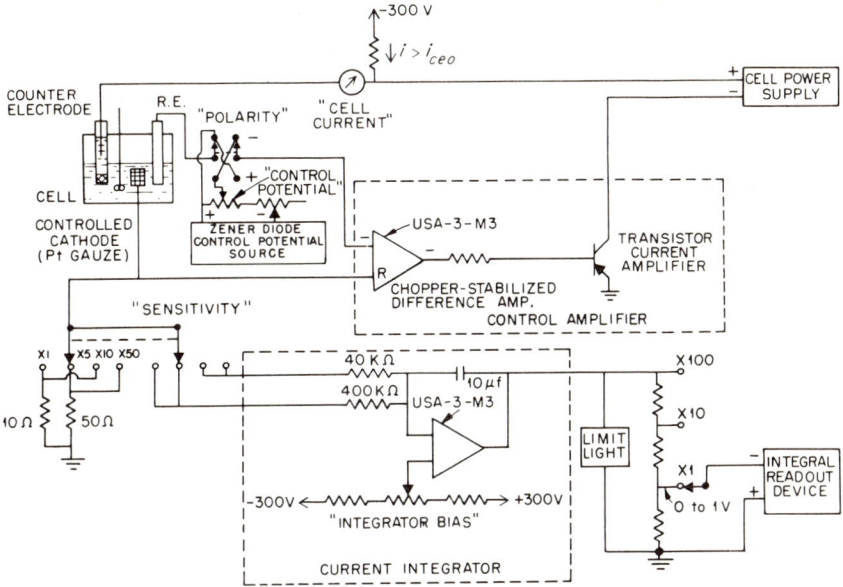

Fig. 29. Block diagram of ORNL model Q-2005-X50 controlled-potential coulometric titrator, reduction mode.

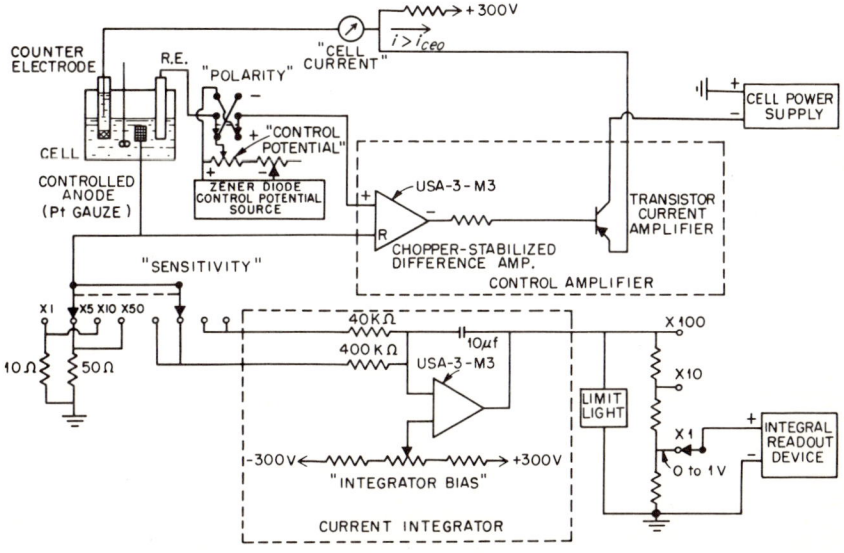

Fig. 30. Block diagram of ORNL model Q-2005-X50 controlled-potential coulometric titrator, oxidation mode.

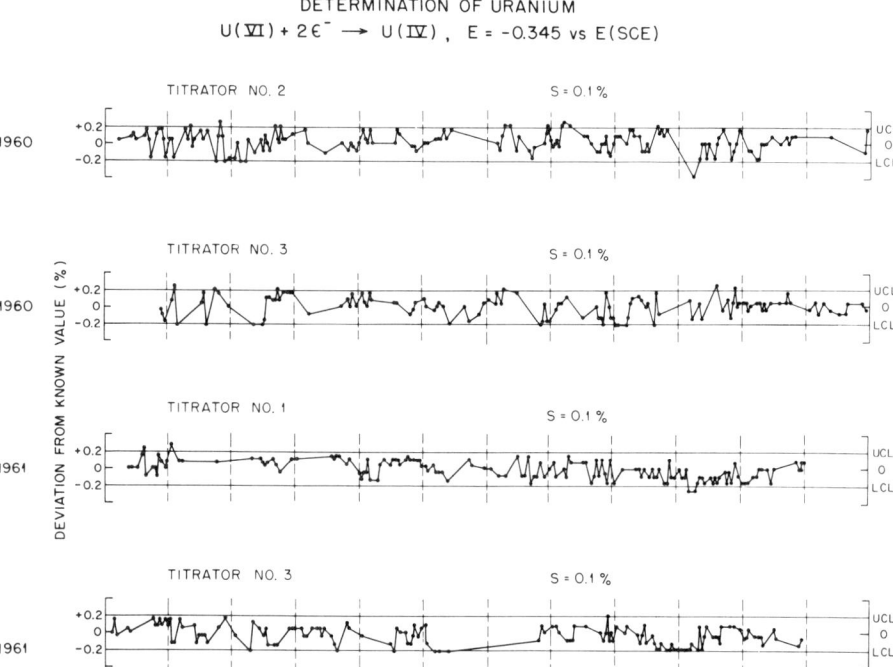

Fig. 31. Control charts, controlled-potential coulometric determination of uranium.

has little validity except in the context of the specific performance characteristics of the methods, parts and devices used for the construction of the respective coulometers. With both methods of integration, performance may be limited by drifts due to the finite temperature coefficients or by calibration changes caused by changes in values with time of critical components of the coulometer. Many published papers attest to the fact that both types of coulometer can be satisfactory. The relative merits are dependent upon the quality of the design (including the selection of the best parts and devices from those available at the time of design) and upon the quality of construction practice in each case.

D. ORNL Controlled-potential Coulometric Titrators

For a number of years at Oak Ridge National Laboratory, ORNL, controlled-potential coulometric determinations of ~ 5 milligrams (~ 40 micro-equivalents) of uranium have been made with a relative standard deviation of 0·1 per cent by means of the ORNL models Q-2005 (188, 195–197) and Q-2005-X50 (188, 195–199) electronic controlled-potential coulometric titrators. Model Q-2005-X50 is commercially available (199). In order to indicate the functional arrangement of the model Q-2005-X50 titrator, block diagrams are given for its operation for reduction reactions, Figure 29, and for oxidation reactions, Figure 30. Control charts are kept for the laboratories that use these instruments. An example, showing that a relative standard deviation of 0·1 per cent is maintained for analyses for uranium, is given in Figure 31. The data in Figure 31 were obtained in 1960–61. Similar data have been obtained throughout the subsequent decade. The ORNL model Q-2564 high-sensitivity coulometric titrator (189, 200–202) was designed primarily for the titration of amounts less than 40 microequivalents. These titrators (both models) are also used at ORNL for the controlled-potential coulometric determination of other materials, including plutonium. The ORNL Master Analytical Manual Method Nos. 1 003029 and 9 003029 (197) has been revised so that it applies both to model Q-2005 and to model Q-2005-X50. The information in the cited Methods (197, 202) includes check out and test procedures and operating instructions for the use of these titrators to analyse samples. As an example, detailed procedures are given (197) for the following tests and checks of the models Q-2005 and Q-2005-X50: 'check for proper grounding, check on the potential control and on the current meter, check on the automatic current cutoff, check on the integrator bias adjustment and extent of drift, check on the overload indicator, in-service oscillation check, test for proper stirring action, calibration of the instrument'. In addition, the following trouble-shooting procedures are given: 'general location of trouble, specific location of trouble in the instrument: power supply, potential control amplifier, cell power supply, transistor, integrating amplifier and other components'. The preparation of the above procedures is an important part of the provision of information for the Instrument User portion of the system. The latter is also undertaken through the preparation of CAPE packages (196, 200) which include a set of electrical and mechanical engineering drawings and parts lists, specifications and photographs that can be used for the fabrication of the instrument. For the titration of less than four microequivalents, the performance of the model Q-2564 titrator is superior to that of the models Q-2005 and Q-2005-X50. With the model Q-2564 titrator, using a smaller volume in a 2 ml capacity cell and by careful attention to procedural details such as the application of suitable blank corrections, it is possible to determine plutonium at a level of 0·05 microequivalent with a relative standard deviation and bias of ~1 per cent (103). In the latter work, it was found that I–Q readout is a useful adjunct to the titrator; initial titration currents were less than the output capability of the titrator and no hangup problems occurred. The maximum output current of the potentiostat of the model Q-2564 titrator is ~8 mA. Later, it was discovered that when the model Q-2564 titrator is used

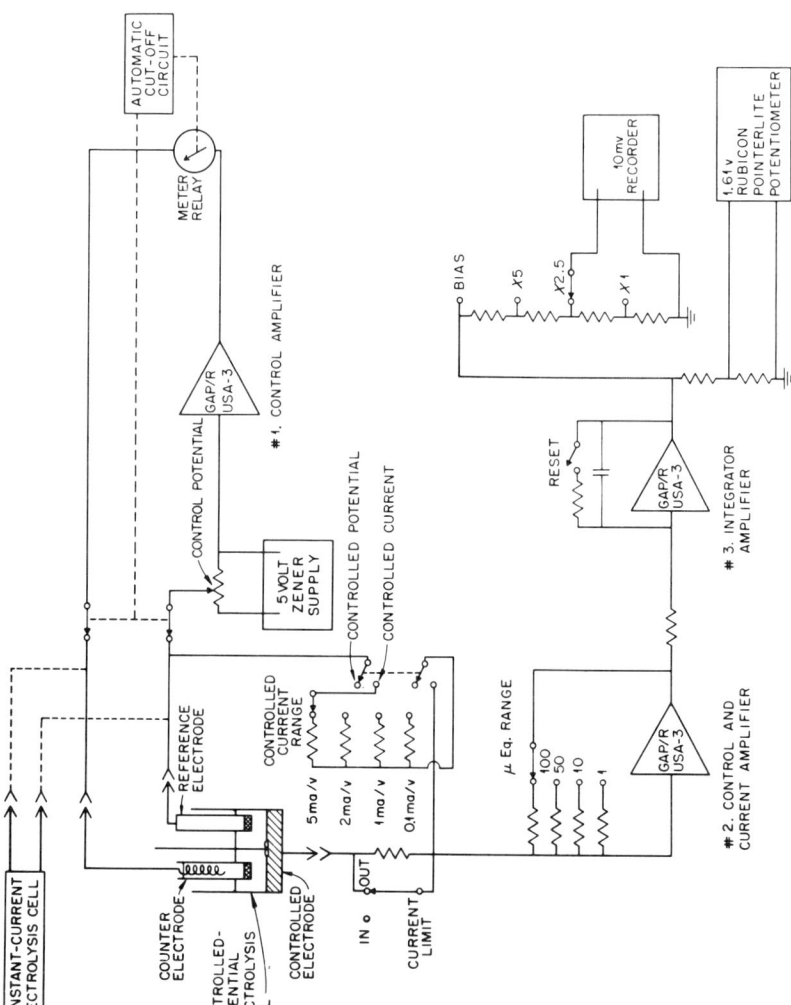

Fig. 32. Block diagram of ORNL model Q-2564 high-sensitivity coulometric titrator.

under conditions such that the potentiostat amplifier is voltage limited during a part of the titration, that titration errors can result during the recovery time (hangup) from voltage limit. By means of a simple circuit modification (201) this source of difficulty has been eliminated. See Section IV-B-4 of this chapter. A block diagram of the model Q-2564 titrator (as modified) is given in Figure 32. Drawings for the construction of the model Q-2564 titrator (200) have been revised to include the above modification and the Master Analytical Manual method (202) has also been revised. The prototype and five other model Q-2564 titrators at ORNL have been field modified. We recommend that this modification be made to other existing titrators of the Q-2564 design. The method of attachment of the I–Q readout mode to the ORNL Q-2564 titrator has been described (201). One of the model Q-2564 titrators has been used in the controlled-current mode (230).

A newer high precision coulometric titrator, model Q-4010, has been designed at ORNL (228, 231). It is taking the place here of models Q-2005-X50 and Q-2564. The potentiostat, current amplifier, meter and an analog coulometer are built in four NIM modules. The performance of the coulometer has been quoted in Section IV-C of this chapter. The titrator output capability is ± 20 V at 200 mA. A feedback circuit is used to limit the cell current to a selected percentage of 200 mA, usually to 90 per cent, by decreasing the control potential. Electrolysis can be terminated automatically when the current falls to a preselected value. Plug-in stabilizing networks are used. A compensating current is available to correct for the continuous, constant background current. Solid-state operational amplifiers are used. Method design studies, including the evaluation of advantages obtained by standard addition and of the relative merits of I–Q readout and of titration to completion, as well as the design of the transducer to optimize Q_F/Q_B, have been scheduled. In 1971, the sixth model Q-4010 was fabricated for use at ORNL for analyses of uranium and plutonium in reactor fuels. These titrators are used for all ORNL accountability analyses for uranium. The titrator can be used either in the controlled-potential or the controlled-current mode.

An automatic, controlled-current, coulometric Karl Fischer titrator has been designed (232). Potentiostatic configurations are used and unique features are provided by means of 14 operational amplifier sub-systems (exclusive of the power supply); etched wiring board construction is used. See Figure 6 of this chapter.

V. DIGITAL SYSTEMS FOR ELECTROCHEMISTRY

Many performance advantages have been obtained since about 1957 through the use of analog operational amplifier sub-systems in measurement systems for electrochemistry, including polarography and coulometry. The use of digital sub-systems is a rapidly growing and important research subject in the area of the design of total systems for measurement. Digital devices, including computers, have already been used for numerous kinds of analytical work, including electrochemistry. Further advantages in system performance are being achieved through the inclusion of digital computers in the total system. Moreover, both analog and digital sub-systems should be considered rather than either type alone. To obtain better information for research and for analysis, instrumentation systems should be designed with optimum mixes of analog *and* digital functions implemented with analog *and* digital sub-systems.

A. Use of Ensemble Averaging to Enhance Sensitivity

The subject of enhancing sensitivity through ensemble averaging (also called signal averaging) was introduced in Section II-F of this chapter. The principles and advantages of enhancing the sensitivity of many kinds of measurements of interest in analytical chemistry through correlation in the frequency domain by filtering, active integration, tuned amplification and synchronous detection together with correlation in the time domain by convolution and by ensemble averaging and by digital filtering with digital instrumentation have been described and exemplary illustrations are given elsewhere in some detail (3). Also, information essential to the effective implementation of correlation in the time and frequency domains is reviewed with illustrations and a selected bibliography. The subjects considered are as follows:

    I.   Description of method
         A. Premises of method
         B. Significance of signal-to-noise ratio S/N
         C. Definition of measurement system
         D. Optimal design of system elements
    II. Principles of S/N optimization and enhancement
    III. Principle and applicability of enhancement of optimized signal-to-noise ratio with an ensemble averaging computer
    IV. Characteristics of signal, S, and of noise, N
    V. Noise
         A. Classification and description of noise time series
         B. Interference noise
         C. White and $1/f$ random noise
         D. Sporadic impulse noise
         E. Noise in flame photometry
    VI. Relative performance of various circuit elements for signal measurement with optimized S/N, and of subsequent information processing systems for further sensitivity enhancement
         A. Basis of retrieval of signal from signal-with-noise

optimize and to enhance S/N and of performance capabilities in exemplary applications

A. Filters
B. Superimposition of replicate measurements
C. Lock-in amplifier system
D. Boxcar correlators
E. Analog correlator
F. Masked correlation spectrometry
G. Fourier transform spectroscopy
H. Ensemble averaging with digital systems
I. Uses of software-programmed digital computers for ensemble averaging
J. Development of hardware-programmed ensemble averagers
K. Uses of adaptations of multichannel analysers for ensemble averaging
L. Uses of hardware-programmed digital computers (including specific purpose ensemble averagers) for ensemble averaging
   References

A specific application of ensemble averaging to substantially increase the sensitivity of flame spectrophotometric measurements has also been presented; other advantages are also cited (4). Rogers and coworkers have demonstrated its utility in gas chromatography (233). Software-programmed general-purpose dedicated-purpose and time-shared digital computers and hardware-programmed specific and general purpose digital computers are available to obtain this advantage. The utilities for polarographic measurements of ensemble averaging and other digital functions are being investigated at ORNL; see Section V-D of this chapter. A hardware-programmed general-purpose digital computer system, Fabri-Tek FT-1064, and, later, a software-programmed Digital Equipment Corporation PDP-8/I dedicated digital computer have been used.

Results detailed in the following papers illustrate the power and usefulness of ensemble averaging in electrochemical investigations and measurements. Perone, Harrar,.Stephens and Anderson demonstrated a ten-fold enhancement in the sensitivity of fast-sweep derivative polarography (234). Goldberg and Bard have used it in simultaneous electrochemical and electron spin resonance measurements (235). Winograd and Kuwana have used it to improve signal-to-noise ratios for internal reflection spectrometry at optically transparent electrodes (236). Zeller and Osteryoung have used ensemble averaging and digital filtering in a computerized electrochemical system for pulse polarography (237). The paper by Savitzky and Golay remains a classic on the subject of digital filtering with simplified least squares procedures (238).

B. Acquisition and Analysis of Digital Data

The development of techniques (including programming) and digital instrumentation and assemblies thereof for the acquisition and analysis of digital data is not a new subject but it remains of interest. Multichannel analysers, signal averagers and signal correlators and digital computers based upon fast and large magnetic core memories and on clock oscillator control of timed steps have been available for about two decades. As better (particularly, larger and faster) digital equipment becomes available, and at lower costs, better digital data acquisition and analysis becomes increasingly feasible. The importance of work that has been done to digitize electro-chemical and other analytical information (via analog-to-digital converters, ADC's and multiplexer and associated interface apparatus) and to perform subsequent software (coded instructions on punched tapes or cards or on magnetic core or tape) or hardware (wired, with plug-in module and switch selected choices) programmed computer computations extends beyond these immediate aims. These techniques are also necessary steps in ensemble averaging and in computer-aided and directed automated analyses. Examples of publications describing work that has been done should be consulted for details; in some cases, the computation (data analysis and transformation) is performed as a separate step or in other cases with a separate time-shared computer and in yet other cases, at the time and site of the experiment itself, with a dedicated computer. Examples of the acquisition and analysis of digital data in electrochemistry are of special interest (239–245). Also, Anson has discussed the subject of cybernation in electrochemistry (246) and Rogers has discussed cybernation for this and other areas of analytical chemistry, including computer-aided and directed analyses (247). The precision and speed with which data can be digitized is becoming vastly improved, as faster high capacity memories and faster cycle times become available. The fact that the subject is of substantial interest is shown, for example, by the title of a report from the 1967 Pittsburgh Conference on Analytical Chemistry and Applied Spectroscopy: 'Analysts Rush to Put Data on Computers' (248) and by the fact that most meetings since that time have featured or included topics on the subject. A paper by Brown and Mueller on the subject of selecting a computer system for on-line applications (though the tabulation of features of equipment then available is now out of date) contains general information and advice that is still pertinent (249). The critical importance of the software aspect of the subject is brought out by Williams and Bailey (250) and it remains of critical importance.

Clynes has pointed out (251) that some ensemble averagers can also be used to precisely digitize and record without a photographic process single measurements of relatively fast wave forms or transients for subsequent readout at the slower speeds of XY plotters or digital printers. Other ensemble averagers, the wave form comparison type, do not provide this useful mode. At present, it is possible to obtain general purpose hardware programmed digital computers which can be used for ensemble averaging and, in addition, whose performance capability for digital data acquisition approaches that of small software programmed general purpose computers but with less versatility and others with interfaces to software programmed digital computers.

Hardware programmed digital computers that can be used for ensemble averaging and also for digital data acquisition and analysis and for other purposes as well use various internal methodological details and principles to sample the signal during the dwell time at each address and have various combinations of built-in features and accessory versatilities. Although the circuitry, construction and the digital and mathematical principles involved are rather advanced, these instruments (particularly, those that provide a normalized output that is always directly related to and calibrated in terms of millivolts of input signal and that include a number of built-in data manipulation and readout capabilities) are as easy to use as a typical high-performance laboratory oscilloscope. Characteristics of these instruments, which differ from model to model, that particularly affect the information about the signal that can be acquired include the following. The memory capacity (number of addresses or words and number of bits that can be stored at each address) affects the resolution or detail with which the signal can be stored and the number of sweeps that can be made in order to enhance by ensemble averaging the signal-to-noise ratio in proportion to the square root of that number. The duration of the minimum available dwell time per address, in accordance with the Sampling Theorem, affects the maximum Fourier component of the signal that can be stored without distortion. The digitizing speed (number of counts, full scale, that can be stored at each address during a particular dwell time, per sweep) and the stability affect the precision (particularly, for the single sweep digital data acquisition mode) with which the signal can be measured. The number and kinds of switch-selected internal transformations of signals stored in memory quarters and halves (fields) and the whole, prior to readout, as well as the number and kinds of readout options and the degree of provision, if any, for the control of the experiment determine the extent to which the instrument approaches the versatility of a software programmed digital computer while featuring the ease of use of a hardware programmed instrument. Examples of the present state of development of hardware programmed instruments are the Fabri-Tek-FT-1064 Instrument Computer and its successors, Nicolet Instrument Corporation Series 1070 and 1080 Signal Averaging and Data Acquisition Systems (252).

At the time of planning of work involving the acquisition, processing and application of analytical data, it is necessary to consider and compare the implementations available at that moment. Implementations include:

assembly of an individual analog-to-digital converter and other interface components whose output can be batch processed or as an interface unit; use of programmable desk calculators; use of a dedicated purpose software programmed computer; use of a time-sharing system; use of an on-site, real-time, hardware-programmed digital computer; use of hierarchical systems. Also, it is impossible to meaningfully select among the options except in the context of specific needs and opportunities and capabilities.

Substantial guidance in the above selection problem is contained in the following group of papers and reviews. Also see references in Section V-C of this chapter. In a review of automated techniques in pharmaceutical analysis, Kuzel, Roudebush and Stevenson present much information about digital systems in analytical chemistry (253). Smith has reviewed applications of on-line digital computers to chemical instrumentation (254). Childs, Hallman and Perrin have reviewed applications of digital computers in analytical chemistry (255). Perone has presented an introduction to digital computers and the fundamentals of on-line computer applications in the chemistry laboratory (256). Venkataraghavan, Klimowski and McLafferty have described uses and relative advantages of different kinds of digital computers for chemical research, with applications to high resolution mass spectrometry, and comments about interfacing problems (257). Ramaley and Wilson have described a general purpose system for computer data acquisition and control (258) and a high-speed autoranging amplifier for digital data acquisition (259). Perone has described his well-known laboratory course on digital computers in chemical instrumentation (260). Wiberg has discussed the small computer in the laboratory (261). Dewey has described principles, functions and advantages of the small dedicated computer in the laboratory (262). Margoshes has discussed advantages of making the computer a part of analytical instrumentation (263). Eggert, Hicks and Davis have described a computer system which will completely perform and control a kinetic experiment by means of a time-sharing system (264). Perone has surveyed computer applications in the chemistry laboratory (265). Dessy and Larsen have discussed the impact of minicomputers on the laboratory and have described how several systems have been designed to solve specific problems and needs and have presented the trade-off and option problems involved in the design or purchase of interfaces and computers (266). Frazer has described the efforts of ASTM Committee E-31 to establish sound guidelines to assist scientists and engineers in taking a systematic approach to the complexities of laboratory automation (267). Schroeder and Walter discuss the time-shared computer and chemical laboratory applications (268). Klopfenstein has described principles and practices of laboratory automation with hierarchical computers (269).

A digital computer system can be used to control or to automate experiments, and, at the same time, to acquire and process data from the experiments. The system output can be interfaced to another larger, central digital computer, perhaps provided with more readout options. In addition to its use for the enhancement of the signal-to-noise ratio by ensemble averaging and digital filtering, it can be used as an excellent laboratory test instrument to collect data essential to the rational design and testing of components, transducers and circuits of other measurement systems. The output of a portion of an instrument can be quantitatively compared to the input. For example those of a voltage follower or a precision peak follower or a sample and hold circuit or an amplifier can be compared, or, the output of a derivative computing circuit can be integrated and compared with the input. The system can be used to obtain the logarithmic or the Nth integral transform of a measurement. The response (single or averaged observations) of instrument circuits to standard test signals can be obtained and then read out to slower recording devices. During the design of a measurement system, it can be used to establish which measurement conditions are optimum.

The related subject of the advantageous use of an optimum mix of digital functions and devices and analog functions and devices to design instrumentation for electrochemistry has received appropriate attention. Bezman and McKinney demonstrated advantages, particularly in the design of the ramp and cyclic scan generator and in its inherent compatibility with computer interrogation and control, in their design of a controlled-potential polarograph (270). Myers and Shain have described their use of a combination of operational amplifiers, diode gates and flip-flops to construct a versatile signal generator for electroanalytical experiments, especially cyclic voltammetry (271). Jones and Perone, following their determination and design of an optimum set of experimental parameters and techniques for stationary electrode polarography through experiments with a dedicated computer, have then hardware implemented them with integrated circuit modules (272).

C. Computer Aided and Directed Automated Analyses

Computer programs can include instructions for executing steps of timing, sequencing and sampling and data identification, data storage steps and data processing steps. The latter are instructions for executing functions consisting of mathematical operations such as ensemble averaging and digital filtering and advantageous data transformations and output presentations. These are all very important and useful conventional digital functions. Another important digital function is that of making measurements and operations feasible that would otherwise be too tedious and formidable; examples include high resolution mass spectrometry, Fourier transform spectrometry and the automated many-step chemical synthesis of complex materials. A very basic fact is that computer programs can provide particularly powerful digital functions consisting of a collection of sets of logical operations. A set in the collection is selected for execution by branching to it in accordance with which is true among a group of true or false comparison queries. This general capability of computer programs enables the invention, preparation and exploitation of additional digital functions that are of great value in measurement systems for research and for analysis. Digital computer sub-systems, as an integral component of measurement systems, whether dedicated, time-shared, or otherwise, enable the systems to obtain information of higher quality through digital acquisition followed by processing with steps like ensemble averaging and digital filtering, through advantageous data transformations, through adaptive processes, through interaction with the experimenter and the results of the experiment in progress, and through graphic presentations which greatly assist interpretation. Incorporated digital computers enable the control of analytical experiments and analyses and, in general, provide better answers and information of higher quality through digital functions in conjunction with analog functions provided in other parts of the system. The digital computer, in contributing to the capabilities of the measurement system, derives its major virtues both from the circumstance that it is a super calculator and the fact that it is much more than a super calculator. Nevertheless, only those well-defined problems for which programs can be written can be solved. A vital area which must be developed and applied more widely by means of digital functions is that of measurement system performance diagnosis and monitoring and the programmed testing and documenting of the validity and reliability of analytical answers, especially for research and for important analyses.

In computer aided and directed automated analyses, in an ultimate implementation, not just the data acquisition and data handling steps but all steps of the analysis are automated. With this type of measurement system, there are resemblances between automated analyses in an analytical chemistry laboratory and continuous in-line or on-stream analyses for industrial processes. One difference in these two application areas has to do with whether the sample is brought to the measurement system and the results are telemetered to the customer or whether the system is not only at the site of the origination of the samples but, ultimately, also controls the characteristics of the samples to be within set limits. There are two application areas in the analytical chemistry laboratory: computer aided and directed automated analyses are not only of interest as a means of increasing the efficiency or quality of measurements for service sample analyses but also for analytical measurements for research purposes. From the system design viewpoint, the immediate incentive for the design work can arise from any of these three application areas but the design work should draw upon the state of the art in all of these areas. With ultimate computer aided and directed automated analysis systems for service or research analytical laboratories, in addition to the automation of the digital data acquisition and data transformation steps, the sampling steps and the pretreatment steps are automated, e.g. sparging in polarography and pretitration in coulometry, the specific desired information (e.g. concentrations or reaction parameters) is automatically calculated and optimally presented and the quality and validity of the answers from the system are self-checked by the system itself at suitable, programmed intervals through analysing standards. The standards are, respectively, similar to the samples (e.g. standard reference materials) or known to obey the theoretical equations and assumptions involved in the calculations. A report published by Hasler ('Cybernation in Spectrochemistry') includes a good summary (p. 34A) of the self-testing functions generally possible in analysis applications of these systems (273). With a dedicated purpose computer all of these analytical steps are accomplished at the time and site of the experiment. A dedicated computer of adequate size is also advantageous for research work because it is more feasible to modify the programming as the experimental work advances.

A potential advantage of computer aided and directed automated analyses is analyses or measurements having better precision, since the computer repeats all of the analytical steps more exactly than may be done by human analysts. A second and very important advantage, both for service and for research areas, is the refinement with which calculations can be made from the data if the computer capacity is large enough; for example correlation steps such as convolution of data with theoretical equations and ensemble averaging and digital filtering, or all, can be readily included. It is of advantage to fully automate reaction rate methods; such methods are very important in areas such as clinical analyses and fast and low cost but high quality, reliable automated analyses are required for any substantial medical screening program. Another application area of special interest is any case where many samples are to be analysed and the range of characteristics of the samples is such that the required method details are well defined for all samples. Automation of this type of work is inherently appropriate and automation can provide time and cost savings in this case in addition to the other advantages.

A question of obvious immediate concern is the relative accuracy and precision with which data can be acquired and transformed by analog and by digital methods. Analog methods (e.g. by means of operational amplifier sub-systems) are readily capable of performing with precision and error as good as 0·1 per cent. (With careful design and use, somewhat better results have been obtained.) In the case of digital methods, there is an inherent conflict between the speed with which each individual data point must be digitized and the number of bits (full scale) that can be put into each memory address for each data point. Also, the data acquisition and processing speed may be limited by the fastest available turnaround or cycle time or jeopardized by uncertain priorities. The question as to the relative accuracy and precision of analog and of digital methods can be answered only in terms of the performance of the specific digital equipment that is available and that can be afforded. Ultimately, at least, digital methods should be capable of doing better than 0·1 per cent, but much available equipment cannot, except for slower speed work or with larger computers.

The meaning of the phrase 'small dedicated computer' appears to be: A computer (and required peripheral equipment) whose cost and performance are such that it could be purchased and used at the site of the experiment for a specific task. Correspondingly, a 'large general purpose computer' is a computer which has more capabilities but which costs too much to be purchased and used at the site of the experiment.

The advantages of small computers are that it is easier to buy them and that they are indeed adequate for many system applications. The disadvantage of small computers is that they are small computers. They tend to appear smaller after purchase, when first put into use. Typically, about 75 per cent of a 4 K-word core memory capacity is used up for the programming steps, even when machine program language is used, limiting the operations that can be performed and leaving, sometimes, a minimal capacity for data and for data manipulation. Also, the speed of acquisition and the number of bits per data point may be serious limitations. These machine languages use up less of the memory capacity than the much easier to use higher level languages that can be used economically with larger computers. The immediate cost of small computers (and required peripheral equipment) may seem relatively small as compared to that of larger computers, but the savings may be more apparent than real because of the resulting performance limitations and the difficulties and expense of writing software that is cleverly designed to require a minimum allotment of the memory capacity. Expensive, complex software must be changed with every change in the requirements of the experiment. However, useful work has been and will be done with systems incorporating small computers. The versatility of such computers, within the limitations of their capacity and peripheral accessories, and as exploited by the ingenuity of the experimenter, is a prominent feature.

The advantage of using a time-shared computer instead of using a smaller computer incorporated at the site of the system is presumably that a larger computer of superior performance is available than could be purchased for use on site as a dedicated purpose computer. Where a substantial number of instruments (e.g. clinical analysis instruments, laboratory gas chromatographs or industrial process instruments) are to perform many sets of specified and clearly defined analyses on similar samples, and where priority problems are solved, time shared systems are very successful. A number of such systems are on the market. The disadvantages of time-shared systems are that expensive interface equipment must be at the site of the experiment and expensive software is required so that the total system cost is far from directly proportional to the cost of the computer element of the system, the quality of the data as received at the computer may be degraded by the transmission thereof, and the software programming is relatively much more complex and expensive. If experimental research work is to be done with the system, the required changes from time-to-time of programs or of sub-programs are difficult as compared to the case with a dedicated computer. Moreover, in some research and analysis applications, the data are lost for ever unless the computer is available at the moment that the data originates unless a priority interrupt or a local data memory store is available. The characteristics of dedicated and time-shared systems have been compared and those of the time-shared system at the University of Illinois have been discussed by Secrest (274).

Summary descriptions of the system design information gain through experience in the substantial program in this area being carried on at Lawrence Radiation Laboratory, Livermore, California, discussions in some detail of the relative merits of several different approaches to the subject, and examples of applications there and elsewhere have been published by Frazer (275, 276). Frazer points out (275) that such systems can add '... new dimensions to experimentation ...' and '... accelerate scientific advancement'. He also stresses the fact that '... the creation of sophisticated software is an expensive proposition. However, the cost of hardware components is rapidly decreasing. Therefore, examine carefully the possible hardware/software trade-offs and their ability to answer to your needs. Third, the system should provide a maximum of freedom for the imaginative scientist. Any system that severely limits the creative, hardworking scientist is to be avoided'. Other information about this program at Livermore and about similar programs elsewhere is available in a report from the 21st annual Analytical Chemistry Summer Symposium (on computer assisted analytical chemistry) (277).

The work in this area at Lawrence Radiation Laboratory, Livermore includes a semiautomated electrochemical measurement system. A dedicated computer (Digital Equipment Corporation model PDP-8/S) has been interfaced and programmed to operate a controlled-potential differential fast-sweep derivative polarograph (278). The programming of the polarograph includes automatic peak detection and the enhancement by means of ensemble averaging of the signal-to-noise ratio of the data obtained. It was found that ensemble averaging improved the analytical sensitivity by at least an order of magnitude (234).

The crucial importance of software has been emphasized repeatedly. Anderson has surveyed the relative advantages of selecting various levels of programming languages for laboratory data acquisition and experiment control and described the early approach at Lawrence Radiation Laboratory, Livermore, toward the development of a conversational language for the laboratory based on modifications to FOCAL (279). In a more recent paper, Anderson up-dates this survey and includes more information about the relative advantages of various language levels, compares some of the compilers and interpretive languages that are more popular with chemists, presents criteria for language choice, and discusses several trends that have emerged during the past few years that have a strong effect upon programming languages (280).

The information in the following group of papers indicates the scope of what has been accomplished with computer aided and directed automated analyses, with emphasis on the area of electrochemistry. Lauer, Abel and Anson have described an electrochemical data acquisition and analysis system based on a digital computer (244). Lauer and Osteryoung have described a general purpose laboratory data acquisition and control system built around a small digital computer; examples of use are in electrochemistry (281). Perone, Jones and Gutknecht used a small digital computer on-line to optimize stationary electrode polarographic measurements in real time; the computer was able to modify the course of the experiment during the experiment based on the nature of the continuously monitored output (i.e. the linear voltage sweep was temporarily interrupted after each reduction step in order to dissipate the reducible species in the diffusion layer and minimize interference with succeeding reduction steps) (282). Hicks, Eggert and Toren discussed the application of an on-line computer to the automation of analytical experiments and designed a system which made it feasible to use a new graphical approach to routinely analyse data from enzyme-substrate studies (283). Huebert and Smith have designed a system which with the aid of a PDP-8/S computer provides on-line experiment control, data acquisition and data analysis for fundamental harmonic ac polarography (284). Keller and Osteryoung have described the application of a computerized electrochemical system to pulse polarography at a hanging mercury drop electrode (285). Frazer, Carlson, Kray, Bertoglio and Perone have described an interactive computer system used for simple instrument control, data acquisition and interactive data reduction so that the experimenter can impose his experienced judgment on the net processing procedure by interacting with the computer during data reduction and applications to spark-source mass spectrometry and gas chromatography are illustrated (286). Perone, Frazer and Kray have illustrated the application of a computerized interactive processing system for the analysis of voltammetric electrochemical data (287). Toren, Carey, Sherry and Davis have described LABTRAN, a new interpretive language for programming chemical experiments (288).

Several research and development programs in the Analytical Chemistry Division at Oak Ridge National Laboratory involve utilizations of advantages of small digital computers in measurement systems and computer aided and directed automated analyses. Examples of this work further illustrate the scope of what has been accomplished in this area. To a rapid photo-metric analyser (289) to which design improvements for analytical applications have been made (Heath model 700 monochromator; parallel-mixing transfer disk; light pipe), a 12-bit 4 K-word memory Digital Equipment Corporation PDP-8/I computer with one 32 K-word disk and an XY recorder has been interfaced. A versatile program to operate this system and to collect and process the data including digital averaging and linear and parabolic regression curve fitting and to printout the data for several applications, FOCG (a modification of 4 K FOCAL 1969 for GeMSAEC) has been written by Myron T. Kelley. One of these studies is a demonstration of the potentialities of this system for analyses needed for environmental evaluations: the determination of orthophosphate by the molybdenum-blue method (290, 291). The programming concepts of FOCG for this use have been described; the disk is used for both program and data storage (292). Programming concepts used in other modifications of FOCAL with 8 K-word PDP-8/I computers with two 32 K-word disks and a Tektronix type 611 storage oscilloscope with alphanumeric readout for the control of rapid photometric analysers in other applications including routine clinical analyses in the ORNL Health Division are also described (292). Programming in a conversational, interpretive language facilitates the development of analytical methods. Modifications to FOCG that facilitate the use of the 4 K system for chemi- and bioluminescent analyses (some investigations were made of the firefly luciferin/luciferase assay for adenosine triphosphate) have been described as well as another version of FOCAL, FODK, which is designed for chemical applications (293). A Tennecomp TP-1371 cartridge tape unit has been incorporated in the 4 K system and used with FOCG (293). A Nuclear Data computer with a 12-bit 8 K-word memory has been acquired primarily for the analysis of X-ray spectra from the scanning electron microscope (294). Also a conversational, interpretive calculating language for the ND-812 computer, ORCAL, has been written; many programs written in FOCAL can be run with no change with the ND-812 provided with the ORCAL interpreter, and some others with minor modifications (295). ORCAL includes $X-Y$ plotter and oscilloscope display functions and a function for reading in data tapes via the teletype tape reader. This report includes a brief description of ORCAL, some instructions for programming, a list of the error diagnostics, a complete listing of the mnemonic and octal coding, and some suggestions on how the user might be able to adapt ORCAL for his own needs (295). Large quantities of data that require considerable data reduction are generated in the laboratories of the Tobacco Smoke Program. FOCAL programs for an 8 K PDP-8/I were developed to computer assist in handling the data (296). Other FOCAL programs for the 8 K PDP-8/I computer facilitate data processing and plotting from analyses by the General Analyses Laboratory (296). A demonstration of the feasibility of determining the U(IV)/U(III) ratio in molten fluoride salt at about 500°C by a computer controlled first derivative cyclic voltammetric method has been made (297). This analysis is very important because the monitoring of the U(III) content in molten fluoride salts used in molten salt reactors indicates that conditions are correct to minimize corrosion (298). The first successful in-line analysis of a constituent (i.e. U(III)) in a flowing molten fluoride salt stream was put in operation in 1971 (299). The ORNL model Q-2943 cyclic voltammeter (the prototype of this voltammeter has been described (300)) has been redesigned to reduce ground loop and other noise pickup in high electrical noise environments: Existing circuitry was rearranged so that the counter electrode could be connected to power ground and the metal framework and container of the engineering equipment. This monitor model, Q-2943A, is interfaced to an 8 K PDP-8/I computer. Programming is in FOCAL.

D. Digital DC Polarography

Some work in digital dc polarography has been done, other work is in progress, and more is planned at ORNL. A brief description of this work is presented here with the hope that it may serve to stimulate ideas on the general subject of digital polarography. Wave forms of interest in digital dc polarography include single and ensemble averaged entire and sampled current–time curves at constant potential and single and ensemble averaged regular and first, second and third derivative dc polarograms.

The digital equipment that has been used so far for this purpose is a Fabri-Tek FT-1064 Instrument Computer and, later, a Digital Equipment Corporation 8 K-word PDP-8/I computer. Other digital data acquisition and processing equipment, such as that cited or mentioned in more detail in Sections V-A, V-B and V-C of this chapter, could also be used.

In Section V-B of this chapter it was pointed out that applications of a software programmable computer or a wired program computer such as the Fabri-Tek FT-1064 Instrument Computer include its use as an excellent laboratory test instrument to collect data essential to the rational design and testing of elements of a measurement system. The digitally integrated output of a first derivative active analog sub-system has been compared to the actual polarographic signal (average current regular dc polarogram) fed into the analog sub-system. The difference between these, as a function of E_{DME}, was read out as such. Similarly, the output of a second derivative analog sub-system has been integrated and compared to the actual polarographic input. Extensive test measurements were not made because the analog derivative circuits had already been tested sufficiently by methods described in Section III-A-2 of this chapter. These digital tests were done with the FT-1064 only to demonstrate that clearly interpretable tests could have been more easily made with a digital system than by more indirect means, had one been available to us at an earlier time. In work done to learn the optimum conditions for operating the DME in the ORNL model Q-2942 electromechanical drop time controller, the FT-1064 was found to be extremely useful as a test instrument, better than an oscilloscope fitted with a camera, for example. The effect was easily learned of various operating conditions such as mechanical hammering intensity and adjustments and average mercury flow rate upon the form and the reproducibility of the current–time curves and upon the form of controlled curves as compared to that of uncontrolled curves.

The utility for polarographic measurements (with regard to precision and with regard to sensitivity) of ensemble averaging is being investigated. Ensemble averaging discriminates with greatest efficiency against random noise.

Mark and Reilley have discussed some applications of analog (oscilloscopic) measurements of current–time curves (301).

It is probable that the equivalent of measurements made by sophisticated polarographic instruments (e.g. pulse polarographs) could be made as well or better with a simple polarograph circuit in conjunction with a small dedicated purpose digital computer.

An area of digital dc polarography is timed current sampling. Much attention has been directed in work in designing measurement systems for polarography to the development of various methods of compensating for or minimizing the observed part of the capacitance current, i_c. Capacitance current is ordinarily the largest part of the residual current. Polarographic methods that involve current sampling for a portion of time at or near the end of the DME drop life include dc methods (e.g. maximum current dc polarography and Tast polarography), fast scan methods (e.g. oscilloscopic polarography) and certain ac methods (e.g. square wave, pulse, radio frequency and high level faradaic rectification polarography). Measurement during only a selected latter part of the drop life enables measurement with a quasi-constant electrode area. In ac methods such as square wave and pulse polarography, time is also allowed for the decay prior to sampling of the greatest part of the charging current associated with the application of the ac modulation to the dc voltage. The technique of timed current sampling exploits the fact that $i_d \sim t^{1/6}$ whereas $i_c \sim t^{-1/3}$. Plots versus time (for a concentration of $\sim 10^{-5}$ M) of i_d and of i_c are shown in Figure 6 of a review paper published by Nürnberg and Wolff (90). The highest ratio of i_d to i_c is at the end of the drop life. From Figure 6 of (90) it would appear that the drop life should be at least 1 second in order to take a substantial advantage of the variation of this ratio with time. Drop times as long as 5 seconds are disadvantageous in derivative dc polarography because the degree of filtering that is then required, even with selective filtering (current-averaging) methods such as those based upon parallel-T filters, is too great to allow a large advantage from the dependence of the magnitude of the Nth derivative signal upon the Nth power of the scan rate. A drop time of 0·5 second has been found to be optimum for first and second derivative dc polarography, which are other highly effective methods of minimizing the effect of i_c upon the sensitivity of polarographic measurements. However, the sensitivity that can be attained in all methods of polarography depends upon the signal-to-noise ratio, S/N, rather than only upon the ratio of faradaic to residual current. Thus, a pertinent and direct question to ask is how does S/N vary during drop life. In the case where current averaging is used, the question is whether the S/N is higher when the current is averaged during the entire drop life, as is now done in the ORNL polarographs, or if, for example, all but the first and/or very last portions of the current time epochs were averaged. It is planned to investigate the S/N at various intervals during the entire drop life by means of digital equipment. Investigations of the optimum design of measurement systems are aided by digital equipment in more ways than in acquiring and/or transforming the data, as illustrated by the following specific example. Any desired portion of the current–time curves stored in memory in a digital sub-system can be used, or the results obtained through the study with the digital equipment would show what timing values would be optimum for simple circuits to be constructed specifically for timed current sampling. It is important to note that the value of optimum timed current sampling can be learned without the necessity of building circuits for the investigation itself.

Other planned work in digital dc polarography includes the following. For investigation of the question of how to obtain higher precision and sensitivity in polarographic measurements, the precisions and sensitivities obtained through utilizing the digital integral of all or a portion of a single or ensemble averaged current–time curve instead of the analog averaged value of the diffusion current or the derivative peak heights will be of interest. Here, again, with digital equipment, it is not necessary to construct sub-systems until it is shown that the use of such sub-systems would be advantageous, and, also, what performance parameters are to be provided in the sub-systems. With the ORNL controlled-potential differential dc polarograph and digital equipment, with the usual solutions having a low specific resistance in the cells, it should be possible to compare observed iR_{inner} values against those calculated by means of the various methods so far proposed. A re-investigation is planned of the practical values (with regard to resolution, sensitivity and precision) of third derivative dc polarography at relatively fast scan rates. The concentrations of the constituents whose first derivative polarographic waves are severely overlapped have been obtained with simultaneous equations; the proportionality factors were evaluated through standard addition in the matrix (302). However, the reported results were obtained with single measurements and at ten-fold slower scan rates than could now be used with more recently designed ORNL dc polarographs (e.g. models Q-1988-FES and Q-2792) with a 0·5 second DME drop time. It is probable that concentrations could be calculated with much better precision and accuracy at the higher scan rates and if, through digital instrumentation, ensemble averaged values were available; the procedure could also be automated.

The following are two illustrations of polarographic data processing options that are available with digital systems. A first or second derivative polarogram as obtained in a single measurement or the average of several measurements of the supporting electrolyte solution can be stored. Then, can be stored the corresponding singly measured or ensemble averaged polarograms for the sample and for the sample plus a standard addition. It is then possible to output the following: the sample polarogram corrected for background and the polarogram that corresponds to the standard addition alone. The background correction is that for the background component of the polarogram obtained in the presence of the sample matrix. It has been noted that the slope of a residual current versus potential curve (i.e. a blank first derivative polarogram) is much more reproducible than the magnitude itself (i.e. a blank regular polarogram). However, if necessary, a stored blank curve can be translated in the memory to that position where it coincides with that of the stored foot of the wave(s), before the subtraction of the blank portion is executed. Moreover, instead of first or second derivative polarograms, i–t curves at a fixed potential in the limiting current region could be stored. The latter portion of these curves, with drop times of several seconds, would be relatively free of the charging current component. If the standard additions could be made through adding a known weight of species, no volume corrections would be needed. If a concentrated standard solution were used but a dropping bottle were used as a weight burette, perhaps the volume correction could be caused to be negligible. As the second illustration, two sample preparations in the supporting electrolyte (with maximum suppressor) could be made up so that the ratio of concentrations is known for the species whose concentration is to be measured. In the absence of overlapping by an adjacent wave (the possibility of overlapping would be much less for first derivative than for regular polarograms), if the ratio were two, the difference between stored polarograms for these preparations would be the polarogram for the species at the lower concentration, with an in-sample-matrix background correction. Then, by subtracting one of these stored polarograms from a third stored polarogram made after a standard addition to that first or second preparation, the necessary calibration factor would be at hand to calculate the concentration of the species of interest and the concentration in the original sample. A linear relationship between diffusion current and concentration is assumed but could be proven through successive standard additions.

With digital equipment, a cross correlation step can be included in the programming. Acquired experimental data can be convoluted with stored theoretical equations. It seems probable that a valuable method of minimizing the effect of undesired responses including residual current and noise upon attainable sensitivity would be to cross correlate single or ensemble averaged observed wave forms with the applicable equations. The usual deviations between observed and calculated wave forms should not detract from the utility of this method for the measurement of very low concentrations. An investigation with digital equipment could demonstrate the value of convolution for this purpose. Following this evaluation, if adequate performance could be obtained at less expense, analog cross correlators could be purchased or designed and built. Ensemble averaging followed by convolution has been used to increase the sensitivity of other kinds of measurements. For example, Pickering and Eckstrom (303) solved the problem of measuring the small infrared absorptions resulting from the adsorption of molecules on the surfaces of nickel and rhodium catalyst film mirrors by using several techniques to enhance the signal-to-noise ratio of the infrared absorption spectra. In the spectrophotometer, light beam chopping and tuned amplification was used and raw signal-plus-noise data were read out digitally via an ADC onto punched cards. The cards were processed by a computer to (a) combine replicate scans by point by point averaging for both sample and background curves, (b) perform point by point subtraction of the averaged sample and background curves to obtain difference curves, (c) obtain a further enhancement of signal-to-noise ratio by using cross correlation with spectrometer response functions and (d) deliver the results on punched cards for subsequent machine plotting.

The interfacing of a polarograph to an 8 K-word PDP-8/I computer and the FOCAL programming of data acquisition and processing have been described briefly (304, 305). A later demonstration of polarographic data acquisition and processing has been reported (306).

Investigations have been planned to explore possibilities of attaining long-term reproducibility of 0·1 per cent for concentration measurement by digital dc polarography. One of the experimental problems that limits the long-term polarographic reproducibility to more like 1 per cent is the difficulty of controlling temperature at the reaction site to the necessary extent. With digital polarography, it should be possible to experimentally thermostat the jacketed solution and DME standpipe to 0·1°C, which is not difficult, to add a temperature sensor near the DME orifice, and to digitally acquire and store ensemble averaged entire or sampled i–t curves and/or derivative dc polarograms along with concurrent temperature readings. Temperature corrections could be made with nominal or exact temperature coefficients of the polarographic diffusion coefficient, D, as a part of subsequent data processing. This should be much easier and more practical than exactly controlling the temperature near the reaction site. It has been pointed out in Section III-G of this chapter that the average mercury flow rate, $\bar{m}$, can be directly controlled to a constant value, but only with some experimental complexity. With digital polarography, with minimum time used for sample measurement and DME calibration, with a clean DME, $\bar{m}$ may not require positive control.

References

1. D. J. Fisher, 'Circumventions of chemical measurement difficulties through systems design,' Abstracts of Papers, 5B–6B, 147th Meeting, American Chemical Society, April, 1964.
2. D. J. Fisher, 'The circumvention of chemical measurement difficulties through the design of appropriate instrumentation systems: A design philosophy, with illustrations,' Abstracts of Papers, ANAL. 009, 157th Meeting, American Chemical Society, April, 1969.
3. D. J. Fisher, 'A method for the design of instrument systems for measurements of highest sensitivity,' *Chem. Instrum.*, **2**, 1–50 (1969).
4. D. J. Fisher, R. W. Stelzner and H. C. Jones, 'The use of ensemble averaging to increase the sensitivity of measurements with flame photometers,' *Chem. Instrum.*, **2**, 51 (1969).
5. D. J. Fisher, W. L. Belew and M. T. Kelley, 'Recent developments in dc polarography,' pp. 89–134 *in* G. J. Hills, Ed., *Polarography 1964*, Vol 1, Macmillan, London, 1966; pp. 112–128.
6. M. T. Kelley, H. C. Jones and D. J. Fisher, *Anal. Chem.*, **31**, 1475 (1959).
7. M. T. Kelley, D. J. Fisher and H. C. Jones, *Anal. Chem.*, **32**, 1262 (1960).
8. M. T. Kelley, D. J. Fisher, W. D. Cooke and H. C. Jones, 'Controlled-potential and derivative polarography', pp. 158–182 *in* I. S. Longmuir, Ed., *Advances in Polarography*, Vol. 1, Pergamon Press, Oxford, 1960.
9. E. J. Bair, *Introduction to Chemical Instrumentation*, McGraw-Hill Book Co., New York, 1962, pp. 325–334.
10. W. D. Shults, D. J. Fisher, H. C. Jones, M. T. Kelley and W. B. Schaap, *Z. Anal. Chem.*, **224**, 1 (1967).
11. H. C. Jones, W. L. Belew, R. W. Stelzner, T. R. Mueller and D. J. Fisher, *Anal. Chem.*, **41**, 772 (1969).
12. R. W. Stelzner, *Chem. Instrum.*, **2**, 213 (1969).
13. D. J. Fisher, W. L. Belew and M. T. Kelley, 'Recent developments in dc polarography,' pp. 89–134 *in* G. J. Hills, Ed., *Polarography 1964*, Vol. 1, Macmillan, London, 1966; pp. 92–94.
14. W. L. Belew, D. J. Fisher, M. T. Kelley and J. A. Dean, *Chem. Instrum.*, **2**, 297 (1970).
15. M. T. Kelley and D. J. Fisher, *Anal. Chem.*, **30**, 929 (1958).
16. M. T. Kelley and D. J. Fisher, *Anal. Chem.*, **28**, 1130 (1956).
17. D. J. Fisher, W. L. Belew and M. T. Kelley, *Chem. Instrum.*, **1**, 181 (1968).
18. D. J. Fisher, W. L. Belew and M. T. Kelley, *Chem. Instrum.*, **1**, 225 (1968).
19. W. D. Shults and W. B. Schaap, *Z. Anal. Chem.*, **224**, 22 (1967).
20. W. D. Shults, D. J. Fisher and W. B. Schaap, *Anal. Chem.*, **39**, 1379 (1967).
21. W. D. Shults and W. B. Schaap, *Anal Chem.*, **39**, 1384 (1967).
22. W. D. Shults, D. J. Fisher and W. B. Schaap, *Chem. Instrum.*, **1**, 7 (1968).
23. D. J. Fisher and H. C. Jones, 'ORNL test specification ST-175A,' *in CAPE-1001*,

U.S.A.E.C. Engineering Materials List, TID-4100, Suppl. 20, Feb., 1964; available from Clearinghouse for Federal Scientific and Technical Information, Springfield, Va. 22151.

24. H. C. Jones and W. L. Belew, 'Polarograph-voltammeter, ORNL model Q-2792, controlled-potential, dc,' method nos. 1 003043 and 9 003043, 3-1-68, *ORNL Master Analytical Manual*; available from M. T. Kelley, Director, Analytical Chemistry Division, Oak Ridge National Laboratory, P.O. Box X, Oak Ridge, Tenn. 37830.

25. L. Němec, *Collection Czech. Chem. Comm.*, **25**, 3085 (1960). English translation: M. T. Kelley, 'Influence of the recording arrangement on the form of polarographic curves,' ORNL-tr-555 (Aug. 1, 1961).

26. M. T. Kelley, H. C. Jones and D. J. Fisher, *Anal. Chem.*, **31**, 1475 (1959); p. 1483.

27. R. W. Schmid and C. N. Reilley, *J. Am. Chem. Soc.*, **80**, 2087 (1958).

28. W. L. Belew and H. P. Raaen, *J. Electroanal. Chem.*, **8**, 475 (1964).

29. J. L. Jones and H. A. Fritsche, *Anal. Chim. Acta*, **56**, 97 (1971).

30. W. L. Belew, D. J. Fisher, H. C. Jones and M. T. Kelley, *Anal. Chem.*, **41**, 779 (1969).

31. W. B. Schaap and P. S. McKinney, pp. 197–214, *in* G. J. Hills, Ed., *Polarography 1964*, Vol. 1, Macmillan, London, 1966.

32. W. B. Schaap and P. S. McKinney, *Anal. Chem.*, **36**, 1251 (1964).

33. M. T. Kelley, H. C. Jones and D. J. Fisher, *Anal. Chem.*, **31**, 1475 (1959); p. 1481.

34. W. D. Cooke, M. T. Kelley and D. J. Fisher, *Anal. Chem.*, **33**, 1209 (1961); p. 1210.

35. D. J. Fisher, W. L. Belew and M. T. Kelley, 'A simple quasi-reference electrode. Applications in controlled-potential polarography and voltammetry and in chronopotentiometry,' pp. 1043–1059 *in* G. J. Hills, Ed., *Polarography 1964*, Vol. 2, Macmillan, London, 1966.

36. D. J. Fisher, W. L. Belew and M. T. Kelley, pp. 89–134 *in* G. J. Hills, Ed., *Polarography 1964*, Vol. 1, Macmillan, London, 1966; pp. 93–94.

37. W. B. Schaap and P. S. McKinney, pp. 197–214 *in* G. J. Hills, Ed., *Polarography 1964*, Vol. 1, Macmillan, London, 1966; pp. 202–203.

38. M. T. Kelley, H. C. Jones and D. J. Fisher, *Anal. Chem.*, **31**, 1475 (1959); Figure 9.

39. W. L. Belew, D. J. Fisher, M. T. Kelley and J. A. Dean, *Chem. Instrum.*, **2**, 297 (1970); Figure 4.

40. W. E. Thomas, Jr. and W. B. Schaap, *Anal. Chem.*, **41**, 136 (1969).

41. D. J. Fisher, W. L. Belew and M. T. Kelley, *Chem. Instrum.*, **1**, 225 (1968); p. 228.

42. D. J. Fisher, W. L. Belew and M. T. Kelley, pp. 89–134 *in* G. J. Hills, Ed., *Polarography 1964*, Vol. 1, Macmillan, London, 1966; pp. 121–125.

43. M. T. Kelley, H. C. Jones and D. J. Fisher, *Anal. Chem.*, **31**, 1475 (1959); pp. 1479–1481.

44. M. T. Kelley, D. J. Fisher, W. D. Cooke and H. C. Jones, pp. 158–182 *in* I. S. Longmuir, Ed., *Advances in Polarography*, Vol. 1, Pergamon Press, Oxford, 1960; p. 171.

45. D. J. Fisher, W. L. Belew and M. T. Kelley, *Chem. Instrum.*, **1**, 181 (1968); pp. 190–192, 195–206, 212–213.

46. M. T. Kelley, H. C. Jones and D. J. Fisher, *Anal. Chem.*, **31**, 1475 (1959); p. 1484.

47. M. T. Kelley, D. J. Fisher, W. D. Cooke and H. C. Jones, pp. 158–182, *in* I. S. Longmuir, Ed., *Advances in Polarography*, Vol. 1, Pergamon Press, Oxford, 1960; Table 2, p. 173.

48. M. T. Kelley, D. J. Fisher, H. C. Jones, W. L. Maddox and R. W. Stelzner, 'Applications of commercial operational amplifiers in instrumentation for chemical analyses,' Table V, Preprint No. NY60-52, Instrument Society of America, Pittsburgh, Pa. (1960).

49. D. J. Fisher, W. L. Belew and M. T. Kelley, pp. 89–134 *in* G. J. Hills, Ed., *Polarography 1964*, Vol. 1, Macmillan, London, 1966; Figure 9 p. 123 and pp. 113–114.
50. D. J. Fisher, W. L. Belew and M. T. Kelley, pp. 89–134 *in* G. J. Hills, Ed., *Polarography 1964*, Vol. 1, Macmillan, London, 1966; Figure 1 p. 104 and Figure 3 p. 107.
51. D. J. Fisher, W. L. Belew and M. T. Kelley, pp. 89–134 *in* G. J. Hills, Ed., *Polarography 1964*, Vol. 1, Macmillan, London, 1966; p. 115.
52. D. J. Fisher, W. L. Belew and M. T. Kelley, *Chem. Instrum.*, **1**, 181 (1968); pp. 198–202.
53. D. J. Fisher, W. L. Belew and M. T. Kelley, *Chem. Instrum.*, **1**, 181 (1968); pp. 192–195 and p. 208.
54. D. J. Fisher, W. L. Belew and M. T. Kelley, pp. 89–134 *in* G. J. Hills, Ed., *Polarography 1964*, Vol. 1, Macmillan, London, 1966; Figure 8 p. 122.
55. R. W. Stelzner, *Chem. Instrum.*, **2**, 213 (1969); Figure 11.
56. D. J. Fisher, W. L. Belew and M. T. Kelley, pp. 89–134 *in* G. J. Hills, Ed., *Polarography 1964*, Vol. 1, Macmillan, London, 1966; pp. 115–118.
57. D. J. Fisher, W. L. Belew and M. T. Kelley, *Chem. Instrum.*, **1**, 181 (1968); pp. 206–211.
58. M. T. Kelley, H. C. Jones and D. J. Fisher, *Anal. Chem.*, **31**, 1475 (1959); p. 1477.
59. H. P. Raaen, *Anal. Chem.*, **34**, 1714 (1962).
60. Helen P. Raaen, R. J. Fox and V. E. Walker, 'Fabrication and assembly of a Teflon dropping-mercury electrode,' *U.S.A.E.C. Report ORNL-3344*, Nov. 30, 1962.
61. H. P. Raaen, *Anal. Chem.*, **36**, 2420 (1964).
62. Helen P. Raaen and Ralph L. Clark, 'Fabrication of the vertical-orifice Teflon capillary of a Teflon dropping-mercury electrode,' *U.S.A.E.C. Report ORNL-3654*, August 1964.
63. H. P. Raaen, *Anal. Chem.*, **37**, 677 (1965).
64. H. P. Raaen, *Anal. Chem.*, **37**, 1355 (1965).
65. H. P. Raaen, 'Instrumentation for polarography of glass-corroding media,' pp. 219–228 *in* Instrument Society of America, Analysis Instrumentation Division, *Analysis Instrumentation—1965*, Plenum Press, New York, 1966.
66. H. P. Raaen, *Chem. Instrum.*, **1**, 287 (1969).
67. H. P. Raaen, *Anal. Chim. Acta*, **44**, 205 (1969).
68. A. M. Bond and T. A. O'Donnell, *Anal. Chem.*, **44**, 590 (1972).
69. D. J. Fisher, W. L. Belew and M. T. Kelley, pp. 89–134 *in* G. J. Hills, Ed., *Polarography 1964*, Vol. 1, Macmillan, London, 1966; p. 103.
70. D. J. Fisher, W. L. Belew and M. T. Kelley, pp. 89–134 *in* G. J. Hills, Ed., *Polarography 1964*, Vol. 1, Macmillan, London, 1966; pp. 100–111.
71. W. D. Cooke, M. T. Kelley and D. J. Fisher, *Anal. Chem.*, **33**, 1209 (1961).
72. Louis Meites, *Polarographic Techniques*, 2nd ed., Interscience Publishers, New York, 1965, pp. 75–76.
73. R. De Levie, *J. Electroanal. Chem.*, **9**, 117 (1965).
74. J. Koryta, L. Němec, J. Pivoňka and L. Pospíšil, *J. Electroanal. Chem.*, **20**, 327 (1969).
75. I. Smoler, *Collect. Czech. Chem. Commun.*, **19**, 238 (1954).
76. R. Kalvoda and I. Smoler, *Zavod. Lab.*, **27**, 549 (1961); *Ind. Lab.*, **27**, 560 (1961).
77. J. Kůta and I. Smoler, pp. 43–63 *in* P. Zuman and I. M. Kolthoff, Eds., *Progress in Polarography*, Vol. 1, Interscience, New York, 1962.
78. I. Smoler, *J. Electroanal. Chem.*, **6**, 465 (1963).
79. Private Communication dated December 11, 1969 from Pascal A. Tarantino, Chemical Research Laboratory, Department of the Army, Edgewood Arsenal, Maryland 21010.

80. J. B. Flato, 'A new polyfunctional electrochemical instrument,' a paper presented at the 19th Pittsburgh Conference on Analytical Chemistry and Applied Spectroscopy, March 1968, Technical Notes T-193, Princeton Applied Research Corp., Princeton, N.J. 08540.
81. T. Kugo, Y. Umezawa and Sh. Fujiwara, *Chem. Instrum.*, **2**, 189 (1969).
82. D. J. Fisher, W. L. Belew and M. T. Kelley, *Chem. Instrum.*, **1**, 181 (1968); p. 193 and p. 218.
83. D. J. Fisher, W. L. Belew and M. T. Kelley, pp. 89–134 *in* G. J. Hills, Ed., *Polarography 1964*, Vol. 1, Macmillan, London, 1966; p. 111.
84. D. J. Fisher, W. L. Belew and M. T. Kelley, *Chem. Instrum.*, **1**, 181 (1968); p. 184 and p. 187.
85. D. J. Fisher, W. L. Belew and M. T. Kelley, pp. 1043–1059 *in* G. J. Hills, Ed., *Polarography 1964*, Vol. 2, Macmillan, London, 1966; pp. 1050–1051 and Figure 1.
86. M. T. Kelley, W. L. Belew, G. V. Pierce, W. D. Shults, H. C. Jones and D. J. Fisher, *Microchem. J.*, **10**, 315 (1966); pp. 316–318 and Figure 1.
87. W. D. Shults, D. J. Fisher, H. C. Jones, M. T. Kelley and W. B. Schaap, *Z. Anal. Chem.*, **224**, 1 (1967); pp. 8–10 and Figure 2.
88. M. T. Kelley, D. J. Fisher, H. C. Jones, W. L. Maddox and R. W. Stelzner, 'Applications of commercial operational amplifiers in instrumentation for chemical analyses,' Preprint Number NY60-52, Instrument Society of America, Pittsburgh, Pa., 1960.
89. M. T. Kelley, D. J. Fisher, H. C. Jones, W. L. Maddox and R. W. Stelzner, Abstracts of Papers, 41B, 101, 144th ACS National Meeting, Los Angeles, California, March 31–April 5, 1963.
90. H. W. Nürnberg and G. Wolff, *Chem. Ing. Techn.*, **37**, 977 (1965); English translation: Scientific Translation Service, 'State of polarographic methods and their instrumentation. Part I. DC methods,' ORNL-tr-966 (1965).
91. D. J. Fisher, W. L. Belew and M. T. Kelley, *Chem. Instrum.*, **1**, 181 (1968); Figure 1.
92. W. D. Shults, D. J. Fisher, H. C. Jones, M. T. Kelley and W. B. Schaap, *Z. Anal. Chem.*, **224**, 1 (1967); Figure 1.
93. M. T. Kelley, D. J. Fisher, W. D. Cooke and H. C. Jones, 'Controlled-potential and derivative polarography,' pp. 158–182 *in* I. S. Longmuir, Ed., *Advances in Polarography*, Vol. 1, Pergamon Press, Oxford, 1960; Figure 3, p. 169.
94. D. J. Fisher, W. L. Belew and M. T. Kelley, pp. 89–134 *in* G. J. Hills, Ed., *Polarography 1964*, Vol. 1, Macmillan, London, 1966; pp. 97–99.
95. J. J. Lingane and R. Williams, *J. Am. Chem. Soc.*, **74**, 790 (1952).
96. D. J. Fisher, W. L. Belew and M. T. Kelley, *Chem. Instrum.*, **1**, 181 (1968); pp. 188–189 and pp. 192–193.
97. D. J. Fisher, W. L. Belew and M. T. Kelley, *Chem. Instrum.*, **1**, 181 (1968); pp. 193–195.
98. R. W. Stelzner, *Chem. Instrum.*, **2**, 213 (1969); pp. 236–238 and p. 242.
99. F. B. Stephens and J. E. Harrar, *Chem. Instrum.*, **1**, 169 (1968); pp. 171–172.
100. W. B. Schaap and P. S. McKinney, *Anal. Chem.*, **36**, 1251 (1964); p. 1253.
101. D. L. Manning, J. M. Dale and G. Mamantov, pp. 1143–1151 *in* G. J. Hills, Ed., *Polarography 1964*, Vol. 2, Macmillan, London, 1966.
102. W. L. Belew, D. J. Fisher, M. T. Kelley and J. A. Dean, *Microchem. J.*, **10**, 301 (1966).
103. M. T. Kelley, W. L. Belew, G. V. Pierce, W. D. Shults, H. C. Jones and D. J. Fisher, *Microchem. J.*, **10**, 315 (1966).
104. G. Mamantov and D. L. Manning, *Anal. Chem.*, **38**, 1494 (1966).
105. J. P. Young, G. Mamantov and F. L. Whiting, *J. Phys. Chem.*, **71**, 782 (1967).

106. D. L. Manning and G. Mamantov, *J. Electroanal. Chem.*, **17**, 137 (1968).
107. G. Mamantov and D. L. Manning, *J. Electroanal. Chem.*, **18**, 309 (1968).
108. W. D. Cooke, M. T. Kelley and D. J. Fisher, *Anal. Chem.*, **33**, 1209 (1961); pp. 1214–1215.
109. D. J. Fisher, W. L. Belew and M. T. Kelley, *Chem. Instrum.*, **1**, 181 (1968); p. 212.
110. D. J. Fisher and P. F. Thomason, *Anal. Chem.*, **28**, 1285 (1956).
111. D. J. Fisher, W. L. Belew and M. T. Kelley, pp. 89–134 *in* G. J. Hills, Ed., *Polarography 1964*, Vol. 1, Macmillan, London, 1966; pp. 128–132.
112. D. J. Fisher, W. L. Belew and M. T. Kelley, pp. 89–134 *in* G. J. Hills, Ed., *Polarography 1964*, Vol. 1, Macmillan, London, 1966; pp. 103–108 and Figure 1.
113. D. J. Fisher, W. L. Belew and M. T. Kelley, pp. 89–134 *in* G. J. Hills, Ed., *Polarography 1964*, Vol. 1, Macmillan, London, 1966; Figure 2, p. 107.
114. D. J. Fisher, W. L. Belew and M. T. Kelley, pp. 89–134 *in* G. J. Hills, Ed., *Polarography 1964*, Vol. 1, Macmillan, London, 1966; pp. 114–115 and Figure 9, pp. 123–124.
115. D. J. Fisher, W. L. Belew and M. T. Kelley, pp. 89–134 *in* G. J. Hills, Ed., *Polarography 1964*, Vol. 1, Macmillan, London, 1966; pp. 124–128.
116. D. J. Fisher, W. L. Belew and M. T. Kelley, *Chem. Instrum.*, **1**, 181 (1968); pp. 206–208, Table 4 and Figure 6.
117. A. Frisque, V. W. Meloche and I. Shain, *Anal. Chem.*, **26**, 471 (1954).
118. D. J. Fisher, W. L. Belew and M. T. Kelley, *Chem. Instrum.*, **1**, 181 (1968); pp. 214–215.
119. D. J. Fisher, W. L. Belew and M.,T. Kelley, pp. 89–134 *in* G. J. Hills, Ed., *Polarography 1964*, Vol. 1, Macmillan, London, 1966; pp. 113–121 and Figures 4, 5 and 6 and Table 1.
120. O. H. Müller, *The Polarographic Method of Analysis*, 2nd ed., Chemical Education Publishing Co., Easton, Pa., 1951, pp. 153–156.
121. D. J. Fisher, W. L. Belew and M. T. Kelley, pp. 89–134 *in* G. J. Hills, Ed., *Polarography 1964*, Vol. 1, Macmillan, London, 1966; Figure 7, p. 120.
122. A.-W. Elbel, *Z. Anal. Chem.*, **173**, 70 (1960).
123. K. Kronenberger and W. Nickels, *Z. Anal. Chem.*, **186**, 79 (1962).
124. P. O. Kane, *J. Polarographic Society*, **VIII**, 10 (1962).
125. M. T. Kelley, D. J. Fisher, W. D. Cooke and H. C. Jones, pp. 158–182 *in* I. S. Longmuir, Ed., *Advances in Polarography*, Vol. 1, Pergamon Press, Oxford, 1960; Figures 7 and 8 and pp. 177–179.
126. D. J. Fisher, W. L. Belew and M. T. Kelley, *Chem. Instrum.*, **1**, 181 (1968); Figure 4.
127. H. C. Jones, W. L. Belew, R. W. Stelzner, T. R. Mueller and D. J. Fisher, *Anal. Chem.*, **41**, 772 (1969); Figure 8.
128. D. J. Fisher, W. L. Belew and M. T. Kelley, *Chem. Instrum.*, **1**, 181 (1968); pp. 215–216.
129. D. J. Fisher, W. L. Belew and M. T. Kelley, pp. 89–134 *in* G. J. Hills, Ed., *Polarography 1964*, Vol. 1, Macmillan, London, 1966; equation 8, p. 103.
130. D. J. Fisher, W. L. Belew and M. T. Kelley, pp. 89–134 *in* G. J. Hills, Ed., *Polarography 1964*, Vol. 1, Macmillan, London, 1966; Figure 2, p. 106.
131. W. D. Cooke, M. T. Kelley and D. J. Fisher, *Anal. Chem.*, **33**, 1209 (1961); Figures 9 and 10, p. 1215.
132. W. D. Cooke, M. T. Kelley and D. J. Fisher, *Anal. Chem.*, **33**, 1209 (1961); Figure 7, p. 1214.
133. D. J. Fisher, W. L. Belew and M. T. Kelley, *Chem. Instrum.*, **1**, 181 (1968); Table 5, p. 210.

134. D. J. Fisher, W. L. Belew and M. T. Kelley, *Chem. Instrum.*, **1**, 181 (1968); Figure 6, p. 207.
135. D. J. Fisher, W. L. Belew and M. T. Kelley, pp. 89–134 *in* G. J. Hills, Ed., *Polarography 1964*, Vol. 1, Macmillan, London, 1966; Figure 3, p. 107.
136. W. D. Shults, D. J. Fisher and W. B. Schaap, *Chem. Instrum.*, **1**, 7 (1968); Figure 3, p. 13.
137. W. D. Shults, D. J. Fisher and W. B. Schaap, *Chem. Instrum.*, **1**, 7 (1968); Figure 2, p. 12.
138. H. C. Jones, W. L. Belew, R. W. Stelzner, T. R. Mueller and D. J. Fisher, *Anal. Chem.*, **41**, 772 (1969); Figure 4.
139. W. L. Belew, D. J. Fisher, M. T. Kelley and J. A. Dean, *Microchem. J.*, **10**, 301 (1966); Figure 2, p. 312.
140. D. J. Fisher, W. L. Belew and M. T. Kelley, *Chem. Instrum.*, **1**, 181 (1968); pp. 216–217.
141. D. J. Fisher, W. L. Belew and M. T. Kelley, *Chem. Instrum.*, **1**, 181 (1968); pp. 217–218.
142. J. Heyrovský, *Analyst*, **72**, 229 (1947); p. 232.
143. J. K. Taylor, *J. Assoc. Offic. Agr. Chemists*, **47**, 21 (1964); p. 23.
144. H. M. Davis and H. I. Shalgosky, pp. 618–627 *in* I. S. Longmuir, Ed., *Advances in Polarography*, Vol. 2, Pergamon Press, Oxford, 1960; p. 626.
145. H. I. Shalgosky and J. Watling, *Anal. Chim. Acta*, **26**, 66 (1962).
146. H. C. Jones, W. L. Belew, R. W. Stelzner, T. R. Mueller and D. J. Fisher, *Anal. Chem.*, **41**, 772 (1969); Figure 5.
147. D. J. Fisher, D. Thiele, W. D. Shults and W. L. Belew, 'Comparative amperometric dc polarography: a new high-precision and high-accuracy method,' U.S.A.E.C. Report ORNL-3889, January 1966, pp. 3–4; available from Clearinghouse for Federal Scientific and Technical Information, Springfield, Va. 22151.
148. W. D. Shults, D. J. Fisher, H. C. Jones, M. T. Kelley and W. B. Schaap, *Z. Anal. Chem.*, **224**, 1 (1967); pp. 19–20.
149. W. L. Belew, D. J. Fisher, M. T. Kelley, R. W. Stelzner and E. S. Wolfe, 'Investigation of precision attainable in single-cell first-derivative dc polarography,' U.S.A.E.C. Report ORNL-4196, January 1968, pp. 7–8; available from Clearinghouse for Federal Scientific and Technical Information, Springfield, Va. 22151.
150. D. J. Fisher and W. L. Maddox, 'Apparatus for precise control of mercury flow rate of dropping-mercury electrodes (D.M.E.'s),' U.S.A.E.C. Report ORNL-4343, December 1968, p. 2; available from Clearinghouse for Federal Scientific and Technical Information, Springfield, Va. 22151.
151. J. J. Lingane, *Anal. Chim. Acta*, **44**, 411 (1969).
152. D. J. Fisher, W. L. Belew, W. L. Maddox and M. T. Kelley 'Polarographic mercury-flow-rate controller,' U.S.A.E.C. Report ORNL-4466, Jan. 1970, pp. 1–2; available from Clearinghouse for Federal Scientific and Technical Information, Springfield, Va. 22151.
153. M. T. Kelley, D. J. Fisher, W. L. Maddox and W. L. Belew, 'High-precision dc polarography,' IUPAC International Symposium on Analytical Chemistry, The University of Birmingham, Birmingham, England, July 21–25, 1969.
154. M. T. Kelley and H. H. Miller, *Anal. Chem.*, **24**, 1895 (1952).
155. D. J. Fisher, 'Polarograph, ORNL Model Q-1338, high-sensitivity, derivative, recording,' ORNL Master Analytical Manual, Method nos. 1 003041 and 9 003041 (2-14-57); single copies available from M. T. Kelley, Director, Analytical Chemistry Division, Oak Ridge National Laboratory, Oak Ridge, Tenn. 37830.
156. D. J. Fisher, 'Polarograph, ORNL Model Q-1673, high-sensitivity, diode filter,

derivative, recording,' ORNL Master Analytical Manual, Method nos. 1 003042 and 9 003042 (2-13-57); single copies available from M. T. Kelley, Director, Analytical Chemistry Division, Oak Ridge National Laboratory, Oak Ridge, Tenn. 37830.

157. M. T. Kelley and D. J. Fisher, *Anal. Chem.*, **30**, 929 (1958); Figures 8, 10, 11, 12 and 13.

158. U.S.A.E.C. Engineering Materials List, 'Controlled-potential and derivative polarograph, Q-1988A,' CAPE-1001 (TID-4100, Supplement 20, Feb. 1964); available from Clearinghouse for Federal Scientific and Technical Information, Springfield, Va. 22151.

159. Indiana Instrument and Chemical Corporation, Bloomington, Indiana 47401, *Anal. Chem.*, **42**, 227 LG (1970); **43**, 200 LG (1971).

160. NUMEC Instruments and Controls Corporation, Monroeville, Pa., *Anal. Chem.*, **40** (9), 202 LG (1968).

161. D. J. Fisher, W. L. Belew and M. T. Kelley, *Chem. Instrum.*, **1**, 181 (1968); pp. 182-183.

162. J. L. Jones, *Anal. Letters*, **1**, 969 (1968).

163. J. L. Hanley and R. T. Iwamoto, *J. Electroanal. Chem.*, **24**, 271 (1970).

164. L. P. Rigdon and J. E. Harrar, *Anal. Chem.*, **40**, 1641 (1968).

165. J. J. Donahue and J. W. Olver, *Anal. Chem.*, **41**, 753 (1969).

166. J. E. Harrar and L. P. Rigdon, *Anal. Chem.*, **41**, 758 (1969).

167. G. Baumgärtel, D. Thiele and S. Radek, 'In-line polarographie zur Uranbestimmung in Prozessabfall-lösungen,' Kernforschungszentrum Karlsruhe Report KFK-1144, July 1970.

168. P. A. Tarantino and S. Sass, *J. Electrochem. Soc.*, **116**, 430 (1969).

169. J. Jordan, *J. Electroanal. Chem.*, **29**, 127 (1971).

170. H. E. Zittel and F. J. Miller, *Anal. Chim. Acta*, **37**, 141 (1967).

171. H. E. Zittel and T. M. Florence, *Anal. Chem.*, **39**, 320 (1967).

172. H. E. Zittel and T. M. Florence, *Anal. Chem.*, **39**, 355 (1967).

173. H. E. Zittel and T. M. Florence, *Anal. Chim. Acta*, **40**, 27 (1968).

174. T. M. Florence and W. L. Belew, *J. Electroanal. Chem.*, **21**, 157 (1969).

175. T. M. Florence, *J. Electroanal. Chem.*, **26**, 293 (1970).

176. T. M. Florence, *J. Electroanal. Chem.*, **27**, 273 (1970).

177. E. B. Sandell and T. S. West, *Pure Appl. Chem.*, **18**, 429 (1969); definition 26, p. 435.

178. T. W. Richards and G. W. Heimriod, *Proc. Amer. Acad. Arts Sci.*, **37**, 415 (1902); p. 415 and p. 443.

179. F. Haber, *Z. Physik. Chem.*, **32**, 193 (1900); pp. 208-213.

180. J. E. Harrar and I. Shain, *Anal. Chem.*, **38**, 1148 (1966).

181. R. W. Stelzner, M. T. Kelley and D. J. Fisher, 'Reference electrode placement in controlled-potential coulometry,' U.S.A.E.C. Report ORNL-3537, pp. 9–12 (issued Feb. 10, 1964).

182. T. R. Mueller, 'Significance of placement of the reference electrode in controlled-potential coulometry,' U.S.A.E.C. Report ORNL-3750, p. 5 (issued Jan. 1965).

183. J. J. Lingane, *Electroanalytical Chemistry*, 2nd ed., Interscience Publishers, New York, 1958, pp. 365-366.

184. P. Delahay, *New Instrumental Methods in Electrochemistry*, Interscience Publishers, New York, 1954, pp. 391-394.

185. W. D. Shults, 'Coulometric methods,' Chapter 23 *in* F. J. Welcher, Ed., *Standard Methods of Chemical Analysis, Vol. 3, Instrumental Methods, Part A*, 6th ed., D. Van Nostrand Company, New York, 1966, p. 464.

186. J. E. Mumby and S. P. Perone, *Chem. Instrum.*, **3**, 191 (1971).

187. R. R. Schroeder and I. Shain, *Chem. Instrum.*, **1**, 233 (1969).
188. M. T. Kelley, H. C. Jones and D. J. Fisher, *Anal. Chem.*, **31**, 956 (1959).
189. H. C. Jones, W. D. Shults and J. M. Dale, *Anal. Chem.*, **37**, 680 (1965).
190. G. L. Booman, *Anal. Chem.*, **29**, 213 (1957).
191. R. C. Propst, *Anal. Chem.*, **40**, 244 (1968).
192. J. J. Lingane, *Electroanalytical Chemistry*, 2nd ed., Interscience Publishers, New York, 1958, p. 366.
193. W. D. Shults, Oak Ridge National Laboratory, Oak Ridge, Tennessee, personal communications, October 8, 1964 and March 8, 1966.
194. J. E. Harrar and E. Behrin, *Anal. Chem.*, **39**, 1230 (1967).
195. M. T. Kelley, H. C. Jones and D. J. Fisher, *Anal. Chem.*, **31**, 488 (1959).
196. U.S.A.E.C. Engineering Materials List (TID-4100): CAPE-785, 'Automatic coulometric titrator,' (August 1961), available from Clearinghouse for Federal Scientific and Technical Information, U.S. Dept. of Commerce, 5285 Port Royal Road, Springfield, Va. 22151.
197. H. C. Jones, 'Automatic coulometric titrator, ORNL model Q-2005, electronic, controlled-potential,' Method Nos. 1 003029 and 9 003029 (revised 8-30-65), Oak Ridge National Laboratory Master Analytical Manual. Copies of this method in limited numbers can be obtained on request from Director, Analytical Chemistry Division, Oak Ridge National Laboratory, P.O. Box X, Oak Ridge, Tennessee 37830.
198. M. T. Kelley, H. C. Jones and D. J. Fisher, *Talanta*, **6**, 185 (1960).
199. Indiana Instrument and Chemical Corporation, Bloomington, Indiana 47401, *Anal. Chem.*, **42**, 227 LG (1970); **43**, 124 LG (1971).
200. U.S.A.E.C. Engineering Materials List (TID-4100): CAPE-1196, 'High-sensitivity coulometric titrator /Q-2564/,' (revision, 1968), available from Clearinghouse for Federal Scientific and Technical Information, U.S. Dept. of Commerce, 5285 Port Royal Road, Springfield, Va. 22151.
201. T. R. Mueller, H. C. Jones, D. J. Fisher and R. W. Stelzner, 'Modifications of the ORNL model Q-2564 high-sensitivity coulometric titrator,' ORNL-TM-2175, March 18, 1968.
202. H. C. Jones, 'Coulometric titrator, controlled-potential, high-sensitivity, ORNL model Q-2564,' Method Nos. 1 003025 and 9 003025 (revision, 3-1-68), Oak Ridge National Laboratory Master Analytical Manual. Copies of this method in limited numbers can be obtained on request from Director, Analytical Chemistry Division, Oak Ridge National Laboratory, P.O. Box X, Oak Ridge, Tennessee 37830.
203. J. G. Truxal, Ed., *Control Engineers' Handbook*, McGraw-Hill Book Co., New York, 1958, Section 3.4 and Figures 3.7, 3.8 and 3.9.
204. J. G. Truxal, Ed., *Control Engineers' Handbook*, McGraw-Hill Book Co., New York, 1958, Sections 2.3–2.10.
205. H. Chestnut and R. W. Mayer, *Servomechanisms and Regulating System Design*, John Wiley and Sons, New York, 1959.
206. J. E. Harrar, F. B. Stephens and R. E. Pechacek, *Anal. Chem.*, **34**, 1036 (1962).
207. C. G. Enke and R. A. Baxter, *J. Chem. Educ.*, **41**, 202 (1964).
208. J. Schoen and K.-E. Staubach, *Regelungstechnik*, **2**, 157 (1954); English translation: W. H. Everhardy, National Institutes of Health, Bethesda, Maryland, May 13, 1955.
209. H. Gerischer and K.-E. Staubach, *Z. Elecktrochem.*, **61**, 789 (1957); English translation: M. T. Kelley, Oak Ridge National Laboratory, Oak Ridge, Tennessee, ORNL-tr-573, Jan. 7, 1965.

210. M. Breiter and F. G. Will, *Z. Elecktrochem.*, **61**, 1177 (1957); English translation: M. T. Kelley, Oak Ridge National Laboratory, Oak Ridge, Tennessee, ORNL-tr-574, Jan., 1964.

211. F. G. Will, *Z. Elektrochem.*, **63**, 484 (1959); English translation: Associated Technical Services, Inc., East Orange, New Jersey, 1964.

212. F. G. Will, *Z. Elektrochem.*, **63**, 689 (1959); English translation: M. T. Kelley, Oak Ridge National Laboratory, Oak Ridge, Tennessee, ORNL-tr-563, Jan. 1964.

213. A. Bewick, A. Bewick, M. Fleischmann and M. Liler, *Electrochim. Acta*, **1**, 83 (1959).

214. A. Bewick and M. Fleischmann, *Electrochim. Acta*, **8**, 89 (1963).

215. E. V. Bohn, *The Transform Analysis of Linear Systems*, Addison-Wesley Publishing Co., Reading, Mass., 1963, Section 6-7, 'Compensating networks'.

216. S. R. Gossmann, 'A transfer function for d.c. operational amplifiers with generalized external networks,' U.S.A.E.C. Report IDO-16809, Feb. 25, 1963.

217. R. H. Brown, 'The design of a potentiostat for electrochemical research,' U.S.A.E.C. Report IDO-16852, May, 1963.

218. G. L. Booman and W. B. Holbrook, *Anal. Chem.*, **35**, 1793 (1963).

219. G. L. Booman and W. B. Holbrook, *Anal. Chem.*, **37**, 795 (1965).

220. I. Shain, J. E. Harrar and G. L. Booman, *Anal. Chem.*, **37**, 1768 (1965).

221. D. T. Pence and G. L. Booman, *Anal. Chem.*, **38**, 1112 (1966).

222. A. Bewick and M. Fleischmann, *Electrochim. Acta*, **11**, 1397 (1966).

223. E. R. Brown, D. E. Smith and G. L. Booman, *Anal. Chem.*, **40**, 1411 (1968).

224. H. Chestnut and R. W. Mayer, *Servomechanisms and Regulating System Design*, Vol. 1, John Wiley and Sons, New York, 1951, pp. 297–299 and p. 302.

225. B. M. Oliver, 'Square wave and pulse testing of linear systems,' Application Note No. 17, Hewlett–Packard Co., Palo Alto, Calif., June 1, 1961.

226. O. I. Elgerd, *Control Systems Theory*, McGraw-Hill Book Co., New York, 1967, pp. 39–41 and 46–57.

227. B. M. Oliver and J. M. Cage, *Electronic Measurements and Instrumentation*, McGraw-Hill Book Co., New York, 1971.

228. T. R. Mueller, 'Coulometric titrator (ORNL Model Q-4010): prototype and instruments for analytical services,' ORNL-4466, p. 2, Jan. 1970.

229. M-T Electronics Co., San Leandro, Calif. 94579, *Anal. Chem.*, **44**, (2), 150 A (1972).

230. J. R. Stokely and W. D. Shults, *Anal. Chim. Acta*, **45**, 528 (1969).

231. T. R. Mueller, H. C. Jones and D. J. Fisher, 'An integrated coulometric titration system,' Abstract for Paper 6, 13th Conference on Analytical Chemistry in Nuclear Technology, Sept. 30–Oct. 2, 1969, Gatlinburg, Tenn. 37738, in *Schedule of Events and Abstracts of Papers.*

232. D. J. Fisher and T. R. Mueller, 'Automatic, controlled-current, coulometric Karl Fischer titrator,' U.S.A.E.C. Report ORNL-4749, pp. 43–45, Jan. 1972. Patent applied for.

233. L. J. Lorenz, R. A. Culp and L. B. Rogers, *Anal. Chem.*, **42**, 979 (1970).

234. S. P. Perone, J. E. Harrar, F. B. Stephens and R. E. Anderson, *Anal. Chem.*, **40**, 899 (1968).

235. I. B. Goldberg and A. J. Bard, *J. Phys. Chem.*, **75**, 3281 (1971).

236. N. Winograd and T. Kuwana, *Anal. Chem.*, **43**, 252 (1971).

237. H. E. Keller and R. A. Osteryoung, *Anal. Chem.*, **43**, 342 (1971).

238. A. Savitzky and M. J. E. Golay, *Anal. Chem.*, **36**, 1627 (1964).

239. M. W. Breiter, *J. Electrochem. Soc.*, **112**, 845 (1965).

240. M. W. Breiter, *J. Electrochem. Soc.*, **113**, 1071 (1966).

241. E. R. Brown, D. E. Smith and D. D. DeFord, *Anal. Chem.*, **38**, 1130 (1966).

242. G. Lauer and R. A. Osteryoung, *Anal. Chem.*, **38**, 1137 (1966).
243. G. L. Booman, *Anal. Chem.*, **38**, 1141 (1966).
244. G. Lauer, R. Abel and F. C. Anson, *Anal. Chem.*, **39**, 765 (1967).
245. G. Lauer and R. A. Osteryoung, *Anal. Chem.*, **39**, 1866 (1967).
246. F. C. Anson, 'Electrochemistry,' *in* H. Eyring, Ed., and C. J. Christensen and H. S. Johnston, Assoc. Eds., *Annual Review of Physical Chemistry*, Vol. 19, Annual Reviews, Inc., Palo Alto, Calif., 1968, pp. 89–90.
247. L. B. Rogers, *J. Chem. Educ.*, **45**, 463 (1968).
248. Anon., *Chem. Eng. News*, **45** (13), 102 (1967).
249. R. M. Brown and E. H. Mueller, *Nucleonics*, **25** (3), 48 (1967).
250. T. J. Williams and S. J. Bailey, *Contr. Eng.*, **14** (10), 65 (1967).
251. M. Clynes, *Instr. Control Systems*, **35** (8), 87 (1962).
252. Nicolet Instrument Corporation, 5225 Verona Road, Madison, Wisconsin 53711.
253. N. R. Kuzel, H. E. Roudebush and C. E. Stevenson, *J. Pharm. Sci.*, **58**, 381 (1969).
254. D. E. Smith, *J. Ass. Offic. Anal. Chem.*, **52**, 206 (1969).
255. C. W. Childs, P. S. Hallman and D. D. Perrin, *Talanta*, **16**, 629, 1119 (1969).
256. S. P. Perone, *J. Chromatogr. Sci.*, **7**, 714 (1969).
257. R. Venkataraghavan, R. J. Klimowski and F. W. McLafferty, *Accounts of Chemical Research*, **3**, 158 (1970).
258. L. Ramaley and G. S. Wilson, *Anal. Chem.*, **42**, 606 (1970).
259. G. S. Wilson and L. Ramaley, *Anal. Chem.*, **42**, 611 (1970).
260. S. P. Perone, *J. Chem. Educ.*, **47**, 105 (1970).
261. K. B. Wiberg, *J. Chem. Educ.*, **47**, 113 (1970).
262. B. Dewey III, 'Dedicated computer in the laboratory,' Chapter 2, *in* C. H. Orr and J. A. Norris, Ed., *Computers in Analytical Chemistry*, Vol. 4 in *Progress in Analytical Chemistry*, Plenum Press, New York, 1970.
263. M. Marghoshes, *Anal. Chem.*, **43**, (4), 101A (1971).
264. A. A. Eggert, G. P. Hicks and J. E. Davis, *Anal. Chem.*, **43**, 736 (1971).
265. S. P. Perone, *Anal. Chem.*, **43**, 1288 (1971).
266. R. E. Dessy and D. G. Larsen, *Chem. Eng. News*, **49** (52), 42 (Dec. 20, 1971).
267. J. W. Frazer, *MTRSA*, **12** (2), 8 (1972).
268. D. L. Schroeder and H. F. Walter, *J. Chromatogr. Sci.*, **10**, 14 (1972).
269. C. E. Klopfenstein, *J. Chromatogr. Sci.*, **10**, 22 (1972).
270. R. Bezman and P. S. McKinney, *Anal. Chem.*, **41**, 1560 (1969).
271. R. L. Myers and I. Shain, *Chem. Instrum.*, **2**, 203 (1969).
272. D. O. Jones and S. P. Perone, *Anal. Chem.*, **42**, 1151 (1970).
273. M. F. Hasler, *Anal. Chem.*, **39** (6), 26A (1967).
274. D. Secrest, *Ind. Eng. Chem.*, **60** (6), 74 (1968).
275. J. W. Frazer, *Anal. Chem.*, **40** (8), 26A (1968).
276. J. W. Frazer, *Chem. Instrum.*, **2**, 271 (1970).
277. Anon., 'Computers play larger role in analyses,' *Chem. Eng. News*, **46** (28), 36 (1968).
278. F. B. Stephens and J. E. Harrar, *Chem. Instrum.*, **1**, 169 (1968).
279. R. E. Anderson, *J. Chromatogr. Sci.*, **7**, 725 (1969).
280. R. E. Anderson, *J. Chromatogr. Sci.*, **10**, 8 (1972).
281. G. Lauer and R. A. Osteryoung, *Anal. Chem.*, **40** (10), 30A (1968).
282. S. P. Perone, D. O. Jones and W. F. Gutknecht, *Anal. Chem.*, **41**, 1154 (1969).
283. G. P. Hicks, A. A. Eggert and E. C. Toren, Jr., *Anal. Chem.*, **42**, 729 (1970).
284. B. J. Huebert and D. E. Smith, *J. Electroanal. Chem.*, **31**, 333 (1971).
285. H. E. Keller and R. A. Osteryoung, *Anal. Chem.*, **43**, 342 (1971).

286. J. W. Frazer, L. R. Carlson, A. M. Kray, M. R. Bertoglio and S. P. Perone, *Anal. Chem.*, **43**, 1479 (1971).
287. S. P. Perone, J. W. Frazer and A. Kray, *Anal. Chem.*, **43**, 1485 (1971).
288. E. C. Toren, R. N. Carey, A. E. Sherry and J. E. Davis, *Anal. Chem.*, **44**, 339 (1972).
289. N. G. Anderson, *Anal. Biochem.*, **28**, 545 (1969).
290. M. T. Kelley, W. D. Shults, R. L. Coleman and J. A. Dean, 'Application of the GeMSAEC rapid photometric analyzer to inorganic microanalysis,' pp. 123–128, Microtechniques, Proceedings 6th International Symposium, Graz, Austria, September 1970, Wiener Medizinsche Akademie Verlag, Wien, Austria, 1970.
291. R. L. Coleman, J. A. Dean, W. D. Shults and M. T. Kelley, *Anal. Letters*, **4**, 169 (1971).
292. M. T. Kelley and J. M. Jansen, *Clin. Chem.*, **17**, 701 (1971).
293. M. T. Kelley, 'Computer analytical systems: FOCAL systems for the GeMSAEC and other analytical instruments; program and data storage,' pp. 40–41, U.S.A.E.C. Report ORNL-4749, Jan. 1972.
294. M. T. Kelley, 'Computer analytical systems: programming for the ND-812,' p. 42, U.S.A.E.C. Report ORNL-4749, Jan. 1972.
295. M. T. Kelley, 'ORCAL—A conversational calculating language for the ND-812 computer,' U.S.A.E.C. Report ORNL-TM-3697, Feb. 1972.
296. R. W. Stelzner, 'Computer analytical systems: PDP-8/I programming,' p. 42, U.S.A.E.C. Report ORNL-4749, Jan. 1972.
297. M. T. Kelley, R. W. Stelzner and D. L. Manning, 'Determination of U(IV)/U(III) ratios in molten fluoride salt for the reactor projects group,' pp. 9–10, U.S.A.E.C. Report ORNL-4466, Jan. 1970.
298. H. W. Jenkins, G. Mamantov, D. L. Manning and J. P. Young, *J. Electrochem. Soc.*, **116**, 1712 (1969).
299. A. S. Meyer and J. M. Dale, 'Analytical developments for the molten-salt reactor program: in-line chemical analysis of molten fluoride salt streams,' pp. 10–11, U.S.A.E.C. Report ORNL-4749, Jan. 1972.
300. T. R. Mueller and H. C. Jones, *Chem. Instrum.*, **2**, 65 (1969).
301. H. B. Mark, Jr. and C. N. Reilley, *J. Electroanal. Chem.*, **3**, 54 (1962).
302. D. J. Fisher, W. L. Belew and M. T. Kelley, 'Recent developments in D.C. polarography,' pp. 89–134 *in* G. J. Hills, Ed., *Polarography 1964*, Vol. 1, Macmillan, London, 1966, pp. 121–128.
303. H. L. Pickering and H. C. Eckstrom, *J. Phys. Chem.*, **63**, 512 (1959).
304. R. W. Stelzner, D. J. Fisher and M. T. Kelley, 'Dedicated digital computer polarographic systems,' Abstract to paper 13, Session 3, 'Application of digital computers in analytical chemistry,' 14th Conference on Analytical Chemistry in Nuclear Technology, Oct. 13–15, 1970, Gatlinburg, Tenn. 37738, 'Schedule of Events and Abstracts of Papers'.
305. R. W. Stelzner, D. J. Fisher and M. T. Kelley, 'Computer analytical systems: digital polarography, pp. 39–40, U.S.A.E.C. Report ORNL-4636, Jan. 1971.
306. R. W. Stelzner, D. J. Fisher and M. T. Kelley, 'Computer analytical systems: computer-electrochemical systems,' p. 42, U.S.A.E.C. Report ORNL-4749, Jan. 1972.

CHAPTER II

Applications of Polarography and Coulometry in Actinide Analysis

G. W. C. MILNER AND G. PHILLIPS, *Actinide Analysis Group,*
Atomic Energy Research Establishment, Harwell, Didcot, Berkshire

I. INTRODUCTION

Electrochemical techniques are playing a very important role in the analytical chemistry of the actinide elements, and this is understandable in view of the many valency states of some of these elements in solution. Information on the valencies of the actinides is included in Table I. It will be seen that the earlier actinides show the greatest number of valencies and that the stable trivalent state, which is characteristic of the lanthanides, does not predominate until americium is reached in the actinide series. The many valencies for uranium, neptunium and plutonium provide excellent scope for the development of electroanalytical methods, and this accounts for the many polarographic and coulometric methods so far developed for the determination of these elements.

TABLE I

The Valency States of the Actinide Elements

Ac	Th	Pa	U	Np	Pu	Am	Cm	Bk	Cf
3	(3)	(3)	3	3	3	3	3	3	3
	4	4	4	4	4	(4)	(4)	(4)	
		5	5	5	5	5			
			6	6	6	6			
				7^a	7^a				

[a] Reported by B. N. Spitsin et al., Doklady Nauk U.S.S.R., **181**, 128 (1968).

Some indication of the possible scope of polarographic and coulometric methods in actinide analysis can be obtained from a consideration of the E_0' values for the reversible couples summarized in Table II. Values for irreversible couples are not given because these couples are not very useful in coulometry and only of limited value in polarography. Several of the values in Table II were obtained by Stromatt and coworkers (1) and by Shults (2) using a controlled-potential coulometer with a platinum working electrode. For the U(IV/VI) and Np(III/IV) couples, however, it was necessary to replace the platinum electrode with a mercury pool electrode because of the higher over voltage of hydrogen on mercury. Information on the Fe(II/III) couple is included in Table II because the E_0' values for this couple

TABLE II

Potentials of Electrochemical Reactions in Various Media

Solution composition	E_0' values in volts versus SCE				
	Fe(II/III)	Pu(III/IV)	Np(V/VI)	Np(III/IV)	U(IV/VI)
HClO$_4$ (N)	+0·47	+0·72	+0·90	−0·10	+0·08
HCl (N)	+0·45	+0·72	—	−0·10	−0·21^a
HNO$_3$ (N)	+0·50	+0·60	+0·91	−0·13	—
H$_2$SO$_4$ (N)	+0·45	+0·49	+0·83	−0·29	−0·19^b

^{a} $E_{1/2}$ data for 2N HCl solution. ^{b} $E_{1/2}$ data.

are in the same potential range as those for some of the actinides, and, moreover, iron must be considered because it is a common impurity of most materials.

It will be seen from Table II that the E_0' values for the plutonium, neptunium and iron couples show little variation in hydrochloric, perchloric and nitric acid solutions. However, the values are generally less positive in sulphuric acid solutions, and this effect is most noticeable in the case of plutonium due to the strong complexing of tetravalent plutonium by sulphate ions. In fact the E_0' values for the plutonium and iron couples in this medium are so close together as to prevent the direct coulometric determination of one element in the presence of the other. It is possible to draw the general conclusion from Table II that controlled-potential coulometry should be a suitable technique for the determination of uranium, neptunium and plutonium individually. When these elements are present together or with iron as an impurity, however, the position is less satisfactory because the separation of the E_0' values is not great enough in some cases to give the complete non-interference of one element on the determination of another. A platinum working electrode should be satisfactory for all determinations except those based on the U(IV/VI) and Np(III/IV) couples. A stirred mercury pool is necessary for these particular couples, and unfortunately this type of electrode is less convenient to use experimentally than the platinum electrode. However, good results can be obtained with it on exercising certain precautions.

The potentials of the U(IV/VI) and Np(III/IV) couples are very suitable for use in polarography, and it should be possible to analyse mineral acid solutions of these elements directly with a dropping mercury electrode. The position is less satisfactory for the polarography of the Pu(III/IV) and Fe(II/III) couples in mineral acid solutions. The potentials of both these couples are too positive for the direct application of the dropping mercury electrode. An accepted technique for overcoming this kind of difficulty involves the addition of a complexing agent to form stable complexes with the Pu(IV) and Fe(III) ions in solution. By this means the polarographic steps for the reduction of these species can be moved into a voltage region suitable for the application of the dropping mercury electrode.

II. URANIUM

A. DC and Cathode-ray Polarography

1. Mineral Acid Supporting Electrolytes

(a) *Hydrochloric acid.* The reduction of uranyl ions at the dropping mercury electrode from acid supporting electrolytes gives a polarographic step which is very suitable for the determination of this element. Harris and Kolthoff (3) studied this reduction from hydrochloric acid solutions, and they identified the reactions taking place at the dropping mercury electrode. In 0·01 to 0·2 M hydrochloric acid solutions uranyl ions give a one-electron step with a half-wave potential of $-0·18$ volts versus SCE, and this is followed by a composite step with an apparent half-wave potential of $-0·93$ volts versus SCE. Although the height of the second step is approximately twice that of the first, this step is not very suitable for analytical purposes.

On increasing the molarity of the hydrochloric acid Harris and Kolthoff (4) observed that the height of the first step increased with acidity, and eventually it reached a height corresponding to a two-electron process. This increase in height occurred at the expense of the height of the second composite step. The explanation for this is that the uranyl ions undergo a one-electron reduction which is a fast reaction:

$$UO_2^{2+} + e \rightleftharpoons UO_2^{+} \tag{1}$$

The UO_2^{+} ions formed at the dropping mercury electrode are unstable and undergo a disproportionation reaction as follows:

$$2 UO_2^{+} + 4 H^{+} \rightarrow UO_2^{2+} + U^{4+} + 2 H_2O \tag{2}$$

The rate of this disproportionation reaction is slow, but it improves on increasing the concentration of hydrogen ions. Under strongly acid conditions this reaction occurs so quickly that, at the potential corresponding to the reaction in equation (1), the uranyl ions are reduced completely to the tetravalent state according to the following equation:

$$UO_2^{2+} + 4 H^{+} + 2 e \rightarrow U^{4+} + 2 H_2O \tag{3}$$

A second step at about $-0·9$ volts versus SCE is obtained for the reduction of tetravalent uranium according to the reaction,

$$U^{4+} + e \rightleftharpoons U^{3+} \tag{4}$$

Although this step results from a reversible reduction, it is of little use analytically because its plateau slopes into the step for hydrogen. For analytical work it is desired to use, wherever possible, the hydrochloric acid solution conditions which give the two-electron reduction of the uranyl ions. Several elements also give polarographic steps in the neighbourhood of the uranium step, and thus cause interference unless a preliminary separation of the uranium is carried out. This behaviour has resulted in later workers investigating the suitability of other supporting electrolytes for uranium.

(b) *Sulphuric acid.* Other mineral acid supporting electrolytes have been shown to be applicable to the determination of uranium. Sheel and Watters (5) observed a step at about -0.20 volts versus SCE for the reduction of uranyl ions from sulphuric acid solutions. They found that the diffusion current is dependent upon the concentration of sulphate in solution, and satisfactory results are obtained from solutions between 1.0 and 2.0 N in sulphuric acid. Shalgosky (6), however, has reported the need for some caution in the use of sulphuric acid solutions under certain conditions. He found that low recoveries could be obtained on heating uranium in concentrated sulphuric acid before polarography.

A second step at about -1.0 volts versus SCE from 0.1 N sulphuric acid solutions has been shown by Heal (7) to be a composite step. It consists of an irregularly-shaped step at -0.9 V followed by a normal step at -1.06 V.

(c) *Perchloric acid.* Zittel and coworkers (8) have studied the suitability of the first step for the reduction of uranyl ions from perchloric acid solutions for quantitative analysis; these investigations being limited to solutions from 0.2 to 2.2 N in perchloric acid. These workers confirmed the earlier findings of Harris and Kolthoff (3) relating to a deviation from linearity on plotting step heights against concentration. They showed that this deviation is present for acid concentrations above 0.2 N and that the deviation becomes greater with increase in the uranium concentration. However, Zittel and coworkers concluded that for concentrations of uranium less than 200 μg per ml the deviation is only slight and the ratio of i_d/C may be considered to be virtually constant.

In this study some consideration was given to the possible effects of certain anions on the height of the uranium step from a 0.1 N $HClO_4$ solution. The anions included sulphate, chloride, nitrate and fluoride. In the case of the first three anions they were added in the form of their mineral acids and their sodium salts. Sulphate increased the step height appreciably, whereas the effects of chloride and nitrate were very much smaller. Fortunately, a plateau region occurs where changes in sulphate concentration do not affect the step height very much, and these workers recommend that the sulphate ion concentration should be controlled above a minimum of 0.4 N in 0.1 N $HClO_4$ to ensure an essentially constant effect.

The addition of even a very small amount of hydrofluoric acid to a perchloric acid solution containing uranyl ions causes a drastic effect on the polarographic results. Two steps are produced instead of the usual single step at about -0.2 volts. Fortunately, this interference can be eliminated by the addition of aluminium ions to complex the fluoride ions.

Perchloric acid solutions have been examined by Elliott and Foreman (9) for the determination of uranium in plant solutions from the processing of nuclear fuels. They obtained a well-defined step for uranium at -0.18 volts versus SCE, the height of which was unaffected by variations of perchloric acid concentration in the range 0.4 to 1.2 M. In the analysis of plant solutions, suitable aliquots were evaporated to fumes of perchloric acid so as to remove nitric acid and destroy any organic matter from solvents used in the separation processes. Excessive losses of perchloric acid were prevented by carrying out oxidations in Erlenmeyer flasks. These workers finally used a 1 M $HClO_4$ + 0.1 M hydrazine perchlorate supporting electrolyte for this particular determination. The hydrazine perchlorate reduces some potentially-interfering elements by chemical action, and under these conditions molybdenum is the only element to cause difficulty.

(d) *Nitric acid.* Another polarographic method has been developed by Mashall and Kindler (10) for the determination of uranium in plant solutions coming from a separation plant employing TBP-kerosine. These workers studied the determination of uranium directly in a nitric acid supporting electrolyte saturated with TBP-kerosine, and they observed a one-electron step with a half-wave potential of -0.17 volts versus SCE from 1.6 N HNO_3 solutions. The diffusion current measured at -0.35 volts was found to be proportional to the concentration of uranium in the range from 1 to 7×10^{-4} M. Interference from nitrite, Fe(III), Ce(IV) and W(IV) was eliminated by the addition of 0.5 g of ascorbic acid to the final solution. Also interference from Mo(VI) and V(V) was overcome by precipitation of these elements with N-benzoyl-N-phenylhydroxylamine. The solid reagent was added to an aliquot of sample with stirring at $60°C$, and after de-aeration, the diffusion current was measured without filtration.

(e) *Phosphoric acid.* Zittel and coworkers (8) report that the reduction of uranyl ions in a 0.2825 N H_3PO_4 solution produces a polarographic step at about -0.2 volts versus SCE. Unfortunately, this step is not well-defined and exact measurements of the half-wave potential and the diffusion current are difficult to carry out. On recording a derivative polarogram, these workers obtained two separate peaks for this ill-defined step. The addition of more phosphoric acid to increase the concentration of the supporting electrolyte caused the two peaks to begin to merge, but there was still some evidence of two peaks at a concentration of 2.8 N H_3PO_4. However, Zittel and coworkers found that the addition of perchloric acid in an amount sufficient to make the ratio of the equivalents of H_3PO_4 to $HClO_4$ equal to 30 to 1 caused the double peak to merge into a single peak. The height of this peak was found to be linearly proportional to the concentration of uranium from 10 to 200 μg per ml in solutions of composition 0·5652 N H_3PO_4 + 0·054 N $HClO_4$.

In further work to understand this system, these workers investigated the effects of adding phosphoric acid and sodium phosphate to a perchloric acid supporting electrolyte containing uranium on the first step for the reduction of uranyl ions. The addition of these reagents resulted in an increase in the height of the uranium step, and this effect was caused at least partially by the phosphate ions and not entirely by the change in the hydrogen ion concentration. Fortunately, a plateau region was found in which changes in the phosphate concentration have little effect on the step height, and this occurred for solutions between the limits of approximately 0·4 N and 2 N H_3PO_4. From this work it is clearly possible to take a phosphoric acid solution and to obtain a satisfactory quantitative polarographic step for uranium by adding a controlled amount of perchloric acid to it.

2. Analytical Applications

(a) *Analysis of UO_2 for the ratio of oxygen to uranium.*   The oxides of uranium are non-stoichiometric, and hence it is necessary to know the ratio of oxygen to uranium dioxide used in the production of nuclear fuels. Uranium dioxide has a tendency to absorb oxygen and to retain it within the lattice as oxide ions. The consequent transfer of electrons from U^{4+} to oxygen thereby raises the effective valency of the uranium, and the determination of uranium in valency states greater than four serves as a measure of the excess oxygen and hence of the overall stoichiometry.

Some of the methods described for the determination of oxygen to uranium ratios are based on the determination of the oxidation states of uranium in UO_2, and a polarographic method originally proposed by Burd and Goward (11) is an important method for the analysis of hyperstoichiometric oxides. The uranium dioxide is dissolved in phosphoric acid under reflux, as this causes no change in the oxidation states. This solution is diluted with (1 + 9) sulphuric acid, and the polarographic step for the reduction of any uranyl ions is recorded starting from zero applied voltage. This method is used routinely in the U.K.A.E.A. by employing either a pen-recording dc polarograph or a cathode-ray polarograph (12).

Shalgosky, Smart and Watling (13) have improved the precision of this determination by employing comparative polarography with the differential cathode-ray polarograph; they succeeded in determining the U(VI) content with a precision of 0·1 per cent. With samples having oxygen/uranium ratios of 2·01 or higher, the standard deviation of this method is ±0·0001 in terms of O/U ratio. However, stringent precautions against air oxidation are necessary for samples with lower ratios, and a standard deviation of ± 0.0002 is reported for samples with oxygen to uranium ratios of 2.001.

Shalgosky and coworkers applied their method to samples of uranium dioxide in the form of pellets. Portions weighing about 0.25 g were obtained by using a steel chisel to break each pellet (with $O/U > 2.01$) into about eight pieces. Each piece took from 1 to 2 hours to dissolve in 5 ml of concentrated phosphoric acid on gentle heating. Then 35 ml of $2 M$ sulphuric acid were added, and, after cooling, the solution was diluted to 50 ml with $2 M$ H_2SO_4. An aliquot of this solution containing between 10 and 60 μg/ml of U(VI) was taken for analysis because calibration graphs of differential peak height against differential concentration were shown to be linear in this range. With near-stoichiometric material the pellets were crushed in a percussion mortar before weighing. The processes of crushing, weighing and immersing in acid were carried out in a glove box filled with purified nitrogen since every possible precaution must be taken against oxidation of the samples at all stages of the analysis.

Sipos and Branica (14) have extended the polarographic method for O/U ratio measurements to include the determination of the total uranium content. This is achieved by oxidizing the uranium completely to the U(VI) state by the addition of ceric ions, and then recording the polarogram. The O/U ratio is calculated from the limiting currents and dilutions according to the following expression:

$$\text{oxygen/uranium ratio for } UO_2 = 2.0000 + \frac{h_1 V_1}{h_2 V_2}$$

where
  h_1 = height of U(VI) step in the dissolved sample solution
  h_2 = height of U(VI) step in the oxidized sample solution
  V_1 = volume of sample taken for determination of total uranium
  V_2 = volume of volumetric flask used for the dilution of sample solution
With this method it is not important to know the weight of the sample accurately, but it should be between 0.1 and 50 mg if the O/U ratio of the sample is between 2.5 and 2.001. The coefficient of variation for the proposed method is about 0.6 per cent.

(b) *The analysis of plutonium metal for traces of uranium.* Plock and Vasquez (15) have used a chloride supporting electrolyte for the determination of trace amounts (> 10 p.p.m.) of uranium in plutonium metal by cathode-ray polarography. From a $2 M$ hydroxylamine hydrochloride supporting electrolyte they obtained a reversible peak corresponding to a one-electron reduction with a peak potential of -0.167 volt versus mercury pool electrode. For this analysis 2 g of plutonium metal is dissolved in 5 ml of $6 M$ HCl, and then any uranium is oxidized with nitric acid (3 ml). The resulting solution is evaporated to dryness in a water bath, then 2–3 ml of 1 M HCl are added and the solution is again evaporated just to dryness. The warm crystals are next dissolved in 4 ml of 5 M hydroxylamine hydrochloride and the solution is warmed on a hot plate for 10 minutes. After cooling and diluting to 10 ml with water, the peak for uranium is recorded.

Investigation of the behaviour of other elements in the supporting electrolyte showed that antimony, copper, molybdenum and titanium could cause interference, if present in the sample. Interference from molybdenum is eliminated by heating the solution in the presence of hydroxylamine hydrochloride. However, difficulty from copper and antimony require electrolysis using a mercury cathode, and an anion exchange separation is needed to separate titanium from uranium.

(c) *The analysis of uranium tetrafluoride for U(VI) content.*   The original method of Burd and Goward (11) for U(VI) in UO$_2$ has been applied by Takani and Morimoto (16) to the determination of U(VI) in uranium tetrafluoride. The sample (0·5 g) is dissolved in 85 per cent phosphoric acid (10 ml) within 4 minutes. Then two drops of 0·02 per cent methyl red are added and the solution is diluted to 50 ml with 1 M sulphuric acid before recording the polarogram. There is no interference from either copper or molybdenum when present in amounts less than 1 p.p.m.

3. Complexing Supporting Electrolytes

There are several elements which can cause interference with the step for uranium from mineral acid supporting electrolytes, and these include copper, molybdenum, titanium, antimony and others. Some improvement in selectivity has been achieved by the inclusion of a complexing agent in the supporting electrolyte, and this behaviour has been clarified by the work of Shalgosky (6). He concluded that the addition of a complexing agent such as oxalic acid, tartaric acid, citric acid, etc., to uranium in a sulphate solution assists mainly in the separation of interfering steps and hardly affects the diffusion current or the mechanism of the reaction at the dropping mercury electrode. Other workers have improved the specificity still further by reducing some constituents to lower valency states before the addition of a complexing agent. Details of the effects of some of the more useful complexing agents are now considered in detail.

(a) *Tartrate.*   An acid–tartrate supporting electrolyte was first suggested by Lewis and Overton (17) for the determination of uranium. However, several years elapsed before Shalgosky (6) carried out a full investigation of this electrolyte. He studied H$_2$SO$_4$–tartrate solutions, and he encountered difficulties on evaporating the uranium solution to fumes of H$_2$SO$_4$ to remove nitrate ions before adding a solution containing 12 g of sodium tartrate, 0·293 g of sodium chloride and 10 ml of proteose peptone per 100 ml. The step height for uranium from this solution was several per cent lower than that from a solution which had not been evaporated to fumes of H_2SO_4. The use of H_2SO_4 for the removal of nitrate ions was therefore abandoned in favour of $HClO_4$, which is known to be a non-complexing medium for uranium. However, since the concentration of $HClO_4$ in the final solution is important, the recommended procedure involved evaporating the solution down to a final volume of 0·1 ml of $HClO_4$, followed by the addition of 0·4 ml 60 per cent $HClO_4$ on cooling. This solution is then transferred to a 5 ml flask with washing, and 0·5 ml of the tartrate stock solution is added. After 2 minutes, 1·0 ml of 5 M H_2SO_4 is added and the solution is diluted to volume with water. The step at $-0·35$ volts versus the mercury pool is quantitative for uranium in the concentration range from 20 to 200 μg per ml. Molybdenum gives a step at $-0·38$ volts which interferes with the uranium step, but those for Bi ($-0·17$ volts), Cu ($-0·18$ volts) and Pb ($-0·55$ volts) do not interfere. Shalgosky recommended the acid–tartrate supporting electrolyte in preference to the acid–oxalate medium proposed by Legge (18) because fewer elements caused interference. This electrolyte is used in a rapid method for the determination of the uranium content of ores, covering a wide range of contents (19).

(b) *Malonic acid.*   Plock and Miner (20) have investigated a supporting electrolyte containing malonic acid for the polarographic determination of uranium. Initial experiments were carried out with solution 0·5 M in di-sodium malonate and 0·5 M in sodium hydrogen malonate. In solutions of low pH (pH 1·1), there is very little complexation of the uranium by the malonic acid. However, the situation changes as the pH is increased, as shown by the fact that the polarographic step moves to more negative potentials. The plot of $E_{1/2}$ versus pH can be divided into straight line portions corresponding to the pH regions 1·0 to 3·3, 3·3 to 4·8 and 4·8 to 6·5. In the first region the shift of $E_{1/2}$ with change of pH may be represented by the equation

$$E_{1/2} = -0·104 - 0·081 \text{ pH}$$

whereas for the pH range 3·3 to 4·8 the equation is $E_{1/2} = -0·250 - 0·037$ pH. In the pH range 4·8 to 6·5 the slope of the curve is approximately zero which indicates that above 4·8 the half-wave potential is not a function of pH.

These workers also observed some variation of step height with pH. Between pH 1·0 and 2·0 the diffusion current increases with increase of pH, but then it becomes constant between pH 2·0 and 3·0. The increase is interpreted as an increase in the rate of disproportionation of U(V). Above pH 3·0 the diffusion current decreases with increase of pH which indicates that there is a lessening of the complexation of uranium with malonate owing to competition from hydroxyl ions.

The uranium step is independent of the concentration of malonate for values greater than 0·1 M, and from this Plock and Miner recommend solutions 1 M malonate with a pH of 2·5 for the quantitative determination of uranium. However, under these conditions they found that antimony, chromium, copper, iron, lead, nickel and titanium interfere. Each of these elements can be removed completely by mercury cathode electrolysis, with the exception of antimony and titanium. The antimony is only partially removed by the method, but the ratio of antimony to uranium can be lowered sufficiently to prevent interference. In the case of titanium, however, an anion exchange separation with Dowex 1 is necessary for its removal. Uranium remains on the column whilst titanium passes through.

Plock and Miner used the step for uranium from the malonic acid supporting electrolyte in a method for the determination of uranium contents greater than 225 p.p.m. in plutonium metal. The plutonium metal sample (1 g or less) is weighed directly into a tared 50 ml volumetric flask and dissolved in 5 ml 6 M hydrochloric acid. Then 3 ml of nitric acid are added, and the solution is evaporated to dryness on a hot plate. To the warm crystals 10 ml of 5 M hydroxylamine hydrochloride are cautiously added, followed by 7·5 ml of 6 M hydrochloric acid. The solution is warmed on a hot plate for 10 minutes to ensure the complete reduction of plutonium to Pu(III). After cooling, the solution is made 0·5 M in disodium malonate and 0·5 M in sodium hydrogen malonate by slowly adding these reagents with continuous swirling of the volumetric flask. The solution is diluted to volume with water. An aliquot of this final solution (pH 2·5) is de-aerated with nitrogen for 10 minutes before recording the polarogram. If interfering elements are shown to be present, these are removed by the methods already described.

(c) *Ethylenediamine tetra-acetic acid* (EDTA).   Uranyl ions are one of the few electroactive species which are only weakly complexed by EDTA, and Pribil and Blazek (21) were the first to take advantage of this fact in the polarography of uranium. They prevented the interference of many element in the polarographic determination of uranium by preferentially complexing these elements with EDTA. They used a supporting electrolyte consisting of 1 M ammonium carbonate $+0·1 M$ EDTA for the determination of uranium directly in the presence of other elements, including Ni, Co, Mn, Zn, Cr and Al. The step for uranium occurs at $-0·8$ volts versus SCE. Copper gives a step at $-0·45$ volts versus SCE which does not cause difficulties except for solutions with unfavourable ratios of copper to uranium. Although titanium is precipitated from solution, small amounts of this element do not cause interference. The main difficulty comes from lead which gives a step coinciding with the uranium step.

In more recent work Auerbach and Kissel (22) have studied an acetate–EDTA supporting electrolyte for uranium, and they found solution conditions suitable for the direct determination of the uranium content of the bismuth alloy used as a fuel in the liquid metal homogeneous reactor. This alloy contains 99·8 per cent Bi, 0·035 per cent Zr, 0·025 per cent Mg and 0·10 per cent U. In this study, these workers used an incremental polarograph giving derivative peaks, and they obtained a peak for uranium from a 0·1 M acetate buffer (0·1 M HAc + 0·1 M NaAc) containing either 0·001 M EDTA or 0·01 M EDTA with an $E_{1/2}$ value of $-0·31$ or $-0·30$ volts respectively. The relationship between peak height and concentration was found to be linear over a range of uranium concentrations from 5×10^{-6} to 8×10^{-4} M. Unfortunately copper gives a peak near to the uranium peak, and complete overlapping is obtained from base solutions containing 0·01 M EDTA. Bismuth gives a polarogram with two separate peaks occurring at potentials more negative than $-0·3$ volts, but the separation between the uranium peak and the first bismuth peak is more than 200 mV. The best resolution for the separation of the peaks for these two elements is obtained from the 0·01 M EDTA + 0·1 M acetate buffer (pH 4·6) supporting electrolyte. Under these conditions it is possible to determine directly a concentration of 2 $\times$ 10^{-6} M uranium in the presence of 0·01 M bismuth. Molybdenum gives two fairly-well resolved peaks between $-0·55$ and $-0·75$ volts, and small concentrations of this element do not cause any interference. However, 10^{-2} M concentrations of molybdenum cause difficulties because of the occurrence of a drawn out wave at $-0·26$ volts just in front of the uranium peak. If necessary, this difficulty can be resolved by increasing the concentration of EDTA to 0·05 M and then the wave starts at $-0·33$ volts. Under these conditions it is possible to determine 5×10^{-6} M concentrations of uranium in the presence of 10^{-2} M molybdenum.

4. Non-aqueous Supporting Electrolytes

With some analytical problems it is necessary to separate and concentrate the uranium from a sample solution before completing the determination polarographically, and in 1956 Fisher and Thomason (23) appreciated the advantages of being able to determine the concentration of uranium directly in the organic extract. These workers carried out a study with a variety of organic solvents, and they obtained satisfactory results on using TBP in isopropyl ether as the extractant for uranium. After separation, this extract was diluted with a twenty-fold excess of a suitable medium containing a supporting electrolyte. Glacial acetic acid containing lithium perchlorate was found to be satisfactory for this purpose, and uranium gave a satisfactory step in the final solution. However, the following problems needed to be overcome before this technique could be accepted: these included difficulties resulting from the relatively high specific resistance of the organic extract and also from the poor performance of the conventional saturated calomel electrode. The subsequent development of a controlled-potential polarograph at the Oak Ridge National Laboratory has helped to overcome these difficulties satisfactorily.

In more recent work Belew, Fisher, Kelley and Dean (24) have concentrated on the use of tri-n-octyl phosphine oxide (TOPO) in cyclohexane as the extractant. The extraction coefficient for uranium with TOPO is high, and in addition this reagent is not reduced at the dropping mercury electrode. Sample solutions 0·5 N in nitric acid are extracted with 0·1 M solutions of TOPO in cyclohexane to obtain solution conditions giving the highest possible extraction coefficient. An equilibration time of 5 minutes is satisfactory, and after separation of the phases, an aliquot of the organic phase is pipetted into the polarographic cell. An equal volume of ethanol containing 0·2 M lithium perchlorate is added to the cell, and the cell solution is then de-gassed with argon for 10 minutes. On recording the polarogram from -0.05 volts versus SCE, a well-defined step is obtained for uranium with an $E_{1/2}$ value of -0.295 volts versus SCE; this step also corresponds with the one-electron reduction of U(VI) to U(V). The height of this step and of the first derivative peak is linear with concentration over the range 0·24 to 25·4 μg uranium per ml in the cell. The determination of concentrations greater than 24 μg per ml are best achieved by diluting the solution to a cell concentration below 24 μg per ml. A Beckman fibre-tip saturated calomel electrode is satisfactory as the reference electrode, provided that the contact time with the solution in the cell is no longer than 15 minutes.

It is claimed that this procedure is especially suitable for the analysis of waste streams containing trace amounts of uranium. Concentrations of uranium as low as 2·4 μg in 250 ml of sample can be determined on using a 5 ml polarographic cell.

B. AC Polarography

1. Square-wave Polarograph

The suitability of the square-wave polarograph for the determination of small concentrations of uranium has been investigated by Milner and Nunn (25). Initially they examined the reduction of uranyl ions from mineral acid supporting electrolytes. A well-defined peak at about -0.20 volts versus SCE was obtained for the reduction of this element from several mineral acid solutions, including HCl, HNO_3, H_2SO_4 and $HClO_4$. The height of the uranium peak proved to be very dependent upon the final acidity in the case of HCl, HNO_3 and H_2SO_4. In the latter supporting electrolyte, a much lower sensitivity was also obtained. The height and $E_{1/2}$ value of the uranium peak from perchloric acid solutions (≤ 3 M) proved to be uninfluenced by variations in the acidity, and this electrolyte was the only one that did not require careful control of the final acidity. In 1 M $HClO_4$

the peak height is linear with concentration over the range from 1 to at least 2500 μg per ml of final solution. Small peaks are obtained for concentrations of uranium less than 1 μg per ml, but these are not suitable for quantitative determinations.

Some interference can result with this step for uranium from the presence of other elements in the same solution. For example bismuth gives a peak occurring very near to that for uranium, and the resolution of the peaks is poor for equal concentrations of these elements. A peak for copper just precedes the uranium peak, and there is negligible interference except for ratios of copper to uranium exceeding 1:1. The lead peak is more negative and fairly well-separated from the uranium peak, and so lead is less troublesome than copper. Molybdenum is the main interfering element since it gives a peak with a half-wave potential coinciding with the value for the uranium peak. In view of these interferences, the peak for uranium is best applied to analyses involving a preliminary separation of the uranium from other constituents by such techniques as solvent extraction or cellulose chromatography. Any organic matter remaining in solution after the separation process can be easily destroyed by wet oxidation before finally evaporating the solution to fumes of perchloric acid and then recording the polarogram.

Milner and Nunn investigated the possibility of adding a suitable complexing agent to the perchloric acid solution so as to reduce the number of interfering elements, and they found a tartrate supporting electrolyte consisting of 2 M HClO$_4$ + 0.08 M sodium tartrate + 0.005 M sodium chloride to be the most useful of those investigated. Interference from bismuth, copper and iron only occurs for high ratios of these elements to uranium, and elements causing no interference include zinc, nickel, cadmium, indium, cobalt, and manganese. Lead and chromium give peaks at about -0.43 volts versus SCE and these elements cause no interference except for high concentrations with respect to the uranium concentration. The only serious difficulty comes from molybdenum which gives a peak at -0.19 volts versus SCE.

These workers succeeded in using this peak for uranium in a method for the direct determination of the uranium content of certain types of mineral samples. In this method 100 mg of sample is taken into solution by sintering with sodium peroxide and extracting the cooled sinter with hydrochloric acid. Silica is then dehydrated by evaporating the solution to dryness, and it is removed by filtration. After diluting the filtrate to 100 ml, a 5 ml aliquot is taken for analysis and 1 ml of 10 N HClO$_4$ is added. This solution is evaporated just to fumes of perchloric acid, and then 0.50 ml of the tartrate supporting electrolyte is added before making the volume up to 5 ml with water. The polarogram is recorded on this solution starting at a potential of -0.1 volts versus SCE. This technique proved very suitable for the analysis of monazite, monazite–dunite, pitchblende–dunite and quartz–iron oxide standard samples containing amounts of uranium in the range from 0·02 to 0·30 per cent. Results for uranium by the polarographic method were in good agreement with values obtained by activation analysis.

2. Pulse Polarograph

In experiments to determine the uranium contents of sea water, Milner, Wilson, Barnett and Smales (26) examined the suitability of a perchloric acid–tartrate supporting electrolyte for the determination of very low concentrations of uranium by the pulse polarograph. Since the pulse polarograph is more sensitive to irreversible reductions than the square-wave polarograph, the behaviour of some elements could be different from that reported previously (25). No polarographic peaks were observed for VO^{2+}, Cr^{3+}, Mn^{2+}, Ni^{2+}, AsO_2^-, TiO^+ and Zn^{2+}; the reductions of these elements being either too irreversible or occurring at too negative a potential to be separated from the background response. Fe^{3+} and Cd^{2+} gave peaks which did not interfere. The responses due to Pb^{2+} and Cu^{2+} ions, though occurring nearer to the uranium peak, did not produce an error greater than 1 per cent in the determination of 1 μg per ml concentrations of uranium if the copper and lead concentrations were ≤ 1 μg per ml. The separation of the peaks due to copper and uranium is only satisfactory, however, if the chloride ion concentration is low. Large concentrations of these ions shift the copper peak, presumably by complexing the hydrated copper ions, such that in a solution 1·0 N in chloride ions the peaks for copper and uranium are not resolved. As with square-wave polarography the most serious interference is given by molybdate ions. The polarogram for this element consists of a small peak at -0.18 volts followed by a large peak at -0.44 volts versus Hg pool, and these peaks are linked by a saddle which is some distance above the recorder base line. In using this uranium peak for analysis, therefore, it is important to separate the uranium chemically so that no molybdenum and non-interfering amounts of copper, lead and chloride ions appear in the final solution.

For the analysis of sea water, Milner and coworkers first separated and concentrated the uranium from 4 litres of sea water by solvent extraction with di-(2-ethylhexyl)-phosphoric acid in carbon tetrachloride. A further purification of the uranium was next carried out by the extraction of uranyl nitrate into ethyl acetate, and the uranium concentration was then determined by employing the peak from the $HClO_4$–tartrate base solution. Uranium-237 tracer was employed as a radioactive tracer to measure the percentage recovery of uranium in the chemical separation process.

C. The Coulometric Determination of Uranium

Volumetric methods for the determination of uranium based upon the U(VI/IV) couple have been widely reported (27). Although direct reductimetric determination of the uranyl ion is possible using standard solutions of chromous and titanous ion, this is rarely attempted due to the practical difficulties of working with standard reducing agents of that nature. The more usual procedure is to carry out a reduction of U(VI) to U(IV) using a column of amalgamated zinc, cadmium, lead or silver followed by oxidation of the U(IV) produced with standard ceric or dichromate solution. Determinations based upon this procedure have to be carried out with care to ensure complete reduction of U(VI) to U(IV) and oxidation of any U(III) produced, and to avoid interferences due to the wide potential span employed. The method is consequently best applied to pure solutions of uranium. Coulometric techniques have been developed with a view to avoiding these difficulties, either by the generation of unstable reductants in situ by constant-current electrolysis, or by the avoidance of interfering reactions by using controlled-potential electrolysis. The merits of the various applications of these two techniques are discussed in the following section.

1. Controlled-potential Methods

The first successful coulometer combining a completely electronic potentiostat and integrator was designed for the coulometric determination of U(VI) at controlled-potential (28). Guided by polarographic information, quantitative reduction of U(VI) to U(IV) either in molar sulphuric acid or in 1 M potassium nitrate $+0.1$ M aluminium sulphate at pH 4·5 was shown to be feasible. Attempts to use the one-electron change of U(VI) to U(V) in dilute acid or buffered acetate solution were unsatisfactory, but, in the solution conditions recommended, disproportionation of U(V) to U(VI) and U(IV) was sufficiently rapid to enable complete reduction to U(IV) to be achieved in 15 minutes or less. In molar sulphuric acid the reduction potential employed was -0.25 volts versus the silver–silver chloride-saturated potassium chloride reference electrode and accurate results were obtained at the 75 and 7·5 mg level of uranium with a precision of 0·03 per cent (coefficient of variation) (29). In potassium nitrate–aluminium sulphate medium the reduction potential employed was -0.60 volts with a pre-reduction at -0.20 volts versus the silver–silver chloride-saturated potassium chloride reference electrode. The technique of prereduction at a controlled-potential eliminates the interference from ions reducing at more positive potentials than that of uranium, and the use of a complexing medium moves the reduction potential for the uranyl ion in a negative direction and further away from that for the reduction of Cu(II) ions. In this medium accurate recoveries were obtained for 0·75 and 0·075 mg quantities of uranium (precisions of 0·06 and 0·4 per cent respectively) in the presence of nitric acid (700:1), Hg(II) (6:1), Cu(II) (0·8:1), Fe(III) (0·5:1) and Ce(IV) (0·1:1). At the 7·5 μg level of uranium the precision obtained was 2·2 per cent (coefficient of variation). Interference from molybdenum and ruthenium would not be eliminated without a separation. Further work by these authors (30) reported a controlled-potential coulometric determination of uranium in solutions of irradiated nuclear fuel specimens. The uranium was separated from the fission product elements by extraction of the tetrapropyl-ammonium uranyl trinitrate complex into methyl isobutyl ketone. Entrained organic material and coextracted ruthenium were eliminated by means of a sodium bisulphate–perchloric acid fusion at 600°C. The coulometric determination of the uranium was carried out in potassium nitrate–aluminium sulphate solution at pH 4·6 with a reproducibility of 0·15 per cent (coefficient of variation).

A similar, but slightly simpler, controlled-potential coulometer reported by Kelley and coworkers (31) has been applied to the determination of uranium in homogeneous reactor fuel solutions (32). These authors confirmed that the presence of minor amounts (<1 per cent) of fission product elements, including Tc, Ag, Eu, Pd, Ce(IV), I, Rb, Te, Rh, Nb, La and Cs, caused no interference in the controlled-potential coulometric reduction of U(VI) in molar sulphuric acid solution at a potential of $-0·30$ volts versus the silver–silver chloride reference electrode. A precision of 0·1–0·2 per cent (coefficient of variation) was obtained on 5 mg amounts of uranium. Interference was noted from elements arising from corrosion of the reactor components, namely Cu(II) (0·06:1), Mo(VI) (0·1:1) and Cr(VI) (0·1:1). Interference from copper at the ratio quoted was avoided by reducing Cu(II) to metal at 0·0 volts versus the silver–silver chloride electrode followed by reduction of U(VI) to U(IV) at $-0·30$ volts. Interference from molybdenum and chromium was eliminated by solvent extraction with α-benzoinoxime/chloroform solution. Later work in the same laboratory (33) reported that uranium in the presence of widely varying ratios of copper should be determined by taking advantage of the fact that the copper reduction at a mercury electrode is reversible, whereas the uranium reduction is irreversible in sulphuric acid electrolyte. Two electrolyses were therefore required to determine the uranium—a coulometric reduction at $-0·3$ volts during which U(VI) is reduced to U(IV) and Cu(II) to Cu(Hg), followed by an oxidation at $+0·175$ volts during which Cu(Hg) is reoxidized to Cu(II). Uranium as U(IV) is not oxidized at this potential and can be determined from the difference in the number of coulombs consumed in the two steps. Both controlled-potential coulometers (28, 31) considered so far are described as absolute instruments and the quantity of material electrolysed is calculated from the number of coulombs consumed and Faraday's law. One paper, however (32), reports a discrepancy of about 1·5 per cent between instrumental calibration and chemical standardization at the 5 mg level.

The irreversible nature of the U(VI/IV) reduction has been used to advantage for determining the stoichiometry of uranium dioxide in a method reported by Stromatt and Connally (34). This determination is based upon the fact that dissolution of hyperstoichiometric UO_2 in 85 per cent phosphoric acid under inert atmosphere does not oxidize U(IV), and the small quantity of U(VI) present can be determined coulometrically. The coulometric reduction is carried out at a potential of $-0·38$ volts versus the SCE in a solution 1 M in sulphuric acid and 0·8 M in phosphoric acid. Total uranium is then determined by adding excess of ceric sulphate solution, prereducing at $+0·05$ volts to remove Ce(IV) ion and then reducing at $-0·38$ volts versus the SCE. The precision of the determination of the U(VI) component was 4 per cent at $UO_{2.0074}$ and 0·13 per cent at $UO_{2.6677}$. At the mercury pool no significant oxidation of U(IV) occurs below about $+0·20$ volts versus the SCE. Any attempt to use higher potentials on the mercury pool results in the dissolution of mercury. Oxidation of U(IV) to U(VI) can be achieved at the platinum electrode using potentials appreciably higher than $+0·5$ volts versus the silver–silver chloride-saturated potassium chloride electrode (35). The rate of oxidation increases with higher positive potentials and reasonably short titration times can be achieved at $+1·4$ volts. At $+1·5$ volts oxidation of the solvent occurs at a rate giving an unacceptably large blank. The authors recommend pretreatment of the platinum electrode at a potential of $+1·5$ volts in order to produce an oxide layer on the surface. This pretreatment results in lower background currents at the working potential of $+1·4$ volts and a correction of less than 3 per cent for the determination of 1 mg amounts of uranium. The determination was carried out in phosphoric, sulphuric, perchloric and nitric acids, and a mean recovery of 100·3 per cent with a coefficient of variation of 0·3 per cent was obtained.

Complexing solutions are commonly employed in polarographic work in order to influence half-wave potential values, and so avoid chemical separations of polarographically interfering elements. The same technique can be applied to controlled-potential coulometric determinations, and Zittel and coworkers (36) have developed a method for the determination of uranium employing a sodium tripolyphosphate medium in order to avoid interference from molybdenum. In this medium, at between pH 7·5 and 9·5, U(VI) is reduced to U(IV) at $-1·40$ volts versus the SCE using a mercury electrode, whereas Mo(VI) is not electroactive and causes no interference for ratios of Mo to U of not more than 1 : 7. Interferences from small quantities of nickel, iron, chromium and copper are eliminated by the pretitration procedure. Ruthenium and nitrate ions cause interference with the U(VI) reduction and must be eliminated by fuming with perchloric and sulphuric acids. The precision for the determination of 5 mg amounts of uranium in the presence of molybdenum is 0·7 per cent or better (coefficient of variation). The tripolyphosphate medium is also suitable for the oxidation of U(IV) to U(VI) at a potential of −0·1 volts versus the SCE at the mercury pool (37). A similar approach to the above, but using a fluoride medium, has been described by Mountcastle and coworkers (38). The potential required for the quantitative reduction of U(VI) to U(IV) is dependent on the concentration of sodium fluoride. With 0·75 M NaF at pH 6 a potential of −1·00 volts versus the SCE is required for quantitative recovery. Minor quantities of sulphate, chloride and nitrate can be tolerated, but nitrite must be removed with sulphamic acid. Relatively large ratios (6:1) of iron as Fe(III) cause no interference on employing a prereduction step at −0·30 volts versus the SCE. No interference from minor quantities of aluminium, chromium, copper, molybdenum and nickel was detected and the method was applied to a number of synthetic reactor-fuel studies. The effects of radiation were not investigated, but careful control of the pH was necessary to avoid the worst effects of fluoride ion corrosion on the cell and electrodes. In this medium and in complexing solutions generally, the controlled-potential coulometric reduction of U(VI) to U(IV) does not proceed as rapidly as in dilute sulphuric acid solution. The controlled-potential coulometric determination of uranium in the presence of niobium has been reported (39). The titration is carried out in 0·5 M sulphuric acid solution containing oxalic acid to complex the niobium. The oxalic acid has no effect on the reducing potential of U(VI) to U(IV), which is −0·325 volts versus the SCE, and the method is applicable to samples in which the Nb:U ratio is not greater than 3:1. At higher ratios niobium tends to precipitate, and it interferes with the uranium determination by giving high results. A solvent extraction procedure using tri-n-octylphosphine oxide is recommended to overcome this difficulty.

A thorough study of the controlled-potential coulometric reduction of U(VI) in 0·5 M sulphuric acid solution has been carried out by Jones and coworkers (40), as part of the evaluation of a coulometer designed for work with small amounts (0·010 to 100 μ equivalents) of materials. These authors found that, at the 8 mg level, a uranium standard prepared from U_3O_8 gave a slight positive bias of ∼0·5 per cent when analysed coulometrically, whereas one prepared from uranium metal did not. They also reported a positive bias which increased as the quantity of uranium taken for the coulometric measurement was decreased. This latter effect could be eliminated by carrying out the pretitration and titration at +0·160 and −0·200 volts versus the SCE respectively, rather than at +0·075 and −0·325 volts. Both these findings are in line with our experiences at A.E.R.E., and the importance of background current evaluation at the μ equivalent level is emphasized in this paper.

It is generally accepted that controlled-potential coulometry is a useful technique for the determination of uranium particularly in sulphuric acid medium, and is subject to few interferences. Some care is needed to establish correlation between instrumental calibration and chemical standards, and the precision of the method is limited to about 0·1 per cent (coefficient of variation). Higher precision (0·006 per cent) has been obtained (41) by applying the technique differentially, but this technique is not generally applicable and is only suitable for the intercomparison of uranium standards. The difficulties associated with controlled-potential coulometry, namely background current effects and the integration of an electrolysis current which varies over several orders of magnitude, can very largely be avoided by constant-current techniques. This technique is, however, more subject to interferences from other elements and is dealt with in the next section.

2. Constant-current Methods

The difficulties encountered in the determination of uranium by constant-current coulometric methods based upon the $U(VI)/U(IV)$ couples arise largely from the fact that this couple is thermodynamically irreversible. Hence, attempts to reduce $U(VI)$ with an electrogenerated reductant or conversely to oxidize $U(IV)$ with an electrogenerated oxidant usually suffer from sluggish end-point behaviour. Most analytical procedures therefore are designed with a view to avoiding this difficulty by the addition of excess of a standard reagent followed by a back titration with an electrogenerated reagent.

Conditions under which Ti^{3+} can be generated with 100 per cent current efficiency have been established by Lingane (42) and this reagent has been applied to the determination of uranium (43). In a slightly modified procedure, Taylor and Marinenko (44) reduce $U(VI)$ to $U(IV)$ with Ti^{3+} generated from 1 M titanyl solution in 9 M sulphuric acid with a current density of 2·5 m amp cm^{-2} at the platinum electrode. These workers acknowledge that the reaction is slow and accordingly they found it advisable to over-titrate and then to wait until equilibrium had been established. An amperometric end-point detection system is used, and the end-point is found by generating additional titanous ion in small increments and measuring the slope of the over-titration curve. By this means they obtained a precision of 0·006 per cent (coefficient of variation) on 1 g samples of pure uranium metal, but unfortunately they found interference from any ions that are reducible by Ti^{3+}.

The reduction of uranyl ions on columns of metallic reductors, such as zinc, cadmium, lead and silver, is a relatively rapid process, and Furman and coworkers (45) used this technique to prepare solutions of $U(IV)$. Quantities of uranium from milligram to microgram amounts were reduced to $U(IV)$

on a cadmium column, which has the advantage of producing little or no U(III), and the reduced solutions were run into excess of a solution of ferric ions. The ferrous ion produced was then determined by titrating to an amperometric end-point with electrogenerated ceric ions produced at constant current. The errors ranged from about ± 2 per cent at the mg level to ± 8 per cent at the $5 \mu g$ level. The method is only applicable to pure uranium solutions since it is subject to interference from any element which can be reduced on the cadmium column and subsequently oxidized by ceric ion. It is known (46) that the current efficiency for the electrogeneration of ceric ion is less than 100 per cent, but overall titration efficiency can still be acceptable by such an approach bearing in mind that ferrous ions can be electro-oxidized directly to ferric at the same time. A similar procedure has been described by Carson (47). In this case the uranium is reduced with a lead column and the U(IV) produced is titrated with electrogenerated bromine. The precision and accuracy of the method ranged from ± 0.3 per cent for 7 mg and ± 6 per cent for 0.03 mg amounts of uranium. Carson, however, found it necessary to carry out the titration at 90–95°C in order to increase the rate of reaction.

The need for close control of nuclear materials has lead to a demand for analytical methods for uranium which are free from bias and give a precision of 0.01 per cent or better. Work of this accuracy and precision is an obvious application for the technique of constant-current coulometry, which depends upon the two physical standards of time and electrical current. These standards can both be measured in absolute terms and with a precision of 1 part in 10^5 or better. Malinowski (48) has described an analytical procedure in which U(VI) is reduced to U(IV) in hydrochloric acid solution by means of metallic aluminium. Any U(III) produced is oxidized to U(IV) on the addition of concentrated phosphoric acid. A weighed excess of potassium dichromate is then added, and the amount in excess is determined by constant-current coulometric generation of ferrous ions. The method therefore depends for its accuracy upon potassium dichromate as a primary standard and on an accurate and precise determination of the slight excess by coulometric means. The precision of the determination of very pure uranium by this method is better than ± 0.01 per cent. Goode and coworkers (49) have also described an analytical procedure for the high precision determination of uranium. This is based upon reduction of U(VI) with an excess of a titanous solution. The excess titanous ion is destroyed with a nitric acid solution containing sulphamate, and the uranium (IV) is reacted with ferric ion. The ferrous ion is then determined by titration with electrogenerated ceric ion. The electrolyte concentration and current density at the gold working electrode are selected to ensure high overall titration efficiency. Under the conditions employed, 80 per cent of the titre resulted from direct electro-oxidation of ferrous ion, and the remaining 20 per cent from electro-generation of ceric ion at an efficiency of 99·66 per cent. There could conceivably have been a bias of $+0·06$ per cent but in fact no such bias was observed, and a precision of $\pm 0·004$ per cent was obtained with an accuracy of better than 0·01 per cent by comparison with alternative methods.

II. PLUTONIUM

A. Polarography

The $Pu^{4+}-Pu^{3+}$ couple behaves reversibly at an electrode in non-complexing support electrolytes with a formal potential of $+0·9818$ volts versus NHE. Koyama (50) attempted to make use of this for the determination of plutonium by square-wave polarography using a stationary platinum electrode. He succeeded in obtaining peaks for the reduction of Pu^{4+} ions in 1 M HCl and 2 M HNO_3 solutions for concentrations below 10^{-5} moles per litre, the peaks occurring at $+0·71$ and $+0·66$ volts versus SCE, respectively. Unfortunately the peaks were not well-defined due to noise from a preamplifier in the polarograph. In addition, interference from a peak at $+0·45$ volts versus SCE for the reduction of ferric iron was more pronounced in nitric acid than in hydrochloric acid solutions. A further problem arose from the production of an oxide layer on the surface of the platinum electrode, particularly from nitric acid solutions, and this resulted in a high residual current which virtually masked the plutonium peak. This work confirmed that a platinum electrode is not very suitable for the polarographic determination of plutonium.

Later workers have studied methods in which the supporting electrolyte contains a complexing agent to complex the tetravalent plutonium preferentially. By this means the formal potential of the $Pu^{4+}-Pu^{3+}$ couple is moved to a less positive potential and into a region more suitable for the use of the dropping mercury electrode. Scott and Peekema (51) provided experimental results on the extent of this movement in oxalate, acetate, citrate and tartrate solutions from coulometric measurements of E_0' values for the $Pu^{4+}-Pu^{3+}$ couple. An oxalate supporting electrolyte may be unreliable for the polarographic determination of plutonium, since both trivalent and tetravalent plutonium can form insoluble oxalate precipitates. However, Nebel and Schwabe (52) showed that tetravalent plutonium produces a satisfactory step from aqueous acetate solutions. These workers used acetate buffer solutions prepared from sodium acetate and acetic acid, and they observed a step in the region of $-0·2$ volts versus SCE caused by the reduction of the plutonium to the trivalent state at the dropping mercury electrode. However, the exact value of the half-wave potential for the step was found to be dependent upon the acetate concentration of the solution as shown by the results in Table III for a plutonium concentration of $\sim 5 \times 10^{-4}$ M.

TABLE III

Dependence of $E_{1/2}$ Values on Acetate Concentration

pH	Acetate concentration (M)	$E_{1/2}$ volts
2·02	3·3 × 10^{-4}	−0·046
2·46	1·00 × 10^{-4}	−0·120
4·25	3·6 × 10^{-3}	−0·216
4·32	1·0 × 10^{-2}	−0·222
3·95	3·37 × 10^{-2}	−0·214
4·24	0·07	−0·226
4·58	0·1	−0·234
4·88	0·193	−0·276
5·0	0·274	−0·297

The temperature dependence of the half-wave potential was studied at the following three acetate concentrations—0·034, 0·055 and 0·214 M. For the temperature range of 25 to 45°C, the temperature coefficient for the half-wave potentials was positive in all three cases with values lying between +0·5 and +0·8 mV per degree centigrade. Nebel and Schwabe found the height of the step to be proportional to the plutonium concentration over the concentration range 0·05 to 1·3 × 10^{-3} M.

In later work Milner and Wood (53) chose a citrate supporting electrolyte for a study by square-wave polarography. A citrate solution, 1 M citric acid +0·1 M in aluminium sulphate adjusted to pH 4·5 with potassium hydroxide, was found to give a satisfactory peak for tetravalent plutonium with an $E_{1/2}$ value of −0·42 volts versus a mercury pool anode, the width of this peak at half peak height being consistent with a reversible one-electron process. The peak height proved to be proportional to concentration for the range 4 × 10^{-5} to 4 × 10^{-4} M on using a gain of 3 with a square-wave amplitude of 8 mV, and also for the range 4 × 10^{-6} to 4 × 10^{-5} M on using a gain of 4 with a square-wave amplitude of 32 mV. The peak for plutonium concentrations less than 10^{-6} M is not well-defined due to the occurrence of a high background current. The peak height is independent of pH in the range from pH 4·0 to 4·8, but outside this range reduced peak heights are obtained.

For quantitative determinations the adjustment of the plutonium to the tetravalent state needs to be carried out very carefully by chemical means, according to the following instructions: Add 1 ml 1·8 M H_2SO_4 to a suitable aliquot of plutonium sample solution in hydrochloric acid and evaporate carefully to dryness. Then dissolve the residue in 1 ml 1·8 M H_2SO_4 and evaporate just to fumes of sulphuric acid. Cool, dilute to 1 ml with water and transfer to a stoppered tube of 10 ml capacity with a spitzer. Place a small piece of platinum foil into the tube, then add Specpure magnesium metal (∼ 10 mg) and allow to dissolve. Repeat this process with two further 10 mg pieces of magnesium metal added separately after the complete dissolution of the previous piece. On completion of the reduction process, cool the solution to room temperature and then add 1 ml HNO_3 (specific gravity 1·42) followed by 0·2 ml 5 M $NaNO_2$. Immerse the tube in a boiling water-bath for 5 minutes. Cool, add a further 0·1 ml 5 M $NaNO_2$ and heat for 5 minutes. Cool, add 4 ml of citrate supporting electrolyte and adjust the solution to pH 4·5 by the addition of 9·6 M KOH ($\sim$ 2 ml required). Dilute with water to 10 ml in a volumetric flask and mix. Then transfer a small aliquot to a polarographic cell, de-aerate with nitrogen and record the polarogram starting from $-0·1$ volts versus the mercury pool anode.

In a study of the behaviour of other elements, the method was slightly modified to include the addition of 0·25 ml 20 per cent w/v sulphamic acid solution after the adjustment of the pH. The sulphamic acid removes the effects of any excess nitrite, which gives a peak at about $-0·9$ volts versus the mercury pool. Although this peak does not interfere with the plutonium peak, its elimination enables an applied voltage from zero to $-1·6$ volts to be used without difficulty. Uranium gives a peak at $-0·75$ volts which is sufficiently well-separated to cause no difficulty. Small amounts of copper do not interfere, but some masking of the plutonium peak occurs with high ratios of copper to plutonium. The greatest interference comes from a double peak at $-0·30$ and $-0·53$ volts, respectively, produced by the reduction of ferric iron.

Although trivalent plutonium is not a stable species and is easily oxidized by air to the tetravalent state, it should be possible to obtain a peak for this species with the square polarograph because of the reversibility of the $Pu^{4+}-Pu^{3+}$ couple. The existence of this peak was confirmed experimentally by reducing 200 μg of plutonium in 1 ml 1·8 M H_2SO_4 to the trivalent state with Specpure magnesium metal added in three separate 10 mg pieces. This reduction process was followed by the addition of 4 ml citrate supporting electrolyte and 9·6 M KOH to adjust the pH in a final volume of 10 ml. The peak height is independent of pH in the range 4·4 to 4·6, but for most analytical purposes the pH range from 4·3 to 4·7 should be satisfactory. Even so, this range is only half that available for the determination of tetravalent plutonium. At pH 4·5 Milner and Wood obtained a linear calibration graph for concentrations of trivalent plutonium from 4×10^{-5} to $3·35 \times 10^{-4}$ M with a slope equal to that for the graph for tetravalent plutonium. Unfortunately the Pu (III)–citrate complex is not very stable and is slowly changed to the Pu (IV) complex, as shown by spectrophotometric studies. This factor, together with the limited pH range for the trivalent complex, has resulted in the peak for the tetravalent complex being preferred for analytical determinations.

In analytical applications some preliminary separation of plutonium is often desirable because of the interference from an iron peak and the closeness of peaks from some other elements. An anion exchange procedure has been developed for the separation of microgram amounts (20–1000 μg) of plutonium without seriously contaminating the final solution with organic degradation products. This procedure is based on the retention by the resin of tetravalent plutonium as $Pu(NO_3)_6^{2-}$ ions from 8 M HNO_3 solutions. Columns of Deacidite FF (150–200 mesh), 3 cm in height and 0·5 ml in volume, are suitable for retaining amounts of plutonium up to a maximum of 4 mg, and washing with 25 ml of 8 M HNO_3 is satisfactory for the removal of those elements with low K_D values. The plutonium is then readily recovered from the column in 10 ml of 1·5 M H_2SO_4, and this solution is evaporated to fumes of sulphuric acid to facilitate the wet oxidation of organic matter with concentrated nitric acid before the determination is completed polarographically. Milner and Wood applied this method satisfactorily to the determination of the plutonium content of U–Pu alloys containing from 0·25 to 1·0 per cent of plutonium.

B. The Coulometric Determination of Plutonium

The importance of coulometric methods in the analytical chemistry of plutonium arises partly from the variety of valency states available in solution, and also from the need to obtain accurate and precise analyses from small samples with the minimum of manipulation. No less than four valency states, Pu^{3+}, Pu^{4+}, PuO_2^+ and PuO_2^{2+} can exist simultaneously in equilibrium in acid solutions and the effect of complexing anions on the system must be taken into account in applying coulometric methods. In spite of this complexity coulometric methods have been widely reported in the plutonium analytical literature (54).

1. Controlled-potential Methods

(a) *Primary methods.* Controlled-potential coulometric methods for the determination of plutonium can either be applied directly using the reversible Pu(III/IV) couple (55–57) or indirectly using ferrous iron as an intermediate reagent following oxidation of the plutonium to Pu(VI) chemically. The direct determination of plutonium using the reversible Pu(III/IV) couple can be carried out in a wide variety of supporting media and relevant E_0' values are given in Table IV. If the plutonium is present in either the ter- or quadrivalent state the determination can be carried out satisfactorily in any mineral acid medium. If, however, significant amounts of hexavalent plutonium are present, then it is essential to use a sulphuric acid medium in order to ensure quantitative reduction of Pu(VI) to Pu(III) in the pre-electrolysis step. It is also desirable, when using a platinum gauze electrode, that the solution should be free from organic material. The present authors (58) have found that plutonium solutions prepared by fusion with ammonium hydrogen sulphate and extraction with molar sulphuric acid are superior to solutions prepared by other methods.

The main difficulty encountered in the determination of plutonium based upon the Pu(III/IV) couple is interference from iron due to the proximity of the Fe(II/III) couple. (See Table IV.) From these values it can be concluded that a sulphuric acid medium is unsuitable for the direct determination of plutonium in the presence of iron, owing to the nearness of the E_0' values. The greatest separation for mineral acid conditions is obtained in 5N hydrochloric acid solutions, but this is insufficient to prevent interference completely. When a potential span is used for 99·9 per cent oxidation of tervalent plutonium in 5N hydrochloric acid (i.e. $E_0' + 0·18$ volts), it can be calculated from the Nernst equation that 1 per cent of the iron is also oxidized. Whether this results in a significant error in the plutonium determination depends on the ratio of iron to plutonium, but at equal weight concentrations the interference is approximately 4 per cent. It is possible to make a correction for the contribution of the iron to the determination, if the iron content of the sample is known. Alternatively, the interference from iron can be decreased considerably by the use of a shorter potential span, so that a known fraction of the total plutonium is reduced at the

TABLE IV

Redox Potentials of the $Pu^{4+}-Pu^{3+}$ and $Fe^{3+}-Fe^{2+}$ Couples in Different Media

Medium	Molarity M	Observed E_0' versus SCE	
		Plutonium volt	Iron volt
Perchloric acid	1	0·705	0·470
	2	0·714	0·492
	5	0·720	0·516
Hydrochloric acid	1	0·707	0·450
	2	0·706	0·436
	5	0·710	0·400
Sulphuric acid	1	0·500	0·420
	2	0·500	0·420
Nitric acid	1	0·700	0·480
Sulphuric acid plus potassium pyrophosphate	1 ⎱ 0·01 ⎰	0·485	—
Sulphuric acid plus 2,2'-bipyridyl or 1,10-o-phenanthroline	pH 2 to 3	0·420	0·820

electrode. Both of these methods depend on the exact reproducibility of E_0' values from solution to solution.

The possibility of using complexing agents for improving the separation of the E_0' values for the Fe^{3+}–Fe^{2+} and Pu^{4+}–Pu^{3+} couples has been studied (59). Several reagents, including pyrophosphoric acid, citric acid and tartaric acid, are suitable for complexing Fe^{3+} and Pu^{4+} ions, thereby causing some change in the E_0' values. Unfortunately, both couples were moved in the same negative direction, and negligible improvements in the separations were effected. A better approach was obtained from the use of the reagents 2,2'-bipyridyl and 1,10-o-phenanthroline. These reagents form complexes with ferrous iron in solutions of pH 1.9 and higher, and, as a result, the E_0' value of the iron couple was found to move in a positive direction, whereas that of the plutonium couple was unaffected. A difference in the E_0' value of 0.40 volts was obtained under these conditions, as compared with 0.31 volts in hydrochloric acid solutions. The interference from iron in the coulometric determination of plutonium should be reduced by this means by about two orders of magnitude, and for equimolar solutions the interference should be small (approximately 0.04 per cent). It was found in practice, however, that attempts to determine plutonium in such a medium using a platinum gauze electrode resulted in low recoveries. This was attributed to poisoning of the platinum working electrode by the organic reagent. This difficulty can now be overcome by using a gold working electrode (60).

(b) *Secondary methods.*   A secondary procedure has been proposed by Shults (61) to enable the coulometric determination of plutonium to be carried out in the presence of iron. The method is based on the irreversible PuO_2^{2+}–Pu^{4+} couple in conjunction with electrogenerated ferrous iron, and is dependent on a suitable chemical step for oxidizing plutonium to the hexavalent state. There are two possible methods available for the quantitative oxidation of plutonium that do not give an excess of oxidant in solution. The oxidizing agent is either argentic oxide or fuming perchloric acid, and the coulometric titration is carried out in two steps. The hexavalent plutonium solution is added to a preconditioned ferric sulphate solution in the coulometer cell, and electrolysis is then carried out at +0.22 volt versus the SCE. Ferrous iron is produced, and this in turn reduces PuO_2^{2+} ions to the Pu^{4+} state. The potential of the working electrode is such that the Pu^{4+} ions are reduced electrochemically to the Pu^{3+} state. The electrolysis is continued until the plutonium is completely reduced to the tervalent state, and a small excess of ferrous iron remains in solution. At this point, which demands an approximate knowledge of the plutonium content, the electrolysis is stopped, and the coulometer reading, Q_R, is recorded. An oxidation is then carried out to completion at +0.70 volts versus the SCE, and the second coulometer reading, Q_O, is noted. In this step, the excess of ferrous iron is oxidized to ferric iron, and the tervalent plutonium is oxidized to the quadrivalent state. The over-all reaction is the reduction of PuO_2^{2+} as follows: $PuO_2^{2+} \rightarrow Pu^{4+}$, and the plutonium content is calculated from the difference reading, $Q_R - Q_O$.

A combination of both primary and secondary controlled-potential coulometric methods has been described by Shults (61) for the analysis of plutonium solutions for valency states. A portion of the sample solution is added to a supporting electrolyte containing a known amount of ferric iron in $1 M$ perchloric acid. The following controlled-potential coulometric titration steps are then applied:

1. Oxidation at $+0.895$ volts versus SCE. This measures Pu(III) by conversion to Pu(IV).

2. Reduction at $+0.285$ volts versus SCE. This converts Pu(VI) and Pu(IV) to Pu(III) and Fe(III) to Fe(II).

3. Oxidation at $+0.895$ volts versus SCE. This converts Pu(III) to Pu(IV) and Fe(II) to Fe(III).

Pu(III) is measured in the first step. Pu(VI) is measured by the difference between the third and the second step. Total plutonium is obtained from the third step after due allowance for the quantity of iron added. Pu(IV) can then be calculated by difference.

The electrolysis cells described in the literature and applied to the controlled-potential coulometric determination of plutonium (55–57) are normally capable of completing a titration in about 15–20 minutes. In order to reduce this time to a minimum, Goode and Herrington (74) have described a high-speed coulometric cell with a large ratio of electrode area to volume of electrolyte. This cell incorporates a tightly wound spiral of platinum or gold mesh which, together with efficient stirring, ensures titration times of two minutes or less. Gold was found to be superior to platinum for the electrode material in that it gives rise to lower and more reproducible blanks. This effect was ascribed to the formation of an oxide film on the platinum. These authors obtained recoveries of 99·99 per cent with a coefficient of variation of 0·05 per cent on 5 mg samples of plutonium in the electrolysis cell described.

(c) *Constant-current methods.* Application of the constant-current coulometric technique to the determination of plutonium has not been widespread probably due to the greater specificity and convenience of controlled-potential methods. Carson (62) described a method in which the plutonium is oxidized to the hexavalent state by heating with potassium permanganate. The excess oxidant is removed with formaldehyde and the latter destroyed by heating with nitric acid. The hexavalent plutonium is then determined by means of electrogenerated ferrous iron. This method has the merit of being applicable in the presence of iron and chromium.

III. NEPTUNIUM

A. Polarography

The formal potentials for neptunium ions in $1\ M\ HClO_4$ have been established, and the following are the accepted values in volts versus the normal hydrogen electrode (63):

$$Np^{3+} \frac{+0.155\ V}{\text{reversible}} Np^{4+} \frac{+0.739\ V}{\text{irreversible}} NpO_2^{+} \frac{+1.137\ V}{\text{reversible}} NpO_2^{2+}$$

There are two reversible systems to chose from for the electrochemical determination of neptunium. The Np^{4+}/Np^{3+} couple is suitable for the mercury electrodes, but it is too near to the reduction of hydrogen for the use of a platinum electrode. The NpO_2^{2+}/NpO_2^{+} couple, on the other hand, can only be used with a platinum electrode. In studying these couples for possible application to the determination of neptunium, $E_0{}'$ values in volts versus SCE have been determined in different mineral acid supporting electrolytes, and the results (64) are summarized as shown in Table V. From these values it is not surprising that the early work on the polarographic determination of neptunium using the dropping mercury electrode was confined to the Np^{4+}/Np^{3+} couple.

TABLE V

Supporting electrolyte	$E_0{}'$ values in volts versus SCE	
	NpO_2^{2+}/NpO_2^{+}	Np^{4+}/Np^{3+}
$HClO_4$ (N)	+0.90	−0.10
HCl (N)	—	−0.10
HNO_3 (N)	+0.91	−0.13
H_2SO_4 (N)	+0.83	−0.29

1. Np^{4+}/Np^{3+} Couple

Hindman and Kirtchevsky (65) were the first to report on the existence of a suitable step for the reduction of tetravalent neptunium in a chloride supporting electrolyte, and Slee, Phillips and Jenkins (66) confirmed this finding in studies with the square-wave polarograph. This reduction corresponds to a reversible one-electron change; the $E_{1/2}$ value being −0.1 volts versus SCE. The peak with the square-wave polarograph is not completely defined owing to slight interference from an anodic peak from the oxidation of chloride ions. Adjustment of the neptunium to the tetravalent state is readily achieved by using either hydroxylamine hydrochloride or ferrous chloride as the reductant. However, it is necessary to record the polarogram within a few hours of the valency adjustment since oxidation of tetravalent neptunium takes place slowly. Several other elements give peaks in the region of -0.1 volts from a chloride base solution, and the presence of any one of these could cause difficulty in the determination of neptunium. For this reason, Slee, Phillips and Jenkins investigated supporting electrolytes containing complexing agents to complex the neptunium and so move its peak to a more negative potential. Ethylenediamine tetra-acetic acid (EDTA) proved to be satisfactory for this purpose, and the supporting electrolyte finally chosen consisted of a solution 0.1 M in EDTA and 1.0 M in NH_2OH. HCl, adjusted to pH 6. The peak for neptunium occurs at about -0.8 volts versus a pool anode from this medium, and it is well away from that caused by the anodic oxidation of chloride ions. Complete reduction of the neptunium to the tetravalent state is not essential before the addition of EDTA, since this state is stabilized by complex formation. In practice this is an advantage, particularly if hydroxylamine hydrochloride is used for the reduction process.

Variations in the pH of the final solution between 2 and 7 were found to have little effect on the height of the neptunium peak. However, $E_{1/2}$ values varied with pH in accordance with the following:

pH	0.26	0.35	0.97	1.83
$E_{1/2}$ volts versus SCE	-0.366	-0.381	-0.457	-0.576
	4.26	4.95	5.38	6.00
	-0.728	-0.772	-0.797	-0.835

On using solutions of pH 6, well-defined polarograms are obtained with peak heights proportional to concentration in the ranges 20 to 110 μg and 10 to 30 μg of neptunium per ml, respectively. Peak heights are reproducible to within ± 1 per cent. In this solution the peak for neptunium is separated from those due to the following elements: cadmium (-1.4 volts), lead (-1.2 volts), copper (-0.5 volts), uranyl (-0.3 volts) and ferric iron (-0.2 volts). Tetravalent plutonium at a concentration of 50 μg per ml produces no peak over the voltage range from -0.2 to -1.4 volts.

Slee, Phillips and Jenkins succeeded in using this peak for the determination of the neptunium content of plutonium metal. ^{237}Np is formed by either a double neutron capture in ^{235}U or by a $(n, 2n)$ reaction in ^{238}U in a thermal nuclear reactor. The chemical processes employed to separate plutonium from uranium and fission products in the irradiated fuels do not separate the plutonium from neptunium. Therefore it is important to know the neptunium content of the resultant plutonium metal produced by this route, and also to be able to follow any decrease in neptunium content during zone refining experiments to purify the plutonium. In the method developed for this determination, the sample (300 mg) is dissolved in hydrochloric acid in the presence of hydroxylamine hydrochloride to adjust plutonium to the trivalent state. Ferrous chloride is added to the solution to reduce neptunium to the tetravalent state, and then this element is separated by solvent extraction into TTA-xylene. Any entrained plutonium is removed from the combined organic layers by washing with 1 M HCl, and the neptunium is finally back extracted into 10 M HNO$_3$. Any organic matter in this extract is destroyed by a wet oxidation process involving the addition of perchloric acid (3 ml), followed by evaporation to fumes of this acid with small additions of concentrated nitric acid. After heating to remove the perchloric acid, the residue is dissolved in 1 ml HCl (specific gravity 1·18) and 1 ml of water to obtain a clear solution. Then 2 ml of 5 M hydroxylamine hydrochloride are added, and the solution is evaporated slowly, without boiling, to reduce the volume to about 0·5 ml. A 0·5 ml aliquot of 1 M EDTA solution is next added, and the solution is adjusted to between pH 5·5 and 6·5 with ammonia solution (specific gravity 0·88). After diluting to 5 ml in a volumetric flask, 2 ml are transferred to the polarographic cell and de-aerated with nitrogen. The peak for neptunium is then recorded at $-0·8$ volts versus the mercury-pool anode by employing a suitable sensitivity of the square-wave polarograph. The neptunium content of the solution is determined by a standard addition technique involving the addition of 0·2 ml of a standard neptunium solution, prepared in EDTA, such that the original peak is approximately doubled in height. These workers found their method to be applicable to samples containing from 10 to 50 p.p.m. of neptunium. The precision (2σ) of the determination at the 500 p.p.m. level is ± 2 per cent, and at the 25 p.p.m. level it is ± 10 per cent.

In view of recent developments with anion-exchange resins in plutonium separations, it might be possible to replace the TTA-xylene extraction method employed by Slee and coworkers by a more convenient method using the anion exchanger, Deacidite FF. Trivalent plutonium is not absorbed by this resin from hydrochloric acid solutions, whereas tetravalent neptunium is completely absorbed from 5–12 M HCl solutions. After dissolving a plutonium metal sample in hydrochloric acid and adjusting the valencies chemically, the passage of this solution through a column of Deacidite FF should result in the plutonium passing through and the neptunium being retained. It would be necessary to wash the column with hydrochloric acid before recovering the neptunium in dilute hydrochloric acid (0·1 M). Any organic matter in this solution should be destroyed by wet oxidation involving the use of perchloric and nitric acids before completing the determination by polarography.

2. NpO_2^{2+}/NpO_2^+ Couple

Plock (67) has investigated materials suitable for use as an indicator electrode in the polarographic determination of neptunium based on the NpO_2^{2+}/NpO_2^+ couple. Glassy carbon proved to be satisfactory, and this worker prepared an electrode by cutting a disc from a glassy carbon plate. He sealed the disc into the end of a glass tube with an epoxy resin, and made electrical contact by pouring mercury into the tube and inserting a copper wire into the mercury.

In studying the reduction of NpO_2^{2+} at this electrode, Plock used a solution prepared by dissolving neptunium dioxide in concentrated nitric acid. After evaporating this solution to incipient dryness, the salt was dissolved in 2·5 M HNO_3 and finally diluted to a suitable volume with water. Measured aliquots of this solution were made 1 M in $HClO_4$ in a volume of 25 ml, and then after degassing the polarograms were recorded. Under these conditions a peak corresponding to a one-electron reduction was obtained at about +0·88 volts, the peak height being proportional to concentration over the range 2·25 × 10^{-4} to 2·25 × 10^{-3} M. A peak potential of +0·886 ± 0·017 volts versus SCE resulted for solutions of varying perchloric acid concentrations in the range 0·05 to 4·0 M.

Nitric acid has no effect on the above peak for concentrations between 0·10 and 4·0 M. However, sulphuric acid in the same concentration range causes the height of the peak to decrease with increase in acid concentration. The neptunium peak potential also becomes more negative with increase in sulphuric acid concentration, and this indicates the formation of a sulphate anion complex. A behaviour similar to this occurs in the presence of phosphoric acid. Even at the positive potential for this peak, there are a few interferences which include $Cr_2O_7^{2-}$, MnO_4^- and Cl^-, but Ce^{4+}, Fe^{3+}, Hg^{2+}, Ni^{2+}, Co^{2+}, Pb^{2+}, Ag^+ and other elements do not interfere.

In a study of the polarographic behaviour of NpO_2^+ ions, portions of the neptunium starting solution were electrolysed by controlled-potential electrolysis to reduce the neptunium to the quinquevalent state in a final acidity of 0·5 M HNO_3. Suitable aliquots were then made 1 M in perchloric acid before polarographic examination. A satisfactory peak resulted from the one-electron oxidation of NpO_2^+ ions under these conditions, the peak height being proportional to concentration over the range 2·27 × 10^{-4} to 2·27 × 10^{-3} M. No peak was obtained on the addition of phosphoric acid, and Plock concluded that the electrode reaction was reversible in HNO_3, $HClO_4$ and H_2SO_4 solutions but not in H_3PO_4 solutions. Ce^{4+}, $Cr_2O_7^{2-}$ and MnO_4^- were found to interfere with this neptunium peak, and chloride ions caused interference for concentrations exceeding 0·1 M.

B. The Coulometric Determination of Neptunium

1. Controlled-potential Methods

Stromatt (68) has reported a method for the determination of neptunium in 1 N sulphuric acid based upon chemical oxidation of neptunium to Np(VI) with Ce(IV), electrolytic reduction of Np(VI) and excess Ce(IV) to Np(V) and Ce(III), followed by a controlled-potential coulometric oxidation of Np(V) to Np(VI) at a potential of $+1.02$ volts versus the SCE. This procedure is applicable to solutions containing neptunium in any valency state and relies upon chemical oxidation of the lower valency states to the higher. The electrolytic reduction of Np(VI) and excess Ce(IV) is carried out at controlled-potential in the region of ≤ 0.66 volts versus the SCE. This potential may be varied slightly in order to enable the neptunium determination to be carried out in the presence of plutonium, (E_0' for Pu(III/IV) = $+0.502$ volts versus the SCE). The authors calculate from the Nernst equation that in the presence of a hundred-fold excess of plutonium interference can be avoided by reducing at a potential of $+0.75$ volts versus the SCE. This will reduce 97 per cent of the neptunium and a corresponding correction is necessary. The potential of $+1.02$ volts used for the final determination is not high enough to oxidize any Ce^{3+} to Ce^{4+}. Phillips and Milner (69) have used a slight variation of the above procedure in which the ceric ion is generated in situ by electrolytic oxidation at an applied potential of $+1.6$ volts versus the SCE for 15 minutes.

The method described by Stromatt (70) can be used to determine the valency states of neptunium in solution in the following manner:

1. Introduce sample. Reduce Np(VI) to Np(V) at the controlled-potential electrode. This gives the Np(VI) concentration.

2. Oxidize Np(V) to Np(VI) at the controlled-potential electrode. This step minus 1 gives Np(V) concentration.

3. Oxidize chemically with Ce(IV). Electrolytically reduce Np(VI) and excess Ce(IV) to Np(V) and Ce(III), respectively.

4. Oxidize Np(V) to Np(VI) at the controlled-potential electrode. This step gives total neptunium.

Fulda (71) recommended the use of a gold electrode rather than platinum for the controlled-potential coulometric determination of neptunium. It is claimed that a gold electrode avoids the background current arising from the $Pt/Pt(OH)_2$ reaction at the high positive potential used to oxidize Np(V) to Np(VI).

The Np(III/IV) couple with an E_0' of -0.29 volts versus the SCE in 0.5 M H_2SO_4 (64) does not appear to have been used in the controlled-potential coulometric determination of neptunium. At that potential a mercury pool electrode is necessary and uranium would interfere.

IV. AMERICIUM

A. Coulometry

Formal potentials for americium ions in acid solutions have been reported, and the following are the accepted values in volts versus the normal hydrogen electrode:

$$E_0, \text{ volts versus NHE}$$

$AmO_2^{2+} + e \rightarrow AmO_2^+$	$+1.64$
$AmO_2^+ + 4H^+ + e \rightarrow Am^{4+} + 2H_2O$	$+1.26$
$Am^{4+} + e \rightarrow Am^{3+}$	$+2.18$
$Am^{3+} + 3e \rightarrow Am$	-2.32

The Am(VI/V) couple is reversible and it is the only one likely to be suitable for the determination of this element by controlled-potential coulometry. Americium usually occurs in solution in the trivalent state, and no direct evidence has been found for the existence of the tetravalent ions in solution. However, higher valency states can be produced by oxidation of Am^{3+} ions by using suitable chemical oxidants or by anodic oxidation. Ammonium persulphate is very suitable for the formation of AmO_2^{2+} ions in solution. Unfortunately, however, the AmO_2^{2+} ions are reduced slowly to the AmO_2^+ state at a constant rate of a few per cent per hour. This reduction probably results from the formation of hydrogen peroxide by the radiolytic decomposition of water by the intense alpha activity of ^{241}Am, since several reducing agents, including hydrogen peroxide, hydrazine and iodide ions, are known to reduce AmO_2^{2+} ions to the trivalent state. This behaviour is likely to cause some difficulty in the development of a satisfactory coulometric method based on the Am(VI/V) couple. In addition the AmO_2^+ ions produced in this determination undergo disproportionation to the Am^{3+} and AmO_2^{2+} states, but fortunately this reaction is slow.

The first attempt to use controlled-potential coulometry for the direct determination of americium was made by Koehly (72). He studied a procedure based on the anodic oxidation of trivalent americium to the AmO_2^{2+} state, followed by the coulometric reduction of these ions to AmO_2^+. Koehly found it preferable to work in sulphate solutions of low acidity prepared by dissolving the oxide (AmO_2) in warm sulphuric acid. The composition of the solution for the anodic oxidation of the americium was $2 M (NH_4)_2SO_4 + 0.1 M H_2SO_4$, and under these conditions the oxidation of the trivalent ions followed the reaction:

$$Am^{3+} + 2H_2O - 3e \rightleftharpoons AmO_2^{2+} + 4H^+$$

On using an anodic potential between $+1.9$ and $+2.0$ volts versus NHE, at least 99 per cent of the americium was oxidized, but the time taken to obtain this was a function of the age of the americium solution. Freshly prepared solutions could be completely oxidized in 80 minutes, but older solutions took longer and a two-month old solution failed to give complete oxidation after 120 minutes. This behaviour is explained by the accumulation of hydrogen peroxide in solution from the radiolysis of the water. The coulometric reduction of AmO_2^{2+} to AmO_2^+ begins as soon as the cathodic potential falls below $+1.55$ volts, and on controlling the potential at $+1.3$ volts, the reduction is complete in 20 minutes on using a 25 ml sample in a cell with an electrode surface area of at least 30 square centimetres.

In quantitative determinations Koehly encountered some difficulty from a high residual current. The resulting error could be reduced, however, by pre-electrolysing a quantity of supporting electrolyte at $+1.3$ volts before adding the oxidized americium solution to it. Experimental details of the recommended procedure are as follows:

Introduce an aliquot of the sample solution into the electrolysis cell. Evaporate to dryness, and then dissolve in 1 ml of 1 M H_2SO_4 by warming. Add 10 ml of a 2 M $(NH_4)_2SO_4$ solution. Whilst stirring oxidize for 2 hours with an anodic potential of $+1.9$ to $+2.0$ volts under thermostatically controlled conditions so as to prevent changes in concentration by evaporation. Pipette 5 ml of this oxidized solution, and introduce it into a cell containing 20 ml of supporting electrolyte previously electrolysed at $+1.3$ V for 20 minutes. Proceed to the coulometric reduction whilst stirring and using a cathodic potential of $+1.3$ volts.

On using the above method for quantitative analysis, Koehler found a systematic error which varied between -1.60 and -1.76 per cent for americium concentrations between 3.125×10^{-3} and $1.25 \times 10^{-4} M$. He also reported the limit of concentration for a precise determination to lie between 1.25×10^{-4} and $6.25 \times 10^{-5} M$. At the $6.25 \times 10^{-5} M$ level the interference from the residual current definitely becomes too significant. The method is very selective for americium since very few couples have a redox potential as high as the value for Am(VI/V), and, of those that have, the majority are irreversible. Cerium is the only element with a redox potential very close to that for the americium couple and thereby causing serious interference. Curium interferes by reason of the effect of its intense α activity causing some reduction of the AmO_2^{2+} ions. The α activities of ^{242}Cm and ^{244}Cm are many times higher than that of ^{241}Am. Fortunately both of these elements can be separated from americium by coprecipitation of their fluorides with lanthanum fluoride.

In later work Stokely and Shults (73) preferred to use chemical oxidation with ammonium persulphate in dilute acid to effect the oxidation of Am^{3+}

to $AmO_2{}^{2+}$. Any excess of persulphate was destroyed by heating the solution, and complete oxidation of the americium was ensured by a brief electrolytic oxidation just before the commencement of the coulometric reduction at $+1.05$ volts versus SCE. Reduction is complete in less than 10 minutes, and in this time there is little influence from the disproportionation of the $AmO_2{}^+$ ions. However, a small amount of $AmO_2{}^{2+}$ is lost by reduction with hydrogen peroxide, and the error from this cause is about 0·5 per cent. Stokely and Shults applied a correction for this error. This method is reported to give very precise results for the determination of americium at the milligram level. However, the accuracy of the method has not been evaluated because of the lack of satisfactory materials for preparing standard solutions, and the non-availability of other accurate methods of analysis for americium.

References

1. R. W. Stromatt et al., Hanford Atomic Products Operation, Richland, Washington, U.S.A. Report HW.58212 (1958).
2. W. D. Shults, *Talanta*, **10**, 833 (1963).
3. W. E. Harris and I. M. Kolthoff, *J.A.C.S.*, **67**, 1484 (1945).
4. W. E. Harris and I. M. Kolthoff, *J.A.C.S.*, **68**, 1175 (1946).
5. S. W. Sheel and J. I. Watters, U.S. Atomic Energy Commission, Report CC.2771 (1945).
6. H. I. Shalgosky, Atomic Energy Research Establishment, Harwell, Report C/R.1869 (1956).
7. H. G. Heal, *Nature*, **157**, 225 (1946).
8. H. E. Zittel, M. T. Kelley, F. L. Conover and G. R. Wilson, *Advances in Polarography*, Pergamon Press, London, Vol. 2 (1960), p. 530.
9. F. Elliott and J. K. Foreman, *Advances in Polarography*, Pergamon Press, London, Vol. 2 (1960), p. 538.
10. J. Mashall and J. Kindler, *Anal. Chim. Acta*, **31**, 490 (1964).
11. R. M. Burd and G. W. Goward, U.S.A.E.C. Report WAPD-205 (1959).
12. U.K. Atomic Energy Authority, P.G. Report 586 (S), (1964), HMSO.
13. H. I. Shalgosky, R. C. Smart and J. Watling, Atomic Energy Research Establishment, Harwell, Report R.4270 (1964) HMSO.
14. L. Sipos and M. Branica, *J. Polarographic Society*, **14**, 3 (1968).
15. C. E. Plock and J. Vasquez, *Anal. Chim. Acta*, **42**, 101 (1968).
16. S. Takani and Y. Morimoto, *Japan Analyst*, **12**, 1189 (1963).
17. J. A. Lewis and K. C. Overton, D.S.I.R., Chemical Research Laboratory, Report CRL/AE41 (1949).
18. D. I. Legge, *Anal. Chem.*, **26**, 1617 (1954).
19. F. E. Wild, Atomic Energy Research Establishment, Harwell, Report C/R 1868 (1956).
20. C. E. Plock and F. J. Miner, *Anal. Chim. Acta*, **38**, 553 (1967).
21. R. Pribil and A. Blazek, *Coll. Czech. Chem. Commun.*, **16**, 567 (1953).
22. C. Auerbach and G. Kissel, *Talanta*, **11**, 85–91 (1964).
23. D. J. Fisher and P. F. Thomason, *Anal. Chem.*, **28**, 1285 (1956).
24. W. L. Belew, D. J. Fisher, M. T. Kelley and J. A. Dean, *Microchemical Journal*, **10**, 301 (1966).
25. G. W. C. Milner and J. H. Nunn, *Anal. Chim. Acta*, **21**, 266 (1959).

26. G. W. C. Milner, J. D. Wilson, G. A. Barnett and A. A. Smales, *J. Electroanal. Chem.*, **2**, 25–38 (1961).
27. C. J. Rodden, Analysis of Essential Nuclear Reactor Materials, Division of Technical Information, U.S.A.E.C., Washington, D.C. (1964).
28. G. L. Booman, *Anal. Chem.*, **29**, 213 (1957).
29. G. L. Booman, W. B. Holbrook and J. E. Rein, *Anal. Chem.*, **29**, 219 (1957).
30. G. L. Booman and W. B. Holbrook, *Anal. Chem.*, **31**, 10 (1959).
31. M. T. Kelley, H. C. Jones and D. J. Fisher, *Anal. Chem.*, **31**, 488 (1959).
32. L. G. Farrar, P. F. Thomason and M. T. Kelley, *Anal. Chem.*, **30**, 1511 (1958).
33. W. D. Shults and P. F. Thomason, *Anal. Chem.*, **31**, 492 (1959).
34. R. W. Stromatt and R. E. Connally, *Anal. Chem.*, **33**, 345 (1961).
35. C. M. Boyd and O. Menis, *Anal. Chem.*, **33**, 1016 (1961).
36. H. E. Zittel, L. B. Dunlap and P. F. Thomason, *Anal. Chem.*, **33**, 1491 (1961).
37. H. E. Zittel and L. P. Dunlap, *Anal. Chem.*, **35**, 125 (1963).
38. W. R. Mountcastle, L. B. Dunlap and P. F. Thomason, *Anal. Chem.*, **37**, 336 (1965).
39. W. D. Shults and L. B. Dunlap, *Anal. Chem.*, **35**, 921 (1963).
40. H. C. Jones, W. D. Shults and J. M. Dale, *Anal. Chem.*, **37**, 680 (1965).
41. G. C. Goode and J. Herrington, *Anal. Chim. Acta*, **38**, 369 (1967).
42. J. J. Lingane, *Electroanalytical Chemistry*, Interscience Publishers Inc., New York, 1958.
43. J. H. Kennedy and J. J. Lingane, *Anal. Chim. Acta*, **18**, 240 (1958).
44. J. K. Taylor and G. Marinenko, Euratom Report EANDC 42S, Brussels (1965).
45. N. H. Furman, C. E. Bricker and R. V. Dilts, *Anal. Chem.*, **25**, 482 (1953).
46. J. J. Lingane, C. H. Langford and F. C. Anson, *Anal. Chim. Acta*, **16**, 165 (1951).
47. W. N. Carson, *Anal. Chem.*, **25**, 466 (1953).
48. J. Malinowski, *Talanta*, **14**, 263 (1967).
49. G. C. Goode, J. Herrington and W. T. Jones, *Anal. Chim. Acta*, **37**, 445 (1967).
50. K. Koyama, *Anal. Chem.*, **32**, 523 (1960).
51. F. A. Scott and R. M. Peekema, *Progress in Nuclear Energy, Series IX, Analytical Chemistry Volume 1*, Pergamon Press, London, 1959, p. 64.
52. D. Nebel and K. Schwabe, *Z. Physik. Chem.*, Leipzig, **220**, 240 (1962).
53. G. W. C. Milner and A. J. Wood, *J. Electroanal. Chem.*, **7**, 190–295 (1964).
54. H. B. Evans and J. O. Karttunen, U.S.A.E.C., Report ANL 6956 (1964).
55. F. A. Scott and R. M. Peekema, U.S.A.E.C., Report HW.58491 (1958).
56. W. D. Shults, U.S.A.E.C., Report ORNL.2921 (1960).
57. G. W. C. Milner and J. W. Edwards, Atomic Energy Research Establishment, Harwell, Report R.3772 (1960).
58. G. W. C. Milner, A. J. Wood, G. Weldrick and G. Phillips, *Analyst*, **22**, 239 (1967).
59. G. Phillips and G. W. C. Milner, in P. W. Shallis, Ed., *Proceedings of the S.A.C. Conference, Nottingham, 1965*, W. Heffer and Sons Ltd., Cambridge, 1965, p. 240.
60. R. G. Monk, private communication.
61. W. D. Shults, *Talanta*, **10**, 833 (1963).
62. W. N. Carson, *Anal. Chem.*, **29**, 1417 (1957).
63. J. J. Katz and G. T. Seaborg, *The Chemistry of the Actinide Elements*, John Wiley and Sons, New York (1957).
64. G. W. C. Milner, *Proceedings Feigl Anniversary Symposium*, Birmingham University, Elsevier, 1963, pp. 225–262.
65. J. C. Hindman and E. S. Kritchevsky, *J. Am. Chem. Soc.*, **72**, 953 (1950).
66. H. J. Slee, G. Phillips and E. N. Jenkins, *Analyst*, **84**, 596 (1959).
67. C. E. Plock, *J. Electroanal. Chem.*, **18**, 289–293 (1968).
68. R. W. Stromatt, U.S.A.E.C. Report HW.59447 (1959).

69. G. Phillips and G. W. C. Milner, Twelfth Conference on Analytical Chemistry in Nuclear Technology, Gatlinburg (1968).
70. R. W. Stromatt, *Anal. Chem.*, **32**, 134 (1960).
71. M. O. Fulda, U.S.A.E.C. Report DP.673 (1962).
72. G. Koehly, *Anal. Chim. Acta*, **33**, 418 (1965).
73. J. R. Stokely and W. D. Shults, *Chem. Engineering News*, **45** (45), 67 (1967).
74. G. C. Goode and J. Herrington, *Anal. Chim. Acta*, **33**, 413 (1965).

CHAPTER III

Voltammetry in Organic Analysis

PHILIP J. ELVING, *The University of Michigan, Ann Arbor, Michigan*

I. INTRODUCTION

In a review of the applicability of voltammetric or polarographic techniques to organic analysis, the question must first be raised as to the relevancy of such an enterprise. Zuman has published a comprehensive book on *Organic Polarographic Analysis* (1b) and comprehensive, well-documented reviews covering various aspects of the subject areas are published every two years in *Analytical Chemistry*. In addition, the present author has published two reviews (2, 3) on the basic methodology of polarographic analysis, which has not changed appreciably since these were published in 1954 and 1961. There is, consequently, a very real question as to the role which an additional statement on the subject can serve. What can be said about organic polarographic analysis that has not already been well said and critically evaluated?

One justification—perhaps the only one—for the present discussion would be the attempt to describe things as they are rather than as they seem to be. The chemist, experienced in organic analysis, who has had contact with laboratories charged with supporting research, development and production by the analysis of organic samples, will have been struck by the wide discrepancy between the extensive literature on the polarography of organic compounds and the rather sparing use made of polarographic methods in actual service analysis. This lack of relation between research output and meaningful application is a serious situation, which must be carefully considered, hopefully from the viewpoint of being able to chart courses of action which will insure a more extensive use of polarographic techniques in organic analysis *if* such an application is warranted.

One facet of the problem is the reluctance of many of those engaged in studying the polarographic behaviour of organic compounds, to apply the results of their research investigations to the development of analytical methods. This reaction is due to the changing attitudes towards analytical chemistry, whose prestige has been decreasing in spite of its continuing and even increasing importance. The cause of the change has been explored by Liebhafsky (4) and others in terms of the change in the contribution of analysis to chemistry from being to a considerable extent a research activity in itself to being largely a service activity.

The function of *analysis* is to supply information; research in *analytical chemistry* should, therefore, be directed towards the discovery, development and implementation of better ways of obtaining such information. However, unless the implementation to the analysis of real samples is finally made or, at the least, the guidelines for such implementation ascertained and made explicit, the task of the analytical chemist has not been completed. It is important for chemists of all categories to carry on research on electrochemical techniques and on the electrochemical behaviour of organic compounds; the analytical chemist, as analytical chemist, should carry such studies of his (or of others) to the further stage where their analytical significance and utilization are evident.

The relatively limited importance of polarography in organic analysis can be seen by checking research journals such as *Journal of the American Chemical Society, Journal of Organic Chemistry, Biochemistry* and *Journal of Polymer Science* to see how often polarography is cited as an analytical method used in the characterization of organic compounds and polymers, and to compare this with the citations of, for example, nuclear magnetic resonance or absorption spectrophotometry.

One reason for the relatively small use made of polarographic techniques by, for example, organic chemists is their lack of sufficient theoretical background in electrochemistry. Although the latter is no more complicated, even if not yet as unified, as the theoretical aspects of radiant energy absorption, radiant energy absorption is generally taught in undergraduate courses as background for an understanding of chemical binding and structure. There is, though, the complication that electrode reactions are *per se* heterogeneous reactions and, by and large, chemists are not oriented towards such reactions, e.g. the relatively minor attention given in undergraduate and graduate chemistry courses to surface phenomena such as adsorption and heterogeneous kinetics.

An allied aspect of the problem regarding the use of polarography is the fact that standards-setting bodies have been slow to accept methods of organic analysis based on polarography as standard methods. It is not always clear whether the rate-determining step is the general conservatism of such bodies or the inferiority of the polarographic methods compared to methods based on other techniques. Some specific studies bearing on the latter point have been made, e.g. comparison of polarographic and gas chromatographic procedures for the determination of styrene in polystyrene resins (5). Some methods have been accepted, e.g. a method for the determination of fumaric acid (6) 'was studied collaboratively by the Association of Official Analytical Chemists and adapted as official first action by the Association at the 1967 annual meeting. A report was published by Taylor and Blomquist (7). It is noteworthy that this is the first polarographic method to appear in the AOAC Methods of Analysis (Supplement)' (8). The American Society for Testing and Materials approved in 1960 a standard procedure for the determination of the bromine index, i.e. extent of bromination, of aromatic hydrocarbons by coulometric titration (9), and the U.S. Pharmacopeia (10) and National Formulary (11) have accepted polarographic methods as subsequently indicated.

On the other hand, there is increasing interest in the application of electro-chemical approaches, frequently involving polarographic techniques, to problems in such areas as biochemistry, biology and medicine, as indicated by the 1971 Rome Symposium on Biological Aspects of Electrochemistry (11a), organized by G. Milazzo, and the 1973 International Symposium on Bioelectrochemistry at Pont à Mousson, organized by Buvet and Thevenot (251).

The author of the present review has attempted to gather some evidence as to the actual extent to which polarographic techniques are being used in connection with organic analysis, interpreted as being the locus of approaches for obtaining information regarding the nature and behaviour of chemical species. Thirty laboratories, primarily covering a variety of research and production organizations, were asked to summarize the uses made of polarographic techniques for organic analysis in their organizations and to comment on the reasons for the use (or non-use) of polarography. The response was excellent; frequent reference is made in the subsequent discussion to the replies received, which showed a fair degree of agreement in the views expressed. A further discussion of the respondents and their views is given in Section I-C.

The principal objective of the present chapter is, then, to portray as realistically as possible the extent to which polarographic and voltammetric techniques are actually being used in organic analysis, especially from the service aspect.

To clarify the nomenclature situation, both *polarography* and *voltammetry* find fairly precise definition among many electroanalytical chemists, as referring to techniques based on current–potential curves obtained under essential constant-potential conditions at the dropping mercury electrode, DME, and at constant area electrodes, respectively. However, in the present chapter, these two words—when used without modifier—will with their adjectival forms refer in general to techniques which have developed from polarography as originally described by Heyrovsky. These techniques are discussed throughout the present volume; a list of many of them is subsequently given in the sub-section on trace analysis (Table V).

A. Current Trends in Organic Polarography

In considering the role which voltammetric techniques are actually playing in organic analysis, it is worthwhile first indicating the current configuration of activity as seen in the literature on organic polarography since, as will be subsequently discussed, the actual use of polarography on a service basis in organic analysis reflects in a rather interesting fashion the relative emphasis placed on different aspects of organic polarography, when considered as an area of research in itself. The author has discussed current trends in organic polarography (12).

In general, recent activity in organic polarography, which is significant in regard to the continuing development of organic polarographic analysis, has been distinguished by attempts to define in depth the paths followed by the electrode process, including the detailed explication of the phenomena which occur in the electrical double layer, preceding, accompanying and following the electron-transfer process. This has involved, *inter alia*, the development of theoretically derived equations and practical approaches for characterizing: (*a*) the interrelationship and kinetics of the chemical and electrochemical steps involved, (*b*) the presence and effect of adsorption, as a necessary or accompanying phenomenon, (*c*) the actual energy levels and concentrations in the interfacial region and (*d*) the roles of proton and solvent in the electrode reaction.

Examples of the activities just indicated will be found throughout the present chapter.

B. Voltammetry as an Analytical Technique

In evaluating analytical techniques in respect to their relative utility for a given problem, the principal factors generally considered are *selectivity*, *sensitivity* and *speed*, which are available at a given level of *reliability* (accuracy and precision), as demanded by the use to be made of the analytical data, and at a given level of *convenience*, where the latter is determined by the availability of the necessary physical facilities and experienced manpower, existing load on these facilities and manpower, and allied factors. Since many physical methods of analysis operate at a precision level of 1 to 3 per cent relative in many measurement situations and convenience is often determined by the existing situation in the laboratory, it is worthwhile looking at organic polarographic analysis from the viewpoint of the three *S* factors. (It is interesting to note that in many analytical approaches, which depend upon physical and physiochemical techniques and which are dynamic in nature, the precision of measurement for quantitative analysis rarely exceeds one part in a hundred unless special efforts are made. On the other hand, methods of measurement, which depend upon exhaustive reaction of the desired constituent, e.g. titrimetry and coulometry, will often yield results good to one part in several hundred for the same amount of effort.)

By *selectivity* or *specificity* is meant the ability to detect and to determine one substance in the presence of other substances with obviously special reference to those other substances which are commonly associated with the desired constituent and which might be expected to interfere with its identification and determination.

The growing need for selectivity in analysis has resulted in the increasing use of methods of separation and measurement based on the spectrum approach, i.e. on the general technique of subjecting the sample to a probe involving a steadily altering energy, time, space or equivalent gradient in order to obtain a record of the relative intensity of sample response versus the applied gradient; from such a record, the sample composition—qualitative and quantitative—can under favourable conditions be deduced. Examples of such techniques are numerous, e.g. absorption spectrophotometry, polarography and gas chromatography.

Specificity is often the prime requirement in the choice of an analytical technique. To quote one respondent, 'In general, voltammetry as an analytical tool, is employed where it will profitably fulfill the following requirements: (a) specificity, (b) accuracy and precision, (c) sensitivity and (d) simplicity. The degree of compliance to any of the above requirements is, of course, a function of the individual analytical problem at hand. However, the pharmaceutical industry is most acutely concerned with the first requirement of specificity. We are constantly striving to develop test methods which will distinguish as unequivocally as possible the intact "drug" from impurities, degradation products and the like'.

The limited number of electroactive functionalities, while it restricts the overall applicability of voltammetry, does tend to improve its selectivity. The latter can often be markedly improved by careful control of the composition of the test solution (cf. Section II-B). Differential and derivative polarography, apparatus for which are now commercially available, may allow the resolution of overlapping waves sufficiently to permit satisfactory quantitative analysis, although careful calibration may be required.

By *sensitivity* is meant one or both of the senses in which this word is customarily used: *absolute sensitivity*, i.e. the smallest amount of material which will give a satisfactory signal in the particular analytical technique and method being used, and *concentrational sensitivity*, i.e. the lowest concentration of material which will give a satisfactory signal. Normally, the latter in the determination of concentration from limiting current measurement is in the range of 0·1 to 0·01 mM; however, variations such as pulse polarography permit extension of the range by one or more orders of magnitude (cf. Section III-A-1).

The major limitation in the sensitivity of most techniques is the signal-to-noise ratio, which in normal polarography would be, operationally, the ratio of faradaic current (due to constituent being measured) to the residual current (with its faradaic and non-faradaic components) (cf. Meites (13)). This limitation cannot always be overcome by increasing the sample size since the latter procedure may actually decrease selectivity due to decreased resolution of component responses in the form of merging waves. A possible solution to the problem of increasing the signal-to-noise ratio is in the development of circuits which will permit direct measurement of the faradaic current component in situations where the non-faradaic contributions are substantial. Brown, McCord, Smith and DeFord (14) have described such an approach for cyclic voltammetry, and fundamental and higher harmonic alternating current polarography with the DME.

There is no need to define *speed*, which is generally a desirable and often a necessary characteristic in an analytical method. An increase in the speed with which an analysis can be made, may mean either of two things: a decrease in the actual man-time required or a decrease in the total elapsed time. The latter is of primary importance when the analytical information is to be used to control a dynamic process or where the holding of a batch of material until its composition is known represents a problem in either economic or physical terms. Frequently, however, the decrease in total man-time is the important factor. The quest for speed often resolves itself into the problem of adjusting to a complex of factors, some of the more important of which are the urgency with which the analytical information is needed, the cost of equipment for handling a given sample load, the relative man-power cost, the frequency with which peak sample loads occur, and the laboratory space available.

Here, polarography often offers some advantage, especially if moderately dilute aqueous solutions are to be analysed, if thermally unstable material is to be determined or if the number of samples to be analysed does not warrant extensive calibration work. The actual measurement time may be very short, especially if the wave is at potentials more negative than that of the oxygen wave, which is frequently the case, and if it is sufficiently well-formed that the wave-height can be determined by measuring the currents at only two potentials, one preceding the wave and one on the current plateau.

From the convenience standpoint, it must be emphasized that most polarographic measurements of concentration, which would constitute the bulk of normal quantitative analysis, can be made and are being made with simple direct-current polarographs, the commercial strip-chart recording models of which retail for under two thousand dollars.

C. Survey of Organic Analytical Laboratories

To gain perspective on the survey made of laboratories engaged in service organic analysis for both research and production, the nature of the respondents should be indicated: eleven were from companies engaged in the general chemical area including several of the largest American chemical companies and one British company; four were with petroleum and petrochemical companies; eight were from pharmaceutical companies; three were from governmental laboratories (Food and Drug Administration, and one Federal and one state agricultural research laboratory); one was from the organic chemicals division of a steel company and one from a large well regarded consulting laboratory. In many cases, the respondent covered not only his own laboratory, but other laboratories in his company, as well as giving his opinion on the general role—present and future—of polarographic techniques in organic analysis. A relatively large percentage of pharmaceutical laboratories were contacted because of the diverse nature of the compounds with which such laboratories are concerned, the high frequency with which these compounds might be expected to show electroactivity, and the well-known commitment of this industry to the use of sophisticated approaches for the study of organic compounds. The chemists, who gave so generously of their time and experience in writing to the author about organic polarographic analysis, are acknowledged at the end of the chapter.

The manuscript for the present chapter was originally submitted to meet an October, 1968, deadline. In January, 1972, letters were sent to the original respondents, inquiring as to any changes in their views concerning the use of polarography in organic analysis. In addition, the literature for the intervening three to four years was surveyed. As far as the general theses of the present chapter are concerned, the replies and the survey have simply reinforced the arguments set forth in the chapter. However, a number of more recent comments and references have been added.

It should be emphasized that the laboratories polled were with only one or two exceptions located in the United States, and consequently, the results obtained do not reflect the situation which prevails, for example, in Czechoslovakia, where polarographic techniques for obvious reasons have found much greater utilization.

Many of the specific comments made have been incorporated into the subsequent text with or without specific creditation to the writer or to 'a respondent'. Some of the comments quoted have been slightly edited, e.g. by omitting connective and other phrases of no substantive importance; in no case was any significant alteration made. In the present section, some comments have been collected in order to set the proper background for the detailed discussion. For example, one of the letters received concerning the use of polarography in organic analysis indicates the potentially wide range of applicability of polarography and evaluates its possible utility (15):

'I shall only mention topics with which I have been personally involved. These are:

Development and application of analytical methods in 4,4'-bipyridyl/Paraquat processes, in particular the determination of 4,4'-bipyridyl, its mono-N-methyl and its N,N'-dimethyl (Paraquat) derivatives in various plant liquors without prior separation. The methods were developed by application of simple polarographic theory and have become the basis of on-line polarographic analysers.

Determination of carbon tetrachloride in various products, such as chlorinated polymers, trichloroethylene and by-product concentrated hydrochloric acid. In many cases, polarography has replaced gas chromatography because of advantages in speed and/or sensitivity.

Study of the stability and ligand number of metal–organic complexes believed to be important in homogeneous catalysis.

Study of the stability and reduction potentials of various stabilizers used for chlorinated solvents.

Direct determination of drugs and metabolites in a variety of biological fluids. This work, with which I have not personally been concerned for over six years, indicated that only polarography, among the various physical methods of analysis, could be used without any separation for many of the biological fluids. Examples: Ethionamide, Metronidazde, Dimetridazde in serum, plasma, C.S.F. urine, saliva, seminal fluid.

It is my opinion that organic polarography would be much more widely used if it were applied in a systematic fashion and that it has been given an undeserved poor reputation by *ad hoc* application. Also, in my opinion, the next logical step forward in organic polarography, is the development of electrodes for the positive range of potentials which are as reproducible as the dropping electrode.'

1. Utilization of Polarography in Organic Analysis

There is general agreement that, since the introduction of polarography and recognition of its use as an analytical technique, polarography has been extensively investigated as a means for organic analysis. This has been due to polarography being a relatively simple experimental technique, which yields fairly readily interpretable results suitable for quantitative and, often, for qualitative analysis, including the evaluation of certain constants and quantities which are applicable to the study of physical phenomena associated with organic compounds and reactions, e.g. reaction rates, redox stability, structure determination and tautomeric equilibria, and possible development of selective synthesis methods using controlled-potential techniques. From the viewpoint of quantitative analysis, polarography offers the possibility of rapid polycomponent analysis from a single analytical record, e.g. under favourable conditions, mixtures of such closely related compounds as *cis–trans* isomers and members of homologous series can be analysed. In many situations, polarography is the simplest method for measuring some types of organic functional groups. In addition, many polarographically inactive organic compounds can be converted to polarographically measurable species by simple, rapid quantitative chemical reactions.

Given these potentialities for success, it is then surprising how relatively little use is made of polarography. The reasons advanced for the latter are, as might be expected, many and varied. Some deal with the failure of potential users to recognize the unique features of polarography as an analytical technique. Others consider the advantages of other techniques, such as gas chromatography, compared to polarography. Still others point to lack of available people experienced in polarography in their organizations. Typical comments follow, including some which are more optimistic regarding the future use of polarography. One purpose, which may be served by the quotations which are given here and elsewhere in this chapter, is the stimulation of research and development which will further the application of polarography in those analytical situations where its use offers advantage.

(*a*) 'With regard to the state of voltammetry in industrial laboratories, it is safe to say that voltammetric techniques are not widely employed. This is particularly true of plant laboratories (almost all plant labs do not have the equipment) but also may be true of research facilities. The reasons for this state of affairs are numerous and complex but at least one involves the kind of literature which has been published in the past decade or so, in that field. A great deal of emphasis has been placed on theoretical matters and mathematical treatment of data. This has alienated the analytical "man in the street" so effectively that polarography either has to be rammed down his throat and kept there or else he slips over to his trusty VPC unit.'

(*b*) 'Our quality control laboratories employ polarography in a limited number of cases; assay of hydrochlorothiazide (see *U.S. Pharmacopeia*) is one quantitative procedure employed. It has no advantage, in my opinion, however, over ultraviolet spectrophotometric determination. I expect that we will develop additional polarographic procedures for quality control from time to time, but do not expect this technique to be employed as extensively as other methods.'

(*c*) 'Though polarography has proven valuable on the industrial scene, it is not as prevalent as it is on the academic scene. Only about 5 per cent of the expended effort in industry is on polarography. However, it often performs a function where no other approach can help.'

(*d*) 'We do absolutely no work in polarography. It does not appear to be an area of chemistry which has application in a commercial laboratory such as our own.'

(e) 'Unfortunately from the standpoint of the use of polarography in our organization, organic polarography finds little application in service analysis. In what might be called the competition through which analytical methods must pass, polarographic methods have not been able to provide the rapid, reliable or low-cost analysis of organic materials which is achieved by the spectroscopic, chromatographic or some of the colorimetric or titration methods. For special purposes, however, organic polarography can be of considerable value, e.g. the polarographic behaviour of metallo-organic compounds is employed to provide information on the structure of these compounds.'

(f) 'There have been so many new and definitive methods introduced in the past decade or so (like DTA, mass spectroscopy, etc.) that some areas of potential importance have to be neglected and the choice is usually made of that method which is of current interest. Often a polarographic method is used only when no other technique can do the job.'

(g) 'We no longer use the polarograph in our service laboratory (of an agricultural utilization research organization). In fact, we gave our equipment to our dairy group in Washington, D.C.; I do not know if they are using it or not. There is another polarograph in the laboratory and it is not being used at present. It seems that as far as work here is concerned, other methods tend to be preferred, partly because we have no one experienced in this area.'

(h) 'We have often thought that we could make better use of polarography but lack of up-dated equipment and lack of experience has prevented us from exploring this field more fully. The government regulatory agencies seem to prefer other methods for documentary purposes.'

(i) 'Based on my previous experience, I would say that polarography is of very little interest for the determination of organic compounds from an analytical service standpoint. I believe it is more useful as a research tool in organic chemistry.'

(j) 'With respect to organic compounds, the greater portion of the effort in electrochemistry for the past few years has gone toward mechanistic studies of charge-transfer reactions with only infrequent use of voltammetric methods for quantitative analysis.'

(k) 'We have looked into numerous analytical possibilities of a service nature based on application of polarography to organic systems but have concluded some other technique had more potential. In this connection, a major consideration has been the lack of voltammetric instrumentation in our refineries and chemical plants. This situation will undoubtedly continue to prevail until an occasion arises where polarography is the only applicable technique for a particular problem.

Two specific examples will illustrate why we have had so little success in exploiting the technique for service work. The AOAC method for assaying pyrethrum extracts is a tedious, time-consuming and exacting procedure. It was thought that polarography might provide a better control method. However, only the total pyrethrum content could be determined, not the individual types. Consequently, it could not be considered for buyer–seller relations. In the other case, a colorimetric procedure was being used to control the amount of a specific phenolic compound added as an oxidation inhibitor to an oil. This method requires a liquid chromatographic separation of the additive agent to eliminate interference from other substances that give a similar colorimetric reaction. Polarography was investigated to find out if the separation step could be eliminated. It was found, however, that the diffusion current was always larger than could be accounted for by the amount of phenol present.

We have done a little research to have a better understanding when voltammetry might be the preferred technique, which has resulted in publication. Such studies have not, as already implied, been productive in the development of superior methods for control application. Our general conclusion is that organic polarography will probably find its greatest application in research studies and special problems not amenable to control type operations.'

(*l*) 'In general, over the years, I have found organic polarography to be very useful in rendering analytical support for petroleum and chemical research; however, in recent years, very little use has been made of this tool, except for a few routine applications which I shall discuss subsequently. The decreased use of organic polarography can be attributed, perhaps, to a large extent in our case to the changing character of the type of research we conduct. Also, newer instrumental techniques have had an effect. I personally have not worked in the field of organic polarography since 1959, and we have had no professionals in this area since 1961. It is difficult to determine which is the cause and which is the effect. When the demand lessened, we reduced manpower in polarography, allotting it to the more glamorous tool, nuclear magnetic resonance, which is now relied upon heavily by organic chemists. Also, our high resolution mass spectrometer and newer gas chromatographic techniques have contributed also to the diminished demand. I am convinced, however, that if professional manpower were available for organic polarography, it could be as useful as it once was in certain areas.'

(*m*) 'You have probably gathered from the paucity of papers in the literature covering applied voltammetry that the technique is disappearing from the realm of everyday application. Polarography has not been used in our laboratories since about six or eight years ago. Prior to that it was used for only a few specific applications, only a portion of which were organic in nature. Examples include determination of dissolved oxygen, elemental sulphur, carbonyl impurities in diluents, and certain metals.

One might ask the question "Why has voltammetry met with disfavour in everyday organic analysis?" Primarily, I think, other techniques have become available which have advantages (when compared with voltammetry) of greater specificity, simplicity and shorter analysis time. For example techniques such as atomic absorption and X-ray fluorescence spectroscopy and modern emission spectrography (e.g. the Quantometer, which can determine eight or ten elements simultaneously) have completely supplanted polargraphic methods for metals. The advent of gas chromatography, and to some extent other chromatographic techniques such as TLC, has increased our ability to separate organic compounds such that the specificity provided by half-wave potentials becomes antiquated in comparison. Improvements in infrared and mass spectral techniques as well as the advent of NMR has assisted greatly in the area of organic analysis. Particularly, by combining an excellent separation technique such as GC with the above-mentioned spectral methods, even the most complex organic systems can be analysed rather completely (in terms of individual specific compounds) in a short period of time. In short, other techniques are replacing voltammetry in organic analysis.

Potentiometric methods and microcoulometry are probably the only other electrochemical techniques which we use very extensively.'

(*n*) 'To sum up, polarography is used in isolated instances as an analytical method but is not the normal method of choice. Unfortunately, it may be too late to effect a change in attitude on the part of many analytical chemists.'

2. Applications Made of Polarography

To counterbalance the possibly overly pessimistic picture indicated in the previous section, it must be emphasized that polarography has been advantageously used in many situations as indicated by the following remarks.

(*a*) 'The industrial activity and applications of polarography seem to take the form of a sine wave. Currently, I think that there is a revitalization of current interest in this area. In addition to the utilization of polarographic techniques to support structural elucidations, we are now actively using this tool for quantitative and qualitative analysis of organic compounds. For example, the polarography of benzodiazephines is one instance of quantitative analysis. We have found polarography useful in the analysis of hydrazine and hydrazine derivatives. The usual inorganic analysis utilizing polarography, where applicable of course, is still carried out. Our current interest in addition to polarography, is constant potential coulometry of organic compound. I believe that this latter technique offers many advantages to the pharmaceutical industry because of its inherent specificity and precision.'

(*b*) 'I think that the use of polarography in organic analysis in the pharmaceutical industry is increasing. Both the U.S. Pharmacopeia XVII and the

National Formulary XII have sections describing the analytical technique of polarography. Hydrochlorothiazide Tablets (U.S.P.) and Chlorothiazide Tablets (N.F.) are officially assayed by polarography. There are people at the Food and Drug Administration who are pushing this method. I do not have any details, but I feel that companies who did no work in polarography ten years ago are doing something today.'

[This optimistic picture should be counterbalanced by the subsequent comment in the same letter: 'We are not using polarography. Its use in organic analysis has been limited by lack of personnel (other tools such as mass spectra, NMR, IR, etc. absorb the available manpower). Outside the research group, it is not used because of personal idiosyncrasies.']

(c) 'In the first place, the voltammetric approach is used infrequently and little in the determination of organic compounds. The polarographic method plays a much more important role, especially from the service aspect, in inorganic analysis and in metalloorganic analysis. The inorganic analyses are mostly very routine type procedures, well known in the general literature. We regard the voltammetric tool using either the DME or the RPE very useful in metalloorganic analysis since it is sometimes the only approach and often the most convenient method. In this type of analysis we are usually interested only in the quantitative aspects of the various oxidation states of the metal ion in the metalloorganic compound. Existing methods in the general literature are modified in our laboratories to suit the type of compounds.

Purely organic analyses by the voltammetric approach are infrequently done. Conventional techniques (with the DME or RPE) are the only ones pursued. There is a two-fold purpose in using voltammetry in our organization:

1. The investigation of oxidation–reduction characteristics of newly synthesized organic compounds with either the DME or the RPE. The information obtained this way aids us in: (a) determining whether the compound if ever produced commercially could be measured voltammetrically, and (b) predicting how easy it is to reduce or oxidize the compound in a series of organic synthesis steps.

2. The usage of polarography as a means of quantitative determination of a specific organic compound in the presence of others on a semiroutine basis. In developing these methods we depend heavily on the information already published in the literature, notably in the monographs of Schwabe (1a) and Zuman (1b). We have found these books to be very reliable and helpful. Compounds determined by the polarographic route are, for instance, (a) maleic acid and maleic anhydride, (b) terephthalic acid and related alkyl-phthalate esters, and (c) cyclohexanone, methylcyclohexanone and cyclohexanone oxime.

This is about the extent to which voltammetry serves us in our programme of organic analysis. I imagine that our organization in view of our business interests and products (basic materials, monomers, and non-oxidizable and non-reducible polymers) makes relatively little use of electroanalytical techniques.'

(d) 'We have used polarographic techniques extensively for a number of years. The majority of this usage has been in the general areas of non-routine analytical investigations (problem solving), correlation of electro-chemical behaviour with other physical properties and effects, and in studies of specific chemical reactions wherein the polarographic method is employed as a monitoring device. This work has Involved both aqueous and non-aqueous media, with the latter predominating in recent years. The following list summarizes the types of organic compounds which come to mind as having received at least some attention: Activated vinyl compounds; disulfides and mercaptans; aldehydes and aldehyde addition and reaction products; nitro compounds; quinones and hydroquinones; phenylene-diamines; azo and hydrazo; numerous heterocyclic systems.

We also make good use of anodic voltammetry at stationary electrodes. Thus, the above list of compounds could be expanded to include other electrochemically oxidizable moieties, such as a substituted phenol, and compounds in some of the classes noted above for which an electro-chemical oxidation is not observable at a mercury electrode.'

(e) 'I have used organic polarography in the determination of a number of organic functional groups: aromatic sulfones and sulfoxides; carbonyls, especially formaldehyde; peroxides and hydroperoxides; carbonyl sulfide by conversion of the gas to piperidine thiocarbamate; mesityloxide in acetone; aromatic sulfonyl chlorides; organic halides; tetraphenylporphorin; rubeanic acid, chrysene and dithioximide; acrylonitrile; condensed ring aromatics; mercaptans; di- and polysulfides.

In addition, the following inorganic materials have been determined in organic liquids: hydrogen cyanide, elemental oxygen and elemental sulfur.

Polarography has been kept alive in our laboratory because of its ability to determine trace components in the presence of multicomponent materials. However, whenever possible, it has been displayed by optical spectroscopy, which also has a capability of determining trace quantities of materials. Polarography is still used when needed to determine some of the above. It is relied upon heavily, however, for the determination of elemental sulfur, dissolved oxygen in hydrocarbons, hydrogen cyanide, and di- and poly-sulfides.'

(f) 'We do not use polarography to any great extent. We have a few isolated cases such as the detection of $\Delta^{1,4}$ diene steroids in the presence of Δ^4 steroids but, generally speaking, we find other methods preferable. For simple batching assay we do use coulometric methods and find these very useful for hydrochloride salts of pharmaceutical importance.'

(g) One respondent (16) gave interesting statistical data 'concerning the actual use of polarography and related voltammetric techniques in our organization. In our work there are two general types of analytical problems which involve applications of electrochemical techniques. The first, and easiest to describe, is the analysis of samples for which a standardized method is available. Records of this type of work are readily available and a summary of the work done in this area is included in Table I. To give a sense of proportion to these numbers, the Analytical Division of the Hercules Research Center handled 86,749 samples during the two-year period covered by the data, so that polarographic work amounted to 0·53 per cent of the standardized analyses carried out in this period. Approximately 30,000 of the samples were analyzed by instrumental methods, so that polarographic analyses amount to roughly 1·5 per cent of the instrumental work.

TABLE I

Polarographic Analyses Performed at Hercules Research Center During 1966–1967 (16)

Method	No. of Samples
I. Organic analyses	
Acrylamide in polymers	200
Aromatic aldehydes	51
Maleic and fumaric acids in resins and polymers	118
	369
II. Inorganic analyses	
Thallium analyses	31
Platinum group metals	11
Copper ore assays	47
	89

A second type of problem, which is more difficult to pin down statistically, is the use of voltammetric techniques to monitor reactions in the pilot plant and laboratory, and to obtain kinetic data concerning reactions involving an electroactive species as product or reactant. In this type of work, a method is often developed and turned over to a chemist in one of the Research Center's product-oriented divisions. In such a case, no record of the amount of work done is easily available. I can only guess at the amount of work done on this basis, but I would estimate that it is at least equal to the amount of work done on a standardized basis, since we have two Sargent Model XXI polarographs and a Beckman Electroscan available for such work, and they are heavily used.'

In 1972, it was indicated that little change has occurred since 1968 and the information then supplied is still valid. 'However, we have slightly increased the number of inorganic samples analysed by electrochemical methods, predominantly using linear scan and anodic stripping voltammetric techniques.'

(*h*) 'We are probably one of the few industrial laboratories which operates a polarograph full time. This is largely a result of the efforts of W. Jura who has been quite successful in developing non-aqueous polarography for analysis of organic systems. Typical analyses are the determination of monomer content of polymers and also the determination of pesticides both for assay purposes and residue analysis. The polarographic technique offers an advantage over VPC or mass spectrometry in that no heating or volatilization is required, which minimizes the chances of alteration of the sample prior to analysis.'

(*i*) 'Polarography is not the most widely used technique in pharmaceutical analysis; however, in appropriate circumstances, it can provide uniquely valuable data. We, for instance, are particularly interested in methods for determining the stability of tranquilizer phenothiazines such as chloropromazine, trifluoperazine, trimeprazine, etc., in pharmaceutical preparations. Polarographic oxidation with a rotating platinum electrode has proven to be quite useful for this purpose; controlled potential coulometry is also very satisfactory, giving even more reproducible and sensitive results.

The stable oxidation product of tranquilizer phenothiazines is the sulfoxide and on occasion we have a need to determine this in pharmaceutical dosage forms. One of the ways this can be done is by polarographic reduction since it gives a well-defined wave (cf. G. S. Porter, *J. Pharm. Pharmacol.*, **19**, 176 (1967)).

Aside from phenothiazines, we have used polarography with the dropping mercury electrode for determining a number of other compounds including: (*a*) organic peroxides, e.g. benzoyl peroxide, urea peroxide, peroxides in fats and oils, (*b*) amine oxides, e.g. scopolamine *N*-oxide, (*c*) hydrochlorthiazide (USP procedure) and (*d*) mercaptans.

We have also used polarographic procedures for determining stability constants of copper chelates, the oxygen content of aqueous solutions as a measure of reaction rates involving oxygen consumption, and the oxidation–reduction potentials of organic compounds in aqueous and non-aqueous solution as a guide to the potential stability of these compounds to air oxidation during storage alone and in pharmaceutical dosage forms. We have also determined half-wave reduction potentials as a guide for large scale electrolytic reduction of organic compounds.

Other pharmaceutical people have reported the use of polarography in analysing steroids, e.g. testosterones, benzodiazepineoxide tranquilizers such as "Librium", nitrofuranes, etc.'

(j) Another respondent tabulated a provocative list of possible applications of polarography in his laboratory, which is indicative of applications which are being made in other laboratories and which might serve as a summary for this sub-section:

'There has not been much activity in using polarography in organic analysis in this laboratory. We believe it is a weakness in our set-up and we have tried to hire someone experienced in this technique but so far have been unable to do so. Some of the area and examples in which we can foresee its application in our work are:

Food analyses: antioxidants, sugars, acids, ascorbic acid in fruits and beverages.

Pharmaceutical and drug analyses: alkaloids, antibiotics, barbiturates, steroids, hormones, nitro groups, vitamins.

Animal poisons: we do quite a few analyses of animal viscera for poisons. Warfarin is one thing that gives us trouble and polarography might be the answer.

Contaminated water: there are chances here for all sorts of approaches.

Pesticides, both residues and formulations: carbamates, organophosphates, nitro compounds.

Animal feed analysis: medicants in medicated feeds.

However, these are only ideas. I am sorry we have little to offer as far as practical experience is concerned.'

Of prime interest to the present chapter is the discussion elsewhere in this volume by Hoffman and Volke (17) of the application of polarographic techniques in pharmacy. The pharmaceutical industry is—at least in some laboratories—vigorously interested in organic polarography as indicated by the replies received to the author's letter of inquiry and the two days of the 1968 Land O' Lakes Conference on Pharmaceutical Analysis being devoted to 'Electrochemical Methods in Pharmaceutical Analysis', most of which was concerned with polarographic techniques and papers on the polarography of benzodiazepams, barbiturates, phenothiazines and imidazolines. As an example, 'classical polarography was found to be an excellent approach to the specific determination of ethacrynic acid in tablet and cryodessicated pharmaceutical formulations' (17a). As a result of a cooperative study of the polarographic determination of organophosphorus pesticide residues, e.g. parathion, methyl parathion, diazinon and malathion, in non-fatty foods (17b), 'the methods described were adopted as official first action by the Association of Official Analytical Chemists, and the procedures are found in the 11th edition of the AOAC Methods published in 1970' (8). The latter respondent noted in 1972 that 'we still note a steady increase in the use of the polarography in the fields of pesticides, heavy metals and food additives. Again, I strongly recommend that polarography be given a more prominent coverage in the chemistry curricula both at undergraduate and graduate levels.'

D. Literature on Polarographic Organic Analysis

The fairly voluminous literature on organic electrochemistry is largely concerned with polarographic reduction (cf. *Electroanalytical Abstracts*, the Sargent bibliography covering the period of 1922–67 (17c), the bibliographies issued by the Prague and Padua polarography centres, and *Interface*, issued ten times per year by Interface Publications of Lansing, Michigan). The current literature is well summarized in reviews, dealing in whole or in part, with the application of polarography to the analysis of various types of materials, which frequently appear, e.g. the reviews of product areas in *Analytical Chemistry* in a special April review issue in odd-numbered years and the reviews of techniques and of some class or material areas, e.g. biochemical analysis, which similarly appear in even-numbered years. Thus, every other year, formerly Hume (18) and now Nicholson (18a) review polarographic theory, instrumentation and methodology, and Pietrzyk (19) reviews organic polarography; other pertinent biennial *Analytical Chemistry* reviews of voltammetric techniques are cited at appropriate locations. Typical examples of class and materials reviews are those of biochemistry (20–22a), agronomy and biology (22b), insecticides and pesticides (23–25), polymers (26, 27) and proteins (28). A helpful listing of books and review chapters and papers will be found on pages 254–60 of P. Zuman, *Organic Polarographic Analysis* (1b).

Of obvious general importance are such standard reference works as Brezina and Zuman, *Polarography in Medicine, Biochemistry and Pharmacy* (Interscience Publishers, 1958), Clark, *Oxidation–Reduction Potentials of Organic Compounds* (Williams and Wilkins, 1960), Kolthoff and Lingane, *Polarography* (2nd ed., Interscience Publishers, 1952), Müller, *Polarography* in Weissberger and Rossiter, Ed., *Physical Methods of Chemistry*, Volume 1 (Part IIA) (Interscience-Wiley, 1971), Vol. 4 (*Electrical Methods of Analysis*) of Part I of the Kolthoff-Elving *Treatise on Analytical Chemistry* (Interscience Publishers, 1963) and Schwabe, *Polarographie und Chemische Konstitution Organischer Verbindungen* (Akademie Verlag, 1957). A particularly useful reference for voltammetric theory and practice including the application of analysis is Meites' well-known book on polarographic techniques (13). Recently published books include ones on organic electrochemistry by Baizer (248) and by Tomilov and associates (252). Frumkin and Ershler (253) have initiated a series on organic electrochemistry, and Yeager and Salkind (254) one on electrochemical techniques. Reviews of the electrochemistry of biologically important compounds have been published by Burnett and Underwood (255) and by Elving, O'Reilly and Schmakel (256).

Perrin has reviewed the general area of organic electrode processes in a chapter in S. G. Cohen and coworkers, *Progress in Physical Organic Chemistry*, Vol. 3, Interscience Publishers, New York, 1965; Zuman has reviewed 'Physical Organic Polarography' in Vol. 5 (1967) of the same series and has discussed mechanisms of organic electrode processes, correlations of $E_{1/2}$ with structural parameters and application of polarography to the study of organic reactions (29). A number of examples of the use of voltammetry in organic analysis will be found in Purdy's book on electroanalytical methods in biochemistry (30).

The journal and review literature are rich in references which could be cited; many will be mentioned during the course of the chapter. The references subsequently given have been selected primarily to cover illustrative studies and recent critical reviews.

II. SYSTEMATICS OF POLAROGRAPHIC ORGANIC ANALYSIS

The essential features and factors involved in the determination of organic compounds by polarography are summarized in the present section. For further treatment of the principal factors in the methodology employed in polarographic organic analysis, the reader is referred to the books by Meites (13) and Zuman (1b) and the previous chapters by the author (2, 3).

A. Electroactive Functions

Generally, polarographic methods of analysis depend on the electrochemical oxidation or reduction of the desired constituent(s) within the potential range limited by the reduction and oxidation of the solvent and background electrolyte system. In the case of amperometric titrations, obviously only the titrant or back-titrant needs to be electroactive. In some cases, an electrochemically inactive species can be determined in terms of, for example, the effect of its adsorption on the electrode on the capacity current.

Quantitative analysis is usually based on the measurement of the limiting or equivalent current, which is directly related to the concentration of the electroactive species for diffusion-controlled as well as for most kinetic-controlled processes. The half-wave potential, $E_{1/2}$, is frequently a sufficiently good identifying characteristic to be used for the qualitative analysis of samples of limited composition as well as for characterizing the behaviour and properties of the electroactive species as subsequently discussed (Section IV).

If a sample contains several components, electroactive within the potential range under consideration, whose waves are close enough in $E_{1/2}$ to result in their merging, it will be necessary to so alter the composition of the solution as to permit resolution of the waves. If this cannot be done, it will be necessary to effect preliminary physical separation of the species involved or to perform chemical reactions which will so change the nature of the species as to give non-interfering polarographic waves.

Typical examples of the direct polarographic determination of elements in organic matrices are those of the determination of elemental sulphur dissolved in petroleum fractions or of oxygen dissolved in organic liquid samples of all types. A more strictly defined example of organic elemental analysis is seen in the determination of compounds on the basis of functionality, i.e. of differences in the electroactivities of the elements concerned, e.g. the analysis of a mixture of chloro-, bromo- and iodoacetones due to differences in the ease of electroreduction of the different carbon–halogen bonds (31) or of a mixture of dichloro- and trichloroacetic acids because of the differing ease of reduction of a carbon–halogen bond depending on the number of halogen atoms on the carbon (32).

In general, polarography is less versatile than absorption spectrophotometry in respect to responsive functions; however, as mentioned, this very restriction tends to build a certain amount of specificity into polarographic measurements.

Electrochemical reduction of an organic compound involves bond fission of some type, which is equivalent to the conversion of a functional group involving two or more atoms into another functional group. This is in marked contrast to inorganic polarography, where the essential electrode process usually involves transformation of an element from one oxidation state to another and, at most, may involve no more than a change in its immediate atomic environment in terms of oxygen attached to it. Most of the applications of polarography to organic compounds have involved irreversible reductions. Reversible organic reductions have apparently been confined to quinones and a few other functional systems such as the phenylene diamines, which resemble the quinones in forming resonating systems.

Only a very few organic oxidations at the DME have had analytical significance, e.g. determination of ascorbic acid. This is due in large part to the relatively few organic oxidation reactions that can be performed at potentials below that at which mercury oxidizes. The development of techniques for using carbon, platinum and other solid electrodes for observing oxidation reactions opens new possibilities for organic analysis, based on extension of the available potential range to the point where oxygen is evolved (cf. Section II-C-1).

A large variety of organic functional groups can be measured at the mercury and other electrodes, which have been described for organic polarography and voltammetry. Since tabulations of such electroactive reducible groups are readily available (cf. references 1, 2, 3 and 33), a listing is not included; many have been listed in Section I-C-2. Too little systematic work has been done on electrolytic oxidation to allow preparation of a meaningful list of electrooxidizable groups.

The ease of electrochemical reduction or oxidation of a given functional group may be markedly affected by other substituents on the molecule and stereochemical factors, since the ease of introducing or removing an electron into or from a functional group, is obviously dependent on the electron density at the reactive site, which is affected to a greater or lesser degree by the whole molecular constitution and configuration. Thus, $E_{1/2}$ of the functional group is influenced by the rest of the molecule, which allows the specific determination of a molecule in terms of its functional group, e.g. determination of the amounts of cis–trans isomers in a mixture or of aromatic ketones in the presence of aliphatic ones. For example, the other substituents on the molecule may change the $E_{1/2}$ of the nitro group sufficiently to make the polarographic method highly specific. Parathion (O,O-diethyl-O-p-nitrophenyl thiophosphate) can be determined by measurement of the nitro group reduction at an $E_{1/2}$ value of -0.39 volts; the accompanying impurity, p-nitrophenol, is reduced with an $E_{1/2}$ value of -0.68 volts (34).

The variety of organic functional groups, which can be measured polaro-graphically in a single laboratory, is illustrated by the following statement:

'The following are functional group areas where we found polarography useful: (a) Carbonyl compounds (aldehydes mainly); (b) azomethine ($-C=N-$); (c) hydrazines and hydrazides; (d) mercaptans; (e) quinones and hydroquinones (includes polymer inhibitors); (f) antioxidants of all types; (g) hypohalites (organic); (h) peroxides (including polymerization catalysts; (i) thioureas; (j) nitro compounds; (k) azo compounds.

In case (a) we determined traces of aldehydes in ketones by polarography and traces of aldehydes in general. Case (b) is described in the literature. We did several hydroazides and hydrazines, case (c). Case (d) is widely used since the —SH group is easily oxidized. Case (e) applies to things like hydro-quinone, t-butyl catechol, chloranil. Case (f) includes butylated hydroxy-toluene (BHT), butylated hydroxyl anisole (BHA) and propyl gallate. In case (g) we monitored methyl hypochlorite at one time. Case (h) is a very common one since peroxides are used a great deal and also are present in many solvents where they cause undesirable effects; polarography provides a sensitive test if a suitable solvent is found. In case (i) we did monitor thiourea but other thioureas should work. Case (j) is an old one but we did use polarography to monitor the reduction of aromatic nitrocompounds to the amines; it works very well. Case (k) is a corollary of case (j) because the azo compounds are intermediates in the reduction of the nitro compounds.'

1. Production of Electroactivity

The polarographic inertness of such organic functionalities as unconjugated aliphatic hydroxyl and carbonyl groups is generally due to the unavailability of sufficient applied potential to affect their reduction or oxidation. The effective limiting potential for reduction processes in aqueous media is around -2.0 to -2.2 volts (versus the saturated calomel electrode, as are all potentials cited in this paper unless otherwise stated) due to discharge of the usual background electrolyte cations involved. If the latter is a tetra-alkyl ammonium ion, the working potential range can be extended to about -2.6, where direct reduction of water occurs. On the positive side, the limiting potential with mercury electrodes is controlled by oxidation of mercury, being between ca 0.0 or 0.1 volts in the presence of species forming complex or insoluble species with mercury, and 0.4 or 0.5 volts in their absence. With solid non-mercury electrodes, potentials as positive as ca. 1.5 may be reached; the limitation is due to oxygen evolution. In some non-aqueous solvents, a somewhat greater potential range may be available.

An important aspect of polarographic organic analysis is the possibility of converting electrochemically inert compounds into active species, usually by one of the following general procedures, some of which depend upon prior use of chemical reactions involving the constituent whose determination is desired; it is necessary that these reactions proceed with high yield, e.g. 95 per cent or greater, and that either the product formed or the added chemical reagent be polarographically active.

(a) Conversion into an electroactive isomer, e.g. a compound, polarographically inert in acidic solution, may be converted in alkaline medium, due to enolization, into a reducible conjugated unsaturated system.

(b) Addition or formation of an active group, e.g. conversion of benzene to dinitrobenzene by reaction with a nitric acid–sulphuric acid mixture; a catalogue of recent papers using nitration to produce polarographic activity is given by Pietrzyk on page 207R of his 1968 review (19).

The Malaprade reaction has been extensively used in the determination of compounds containing adjacent hydroxyl or similar groups, which are quantitatively oxidized by periodic acid to reducible carbonyl groups, e.g. mixtures of ethylene and propylene glycols can be analysed by oxidizing the glycols with periodate, distilling the formaldehyde and acetaldehyde formed into a suitable solvent, and polarographically examining the resultant solution (35). The wide applicability of the Malaprade reaction in the oxidation of organic functional groups is well summarized by Dryhurst (35a) and by Dyer (36).

The use of chemical reaction to prepare a sample for measurement is further exemplified by analysis of mixtures of ethylene and propylene chlorohydrins via hydrolysis of the latter to the glycols and application of the procedure just outlined (37). Mixtures of ethylene and propylene oxides have been analysed by hydrolysing them to chlorohydrins, further hydrolysing the latter to the glycols, and then going through the periodate oxidation and polarographic measurement.

Mixtures of maleic and fumaric acids and of their diethyl esters can be readily analysed by examining the wave patterns obtained before and after hydrolysis (37a). Before hydrolysis three waves are observed: a composite ill-defined wave due to both esters and two well-defined waves due to each of the acids. After hydrolysis, only the latter two waves are observed and these have increased in amount equal to the respective esters originally present. Aspartic acid can be determined in rather complex samples of protein hydrolyzates by converting the aspartic acid through a simple quantitative reaction with dimethyl sulphate to an adduct which can then be readily converted to a mixture of maleic and fumaric acids (38).

(c) Formation of an electroactive complex in some few instances due to the organic ligand affecting the ease of reduction of the metal ion with which it is complexed. An area, which has been insufficiently explored is the determination of organic compounds at mercury electrodes by the use of anodic waves due to mercury oxidation, analogous to those obtained with halide ion. Thus, a mercaptobenzothiazole which gives a cathodic wave due to the catalytic evolution of hydrogen, gives an anodic wave due to the formation of the mercury(I) compound (39).

(d) Use of catalytic waves. The catalytic wave due to catalysed hydrogen ion discharge has frequently been suggested for quantitative analysis since it is proportional to the concentration of the catalyst. The latter class includes certain types of amino acids, proteins, alkaloids and sulphhydryl compounds. The catalytic wave usually manifests itself as a large maximum on the current-potential curve; the current is often many times the magnitude expected for a diffusion-controlled process. The classic example is the wave given by cystine and cysteine in ammonia buffer solutions containing cobaltous ion, e.g. references 39a and 39b. The topic of catalytic and kinetic waves in polarography is thoroughly reviewed by Mairanovskii (39c). The most interesting and probably most important example of catalytic waves has been the controversial application to the clinical diagnosis and study of cancer first suggested by Brdička; Müller (40), Rappolt (41) and Homolka (41a) have reviewed the Brdička reaction; a group of papers on the Brdička reaction was also presented at a symposium (42).

(e) Consumption of an electroactive reagent with the obvious limitation that such means of measurement are usually not specific in their applicability, except insofar as the chemical reaction used in determining the desired constituent is specific, e.g. determination of carbonyl group by measurement of sulphite consumed.

2. Non-faradaic and Catalytic Analytical Methods

In addition to fairly straightforward methods for organic analysis, based on measurement of a faradaic current due to reduction or oxidation of the desired constituent, other polarographic phenomena have been used to

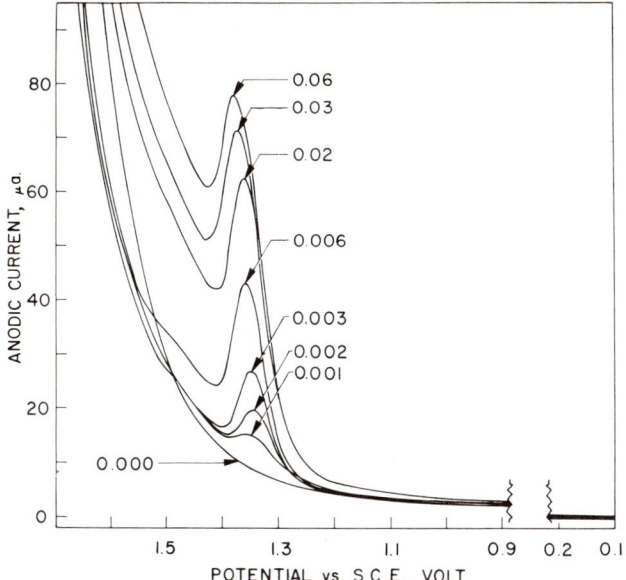

Fig. 1. Voltammograms of $0.25\,M$ K_2SO_4 solutions containing varying amounts of Triton X-100 (wt. per cent indicated on curves). Polarization rate: $200\,\text{mv min}^{-1}$. Temperature: 25°C (45).

some extent for analytical purposes, e.g. use of overpotential phenomena and of tensammetry involving alternating current polarography to determine electrolytically inactive compounds which adsorb on the electrode (43).

Suppression of the oxygen maximum by surface-active substances has long been advocated for determining minor organic constituents. This approach would seem to be especially advantageous in the determination of surface-active substances of unknown nature, e.g. assay of the purity of refined sugar (44).

Although, as mentioned in the previous section, many analytical methods have been based on the catalytic effect of the constituent to be determined in lowering the overpotential of hydrogen evolution on mercury, little or nothing has been done on the similar utilization of the oxygen overpotential. The catalytic evolution of oxygen at the graphite electrode was made the basis of an analytical method, using the fact that in the presence of non-ionic surface-active agents, e.g. the polyoxyalkalene ethers, an anodic wave appears at the graphite electrode at very positive potential (45). This peak current is due to a prior adsorption of surfactant on the electrode surface, which lowers the overpotential of oxygen on graphite (perhaps in water bound to the surfactant) and results in an evolution of oxygen with the concomitant desorption of the surfactant film.

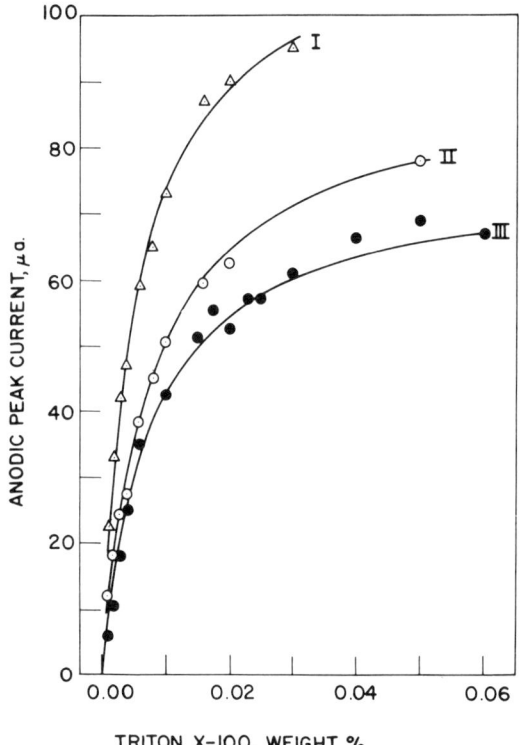

Fig. 2. Peak current–concentration variation for Triton X-100 in various electrolytes. Curves are calculated from equation (1); points are experimental values (corrected for the residual current in the case of curve III). I (triangles): 0·25 M NaOAc–HOAc (pH 4·7); 40°C; calculated constants, $k = 193$ (wt. per cent)$^{-1}$, $q = 114\,\mu$a. II (open circles): 0·25 M NaOAc–HOAc (pH 4·7); 25°C.; calculated constants, $k = 126$ (wt. per cent)$^{-1}$, $q = 90\,\mu$a. III (solid circles): 0·25 M K$_2$SO$_4$ (pH 6·2); 25°C.; calculated constants, $k = 128$ (wt. per cent)$^{-1}$, $q = 76\,\mu$a (45).

Since i_p varies with the amount of surfactant adsorbed, which in turn depends on the surfactant concentration in the solution, i_p can be used to estimate the latter. A series of curves obtained as surfactant concentration is varied as shown in Figure 1. Plots of i_p versus concentration for three such experiments are given in Figure 2, where the points are the experimental values and the lines have been calculated on the basis of a Langmuir adsorption isotherm expressed as

$$i_p = qkC/(1 + kC) \tag{1}$$

where q is a measure of the surface area of the electrode in terms of the current produced, k is a measure of the strength of adsorption and C is the concentration of surfactant. Analytically, the method can be applied, e.g. to the determination of 0·001 to 0·01 wt. per cent Triton X-100 with an accuracy of ± 5 per cent or better; 0·001 per cent Triton corresponds to 10 μg ml^{-1} of voltammetric test solution.

3. Investigation of Electrode Processes

One of the impressive features of polarographic organic analysis has been the stress being placed on fundamental studies of the polarographic behaviour of organic compounds in the belief that such studies provide the optimum basis for the rational development of analytical procedures. Thus, an investigation (46) of the complicated polarographic patterns exhibited by the three chloroacetaldehydes due to the hydration equilibria involved, revealed the possibilities for polarographic determination of the individual aldehydes and analysis of mixtures (47). However, there has been decreasing development (or, at least, publication) of analytical methods based on fundamental studies even though the number and thoroughness of such studies have been steadily increasing.

A great deal of effort is currently being given to the study of organic electrode processes with an ever increasing variety of approaches to elucidating the intimate details of each step in the overall electrode reaction and revealing the interrelationship of these steps. This is quite evident from a consideration of the material presented at three recent symposia: The International Conference of Polarography held in Kyoto (1966) (48), the Faraday Society Discussion on Electrode Reactions of Organic Compounds (1966) (49) and Intermediates in Electrochemical Reactions (1973), and the symposium on investigation of reaction intermediates at electrode surfaces held at the Eastern Analytical Symposium in New York in November, 1968.

Current understanding of the mechanisms of organic electrode processes has been perceptively and critically summarized by Perrin (50), who emphasizes, as others have done, that, because an electrochemical reaction is a heterogeneous one, a thorough description of its mechanism must await a detailed understanding of the structure of the double layer and the surrounding solvent layer. At the present time, descriptions of organic electrode processes all too frequently still consist only of a listing of the products of the electrode reaction with, in some cases, more or less speculative postulation of intermediates formed during the reaction. Only in a few cases is the experimental evidence sufficiently detailed to allow the postulation of structures for the transition states involved.

In general, it is to be expected that helpful information about energy-controlling steps and long- and short-lived intermediates in redox processes will be obtained by more extensive use of electrochemical investigative techniques of increased information output, supplemented by optical and magnetic examination of solutions during and after electrolysis. Thus, it is often possible to separate the electron-transfer step from the subsequent reactions of the reduced species, e.g. a free radical, by use of perturbation techniques.

Hopefully, systematic studies of the electrochemical behaviour of specific organic functional systems in a variety of compounds will aid in defining such important factors as the extent and manner of adsorption and orientation of reactant and products at the electron-exchange surface, the steps in the overall redox reaction including the reversibility of each step and the presence of preceding, accompanying and following chemical reactions, the role of free radical and other intermediate species, and the changes, if any, produced on the electrode surface during the redox reaction, e.g. film formation, as well as evaluating the rate constants for the various homogeneous and heterogeneous reactions involved. It should then be possible, for example, to establish more firmly the sequence of electron-transfer and non-electron-transfer steps in redox processes.

A general mechanistic pattern for organic electrode processes has been outlined (51), which serves to rationalize not only the general course of such processes but also to explain the changes in mechanism with experimental conditions or between members of a homologous series, which occur in some cases, e.g. reduction of a ketone to either a pinacol or a carbinol by selection of pH and applied potential. Application of this viewpoint to organic compounds of biological interest is indicated in a review of the polarographic behaviour of heterocyclic bases and nucleosides and nucleotides (52).

Current research on organic electrode processes seems to emphasize two principal trends. One involves the detailed investigation and reinvestigation as more sophisticated methodologies and experimental techniques become available of a single type of electroactive group; outstanding examples are the extensive work being done on the reduction and oxidation of aromatic hydrocarbons in non-aqueous media and the thorough study of aromatic amines made over a period of years by Adams and his collaborators (97). The other tendency is the attempt to explore the essential features of electrode processes involving functional groups which have not previously been explored. This trend is exemplified by the extensive studies in the last few years of nitrogen-containing heterocyclic organic compounds, which constitute the fundamental building blocks or components of a variety of biologically important compounds, e.g. purines and pyrimidines.

B. Composition of the Test Solution

The composition of the test solution is the critically decisive factor in the success of a polarographic method of analysis for organic compounds in a given sample. The important experimental variables in obtaining well-defined, well-developed waves and in separating such waves for related compounds, are the nature of the solvent, nature and concentration of the background electrolyte, e.g. buffer system, including pH and ionic strength, and maximum suppressors added. These factors have been frequently discussed and need not be further considered except to emphasize some items of particular significance for analysis.

One of the factors contributing to more rational and logical decisions regarding the composition of the polarographic test solution is the increasing volume of investigation on the nature of phenomena occurring in and at the solution-electrode interface or electrical double layer, e.g. in respect to the effects of adsorption on the electrode and of the potential variation in the double layer.

Much of the work in this area is well summarized in the books by Damaskin, Petrii and Batrakov (52a), Delahay (53) and Gileadi (54), and the review papers of Frumkin (55), as well as in the extensive studies of Gierst, Parsons, Mairanovskii and others, e.g. references (56) and (57). The reader is also referred to the lucid reviews of the theory of the electrical double layer by Mohilner (58) and of adsorption in polarography by Reilley and Stumm (59), of the adsorption of organic compounds at the electrode–solution interface, largely as observed by electrocapillary and differential capacity methods by Frumkin and Damaskin (55), of the measurement of adsorption at electrodes by Bauer, Herman and Elving (60) of the theory of charge effects in the double layer by Barlow and MacDonald (61), and of double layer structure and influence on electrode reaction rates by Parsons (62).

1. Solvent

The use of solvents other than water by itself as media for test solutions for organic analysis has long been common because of the practical factors of increased solvent power and decreased solvolysis. This has also been true in polarographic organic analysis, where mixed water–solvent media were commonly employed. However, there has been increasing interest in the electrochemical behaviour of organic compounds in non-aqueous media due to theoretical considerations, e.g. the effect of the solvent on the mass transport process through viscosity and solvation effects, and on the potential–reaction relationship through its participation directly or via derived species in (a) polarization of the reactant molecule, (b) the primary electrochemical reaction, (c) accompanying chemical reactions and (d) the structure of the double layer. Study of the electrochemical behaviour of organic functionalities in non-aqueous solvents is particularly relevant in respect to the factors of proton participation and free-radical formation and reaction.

An example of the use of non-aqueous media for systems, where water interferes, is the investigation of the ease of reduction of maleic anhydride in pyridine in order to compare it with those of maleic and fumaric acids and esters (63); the analytical implications for the analysis of mixtures is obvious.

Selection of a solvent for a polarographic method of analysis must consider not only the availability, purity and solvent powers of the solvent, but also its electrochemical inertness in reference both to oxidation and reduction, its effect on electrocapillary properties, and either its lack of reactivty or its reproducible reactivity with the substance to be measured. Where applicable, water continues to be the favourite solvent. Mixtures of water with an alcohol are frequently used. The organic solvents, which have been extensively used in studies of the polarographic behaviour of organic compounds, such as acetonitrile, dimethylformamide and dimethylsulphoxide, have found only limited use in analysis. Often, solvent mixtures are used to obtain both sample and background electrolyte solubility, e.g. benzene and methanol.

The possible important effect of minute amounts of water in organic solvents needs to be more carefully considered than it usually is. The presence of 0·01 per cent water in a typical solvent corresponds to a 4 or 5 mM water solution, which is many times the concentration of the electroactive species. At the present time, the effect of such residual amounts of water is commonly overlooked except when the presence of a hydrogen ion source is required to complete a postulated reaction scheme.

General aspects of the polarographic behaviour of organic compounds in solvents are summarized in references such as 64 to 71. Analytical aspects of voltammetry in non-aqueous solvents and fused salts (melts) is discussed in the present volume by Badoz-Lambling and Cauquis (72). The overall area of electrochemical behaviour in non-aqueous systems is reviewed by Mann and Barnes (64); this book provides a useful guide to the literature. Non-aqueous solvents for electrochemical use (72a), reference electrodes for aprotic solvents (72b) and acid–base behaviour in aprotic solvents (72c) have also been reviewed.

Among reviews dealing with specific solvents and classes of compounds are ones on electrochemistry in dimethylsulphoxide (73) and propylene carbonate (74), electrochemical reduction and oxidation of aromatic hydrocarbons and related substances, largely in aprotic solvents (75, 75a), and polarography of aromatic compounds in acetonitrile (76), of hydrogen ion in non-aqueous solvents (77) and of organic compounds in non-aqueous media (78, 79). The papers presented at the 1970 Paris Symposium on Electrochemistry in Non-aqueous Media, organized by J. Badoz-Lambling, have been largely published in the *Journal of Electroanalytical Chemistry*.

As far as general references on the properties and behaviour of non-aqueous systems are concerned, in addition to the two standard references by Jander (80) and Audrieth and Kleinberg (81), the books by Charlot and Tremillon (82), Waddington (83), Conway and Barradas (84) and Janz, Kelly and Venkatasetty (85) have been found to be quite helpful. Special attention should be called to the series on chemistry in non-aqueous solvents being edited by Lagowski (86).

(a) *Solvent as a reactive species.* Use of the solvent as a reactive species in developing analytical methods merits more exploitation than has been given to this approach. The possibilities are indicated by the polarographic behaviour of the pyridinium species produced on dissolution of Brønsted and Lewis acids in pyridine as solvent (87, 87a, 88). Such solutions commonly produce a $1e$ polarographic wave per acidic function, whose $E_{1/2}$ is controlled by the equilibria in the coordination system involving the hydrogen-bonded non-ionic pyridine–acid adduct, the ionized but undissociated ion pair and the dissociated pyridinium and acid anion ions (87). When the background electrolyte is $LiClO_4$, all Brønsted acids of aqueous pK_a less than ca. 9, produce identical $E_{1/2}$ values of ca. -1.3 volt versus the normal silver nitrate electrode in pyridine, due to an ion-exchange reaction between the coordination system representing the monosolvated acid, SHA, and the background electrolyte, MB, to produce MA and the common reducible species SHB, where HB is generally a much 'stronger' aqueous acid than HA, e.g. HB is $HClO_4$ (87, 88), which permits a simple determination of the total acidity in the sample (88).

When the background electrolyte consists of large univalent ions, e.g. Et_4NClO_4, parallel linear correlations exist for each type of acid investigated (carboxylic, phenolic and nitrogen heterocyclic (purine)) between $E_{1/2}$ and aqueous pK_a (Figure 3) with a levelling effect for acids of aqueous

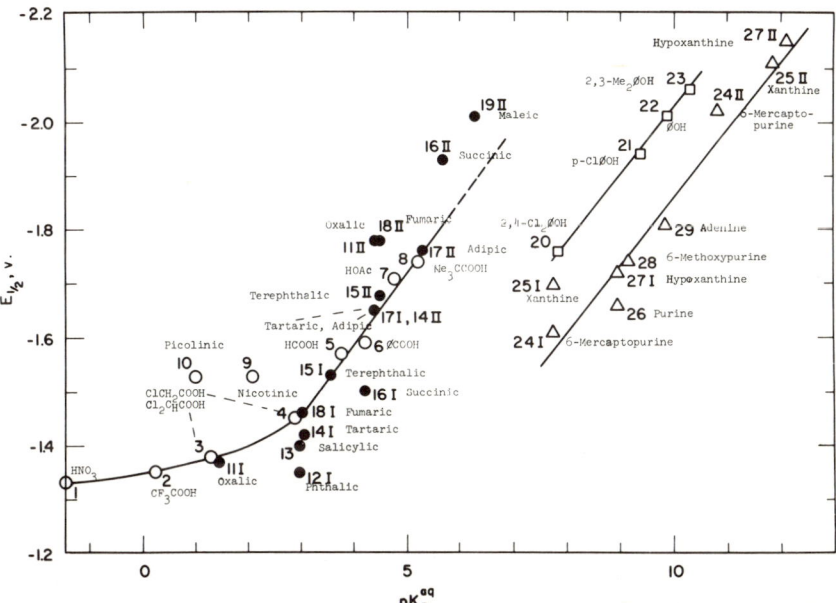

Fig. 3. Relation between half-wave potentials of waves given by solutions of Brønsted acids in pyridine containing $0.1\ M$ tetraethylammonium perchlorate at 25°C and the aqueous pK_a values of the acids. Circles: nitric and monocarboxylic acids; solid circles: salicylic and dicarboxylic acids; squares: phenols; triangles: purines (87).

pK_a less than ca. 3, which is due to the reduction of essentially unpaired pyridinium ion (87). (Deviations from regularity of the $E_{1/2}$–pK_a (aqueous) relationship can be interpreted on the basis of the changes in dielectric constant and solvent basicity and coordination (solvation), as reflected in the increased stability of monoanions of basic acids, and zwitterion phenomena.) This behaviour allows the analysis of acid mixtures by polarography in pyridine, if the aqueous pK_a values differ sufficiently (cf. Figure 4).

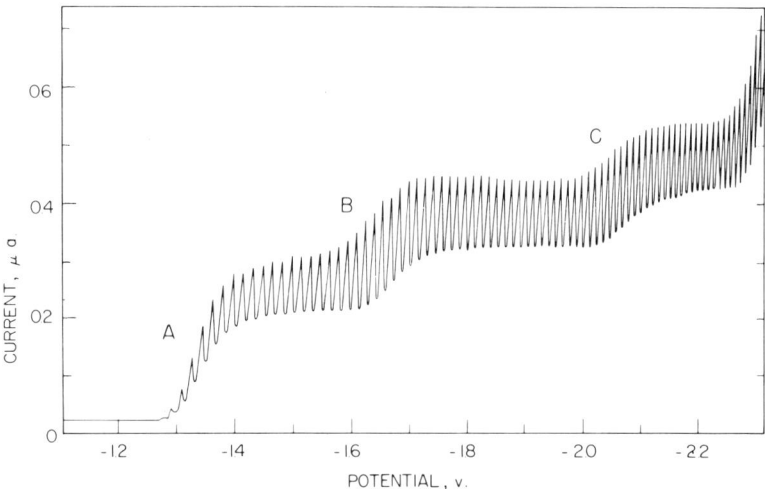

Fig. 4. Polarogram of a mixture of 0·92 mM nitric acid (added as pyridinium nitrate; wave A) and 0·60 mM 6-mercaptopurine (waves B and C) in pyridine (0·1 M in Et$_4$NClO$_4$) (87).

2. Proton Activity

The most readily controllable factor in obtaining specificity of measurement in polarographic organic analysis is usually the pH of the aqueous or partially aqueous test solution. A classical example is the simultaneous determination of maleic and fumaric acids, which depends on the difference over the pH range in variation of $E_{1/2}$ with pH for the two compounds (Figure 5) (89), and which is one of the frequently used polarographic methods in organic analysis as indicated elsewhere in this chapter.

The factors involved in the variation of the $E_{1/2}$ of organic compounds with pH and the analytical utilization of such behaviour has been discussed (90).

An important aspect of the study of organic electrode processes has been the elucidation of pH-dependent processes by investigation of such processes in proton-poor solvents, coupled with the controlled addition of proton donors such as phenol and benzoic acid, e.g. discussion of Hoijtink's pioneering interpretation by Kolthoff (65) and reference 75a.

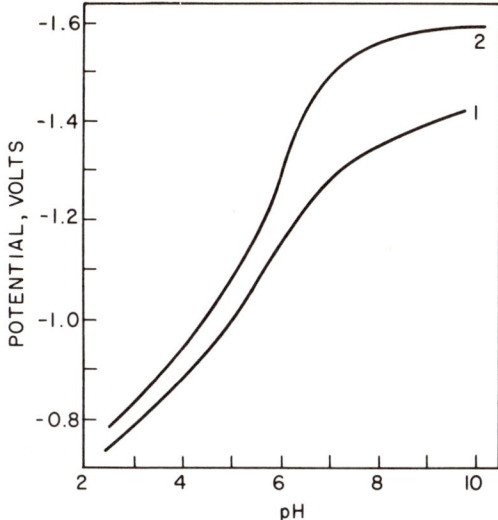

Fig. 5. Variation of polarographic half-wave potentials
with pH : 1, maleic acid ; 2, fumaric acid (89).

(a) *Buffer systems.* Simplicity of buffer system composition is desirable
from the viewpoint both of ease of preparation and of avoidance of interaction between buffer components and the organic species to be measured.
For example boric acid is usually an undesirable constituent due to its
tendency to react with carbonyl and vicinal hydroxyl groups, which often
results in the removal of waves; in some cases, borate can be advantageously
used as a masking agent to inactivate an electroactive compound in order to
be better able to observe the wave due to some other compound. An example
of the use of boric acid to secure favourable polarographic waves is in the
determination of ninhydrin (indantrione hydrate) (91). However, undesirable interactions of ninhydrin with borate buffers are reported in more
recent work of Holleck and Lehmann (91a).

Occasionally, interaction of a solution component with the desired constituent or its reduction product may be quite helpful, e.g. cyclooctateraene
gives well-defined waves proportional to concentration only in the presence
of tetramethylammonium ion (92) and non-reducible carbonyl compounds
may form reducible imines with ammonia or primary amines, e.g. references
93 and 94.

The elementary fact is worth emphasizing that a buffer system, at the usual
concentrations used, generally provides adequate capacity in the pH range
of $(pK_a - 1)$ to $(pK_a + 1)$, where pK_a represents the negative logarithm of
the acid dissociation constant of the active component of the buffer system,
e.g. an acid of dissociation constant 10^{-6} can be used with its sodium or
potassium salt to provide buffering action over the pH range of 5 to 7. In

addition, the buffering system must have sufficient capacity and speed of reaction to keep the hydrogen ion concentration at the interface constant; this is partially assured by making the solution 0·1 M or greater in the buffer system and by avoiding known sluggish buffer systems. (Buffer capacity is obviously a function of the relative concentrations of electroactive species and buffer.)

Even with perfectly adequate buffers, the slowness of dissociation and association of electrode reactants and products may be the cause of the appearance of several waves.

It is, unfortunately, still true that, in the development of analytical procedures, trial and error is often the only means of selecting the optimum buffer. However, this selection can be made more rational by first investigating the behaviour of the electroactive species over a reasonable range of pH. The matter of buffer system selection is obviously connected to that of the background electrolyte, which is discussed next.

In attempting to use a published polarographic method of organic analysis, it is important that the same buffer system as that recommended by the original investigators be used. If a different system is to be used, it should first be examined in reference to the electroactive species involved to see whether it will yield the desired results.

3. Background Electrolyte

The background electrolyte and other solution components, such as surfactants, may markedly affect the observed polarographic behaviour of organic compounds through, among others, the following factors and phenomena: (a) composition, structure and potential gradient in the double layer, (b) competitive adsorption on the electrode of electroactive species, intermediates and products, and other solution components, (c) activities of electroactive species, intermediates and products as affected by ionic strength, equilibrium constants, etc., (d) solvation of such species, (e) their complexation, including ion-pairing, protonation and other Lewis acid–base adduct equilibria and charge-transfer complex formation, (f) proton activity and concentrational stability, e.g. buffering and (g) kinetics of the various equilibria listed as well as of others involving the electroactive species, e.g. its formation from a more stable solution component.

Reference was made in the previous sub-section to the use of proper background electrolyte components to produce more favourable wave patterns, e.g. to produce more readily reduced imines from ketones by using amines in the background electrolyte.

An important potential result of the studies of the polarographic behaviour of organic compounds in non-aqueous media is a clearer picture of the role of background electrolytes. Often, a change in background electrolyte cation results in a marked alteration in observed current–potential relations with the alterations being at times assigned on rather slim evidence to formation of stable ion pairs, insoluble compounds, orientation of reactants in the double layer, etc. It is hoped that continuing fundamental studies may afford more detailed models for electrolyte effects, e.g. in respect to specific solute–solvent interaction. Use of non-aqueous media and proper background electrolytes for analytical methods may possibly result in increased specificity and perhaps even in increased sensitivity.

(a) *Ionic strength.* Ionic strength has been moderately thoroughly studied as a factor in the polarographic behaviour of organic compounds. One of its main roles is its effect on the rates of equilibration of organic systems, which partake in the reaction, such as an acid–anion one, presumably through its influence on the activities of the species involved.

Attention should be given in the development of polarographic analytical procedures to the effect of ionic strength, especially in view of the usual practices in preparing and using buffered solutions. Customary references on the composition of buffer systems are mainly concerned with the pH of the resulting solution with little or no attention being paid to the effective ionic strength. In many buffer systems, e.g. those involving polyprotoic acids, the ionic strength of the solution will change considerably over the recommended optimum pH buffering region.

(b) *Maximum suppressors.* It has been the author's experience that maximum suppressors should be used with the utmost of care in organic polarography. Frequently, troublesome maxima can be effectively removed or minimized by decreasing the concentration of the electroactive species to a value of about 0.1 mM; such concentration decrease is far preferable to the addition of a substance, which may affect other parts of the measurable phenomena in addition to the undesirable maximum. Use of dilute sample solutions has the added advantage of aiding the geometrical separation of adjacent waves.

C. Instrumentation

Probably the most significant development in techniques and apparatus for both research and analysis has been the increasing availability of operational amplifier-based systems for the convenient application of a variety of electrochemical approaches (95); a typical example is the completely transistorized polarographic system described by Bezman and McKinney (95a).

Other developments in instrumentation of particular relevance for the investigation and analytical utilization of organic electrode processes, include (a) three-electrode configurations, largely involving operational amplifier control systems, which have minimized to a considerable extent the problems associated with the large IR drop experienced on the polarography of organic solvent solutions of high resistance, (b) reliable indicating electrodes based on graphite, which have extended the potential range available for studying electrochemical oxidations, (c) cyclic voltammetry for the identification of reversible redox couples and electroactive intermediates, (d) alternating current polarography and (e) optical and magnetic resonance techniques for identifying electrode reaction products and intermediates.

Fisher (96) reviews elsewhere in this volume developments in instrumentation for dc polarography and coulometry.

1. Electrodes

Until relatively recently, study and utilization of the electrochemical oxidation of organic compounds was limited by the unavailability of an electrode, which would not itself be oxidized at the relatively positive potential necessary for investigating organic compounds, whose oxidation usually involves a high activation overpotential. This situation has been changed by the development of reliable indicating electrodes, based on various types of carbon including wax-impregnated spectroscopic carbon rods, mixtures of graphite paste with various organic solvents, pyrolytic graphite and glassy graphite, e.g. Adams' book on electrochemistry at solid electrodes (97) and the discussion of voltammetry with stationary and rotating electrodes by Piekarski and Adams (98). The negative potential range available at graphite is less than that at mercury, although greater than that at platinum; the extended positive range available at graphite, ca. 1·5 volts versus SCE, is a decided advantage.

The theory, methodology and applicability of graphite indicating electrodes have been reviewed (97–99a); Beilby (99b) has prepared a bibliography on carbon electrodes. Figures 6 and 7 show typical voltammograms obtained under different conditions for the analytical oxidation of organic compounds. Oxidation of guanine in $2 M$ sulphuric acid solution provides a highly specific method for its determination in the presence of adenine, cytosine and other naturally occurring purines and pyrimidines; a combination of mercury and graphite electrodes can be used for analysing mixtures of the latter compounds (100, 101). The analytical precision attainable is evident from the i_p/C ratios in Table II; the constancy of $E_{p/2}$ is also quite good (102).

The concordance of the potential data obtained at platinum, mercury and graphite electrodes for a reversible organic system (nitrosobenzene-phenylhydroxylamine) is indicated in Table III (103); if anything, the data seem to be too good when consideration is given to the variation in solvent composition. One advantage of the graphite electrode is evident; it could be used

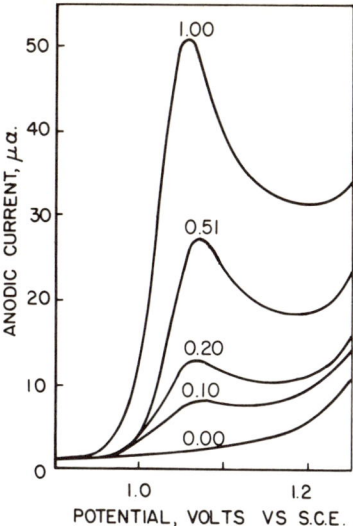

Fig. 6. Voltammograms of guanine in 2 M sulphuric acid at a stationary graphite electrode; millimolar guanine concentrations are indicated (100).

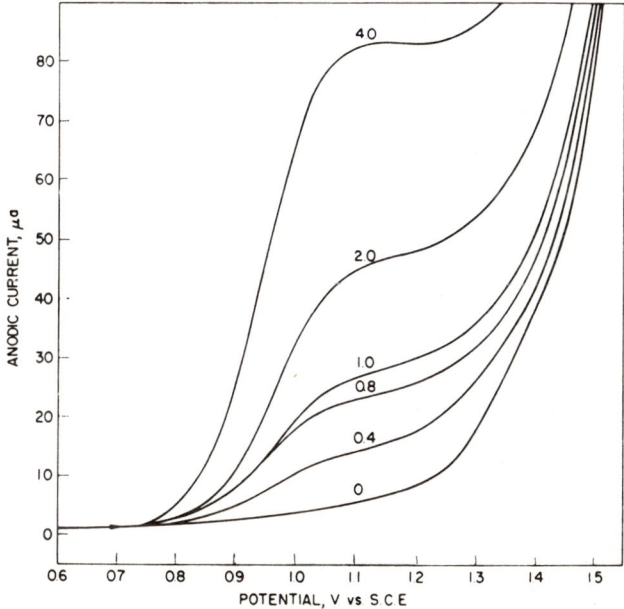

Fig. 7. Voltammograms of sulphanilamide at a stationary graphite electrode with the polarograph highly clamped. Rate of polarization: 200 mv min^{-1}. Background electrolyte: 0·5 M NaOAc + HOAc, pH 4·57. Millimolar concentration as indicated. Lower curve is a voltammogram of the background solution (102).

TABLE II

Anodic Voltammetry of Organic Compounds at the Graphite
Electrode[a] (102)

Concentration mM	i_p/C $\mu a/mM^{-1}$	$E_{p/2}$ volts
Adenine[b]		
1·04	47·6	1·07
0·55	46·7	1·06
0·36	48·6	1·03
Mean and standard deviation:	47·6 ± 1·0	
Ascorbic acid[c]		
4·00	12·7 ± 0·3	0·301 ± 0·011
2·00	12·9 ± 0·2	0·281 ± 0·014
1·00	12·6 ± 0·3	0·279 ± 0·013
0·50	12·4 ± 0·4	0·276 ± 0·014
0·10	12·6 ± 0·0	0·276 ± 0·008
Mean and standard deviation:[d]	12·6 ± 0·2	
Sulphphanilamide[e]		
4·0	20·5 ± 0·7	0·946 ± 0·005
2·0	20·0	0·954
1·0	20·4 ± 0·5	0·942 ± 0·007
0·80	20·8	0·938
0·40	20·0 ± 0·3	0·946 ± 0·009
0·10	16·8	0·948

[a] Rate of polarization: 200 mv min^{-1}. The polarograph was highly damped.

[b] Background electrolyte: 0·5 M NaOAc plus HOAc; pH 5·5.

[c] Background electrolyte: 0·1 M citric acid + 0·09 M Na$_2$HPO$_4$; pH 3·6. Mean and standard deviation of three runs using a different surface of the same graphite electrode for each run.

[d] Mean and standard deviation of the individual means for the different concentration levels.

[e] Supporting electrolyte: 0·5 M NaOAc + HOAc; pH 4·57. Values with standard deviations are the averages of five runs using a different electrode surface of the same graphite electrode for each run.

below pH 4, where oxidation of mercury at the potentials involved prevents use of the DME.

Mention should be made of the hanging mercury drop electrode, originally proposed by Kemula, which is advantageously used for cyclic voltammetry and for stripping analysis.

One respondent expressed a not uncommon attitude towards electrodes:

'There has been considerable interest in our laboratories in electrode systems other than mercury for voltammetric use. Noble metal electrodes have been extensively employed, and, although it cannot be classed as a

TABLE III

Potential Data for Nitrosobenzene–Phenylhydroxylamine Couple at Graphite, Mercury and Platinum Electrodes

Technique and electrode	Solvent	Potential[a] volts	Reference
Potentiometric titration; Pt	50 per cent acetone 0·1 N HCl	$E_c^0 = 0.337$	R. E. Lutz and M. R. Lytton, *J. Org. Chem.*, **2**, 68 (1938)
Polarography; DME	10 per cent EtOH 0·005 per cent gelatine pH 4 to 10	$E_{1/2} = 0.33 - 0.061$ pH	J. W. Smith and D. R. Waller, *Trans. Faraday Soc.*, **46**, 290 (1950)
Polarography; DME	Aqueous pH 6 to 10	$E_{1/2} = 0.339 - 0.0584$ pH	R. M. Elofson and J. G. Atkinson, *Can. J. Chem.*, **34**, 4 (1956)
Polarography; DME	52 per cent EtOH pH 6 to 13	$E_{1/2} = 0.323 - 0.0603$ pH	R. M. Elofson and J. G. Atkinson, *Can. J. Chem.*, **34**, 4 (1956)
Voltammetry PGE[b]	50 per cent EtOH pH 1·6 to 12·5	$E_{p/2} = 0.33 - 0.060$ pH	L. Chuang, I. Fried and P. J. Elving, *Anal. Chem.*, **36**, 2426 (1964)

[a] All potentials are versus aqueous SCE.
[b] Stationary pyrolytic graphite electrode.

noble metal electrode, we have considerable interest in the application of a silver electrode. We have had a high degree of interest in carbon electrodes, especially carbon paste and pyrolytic graphite electrodes. For analytical use, we have had particular success with a pyrolytic graphite electrode. At one time or another we have probably examined most of the electrodes suggested for voltammetric work.'

2. Electrochemical Techniques

Reference has already been made to the use of operational amplifier-based circuits. Mention should be made of several techniques which are becoming increasingly important, e.g. use of perturbation techniques such as cyclic and potential-step voltammetry may permit separation of the electron-transfer step from subsequent reactions of the reduced or oxidized species. Cyclic voltammetry, alternating current (ac) polarography and related techniques are thoroughly discussed by Brown and Large in a recent book chapter (103a). Comparative polarograms produced by dc polarography, ac polarography and cyclic voltammetry are illustrated in Figure 8 (104).

The value of cyclic voltammetry at constant area electrodes in mechanistic studies of electrode processes has been amply demonstrated; it facilitates identification of reversible redox couples and frequently of electroactive intermediates formed chemically or electrolytically, e.g. studies of the quinone-hydroquinone system (105) and of benzophenone (106) in pyridine.

Papers on the theory of stationary electrode polarography by Shain, Nicholson and others, e.g. references 103a and 107 to 110, have provided possible means for quantitatively resolving the electrochemical and associated chemical steps in electrode processes.

There is increasing interest in the use of alternating current or ac polarography and tensammetry in investigating organic electrode processes with some emphasis on the utility of this approach in defining adsorption on the electrode and other phenomena occurring in the double layer. One factor is the observation of ac waves for compounds, which were not expected to undergo reversible electrode processes, e.g. Figure 8, where it is apparent that the second wave is better defined on ac than on dc polarography. The theory of such ac polarographic waves for irreversible systems has been recently presented by Timmer, Sluyters-Rehbach and Sluyters (111, 112) and by Smith and McCord (113).

AC photographic theory and practice are helpfully reviewed by Smith (114), Breyer and Bauer (43, 115) and Sluyters-Rehbach and Sluyters (115a). The use of phase-selective detection (114, 115b to 115e) considerably enlarges the variety and precision of the information obtainable.

The relatively old technique of electrolysis at controlled electrode potential (116) is still generally essential in studying organic electrode processes,

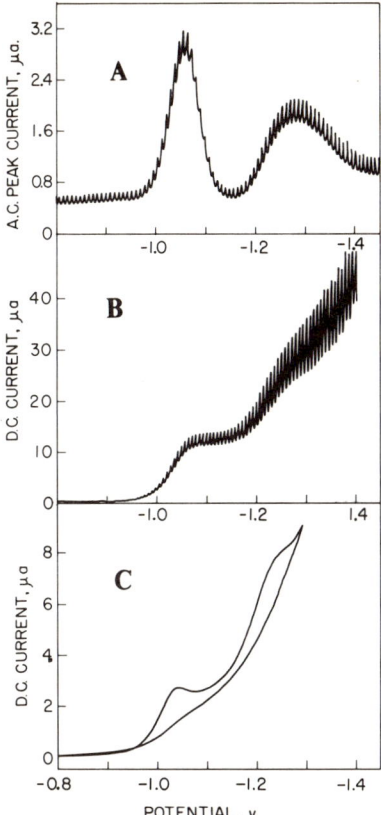

Fig. 8. Response of purine in pH 4·1 aqueous acetate buffer solution (ionic strength: 0·5 M) A: ac polarography at DME; dc scan rate, 2 mv sec^{-1}; ac amplitude, 10 mv peak-to-peak; frequency, 50 Hz; 2 mM purine. B: dc polarography at DME; scan or polarization rate, 2 mv-sec^{-1}; 2 mM purine. C: Cyclic voltammetry at hanging mercury drop electrode; scan rate, 20 mv sec^{-1}; 0·5 mM purine (104).

e.g. in preparing sufficiently large amounts of products to permit their isolation, characterization, identification and determination. The use of preparative electrolysis in connection with the elucidation of organic electrode reaction mechanisms has been discussed by Zuman (117). Controlled-potential electrolysis in analysis and research has been reviewed by Meites (118).

The products of transitory electrolysis at micro electrodes and of exhaustive electrolysis at massive electrodes are generally similar except where chemical rate phenomena can cause differences. Controlled-potential electrolyses have permitted the elucidation of complex electrode processes with unexpected products, e.g. reduction of the solvent when Al(III) solutions in pyridine are electrolysed (119), stereochemistry in the reduction of acetylenedicarboxylic acid (120) and dicarbonium ion intermediates in oxidation of uric acid (121) and adenine (122).

Two increasingly important applications of controlled-potential electrolysis are in the generation of free radicals for study by electron spin resonance (cf. next sub-section) and in the development of electrolytic methods of organic synthesis (cf. Section IV-F).

A considerable number of cells for polarographic examination of flowing solutions have been described (cf. references given in reference 122a and recent papers by Pungor, Nagy and Feher (122b); also references 150 and 151).

The possibilities inherent in the use of chronoamperometry, chronocoulometry and chronopotentiometry for analysis and for investigating organic reactions are thoroughly explored by Murray (122c).

The following quotations from respondents illustrate the wide variety in the extent of electrochemical techniques employed in laboratories, which do use polarography in organic analysis.

(a) 'We use conventional type instruments such as the Sargent XV, Sargent XXI and the Metrohm Polarocord in the analysis of drugs, food additives, etc., when the element or compound of interest is present in amounts large enough to be measured conveniently, i.e. 1 mg or more. Most of my investigations are in the field of pesticide residues and food additives when present in microgram amounts or less. In order to determine these compounds, we required more sensitive type polarographs and have found that the linear sweep type instruments do the job very satisfactorily. In our laboratories, we are using two linear sweep type instruments, the Polarotrace K1000 and Davis Differential CRP A1660.'

(b) 'At present there are seven polarographic instruments in the Research Laboratories. In addition, there are two instruments specifically constructed for chronopotentiometric work, two instruments designed only for cyclic voltammetric work, one instrument specifically reserved for controlled-potential coulometric work and six multifunctional electrochemical instruments. Approximately two-thirds of the instruments in existence are of operational amplifier circuitry; two of these are of fairly recent vintage and were locally constructed employing solid-state operational amplifiers. There are two additional multifunctional instruments presently in the design stage to be locally constructed with solid-state operational amplifier circuitry. Electrochemical cells of a number of different types are employed in this work, some of these for specific applications and thereby are unique in design.'

(c) 'There are currently no standardized analyses run in our division using voltammetric techniques other than dc polarography. We have often made use of chronopotentiometry and linear scan voltammetry at solid electrodes for special studies, and to help in the interpretation of phenomena observed in polarographic studies, but we have no statistical data available on the amount of work done by these techniques. We are not currently equipped for sine wave or square wave ac polarography, and have done no work with these techniques.'

(d) 'In regard to related techniques, we also offer cyclic voltammetry on a routine basis to our organic chemists as a tool for investigating the oxidation or reduction mechanisms of organic molecules. Again, this is primarily performed in non-aqueous solutions. Another service which we nominally consider as routine is a preliminary study to determine the feasibility of electrochemical synthesis of a product or intermediates of interest. The technique employed in this case is normally constant potential electrolyses. In addition, we engage in longer-range research projects designed to study and to aid in the development of electrochemical devices and processes. This work is not generally considered service work but it employs almost all of the electroanalytical techniques currently available. Examples of such techniques which we have used in the recent past include chrono-potentiometry, rotated ring disc electrode, MATR at transparent electrodes, in situ observation of electrode products by ESR and thin-layer chrono-potentiometry.'

3. Optical and Magnetic Resonance Spectroscopy

Adams (123) has reviewed the application of electron spin or paramagnetic resonance techniques in electrochemistry for the detection and identification of radical intermediates in electrode reactions (predominantly organic in nature), as well as for the insufficiently exploited study of the solution interaction of radical ions as redox intermediates; the electrochemical generation of radical ions is also reviewed. In general, it would seem that electron spin resonance will detect free-radical intermediates where the half-lives of these species are sufficiently long. However, where the free radical reacts either chemically or electrochemically, it may not be feasible to detect its presence. Consequently, the technique must be used with caution; there are frequent failures to detect free-radical formation and even to obtain patterns which are interpretable (124).

The coupling of voltammetry with electron spin resonance for the investigation of organic reaction mechanisms is discussed in this volume by Kastening (125).

The use of optical techniques for the detection and identification of intermediates in electrode processes by, for example, spectrum scanning of the solution-electrode interfacial region during electrolysis has included (*a*) ellipsometry to investigate electrode surface phenomena including film formation, (*b*) internal reflectance techniques to detect and identify electrode reaction intermediates in the solution–electrode interfacial region during electrolysis, (*c*) optically transparent electrodes for the concurrent examination of solutions during electrolysis by absorption spectrophotometry or internal reflectance spectroscopy and (*d*) examination of luminescence phenomena which accompany the electrolytic generation of certain free radicals. Kuwana and associates, for example, have applied optically transparent electrodes, made from 'doped' tin oxide coated glasses, to electrochemical studies using both normal transmission (126, 127) and internal reflectance spectroscopy (128, 129); transmission spectroscopy was used to evaluate the homogeneous chemical reaction following the charge-transfer step in the electrochemical oxidation of *o*-tolidine (130).

D. Measurement and Data Conversion

The critical factors involved in the preparation of the analytical test solution have been summarized (1–3), as has been the analysis of the polarographic curve, which in the case of the irreversible, drawn-out waves shown by many organic compounds does often cause problems, especially in the case of closely adjacent waves (1–3, 13).

The large potential (*iR*) drop due to the high resistance of many organic solvent solutions, when an appreciable current flows, results in a considerable distortion of the current–potential curve, which may necessitate the replotting of this curve with correction for the varying *iR* drop, before reliable measurements can be made. However, the availability of automatic correction for the *iR* drop in such high resistance solutions, including the use of three-electrode configurations, has minimized the problem.

If the current steps in the polarographic record are clearly demarcated, satisfactory data can be often obtained by a spot-reading procedure in which the current is read at definite potentials before and after the wave for both test and background solutions in order to obtain the current increments due to electroactive species (cf. application in analysis of mixtures of formaldehyde and acetaldehyde (35)).

Of the commonly used general techniques for converting numerically obtained data to composition (percentage fraction or absolute amount) (cf. reference 2), the *standard series* method finds the widest use in polarography with a linear current–concentration ratio commonly being found for organic compounds between the concentration limits of 0·1 and 1 m*M*. The *standard addition* method has been increasingly used, especially in connection with the determination of infrequently measured compounds. The *internal standard* method finds little application, which is also the situation with the use of a *fundamental law calculation*, here, of course, the Ilkovic equation. The latter, though, finds a number of semiquantitative uses, especially in view of the fact that the diffusion current constant at the DME for many organic compounds in a variety of solvent-background electrolyte media is about two per electron transferred. *Empirical methods* of calibration are occasionally used, e.g. for catalytic waves or repression of the oxygen maximum; such approaches are usually very sensitive to experimental conditions and to small variations in the sample composition.

The accuracy and precision normally obtainable in the polarographic determination of organic compounds is comparable to that obtained with other physical and chemical methods of functional group estimation, i.e. errors are of the order of 1 to 3 relative per cent.

1. Analysis of Mixtures

If the waves for a mixture are far enough apart on the potential axis, e.g. 0·15 to 0·25 volts for typical irreversible electrode processes, their measurement is relatively simple and care need only be taken to correct currents for all waves at the DME after the first for the effect of the electrocapillary curve upon the currents, if the latter are measured at potentials more negative than ca. −1·2 volts, where the decreasing drop-time of the capillary may introduce a significant effect.

Where the waves are of widely differing current magnitudes, the usual techniques of current compensation and change of current recorder sensitivity can be used to make measurement of the waves more precise.

Where waves are close in $E_{1/2}$, resolution can be increased by making the sample concentration in the test solution as low as possible since the potential span covered by the rising portion of the curve depends upon the concentration of the electroactive species; this is of considerable importance where several waves of low slope appear when examining an organic sample, e.g. reference 131. Variation in experimental conditions is often of help, e.g. selection of proper pH, when $d(E_{1/2})/d(pH)$ for the compounds differ.

Awareness of the extent of the shift of $E_{1/2}$ with structure and substitution as well as with pH and ionic strength will help the analytical chemist (a) in deciding whether certain multicomponent mixtures can be analysed and (b) in so altering conditions as to make analysis possible by shifting $E_{1/2}$ values and thereby minimizing interference due to merging or superimposable waves. The fact that the reduction potential of the disulphide group linked to a phenyl group is −0·5 volts and to an alkyl group is −1·25 volts (132), permits both qualitative and quantitative analysis for such compounds.

The use of derivative and differential techniques to increase wave resolution has been frequently described.

Subtractive polarography, which employs two cells, one containing the sample solution and the other having an identical composition except for the absence of the compounds to be measured, has long been suggested for analysis, but has found essentially no acceptance due to the difficult feat of synchronizing two dropping mercury electrodes. The use of a commercial differential instrument, employing drop-rate controllers for this purpose, has been described (133).

A series of studies by Schaap, Shults and collaborators (134–137) explored the analytical possibilities currently available from the use of differential dc polarography; limits of accuracy and precision were improved under optimum conditions to 0·1 per cent.

Israel (138) investigated the use of matrix algebra to resolve overlapping polarographic waves and derivative polarographic curves obtained for multicomponent systems. While concentrations calculated for two-component systems were moderately satisfactory, unsatisfactory results were obtained for a three-component system (mixture of *o*-, *p*- and *m*-nitrobenzoic acids with a $E_{1/2}$ span of 80 mV).

(*a*) *Equilibrium systems.* The occurrence of mixtures of equilibrium forms of a single substance is a potential source of error in analytical methods based on polarography. Typical equilibrium systems include keto–enol, acid–anion, base–cation, lactam–lactim and carbonyl-*gem*-glycol. The half-wave potentials of the two forms may differ considerably or one form may be non-electroactive within the available potential range; cognizance must be taken of such situations if significant analytical data are to be obtained.

If the interconversion rate of one form to the other is very rapid, only the wave of the more readily reducible form will be measured in the usual DME technique. However, if this rate is very slow compared to the rates of reduction, two waves may be obtained, one for each equilibrium form. The wave at less negative potential will represent the equilibrium concentration of the more easily reduced form plus a contribution from that of the other form, which would have been converted in the electrode region to the more easily reduced form during the mercury-drop life-time as the latter form is removed via reduction.

Since the interconversion of forms is like all other kinetic processes quite sensitive to temperature and the composition of the environment, the only safe analytical practice in dealing with an equilibrium system is to measure the sum of the waves due to the participants in the equilibrium. Even this practice is open to error if the diffusion coefficients of the two forms are appreciably different and if appreciable concentrations of the forms coexist.

The possibility of regulating conditions for satisfactory analyses is illustrated by the polarographic determination of oxaluric acid (139), where the undissociated acid is reduced in a pH-dependent process between pH 1 and 5, and the anion is reduced in a pH-independent process above pH 7.

2. Preliminary Separation

A considerable number of suggestions have been made for combining polarographic measurement with various types of physical separation such as extraction or column chromatography, usually preliminary to polarography, e.g. the coupling of polarography with paper chromatography to yield quantitative procedures suitable for quality control of pharmaceutical products (140). The purpose is either to secure greater specificity in the polarographic measurement by removing interfering substances or to permit data to be obtained due to fractionation effects which will allow the multicomponent analysis of a single sample. Strictly speaking, these approaches fall outside of the scope of a discussion of organic polarographic analysis, especially if the separation is preliminary to the polarography, since in many types of analysis the measurement step is much less difficult and time-consuming than the steps needed to obtain a sample, in which the desired constituents can be measured free of interference.

Polarographic observation of the effluent of chromatographic columns as a conjoint approach to analytical problems has been considered in a series of papers by Kemula, e.g. reference 141. In a typical application, residue analysis for parathion and methylparathion on crops has been described by Koen and Huber (141a).

3. Use of Computers

Although it is not specifically connected with organic polarography applied to analysis *per se*, reference must be made to rapidly developing interest in the use of computers not only for handling numerical computations based on data fed to them, but also for programming and otherwise controlling laboratory experimental arrangements, and, to some extent, evaluating and interpreting the resulting data, as well as for controlling more or less completely automated analysis systems, including the taking of proper remedial or other action on process units as samples are analysed (the latter may, of course, be a process stream).

Several recent descriptions of the use of digital data acquisition techniques in electrochemical measurement, e.g. references 142 to 144a, were followed by the logical development of including the digital computer as part of the experimental arrangement, which is facilitated by the availability of small computers, e.g. the general purpose laboratory data acquisition and control system described by Lauer and Osteryoung (145), which can be used with most analytical laboratory instruments; the examples described are centred on polarographically related equipment, e.g. for chronocoulometry (cf. also Lauer, Abel and Anson (146)).

As a step toward the complete automation of voltammetric procedures, Perone, Harrar, Stephens and Anderson (147) described an instrumental arrangement and program for fast-sweep derivative polarography with a controlled short drop-time DME. Subsequently, on-line computer process-ing of voltammetric data, including the numerical deconvolution of over-lapping stationary electrode curves, was described (147a, 147b).

Mohilner and Mohilner (148) have described a digital computer method for the thermodynamic analysis of electrocapillary data; data for the adsorp-tion of pyridine on mercury from aqueous solution were used as an example. Kinetic parameters for rapid electrochemical processes, based on potential-step measurements, are advantageously calculated by computer (148a).

III. QUANTITATIVE ORGANIC ANALYSIS BY VOLTAMMETRY

No attempt will be made to review any significant fraction of the litera-ture dealing with the use of polarographic techniques for determining organic functional groups and organic compounds. Instead, an attempt will be made to survey the principal types of analytical measurement from the viewpoint of their actual and potential utilization in organic chemistry. Consequently, statements may be made which seem at variation with fact, e.g. reference to 'some use is beginning to be made' when the published literature contains descriptions of a large number of suggested procedures for the purpose.

A. Determination of Concentration

Conventional polarography with the DME is still the most commonly used polarographic method for the determination of concentration, usually on a known weight of sample dissolved to a known volume, which permits the calculation of percentage composition by weight after suitable calibra-tion.

Some use is beginning to be made of graphite electrodes for concentra-tion measurement, especially in connection with the oxidation of non-reducible organic compounds. Little or no significant use for quantitative analysis is being made of such phenomena as the one-electron electro-chemical reduction and oxidation of aromatic hydrocarbons in non-aqueous media, on which there is a copious literature.

The general situation seems to be that the organic functionalities, which are susceptible to electrochemical reduction, are sufficiently activated by adjacent structure as to produce chromophores of high absorptivity in the ultraviolet and/or chemical reactivity which readily allows formation of such chromophores. For the analysis of large numbers of similar samples, spectro-photometric absorption is often a more convenient method of analysis than polarography.

Many examples of quantitative analysis have been indicated in the quota-tions previously given, especially in Section I-C-2, and in the general discus-sion. Other examples follow.

'Our Chemicals Division is using polarography routinely for the following determinations: fumaric acid in maleic acid and maleic anhydride; maleic anhydride in crude phthalic anhydride; aldehydes in alcohols; coumarin in phthalic anhydride.'

The reduction of the nitro group, especially, when on an aromatic ring, is frequently advantageously used as the basis of an analytical method (cf. reference to introduction of the nitro group in Section II-A-1). For example, an 'assay based on dc polarography was developed for Ranidazole (1-methyl-5-nitroimidazole-2-methanol carbamate ester) in animal feeds as well as for drug residues in extracts of animal tissues. The requirements of specificity and sensitivity were fulfilled nicely by polarography. In this regard, I refer you to the following references: Dumanovic and coworkers, *J. Pharm. Pharmacol.*, **18**, 507 (1966). Several other organic nitro compounds have been examined by polarography successfully.'

Typical of the flexibility in analysis resulting from the variation of the ease of polarographic reduction with pH is seen in the determination of oxaluric acid (Figure 9) (139). Oxaluric acid (oxalic acid monoureide, carbamyloxamic acid) is polarographically reduced in a two-electron process apparently to glyoxyllic acid monoureide. The reduction occurs by two distinct mechanisms; the first, operative in acid solution (pH 1 to 5), is pH-dependent and involves reduction of the undissociated acid. Above pH 7, the oxalurate anion is reduced in a pH-independent process. Between pH 5 and 6, both processes occur, the heights of the respective waves depending on the extent of dissociation of the acid and the rate of recombination of anion and hydrogen ion. The precision of current measurement over the whole pH range is 1 to 3 per cent. From the viewpoint of quantitative analysis, the pH used can be selected in the first place on the basis of a pH where the oxaluric acid wave does not fall in the same potential region as waves due to other electroactive components in the sample. Where a choice of pH is available, that pH should be selected where the oxaluric acid wave has the most readily measured shape as well as a satisfactory i_1/C ratio, e.g. pH 0.75 would normally be preferable to pH 4.7 in the range of 0.2 to 1 mM oxaluric acid, whereas pH 4.7 might be preferable for concentrations below 0.2 mM because of the more nearly linear current-concentration relationship in that region. The pH region in which the wave is split, i.e. around pH 6, should generally be avoided for quantitative analysis.

A provocative example of what can be done with polarography is provided by the work of Gajan on pesticide residues; he is also currently investigating the use of polarography to determine trace elements in food. Gajan (8) maintains that 'in pesticide residue analysis we feel that polarography is an ideal technique to quickly confirm the identity of components isolated by the various multiple detection methods, e.g. GLC and TLC, now routinely used. Such verification can be made in less than 20 minutes in many cases.

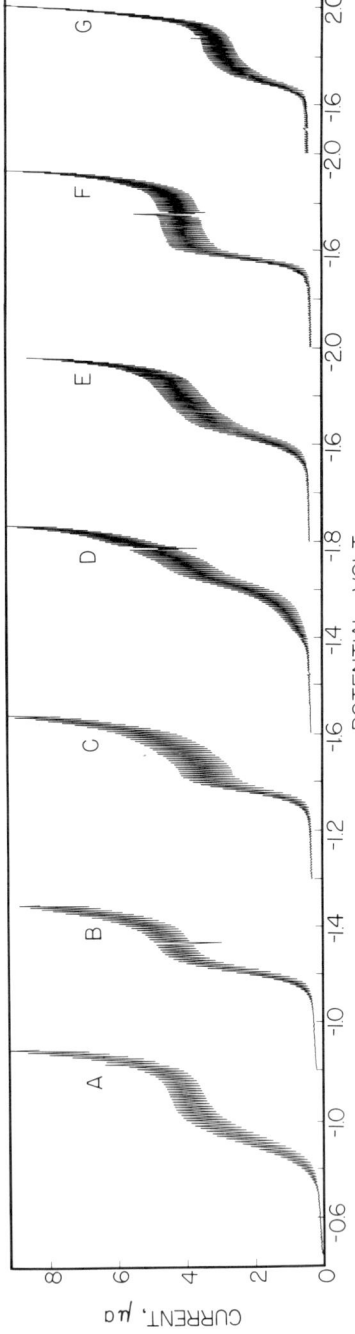

Fig. 9. Polarographic wave pattern for oxaluric acid (0·5 mM solution) over the normal pH range. A: pH 1·2 chloride buffer; B: pH 2·9 chloride buffer; C: pH 4·7 acetate buffer; D: pH 6·1 McIlvaine buffer; E: 7·9 McIlvaine buffer; F: pH 9·2 ammonia buffer; G: pH 11·2 chloride buffer (148b).

Table IV lists some of the pesticides we can determine polarographically.'
However, another investigator, while praising Gajan's work, pointed out
that, for the determination of the insecticides made by his company as plant
residues, he tends to use gas chromatography since he finds the 'clean-up'
problems are less severe with this technique than with polarography (cf.
discussion of polarographic determination of organic pesticides by Tachi
and Senda (149)).

TABLE IV

Pesticides Advantageously Determined Polarographically (8)

Pesticides	Reference
(1) *o-p*-DDT *p-p*-DDT Methoxychlor- Keltane	(1) R. J. Gajan and J. Link, *J. Assoc. Offic. Anal. Chemists*, **47**, 1120 (1964).
(2) Parathion Methyl Parathion EPN Paraoxon	(2) R. J. Gajan, *J. Assoc. Offic. Anal. Chemists*, **46**, 216, 220 (1963).
(3) PCNB (Pentachloronitrobenzene)	(3) A. K. Klein and R. J. Gajan, *J. Assoc. Offic. Anal. Chemists*, **44**, 712 (1961).
(4) Systox thiol Systox thiol sulphoxide Systox thiol sulphone Systox thiono Systox thiono sulphoxide Systox thiono sulphone Thimet Disyston	(4) R. J. Gajan, *J. Assoc. Offic. Anal. Chemists*, **45**, 401 (1962).
(5) Carbaryl	(5) R. J. Gajan, W. R. Benson and J. M. Finocchario, *J. Assoc. Offic. Anal. Chemists*, **48**, 958 (1965). W. R. Benson and R. J. Gajan, *J. Org. Chem.*, **31**, 2498 (1966).
(6) Malathion Guthion Diazinon Dimethoate Trithion Dinitro-*o*-cresol	(6) R. J. Gajan, not yet published.

Blaedel and Laessig (150) list several organic applications of amperometry, coulometry, coulometric titration and mercury electrode polarography, which have been considered or are being used for continuous

analysis (cf. pages 122–125 of reference). Continuous polarographic analysers suitable for continuous control and process regulation have also been reviewed by Novak (151).

1. Trace Analysis

The steadily increasing interest in trace analysis has been evident in polarography, difficult as it may be to define 'trace'. The percentage composition ranges, which are considered to be included, vary with the nature of the sample and the purposes for which it is to be used. In general, trace analysis covers the concentration range from 1 part in 10^4 to one part in 10^8 or even lower and absolute amounts in the microgram and lower range. Because of factors such as the effects of matrix and other components and signal-to-noise ratio in measurement, trace analysis usually demands a fairly sophisticated approach which often involves 'clean-up' and concentration separation steps preliminary to measurement. This situation is evident in the important area of pesticide residue analysis.

The tenuous nature of the semantic (and psychological) problem can be considered by examining the situation in conventional dropping mercury electrode polarography, which is among the less sensitive of polarographic techniques. If we are to measure a compound of formula weight of about 100 and use 10 ml of a test solution, which is 0·5 mM in the electroactive compound (a situation frequently encountered in normal analytical practice), the total amount of compound required for the determination is 0·5 mg. Without going to undue difficulties, the same determination can be made on 1 or 2 ml of test solution, which is 0·1 mM in the electroactive species, thus requiring only 10 to 20 micrograms of the compound. If the amount of sample available is truly a limiting factor, it may be possible to make the polarographic measurement on a single drop of solution (0·03 to 0·05 ml), which is 0·01 mM in electroactive species, for a total sample requirement of 30 to 50 nanograms of compound. The amount of the compound consumed during the polarographic measurement would normally be negligible.

Other electroanalytical techniques derived from polarography (cf. Table V) can be applied to solutions orders of magnitude less concentrated than those indicated and no greater in volume, thus moving the polarographic method down into the ultra–ultramicro range. For example, pulse polarography provides a many-fold increase in sensitivity over classical polarography; it has been applied to the estimation of nanogram quantities of proteins (151a). The limitation of detection of cystine in pure solutions, using its Co(II) activated catalytic hydrogen wave, is 10^{-8} M (39a).

In an investigation of the applicability of pulse polarography to organic analysis, Wolff and Nürnberg (152) recommended the use of derivative pulse methods and concluded that the sensitivity may reach 1 part in 10^9 or below

TABLE V

Quantitative Sensitivity Limits of Some Electroanalytical Methods (153)

Sensitivity limit M	Methods
$10^{-4}–10^{-5}$	AC polarography, chronopotentiometry, thin layer coulometry, potentiometry with metal-specific glass or membrane electrodes
$10^{-5}–10^{-6}$	Classical polarography, coulometry at controlled potential, chronocoulometry, tensammetry, precision null-point potentiometry
$10^{-6}–10^{-7}$	Tast polarography, derivative polarography, square-wave polarography, second harmonic ac polarography, phase-sensitive ac polarography, linear sweep voltammetry, staircase voltammetry, derivative voltammetry, coulostatic analysis, chemical stripping analysis
$10^{-7}–10^{-8}$	Pulse polarography, R.F. polarography, coulometric titrations, amperometry with rotating electrodes, conductivity (aqueous)
$10^{-7}–10^{-9}$	Anodic stripping with hanging mercury drop electrodes
$10^{-9}–10^{-10}$	Anodic stripping with thin film electrodes or solid electrodes

in favourable cases, e.g. nitro compounds. 'Differential pulse polarography has been successfully applied to trace analysis for ronidazole and dinitrocarbanilide, at concentrations down to 10^{-7} M' (152a).

Laitinen (153) and Taylor, Maienthal and Marinenko (154) have reviewed the application of electrochemical methods to the determination of substances at the trace level with emphasis on polarography and voltammetry and derived techniques, which offer special advantages for the measurement of very low concentrations and very small amounts of substances, e.g. ac polarography, chronopotentiometry, coulometry, coulostatic analysis, derivative polarography, differential polarography, pulse polarography, square-wave polarography, stripping analysis, strobe ('tast' or intermittent) polarography, chronocoulometry, cyclic voltammetry, voltage step chronoamperometry, etc., as well as the use of catalytic currents, adsorption effects, etc.

The reader interested in trace analysis is urged to consult these two comprehensive reviews, in which, although the emphasis is on inorganic applications, there are some references to organic applications (cf. in this respect Table 1 (pages 397 to 399) of the Taylor chapter (154), in which methods for the determination of a variety of organic compounds at the microgram and p.p.m. levels are given). The applications to trace analysis of coulometry, polarography and voltammetry, and amperometry have also been discussed by Rogers (155), Delahay (156) and Reilley (157), respectively.

Table V, taken from Laitinen's chapter (153), illustrates the range of concentrational sensitivities of electroanalytical methods.

B. Titration

Rather extensive use is being made in organic analysis of titration methods involving polarographic detection of the equivalence point (*amperometric titration*) and electrochemical generation of titrant (*coulometric titration*) with amperometric or related, e.g. 'dead-stop' or other polarized electrode pair, equivalence point detection; measurement of compounds by direct electrolysis, usually at controlled potential (*coulometry*). in which the titrant is the passage of electrons, finds some use.

1. Amperometric Titration

The simplicity of amperometric titration apparatus and technique makes its use quite advantageous as does the possibility of polarographic determination of inert organic compounds by using electroactive titrants. Presumably, amperometric titration can be applied to any reaction involving an organic compound which is reproducible and which proceeds rapidly enough. In some cases, substances which react slowly with the reagent can be determined by addition of an excess of reagent and titrimetric measurement of the unreacted reagent.

In any amperometric titration involving an organic compound, where the measurement is to be partially or entirely based on the wave due to the organic compound, it is obviously desirable that the basic polarographic behaviour of the compound be thoroughly studied in order to determine the optimum conditions for the titration process. For example, examination of the polarographic behaviour of the familiar organic reagent, cupferron, revealed the possibility of using it as a selective electroactive titrant for polarographically inert zirconium (158, 159).

Among the more useful titrants for organic compounds is potassium bromate, which, on addition to an acidic bromide solution, generates bromine which reacts with unsaturated and phenolic compounds. The first excess of bromine produces a current which can be detected voltammetrically at a rotating platinum microelectrode. Styrene can be determined in 75 per cent methanol solution with an accuracy of 0·5–1·0 per cent; amounts as small as 0·2 mg can be measured (160). Phenol and phenolic derivatives such as 8-quinolinol (8-hydroxyquinoline) and p-aminosalicyclic acid can be determined with an accuracy of 0·5 per cent (161, 162).

Silver nitrate is another useful titrant, especially for sulphhydryl groups in mercaptans and other sulphur-containing compounds. Originally described by Kolthoff and Harris (163), the technique has been applied in a variety of ways, even on a microgram scale (164). Mercaptans can also be titrated with iodine solution (165). Other typical adaptations of organic reagents to amperometric titration are the titration of aromatic diamidines with alizarin sulphonic acid (166) and of couplers such as pyrazolone or a naphthol with a diazotized aromatic amine (167).

The amperometric titration of organic compounds, as well as the more extensive literature on the use of electroactive organic compounds for the titration of inorganic species, is well summarized by Stock in his book on amperometric titration (168); amperometric titrations are also reviewed every other year by Stock (169) with a moderate number of titrations of organic compounds being listed. The possibilities inherent in the more extensive use of organic oxidations at graphite electrodes should allow extension of the application of amperometric titration, e.g. the oxidation of tetraphenylborate at graphite, which was originally applied for the amperometric titration of potassium with tetraphenylborate (170), has been applied to the titration of a wide variety of amines (171).

2. Coulometry and Coulometric Titration

These two techniques based on Faraday's laws of electrolysis and polarographic techniques, offer many advantages for organic analysis, which accounts for their increasing utilization. Coulometry and coulometric titration are reviewed every other year by Bard (172) and by Davis (172a); numerous organic examples are always cited.

Coulometry, involving electrolysis at controlled potential, is based on selective oxidation or reduction of the desired constituent at an electrode and on measurement of the amount of the constituent reacted via measurement of the total current flow with a coulometer, which can conveniently consist of a small series-connected resistor, the potential drop in which is converted to a frequency and counted. As previously mentioned, such an exhaustive method, which is based on chemical reactivity, allows in principle and often in practice a precision of measurement which exceeds by one or more orders of magnitude that possible by measuring a physical property of a system involving the constituent. It is possible by coulometry at controlled potential to determine trichloroacetic acid in the presence of an eight-fold excess of dichloroacetic acid with an accuracy and precision of 0·2 per cent (173).

It was pointed out by a respondent that coulometry, 'where applicable to a given problem, can certainly "hold its own" in regard to precision with almost any other physical-chemical approach to purity determination'. Another respondent stressed that, 'despite the fact that we are not using polarography, we are making extensive use of other electrochemical techniques such as microcoulometry'. Amylase can be satisfactorily determined in serum for clinical analysis by coulometry (173a).

The principles and applicability of controlled-potential electrolysis are summarized by Lingane (174), Meites (116) and Bard and Santhanam (174a). Specific reviews cover the applicability of coulometry to clinical analysis (174b), analysis of drugs (174c) and general industrial analysis (174d). In an interestingly designed rotated mercury coulometric cell, the effect of the background current is eliminated by digital normalization (174e).

Coulometric titration methods, usually based on constant current electrolytic generation of titrant with time required for reaction being the measurement, are more generally applicable to analysis, if only because of the vast existing literature on titration of organic compounds as indicated by Ashworth's monumental compilation on that subject (175). However, actual applications have been largely restricted to silver precipitation and bromination reactions.

A major area of application is in coulometric titration of halogen, employing silver generator electrodes and an amperometric silver electrode endpoint circuit, for the determination of halide ion produced on combustion of organic halogen-containing compounds, e.g. by an oxygen flask method. Commercial units are now available for this titration, as well as for the coulometric titration of mercaptan with electrolytically generated silver(I). As an indication of the achievable accuracy with a commercial titration unit, samples of National Bureau of Standards chlorobenzoic acid (22·65 per cent Cl by theory), when analysed at intervals of several months by different analysts, gave values of 22·66, 22·49, 22·72 and 22·63 per cent.

Another major area of application of coulometric titration involves the generation of bromine for the determination of unsaturation in organic samples; a dead-stop end-point is usually employed. As mentioned previously, this has been accepted as a standard method by the A.S.T.M. (9).

C. Measurement of Reaction Rates

Polarographic measurement has been generally applied for the measurement of three types of reaction kinetics. One type involves moderately slow homogeneous solution reactions, for which polarography is used to monitor bulk solution concentration of electroactive species. For this class of applications, polarography is one of many competing techniques with the possible advantage of being applicable in the millimolar region. Another type of application involves the unique nature of polarography and is applicable to rapid reactions, which can be made to occur at the electrode surface, due to the disturbance of an equilibrium by electrolysis of one component and the resulting reestablishment of equilibrium, e.g. measurement of the recombination rate of hydrogen ion and an acid anion due to electrolysis at the applied potential of only the more readily reduced acid. The third type concerns the rates of heterogeneous reactions involving electron transfer across the solution–electrode interface.

1. Homogeneous Kinetics

Polarography has long been used in homogeneous kinetic studies as a means of measuring changes in concentrations of reactants and/or products either by repeated scanning of the necessary potential range, if the reaction were slow enough or the potential sweep technique used were rapid enough, or by following concentration change amperometrically at a fixed applied potential. Many, probably a majority of the reactions studied have been organic in nature. By 1950, the volume of literature on the use of polarographic measurement in chemical kinetics was sufficient to justify a 167-page review by Semerano (176), covering isomerization, elimination, addition, autooxidation, polymerization and depolymerization, dissociation, decomposition and other types of reactions. Zuman (177) has reviewed the study of the kinetics of homogeneous organic reactions by polarographic measurements. Kies (177a) and Janata and Mark (177b) have reviewed the use of controlled-current coulometry for the investigation of reaction kinetics.

Typical of what can be done with polarography are kinetic studies of the 2-nitro-1-butyl acetate-2-nitro-1-butene-2-nitro-1-butanol reaction sequence (178), the Beckmann rearrangement of 6α-methyl-17α-acetoxyprogesterone-3-oxime (178a) and the reduction of dimedone (178b). Another example is taken from a respondent: 'Thus, we have used polarography rather successfully in a number of degradation kinetic studies. We encounter compounds from time to time whose degradation reactions involve one or more equilibrium reactions (usually hydrolytic). I am always concerned in such cases about upsetting such equilibria, particularly where colorimetric techniques are employed. Polarography in some cases has been very useful in providing in situ data.'

2. Kinetics Near the Electrode

The second type of application, for which new techniques and methodologies are constantly being devised, is the study of the mechanisms and kinetics of chemical reactions, which may precede, accompany or follow electron-transfer processes and which may be studied by their effect on the observed polarographic currents. The classical example is the reduction of the protonated form of a species, in which the current is controlled not only by the bulk concentration and diffusion coefficient of the unprotonated species and by the rate of the charge-transfer process, but also by the rate of protonation of the unprotonated species in the layer next to the electrode as the more readily reduced protonated species is reduced,

$$H^+ + A^- \rightleftharpoons HA \tag{2}$$

The basic theory of such polarographic kinetic currents at the DME has been reviewed by the original investigators of such currents, Brdička, Hanuš and Koutecky (179). The subject of kinetic waves is thoroughly discussed in Mairanovskii's book on the subject (39c). Nürnberg and Wolff (179a) have critically examined the effect of various factors on homogeneous chemical reactions occurring in the diffuse double layer as examined by various polarographic techniques. The determination of the kinetics of fast reactions in solution by electrochemical methods has been reviewed by Bewick, Fleischmann and Hiddleston (180).

Oglesby, Johnson and Reilley (181) measured the kinetics of the benzidine rearrangement by four polarographic techniques (step-current reversal chronopotentiometry, reverse-ramp chronopotentiometry, thin-layer chronopotentiometry and thin-layer potential step electrolysis), based on the reversible electrochemical reduction of azobenzene to hydrazobenzene and the acid-catalysed rearrangement of the latter compound.

The development of the theory of observed current–potential curves for constant area electrodes under the condition of rapid polarization (high values of (dE/dt)) has included extensive work on measurement of the rates of chemical reactions accompanying electron-transfer processes, e.g. references 103a and 107–110. A nomenclature has evolved where E refers to an electron-transfer reaction and C to a chemical reaction. Thus, an ECE mechanism for a single polarographic wave refers to a process in which an electrochemical step involving electron transfer produces a species which then undergoes a chemical reaction to form a product unstable at the potential, at which it is formed, and consequently rapidly reduced in a second electrochemical reaction.

In a typical study using cyclic voltammetry (181a), pyrimidine, which is the model compound for the reduction of the biologically important purine-pyrimidine series, is initially reduced electrochemically in acetonitrile in a one-electron, reversible; diffusion-controlled process to produce an unstable radical anion, which is quickly deactivated via two competitive pathways: (a) fast dimerization to form a pyrimidine anionic dimer (probably the 4,4′ species) with a rate constant of $8 \pm 5 \times 10^5$ l/mole sec (EC process), and (b) proton abstraction from residual water (rate constant of ca. 7 l/mole sec^{-1}), which is rapidly followed by a further one-electron reduction to produce ultimately 3,4-dihydropyrimidine (ECE process). Addition of acid produces a new wave at less negative potential, which is attributed to the one-electron reduction of an N-protonated pyrimidine species to produce a free radical which dimerizes very quickly, probably to the 4,4′ compound;

this wave is very similar to the first one-electron wave of pyrimidine in aqueous media.

Other examples of such measurements involve carbanion protonation as well as dimerization and other free radical reactions (257–260).

Ashley and Reilley (182) have reviewed chemical kinetics in electrochemical processes.

3. Heterogeneous Electrode Kinetics

The third type of polarographically measured kinetics involves determination of the heterogeneous rate constants and transfer coefficients for electrode reactions; measurements on organic species have been relatively scanty. The difficulties in applying the Koutecky approach to organic compounds have been discussed (71, 183–185). Recent developments in electrochemical relaxation and other cyclic techniques offer hope for the revival of interest in the accumulation of adequate data on the subject; Reinmuth (186) and Roe (186a) have reviewed the use of relaxation techniques to study electrochemical reactions (cf. use of non-linear relaxation methods for the study of very fast electrode processes (187) and review of the study of fast electrode processes by relaxation methods (188)).

The standard reference on electrochemical kinetics is Vetter's book (189), an English translation of which is available (190, 191). Damaskin's brief book (192) is a helpful summary of the principles underlying the methods used to investigate the kinetics of fast electrode processes. The kinetics of electrode processes are fully treated in the significant books by Conway (193), Delahay (53) and Reynolds and Lumry (194), the group of stimulating papers on chemical and electrochemical electron-transfer theory by Marcus, e.g. reference 195, and many chapters in the three series on recent developments in electrochemistry edited by Bard (196), Bockris (197) and Delahay and Tobias (198). The effect of adsorption of organic compounds on their electrochemical kinetics has been considered (199).

IV. QUALITATIVE ORGANIC ANALYSIS BY VOLTAMMETRY

'Qualitative organic analysis' is being interpreted in the present context as including not only the identification and characterization of the structure of organic compounds, for which purposes polarographic techniques find at best only limited use, but also as including (a) the use of characteristic potentials, e.g. $E_{1/2}$, as indices to the chemical, physical and biological properties and behaviour of organic compounds and (b) the elucidation of electrode reaction paths or mechanisms as a guide to methods of synthesis and to chemical and biological reaction paths. The extent of the area of application of polarographic data just delineated is steadily growing, e.g. there is increasing reference to the use of polarographic behaviour, particularly $E_{1/2}$

values, to characterize organic, metalloorganic and metal–organic ligand complexes. The evidence is strong that the major contribution of polarographic organic analysis to the practice of chemistry may be in this area.

A. Identification and Characterization

The use of polarography for qualitative analysis in the restricted interpretation of identifying organic compounds is quite limited. Normally, a compound's $E_{1/2}$ is regarded as its identifying characteristic and use has been made of half-wave potentials in this respect when the expected population of compounds is small in number and of approximately known type. This restricted use is related to the nature of the polarographic half-wave potential.

$E_{1/2}$ for a typical organic compound involved in an irreversible electrode process is sensitive to experimental conditions and, as normally recorded, cannot be measured more precisely than ± 0.01 to ± 0.02 volts. Although this precision seems reasonable when compared to the range of 0.0 to -2.0 volts normally available at mercury electrodes, it does not seem as adequate when compared to the more restricted range of a few tenths of a volt in which compounds of the type under consideration might be reduced. In this respect, polarography might be compared with ultraviolet absorption spectrophotometry, since, as mentioned earlier, polarographically reducible compounds frequently absorb in the ultraviolet due to the presence of electronic conjugation. While the wavelength of an absorption maximum may not be measured with any better precision than $E_{1/2}$, the use of the *absorptivity* provides a second handle for a pure compound which is far superior to the limited information conveyed by the limiting current due to the small range of variation in diffusion coefficient and the limited number of integral values of the faradaic n.

Although $E_{1/2}$ is sensitive to experimental conditions, such characteristics as $d(E_{1/2})/d(\text{pH})$ are useful for identification only in special cases.

Occasionally, it is possible to use several features of a polarographic pattern, e.g. number of waves, relative magnitudes, $E_{1/2}$ values, effect of pH, etc., for the identification of a pure compound or its recognition in a mixture.

On the positive side, there is the fact that many electroactive groups are markedly affected in respect to their ease of reduction, i.e. $E_{1/2}$, by substituents on the molecular skeleton, e.g. the variation in ease of reduction of pyrimidines and purines (200) or of a given carbon–halogen bond (201, 202; footnote 1 in reference 203). In aliphatic compounds with a single functionality, the effect may be small, e.g. formaldehyde is more easily reducible than acetaldehyde and the higher saturated aldehydes, whose $E_{1/2}$ values are close together. More extensive treatment of the effect of substituents and structure on ease of reduction will be found in the discussions by Perrin (50) and Zuman (1b, 33).

Characterization of organic compounds is, then, one of the uses to which polarography could be put. However, few organic chemists actually use the technique for this purpose as is readily evident from the literature; this is principally due to two factors. Molecular spectroscopic techniques such as infrared absorption and nuclear magnetic resonance yield more bits of information about the compound, frequently about small discrete component portions of the molecule, which facilitate structure elucidation and resulting compound characterization. The other factor involves the failure to recognize the utility of the half-wave potential as a readily measured characteristic of a molecule as a whole (or at least in terms of the electroactive moiety and the surrounding molecular foliage). The result is that half-wave potentials can rarely be used as effectively as absorption maximum wavelengths in characterizing an active function and its immediate foliage, e.g. in the case of multiple conjugated unsaturation or in the shift of $E_{1/2}$ or λ_{max} with pH.

The use of $E_{1/2}$ for characterizing an organic compound is presently largely limited to members of series, for which quantitative relations between $E_{1/2}$ and some parameter, which is sharply dependent on structure (cf. Section IV-E), have been experimentally obtained; Zuman (204) discusses the use of experimental Hammett–Taft relations between $E_{1/2}$ and *sigma* values for deciding on the presence of certain substituents on an unknown member of an investigated series of known *rho*.

The straight-forward use of polarographic characterization of reaction products in complex mixtures is seen in the recognition of parabanic acid and alloxan among the products of the electrolytic oxidation of uric acid (121). The utility of polarography in the characterization of organic compounds is further illustrated by Abrahamson's studies of heterocyclic compounds (205, 206) and Acker and Hertler's study of 7,7,8,8-tetracyanoquinodimethan (207). Its use in characterizing the structure and chemical binding of metal complexes with organic ligands is evident from Olson's studies of transition metal complexes (208, 209).

B. Determination of Chemical and Physical Properties

The use of polarography to determine oxidation states, redox potentials, thermodynamic quantities related to potentials in the case of reversible systems, equilibrium constants based on concentration measurements, etc., is apparent, e.g. the structure, conformation, orientation and association of the adenine nucleoside–nucleotide series in solution and at the electron-transfer interface have been correlated with their electrochemical reduction (209a). Typical applications are illustrated by the following case histories and by the subsequent discussions on the polarographic determination of structure, of reaction paths and of correlations with other properties and reaction behaviour.

(a) 'We have used polarography and solid electrode voltammetry some-what routinely to obtain information about oxidizable or reducible groups in our physical-chemical studies of new compounds. We employ aqueous and non-aqueous systems (DMF, acetonitrile) in these studies. We feel that such information is of assistance to us in predicting the stability of the compound in contact with oxygen, metals, etc. If the oxidizable or reducible function is involved in hydrolytic or oxidative degradation of the compound, polaro-graphy can provide a suitable method in some cases for kinetic studies of the degradation reaction and for stability studies of the compound in dosage formulations.

We attempt to define the mechanisms of such degradation processes rather carefully in our physical-chemical studies to insure that whatever analytical methods are developed are suitably specific to follow these reactions.'

(b) 'We have obtained some data comparing the case of oxidation of the α-ketol group of a number of steroids with their ease of reaction with blue tetrazolium.

Another interesting investigation carried out in our laboratories con-cerned the reducible properties of a series of 3-keto steroids containing one or more conjugated double bonds, e.g.

(I)   (II)   (III)

This study was carried out in an attempt to develop methods for the deter-mination of I and III as impurities in II. Although distinctly different half-wave potentials were observed, they were too close to provide useful quantita-tive information. I suspect that first derivative curves may have been useful in this case. The problem was solved, however, by gas–liquid chromato-graphy.

We are presently investigating the polarographic properties of N-nitroso derivatives of a number of imidazoline compounds, e.g.

(c) 'In our work in the research department, virtually no use is made of polarography as a quantitative tool in organic analysis. The technique is used primarily as a qualitative tool to establish the oxidative and reductive stability of newly synthesized materials. Sometimes this is useful in patent considerations. In many other cases, it is merely information that is incorporated into publications to describe the physical characteristics of the compound. In these cases, the organic research chemist is interested in the reversibility of the oxidation or reduction, and the number of electrons involved, if reversible. Infrequently, we attempt to scale up and identify the reaction product. Much of the work is in non-aqueous solvents and acetonitrile is frequently the solvent of choice.

In earlier years, attempts were made to demonstrate, by more detailed work, the utility of polarography in characterization of organic materials. This work did excite some enthusiasm for the technique amongst organic chemists, and, during the period of cyanocarbon research, polarography was used to a considerable degree to relate half-wave potentials to strength of acids in a qualitative manner.

However, the current use of polarography in our shop is so limited as to be rather non-existing at the present time.'

(d) 'Our principal interests in polarography, as regards phenothiazines, is as a means of detecting structure-activity relationships. We have found a good correlation between anodic oxidation half-wave potentials and electronegativity of groups substituted on the 2-position in phenothiazine tranquilizers (cf. Kabasakalian and McGlotten, *Anal. Chem.*, **31**, 431 (1959)). Within limited series, there is also a correlation between these $E_{1/2}$ values and biological activity as tranquillizers (to be published). The relationship between resistance to oxidation and biological activity is not unexpected since these compounds may be subject to considerable oxidative attack and destruction during biological transport to the active site. It seems reasonable to presume that the members of a series most resistant to oxidative attack would be active at the lowest concentrations.'

C. Characterization of Chemical Reactions

The polarographic investigation of reactions is exemplified by the study (210) of the alloxan(IV)-alloxantin(V)-dialuric acid(VI) system,

which constitutes a reversible redox system of the semiquinone type,

$$\text{alloxantin} \rightleftharpoons \text{alloxan} + \text{dialuric acid} \qquad (3)$$

Investigation of the oxidation–reduction of the three compounds polarographically, coulometrically and by controlled potential electrolysis over the normal pH range with polarographic, spectrophotometric and chemical examination of the reaction product, served to rectify a number of incorrect statements in the literature concerning the system and the individual compounds. In particular, it was established that, contrary to the earlier literature involving potentiometric studies, alloxantin is only very slightly dissociated.

An interesting example of the application of polarography to the study of interactions in non-aqueous media, such as those involving donor–acceptor charge-transfer complexes, and the characterization of the products, is provided by Peover (211), who was able to detect the formation of intermediate species by polarography which could not be picked up by spectrophotometry (212). The equilibrium constants for the adducts determined polarographically are in agreement with those determined by other techniques, where the latter are also applicable.

Peover has also found linear correlations between the reduction $E_{1/2}$ for quinones and some anhydrides and cyanohydrocarbons, and the energy of the longest wavelength absorption band of the charge-transfer complexes of these compounds with suitable donor molecules (213, 214) (cf. Section IV-E on correlations).

D. Elucidation and Correlation of Reaction Paths

Some mention has been made of the use of polarography to identify intermediate and final products in chemical reactions. A potentially more important area of application involves utilization of the knowledge, which is being accumulated on the reaction paths followed by electrochemical reductions and oxidations, to interpret chemical and biological redox reactions. A great deal of attention, as already indicated, is being given to the investigation and interpretation of organic electrode processes with increasing refinement of the intimate details of the mechanism as developments in methodology and electrical and auxiliary instrumentation make this possible.

Consequently, for example, greater attention should be given by electrochemists to the investigation of the electrochemistry of biologically significant compounds because of the provocative possibility of the correlation of the mechanisms of electrochemical and biological processes, which is related to the fact that electrolytic oxidations and reductions occur under conditions resembling those of enzymatic and other biological transformations:

1. Both types of processes involve heterogeneous electron-transfer reactions, in which the nature of the electrical double layer and adsorptive phenomena are important factors.

2. Both occur in dilute aqueous solution of generally similar pH range with electrolyte compositions of comparable ionic strength.

3. The temperature range involved is also more or less comparable between electrochemical studies at 25° and biological reactions at animal body temperature.

4. The mass transport processes of convection and diffusion are operative and are not too dissimilar for both types of processes.

The desirability of electrochemical study of biological compounds is further indicated by the viewpoints (a) that, since biological processes are frequently associated with electron-transfer reactions, information in respect to the energy levels and reaction paths involved in the latter is of obvious utility, and (b) that polarographically determined half-wave potentials and electrode mechanisms provide such data for aqueous solutions under conditions resembling those frequently encountered in biological systems; polarographic potentials also provide readily measured indices for evaluation of theoretical reactivity and structural approaches for calculating the energy-determining step in electron transfer.

Quantum chemists concerned with correlating the biological, chemical and physical properties of biologically important compounds with structural and other theoretically calculated characteristics of such compounds have used polarography as a means of evaluating some of their calculations (cf. Pullman and Pullman (215)). Since these quantum chemists have given a great deal of effort to correlating calculated energies with chemical and biochemical activity for a variety of molecules, a successful correlation of such calculations with polarographic data will facilitate correlation of chemical and biochemical activity of compounds with their polarographic behaviour. It is clear, however, that, in order validly to correlate theoretically calculated data with redox potentials, the effects of adsorption, electron-transfer reversibility and solvation energy as discussed in Section IV-E, must be considered, since these may seriously alter the potentials associated with the processes and perhaps even the mechanistic route.

Postulation of the degree of correlation that may exist between electrolytic and enzymatic processes must at present be speculative; nevertheless, there is the similarity of conditions under which the two kinds of processes occur. Hopefully, knowledge gained in the investigation of the electrochemical behaviour of compounds of biological interest may allow evaluation of the extent to which correlations can be made. For example the electrolytic oxidation of uric acid (2,6,8-trihydroxypurine) has been shown (121) to proceed by a mechanism analogous to that postulated for the enzymatic oxidation, in which the uric acid is oxidized in a two-electron process to a dicarbonium ion, which then undergoes hydrolysis and rearrangement. Subsequently, adenine (6-aminopurine), which is one of the two principal purines found in nucleic acid and, in the form of adenosine triphosphate (ATP), plays an important role in many metabolic processes, was shown (122) to be electrochemically oxidized in a process involving a total of six electrons per molecule, which proceeds by sequential two-electron two-proton oxidations initially to 2-hydroxyadenine and then rapidly to 2,8-dihydroxyadenine (these processes are presumably of the type suggested for enzymatic purine oxidation). The latter compound is similar to uric acid except for the presence of an amino group in place of an hydroxyl group at the 6-position. Since ease of oxidation of purines increases with the number of oxygenated functions on the molecule, oxidation continues with further removal of two electrons, which results in oxidation of the 4,5-double bond and formation of a dicarbonium ion of the same type as that postulated for the electrochemical oxidation of uric acid. The dicarbonium ion intermediate, being unstable, undergoes further reaction. Thus, the electrochemical oxidation of adenine appears to follow initially the same path as the enzymatic oxidation.

The fact that analogous reaction paths have to be postulated to explain both the enzymatic and the electrochemical oxidation of two key compounds (adenine and uric acid) in biochemical processes, reflects a parallelism between the two types of processes, which supports the notion that study of the electrolytic mechanistic pathways and the associated potentials for the oxidation and reduction of biologically significant compounds may be of service in elucidating their biological behaviour in electron-transfer processes in respect to sites of oxidation and reduction, reaction paths, experimental conditions, effect of substituents and adsorption at the electron-transfer interface.

In this connection, it is worth recalling the past importance of electrical measurements, mainly by potentiometry, in understanding and utilizing the redox behaviour of organic compounds. However, due to the fact that few organic redox systems behave reversibly, valid potentiometric data have been obtained for only a limited number of compounds. Where such data have been systematically gathered and analysed, the results have been fruitful, e.g. the work of Michaelis, Clark and their collaborators, which was largely centred in the 1920's and early 1930's (cf. Clark's excellent book (216)). Since then, the bulk of the studies of the redox behaviour of organic compounds has been based on polarography, generally at the DME. Frequently, $E_{1/2}$ data on organic compounds, where they can be secured, are the only energetic data readily obtainable by electrochemical measurement. For this and other reasons, such data—including their variation with experimental conditions—should become of increasing interest in connection with various studies of the nature and properties of organic compounds.

E. Correlations Involving Polarographic Potentials

It is by now apparent that an increasingly important aspect of the study of organic electrode processes is the correlation of polarographic behaviour with environment, structure and reactivity. An unusual example of such correlation is provided by the following:

'Our laboratory, at one time, worked on the determination of condensed ring aromatics by polarography and I feel that this has much merit for certain applications; however, it suffers from the defect that, when a large number of reducible condensed ring aromatics are present, there is no readily distinguishable top to the wave and an arbitrary voltage had to be selected as the limiting voltage. Despite this drawback, the method did provide useful results in predicting the potential carcinogenicity of a given material.'

Correlation of polarographic $E_{1/2}$ values with numerical structural and reactivity characteristics is usually based on the implicit or explicit postulation that the characteristic $E_{1/2}$ of a compound is a function of electron density and other factors, which are also relatively simply related to some biological, chemical or physical property. The frequently resulting linear relationship between $E_{1/2}$ for a series of more or less closely related compounds and a suitably selected mathematical function of the values of the given property for that series of compounds permits (a) prediction of the magnitude of the property of a compound from its readily measured $E_{1/2}$ and (b) the rapid comparative evaluation of a property based on comparison of $E_{1/2}$ values.

A large variety of experimental and theoretically calculated properties and phenomena have been compared to polarographically determined potentials, e.g. ionization potentials, degree of carcinogenesis, transition energies of charge-transfer complex formation, triplet energies for photochemically excited states, nuclear quadruple resonance frequencies, wavelengths of spectrophotometric absorption maxima and antioxidant ability; the most extensively used correlations have involved various forms of the Hammett *sigma–rho* equation based on polar substituent quantities and the Taft modification; perhaps the potentially most significant have involved quantum mechanically calculated parameters. To quote a recent paper (217), 'One of the most successful applications of the Hückel molecular orbital theory has been in the correlation of polarographic oxidation and reduction potentials'.

The principal advantages in the use of polarographic half-wave potentials in linear free energy or other types of correlations, e.g. in verifying assumptions regarding the structure of a molecule or in evaluating a theoretical approach to calculating the energy level for adding an electron to the compound, are in the ease with which such potentials can be measured, as compared, for example, to rate or equilibrium constants, and the relative precision which is obtainable.

The following examples indicate the wide range of application now being made of correlations based on half-wave potentials. Anodic half-wave potentials were correlated with fungicidal activity for 25 p-phenylenediamines (218).

Hansen, Toren and Young (219–221) in a study of nitro-*p*-terphenyls involving charge-transfer complexes, polarography and electron paramagnetic resonance spectra, suggested that the slope of the linear relation between the polarographic oxidation potentials and the charge-transfer absorption maximum frequencies indicated that the nitro terphenyls interacted with the platinum electrode surface, perhaps by complexing with platinum oxide. The fact that steric strain in two isomers is relieved by completely different mechanisms, was supported principally by electrochemical measurements.

Figure 10 shows the correlation of Taft sigma values (cf. next sub-section) and nuclear magnetic resonance chemical shifts with $E_{1/2}$ for the lower aliphatic aldehydes (222).

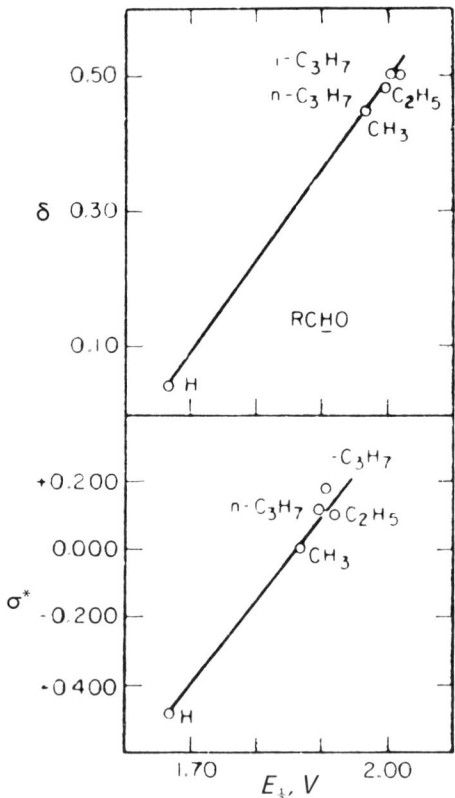

Fig. 10. Correlation of Taft σ^* and nuclear magnetic resonance σ values with $E_{1/2}$ values for aliphatic aldehydes RCHO ($\sigma = 10^5 (H_r - H_c)/H_r$, where H_c is applied magnetic field for H^1 resonance in proton indicated and H_r is that for water) (222).

Three frequently overlooked but important factors must be emphasized. The reliability of correlations of the type discussed depends on calculations and measurements having been made on identical molecular species. It is for this reason among others that the details of the electrochemical redox reaction pathway, including the sites at which electron transfer occurs and their exact chemical nature, need to be elucidated, since only on this basis is it possible to assure that theoretical and experimental data are being compared for the same molecular species.

A second factor is the influence of the degree of reversibility of the electron-transfer process upon observed potential. While the potential observed for a 'completely reversible' electrode process is directly relatable to the standard free energy of the process, that observed for one, which is not so reversible, contains a component due to the necessary energy of activation as well as, perhaps, components due to adsorption and other accompanying phenomena. The assumption is usually made that, while irreversibility and adsorption are complicated factors, they may be expected to operate uniformly in a series of closely-related derivatives; the validity of this assumption is usually judged on the basis of the consistency of the correlation, e.g. reference 223. Thus, $E_{1/2}$ values for overall irreversible polarographic reductions have frequently been correlated with molecular orbital and free energy indices, e.g. references 224 and 225; the usefulness of such correlations has been emphasized, e.g. by Zahradnik and Parkanyi (226).

Finally, account must be taken of the solvation energy contribution to the $E_{1/2}$. Theoretical calculations are frequently based on an idealized gas-phase molecule. Consequently, there has been again the implicit or explicit assumption in many past correlations that the solvation energy term is constant, is inconsequential or varies in a regular fashion for the series of compounds studied. While this assumption may be essentially valid for a series of large aromatic hydrocarbons and their one-electron oxidation or reduction products in an 'inert' organic solvent, its validity would have to be carefully examined, for example, for a series consisting of a purine and its nucleosides and nucleotides in aqueous medium because of the variation in nature, number and strength of hydrogen-bonding donor and acceptor groups in the different compounds and in their redox products.

In general, correlations of calculated molecular orbital and structure-reactivity indices with experimental electron affinity values, e.g. $E_{1/2}$ for reduction, and other properties, e.g. ultraviolet absorption bands, must—for the time being—be considered as semiempirical, since there are no exact methods for calculation of molecular orbital and other parameters which would involve all possible intra- and intermolecular effects. In any event, comparison of calculated values with $E_{1/2}$ for a series of related compounds may provide information on (a) whether effects of the types considered operate with all individual derivatives to comparatively the same extent and

(b) whether differences in $E_{1/2}$ due to structural changes have a counterpart in differences in molecular orbital and other indices, and vice versa. This approach is related to one of the as of yet relatively unexploited applications of $E_{1/2}$-property relations: The use of deviations from the expected linear relationship to shed light on the nature of the property and electrochemical process, e.g. deviations from expected correlations for electrochemical carbon–halogen bond fission in aliphatic compounds could be related to adsorption on the mercury electrode with resulting polarization of the molecules and other stabilizing effects (227).

1. Linear Free Energy Correlations

Zuman (204) has reviewed the correlation of half-wave potentials based on linear free energy relationships of the Hammett–Taft variety in a book, which examines exhaustively the effect of substituents on polarographic behaviour on the basis of polar, resonance and steric phenomena, and the consequent applicability of polarography to the study of mechanisms and structural effects in organic chemical and electrochemical reactions. (The reader is referred to Rosseinsky's informative review (228) of Zuman's excellent book (204).)

The physical basis for using half-wave potentials in linear free energy relations such as the Hammett–Taft equations is the fact that $E_{1/2}$ is a simple function of the logarithm of (a) the heterogeneous rate constant for irreversible electrode processes or (b) the equilibrium constant for reversible electrode processes. Hence, the basic Hammett equation for two members of a related series of compounds can be expressed as

$$(E_{1/2})_i - (E_{1/2})_j = \rho(\sigma_i - \sigma_j) \tag{4}$$

where ρ is a proportionality constant for the series, σ is a substituent constant summing up the effects of the substituents on a compound by which it differs from the selected reference compound, and one of the compounds (usually, the least substituted in respect to the substituents being examined) may be selected as the reference compound of $\sigma = 0$. A plot of $E_{1/2}$ versus σ for the series of compounds should be linear with a slope of ρ if the σ values used correctly assess the factors determining $E_{1/2}$.

Application of the Hammett approach can be illustrated by the data for a series of 6-substituted purines (the naturally occurring purines are, with few exceptions, substituted in that position, including the two major purine constituents of nucleic acids, adenine (6-aminopurine) and guanine (2-amino-6-hydroxypurine), as well as minor constituents such as 2-methyladenine, 6-methylaminopurine and 6-dimethylaminopurine) (229). The parent compound, purine, gives two 2e reduction waves at the DME, as does 6-methylpurine, while adenine (6-aminopurine) and the other 6-substituted purines studied give one $4e$ wave; reduction involves first the 1,6 N=C bond and then the 3,2 N=C bond (229, 230); the protonated purine (at N_1) seems to be an electroactive species (229).

Since the application of free-energy relationships to the polarographic behaviour of 6-substituted six-membered heterocyclic compounds of the purine type had not been previously reported, a trial and error approach was used. A fairly linear correlation is obtained on plotting (Figure 11) $E_{1/2}$ against total polar substituent constant, σ_p, which constant is dependent on the kind and position of the substituent, as well as to some extent on the nature of the aromatic ring, but is presumably independent of reaction and reaction conditions (204). Plots of $E_{1/2}$ against other constants, which are frequently used in modifications of the Hammett equation (204, 231, 232), do not result in reasonably linear correlations.

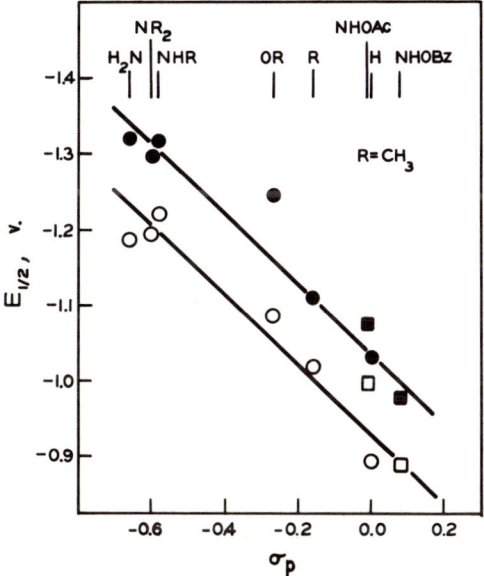

Fig. 11. Variation of $E_{1/2}$ of 6-substituted purine reduction wave with the total polar substituent constant, σ_p; substituents are indicated. $E_{1/2}$ determined in McIlvaine buffer (circles, pH 2·5; solid circles, pH 4·0). Slope, ρ, is 0·46 V at both pH 4·0 and pH 2·5 (229).

In the 6-substituted alkylaminopurines, the group initially reduced polaro-graphically (the 1,6 N=C) is separated by an —NH— group from the substituent. As a rough approximation, the —NH— group may be considered analogous to the —CH_2— group and $E_{1/2}$ may be plotted against polar substituent constant, σ^* (204, 231, 232); a linear relation is obtained (Figure 12).

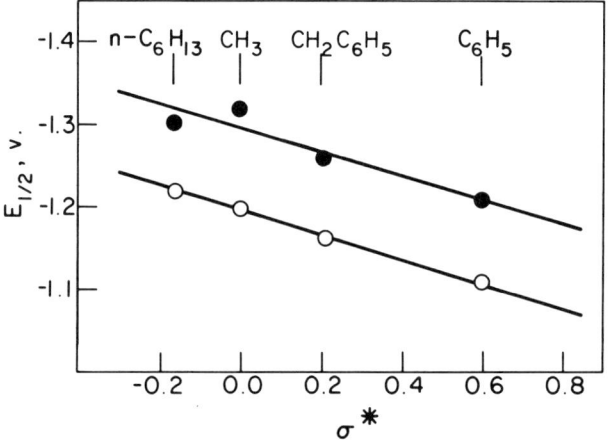

Fig. 12. Variation of $E_{1/2}$ of reduction wave of N'-substituted 6-aminopurines with the polar substituent constant, σ^*; substituents are indicated. $E_{1/2}$ determined in McIlvaine buffer (circles, pH 2·5; solid circles, pH 4·0). Slope, ρ is 0·14 V for pH 4·0 and 0·15 V for pH 2·5 (229).

In agreement with the polarographic behaviour (229), i.e. 6-methylpurine resembles the parental purine in being reduced in two $2e$ waves, while the other 6-substituted purines resemble adenine in being reduced in one $4e$ wave, the correlation of $E_{1/2}$ with total polar substituent constant (Figure 11) places 6-methylpurine closer to purine than to the 6-amino derivatives. The small change in the polarographic reducibility of purine produced by introducing the 6-methyl group also corresponds to small differences in LEMO energies and other electronic indices (Figures 13 to 15).

The two linear relationships just described also support the fact that the mechanisms for the polarographic reduction of the 6-substituted purines studied are essentially the same as those of the parental compounds, purine and adenine. Correlation of $E_{1/2}$ with the polar substituent constants for the 6-alkylaminopurines and with the total polar substituent constants for all 6-substituted purines examined, indicates either that no mesomeric and steric effects due to the substituents are involved or that they are transmitted by an inductive mechanism (204, 231, 232).

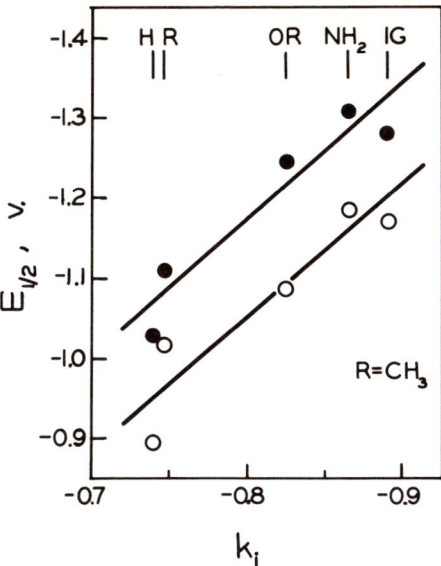

Fig. 13. Variation of $E_{1/2}$ of 6-substituted purines (substituents are indicated) with LEMO energies (k_i). IG = isoguanine. $E_{1/2}$ determined in McIlvaine buffer (circles, pH 2·5; solid circles, pH 4.0) (229).

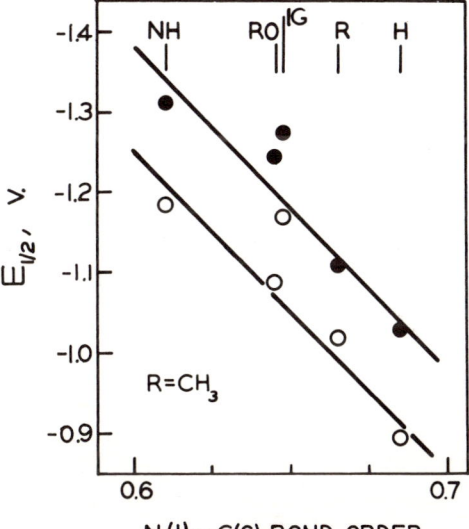

Fig. 14. Variation of $E_{1/2}$ of 6-substituted purines (substituents are indicated) with calculated $N_{(1)}-C_{(6)}$ bond order. IG = isoguanine. $E_{1/2}$ determined in McIlvaine buffer (circles, pH 2·5; solid circles pH 4·0) (229).

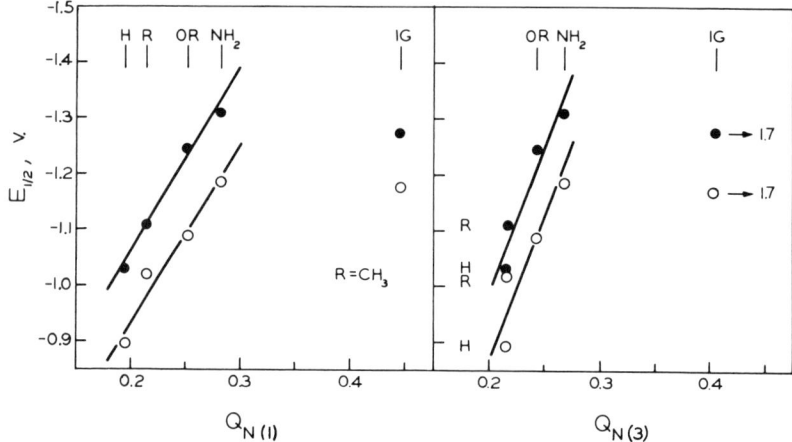

Fig. 15. Variation of $E_{1/2}$ of 6-substituted purines (substituents are indicated) with net negative charges on $N_{(1)}$ and $N_{(3)}$. IG = isoguanine. $E_{1/2}$ determined in McIlvaine buffer circles, pH 2·5; solid circles pH 4·0 (229).

2. Quantum Mechanically Calculated Electronic Indices

Correlations of molecular orbital (MO) calculations with polarographic data have commonly involved the energies for adding an electron to the lowest empty MO (LEMO) or removing one from the highest occupied MO (HOMO), based on calculation of the energies of the molecular orbitals of the mobile or π electrons. These energies are frequently of the form

$$E_i = \alpha + k_i\beta \tag{5}$$

where α is the coulomb integral and β the resonance integral (β is ca. 20 kcal mole^{-1}). Positive k_i values are associated with occupied (bonding) orbitals in the ground state of the molecule; negative k_i values with un-occupied (antibonding) orbitals, which are usually occupied only in the excited state. For a homologous series, the smaller k_i, the greater the electro-negativity and, consequently, the greater the electron acceptor properties of the molecule.

A priori, the optimum approach would seem to involve the use of electro-chemical data based on initial $1e$ processes, since the MO data apply to such processes. Many organic compounds, which give multiple-electron polaro-graphic waves in proton-available media, e.g. aqueous solution, give initial $1e$ waves in non-aqueous media of low proton availability; the latter fact undoubtedly accounts for the commonly better correlations based on $E_{1/2}$ data obtained in non-aqueous media. The validity of this approach is shown by the work of Hoijtink (233), Streitwieser (234, 235), Pysh and Yang (223) and Neikana and Desmond (236), who have described the correlation of the potentials for the oxidation of organic compounds in non-aqueous media at platinum electrodes with a variety of calculated values based on molecular-orbital theory. However, data obtained for multielectron waves in aqueous and non-aqueous media have also been used, e.g. references 224 and 225. In the case of biologically important molecules such as purines, it may for certain applications be preferable to examine data obtained under conditions similar to physiological ones, e.g. in aqueous media of normal pH range.

In addition to correlation of $E_{1/2}$ with LEMO and HOMO values, correlations can be investigated involving other calculated parameters, e.g. between $E_{1/2}$ and bond orders for selected pairs of atoms in the molecules, and between $E_{1/2}$ and net charges on certain atoms in the molecules in the attempt, for example, to determine whether such correlations can be used to identify the electroactive sites in the molecules (cf. Janik and Elving (229)).

The *bond order* (cf. Pullman and Pullman (215) and Streitwieser (234) for discussion of this concept) may be physically associated with the binding power of a bond as the molecular orbital analog of the resonance theory concept of double-bond character, e.g. p values are 1 for the carbon–carbon double bond in ethylene, 0·667 for each carbon–carbon bond in benzene and 0·894 and 0·447 for the double and single carbon–carbon bonds, respectively, in the conventional structural symbolism for 1,3-butadiene. The *net charge* at an atom represents the difference between the charge that the atom would carry in the absence of delocalization and its actual calculated electronic charge (215). The latter, e.g. the total electron density at an atom, is the sum of electron densities contributed by each electron in each MO and is a convenient value, even though it has only approximate physical significance (234).

Correlations of the types just mentioned can be illustrated by data for the reduction of the 6-substituted purines (229), already considered in the case of linear free energy correlations.

When the ease of polarographic reduction ($E_{1/2}$) is compared with electron-acceptor properties (LEMO energy calculated by the Huckel LCAO molecular-orbital approximation method (215)), a fairly linear relationship (Figure 13) is obtained between $E_{1/2}$ and LEMO k_i coefficient for purine, 6-methylpurine, adenine and 6-methoxypurine (data calculated for the hypoxanthine lactim form were used for the latter); it is of interest that ATP (adenosine triphosphate) fits the correlation but AMP (adenosine monophosphate) does not. The deviant behaviour of isoguanine is subsequently discussed.

Substitution in the 6-position influences mostly the $N_{(1)}$–$C_{(6)}$ bond order and the electronic charge distributions on $N_{(1)}$ and $N_{(3)}$, which, in turn, largely determine the reducibility; $E_{1/2}$ correlates with these (Figure 14), but not with the $C_{(2)}$–$N_{(3)}$ bond order and the charges on $C_{(6)}$ and $C_{(2)}$.

These results are in accord with the association of the initial purine and adenine reduction steps with the 1,6 N=C bond, which bond, moreover, seems to be more accessible for reduction than the 2,3 C=N bond, e.g. isoguanine (2-hydroxy-6-aminopurine) produces a well-defined wave while hypoxanthine (6-hydroxypurine) only gives an ill-defined inflection on the background discharge (230) and guanine (2-amino-6-hydroxypurine) is polarographically reducible only at potentials more negative than background discharge by a mechanism different from that for other purine derivatives (230, 237, 238).

As noted, $E_{1/2}$ for the reduction of isoguanine deviates markedly from the linear correlations of $E_{1/2}$ with net negative charge on $N_{(1)}$ and $N_{(3)}$, but only slightly from the linear plots of $E_{1/2}$ versus LEMO energy and $N_{(1)}$–$C_{(6)}$ bond order (Figure 15). This behaviour may be correlated—at least to some extent—with the reduction site and the effect of protonation on the latter. In isoguanine, only the 1,6 N=C bond is reduced since the 2,3 C=N bond is unavailable due to the stability of the keto form. The agreement of isoguanine $E_{1/2}$ with LEMO energy and $N_{(1)}$–$C_{(6)}$ bond order may then be ascribed to the connection of the latter with the reduction site. However, protonation of isoguanine does not seem to play a decisive role in its polarographic reduction (229); even if a protonated species were involved in the reduction, the most probable protonation site, based on comparison of N— charges (215), is $N_{(7)}$ which is not part of the 1,6 N=C reduction site.

Methods of calculation of MO parameters are described by Pullman and Pullman (215), Streitwieser (234) and other books dealing with molecular-orbital calculations for chemists.

The value of reliable experimental redox potentials for organic compounds as tests of theoretical approaches to the electronic properties of molecules is typified by a paper (239), in which calculations for the four nucleic acid bases (guanine, adenine, cytosine and uracil) by a self-consistent-field (SCF) method, as well as by the Hückel molecular orbital (HMO) approximation method, are evaluated by comparing the values obtained for ionization potentials (electron-donor properties) and for electron affinities (electron-acceptor properties) with oxidation potentials determined at the graphite electrode and reduction potentials determined at the DME.

Gleicher and Gleicher (217), using $E_{1/2}$ data obtained by others for the polarographic oxidation of aromatic hydrocarbons and their alkyl derivatives, concluded that the SCF approach has a small but real advantage over the HMO approach in treating alternant and non-alternant systems within the same correlation. Figure 16 shows the impressive agreement obtained by them for 50 compounds.

Zahradnik and Parkanyi (226) have reviewed empirical correlations of $E_{1/2}$ data with HMO characteristics and compared these relationships with

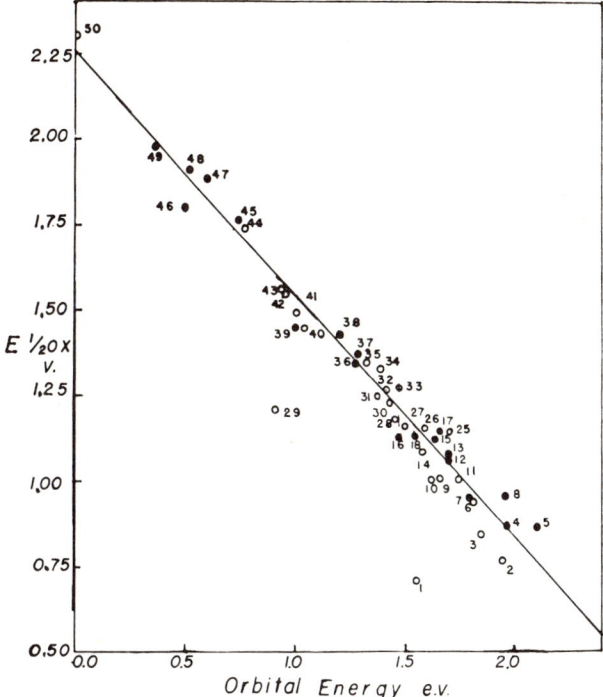

Fig. 16. Plot of polarographic oxidation potentials against the energy of the highest occupied molecular orbital of the neutral molecule as calculated by an SCF approach with constant bond lengths. Points 19–24 have been omitted to preserve clarity: ○, aromatic molecules; ●, alkylaromatic molecules. Numbering given in Table I of reference (217).

analogous dependencies for electronic spectra and ionization potentials. Perrin (50) provides a succinct general summary of $E_{1/2}$ correlation.

F. Electrochemical Synthesis

The potentialities of preparative organic electrochemistry at controlled or uncontrolled electrode potential, by and large, have scarcely been realized despite the large amount of work done. The main advantages are (a) precise control, when operating at controlled potential, of the intensity of reactivity of the reagent, i.e. the electron, (b) resulting possibly high selectivity of reaction resulting in a high yield of pure product and (c) facilitation of product recovery due to the presence of minimum amounts of side products and absence of excess reagent. In addition, electrolytic processes are readily automated and consequently often require a minimum of operator attention, are usable on a micro to macro scale, and can use polarography as a screen to select optimum conditions. As an example of a type of synthesis where the electrochemical approach can be advantageously used, the preparation of bicyclobutanes and other strained cycloalkanes from appropriately disubstituted halides by Rifi (240) might be cited.

The limited use of preparative organic electrochemistry has been largely due to the difficulty of controlling the applied potential, which is generally quite large because of the iR drop due to the appreciable current flow desired in a preparative process; the latter problem is greatly aggravated when non-aqueous solvents have to be used. At present, controlled potential electrolysis is used commercially only for the small-scale production of relatively costly products. 'As pharmaceutical molecules get more complex, reactions involving chemical oxidation and reduction reagents start losing out to the selectivity of electrochemical processes, at least on the small laboratory scale. The related polarographic analysis is soon carried over to competing chemical synthetic work.'

Brown and Harrison (240a) have provocatively reviewed the reactions of cathodically generated radicals and anions, and their potentialities in organic synthesis.

Frequently, it is possible to control the potential by control of experimental conditions and the current drawn. An outstanding example of this approach is the work of Baizer and his collaborators (241), who have extensively investigated both intermolecular and intramolecular electrolytic reductive coupling of unsaturated species and have produced a variety of useful industrial compounds, many of which are not readily accessible by more conventional synthesis employing chemical reagents. Adiponitrile, used in the manufacture of nylon, is now being made on a commercial scale from the electrolytic reductive coupling of acrylonitrile. (Cf. review by Baizer and associates of electrolytic reductive coupling as a synthetic tool (242).)

Two important areas of application of electrosynthesis are (a) the generation of free-radical species for study by electron spin resonance (125) and (b) controlled potential electrolysis and coulometry in elucidating the nature of organic electrode processes. Zuman (117) has reviewed preparative methods used in the elucidation of organic electrode processes.

Chang, Large and Popp (242a) and Barendrecht (242b) recently reviewed important developments in preparative electrolytic organic chemistry, supplementing the reviews of the subject by Wawzonek (243), Swann (244) and Popp and Schultz (245). The review by Popp and Schultz (245), which covers the period of 1940-61, has the useful feature of presenting tabular data on the reduction of a large variety of compounds, including product(s), cathode(s) used and percentage yield, where data are available for the latter. Vijh and Conway (246) have recently reviewed the much investigated Kolbe reaction. Three books stressing synthetic organic electrochemistry have recently appeared (248-250).

V. FUTURE PROSPECTS FOR VOLTAMMETRY IN ORGANIC ANALYSIS

The nature and extent of the future utilization of voltammetric techniques in organic analysis have been repeatedly referred to in the present chapter. All will agree that chemists concerned with organic analysis in all of its varied aspects should be aware of the potentialities of the gamut of voltammetric techniques and should use them when they provide better answers to problems than do other techniques under the conditions and requirements by which analytical approaches are judged.

In summary, it seemed best to the author that he quote two chemists, who are very much involved with the practice of organic analysis and who are concerned as to how it can best be served by voltammetric and polarographic techniques; one is employed by a photographic company whose research stresses chemistry; the other is at a governmental regulatory laboratory.

(a) 'For what it is worth, I might pass on my personal feelings concerning the application of polarographic and voltammetric techniques in future work in the Research Laboratories. In recent years, instances in which we have employed these techniques for the determination of the quantity of the material present in a sample have been decreasing and I expect this trend to continue. We do make considerable application of coulometric stripping analysis, although not for organic compounds at present. We continue to see a need for voltammetric and related techniques for the determination of relatively low concentrations of organics and may indeed expand our efforts in this area. However, quantitative applications of voltammetric techniques probably account for only ten per cent of our total effort at present, and will most likely not expand beyond that point in the immediate future. This trend is, of course, in direct opposition to the general increase in interest in electrochemical techniques in the Research Laboratories. During the past five years the number of persons actively involved in electrochemical work has more than tripled. For the past year or so, a seminar in electrochemistry has been held in the Research Laboratories on a regular basis. Thus, those of us actively engaged in the field have a strong feeling that electrochemical studies are presently, and will continue to be, a vital part of the Company's research program ' (247).

(b) 'I would like to make some general observations on the present and the future of organic polarography. One needs only to note that the number of references cited in the biannual review of organic polarography in the April issues of *Analytical Chemistry* have doubled since 1960.

In 1960 when I first began to concentrate on organic polarography very few laboratories were using the linear sweep type instruments. Although progress has been slow, I know an increasing number of laboratories have been using such instrumentation successfully. A number of industrial firms are using them routinely for quality control work and for research.

I am constantly amazed at the number of chemists at all degree levels who for some reason or another have only a limited knowledge of modern polarographic techniques. Most have had only a fleeting acquaintance with the technique and are unaware of its tremendous possibilities. On the other hand, we have had many foreign visitors in our laboratories who are well versed in the technique and whose laboratories use several polarographs on a routine basis. It seems to me that most university chemistry departments lump polarography under a general heading along with other electroanalytical techniques, stressing the theoretical aspects and at the same time neglecting the practicable applications, e.g. analysis of materials of all types.

I have noted over the years that many investigators will turn to polarography only as a last resort when all other approaches have failed. And when polarography also fails to do the job they will say they tried the method and it did not work. For example a chemist told me at the 1960 ACS meeting in New York that "Polarography is dead!" This approach seems typical of many of our chemists today. We should exploit polarography where it works well, i.e. nitro groups, ketones, organometallics, etc., and I am sure the technique will be well received.

There is hardly a field of organic chemistry that has not been investigated polarographically, i.e. drugs, drug metabolism, pesticides, polymers, biochemistry, etc. I think with the advent of modern instruments and techniques that polarography is and should be a part of any organic research group or analytical or chemical laboratory.' (8)

(c) Reviewing the comparative use of polarographic and voltammetric techniques in organic analysis in 1968 and 1972, caused the respondent (247) previously reported to write 'Be it healthy or not, our current use of electrochemical techniques in organic analysis is much the same as that described in my 1968 letter. I could add that the total number of people engaged in electrochemical work in these Laboratories has increased since 1968, and that new programmes in electrochemistry have been initiated in the elapsed three-plus years. However, our use of electrochemistry for analysis *per se* for organic compounds has not increased, and may in fact have decreased somewhat (this is also true for inorganic analysis).

Perhaps our situation is unique in that we have a full complement of analytical techniques to bring to bear on any given problem. Several of the techniques that can be considered to be alternatives to electrochemical methods are, in general, less expensive in time and manpower, on a per-sample basis, than electrochemical methods even though the capital expenditure for some of them is rather sizeable. We thus tend to apply these other methods first, and, if they succeed, do not utilize electrochemical methods.

This may mean that, through lack of practice, we are overlooking some potential applications of electrochemical methods of special value.'

ACKNOWLEDGMENTS

The author thanks the following chemists (listed alphabetically), who took the time to write to him about the current state of organic polarographic analysis, often in considerable detail but always with an informative and helpful attitude: E. A. Abrahamson, Joseph F. Alicino, T. M. Bednarski, Ralph O. Clark, Edward M. Cohen, Harry V. Drushel, Irving B. Eisdorfer, Raymond J. Gajan, J. Gordon Hanna, Peter Kabasakalian, P. O. Kane, J. H. Karchmer, Alan F. Krivis, R. La Coste, Robert F. Large, D. L. Maricle, L. M. Melnick, Clyde L. Ogg, Edward C. Olson, Carl R. Rehm, Noel Schwartz, Bernard Z. Senkowski, Sidney Siggia, Al Steyermark, George A. Ward, F. T. Weiss, D. H. Wilkens and F. G. Willeboordse. He also thanks the U.S. Atomic Energy Commission, which long supported his work on the polarographic behaviour of organic compounds, and the National Science Foundation, which is currently supporting it.

REFERENCES

1a. K. Schwabe, *Polarographic und Chemische Konstitution organischer Verbindungen*, Akademie-Verlag, Berlin, 1957.
1b. P. Zuman, *Organic Polarographic Analysis*, Pergamon Press, New York, 1964.
2. P. J. Elving, *Organic Analysis*, J. Mitchell, Ed., Vol. II, Interscience Publishers, New York, 1954, pp. 195–236.
3. P. J. Elving, *Progress in Polarography*, P. Zuman, Ed., Interscience Publishers, New York, 1962, pp. 625–648.
4. H. A. Liebhafsky, *Anal. Chem.*, **34**, 23A (1962).
5. E. P. Ragelis and R. J. Gajan, *J. Assoc. Offic. Agr. Chemists*, **45**, 918 (1962).
6. H. R. Smith and R. J. Gajan, *J. Assoc. Offic. Anal. Chemists*, **48**, 769 (1965).
7. J. R. Taylor and V. H. Blomquist, *J. Assoc. Offic. Anal. Chemists*, **51**, 533 (1968).
8. R. J. Gajan, private communication.
9. American Society for Testing and Materials, Standard D-1492-60, 'Method of Test for Bromine Index of Aromatic Hydrocarbons by Coulometric Titration'.
10. U. S. Pharmacopeial Convention, 'The Pharmacopeia of the United States of America', 17th revision, New York, 1965.
11. American Pharmaceutical Association, 'National Formulary xii,' Washington, 1965.
11a. G. Milazzo, P. E. Jones and L. Rampazzo, Ed., *Biological Aspects of Electrochemistry*, Birkhauser, Basel, 1971; *Experientia Suppl.*, **18**, 1–703 (1971).
12. P. J. Elving, *Pure Appl. Chem.*, **15**, 297 (1967).
13. L. Meites, *Polarographic Techniques*, 2nd ed., Interscience Publishers, New York, 1965.
14. E. R. Brown, T. G. McCord, D. E. Smith and D. D. DeFord, *Anal. Chem.*, **38**, 1119 (1966).

15. P. O. Kane, private communication.
16. G. A. Ward, private communication.
17. H. Hoffman and J. Volke, *Electroanalytical Chemistry*, H. W. Nürnberg, Ed., Interscience Publishers, New York, 1974, pp. 287–334 (this volume).
17a. E. M. Cohen, *J. Pharm. Sci.*, **60**, 1702 (1971).
17b. R. J. Gajan, *J. Assoc. Offic. Anal. Chemists*, **52**, 811 (1969).
17c. C. L. Schmitz, E. F. Ewen and S. P. Dodd, Ed., *Bibliography of Polarographic Literature*: 1922–1967, Sargent-Welch Scientific Co., Skokie, 1969.
18. D. M. Hume, *Anal. Chem.*, **40**, 174R (1968).
18a. R. S. Nicholson, *Anal. Chem.*, **42**, 130R (1970); **44**, 478R (1972).
19. D. J. Pietrzyk, *Anal. Chem.*, **40**, 194R (1968); **42**, 139R (1970); **44**, 457R (1972).
20. P. J. Elving, *Abhandl. Deut. Akad. Wiss., Kl. Med.*, **1966**, 635.
21. A. J. D'Eustachio, *Anal. Chem.*, **40**, 19R (1968).
22. W. C. Purdy, *Ann. N.Y. Acad. Sci.*, **137**, 390 (1966).
22a. E. Palecek, *Progress in Nucleic Acid Research and Molecular Biology*, Vol. 9, J. N. Davidson and W. E. Cohn, Eds., Academic Press, New York, 1969.
22b. P. Nangniot, *La Polarographie en Agronomie et en Biologie*, Editions J. Duculot, Gembloux, 1970.
23. K. I. Beynon and K. E. Elgar, *Analyst*, **91**, 143 (1966).
24. A. Calderbank, *Residue Rev.*, **12**, 14 (1966).
25. R. J. Gajan, *J. Assoc. Offic. Agr. Chemists*, **48**, 1001 (1965).
26. E. L. Bush, *J. Polarog. Soc.*, **11**, 41 (1965).
27. J. Ukida, S. Usami and T. Kominami, *Talanta*, **12**, 1163 (1965).
28. R. Brdička, M. Brezina and V. Kalous, *Talanta*, **12**, 1149 (1965).
29. P. Zuman, *Progress in Physical Organic Chemistry*, Vol. V, A. Streitwieser and R. W. Taft, Eds., Interscience Publishers, New York, 1967, pp. 81–206.
30. W. C. Purdy, *Electroanalytical Methods in Biochemistry*, McGraw-Hill Book Co., New York, 1965.
31. P. J. Elving and R. E. Van Atta, *Anal. Chem.*, **27**, 1908 (1955).
32. P. J. Elving and C. S. Tang, *Anal. Chem.*, **23**, 341 (1951).
33. P. Zuman, *Chem. Eng. News*, **46**, No. 12, 94 (1968).
34. R. Buckley and J. K. Taylor, *J. Res. Natl. Bur. Stds.*, **34**, 97 (1945).
35. B. Warshowsky and P. J. Elving, *Ind. Eng. Chem., Anal., Ed.*, **18**, 253 (1946).
35a. G. Dryhurst, *Periodate Oxidation of Diol and Other Functional Groups*, Pergamon Press, London, 1970.
36. J. R. Dyer, *Methods of Biochemical Analysis*, D. Glick, Ed., Vol. 3, Interscience Publishers, New York, 1956, pp. 111–152.
37. W. A. Cannon, *Anal. Chem.*, **22**, 928 (1950).
37a. P. J. Elving, A. J. Martin and I. Rosenthal, *Anal. Chem.*, **25**, 1082 (1953).
38. B. Warshowsky and M. W. Rice, *Anal. Chem.*, **20**, 341 (1948).
39. C. V. Francis, *Anal. Chem.*, **21**, 1238 (1949).
39a. D. D. Gilbert, *Anal. Chem.*, **41**, 1567 (1969).
39b. M. Brezina and V. Gultjoj, *Collec. Czech. Chem. Commun.*, **28**, 181 (1963).
39c. S. G. Mairanovskii, *Catalytic and Kinetic Waves in Polarography*, Plenum Press, New York, 1968.
40. C. H. Müller, *Clin. Chem.*, **11**, 270 (1965).
41. R. Rappolt, *Polarography 1964*, G. J. Hills, Ed., Interscience Publishers, New York, 1966, pp. 977–983.
41a. J. Homolka, *Methods of Biochemical Analysis*, Vol. 19, D. Glick, Ed., Interscience Publishers, New York, 1971, pp. 435–555.

42. H. Berg, Ed., *Elektrochemische Methoden und Prinzipien in der Molecular-Biologie*, Akademie-Verlag, Berlin, 1966, pp. 423–484.
43. B. Breyer and H. H. Bauer, *Alternating Current Polarography and Tensammetry*, Interscience Publishers, New York, 1963.
44. I. Vavruch, *Anal. Chem.*, **22**, 930 (1950).
45. P. J. Elving and D. L. Smith, *Microchemical Techniques*, N. D. Cheronis, Ed., John Wiley and Sons, New York, 1962, pp. 829–842.
46. P. J. Elving and C. E. Bennett, *J. Electrochem. Soc.*, **101**, 520 (1954).
47. P. J. Elving and C. E. Bennett, *Anal. Chem.*, **26**, 1572 (1954).
48. International Union of Pure and Applied Chemistry, *International Congress of Polarography*, Butterworths, London, 1967; also, in *Pure Appl. Chem.*, **15**, No. 2 (1967).
49. Faraday Society, 'Electrode reactions of organic compounds,' Discussions No. 45, 1968.
50. C. L. Perrin, *Progress in Physical Organic Chemistry*, Vol. III, S. G. Cohen *et al.*, Eds., Interscience Publishers, New York, 1963.
51. P. J. Elving and B. Pullman, *Advances in Chemical Physics*, I. Prigogine, Ed., Vol. III, Interscience Publishers, New York, 1961, pp. 1–31.
52. B. Janik and P. J. Elving, *Chem. Rev.*, **68**, 295 (1968).
52a. B. B. Damaskin, O. A. Petrii and V. V. Barakov, *Adsorption of Organic Compounds on Electrodes*, Plenum Press, New York, 1971.
53. P. Delahay, *Double Layer and Electrode Kinetics*, Interscience Publishers, New York, 1965.
54. E. Gileadi, Ed., *Electrosorption*, Plenum Press, New York, 1967.
55. A. N. Frumkin and B. B. Damaskin, *Modern Aspects of Electrochemistry*, J. O'M. Bockris, Ed., Vol. 3, Butterworths, London, 1964, pp. 149–223.
56. R. Parsons, *J. Electroanal. Chem.*, **8**, 93 (1964).
57. S. G. Mairanovskii, *J. Electroanal. Chem.*, **4**, 166 (1962).
58. D. M. Mohilner, *Electroanalytical Chemistry*, A. J. Bard, Ed., Marcel Dekker, New York, 1966, pp. 241–409.
59a. C. N. Reilley and W. Stumm, *Progress in Polarography*, P. Zuman, Ed., Interscience Publishers, New York, 1962, pp. 81–121.
59b. H. W. Nürnberg and M. v. Stackelberg, *J. Electroanal. Chem.*, **4**, 1 (1962).
59c. B. Kastening and L. Holleck, *Talanta*, **12**, 1259 (1965).
60. H. H. Bauer, P. J. Herman and P. J. Elving, *Modern Aspects of Electrochemistry*, Vol. 7, B. E. Conway and J. O'M. Bockris, Eds., Plenum Press, New York, 1972, pp. 143–197.
61. C. A. Barlow and J. R. MacDonald, *Advances in Electrochemistry and Electrochemical Engineering*, Vol. VI, P. Delahay and C. Tobias, Eds., Interscience Publishers, New York, 1967, pp. 1–199.
62. R. Parsons, *Advances in Electrochemistry and Electrochemical Engineering*, Vol. I, P. Delahay and C. Tobias, Eds., Interscience Publishers, New York, 1961, pp. 1–64.
63. R. Takahashi and P. J. Elving, *Electrochim. Acta*, **12**, 213 (1967).
64. C. K. Mann and K. K. Barnes, *Electrochemical Reactions in Nonaqueous Systems*, Marcel Dekker, New York, 1970.
65. I. M. Kolthoff, *Polarography 1964*, G. J. Hills, Ed., Interscience Publishers, New York, pp. 1–24.
66. S. G. Mairanovskii, *Talanta*, **12**, 1299 (1965).
67. S. G. Mairanovskii, *Polarography 1964*, G. J. Hills, Ed., Interscience Publishers, New York, 1966, pp. 719–730.

68. M. E. Peover and J. D. Davies, *Polarography 1964*, G. J. Hills, Ed., Interscience Publishers, New York, pp. 1003–1016.
69. K. Schwabe, *Progress in Polarography*, P. Zuman, Ed., Interscience Publishers, New York, 1962, pp. 333–355.
70. S. Wawzonek, *Talanta*, **12**, 1229 (1965).
71. P. G. Grodzka and P. J. Elving, *J. Electrochem. Soc.*, **110**, 225 (1963).
72. Janine Badoz-Lambling and G. Cauquis, *Electroanalytical Chemistry*, H. W. Nürnberg, Ed., Interscience Publishers, New York, 1974, pp. 335–419 (this volume).
72a. C. K. Mann, *Electroanalytical Chemistry*, Vol. 3, A. J. Bard, Ed., Marcel Dekker, New York, 1969, pp. 57–134.
72b. J. N. Butler, *Advances in Electrochemistry and Electrochemical Engineering*, Vol. 7, P. Delahay and C. Tobias, Eds., Interscience Publishers, 1970, pp. 77–175.
72c. M. M. Davis, *Acid-Base Behaviour in Aprotic Organic Solvents*, U.S. Government Printing office, Washington, 1968.
73. J. N. Butler, *J. Electroanal. Chem.*, **14**, 89 (1967).
74. H. Jasinski, *J. Electroanal. Chem.*, **15**, 89 (1967).
75. M. E. Peover, *Electroanalytical Chemistry*, A. J. Bard, Ed., Vol. 2, Marcel Dekker, New York, 1967, pp. 1–51.
75a. G. J. Hoytink, *Advances in Electrochemistry and Electrochemical Engineering*, Vol. 7, P. Delahay and C. Tobias, Eds., Interscience Publishers, 1970, pp. 221–281.
76. T. A. Gough and M. E. Peover, *Polarography 1964*, G. J. Hills, Ed., Interscience Publishers, New York, pp. 1017–1033.
77. P. J. Elving and M. S. Spritzer, *Talanta*, **12**, 1243 (1965).
78. G. Le Guillanton, *Bull. Soc. Chem. France.*, **1963**, 2359.
79. P. H. Given and M. E. Peover, *Advances in Polarography*, I. S. Longmuir, Ed., Pergamon Press, London, 1960, pp. 948–964.
80. G. Jander, *Die Chemie in Wasserähnlichen Losungsmitteln*, Springer-Verlag, Berlin, 1949.
81. L. F. Audrieth and J. Kleinberg, *Non-Aqueous Solvents*, John Wiley and Sons, New York, 1953.
82. G. Charlot and B. Tremillon, *Les reactions chimiques dans les solvants et les sels fondus*, Gauthier-Villars, Paris, 1963.
83. T. C. Waddington, Ed., *Non-aqueous Solvent Systems*, Academic Press, New York, 1965.
84. B. E. Conway and R. G. Barradas, Eds., *Chemical Physics of Ionic Solutions*, John Wiley and Sons, New York, 1966.
85. G. J. Janz, F. J. Kelly and H. V. Venkatasetty, *A Survey of Non-aqueous Conductance Data*, Rensselaer Polytechnic Institute, Troy, 1962.
86. J. J. Lagowski, Ed., *The Chemistry of Non-aqueous Solvents*, Vol. I: *Principles and Techniques*, Academic Press, New York, 1966.
87. K. Tsuji and P. J. Elving, *Anal. Chem.*, **41**, 286, 1571 (1969).
87a. P. J. Elving, *J. Electroanal. Chem.*, **29**, 55 (1971).
88. M. S. Spritzer, J. M. Costa and P. J. Elving, *Anal. Chem.*, **37**, 211 (1965).
89. P. J. Elving and C. Teitelbaum, *J. Am. Chem. Soc.*, **71**, 3916 (1949).
90. P. J. Elving, *Pure Appl. Chem.*, **7**, 423 (1963).
91. A. K. Vlček, E. Špalek and L. Krátky, *Coll. Czech. Chem. Commun.*, **15**, 340 (1950).
91a. L. Holleck and O. Lehmann, *Coll. Czech. Chem. Commun.*, **30**, 4024 (1965).
92. R. M. Elofson, *Anal. Chem.*, **21**, 917 (1949).
93. M. E. Hall, *Anal. Chem.*, **31**, 2007 (1959).
94. R. E. Van Atta and D. R. Jamieson, *Anal. Chem.*, **31**, 1217 (1959).
95. Symposium on Operational Amplifiers, *Anal. Chem.*, **35**, 1770–1833 (1963).

95a. R. Bezman and P. S. McKinney, *Anal. Chem.*, **41**, 1560 (1969).
96. D. J. Fisher, *Electroanalytical Chemistry*, H. W. Nürnberg, Ed., Interscience Publishers, New York, 1974, pp. 1–158 (this volume).
97. R. N. Adams, *Electrochemistry at Solid Electrodes*, Marcel Dekker, New York, 1969.
98. S. Piekarski and R. N. Adams, *Physical Methods of Chemistry*, A. Weissberger and B. W. Rossiter, Eds., Vol. I, Part IIA, Wiley-Interscience, New York, 1971, pp. 531–589.
99. P. J. Elving, I. Fried and W. R. Turner, *Polarography 1964*, G. J. Hills, Ed., Interscience Publishers, New York, 1966, pp. 277–297.
99a. P. J. Elving and D. L. Smith, *Analytical Chemistry 1962*, Elsevier, Amsterdam, 1963, pp. 204–213.
99b. A. L. Beilby, *Bibliography on the Use and Nature of Carbon as an Electrode Material in Electroanalytical Methods*, Pomona College, Claremont, Calif., 1965.
100. D. L. Smith and P. J. Elving, *Anal. Chem.*, **34**, 930 (1962).
101. P. J. Elving, W. A. Struck and D. L. Smith, *Mises Point Chim. Anal. Org. Pharm. Bromatol.*, **14**, 141 (1965).
102. P. J. Elving and D. L. Smith, *Anal. Chem.*, **32**, 1849 (1960).
103. L. Chuang, I. Fried and P. J. Elving, *Anal. Chem.*, 2462 (1964).
103a. E. R. Brown and R. F. Large, *Physical Methods of Chemistry*, A. Weissberger and B. W. Rossiter, Eds., Vol. I, Part IIA, Wiley-Interscience, New York, 1971, pp. 423–530.
104. G. Dryhurst, M. Rosen and P. J. Elving, *Anal. Chim. Acta*, **42**, 143 (1968).
105. W. R. Turner and P. J. Elving, *J. Electrochem. Soc.*, **112**, 1211 (1965).
106. R. F. Michielli and P. J. Elving, *J. Am. Chem. Soc.*, **90**, 1989 (1968).
107. R. S. Nicholson and I. Shain, *Anal. Chem.*, **36**, 706, 1212 (1964); **37**, 178, 190 (1965).
107a. D. S. Polcyn and I. Shain, *Anal. Chem.*, **38**, 370 (1966).
107b. R. H. Wopschall and I. Shain, *Anal. Chem.*, **39**, 1527 (1967).
107c. R. L. Myers and I. Shain, *Anal. Chem.*, **41**, 980 (1969).
107d. M. H. Hulbert and I. Shain, *Anal. Chem.*, **42**, 162 (1970).
108. R. S. Nicholson, *Anal. Chem.*, **37**, 1351 (1965); **38**, 1406 (1966).
108a. M. L. Olmstead and R. S. Nicholson, *Anal. Chem.*, **38**, 150 (1966); **41**, 862 (1969).
108b. R. S. Nicholson, J. M. Wilson and M. L. Olmstead, *Anal. Chem.*, **38**, 542 (1966).
108c. M. L. Olmstead, R. G. Hamilton and R. S. Nicholson, *Anal. Chem.*, **41**, 260 (1969).
109. G. L. Booman and D. T. Pence, *Anal. Chem.*, **37**, 1366 (1965); **41**, 746 (1969).
109a. J. M. Saveant and E. Vianello, *Electrochim. Acta*, **12**, 1545 (1967).
109b. M. Mastragostino, L. Nadjo and J. M. Saveant, *Electrochim. Acta*, **13**, 721 (1968).
109c. B. Kastening, *Anal. Chem.*, **41**, 1142 (1969).
110. M. S. Shuman, *Anal. Chem.*, **42**, 521 (1970).
111. B. Timmer, M. Sluyters-Rehbach and J. H. Sluyters, *J. Electroanal. Chem.*, **14**, 169 (1967).
112. B. Timmer, M. Sluyters-Rehbach and J. H. Sluyters, *J. Electroanal. Chem.*, **14**, 181 (1967).
113. D. E. Smith and T. G. McCord, *Anal. Chem.*, **40**, 474 (1968).
114. D. E. Smith, *Electroanalytical Chemistry*, A. J. Bard, Ed., Vol. 2, Marcel Dekker, New York, 1966, pp. 21–155.
115. B. Breyer, *Pure Appl. Chem.*, **15**, 313 (1967).

115a. M. Sluyters-Rehbach and J. H. Sluyters, *Electroanalytical Chemistry*, Vol. 4, A. J. Bard, Ed., Marcel Dekker, New York, 1970, pp. 1–128.

115b. R. F. Evilia and A. J. Diefenderfer, *Anal. Chem.*, **39**, 1885 (1967).

115c. E. R. Brown, D. E. Smith and G. L. Booman, *Anal. Chem.*, **40**, 1411 (1968).

115d. R. DeLevie and A. A. Husovsky, *J. Electroanal. Chem.*, **20**, 181 (1969).

115e. P. J. Elving and S. J. Pace, *Experientia Suppl.*, **18**, 35 (1971).

116. L. M. Meites, *Physical Methods of Chemistry*, A. Weissberger and B. W. Rossiter, Eds., Vol. I, Part IIA, Wiley-Interscience, New York, 1971, pp. 645–712.

117. P. Zuman, *J. Polarog. Soc.*, **13**, 53 (1967).

118. L. Meites, *Record Chem. Progress (Kresge-Hooker Lib.)*, **22**, 81 (1961).

119. A. Cisak and P. J. Elving, *Electrochim. Acta*, **10**, 935 (1965).

120. I. Rosenthal, J. R. Hayes, A. J. Martin and P. J. Elving, *J. Am. Chem. Soc.*, **80**, 3050 (1958).

121. W. A. Struck and P. J. Elving, *Biochem.*, **4**, 1343 (1965).

122. G. Dryhurst and P. J. Elving, *J. Electrochem. Soc.*, **115**, 1014 (1968).

122a. E. Scarano, M. G. Bonicelli and M. Forina, *Anal. Chem.*, **42**, 1470 (1970).

122b. E. Pungor, Z. Feher and G. Nagy, *Anal. Chim. Acta*, **51**, 417 (1970); **52**, 417 (1970); *Anal. Chim. Acta*, Proceedings of Second Conference on Applied Physical Chemistry, 1971.

122c. R. W. Murray, *Physical Methods of Chemistry*, A. Weissberger and B. W. Rossiter, Eds., Vol. I, Part IIA, Wiley-Interscience, New York, 1971, pp. 591–644.

123. R. N. Adams, *J. Electroanal. Chem.*, **8**, 151 (1964).

124. R. N. Adams, Fourth International Congress on Polarography, Prague, July, 1966.

125. B. Kastening, *Electroanalytical Chemistry*, H. W. Nürnberg, Ed., Interscience Publishers, New York, 1974, pp. 421–494 (this volume).

126. T. Kuwana, R. K. Darlington and D. W. Leedy, *Anal. Chem.*, **36**, 2023 (1964).

127. J. W. Strojek and T. Kuwana, *J. Electroanal. Chem.*, **16**, 471 (1968).

128. W. N. Hansen, R. A. Osteryoung and T. Kuwana, *J. Am. Chem. Soc.*, **88**, 1062 (1966).

129. W. N. Hansen, R. A. Osteryoung and T. Kuwana, *Anal. Chem.*, **38**, 1810 (1966).

130. J. W. Strojek, T. Kuwana and S. W. Feldberg, *J. Am. Chem. Soc.*, **90**, 1353 (1968).

131. B. Warshowsky, P. J. Elving and J. Mandel, *Anal. Chem.*, **19**, 161 (1947).

132. M. I. Gerber and A. D. Shusharina, *Zh. Anal. Khim.*, **5**, 262 (1950); through *Chem. Abs.*, **44**, 10593 (1950).

133. K. G. Powell and G. F. Reynolds, *Polarography 1964*, G. J. Hills, Ed., Interscience Publishers, New York, 1966, pp. 249–259.

134. W. D. Shults *et al.*, *Z. Anal. Chem.*, **224**, 1 (1967).

135. W. D. Shults, D. J. Fisher and W. B. Schaap, *Anal. Chem.*, **39**, 1379 (1967).

136. W. D. Shults and W. B. Schaap, *Anal. Chem.*, **39**, 1384 (1967).

137. W. D. Shults and W. B. Schaap, *Z. Anal. Chem.*, **224**, 22 (1967).

138. Y. Israel, *Talanta*, **13**, 1113 (1966).

139. G. Dryhurst and P. J. Elving, *Anal. Chem.*, **40**, 492 (1968).

140. H. J. Pazdera, W. H. McMullen, L. L. Ciaccio, S. R. Missan and T. C. Grenfell, *Anal. Chem.*, **29**, 1649 (1957).

141. W. Kemula, *Zh. Anal. Khim.*, **22**, 562 (1967).

141a. J. G. Koen and J. F. K. Huber, *Anal. Chim. Acta*, **51**, 303 (1970).

142. G. L. Booman, *Anal. Chem.*, **38**, 1141 (1966).

143. E. R. Brown, D. E. Smith and D. D. DeFord, *Anal. Chem.*, **38**, 1130 (1966).

144. G. Lauer and R. A. Osteryoung, *Anal. Chem.*, **38**, 1137 (1966).

144a. S. P. Perone, D. O. Jones and W. F. Gutknecht, *Anal. Chem.*, **41**, 1154 (1969).
145. G. Lauer and R. A. Osteryoung, *Anal. Chem.*, **40**, 30A (1968).
146. G. Lauer, R. Abel and F. C. Anson, *Anal. Chem.*, **39**, 765 (1967).
147. S. P. Perone, J. E. Harrar, F. B. Stephens and R. E. Anderson, *Anal. Chem.*, **40**, 899 (1968).
147a. W. F. Gutknecht and S. P. Perone, *Anal. Chem.*, **42**, 906 (1970).
147b. S. P. Perone, J. W. Frazer and A. Kray, *Anal. Chem.*, **43**, 1485 (1971).
148. D. M. Mohilner and P. R. Mohilner, *J. Electrochem. Soc.*, **115**, 261 (1968).
148a. K. Niki, Y. Okuda, T. Tomonari, E. Buck and N. Hackerman, *Electrochim. Acta*, **16**, 487 (1971).
149. I. Tachi and M. Senda, *Progress in Polarography*, P. Zuman, Ed., Interscience Publishers, New York, 1962, pp. 711–725.
150. W. J. Blaedel and R. J. Laessig, *Advances in Analytical Chemistry and Instrumentation*, Vol. 5, C. N. Reilley and F. W. McLafferty, Eds., Interscience Publishers, New York, 1966, pp. 69–168.
151. J. V. A. Novak, *Progress in Polarography*, P. Zuman, Ed., Interscience Publishers, New York, 1962, pp. 569–582.
151a. E. Palecek and Z. Pechan, *Anal. Biochem.*, **42**, 59 (1971).
152. G. Wolff and H. W. Nürnberg, *Z. Anal. Chem.*, **216**, 169 (1966).
152a. E. M. Cohen, private communication.
153. H. A. Laitinen, *Trace Characterization: Chemical and Physical*, W. W. Meinke and B. F. Scribner, Eds., Government Printing Office, Washington, 1967, pp. 75–120.
154. J. K. Taylor, E. J. Maienthal and G. Marienko, *Trace Analysis: Physical Methods*, G. Morrison, Ed., Interscience Publishers, New York, 1965, pp. 377–433.
155. L. B. Rogers, *Trace Analysis*, J. H. Yoe and H. J. Koch, Eds., John Wiley and Sons, New York, 1957, pp. 270–282.
156. P. Delahay, *Trace Analysis*, J. H. Yoe and H. J. Koch, Eds., John Wiley and Sons, New York, 1957, pp. 283–316.
157. C. N. Reilley, *Trace Analysis*, J. H. Yoe and H. J. Koch, Eds., John Wiley and Sons, New York, pp. 317–345.
158. P. J. Elving and E. C. Olson, *J. Am. Chem. Soc.*, **79**, 2797 (1957).
159. E. C. Olson and P. J. Elving, *Anal. Chem.*, **26**, 1747 (1954); **27**, 1817 (1955); **28**, 251, 338 (1956).
160. I. M. Kolthoff and F. A. Bovey, *Anal. Chem.*, **19**, 498 (1947).
161. G. Duyckaerts, *Bull. Soc. Roy. Sci. Liège*, **18**, 152 (1949).
162. A. Liberti, *Atti accad. nazl. Lincei, Rend. classe sci. fis., mat. e nat.*, **8**, 608 (1950); through *Chem. Abs.*, **45**, 68 (1951).
163. I. M. Kolthoff and W. E. Harris, *Ind. Eng. Chem., Anal. Ed.*, **18**, 161 (1946).
164. S. Rosenberg, J. C. Perrone and P. L. Kirk, *Anal. Chem.*, **22**, 1186 (1950).
165. I. M. Kolthoff and W. E. Harris, *Anal. Chem.*, **21**, 963 (1949).
166. J. B. Conn, *Anal. Chem.*, **20**, 585 (1948).
167. R. M. Elofson and P. A. Mecherly, *Anal. Chem.*, **21**, 565 (1949).
168. J. Stock, *Amperometric Titrations*, Interscience Publishers, New York, 1965.
169. J. T. Stock, *Anal. Chem.*, **40**, 392R (1968); **42**, 276R (1970); **44**, 1R (1972).
170. D. L. Smith, D. R. Jamieson and P. J. Elving, *Anal. Chem.*, **32**, 1253 (1960).
171. E. Smith, L. F. Worrel and J. E. Sinsheimer, *Anal. Chem.*, **35**, 58 (1963).
172. A. J. Bard, *Anal. Chem.*, **40**, 64R (1968); **42**, 22R (1970).
172a. D. G. Davis, *Anal. Chem.*, **44**, 79R (1972).
173. T. Meites and L. Meites, *Anal. Chem.*, **27**, 1531 (1955).

173a. J. R. Moody and W. C. Purdy, *Anal. Chim. Acta*, **53**, 31 (1971).

174. J. J. Lingane, *Electroanalytical Chemistry*, 2nd ed., Interscience Publishers, New York, 1958.

174a. A. J. Bard and K. S. V. Santhanam, *Electroanalytical Chemistry*, Vol. 4, A. J. Bard, Ed., Marcel Dekker, New York, 1970, pp. 215–315.

174b. W. C. Purdy, *Z. Anal. Chem.*, **243**, 17 (1968).

174c. G. Patriarche, *Sci. Pharm., Proc.*, **2**, 67 (1966).

174d. W. Buechler, P. Gisske and J. Meier, *Z. Anal. Chem.*, **239**, 289 (1968).

174e. R. C. Clem, F. Jakob, D. H. Anderberg and L. D. Ornelas, *Anal. Chem.*, **43**, 1338 (1971).

175. M. R. F. Ashworth, *Titrimetric Organic Analysis*, Parts I and II, Interscience, Publishers, New York, 1964–65.

176. G. Semerano, *Ric. Sci.*, **22**, Suppl. 197 (1952).

177. P. Zuman, *Advances in Physical Organic Chemistry*, V. Gold, Ed., Vol. 5, Academic Press, New York, 1967, pp. 1–52.

177a. H. L. Kies, *J. Electroanal. Chem.*, **4**, 257 (1962).

177b. J. Janata and H. B. Mark, *Electroanalytical Chemistry*, Vol. 3, A. J. Bard, Ed., Marcel Dekker, New York, 1969, pp. 1–55.

178. W. J. Seagers and P. J. Elving, *J. Am. Chem. Soc.*, **73**, 947 (1951).

178a. A. P. Shroff and C. J. Shaw, *Anal. Chem.*, **43**, 455 (1971).

178b. E. Kariv, J. Hermolin and E. Gileadi, *Electrochim. Acta*, **16**, 1437 (1971).

179. R. Brdička, V. Hanuš and J. Koutecky, *Progress in Polarography*, P. Zuman, Ed., Interscience Publishers, New York, pp. 145–199.

179a. H. W. Nürnberg and G. Wolff, *J. Electroanal. Chem.*, **21**, 99 (1969).

180. A. Bewick, M. Fleischmann and J. N. Hiddleston, *Polarography 1964*, G. J. Hills, Ed., Interscience Publishers, New York, 1966, pp. 57–78.

181. D. M. Oglesby, J. D. Johnson and C. N. Reilley, *Anal. Chem.*, **38**, 385 (1966).

181a. J. E. O'Reilly and P. J. Elving, *J. Am. Chem. Soc.*, **93**, 1871 (1971).

182. J. W. Ashley and C. N. Reilley, *J. Electroanal. Chem.*, **7**, 253 (1964).

183. P. G. Grodzka and P. J. Elving, *J. Electrochem. Soc.*, **110**, 231 (1963).

184. M. Suzuki and P. J. Elving, *J. Phys. Chem.*, **65**, 391 (1961).

185. M. Suzuki and P. J. Elving, *Coll. Czech. Chem. Communs.*, **25**, 3203 (1960).

186. W. H. Reinmuth, *Anal. Chem.*, **40**, 185R (1968).

186a. D. K. Roe, *Anal. Chem.*, **44**, 85R (1972).

187. G. C. Barker, *Polarography 1964*, G. J. Hills, Ed., Interscience Publishers, New York, 1966, pp. 25–47.

188. P. Delahay, *Advances in Electrochemistry and Electrochemical Engineering*, Vol. I, P. Delahay and C. Tobias, Eds., Interscience Publishers, New York, 1961, pp. 233–318.

189. K. J. Vetter, *Elektrochemische Kinetik*, Springer-Verlag, Berlin, 1961.

190. K. J. Vetter, *Electrochemical Kinetics: Theoretical and Practical Aspects*, Academic Press, New York, 1967.

191. K. J. Vetter, *Electrochemical Kinetics: Theoretical Aspects*, Academic Press, New York, 1967.

192. B. B. Damaskin, *Principles of Current Methods for the Study of Electrochemical Reactions*, McGraw-Hill Book Co., New York, 1967.

193. B. E. Conway, *Theory and Principles of Electrode Processes*, Ronald Press, New York, 1965.

194. W. L. Reynolds and R. W. Lumry, *Mechanisms of Electron Transfer*, Ronald Press, New York, 1966.

195. R. A. Marcus, *Ann. Rev. Phys. Chem.*, **15**, 155 (1964).

196. A. J. Bard, Ed., *Electroanalytical Chemistry*, Vols. 1 to 6, Marcel Dekker, New York. 1966–72.
197. J. O'M. Bockris and B. E. Conway, Eds., *Modern Aspects of Electrochemistry*, Vols. 1 to 7, Butterworths, London, and Plenum Press, New York, 1954–1972.
198. P. Delahay and C. W. Tobias, Eds., *Advances in Chemistry and Electrochemical Engineering*, Vols. 1 to 7, Interscience Publishers, New York, 1961–1970.
199. S. G. Mairanovskii, *Electrochim. Acta*, **9**, 803 (1964).
200. P. J. Elving, *Ann. N.Y. Acad. Sci.*, **158**, 124 (1969).
201. P. J. Elving, *Record Chem. Progr.*, **14**, 99 (1953).
202. P. J. Elving, *Ric. Sci.*, **30**, Suppl. 205 (1960).
203. P. J. Elving and J. T. Leone, *J. Am. Chem. Soc.*, **82**, 5076 (1960).
204. P. Zuman, *Substituent Effects in Organic Polarography*, Plenum Press, New York, 1967.
205. E. A. Abrahamson, *J. Am. Chem. Soc.*, **81**, 2692 (1959).
206. E. A. Abrahamson, *J. Am. Chem. Soc.*, **81**, 3919 (1959).
207. D. S. Acker and W. R. Hertler, *J. Am. Chem. Soc.*, **84**, 3370 (1962).
208. D. C. Olson, V. P. Mayweg and G. N. Schrauzer, *J. Am. Chem. Soc.*, **88**, 4876 (1966).
209. D. C. Olson, *Anal. Chem.*, **39**, 1785 (1967).
209a. B. Janik and P. J. Elving, *J. Am. Chem. Soc.*, **92**, 235 (1970).
210. W. A. Struck and P. J. Elving, *J. Am. Chem. Soc.*, **86**, 1229 (1964).
211. M. E. Peover and J. D. Davies, *Polarography 1964*, G. J. Hills, Ed., Interscience Publishers, New York, pp. 1003–1016.
212. M. E. Peover, *Trans. Faraday Soc.*, **60**, 417 (1964).
213. M. E. Peover, *Trans. Faraday Soc.*, **58**, 1656 (1962).
214. M. E. Peover, *Trans. Faraday Soc.*, **58**, 2370 (1962).
215. B. Pullman and A. Pullman, *Quantum Biochemistry*, Interscience Publishers, New York, 1963.
216. W. M. Clark, *Oxidation-Reduction Potentials of Organic Compounds*, Williams and Wilkins, Baltimore, 1960.
217. G. J. Gleicher and M. K. Gleicher, *J. Phys. Chem.*, **71**, 3693 (1967).
218. G. J. Niemann and J. Dekker, *Neth. J. Plant Pathol.*, **72**, 213 (1966).
219. R. L. Hansen, *J. Phys. Chem.*, **70**, 1646 (1966).
220. R. L. Hansen, P. E. Toren and R. H. Young, *J. Phys. Chem.*, **70**, 1653 (1966).
221. R. L. Hansen, R. H. Young, and P. E. Toren, *J. Phys. Chem.*, **70**, 1657 (1966).
222. C. E. Bennett and P. J. Elving, *Coll. Czech. Chem. Commun.*, **25**, 3213 (1960).
223. E. S. Pysh and N. C. Yang, *J. Am. Chem. Soc.*, **85**, 2124 (1963).
224. J. W. Sease, F. G. Burton and S. L. Nickol, *J. Am. Chem. Soc.*, **90**, 2595 (1968).
225. W. W. Hussey and A. J. Diefenderfer, *J. Am. Chem. Soc.*, **89**, 5359 (1967).
226. R. Zahradnik and C. Parkanyi, *Talanta*, **1965**, 1289.
227. I. Rosenthal, C. H. Albright and P. J. Elving, *J. Electrochem. Soc.*, **99**, 227 (1952).
228. D. R. Rosseinsky, *Analyst*, **93**, 62 (1968).
229. B. Janik and P. J. Elving, *J. Electrochem. Soc.*, **116**, 1087 (1969).
230. D. L. Smith and P. J. Elving, *J. Am. Chem. Soc.*, **84**, 1412 (1962).
231. P. Zuman, *Ric. Sci.*, **30**, *Suppl.* 229 (1960).
232. P. Zuman, *Modern Aspects of Polarography*, T. Kambara, Ed., Plenum Press, New York, 1966, pp. 102–116.
233. G. L. Hoijtink, *Rec. Trav. Chim.*, **77**, 555 (1958).
234. A. Streitwieser, *Molecular Orbital Theory for Organic Chemists*, John Wiley and Sons, New York, 1961.
235. A. Streitwieser and C. Perrin, *J. Am. Chem. Soc.*, **86**, 4938 (1964).
236. W. C. Neikana and M. M. Desmond, *J. Am. Chem. Chem. Soc.*, **86**, 4811 (1964).

237. B. Janik and E. Palecek, *Z. Naturforsch.*, **216**, 118 (1966).
238. B. Janik and E. Palecek, *Abhandl. Deut. Akad. Wiss. Berlin, Kl. Med.*, **1966**, 513.
239. H. Berthod, C. Giessner-Prettre and A. Pullman, *Theoret. Chim. Acta*, **5**, 53 (1966).
240. M. R. Rifi, *J. Am. Chem. Soc.*, **89**, 4442 (1967).
240a. O. R. Brown and J. A. Harrison, *J. Electroanal. Chem.*, **31**, 387 (1969).
241. M. M. Baizer, *J. Electrochem. Soc.*, **111**, 215 (1966) and subsequent papers in *J. Electrochem. Soc., J. Org. Chem.* and *Tetrahedron Letters.*
242. M. M. Baizer, J. D. Anderson, J. H. Wagenknecht, R. M. Ort and J. P. Petrovich, *Electrochim. Acta*, **12**, 1377 (1967).
242a. J. Chang, R. F. Large and G. Popp, *Physical Methods of Chemistry*, A. Weissberger and B. W. Rossiter, Eds., Vol. I, Part IIB, Wiley-Interscience, New York, 1971, pp. 1–89.
242b. E. Barendrecht, *Chem. Weekblad*, **67**, No. 21, 14 (1971).
243. S. Wawzonek, *Sci.*, **155**, 39 (1967).
244. S. Swann, *Catalytic, Photochemical and Electrolytic Reactions*, A. Weissberger, Ed., 2nd ed., Interscience Publishers, New York, 1956, pp. 385–523.
245. F. D. Popp and H. P. Schultz, *Chem. Rev.*, **62**, 19 (1962).
246. A. K. Vijh and B. E. Conway, *Chem. Rev.*, **67**, 623 (1967).
247. R. F. Large, private communication.
248. M. M. Baizer, Ed., *Organic Electrochemistry*, Marcel Dekker, New York, 1973.
249. L. Eberson and H. Schafer, *Organic Electrochemistry*, Springer, Berlin, 1972.
250. A. J. Fry, *Synthetic Organic Electrochemistry*, Harper and Row, New York, 1972.
251. Proceedings of Second International Symposium on Bioelectrochemistry (Pont à Mousson, October, 1973), *Bioelectrochem. Bioenergetics*, 1, No. 1/2, 3/4 (1974).
252. A. P. Tomilov, S. G. Mairanovskii, M. Ya. Fioshin and V. A. Smirnov, *Electrochemistry of Organic Compounds*, Halsted Press, New York, 1972.
253. A. N. Frumkin and A. B. Ershler, Eds., *Progress in Electrochemistry of Organic Compounds*, Vol. 1, Plenum Press, New York, 1971.
254. E. Yeager and A. J. Salkind, Eds., *Techniques of Electrochemistry*, Vols. 1 and 2, Interscience Publishers, New York, 1972–73.
255. A. L. Underwood and J. N. Burnett, *Electroanalytical Chemistry*, Vol. 6, A. J. Bard, Ed., Marcel Dekker, New York, 1972, pp. 1–85.
256. P. J. Elving, J. E. O'Reilly and C. O. Schmakel, in *Methods of Biochemical Analysis*, Vol. 21, D. Glick, Ed., Interscience Publishers, New York, 1973, pp. 287–465.
257. J. E. O'Reilly and P. J. Elving, *J. Am. Chem. Soc.*, **94**, 7941 (1972).
258. P. J. Elving, S. J. Pace and J. E. O'Reilly, *J. Am. Chem. Soc.*, **95**, 647 (1973).
259. J. W. Webb, B. Janik and P. J. Elving, *J. Am. Chem. Soc.*, **95**, 991 (1973).
260. K. S. V. Santhanam and P. J. Elving, *J. Am. Chem. Soc.*, **95**, 5482 (1973).

CHAPTER IV

Polarographic Analysis in Pharmacy

H. Hoffmann, *Institute of Pharmacy,*
University of Frankfurt, Federal Republic of Germany, and J. Volke,
'J. Heyrovský Institute' of Physical Chemistry and Electrochemistry,
Czechoslovakian Academy of Sciences, Prague, CSSR

I. INTRODUCTION

In the general setting of the growth in demand for pharmaceutical analysis and the corresponding development and application of new analytical methods, the electrochemical techniques, especially polarography, have gained steadily in importance during recent years. This applies to production control and to legal regulations for the registration of new pharmaceuticals as well as to qualitative and quantitative analysis of drugs in unknown preparations and to the identification of pharmaceuticals in fluids of the human body with respect to toxicology, pharmacology and pharmacological kinetics. The acceptance of the significance of dc polarography is featured by the fact that this method has been taken up by several pharmacopoeas for the analysis of various substances (in Pharmacopoea Bohemo-Slovenica II, 1954, for ascorbic acid, chlortetracycline, chloramphenicol, nicotin-amide, insulin zinc suspension; in USP XVII, 1965, for acetazolamide, dichlorophenamide, hydrochlorothiazide, methazolamide, nitrofurantonily; in DAB 7 DDR, 1967, for insulin). While spectrophotometry is frequently applied in the analysis of pure materials, polarography has significant advantages when pharmaceutical preparations have to be analysed, i.e. if various auxiliary substances are also present in the sample. The levels of sensitivity and precision which can be attained by dc polarography and spectrophotometry are comparable. The typical determination limit for both methods is a concentration of $10^{-5} M$ if the error is to be less than 2 to 3 per cent. The above-mentioned concentration corresponds to a quantity of $5 \mu g$ in 1 ml and under optimal conditions the error may be reduced to ± 1 per cent. Its distinct character as a microanalytical technique therefore makes dc polarography especially suitable for pharmaceutical analysis. Furthermore, several other polarographic techniques have been developed in the past which can be applied to solve special problems and will be described later.

The number of publications dealing with the application of polarography to pharmaceutical problems is steadily increasing. Thus the present survey can only review, without claiming completeness, the development during the past years and demonstrate via selected examples the great possibilities still to be explored in detail in the future.

II. SUBSTANCES DETERMINABLE BY POLAROGRAPHY

The prerequisite for the direct polarographic determination of a sub-stance is that it contains a reducible or oxidizable functional group, i.e. a 'polarographically active group'. In a later section, possibilities will be described by which other substances can also be made suitable for polarography.

Table I provides a survey of all the important polarographically active functional groups relevant to the analysis of pharmaceuticals. The single C—C bond is usually not reduced at the mercury electrode. Exceptions, as for example with the p-phthalic acid mononitrile, which is reduced with the uptake of two electrons and two protons to benzoic acid and CN^-, have no

TABLE I

Bond grouping	Example

C—N

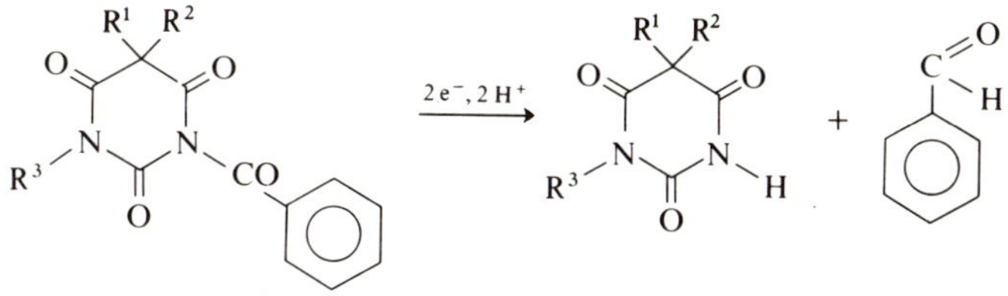

R^1, R^2 barbiturate with R^3–N, N–CO–phenyl $\xrightarrow{2e^-,2H^+}$ barbiturate (R^1, R^2, R^3–N, N–H) + benzaldehyde (C=O, H + phenyl)

C—O

CH_3–N methoxy-O-bridged structure $\xrightarrow{2e^-,H^+}$ CH_3–N structure with OCH_3, O^-

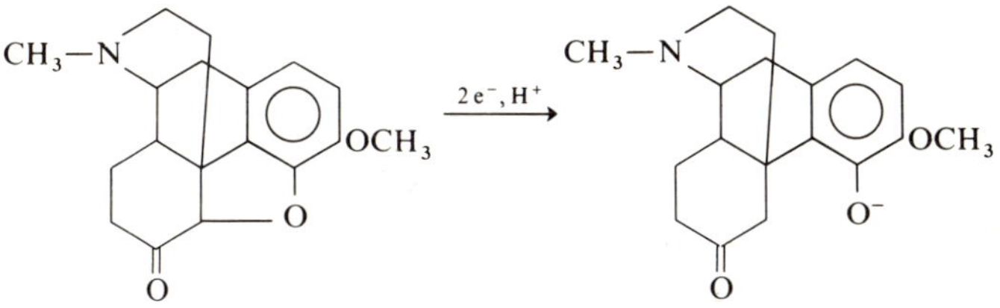

C—Hal

$$R-CH_2-Br \xrightarrow{2e^-,H^+} R-CH_3 + Br^-$$

$$C_6H_6Cl_6 \xrightarrow{6e^-} C_6H_6 + 6\,Cl^-$$

N—N

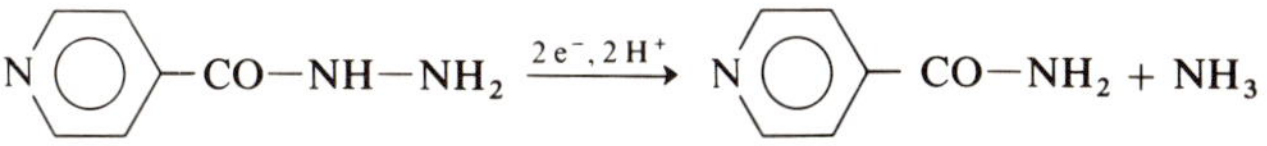

$$N\text{-pyridyl}-CO-NH-NH_2 \xrightarrow{2e^-,2H^+} N\text{-pyridyl}-CO-NH_2 + NH_3$$

O—O

peroxide $>\!\!-\!\!\langle O-O \rangle\!\!-\!\!< \xrightarrow{2e^-,2H^+}$ diol HO—...—OH

S—S

$$HOOC-\underset{NH_2}{CH}-CH_2-S-S-CH_2-\underset{NH_2}{CH}-COOH \xrightarrow{2e^-,2H^+}$$

$$2\ HS-CH_2-\underset{NH_2}{CH}-COOH$$

C=C

$$RO-\langle\ \rangle-CH=CH-\langle\ \rangle-OR \xrightarrow{2e^-,2H^+}$$

$$RO-\langle\ \rangle-CH_2-CH_2-\langle\ \rangle-OR$$

C=O

$$F-\langle\ \rangle-CO-CH_2-CH_2-CH_2-N\overset{\frown}{\underset{\smile}{\ }}N-\langle\ \rangle_{OCH_3}$$

$$=Ar-CO-Alk \xrightarrow{e^-,H^+} Ar-\overset{\cdot}{C}OH-Alk$$

$$Ar-\overset{\cdot}{C}OH-Alk \nearrow \begin{array}{l} Ar-\underset{|}{C}OH-Alk \\ Ar-\overset{|}{C}OH-Alk \end{array}$$

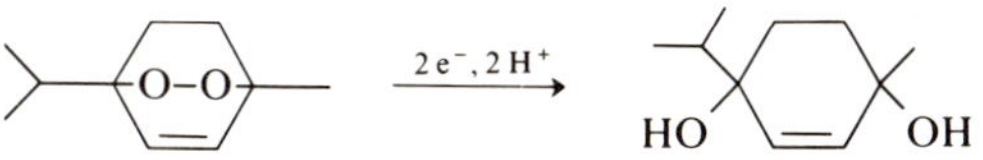

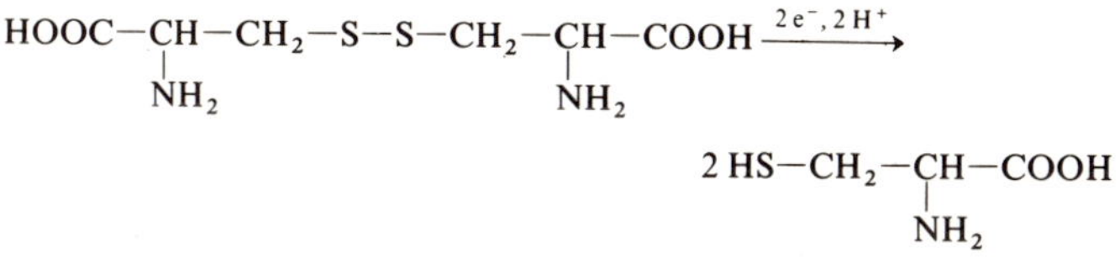

$$Ar-\overset{\cdot}{C}OH-Alk \xrightarrow{e^-,H^+} Ar-CHOH-Alk$$

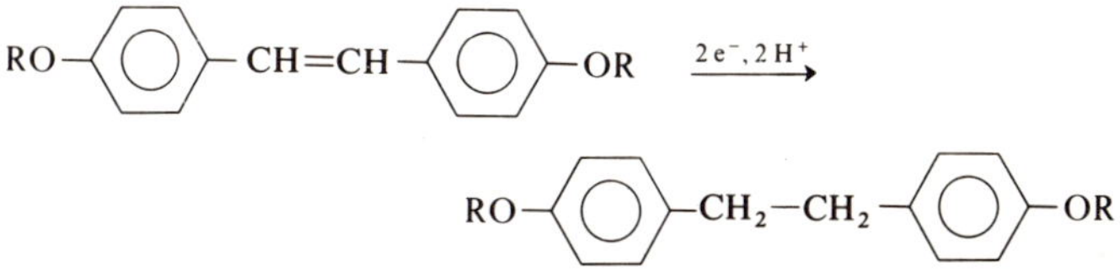

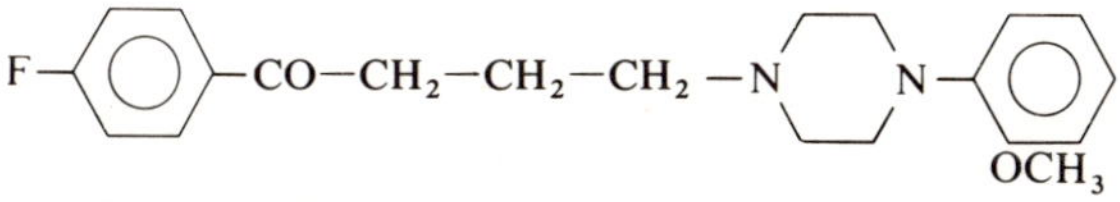

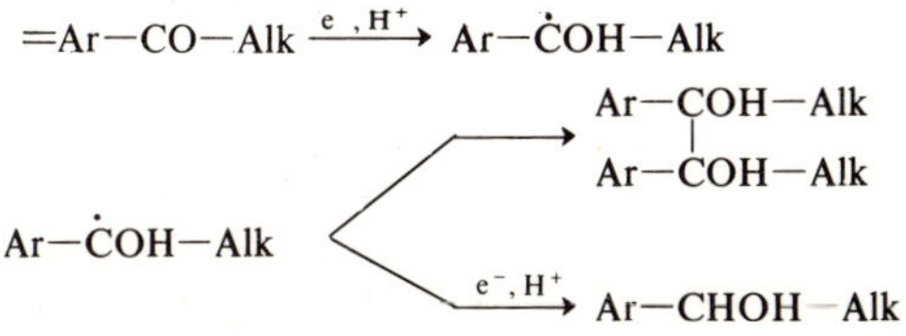

TABLE 1—*contd.*

Bond grouping	Example

C=N

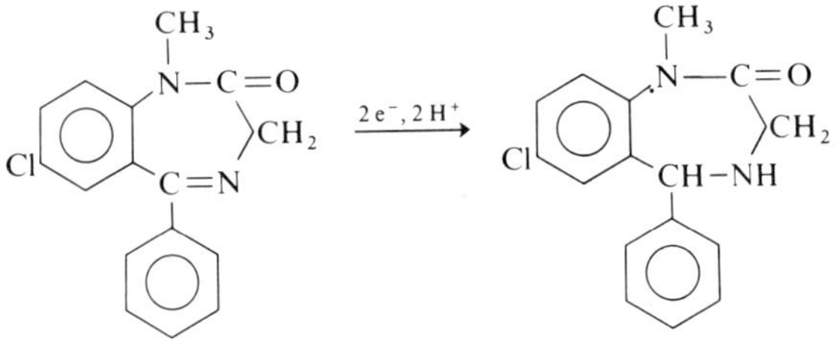

$$\text{(structure: 7-chloro-1-methyl-5-phenyl-1,3-dihydro-2H-1,4-benzodiazepin-2-one, with N—C=O, CH}_2\text{, C=N)} \xrightarrow{2e^-,2H^+} \text{(reduced product with CH—NH)}$$

N=N

$$\text{C}_6\text{H}_5\text{—N=N—(aryl)—NH}_2 \text{ (with NH}_2\text{)} \xrightarrow{2e^-,2H^+} \text{C}_6\text{H}_5\text{—NH—NH—(aryl)—NH}_2 \text{ (with NH}_2\text{)}$$

N → O $\text{CH}_3\text{—(aryl)—CH}_3$ with $\text{NH—CO—CH}_2\text{—N}{=}(\text{C}_2\text{H}_5)_2$ and $\overset{\downarrow}{\underset{\text{O}}{}}$ $\xrightarrow{2e^-,2H^+}$ $\text{CH}_3\text{—(aryl)—CH}_3$ with $\text{NH—CO—CH}_2\text{—N}{=}(\text{C}_2\text{H}_5)_2$

N=O

$$\text{CH}_3(\text{CH}_2)_3\overset{+}{\text{N}}\text{H}\overset{\text{N=O}}{\underset{}{}}\text{—(aryl)—COO—CH}_2\text{CH}_2\text{—N}{=}(\text{CH}_3)_2 \xrightarrow{4e^-,4H^+}$$

$$\text{CH}_3(\text{CH}_2)_3\text{—N}\overset{\overset{+}{\text{NH}_3}}{\underset{}{}}\text{(aryl)—COO—CH}_2\text{CH}_2\text{—N}{=}(\text{CH}_3)_2$$

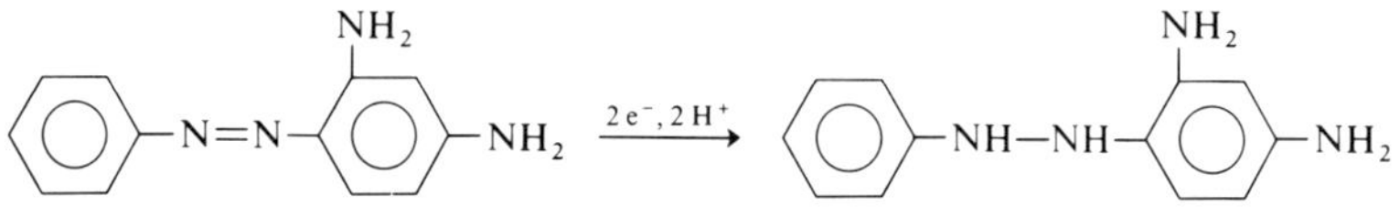

NO₂

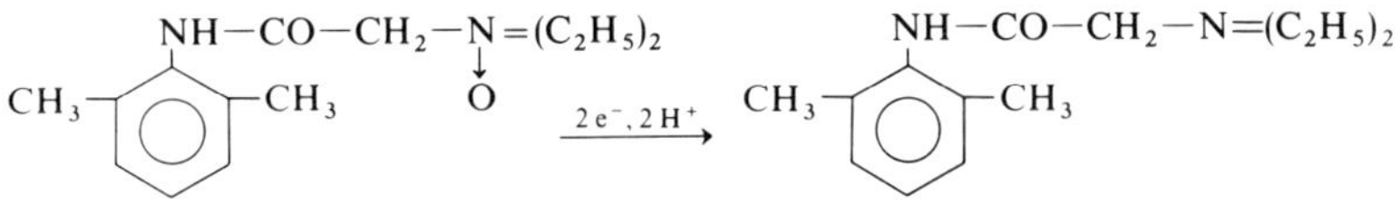

$$\text{O}_2\text{N—(aryl)—CHOH—}\overset{\overset{\text{NH—CO—CHCl}_2}{|}}{\text{CH}}\text{—CH}_2\text{OH} \xrightarrow{4e^-,4H^+}$$

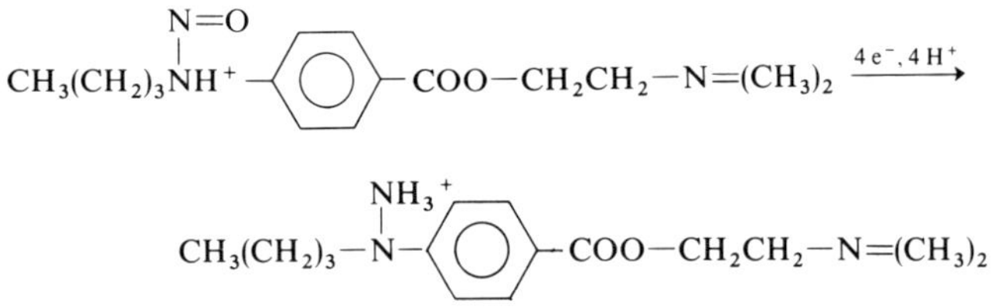

$$\text{HONH—(aryl)—CHOH—}\overset{\overset{\text{NH—CO—CHCl}_2}{|}}{\text{CH}}\text{—CH}_2\text{OH}$$

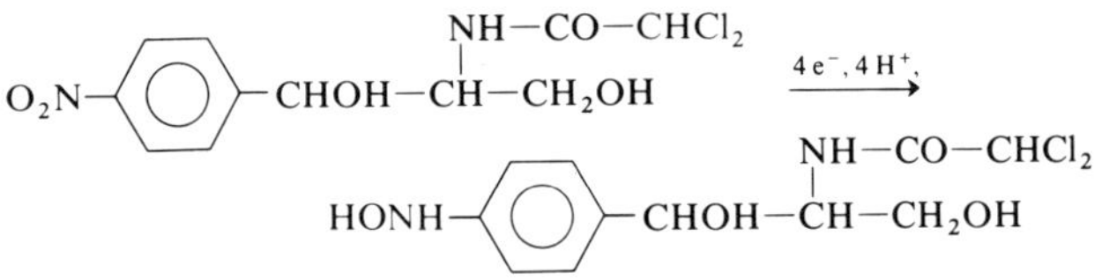

C=N
(heterocyclic)

$$\text{(quinoline with CO—NH—CH}_2\text{CH}_2\text{—N}{=}(\text{C}_2\text{H}_5)_2 \text{ and O—C}_4\text{H}_9) \xrightarrow{2e^-,2H^+}$$

$$\text{(dihydroquinoline with CO—NH—CH}_2\text{CH}_2\text{—N}{=}(\text{C}_2\text{H}_5)_2, \text{ H, N—H, and O—C}_4\text{H}_9)$$

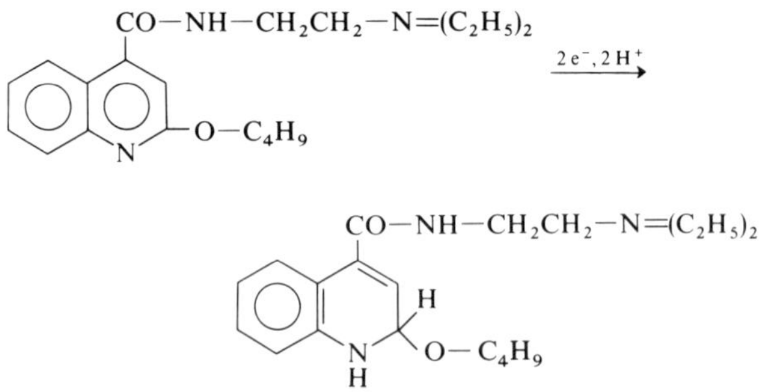

practical significance for pharmaceutical analysis. Except in a few other compounds, a reductive fission of the single C—N bond occurs only in quarternary phenyl- and alkylammonium salts and then only at very negative electrode potentials. Thus these substances are usually suitable supporting electrolytes even for compounds demanding rather negative electrode potentials for reduction. The reduction of the single C—O bond at the mercury electrode is also restricted to a few cases.

A better reducibility, increasing in the sequence Cl < Br < I, is shown by the halides. Polyhalides have a greater reducibility than the monoderivatives. Vicinal and geminal dihalides are reduced without consumption of protons. Under fission of the second C—X bond and release of the second halogen atom a double bond is formed.

Isolated C=C double bonds are non-reducible in the potential range attainable with the mercury electrode. However, the reduction of conjugated double bonds gives defined polarographic steps. A similar behaviour is shown by C≡C triple bonds.

Among the aromatic hydrocarbons, benzene is non-reducible while naphthalene gives a two-electron reduction step and higher polycyclic aromatics may give several steps.

A pronounced dependence on the nature of the substituent is shown in the reduction of the carbonyl group. Aldehydes give at relatively positive potentials (ca −1 volt) a well-defined step, while ketones are reduced at more negative potentials. A conjugation as well as an amino group in the β-position facilitates significantly the reduction of the keto group. The same is true for the thiones. Ketone derivatives, such as imines, oximes or hydrazones, are often applied for indirect ketone determinations due to their convenient reducibility. Amine oxides and nitroso compounds have no great significance as pharmaceuticals, but they are important for the indirect determination of tertiary and secondary amines. Usually the nitro group is reduced in a four-electron step which is followed by a two-electron step. As numerous compounds, non-accessible to polarographic reduction, may be easily nitrated they are frequently determined by polarography after transformation into the respective nitro derivatives. The relatively large height and the usually well-defined form of the step corresponding to the reduction of the nitro group allow the determination of comparatively small quantities (1).

The C=N double bond in heterocycles (pyridines, pyrimidines, purines, quinolines, quinoxalines, isoalloxazines, acridines and similar derivatives) is reducible at the mercury electrode and this forms the basis for a number of polarographic determination methods for drugs containing the above-mentioned heterocyclic structures. The values of the half-wave potentials depend on the nature of the π-electron system, the position of the heteroatoms in the ring and the nature of the substituents (2, 3).

Further reducible groups such as the C≡N triple bond, the N—S, O—S, C—S or S—Hal bonds are of no significance for pharmaceutical analysis.

Determination methods based on the oxidation of substances are rare, as depending on the nature of the supporting electrolyte, the mercury electrode can be applied only up to potentials of $+0.3$ volts versus SCE. As an example of pharmaceutical interest, ascorbic acid is mentioned in this connection. The development of electrodes which can be polarized much further in the anodic direction, i.e. platinum or graphite electrodes, has led to a number of analytical applications. Examples are the direct determination of morphine (4) and of tocopherol (5).

III. INDIRECT METHODS

Besides numerous compounds which are determinable via a polarographically active group, there are a great many substances which cannot be reduced or oxidized at the mercury electrode, although polarography would also be a very advantageous analytical approach for these substances (viz. Section V, p. 305). In these cases an indirect method can frequently be applied. In principle one may characterize several possibilities commonly applied with small variations in pharmaceutical analysis:

1. The introduction of a polarographically active group into an inactive molecule;
2. The defined oxidation of the substance to be determined and the subsequent reduction of the oxidation product;
3. The amperometric titration;
4. Methods based on catalytic waves.

Among the methods mentioned previously nitration plays a dominant role as it is frequently a rapid and quantitative chemical reaction and usually gives well-defined final products. However, even if several products are formed nitration may be applied if experimental conditions are selected which secure reproducible values. Excess of the nitration mixture does not interfere with the determination if the pH is adjusted to give an alkaline medium and consequently the experimental effort to be taken is made minimal.

A survey of the possible applications of this approach is given in reference 6. Recent investigations deal mainly with studies on the influence of pharmaceuticals (7), of tablet additives (8–11) and further auxiliary materials (12, 13), on the application of nitration in eluates from column chromatography separations (14) and on the possible determination of substances suitable for nitration in blood, urine and tissues (15–17).

Besides the direct nitration with HNO_3/H_2SO_4 mixtures the nitration catalysed by nitrous acid has gained in importance. Lund (18) has shown in

1958 that if morphine is nitrated under conditions first proposed by Bagges-gar-Rasmussen (19) nitromorphine rather than nitrosomorphine is formed. Studies of the kinetics and chemistry of the formation of nitrophenacetin (20) or of nitro-p-hydroxybenzoic acid esters (21) also furnished evidence that for reactive aromatics the nitro derivatives are formed by nitrosating nitration. The reaction proceeds under mild conditions. Thus, other substances and auxiliary materials frequently remain unchanged, as has been shown for the nitration of emetin (22), phenacetin (7, 23), phenatone (24) and chlorogenic acid (25). However, in the presence of sugars and other easily oxidizable substances, a nitration should not be considered before these substances have been separated. In the presence of basic compounds, as for instance lidocaine and thiamine, an enhancement of reaction rate and the formation of dinitro products has been observed (21). Thus a prerequisite for every case is the elucidation of the optimal conditions. Even after isolation of the substance to be determined from cerebral substances (26), tablets (27, 28), syrups (12), plant material (29, 30) or sorbents from thin layer chromatography plates (31), the nitration method provides the possibilities of detecting very small quantities. In testing the capacity of an ion exchanger with morphine, the time for the morphine determination, being 3 to 3·5 hours if colorimetric measurements were made, could be reduced to 40 to 50 minutes if indirect polarographic analysis was applied (32).

Secondary amines also become accessible via nitrosating. The N-nitroso derivative formed quantitatively gives in acid buffers, where it is reduced to $R_2H\text{-}NH_3{}^+$, a four-electron step, while in alkaline solutions the reduction goes in a two-electron step to $R_2HN^+ + N_2O$. In this way, concentrations down to $10^{-5}\,M$ become determinable. This method is particularly suitable for the analysis of local anaesthetics of the type of the basic esters of N-alkyl-p-aminobenzoic acid (viz. tetracaine) (33).

An important and frequently applied indirect polarographic method is the determination of ketones after transformation into the corresponding ketimines. The equilibrium of the reaction:

$$\diagdown \!\!\!\!\!\diagup \!\! C{=}O + H_2N{-}R \rightleftharpoons \diagdown \!\!\!\!\!\diagup \!\! C{=}N{-}R$$

adjusts rapidly and lies almost completely on the ketimine side. Non-conjugated ketones may be reduced in principle directly at the mercury electrode. However, the reduction occurs at rather negative potentials (acetone $-2·2$ volts versus SCE) and consequently the corresponding steps are frequently ill defined. On the contrary the ketimines are more easily reduced (acetonimine $-1·52$ volts versus SCE) and give steps which can be evaluated well.

In the same way as the ketimines, the hydrazones also offer good possibilities for indirect polarographic analysis of ketones. The Girard T or D reagents trimethylglycylhydrazide or dimethylglycylhydrazide have been proved successful for a number of ketosteroids. This method well known for a long time (the first studies were made by Wolfe (34) in 1940) has become in the meantime a standard procedure for the indirect polarographic determination of ketones (viz. references 6 and 35).

In oscillopolarographic studies Kitaev and collaborators (36) found for some phenylhydrazones, semi- and thiosemicarbazones, besides the peak related to the reduction of the C=N double bond at −0·15 volts (SCE), a further kinetically controlled peak.

The reciprocal of the outlined method for ketones is the determination of amino acids and mixtures of them after addition of formaldehyde (37).

Other preliminary chemical reactions which serve to introduce a polarographically active group have no general importance in pharmaceutical analysis but may lead to good results in special cases as, for instance, in the determination of citric acid as dibromoacetic acid (38). Unsaturated alcohols such as phytol (39) or allyl alcohol (309) may be determined indirectly by polarography after reaction with HNO_2. Thymol and cavacrol have been polarographed after reaction with 2,4-dinitrofluorobenzene and other hydroxy compounds after formation of their dinitrobenzoate (40, 41) derivatives. Another possibility for indirect polarographic determinations is offered by the formation of salts or complexes with inorganic ions. In this way the barbiturates have been determined as their mercury compounds (42, 43), aminophenazone after reaction with an excess of cadmium iodide (44) and amino acids (45) or various alkaloids (viz. Solanidine and solanine (46)) via their copper complexes.

Some substances may be converted into polarographically active compounds by adjustment of the pH to a suitable value. 2,5-Bis-aminooxymethyl-3,6-diketopiperazine, an impurity of cycloserine, forms quantitatively in alkaline solution 2,5-dimethylene-3,6-diketopiperazine, which gives a well-defined two-electron step (47). Thioxolone is converted in alkaline solution into 2,2'-dithiodiresorcinol (48) and carnitine becomes determinable after transformation into crotonobetaine (49). A method for the analysis of cyclamate is based on its reaction with HNO_2 in acetic acid solution leading to the formation of sulphate ions which are precipitated with lead ions (50).

For the oxidation of polarographically inactive compounds to products which may be subsequently reduced at the electrode, various oxidants have been used. The best known method is the periodate oxidation of alcohols with several hydroxyl functions according to Malaprade. While formic acid is usually titrated if large amounts of alcohol are present in the sample, one determines small alcohol quantities via the formaldehyde. The colorimetric determination with chromotropic acid demands that the formaldehyde is distilled off while by polarography a direct determination of the formaldehyde in the sample solution in the presence of periodate and iodate is possible, as these anions are reduced at potentials sufficiently distant from the reduction of the formaldehyde and thus a good separation of the respective steps is secured. As the oxidation is not interrupted for the recording of the polarograms this approach is very suitable for kinetic studies of the Malaprade reaction (viz. the survey in reference 35). After oxidation with iodate, epinephrine and norepinephrine give quantitatively iodoadrenochrome and iodonoradrenochrome, respectively, which are suitable for polarographic determination (51,52). Chlorpromazine can be oxidized with bromine (53, 54) leading probably to the corresponding divalent cation (55) which is sufficiently stable. The bromine excess is easily eliminated by passing nitrogen through the solution. The subsequent reduction at the mercury electrode permits the determination of 0·1 to 5 μg ml^{-1} with good accuracy in aqueous solution as well as in urine. In the same way, chlorpromazine has been determined in presence of its directly reducible sulphoxide. After recording the step due to the sulphoxide one oxidizes and the resulting step corresponds to the sum of both substances.

A preliminary oxidation also makes possible the polarographic determination of tertiary amines (56, 57). They are not directly accessible to polarography if the molecule contains no additional active functional groups. A number of amines, especially alkaloids, cause catalytic hydrogen waves. However, the application of those waves for analytical purposes is rather limited. The main reason for this limited application is that the height of catalytic hydrogen waves is not a linear function of the concentration of the catalyst and shows, furthermore, a strong dependence on the pH value and on the buffer capacity of the supporting electrolyte. Via the oxidation of tertiary amines, with hydrogen peroxide to the corresponding amineoxides in a quantitative manner, the great group of the aliphatic, cycloaliphatic and aromatic amines becomes accessible to polarography. The excess of oxidant is easily decomposed with catalase. No interfering side reactions have been observed in the polarography of a number of pharmaceuticals.

A further possibility for indirect polarographic determinations consists of the application of polarometric titrations. With this method, introduced by Heyrovský for the determination of lead with sulphate ions, a high selectivity is attained by suitable selection of the titrant. Cysteine, some alkaloids and thiobarbituric acids can be titrated with silver nitrate. Mercurimetric methods, e.g. the use of mercury perchlorate, have been applied for the determination of gluthatione, salicylates, phenazone, 'Irgapyrine' and barbiturates. A number of basic nitro compounds gives with BiI_4^- ions and similar complexes rather insoluble precipitates and the excess of BiI_4^- then causes a current increase in the titration curve. Substances reacting with heteropolyacids (viz. $SiO_2 . 12WO_3$) may also be determined in a similar way. Platinum electrodes have been used for the polarometric titration of tetracaine with potassium bromate (58) and of tocopherols with Ce(IV) (5). A comprehensive survey of the possibilities is, for instance, given by Zuman (35). Recent developments in the field are reviewed every two years in reference 59. The method has further been applied successfully for the analysis of inorganic ions such as sulphate (60), calcium (61) and bismuth (62) in pharmaceuticals or in serum.

Coulometric titrations, where the titrant is generated by electrolysis, also belong to the indirect methods. These methods, compared with the conventional titrations, have the advantage that the result remains unaffected by uncertainties in the concentration of the titrant solution, a point of particular significance for micro determinations. Examples are the coulometric titration of phenothiazines (63–66) and of barbiturates (67).

Finally the possibility of basing the determination of substances on their surface activity is mentioned. The quantitative determination of small amounts of pilocarpine in the liquor of the eye chamber becomes possible via the suppression of the oxygen maximum (68), because the potential of maximum termination is related to the pilocarpine concentration.

Detergents, such as ammoniumbutylnaphthalenesulphonate (69) or polyethylene-glycolether (70), cause, for the step of nitrobenzoic sulphonic acid, a fission into two steps which is proportional to detergent concentration in the range 0·01 to 0·1 g l^{-1}.

IV. INSTRUMENTATION

In this section only a brief outline description of the respective techniques will be given. For a more detailed discussion of this topic which is beyond the scope of this contribution, reference should be made to the corresponding literature (71–74).

A. DC Polarography

Conventional dc polarography is still the most frequently and widely applied method in pharmaceutical analysis. The commercial recording instruments (for a survey viz. Nürnberg and Wolff (75)) offer a degree of accuracy, reproducibility and operating simplicity quite sufficient for most problems. The recorded polarogram is also a permanent documentation of the analytical result. The determination limit of 10^{-5} to 10^{-6} M is caused by the magnitude of the capacity current which reaches at this concentration level the magnitude of the faradaic current and can thus not be exceeded in conventional dc polarography for fundamental reasons. Compensation of the capacity current possible with all instruments leads to less distorted polarograms but cannot increase the sensitivity significantly. For precise measurements the recording of the current–potential curve of the respective blank supporting electrolyte solution is strongly recommended. Among the versions of polarography which have been further developed, the techniques with controlled drop time have gained particular importance. They make it possible for the drop time to be restricted to 0·1 to 1 sec, compared with the 3 to 5 sec drop time of the conventional dropping capillary, with much higher rates for the potential scan and the current registration. Thus only $\frac{1}{5}$ to $\frac{1}{10}$ of the time for the recording of a conventional dc polarogram is now demanded (rapid polarography). This advantage is particularly important for production control. Due to the short drop time each drop is polarized by a quasi-stationary potential despite the rapid potential scan. Simultaneously the current oscillations become smaller which permits damping to be reduced and thus undistorted polarograms are obtained.

Controlled drop time and a predetermined measuring interval in the final stage of the drop life is applied in Tast polarography. While the capacity current (i_c) decreases with the growth of the drop, the faradaic current (i_F) increases. Thus the ratio i_F/i_c is near the end of the drop life larger than at the beginning and in this way a sensitivity increase of a factor 1·5 is attained. Furthermore a higher registration rate and a smaller damping degree becomes possible. A simple device which can be attached to recording polarographs is described by Kane (76). The application of Tast polarography has been shown to be advantageous for the recording of derivative polarograms. As the larger current alterations at the beginning of the drop life and at its detachment are not recorded the oscillations become for the first derivative of a Tast polarogram smaller than those for a conventional derivative polarogram, therefore facilitating the evaluation. Due to the faster recording at shorter drop times, the reduction in sensitivity is smaller than in conventional derivative polarography. With this technique, even the precise determination of those substances which give in dc polarography insufficiently separated steps or which show due to the nearness of their step to that of the supporting electrolyte increasing limiting currents becomes possible.

If one has solutions of poor conductivity or reference electrodes with high resistance, the application of conventional polarographs where the polarizing voltage is adjusted by a voltage divider, is not possible. Compensation for the iR drop may be achieved with the aid of a second reference electrode. By introducing an impedance transformer, the potential difference between both reference electrodes measured with a high resistance is then superimposed on the polarization voltage with a low resistance. These devices are available as attachments to polarographs. More recent polarographs are constructed usually under application of operational amplifiers. The potential between working electrode and auxiliary electrode is adjusted in such a manner that the apparent voltage between both these electrodes remains constant within several millivolts. As practically no current flows between these electrodes, the solution may for instance be separated from the reference electrode by a high resistance diaphragm. The anode chamber may also be separated by a diaphragm, thus excluding in practice any contamination of the solution, a considerable advantage for the analysis of series.

Although the exact adjustment or determination of the real potential difference over the interface mercury/solution is difficult in supporting electrolytes with a rather high resistance (the probe of the reference electrode has to be as near as possible to the dropping mercury electrode) the three-electrode technique becomes the chosen method for quantitative analysis in many cases, particularly when the precise knowledge of the true half-wave potential is of minor importance. Some polarographs contain a potentiostat by which the apparent potential may be controlled for currents up to 300 milliamps and for potential differences between the working electrode and auxiliary electrode up to 100–150 volts. With these devices potential-controlled electrolysis at large electrodes is possible, which is an important aspect for the subsequent study of reaction products. For technical details the corresponding review in references 73, 75 and 77 should be consulted.

New possibilities for the application of polarographic techniques have been opened by the inclusion of data processing units (78, 79, 302–306). The high flexibility obtained in this way, both with respect to the selection of various parameters and with regard to the high registration rate, limits computer applications not only to dc polarography. On the contrary it is a domain for the rapid ac and pulse techniques. The possibilities of comparing computed and experimentally obtained currents in the form of graphical representations, to adjust various time functions for the alteration of experimental parameters or to simulate the influence of instrumental and cell parameters, are of paramount significance in the study of electrochemical reactions. For analytical purposes the following points are of most interest: improvement of the signal-to-noise ratio via repetitive registration; subtraction of current components not related to the substance to be determined; general data processing and smoothing including error calculation; documentation of results.

B. Inverse Polarography

The inverse polarography (80) (anodic stripping), where an electrolytic preconcentration of the substance to be determined is made before the recording of the polarogram, remains usually restricted to inorganic substances. At a stationary electrode (hanging drop in the form of the Kemula electrode, a noble metal electrode or a carbon paste electrode) the substance is deposited for a given time interval at a predetermined potential and with steady stirring of the solution. During the subsequent polarization in the anodic direction the deposited material is then redissolved. By this technique polarographic sensitivity is enhanced very significantly. Concentrations down to $10^{-10} M$ become determinable. Suitable electrodes and instrumental attachments for the conventional polarographs are commercially available from various companies. Pharmaceutical applications of inverse polarography are focused on determinations of metal traces in pharmaceuticals (81) or in blood, urine and tissues (82). Attempts to determine organic depolarizers by inverse polarography have, as yet, not been very successful.

C. Cathode Ray Polarography

In cathode ray polarography, the electrode is polarized over its whole potential range once (single sweep) or several times (multi sweep) during the final part of each drop life. The delay time (typically 3 sec) from the beginning of the drop life before the sweep, the sweep rate (typically 0·5 volts sec^{-1}) and the starting potential of the sweep may be adapted to the respective problem. As the sweep is restricted to the final part of the drop life, a somewhat improved ratio of i_F/i_c is obtained. In analogy to Tast polarography the i_c component related the growth of the drop surface remains rather small. The i_c component due to the high sweep rate v is proportional to v and thus easily compensated. The synchronization between the polarization and the drop time is achieved via an electronically controlled knocking-off device. The current–voltage curves recorded with an oscillograph give for reversible processes a peak which becomes flatter with the higher the degree of irreversibility. The peak height is proportional to the depolarizer concentration (307). Besides the ten times higher sensitivity (detection limit about $10^{-7} M$) compared with conventional dc polarography the further advantages of the method consist of the rapidity of recording and the absence of oscillations caused by the drop detachment. The registration of the first derivative results in a better resolution than in conventional dc polarography permitting the separation of reversible processes of equal quantities even if the peak potential difference is only 30 to 35 millivolts. The current of a more easily reducible excess component may be compensated enabling a determination to be made even at an excess ratio of 2000 : 1. If necessary an excess substance may be depleted by electrolysis at a more positive potential than the reduction of the substance to be determined. In this way, even at very unfavourable excess ratios good results may be obtained.

A further increase in sensitivity is achieved with differential cathode ray polarography. In this case two equal and synchronized capillaries are used of which one dips into the blank and the other into the sample solution. As only the difference between the currents flowing through both cells is recorded the determination of very small amounts even in very complicated matrices becomes possible. When the current difference is recorded between two equal cells containing the same sample solution, the first derivative of the current–voltage curve is obtained if both electrodes are polarized with a small time difference of 10 to 50 millisec, or if a time difference for the signals is produced via an RC link. The combination of both versions gives the second derivative. A disadvantage, which should be mentioned is that the precision of cathode ray polarography is lower than that of conventional dc polariography and that the peak height depends more sensitively on the experimental conditions, particularly the composition of the solution. Thus, cathode ray polarography is applied if traces have to be determined with reduced precision, for instance in the purity control of inorganic substances and pharmaceuticals or for the determination of micro amounts of pharmaceutical mixtures where the substance to be determined is not the component of predominant pharmaceutical efficiency.

The detection limit for a number of pharamaceuticals (3-ketosteroids, nitrofurazone, nitrofurantoin, chloramphenicol, glyceryl trinitrate and similar compounds) is 5 μg ml^{-1}, where in some cases the composition of the supporting electrolyte is of major importance (83). The detection of tetracycline in urine has been possible down to 0·1 μg ml^{-1} (84).

Further examples are the determination of chlorpromazine in the presence of the corresponding sulphoxide (53) and in urine (54) (between 0·1 and 5 μg ml^{-1}), of epinephrine and norepinephrine in aqueous solution (85), of pigments from bacteria cultures (86), of pesticide residues (87, 88) and of isomeric hexachlorocyclohexanes (89). The quantitative determination of very small amounts (0·1 to 0·01 μg ml^{-1}) folic acid has been achieved with cathode ray polarography at the hanging drop using $CdCl_2$ as a kind of 'internal standard' for the definition of the base line (90). The recording of oscillograms in some tenths of a second has been described for the study of rapid processes at a vibrating mercury drop electrode (91). The registration of the second derivative has been applied for the discrimination of very similar 17-ketosteroids (92).

In the multi-sweep version the electrode potential is altered repetitively with a given rate between two limiting values within a predetermined time interval. The direction of polarization may be selected with most of the instruments. Usually the cathodic as well as the anodic currents are recorded (cyclic voltammetry). In this way, important information on the reaction sequence at the electrode may be obtained. From the numerous applications, the study on the oxidation of epinephrine by Adams (93) is mentioned. In 1 M H_2SO_4 a cathodic peak is obtained. The corresponding anodic peak appears nearly at the same potential. The recorded currents correspond to the quasi-reversible redox reaction of protonated epinephrine and the epinephrine-o-quinone in the open-chain form. With increasing pH the peak potential difference increases steadily, i.e. the electron-transfer step becomes slower. At pH 3 an additional cathodic and anodic peak are then observed which are related to the redox couple adrenochrome/leucoadrenochrome. The leucoadrenochrome has been formed from epinephrine-quinone by intramolecular 1,4-addition.

D. Coulometry, Controlled Potential Electrolysis

Among the techniques related to dc polarography, coulometry has a certain significance for pharmaceutical analysis. As the electrical charge Q consumed in an electrode reaction is proportional to the number n of electrons transferred per molecule and to the concentration c of the electroactive substance, n or c may be determined via the measurement of Q. For the fundamental theory and technical details reference is made to the corresponding literature (94–96). The determination of n is of special interest in the development of new analytical methods and the necessary elucidation of the involved electrode processes or in the field of structure investigations. Then coulometry is usually applied at large area mercury pool electrodes at constant potential and in stirred solution. The determination of Q is nearly always performed with electronic integrators. In the absence of subsequent chemical reactions and if the electrode potential can be adjusted in such a way that further electrochemical processes are excluded, accurate results are obtained. However, a certain amount of substance (about $10^{-4}\,M$) and a potentiostat applicable up to about 200 milliamps are prerequisites.

Coulometry at the dropping mercury electrode or other solid microelectrodes may be performed with a polarograph and significantly smaller amounts of substance (down to $10^{-6}\,M$). Errors up to ± 10 per cent have to be accepted.

Concentration determinations are possible at larger electrodes or by microcoulometry. As factors like temperature, solvent viscosity or capillary characteristics remain unimportant accuracy may be better than in dc polarography. Resolution is somewhat poorer. More easily reducible substances can be separated even if they are present in greater excess by preelectrolysis. A higher precision, however, can only be achieved by an increased time expense and, especially for microcoulometry, by increased experimental carefulness. For these reasons the application of coulometry in pharmaceutical analysis is restricted to those cases in which other techniques fail. In these cases, however, good results can be obtained as was shown in the determination of dipyridamole (97) and of very low amounts of water (98).

E. AC Polarography

In ac polarography according to Breyer (viz. reference 99), the continuously increasing dc voltage is superimposed by an ac voltage (in general

50 cps) of small amplitude (10 to 50 millivolts), Peaks are observed in the corresponding ac current versus dc voltage curves, resulting from either electrochemical reactions (faradaic peaks) or changes of the double layer by adsorption, desorption or by structural changes (tensammetric peaks). Well-shaped faradaic peaks are only obtained for reversible electrode reaction, while increasing irreversibility results in flattening of the curves; hence oxygen, showing only a small current increase, need not be removed in many cases (100) in the author's opinion. A particular advantage of ac polarography results from a considerably increased resolution. With $n = 2$ only 45 millivolts difference of peak potentials are sufficient for the separation of two current peaks of approximately equal heights. Though larger differences are necessary for irreversible reactions, the separation is at any rate better than in a dc polarogram. Moreover, the significance of Breyer's ac polarography results from the possibility of determining by tensammetric peaks compounds not undergoing reduction or oxidation, if they cause a change of the double layer capacity. This is also a simple and important proof of adsorption processes, the knowledge of which may also be of interest for the practical application of classic dc polarography, because surfactants may affect the shape of current–voltage curves to a considerable extent (101). Surprisingly, only little application has been made so far of ac polarography in pharmaceutical analysis in spite of its advantages; this may be caused to some extent by the need for additional instruments. It has been shown in some work that the proficiency of this technique with respect to resolution and sensitivity is particularly suitable in solving problems of drug analysis, e.g. for the determination of ketosteroids (102, 103), various antibiotics (104), thiomersal (105), anethole trithione (106–108) or barbiturates (42).

These considerations apply even more to other polarographic ac techniques such as pulse polarography and high level faradaic rectification (109), square-wave and radio-frequency polarography as well as second-order techniques making use of higher harmonics (see reference 110). These techniques have rarely, if at all, been used for specific pharmaceutical problems, which are of more general chemical interest and are, therefore, beyond the scope of this chapter.

F. Oscillopolarography with Constant AC Amplitude

In this technique (viz. references 111 and 112), the DME is polarized by an alternating current of given constant amplitude (0.05 to 5 milliamps) and frequency (in general 50 cps) and the resulting dE/dt versus E curve is recorded with an oscillograph. This curve being in principle of elliptic shape in the presence of the supporting electrolyte only shows characteristic incisions in the presence of electroactive substances. The essential advantage of this method results from the fact that even substances not undergoing reduction or oxidation at the electrode give rise to the so-called capacitive incisions, if they affect by adsorption, desorption or structural changes the double layer capacity. The position and the relative magnitudes of the incisions resulting in oscillopolarograms are characteristic for a given compound and its quantity. Therefore, it is possible to determine a series of substances simultaneously which are not accessible by other electrochemical techniques. Some of the various applications may be mentioned: vitamins A and D (113), lysine (114), inorganic polyphosphates (115), heparin, hyaluronidase and dextran sulphate (116), reserpine and yohimbine (117), thalidomide (118), barbiturates and xanthine derivatives (43), phenothiazines (119), diphenylhydramine and tripelennamine (120), ataractics of different structure (121) and anthraquinones (122). Dusinský and Faith (123) presented a scheme by which 134 organic drugs could be identified. Even the separate determination of very similar compounds, like different tetracyclines, is possible (124, 125).

Because the quantity of dE/dt is inversely proportional to the concentration and a calibration curve, therefore, is not a straight line, only a comparatively narrow range of concentrations is accessible for content determinations. In general this range is between 10^{-3} and 10^{-4} M. This drawback is balanced by the simplicity of the method by which it is very suitable for routine production control. Oscillopolarography permits the determination of the stability of drugs or dispersions (see Section VII), or may be applied to the study of conversions (126) in a simple way.

G. Electrodes

The 'standard' electrode of polarography is the dropping mercury electrode (DME). For most problems this electrode is the most suitable because of the continuously renewing surface, making it insensitive to contaminations, and due to the large hydrogen overvoltage. Among several modifications of the usual thick-walled straight capillary, Smoler's suggestion with the plane of the capillary tip being bent through 45° has seen practical application. It offers special advantages if the ordinary drop growth may be subject to disturbances caused by the presence of surfactants, because irregularities are observed to a considerably lesser extent with this type of capillary. For fundamental investigation it is of importance in that the depletion effect of the solution in the vicinity of the electrode can be neglected with the bent capillary.

Besides the dropping mercury electrode which, because of the dissolution of mercury cannot be polarized to potentials more positive than ca. $+0.3$ volts versus SCE, rotating platinum electrodes are applied especially for oxidations. A brief discussion of the properties and application ranges of this electrode with several references is given for instance by Adams (93). It has been applied for the determination of amines (127–129, 313), methionine (130), morphine and its derivatives (4, 131) and aminophenazone (132). Platinum electrodes are suitable also for end-point determinations in amperometric titrations (58, 60, 62, 133) as well as for kinetic studies (71, 134). In aqueous media their sensitivity against oxygen and contamination of the electrode surface must be taken into consideration, because the reproducibility of the results is affected under such conditions. These interferences occur to a much smaller degree in non-aqueous solvents.

Other electrode materials like graphite, boron carbide and tungsten as well as carbon paste electrodes from graphite and nujol or bromonaphthalene have not so far been applied practically in pharmaceutical analysis.

H. Solvents

The main applications of polarographic techniques in pharmacy (i.e. the determination of contents and the control of purities) require that polarograms are obtained from which the desired quantities can easily be deduced. This is an essential aspect in the choice of the solvent. The least number of complications occurs if measurements can be made in aqueous solutions. Even the increased cell resistance upon using non-aqueous solvents may cause a distortion of the curves, if the iR-drop is not eliminated by additional instrumental appliances. With organic compounds only poorly soluble in water one would prefer, therefore, to use water-like solvent mixtures by dissolving the compound under investigation in an organic solvent which is then diluted with a suitable buffer solution. Preferably lower alcohols have been used (135–138, 197) which show sufficient dissolving properties for supporting electrolytes, especially lithium chloride or tetraalkylammonium salts. More recently increasing application has been made of dimethylformamide (DMF) (139–145); this shows exceptionally good dissolving properties. DMF additions up to 40 per cent result in virtually no distortion of the curves; only a decrease of diffusion coefficients takes place due to an increased viscosity. Interference caused by the adsorption of the depolarizer or the reaction product can frequently be avoided by the addition of DMF. An addition of DMF may also avoid an adsorption at insoluble concomitants occasionally present. The purity of commercially available DMF is in general not sufficient; it has to be purified by distillation or by column chromatography (146). Pure DMF (partially diluted with added amounts of water) has been used as a solvent with tetraalkylammonium salts as supporting electrolytes for the determination of fat-soluble vitamins (147–151) and of ketosteroids (152). Application has also been made of other solvents such as acetonitrile (97, 153–155), tetrahydrofuran (156) and dimethylsulphoxide (98).

Possibilities for the quantitative determination of pharmaceuticals in non-aqueous solvents have been tested with 24 compounds (155). A monograph (157) which appeared in 1970 deals exclusively with electrochemical reactions in non-aqueous solvents. Besides an introduction to experimental problems the behaviour of a great number of substances (classified into substance classes) is described. Therefore special reference to this monograph is made for details and an extensive survey of the literature.

V. SAMPLE PRETREATMENT

A. Solutions

The determination of the contents of solutions, drops, syrups, eye drops, injections or tinctures, can easily be done if interfering concomitants are absent. In these cases the solution is diluted with a suitable solvent and filled up with the supporting electrolyte solution to a definite volume (104, 136, 158–162). Indirect determinations (7, 11, 13, 51, 53) can also be carried out in this way. When larger amounts of additives are present, especially in the case of suspensions, then the method of standard addition (12) can be applied or a calibration curve can be established making use of a solution which contains simultaneously all additives present in the sample (145). In the presence of interfering ingredients, an extraction is necessary even for solutions and for oils, if the oil is not soluble in a solvent suitable for the polarographic determination (27).

B. Tablets

Tablets are powdered and the active ingredient is dissolved for analysis in a suitable solvent (47, 53, 104, 132, 136, 161, 163–170). Polarograms can be recorded in many cases immediately after dilution of the suspension thus obtained; the readsorption of the compound under investigation on the additives can be avoided by proper dilution or by selecting a suitable solvent such as DMF. Filtration prior to the addition of the supporting electrolyte is recommended by other instructions. This procedure is necessary especially if a reaction is to be expected between the unsolved additives and the buffer solution. If the extraction can only be carried out with a solvent immiscible with the supporting electrolyte, then a reextraction must be carried out or the solvent has to be evaporated and the residue is redissolved in a suitable medium. Special attention has to be paid to strongly surface active additives in tablets. While starch, talcum, magnesium stearate and lactose in general do not interfere, a distortion of waves corresponding to irreversible reactions and especially, of course, of adsorption controlled waves (42, 171) will be effected by gelatin, polyvinylpyrrolidone, methyl cellulose and similar compounds. However, satisfactory results can also be obtained in these cases by applying the method of standard addition (172, 173, 246). If interfering concomitants cannot be separated by a suitable choice of the solvent, then a chromatographic separation is necessary (see Section V-E, p. 307). As in the case of solutions, techniques of indirect determinations may be applied to tablets as well (8, 11, 47, 50, 53). The determination of inorganic ions, necessary, for instance, in the analysis of vitamin–mineral salt combinations, can best be carried out after ashing the tablets (174, 175).

C. Ointments, Creams, Suppositories and Similar Preparations

For a content determination of these medicaments, a separation of the fatty ingredients is always necessary. For this purpose, a suitable quantity of the preparation can be dissolved in a non-polar solvent (petroleum ether, hexane, ether, etc.), with heating if necessary, and the active ingredient is extracted by shaking with an acid or buffer solution of suitable pH value, if the substance shows sufficient solubility in the corresponding aqueous medium. Polarograms of the solutions thus obtained can immediately be recorded after making up to a certain volume. The procedure has been applied, for instance, to preparations containing chloramphenicol (161), tetracycline (104), hydrocortisone (176) or zinc oxide (177). In the case of compounds showing good water solubility and of non-emulsifying fats, a repetitive grinding of the ointment or paste with the supporting electrolyte solution is frequently sufficient (178). In the case of ingredients not soluble in water, extraction with alcohol from the molten substance is possible (179, 180). Creams can be separated by shaking with chloroform or ethyl acetate. The solution containing the fats and the active ingredient is then evaporated and the active ingredient is extracted from the residue with warm alcohol (161). Furthermore, the emulsion can be broken by warming and the addition of water and a large amount of hard paraffin. After cooling, the fats are deposited in solid form, while the active ingredient remains dissolved in the aqueous phase (181). The separation of monostearins from ointments becomes possible via the formation of urea inclusion compounds (182).

For the analysis of suppositories, based on polyethylene or gelatin, these are dissolved and extracted with water as far as possible (136, 183). Due to the high viscosity of the aqueous solution, an immediate content determination is only possible, if the quantity of the ingredients present permits a high dilution.

In spite of the necessity to isolate the active ingredient, polarography exhibits decisive advantages as compared with spectrophotometry in the content determination of fatty medicines, because the complete separation of the fats, which can only be carried out with large difficulties, is not required. The recording of polarograms is not disturbed, although the solutions obtained frequently are turbid.

D. Plant Contents and Galenical Preparations

The determination of plant contents is in general preceded by an extraction. A polarographic analysis of the contents of essential oils can be carried out in most cases by dissolving the oil in ethanol and dilution with a suitable aqueous buffer solution, e.g. for the determination of citral and citronellal in citrus oil or lemon grass oil, of anisaldehyde in the oil from star anise, anise oil or fennel oil, of cinnamic aldehyde in cinnamon oil and of carvone in dill oil. Other substances can directly be extracted with the solvent from the plant material. In this way, however, mixtures are obtained in most cases permitting only the determination of the total content of, for example, quinine and quinidine in Peruvian bark (184) or of the sum of anthraquinones in frangula bark and the bark from Oreoherzogia fallax (122). A separation prior to the polarographic analysis is necessary in these cases, if the contents of individual components are to be determined.

E. Techniques with Preliminary Chromatographic Separation

If a content determination cannot be carried out by one of the techniques described in the previous sections, because interfering ingredients or concomitants are transferred to the sample solution, then a further separation becomes necessary. Chromatographic techniques are particularly suitable for this purpose. A polarographic determination of substances in the effluent may be preferred compared with spectrophotometry, because the choice of the solvent is not restricted in the same manner as it is in the photometric determination and because particles of the solid adsorbent which get into the effluent do not cause any disturbance. This is particularly true for the evaluation of paper and thin-layer chromatograms. No general rules can be given for the choice of the most suitable separation technique, as this depends on the respective problem. Reference is made in this respect to the handbooks of chromatography. Some examples may, however, illustrate the possibilities of application. Ion exchangers have been applied in tracing chlorpromazine in urine (54), morphine (32) or for various residues of pesticides (185). Column chromatographic separation (the solid adsorbent given in parantheses) has been applied for the determination of oestradiol-benzoate (14) (Al_2O_3), adenine in the presence of vitamins B_1 and B_{12} (186) (silica gel), merobromine (187) (celite), chlorogenic acid in plant material(29) (polyamide) and furazolidone in the presence of syn- and anti-nitrofurald-oxime (188) (complexed $Ni(4\text{-picoline})_4 . (SCN)_2$ and $Ni(4\text{-picoline}_3 . (3\text{-picoline}) . (SCN)_2)$. Besides adsorption chromatography, column distribution chromatography has also been applied successfully for the separation of effective substances from capsules, tablets and solutions (189).

Paper chromatographic separations with subsequent elution have been applied, for example, for the determination of various furocoumarins (atha-manthine, pastinacine, xanthotoxine and bergaptene) in drugs (190), of khelline in the presence of khellolglucosides and coumarones (191) in tablets and for the tracing of chlorogenic acid (29). The last-mentioned example refers to an indirect method in which elution from the paper and nitrosation proceeds simultaneously. A special success was the polarographic evaluation of thin-layer chromatograms introduced in 1965 (141, 143, 192). While the elution of spots on thin-layer plates with subsequent photometric determination, introduced earlier, frequently results in the loss of 5 to 10 per cent, because the adsorbent must be separated, the polarographic determination can be carried out in the presence of the adsorbent; readsorption can be avoided by the addition of ca. 20 per cent DMF to the electrolyte solution. The detection of the spots is done best under u.v. light on plates impregnated with fluorescent pigments (e.g. zinc silicates). Interference of zinc ions from the fluorescent pigments which could occur in acid solutions, can be avoided in the presence of EDTA if reduction of the substance to be determined in acid solution occurs not too close to the hydrogen evolution. As well as silica gel, other adsorbents like Al_2O_3, magnesium silicate and polyamide are suitable. This method, proved for the tracing of 1,4-benzodiazepines, has also been applied to the determination of morphine in opium (31), quinoline in the presence of isoquinoline (193), various corticosteroids (194–196) and of insecticides like DDT and gammexane (197). Tetrahydroquinone and 2-ethyl-anthraquinone, not being separated in chromatography, have been eluated together and, because of their different half-wave potentials, determined quantitatively after separation of the remaining concomitants (198). The application of Smoler's capillary (bent through 45°) proved to be advantageous for polarography in the presence of the adsorbent (195).

In chromato-polarography (see references 35 and 199) introduced by Kemula (200), the concentration of a polarographically active substance is continuously determined in the effluent of a column. The dropping mercury electrode is polarized in this case to the potential of the limiting current of the respective compound. The rate of elution and the size of the polarographic vessel passed by the effluent must be chosen such as to avoid disturbances of a purely diffusion controlled current by convection. Chromatographic as well as polarographic properties of the compound under investigation have to be taken into account in the selection of the eluating medium. The solvent has to dissolve a sufficient amount of supporting electrolyte which, however, must not be adsorbed by the stationary phase of the column. Oxygen can be removed either by addition of sulphite or by deaerating the solution prior to the introduction into the column. A separation of brucine and strychnine communicated by Kemula (201, 202) making use of swollen rubber with benzene as the stationary phase is of pharmaceutical interest. The appearance of the alkaloids in the effluent is indicated by a catalytic hydrogen wave. Alkaloids from haplophyllum have also been determined by chromato-polarography (203).

VI. POLAROGRAPHY OF PHARMACEUTICALS

There has been a considerable increase in the application of polarographic techniques in pharmaceutical analysis since Brezina's and Zuman's comprehensive book *Die Polarographie in der Medizin, Biochemie und Pharmazie* (6) was published. This fact has been reflected also by the reviews reporting every two years on the progress made in pharmaceutical analysis (204). An attempt is made in the following to point out this development guided by a selection of some important work from the past years. This involves further applications of known techniques as well as analysis methods for new drugs. The classification of the substances according to their medical indication may not be the most logical from the electrochemical viewpoint. However, it turns out that substances having similar medical effects frequently also have related structures. Furthermore the width of application possibilities of polarographic methods is to be presented under the medical aspects with which the pharmaceutical analyst rather frequently is confronted.

Sympathomimetics of the phenylalkylamine type do not undergo direct polarographic reduction. Epinephrine and norepinephrine, however, are oxidized by periodate (51) or potassium tetrahydromercurate(II) (205) to iodoadrenochrome and iodonoradrenochrome, respectively. The oxidation proceeds quantitatively within a few minutes and well-developed waves are obtained in acetate buffer solution. In this way, epinephrine has been determined in injection solutions containing 2 to 20 μg ml^{-1} (51). The influence of copper and iron ions on the autooxidation of epinephrine has been followed in this work. A chromatographic separation is necessary for the analysis of epinephrine in biological material. Other sympathomimetics not undergoing oxidation do not interfere with this determination. Other authors (85) have determined epinephrine and norepinephrine by anodic oxidation; the indirect method, however, seems to be preferable because at the DME the anodic waves are not well shaped. Studies of the stability of epinephrine can be carried out due to the good suitability of the primary oxidation product of epinephrine, adrenochrome. For polarographic determination, ephedrine and related substances undergo quantitative nitrosation by nitrite in acetate buffer at 80°C within 15 minutes (27); the polarogram is recorded after the solution has been diluted. Ephedrine contained in tablets, syrups, oil solutions or suppositories is extracted with an organic solvent and subsequently reextracted with hydrochloric acid.

Some of the usual local anaesthetics are directly reduced at the dropping mercury electrode. Among these are the basic ketones propipocaine and dyclonine showing waves at rather negative potentials ($E_{1/2}$ ca. $-1\cdot3$ volts at pH 5·8) the wave heights being proportional to the concentration (33). Quinoline and isoquinoline derivatives such as ethylhydrocupreine, iso-amylhydrocupreine and cinchocaine, are reduced under uptake of two electrons per molecule, while the number of electrons is four in the case of quinisocaine. With the exception of cinchocaine, half-wave potentials are comparatively negative, thus allowing satisfactory content determinations only in a concentration range between 2×10^{-4} and $10^{-2} M$ due to the ascending character of the limiting currents. The wave of cinchocain ($E_{1/2} = -1\cdot02$ volts) exhibits a limiting current parallel to the residual current thus allowing its determination in $10^{-5} M$ solutions with proper accuracy (33). N-substituted basic esters of p-aminobenzoic acid can indirectly be determined by the formation of the corresponding N-nitroso compounds. Numerous compounds of this group (tetracaine, hydroxytetracaine, paridocaine, benzonatate and their homologues and isomers) have been studied (33). Nitrosation proceeds in acid solutions within 10 minutes; polarograms can be recorded immediately after making the solutions alkaline and filling up to a certain volume. Substances with a primary amino group, such as butethamine, interfere with this determination by the formation of azo compounds; also even small amounts of adrenaline interfere, whereas strongly surface-active substances result in a distortion of the waves. Local anaesthetics of tertiary amine type, such as lidocaine or stadacaine (diethylamino-p-butoxybenzoate), can be determined quantitatively after transformation into the corresponding amineoxides with H_2O_2 (56, 57).

Indirect polarographic determinations have been elaborated for the frequently used analgesics and antipyretics phenacetin and phenazone, involving nitrosating nitration (7, 23, 24). Polarograms can be recorded in the nitration solution immediately after dilution with a buffer solution. Numerous drugs frequently combined with these substances either do not interfere or can be separated by column chromatography. Aminophenazone has been determined by anodic oxidation at a rotating platinum electrode (132) the error at concentrations between 10^{-5} and $10^{-3} M$ being ± 3 per cent. Also in this case various substances and additives being present in combinations of analgesics at a ten- to twenty-fold excess do not interfere. For aminophenazone an indirect method based on the formation of the CdI_2-complex and the polarographic determination of the Cd^{2+} excess has been described (44). In the polarographic analysis of the narcotic analgesics, the determination of morphine mentioned above (Section III) which is occasionally carried out subsequent to thin-layer chromatographic separation, is of special importance. It is suitable for content determinations of drug dispersions as well as for the determination of small amounts of morphine in organs; thus, a reliable quantitative indication of 0·4 to 1 μg in

5 g of rat brain has been carried out (26). Besides, a morphine determination by oxidation at a rotating platinum electrode in 2 M sulphuric acid has been communicated (131). In this case morphine, pseudomorphine, dihydromorphinone, dihydromorphine, nalorphine and apomorphine exhibit anodic waves, whereas derivatives substituted on the oxygen atom (ethylmorphine, diacetylmorphine, benzylmorphine, codeine and thebaine) do not undergo oxidation. The keto derivatives of morphine, however, are immediately reduced at the dropping mercury electrode. If conjugation of the keto group is absent, then reduction takes place at rather negative potentials, whereas conjugated carbonyl groups are reduced more easily. This permits a separate determination of hydroxycodeinone and dihydrohydroxycodeinone in mixtures (206). For more recently developed narcotic analgesics derived from morphine, indirect methods are again applicable in the first line because these compounds do not show polarographic activity by themselves. Phenadoxone (11) and methadone (16, 17) have been nitrated at 100°C with HNO_3/H_2SO_4, and the polarograms of the nitro derivatives have been recorded after making the solutions alkaline. This method can be applied successfully for the determination of small amounts of the analgesics in tablets, solutions, blood, urine and in the liver. Ketobemidone and similar substances can be determined quantitatively after conversion to the corresponding N-oxides (56).

Dihydroethaverine may be mentioned among those more recently developed spasmolytics which can be determined by polarography. In this compound, which is an isoquinoline derivative like many substances with spasmolytic activity, the $C=N$ double bond within the heterocycle is responsible for the polarographic activity. The corresponding tetrahydro derivative cannot be reduced. Dihydroethaverine has been determined in the presence of nicotinic acid, homatropine methylbromide, isopropamide and quietidine (167). Oxyphencyclimine can be determined in weakly alkaline buffer solutions at concentrations of 12 to 60 μg ml^{-1}; the half-wave potential, however, is rather negative ($E_{1/2} = -1.53$ volts at pH 7 to 9). In a study on the stability of 0.1 to 2 per cent solutions of papaverine, concentrations of 0.002 to 0.01 per cent papaveraldine have been determined (159). Khelline derivatives are of special interest among the spasmolytically active chromones. A well-shaped wave (207) is obtained in 0.01 M Et$_4$NI with 7 per cent Na$_2$SO$_3$, the error being ± 2 per cent. The determination of khelline in the presence of chemically similar furanochromones in plant material has been carried out following extraction and paper chromatographic separation (191). Dipyridamole, acting in a similar manner to khelline mainly as a coronary dilatator, can be oxidized in acetonitrile to a stable radical which undergoes further oxidation in a second wave. A microcoulometric technique has been developed for the content determination of this substance (97). 2-Methyl-3-piperid-1-yl-3-p-tolyl-1-propanone (mydetone, mydocalm) acting specifically in the case of muscle spasma can also be traced polarographically (208). The reducibility of β-amino ketones is more facile than that of ordinary ketones (209), the effect being more pronounced in the presence of a phenyl group. Well-shaped waves are, therefore, obtained at half-wave potentials shifted to comparatively positive potentials.

Among the hypnotics, barbiturates still maintain a dominating position. Apart from the well-known determination method by anodic oxidation of the mercury salts formed, two studies may be mentioned here. Cohen (42) studied the polarographic behaviour of barbiturates making use of ac polarography according to Breyer; beside the anodic wave he found an adsorption wave. He also studied the influence of the ionic strength and of several surface active additives in tablets. It has been supposed that the formation of linear polymers which cannot take place with N-methyl barbiturates, is required for polarographic activity. In the coulometric titration (with Hg^{2+} as the electrochemically produced titrant) polymers with continuous $Hg-N$ bonds are formed (67) as well. An oscillopolarographic determination of barbiturates, usually not showing an incision on the dE/dt versus E curves, becomes possible upon addition of phenylmercury acetate or salicylate (43); a complex is formed in this case from 2 moles of phenyl-mercury and one mole of barbiturate. This determination, however, is not very specific, because all barbiturates, except the N-methyl compounds, as well as other compounds like theobromine and theophylline are active under these conditions. Moreover, the polarographic behaviour of thiobarbiturates has been studied (210, 211); their polarographic activity results also from the formation of mercury salts. A linear relation between concentration and wave height has not been found. For N-benzoyl barbituric acids the exocyclic $C-N$ bond is reduced while the further reduction of the formed benzalde-hyde leads to benzyl alcohol (212). The quinazolone derivatives methaqua-lone and ethinazone may be mentioned among the more recently developed hypnotics; a content determination in tablets can be carried out after dissolution in ethanol and dilution with the supporting electrolyte solution without further operations (213).

Among the tranquillizers, phenothiazines and 1,4-benzodiazepines have obtained special importance. While phenothiazines cannot directly be reduced at the DME, their oxidation is possible at the rotating platinum electrode (214). After oxidation with bromine, a cathodic step at the DME is obtained (53). The excess of bromine is removed during deaeration of the solution with nitrogen. By cathode ray polarography, the determination can be made within 6 minutes, and quantities of less than $1 \mu g \ ml^{-1}$ can be traced. The method has been applied to the determination of the chlorpro-mazine content of injections, tablets and syrups as well as to the stability control of these preparations when exposed to the light and the air; the products of decomposition do not interfere. Phenothiazine sulphoxides and sulphones exhibit reduction waves in tetraalkylammonium salt solutions and can thus directly be determined (215). As as already been mentioned, phenothiazine sulphoxides can be determined as well as the unchanged phenothiazines (53). This method is of special interest for investigations of metabolic processes; interfering constituents of the urine have been separated by ion exchangers (54). Successful use has been made of oscillopolarography with alternating current for the identification and the toxicologic tracing of phenothiazines (121). Although the differentiation of compounds having very similar structures is difficult, no problems arise in a reliable indication upon making use of thin-layer chromatographic preseparations. For control measurements during the synthesis of phenothiazine derivatives, the polarographic determination of the precursor diphenylamine proved successful (216); the polarographic activity of this latter compound is reached by the formation of a molecular complex with mercury acetate. This reaction proceeds within $7\frac{1}{2}$ hours at 50°C in 10 per cent acetic acid solution in methanol. The coulometric titration of a number of phenothiazine derivatives is possible with various electrochemically produced titrants (63–65).

The polarographic behaviour of 1,4-benzodiazepine has been examined in detail. Chlordiazepoxide shows three steps (144, 217, 218), of which the first two lend themselves readily to a quantitative evaluation of the polarograms. The two-electron reduction corresponding to the third step results in a contraction of the diazepine ring (219); this step is strongly dependent upon the experimental conditions. The derivatives of chlordiazepoxide, diazepam (139, 140), nitrazepam (142), oxazepam (166, 220) and medazepam (221) can also readily be determined polarographically, since they all contain the $C=N$ double bond, which is reduced at fairly positive potentials and shows well-developed steps. For nitrazepam, the four-electron reduction step of the nitro group may be used for the determination. It is possible, for example, to determine this substance with satisfactory accuracy, in tablets in the presence of a hundred-fold excess of the inert tablet additives, without prior isolation of the effective substance (142). Combination of thin-layer chromatography with polarography allows the determination of extremely small amounts of benzodiazepine in the body. In comparison, extraction from blood or urine, followed by back extraction into 1 M HCl with subsequent polarographic analysis showed a yield of only about 90 per cent (222). The p-fluorobutyrophenones triphenperidol, fluranisone and haloperidol give reduction steps due to the aromatic carbonyl group (223). A simultaneous analysis of the three substances is not possible because of the almost equal half-wave potentials.

The analysis of antihistamines of the diphenhydramine- or tripelennamine-type is possible only with indirect polarographic methods (56, 120). The phthalazo derivatives, recently introduced, are reducible forming 3,4-dihydro compounds and produce well-developed steps in acid or neutral solutions, enabling their analysis in a concentration range of 5×10^{-6} to $10^{-3} M$, with a variation of 2 per cent (224).

The antibiotics represent a large group of compounds that lend themselves readily to polarographic analysis. Penicillin itself is not polarographically active, although oscillopolarography has attained some importance here for purity control and in following the breakdown of penicillin in various media or under the action of penicillinase (225). Reduction steps at rather negative potentials are obtained for cephaloridin derivatives (cephalosporin C, cephalotin, cephaloridine). The step of cephaloridine is the most positive and is thus the best to be evaluated for quantitative purposes (226). The purity of cephaloridine has also been controlled by oscillopolarography $(dE/dt = f(E))$ (227). An attempt to determine the purity of penicillin by a catalytic step in media containing Co^{2+}, led to results no better than can be obtained by spectrophotometry (228).

Polarography was used early for quality control of chloramphenicol, especially for the determination of the toxic keto derivatives (cf. reference 6). The simultaneous determination of p-nitroacetamino-3-hydroxypropiophenone and 1-p-nitrophenyl-2-amino-1,3-propandiol is possible (229). More recent work is concerned with the determination in various preparations (161, 181). While capsules, eye- and ear-drops and preparations for the manufacture of aqueous or injection solutions require merely dilution with isopropanol, extraction of the active substance is required for lipid preparations. After separation of the lipid, the alcoholic solution of chloramphenicol may then also be polarographed, after suitable dilution with buffer. For oscillopolarographic determinations of chloramphenicol, the limit of detection has been given (83) as $0.2\ \mu g\ ml^{-1}$. Of great relevance is the fact that, due to its third step in dc polarography, and the second peak in oscillopolarography, chloramphenicol can be readily distinguished from nitrofurane derivatives.

Tetracyclines including half-synthetic derivatives (312) may also be determined polarographically, but the steps are not very predictable and the experimental conditions must be held extremely constant. Some time ago, the polarographic behaviour of oxytetracycline was once again examined in detail. It appears that the two steps observed in acid media correspond to a total of four electrons. At higher pH values, a third, middle, step develops in close proximity to the first step, so that six electrons are now transferred. Increasing the pH even higher, one observes that the first two steps join into one; in strong alkali a step height reduction begins, which gradually results in a long drawn out four-electron step (230).

The quantitative determination of oxytetracycline, α- and β-terramycine and terrinolide is affected by the breakdown products of these antibiotics. The use of ac polarography according to Breyer has proved advantageous in the analysis of tetracyclines, because of the increased sensitivity and better resolution of the peaks as compared with conventional dc polarography (104). With ac polarography, it is possible to analyse oxytetracycline and chlortetracycline together. The determination in tablets, capsules, syrups or injections has been described in this connection. Oscillopolarography is useful for distinguishing the three tetracyclines (tetracycline, oxy- and chlortetracycline (124), as well as for following their breakdown in solution (125). While the sensitivity of dc and ac polarography allows the detection of about $20 \mu g \, ml^{-1}$, it was possible to reach a detection limit of $0.1 \mu g \, ml^{-1}$ for pure solutions, and $5 \mu g \, ml^{-1}$ for urine with differential cathode ray polarography (84). In solutions, containing cells of *Escherichia coli*, the oxytetracyline concentration could be determined above $6 \mu g \, ml^{-1}$ (231).

Streptomycin, due to the aldehyde group in the 3-methylpentose, shows a kinetic double step, whose height is strongly dependent upon pH and temperature. At higher pH, only the second part of the step is independent of the reservoir height, and in $0.1 M$ NaOH a single step is obtained, whose temperature coefficient is little above that for a purely diffusion controlled step. This step is generally used for analysis (232, 233). Pure, crystalline streptomycin trihydrochloride, in 90 per cent methanol and LiCl, shows only a single step immediately after preparation of the solution. After a short time a second step appears in the polarogram, which gradually leads to the double step already mentioned above, which we obtain also from amorphous streptomycin. Because of these properties, the tautomeric transformations of streptomycin were examined (234). Novobiocin (233) is reduced in 20 per cent alcohol in a double step, owing to the reduction of the hydroxy-coumarin group. Other antibiotics, which are themselves not polarographically active, were determined indirectly by the lowering of the oxygen maximum (polymyxin B, fungicin, paramomycin), or by the production of a catalytic hydrogen evolution step (erythromycin) (233).

The quantitative determination of nitrofuran derivatives, such as nitrofural, nitrofurantoin, furazolidone, furmethonol and thiofuradene, is based on the polarographic activity of the nitro group. The four-electron step allows the detection of very small amounts (down to about $1 \mu g \, ml^{-1}$), even in preparations (145, 169, 235), in the presence of other active substances (168), and in urine (236). The simultaneous determination of different nitrofuran derivatives, however, is difficult. By the use of an electrolyte containing tetrahydrofurfuryl alcohol it was possible to determine (83) nitrofuran and nitrofurantoin by cathode ray polarography, since their peak potentials differ by 0.3 volts under these conditions. For the weakly soluble nitrofurans, acetone (169), methylcellosolve (168) and DMF (145) have been used as solvents. The nitroimidazoles behave in a similar manner to the nitrofurans, and polarographic analysis has been used for pharmaceutical preparations (169) as well as for the simultaneous detection of different isomers during synthesis control (237). Actinomycin, due to its phenoxazone-2-derivative structure, is reduced reversibly in acid buffers. At concentrations above 5×10^{-5} M of this non-protonated molecule, a dependence upon concentration of the half-wave potential is observed, because the reduction rate is slowed down by self adsorption of the actinomycin (238). Griseofulvin shows in the middle pH-range, rather high maxima, which can be suppressed by the addition of gelatin. The use of higher alcohol concentrations inhibits the formation of the maxima, but adversely affects the analytical value of the polarographic steps.

Sulphonamides can be determined in solutions of tetraalkylammonium salts, due to the hydrogen reduction step of the acid proton of the nitrogen. Although this technique is not very specific, it nevertheless permits the detection of these substances in a simple manner, especially in preparations of known composition, containing no other acid components. If the sulphonamide contains other reducible groups as well, then the polarographic analysis will be more specific. Salicylazosulphapyridine, with a half-wave potential of -0.25 volts, for example, can be determined with good accuracy at pH 5·2. The step height is lowered by human albumin, and, interestingly, this effect can be largely eliminated by the addition of 1 mg ml^{-1} of phenylbutazone (239). Acetazolamide is one of the most frequently polarographically analysed sulphonamides; in fact, the USP XVIII prescribes this method for acetazolamide tablets. In comparison with other methods of analysis (240), polarography proved to be among the best. In acetonitrile tertiary benzene- and p-toluene-sulphonamides are reduced under uptake of two electrons per molecule to the corresponding sulphinates and amine anions R_2N^-. The latter react in the absence of other proton donors with the solvent. However, if primary or secondary sulphonamides are present, their released protons react with the amine anions. The formed sulphonamide anion is not further reducible and this fact leads virtually to the consumption of one electron per molecule (241). The p-aminobenzene sulphonamides, important in therapy, have not yet been included in these investigations.

An important role is played by the isoniazide derivatives among the tuberculostatics. The known reduction of these substances in two two-electron steps was used repeatedly for the analysis of isoniazide and similar compounds (242, 243). Of the more recent work, the determination of isoniazide alongside 4-aminosalicylic acid deserves mention (244). Even in the presence of a 25-fold excess of aminosalicylic acid in 0·1 N HCl, the accurate determination of isoniazide was possible by the method of standard addition. The total concentration of aminosalicylic acid was subsequently determined spectrophotometrically. A process for the anodic oxidation of these compounds has also been described. In phosphate buffer of pH 11·2, isonicotinic acid isopropylhydrazide, p-aminosalicylhydrazide, N-benzyl-(isonicotinylhydrazino)-propionamide and glucuronolactone-isonicotinylhydra-zide gave anodic waves at potentials between -0.08 and -0.18 volts versus SCE, whose heights are proportional to concentration and which can be used for the determination of these substances in tablets (245). Hydrazones of isonicotinic acid (163, 246), e.g. oxopimelinic acid isonicotinylhydrazone, are reduced at the azomethine group. The electrode process is here very much influenced by substituents. A pH-dependent splitting of the step can be observed in the case of aromatic ketones or aldehydes, but if the aromatic ring is not directly bound to the carbonyl group forming the hydrazone, a single step is obtained (246). The determination in tablets was possible by the method of standard addition and by the use of a calibration curve. The thiosemicarbazones contain an equally well reducible $CH=N$ group, but the $C=S$ double bond and the unconjugated $C=NH$ bond in ambazone are not reducible. Since we obtain a four-electron step in acid solutions of thiacetazone and ethylureidobenzaldehydethiosemicarbazone, further reduc-tion must be taking place simultaneously, and this has been attributed to the carbonyl group (242).

Polarographic analysis of vitamins has attained large proportions. Since this field is well documented, with full literature citation in two newer monographs (247, 248), merely the points of general importance will be made here and some more recent work is mentioned. Carotins, after iodination, may be anodically determined (247). The possibility of polarographic analysis of vitamin A was recognized and exploited recently. In dioxan or DMF, with tetraalkylammonium salts, e.g. Et_4NI, vitamin A acetate shows three reduction steps at -1.9, -2.34 and -2.73 volts versus SCE (113, 151). The first is suitable for quantitative analysis while the third merges with the supporting electrolyte reduction. The influence of various proton donors and the mechanism of the reduction has been investigated by the same authors (148). The three steps of retinol and retinol acetate correspond to three separate uptakes of a pair of electrons and protons. The heights of the second and third steps decrease with decreasing concentration of proton donors, and the steps become kinetically controlled. Vitamin A aldehyde also shows, in the presence of proton donors, three steps, of which the first two are one-electron steps and probably correspond to the formation of radicals. The two steps merge into one at sufficiently high proton concentra-tions. Retinoic acid behaves similarly, but steps 1 and 2 are irreversible. Controlled-potential electrolysis indicated that the result of reduction is an elimination in the side chain; the mechanism has been discussed (148). The polarographic properties of vitamin A were used for the determination of the pure crystalline substance and its concentration in oily solutions and concentrates, in fish oils, granulates and tablets (160), and it was possible to distinguish between vitamin A esters and the corresponding alcohols. Since vitamin D_2 produces only a single step at -2.62 volts versus SCE, a simultaneous determination of both substances is possible. Oscillopolaro-grams, with alternating current of vitamin A emulsions or solutions, show, in 0·1 M HCl, two pairs of cathodic and anodic indentations, whose depths are dependent on concentration and can be used for quantitative analysis. The indentations can be attributed to adsorption and desorption of the vitamin (113).

Thiamine produces a catalytic step (with or without Co^{2+}) in unbuffered solutions, a reduction step in alkaline solution and an anodic step which probably results from the formation of the mercury salt of the thiol form. All three steps have been used for quantitative determination, but the anodic one is the best. The separation of thiamine as a Reinecke salt permits its determination in multivitamin preparations (249). Riboflavin is reduced reversibly at the DME. The rather positive half-wave potential ($E_{1/2} = -0·47$ volts, pH 7·5) means that, generally, other vitamins do not interfere so that the polarographic method can be used for the determination of riboflavin in multivitamin preparations. A comparison with other analytical methods for vitamin combinations leads to the conclusion that polarography is the best method for the simultaneous determination of riboflavin, thiamine and nicotinamide (250). Due to the effect on catalytic hydrogen evolution, riboflavin, flavin adenine dinucleotide and flavin mononucleotide can be determined in extremely small concentrations (down to $10^{-8} M$) (251). This method is restricted, of course, to solutions containing no other catalyti-cally active substances. Within the vitamin B_6 group, only pyridoxal and pyridoxal-5-phosphate are polarographically active. Half-wave potentials of $-1·47$ volts were found in solutions of LiCl, with gelatin present. Since the step begins near the electrolyte decomposition, other methods are more accurate but polarography may nevertheless often be preferred in cases where less interference by other vitamins is encountered than in other methods (232).

It has been reported that the dc polarographic determination of folic acid is possible only at relatively high concentrations since the range in which the step ($-0·9$ volts, pH 9·4) is proportional to concentration lies between 0·2 and 1 mg ml^{-1} folic acid (248). In more recent papers the con-centration range in which quantitative analysis is possible with an error of $\pm 0·5$ to 2 per cent is stated as 0·04 to 0·1 mg ml^{-1} (253, 254). The determina-tion of very small amounts of folic acid by cathode ray polarography has already been mentioned (90) (Section IV-C, p. 299).

Vitamin B_{12} may be determined either by the resulting catalytic hydrogen wave, or by determination of the cobalt (255, 256). Adsorption effects interfere with the quantitative analysis, so that the polarographic determina-tion has less importance. Polarographic analysis of vitamin B_{15} (pangamic acid) has been suggested (257).

Nicotinic acid is reduced as the undissociated form at the DME. Since the anion is not polarographically active, the composition of the electrolyte, with respect to pH and buffer capacity, must be carefully controlled. The method of standard addition is generally used. For the determination of nicotinamide, its catalytic hydrogen step is used. By the variation in height of the steps of nicotinic acid and nicotinamide, at different pH values, it is possible to analyse these two substances together. Of the more recent work on the polarographic determination of the vitamin B_6 group, that work concerning the determination in various multivitamin preparations is considered here of greatest importance. In this context the analyses of tablets containing vitamins B_1, B_2, PP and nicotinamide (258), of thiamine, riboflavin, pyridoxal and nicotinamide (165), of vitamin B_1, B_2, nicotinic acid and nicotinamide as well as riboflavin and folic acid in the presence of larger amounts of vitamin C (173, 253, 259), further the detection of adenine in the presence of vitamins B_1 and B_{12} (186), have been described.

For the quantitative analysis of vitamin C, its oxidation at the DME in acetate buffer is an excellent and often used technique. The anodic step is rarely subject to interference by other substances, and the analysis is thus very specific and can be employed for many preparations. It is especially useful for the examination of the stability of vitamin C preparations (cf. Section VII, p. 322).

Similarly to vitamin A, the polarographic behaviour of vitamin D_2 has only recently been examined (113, 147, 149, 153, 260). Suitable solvents here are DMF or mixtures of benzene and acetonitrile. A single step is obtained ($E_{1/2} = -2·62$ volts versus SCE), due to reduction of a double bond in the conjugated side chain. Accordingly, other substances related to vitamin D_2 show steps at similar potentials. The simultaneous determination of vitamin A and vitamin D is possible (113); that of transergocalciferol and tachysterol, during the control of irradiation of ergosterol, was made possible by condensation of the tachysterol with maleic acid anhydride.

Tocopherols can be determined by the anodic oxidation of the 6-hydroxy group of the chromane residue to the respective quinone (5). The two-electron step is not well developed, and can in general only be safely used for the analysis of pure substances or after suitable extraction of the vitamin. An indirect method (170) is applied, therefore, for the quantitative determination of tocopherols in natural substrates and in pharmaceuticals based upon the determination of the quinone formed by the action of Ce^{4+}. It was possible to distinguish between the mono-, di- and tritocopherylquinones (247). An amperometric titration based on the same principle is possible (247, 261). During the synthesis of tocopherol, polarographic control has been used for the detection of 5-bromo-3,6-dinitropseudocumene, trimethylbenzoquinone, trimethylhydroquinone and phytol (39, 262). Phytol was determined here by polarography, after reaction with $NaNO_2$ in acetic acid solution. The determination of tocopheronolactone in urine was possible also after isolation and hydrolysis (263).

The polarographic analysis of the K vitamins, which are derivatives of 1-methyl-1,4-naphthoquinone or of 2-methyl-1,4-naphthohydroquinone, is based on the reduction or oxidation of this part of the molecule. In many cases, this is the most suitable method since breakdown products and $NaSO_3$, used as a stabilizer, can be detected besides the substance itself. Accompanying substances may lead to a slight deformation of the steps, but this difficulty was overcome well in such cases by the standard addition method (172). The reader is referred for details to the monographs mentioned above (6, 247, 248). Only one other investigation, which bears on the polarographic behaviour of vitamin K_3 in acetonitrile, is mentioned here and where two steps were found as expected, corresponding to the reduction of the quinone to the semiquinone and to the hydroquinone (153). By microcoulometric titration with Ce(IV) good results are also obtained (264). Finally the water-soluble vitamin P (4-methylesculetin-6,7-disulphonic acid) is to be mentioned (311). Due to the coumarin structure this compound gives a one-electron step at -1.6 volts (SCE) which permits a rather selective determination from various pharmaceutical preparations.

The polarographic determination of the steroids has been described in detail by Brezina and Zuman (6, 35). Those ketosteroids conjugated with one or more double bonds are directly reducible, and conjugated ketones can be indirectly determined after conversion to the corresponding hydrazone or similar derivatives or to hydroxy compounds by treatment with 2,4-dinitrofluorobenzene or the respective chloro derivative. The reader is referred here only to the analysis of some preparations (138, 171,.176, 179, 265), as well as to the use of ac polarography (102, 103), of cathode ray polarography (83) and of separation by thin-layer chromatography followed by polarographic evaluation (194–196), which allows the detection of minute amounts. More recent work reports on the polarographic behaviour of unsaturated ketosteroids in DMF and acetonitrile. Due to a keto–enol tautomerism the reduction mechanism of these compounds depends significantly on the composition of the solvent (152, 266).

Among the anticoagulants, derivatives of hydroxycoumarin, the furocoumarins and the substituted 1,3-indandiones have attained an important role in the last few years. 4-Hydroxycoumarins are not reducible in aqueous solution with NH_4Cl as electrolyte, because they form anions stabilized by resonance. For this reason, an indirect method for warfarin, by nitration, has been suggested (13). However, a reduction takes place in acetonitrile or in 50 per cent ethanol with 0.1 N Bu_4NI (154, 267). The half-wave potentials for 4-hydroxycoumarin and warfarin are -1.26 and -1.31 volts in acetonitrile and -1.60 and -1.65 volts in 50 per cent ethanol, respectively.

The possibility that warfarin exists in an equilibrium with a cyclic hemiacetal, which could cause a difficult reduction, was eliminated by a comparison with similar cyclic compounds, since these were reduced at even more negative potentials. Substances which contain a further polarographically active group, such as acenocoumarol are accessible to direct analysis in aqueous solution (268). The polarographic behaviour of furocoumarin is similar to that of coumarin itself, i.e. one obtains a one-electron step, dependent upon concentration (269). Accordingly, a series of furocoumarins (athamantin, pastinacin, bergapten, xanthotoxin) in pharmaceutical preparations were determined (136, 190). The reduction of the two keto groups of 1,3-indandiones in acid solution takes place in two one-electron, diffusion controlled steps of pH-dependent $E_{1/2}$. In alkaline solution, only one one-electron step is obtained, which is better suited for the quantitative evaluation of the polarograms. The half-wave potentials of variously substituted compounds lie around -1.41 volts. The carboxyl group in 2-acylindandione-1,3 is not affected at the DME. In this manner, a series of compounds, the pure substances as well as in tablets and (after extraction) in blood serum, was determined (164, 270–272).

In the analysis of antiseptics, polarographic determination is focused on mercury-containing preparations. The polarographic behaviour of phenyl-mercury compounds has been examined earlier (273). Phenylmercury is formed with the uptake of one electron and the adsorbed substance is reduced further in a second one-electron step. This step is generally made use of for the quantitative evaluation of polarograms (183, 187, 274–276). Due to the adsorption effects, irregularities are seen in the drop times when using cathode ray polarography, but these can be eliminated by addition of gelatin (187), or by substantial dilution (0·4 to 20 μg ml^{-1}) (277). Here the following determinations are again mentioned: the polarographic determination of paraformaldehyde in the presence of eugenol and propipocaine (178), of peracids (performic, peracetic, perpropionic acids) in the presence of naphthol and H_2O_2 (162) and the indirect determination of thymol and carvacrol (after reaction with 2,4-dinitrofluorobenzene (41)) as well as of tioxolone (after conversion to 2,2′-dithioresorcinol (48)).

Cytostatics form a large and chemically very differentiated group, and here again only a few examples will be mentioned where successful polarographic analysis has been carried out. 6-Mercaptopurine produces an anodic step (278), which was used in the analysis of the pure substance or during synthesis control of 6(4-carboxybutyl)-thiopurine, where the above-mentioned substance may appear as an impurity (279). The ethyleneiminobenzoquinones are polarographically active due to their quinone structure (280). Other ethyleneimines, such as TEPA, tretamine or ethyleneiminohydroxybutene, produce only catalytic steps, which may be useful for analysis eventually (280). 1,6-Bis(ethyleneimino)-1,6-dideoxy-3,4-isopropylidenemannitol is reduced in weakly acid buffer under uptake of four electrons; determinations have been carried out in solutions of concentrations from 10^{-4} to 10^{-7} M (281). With respect to the large group of heterocyclic aromatic N-oxides, which are all reducible at the DME, the reader is referred to the exhaustive monograph of Ochiai (282). Of the unsaturated halogenated compounds β-4-methoxybenzoyl-β-bromoacrylic acid is mentioned here; this acid was determined polarographically simultaneously with a raw material needed for its manufacture (283). The N-alkyl-N-nitrosoureas, used in the treatment of leukemia, have also been polarographically examined in detail (284).

The alkaloids assume a special position in the polarographic literature and, despite their very heterogeneous chemical structure and physiological activity, they are often treated together. Direct polarographic analysis is possible especially for those compounds that possess either a conjugated carbonyl group or a reducible C=N double bond in an N-heterocyclic ring.

Next to this, the N-oxides play a certain role. Reviews have been published (6, 308, 285). Recent work has dealt, among others, with the polarographic determination of quinolizidine alkaloids (sparteine, etc.) (286), nuphleine (287), cytisine (288), lobeline and its derivatives (158, 289), rauwolfia alkaloids (290), glyco-alkaloids from Solanum (46) as well as alkaloids from Haplophyllum (203) and Cinchona (184, 310).

VII. FURTHER APPLICATIONS OF POLAROGRAPHIC METHODS

Among the numerous further possibilities for the application of polarography (viz. reference 35), e.g. the determination of equilibrium constants and reaction rates, and the elucidation of structure and reaction mechanism coupled with the actual electrode process, the greatest significance for pharmacy has been attained by stability studies and studies of the connection between polarographic properties and physiological activity.

While generally, in other methods of analysis, interference is often encountered from breakdown products, the use of polarography for continuous measurement is often possible without prior separation of single constituents. As examples of a series of such cases we may mention here the study of the effect of copper ions (292) on the stability of vitamin C and tests of KCNS, $K_4Fe(CN)_6$, $K_3Fe(CN)_6$ and EDTA (293) as possible stabilizers. It is possible here to follow at the same time the concentrations of oxygen and hydrogen peroxide, which are important for the reaction, from one and the same polarogram (292, 294). The decomposition of L-epinephrine can be followed either by the step due to adrenochrome or by the indirect determination of epinephrine (51) (cf. Section III, p. 295). The conditions favourable to the stability of thyroxine derivatives (295) were found by polarography, and also the decomposition of oxopimelic acid isonicotinylhydrazone in alkaline solution (163). The course of the oxidation of phenothiazine by peroxidase in propylene glycol and its dependence on water concentration was also followed polarographically (296). The elution of zinc ions from rubber stoppers and the formation of complexes between zinc and thiomerosal was proved by polarography (275).

Because of the ease and speed of the application of ac oscillopolarography for analysis, this method has been repeatedly employed in stability studies of pharmaceutical preparations. Changes in solutions of alkaloids (morphine, papaverine, sparteine, reserpine and emetine (297), and also reserpine, serpentine and sarpagine (290)) were examined in this manner. Lobeline was detected as well as its decomposition products lobelanine, lobelaninidine and the corresponding acetophenone derivatives. The decomposition of various tetracyclines (125) as well as the enzymatic hydrolysis of penicillin derivatives (225) and cephaloridines (227) was examined oscillopolarographically.

A connection between polarographic behaviour and physiological activity is especially conclusive in such cases where the mechanism of the activity rests upon the formation of radicals or complexes, whose properties can be detected polarographically. Reference is made to previous work (6) on the connection between $E_{1/2}$ and physiological efficiency of acridines. The half-wave potentials of the most efficient of the acridines was more negative than -0.4 volts and distribution phenomena also play a decisive role. During a study of bipyridilium salts, it was possible to show that the formation of a radical was responsible for the herbicidal activity of these compounds. Beyond this, it may be concluded from a comparison of the polarographic behaviour of the mostly inactive benzylviologen and the active morphamquat, that solubility and adsorbability of the compounds and the ease of further oxidation of the radical are crucial for their efficiency (298). Tests of various phenylenediamines on powdery mildew (Sphaerotheca fuliginea) and rust (Uromyces appendiculatus) and comparison with the polarographic half-wave potentials showed that, probably, the semiquinone appearing as the oxidation product is responsible for the effect (299). A linear relationship was found between half-wave potentials of some N-alkyl-N-nitroso-ureas and the rate constants of OH^--catalysed solvolysis. Although no relation could be found between the available data and the antileukemic and antivirial activity of these compounds, their instability seems to play a role in their effect (284). The antimicrobacterial activity and the polarographic properties of quinolinol metal complexes suggested that the oxidation potential of these compounds is of importance for the interaction with enzyme systems (300). Some attention was given also to a connection between half-wave potential and physiological activity during a study of ethyleneimine- and chlormethine-derivatives of benzoquinone. The authors (301) take the view that, besides the cytostatically effective substituents, the altered redox potential is responsible for the activity; the cytostatic effect increases when the half-wave potential is more positive than -0.5 volts versus NCE.

VIII. FINAL REMARKS

The foregoing survey shows that polarographic analysis of pharmaceuticals has decisive advantages, which can be summarized as follows:

1. Since a large number of substances applied in therapy are also electro-chemically active, the usefulness of the method is not restricted to merely a few substances. Furthermore, the field of application is extended by indirect methods. The simultaneous determination of several substances is possible, even when some component is in large excess relative to the others.

2. For the analysis of pharmaceutical products, the separation of the excipients is in many cases not necessary and this simplifies the preparation of the samples.

3. Turbid and coloured solutions, which make other methods impossible, can be analysed.

4. The sensitivity of polarography is sufficiently high to enable the deter-mination of small amounts of active substances or traces of (perhaps toxic) impurities.

5. Only small volumes of samples are necessary, and these are later still available for further examination.

6. The method makes possible continuous control, which may be of importance during production surveillance.

In conclusion for the polarographic study and determination of new substances with unknown electrochemical properties the following rules are recommended:

1. The solubility in water or, if necessary, in non-aqueous solvents, is to be tested. The stock solution used should usually have a concentration between 10^{-2} and 10^{-3} moles litre^{-1}. The substance to be examined must not be precipitated by dilution with the supporting electrolyte. On the other hand, in cases where the non-aqueous solvent is used it must not lead to precipitation of salts from the stock solution (this danger exists, for example, with phosphate buffers in ethanolic solutions). The solubility in various solvents also gives useful information for later extraction procedures, should this be necessary.

2. The next step is to examine the polarographic behaviour in relation to the pH of the (buffered) solution, and to find out the optimum composition of the supporting electrolyte. Together with this, one should test the stability of the substances in the various media.

3. The most important finding for an analytical method is whether and in which concentration range the step height is proportional to the concentration of the substance to be determined. The analysis should be carried out in a range where the limiting current is neither pH- nor time-dependent. If at all possible, diffusion-controlled steps should be used for quantitative determination.

4. It is important to test whether the development of the polarographic step is influenced by other substances also present in the samples. Special attention should be paid to surface active substances, which may, especially for irreversible electrode processes, shift the half-wave potential and may also alter the value of the limiting current. In some cases different concentrations of electrolyte may lead to different steps, as, for example, when chemical reactions occur with the substance to be determined (borates with hydroxyketones and primary amines with ketones) or when adsorption effects influence the rate of the reaction. If the composition of the solution is unknown concentration determinations by the standard addition method are to be preferred to the application of calibration curves.

5. For the qualitative characterization of a substance, the half-wave potential/pH relationship and the number n of electrons transferred per molecule, obtained from the step height, are useful.

6. After these studies, using the pure substance, a procedure may be worked out to separate interfering components from the sample. By the application of techniques with higher resolution capability (e.g. derivative, ac and cathode ray polarography), however, such separations can be avoided in many cases.

REFERENCES

1. H. W. Nürnberg and G. Wolff, *Z. Anal. Chem.*, **216**, 169 (1966).
2. J. Volke, *Talanta*, **12**, 1081 (1965).
3. B. Janik and P. J. Elving, *Chem. Rev.*, **68**, 295 (1968).
4. A. Rashid and R. Kalvoda, *Cesk Farm.*, **20**, 143 (1971), (*Chem. Abstr.*, **75**, 52856b (1971)).
5. L. Lucarini, M. Cospito, and G. Raspi, *Farmaco, Ed. Prat.*, **25**, 39 (1970), (*Chem. Abstr.*, **73**, 7306v (1970)).
6. M. Brezina and P. Zuman, *Die Polarographie in der Medizin, Biochemie und Pharmazie*, AVG Leipzig 1956; *Polarography in the Medicine Biochemistry and Pharmacy*, Interscience Publishers, New York, 1958.
7. G. Brockelt, *Pharmazie*, **20**, 136 (1965).
8. G. Bozsai and G. Vastagh, *Pharm. Zentralhalle*, **103**, 403 (1964).
9. A. Danek and H. Strozik, *Dissert. Pharm.*, **18**, 519 (1966), (*Chem. Abstr.*, **67**, 47115q (1967)).
10. A. Danek, A. Madej and W. Ztark, *Dissert. Pharm.*, **17**, 329 (1965), (*Chem. Abstr.*, **64**, 11030g (1966)).

11. M. Skóra-Zietek, *Acta Polon. Pharm.*, **23**, 123 (1966), (*Chem. Abstr.*, **65**, 10426h (1966)).
12. U. Rutkowska, *Biul. Inst. Roslin. Leczniczych*, **9**, 94 (1963), (*Chem. Abstr.*, **61**, 2907c (1964)).
13. M. Skóra-Zietek and J. Fidelus, *Mikrochim. ichnoanalyt. Acta*, **1965**, 988, (*Anal. Abstr.*, **14**, 2303 (1967)).
14. H. Sachweh, G. Seidenglanz and J. Richter, *Arzneimittelstandardisierung*, **7**, 697 (1966), (*Chem. Abstr.* **67**, 25428s (1967)).
15. J. Jindřichová, V. Vortel, A. Fingerland, K. Jindrák and L. Chrobák, *Vnitřni Lékař*, **11**, 995 (1965), (*Chem. Abstr.*, **64**, 8837b (1966)).
16. M. Skóra, *Dissert. Pharm.*, **15**, 433 (1963), (*Chem. Abstr.*, **61**, 2910b (1964)).
17. M. Skóra-Zietek, *Dissert. Pharm.*, **17**, 301 (1965), (*Chem. Abstr.*, **65**, 2607b (1967)).
18. H. Lund, *Acta Chim. Scand.*, **12**, 1444 (1958).
19. H. Baggesgaard-Rasmussen, C. Hahn and K. Ilver, *Dansk Tidskr. Farm.*, **19**, 41 (1945).
20. H. Oelschläger, H. Hoffman and U. Matthiesen, *Arch. Pharmaz.*, **302**, 43 (1969).
21. S. Tammilehto and M. Perala, *Pharm. Acta Helv.*, **46**, 351 (1971).
22. A. De Marco and E. Mecarelli, *Boll. Chim. Farm.*, **109**, 516 (1970).
23. C. Toporski, Dissertation, Frankfurt/Main 1968 (c.f. 195).
24. H. Oelschläger and D. Hamel, *Arch. Pharmaz.*, **302**, 847 (1969).
25. J. Davidek and Yu. Sil'yanova, *Biokhimiya*, **30**, 927 (1965), (*Chem. Abstr.*, **64**, 5757e (1966)).
26. T. Johannesson and K. Milters, *Radiometer News*, **5**, 3 (1964).
27. A. De Marco and E. Mecarelli, *Farmaco, Ed. Prat.*, **22**, 795 (1967), (*Chem. Abstr.*, **68**, 43202p (1968)).
28. B. Zsadon and T. Paal, *Herba Hung.*, **8**, 157 (1969), (*Chem. Abstr.*, **72**, 103784z (1970)).
29. J. Davidek and Yu. Sil'yanova, *Sb. Vyz. Skoly Chem.-Technol.Praze, Potravin. Technol.*, **10**, 55 (1966), (*Chem. Abstr.*, **66**, 26491y (1967)).
30. R. J. Gajan, W. R. Benson and J. M. Finocchiaro, *J. Assoc. Offic. Agric. Chem.*, **48**, 958 (1965).
31. D. Heusser and E. Jackwerth, *Deutsch. Apotheker Ztg.*, **105**, 107 (1965).
32. Yu. E. Orlov, Yu. I. Ignatov and Yu. V. Shostenko, *Med. Prom. SSSR*, **18**, 44 (1964), (*Chem. Abstr.*, **60**, 15681d (1964)).
33. H. Burghardt, H. Jäger and M. v. Stackelberg, *J. Electroanal. Chem.*, **17**, 191 (1968).
34. J. K. Wolfe, E. B. Hershberg and L. F. Fieser, *J. Biol. Chem.*, **136**, 653 (1940).
35. P. Zuman, *Organic Polarographic Analysis*, Pergamon Press, Oxford, 1964.
36. Yu. P. Kitaev, G. K. Budnikov, V. A. Mikhailov and T. V. Troepol'skaya, *Izv. Akad. Nauk SSSR. Ser. Khim.*, **1967**, 292, (*Chem. Abstr.*, **66**, 121541y (1967)).
37. B. P. Zhantalai, *Biokhimiya*, **29**, 1009 (1964), (*Chem. Abstr.*, **62**, 7590c (1965)).
38. J. Mohay–Farkas, E. Szepesvary–Rath and E. Pungor, *Acta Pharm. Hung.*, **40**, 205 (1970), (*Chem. Abstr.*, **73**, 134027e (1970)).
39. G. P. Tikhomirova and S. L. Belen'kaya, *Ukr. Khim. Zh.*, **31**, 954 (1965), (*Chem. Abstr.*, **64**, 3285g (1966)).
40. W. Fürst, *Pharm. Zentralhalle*, **107**, 184 (1968).
41. W. Poethke and H. Köhne, *Pharmazie*, **22**, 639 (1967).
42. E. M. Cohen, *Dissert. Abstr.*, **26**, 108 (1965).
43. W. Fürst, *Pharm. Zentralhalle*, **103**, 341 (1964).
44. M. M. Marcu, *An. Stiint. Univ. 'Al. I. Cuza' Iasi, Sect. I C*, **1967**, 13 (1), 45, (*Chem. Abstr.*, **69**, 12972f (1968)).

45. K. Lindner, *Elelmiszervizsgalati Kozlemeny*, **10**, 3 (1964), (*Chem. Abstr.*, **64**, 7267g (1966)).
46. T. Pierzchalski, *Chem. Anal.* (*Warsaw*), **8**, 443 (1963), (*Anal. Abstr.*, **11**, 3322 (1964)).
47. E. Svatek and J. Vachek, *Czech. Pat.* 115 980, 15 September, 1965, (*Chem. Abstr.*, **65**, P 577f (1966)).
48. A. Sugii and Y. Kabasawa, *Yakugaku Zasshi*, **84**, 1138 (1964), (*Chem. Abstr.*, **62**, 5145c (1965)).
49. E. Strack and W. Z. Kunz, *Z. Physiol. Chem.*, **333**, 46 (1964).
50. Y. Kurayuki, Y. Miznoya and H. Kojima, *J. Pharm. Sci. Japan*, **86**, 890 (1966), (*Anal. Abstr.*, **15**, 1064 (1968)).
51. M. Dezelić, M. Trkovnik, R. Popovic and D. Dimitrijević, *Acta Pharm. Jugoslav.*, **17**, 81 (1967), (*Chem. Abstr.* **67**, 94042y (1967)).
52. J. Henderson and A. S. Freedberg, *Anal. Chem.*, **27**, 1064 (1955).
53. G. S. Porter, *J. Pharm. Pharmacol.*, *Suppl.*, **16**, 24S (1964).
54. G. S. Porter and J. Beresford, *J. Pharm. Pharmacol.*, **18**, 223 (1966).
55. G. Dušinský, *Pharmazie*, **13**, 478 (1958).
56. H. Hoffmann, *Arch. Pharmaz.*, **304**, 614, 741u. 849 (1971); **305**, 254 (1972).
57. H. Oelschläger and H. Hoffmann, *Arch. Pharmaz.*, **299**, 1025 (1966).
58. G. Bozsai-Bolda and L. Mosonyi-Sata, *Gyogyszereszet*, **1970**, 181, (*Chem. Abstr.*, **73**, 69892s (1970)).
59. J. T. Stock, *Anal. Chem.*, **42**, 276 R (1970); **40**, 392 R (1968) and earlier reviews.
60. M. K. Abramov and I. L. Teodorovich, *Aptechn. Delo*, **13**, 66 (1964), (*Chem. Abstr.*, **61**, 8136c (1964)).
61. M. K. Abramov and Yu. G. Semina, *Aptechn. Delo*, **13**, 58 (1964), (*Chem. Abstr.*, **61**, 8136d (1964)).
62. M. K. Abramov, *Aptechn. Delo*, **14**, 54 (1965), (*Chem. Abstr.*, **64**, 1903g (1966)).
63. G. J. Patriarche and J. J. Lingane, *Ann. Pharm. Fr.*, **28**, 511 (1970).
64. G. J. Patriarche and J. J. Lingane, *Anal. Chim. Acta*, **49**, 25 (1970).
65. G. J. Patriarche and J. J. Lingane, *J. Pharm. Belg.*, **25**, 57 (1970).
66. G. J. Patriarche, *Mikrochim. Acta*, **1970**, 950.
67. J. R. Monforte and W. C. Purdy, *Anal. Chim. Acta.*, **52**, 25 (1970).
68. O. Hockwin, *Z. Anal. Chem.*, **216**, 255 (1966).
69. F. Peter, A. Lorinc and I. Palko, *Magy. Kem. Folyoirat*, **72**, 101 (1966), (*Chem. Abstr.*, **64**, 17872g (1966)).
70. F. Peter and I. Ozvald, *Magy. Kem. Folyoirat*, **71**, 490 (1965), (*Chem. Abstr.*, **64**, 9959g (1966)).
71. R. N. Adams, *Electrochemistry at Solid Electrodes*, Marcel Dekker Inc., New York, 1969.
72. P. Delahay, *New Instrumental Methods in Electrochemistry*, Wiley, New York, 1954.
73. D. N. Hume, *Anal. Chem.*, **36**, 200 R (1964); **38**, 261 R (1966); **40**, 174 R (1968).
74. R. S. Nicholson, *Anal. Chem.*, **42**, 130 R (1970).
75. H. W. Nürnberg and G. Wolff, *Chem. Ing. Technik*, **37**, 977, (1965).
76. P. O. Kane, *J. Electroanal. Chem.*, **11**, 276 (1966).
77. G. W. Ewing, *J. Chem. Ed.*, **46**, A 717 (1969).
78. R. G. Clem and W. W. Goldsworthy, *Anal. Chem.*, **43**, 918 (1971).
79. H. E. Keller and R. A. Osteryoung, *Anal. Chem.*, **43**, 342 (1971).
80. R. Neeb, *Inverse Polarographie und Voltammetrie*, Verlag Chemie, Weinheim, 1969.
81. D. Monnier, E. Martin and W. Haerdi, *Anal. Chim. Acta*, **34**, 346 (1966).
82. C. L. Newberry and G. D. Christian, *J. Electroanal. Chem.*, **9**, 468 (1965).

83. J. S. Hetman, *Abhandl. Deutsch. Akad. Wiss. Berlin, Kl. Chem., Geol., Biol.,* **1964**, (1), 169.
84. J. S. Hetman, *Abhandl. Deutsch. Akad., Wiss. Berlin, Kl. Chem., Geol., Biol.,* **1964**, (1), 174.
85. W. Holobut and A. Kolataj, *Acta Physiol. Polon.,* **16**, 903 (1965), (*Chem. Abstr.,* **64**, 17357e (1966)).
86. G. C. Whitnack and G. Soli, *J. Electroanal. Chem.,* **12**, 60 (1966).
87. D. O. Eberle and F. A. Gunther, *J. Assoc. Offic. Agric. Chem.,* **48**, 927 (1965).
88. R. J. Gajan, *J. Assoc. Offic. Agric. Chem.,* **48**, 1027 (1965).
89. A. Sugii, Y. Kabasawa and H. Morita, *Nippon Daigaku Yakugaku Kenkyu Hokoku,* **9**, 1 (1968), (*Chem. Abstr.,* **73**, 59337a (1970)).
90. A. H. I. Ben-Bassat, G. Frydman-Kupfer and M. Ben-Bassat, *Polarography* 1964, G. J. Hills, Ed., Vol. 2, Macmillan, London, 1964, p. 993.
91. V. G. Mairanovskii, *Zavodsk. Lab.,* **31**, 1187 (1965), (*Chem. Abstr.,* **64**, 3049c (1966)).
92. R. C. Rooney, *Chem. Ind. London,* **1966**, 881.
93. R. N. Adams, *J. Pharm. Sci.,* **58**, 1171 (1969).
94. A. J. Bard, *Anal. Chem.,* **42**, 22 R (1970); *Anal. Chem.,* **40**, 64 R (1968); and earlier reviews.
95. A. J. Bard and K. S. V. Sanathanam, *Electroanalytical Chemistry,* A. J. Bard, Ed., Vol. 4, Chap. 3, Marcel Dekker Inc., New York, 1970.
96. L. Meites, *Polarographic Techniques,* Interscience Publishers, New York, 1965, p. 523 f.
97. L. Ladanyi, G. Fauvelot, G. Marchon and J. Badoz-Lambling, *Bull. Soc. Chim. France,* **1967**, 1846.
98. M. R. Lindbeck and H. Freund, *Anal. Chem.,* **37**, 1647 (1965).
99. B. Breyer and H. H. Bauer, *Alternating Current Polarography and Tensammetry,* Interscience Publishers, New York, 1963.
100. A. M. Bond and J. H. Canterford, *Anal. Chem.,* **43**, 228 (1971).
101. B. Kastening and L. Holleck, *Talanta,* **12**, 1259 (1965).
102. J. L. Spahr and A. M. Knevel, *J. Pharm. Sci.,* **55**, 1020 (1966).
103. J. L. Spahr, *Dissert. Abstr.,* **27 B**, 4269 (1967).
104. M. E. Caplis, H. S. Ragheb and E. D. Schall, *J. Pharm. Sci.,* **54**, 694 (1965).
105. T. Omura, S. Morishita and Y. Ueda, *Bunseki Kagaku,* **19**, 941 (1970), (*Chem. Abstr.,* **73**, 80549f (1970)).
106. A. Sugii and Y. Kabasawa, *Nippon Diagaku Yakugaku Kenkyu Hokoku,* **10**, 36 (1969), (*Chem. Abstr.,* **73**, 69928h (1970)).
107. A. Sugii, Y. Kabasawa and T. Akamatsu, *Yakugaku Zasshi,* **88**, 1371, (1968), (*Chem. Abstr.,* **70**, 71120v (1969)).
108. A. Sugii and Y. Kabasawa, *Yakugaku Zasshi,* **90**, 491 (1970), (*Chem. Abstr.,* **73**, 28991e (1970)).
109. H. W. Nürnberg, *Fortschr. Chem. Forschg,* **8**, 241 (1967).
110. H. W. Nürnberg and G. Wolff, *Chem. Ing. Technik,* **38**, 160 (1966).
111. J. Heyrovsky and R. Kalvoda, *Oszillographische Polarographie mit Wechelstrom,* Akademie Verlag, Berlin, 1960.
112. R. Kalvoda, *Die Technik der oszillopolarographischen Messungen,* Steinkopff, Dresden, 1965.
113. V. G. Mairanovskii and G. I. Samokhvalov, *Zh. Vses. Khim. Obshchestva im. D. I. Mendeleeva,* **9**, 358 (1964), (*Chem. Abstr.,* **61**, 8132e (1964)).
114. V. Bulant, *Czech. Pat.* 118 188, 15 April, 1966, (*Chem. Abstr.,* **66**, 22297s (1967)).
115. J. Boháček and Ch. Sing, *J. Electroanal. Chem.,* **7**, 222 (1964).
116. J. Boháček and Ch. Sing, *Analyt. Biochem.* **15**, 1 (1966).

117. G. Dušinský, *Abhandl. Deutsch. Akad. Wiss. Berlin, Kl. Chem., Geol., Biol.*, **1964** (1) 176.
118. J. S. Hetman, *Chem. Zvesti*, **18**, 422 (1964), (*Chem. Abstr.*, **61**, 10538d (1964)).
119. J. Blazek, *Pharmazie*, **22**, 129 (1967).
120. E. V. Cienfuegos, *Anales Fac. Quim. Farm. Univ. Chile*, **16**, 177 (1964), (*Chem. Abstr.*, **64**, 11030e (1966)).
121. I. Hynie, J. Prokes and K. Kacl, *Česk Farm.*, **14**, 466 (1965), (*Chem. Abstr.*, **64**, 14025 (1966)).
122. W. Poethke and H. Beherendt, *Pharm. Zentralhalle*, **104**, 4 (1965).
123. G. Dušinský and L. Faith, *Pharmazie*, **22**, 475 (1967).
124. L. Faith, *Farm. Obzor*, **32**, 53 (1963), (*Chem. Abstr.*, **64**, 15676d (1966)).
125. V. Parrák, J. Ruzickova and P. Antolik, *Pharmazie*, **21**, 418 (1966).
126. G. Palyi, Z. Balthazar and A. Merenyi, *Magy. Kem. Foly.*, **73**, 103 (1967), (*Chem. Abstr.* **67**, 74447f (1967)).
127. K. K. Barnes and C. K. Mann, *J. Org. Chem.*, **32**, 1474 (1967).
128. V. Dvořák, I. Němec and J. Zýka, *Microchem. J.*, **12**, 99, 324, 350 (1967).
129. S. Wawzonek and T. W. McIntyre, *J. Electrochem. Soc.*, **114**, 1025 (1967).
130. D. Kyriacou, *Nature*, **211**, 519 (1966).
131. H. P. Deys, *Pharm. Weekblad*, **99**, 737 (1964).
132. S. V. Lugovoi and J. P. Ryazanov, *Zh. Anal. Khim.*, **22**, 1093, (1967), (*Chem. Abstr.*, **67**, 111474k (1967)).
133. I. Bozsai and M. Mosonyi, *Acta Pharm. Hung.*, **34**, 246 (1964), (*Chem. Abstr.*, **62**, 12982c (1965)).
134. P. A. Malachesky, L. S. Marcoux and R. N. Adams, *J. Phys. Chem.*, **70**, 4068 (1966).
135. V. D. Bezuglyi and T. M. Rapota, *Elektrokhimiya*, **2**. 50 (1966), (*Chem. Abstr.*, **64**, 15389a (1966)).
136. Yu. E. Orlov and N. P. Dzyuba, *Med. Prom. SSSR*, **19**, 44 (1965), (*Chem. Abstr.*, 1659f (1965)).
137. P. N. Smirnov, *Med. Prom. SSSR*, **19**, 53 (1965), (*Chem. Abstr.*, **63**, 7285g (1965)).
138. N. E. Vorob'ev and N. P. Dzyuba, *Farmatsevt, Zh. (Kiev)*, **19**, 18 (1964), (*Chem. Abstr.*, **61**, 6862d (1964)).
139. H. Oelschläger, J. Volke and H. Hoffmann, *Collect. Czechoslov. Chem. Commun.*, **31**, 1264 (1966).
140. H. Oelschläger, J. Volke and E. Kurek, *Arch. Pharmaz.*, **297**, 431 (1964).
141. H. Oelschläger, J. Volke and G. T. Lim, *Arch. Pharmaz.*, **298**, 213 (1965).
142. H. Oelschläger, J. Volke, G. T. Lim and U. Frank, *Arzneimittelforsch.*, **16**, 82 (1966).
143. H. Oelschläger, J. Volke and G. T. Lim, *Arzneimittelforsch.*, 17, 637 (1967).
144. H. Oelschläger, J. Volke, H. Hoffmann and E. Kurek, *Arch. Pharmaz*, **300**, 250 (1967).
145. J. Pasich and M. Lehmann, *Farm. Polska*, **20**, 731 (1964), (*Chem. Abstr.*, **62**, 11631d (1965)).
146. H. J. Ferrari and J. G. Heider, *Microchem. J.*, **7**, 194 (1963).
147. V. G. Mairanovskii and G. I. Samokhvalov, *Elektrokhimiya*, **1**, 996 (1965), (*Chem. Abstr.*, **63**, 11253c (1965)).
148. V. G. Mairanovskii and G. I. Samokhvalov, *Elektrokhimiya*, **2**, 62 (1966), (*Chem. Abstr.*, **64**, 15389e (1966)).
149. V. G. Mairanovskii, N. A. Bogoslovskii and G. I. Samokhvalov, *Med. Prom. SSSR*, **20**, 39 (1966), (*Chem. Abstr.*, **64**, 14025a (1966)).
150. V. G. Mairanovskii, L. A. Vakulova and G. I. Samokhvalov, *Elektrokhimiya*, **3**, 23 (1967), (*Chem. Abstr.*, **66**, 91183p (1967)).
151. I. E. Valashek and G. I. Mairanovskii, *Khim.-Farm. Zh.*, **2**, 51 (1968), (Chem. *Abstr.*, **70**, 50499k (1969)).

152. N. Shinriki and T. Nambara, *Yakugaku Zasshi*, **91**, 611 (1971), (*Chem. Abstr.*, **75**, 70700a (1971)).
153. R. Takahashi and I. Tachi, *Abhandl. Deutsch. Akad. Wiss. Berlin, Kl. Med.*, **1966**, (4), 589, (*Chem. Abstr.*, **66**, 108293z (1967)).
154. S. Wawzonek and T. W. McIntyre, *J. Electroanal. Chem.*, **12**, 544 (1966).
155. A. L. Woodson and D. E. Smith, *Anal. Chem.*, **42**, 242 (1970).
156. J. Hakl, *Chem. Listy*, **61**, 536 (1967), (*Chem. Abstr.*, **66**, 111002j (1967)).
157. Ch. K. Mann and K. K. Barnes, *Electrochemical Reactions in Nonaqueous Systems*, Marcel Dekker Inc., New York, 1970.
158. O. Mohelska, F. Machovicova, V. Parrak and E. Radejova, *Sci. Pharm., Proc.*, 25th 1965, **2**, 249, Butterworths, London, 1966, (*Chem. Abstr.*, **69**, 99337z (1968)).
159. E. Pawelczyk and H. Tadeusz, *Chem. Anal. (Warsaw)*, **13**, 617 (1968), (*Chem. Abstr.* **69**, 99427d (1968)).
160. I. A. Solunina and V. A. Devyatnin, *Prikl. Biokhim. i Mikrobiol.*, **1**, 544 (1965), (*Chem. Abstr.*, **64**, 7285g (1966)).
161. A. F. Summa, *J. Pharm. Sci.*, **54**, 442 (1965).
162. V. Žikeš, Česk. *Hyg.*, **12**, 162 (1967), (*Chem. Abstr.*, **67**, 36385b (1967)).
163. G. Allesandro and E. Mecarelli, *Bull. Chim. Farm.*, **103**, 427 (1964).
164. A. Danek, *Dissert. Pharm.*, **16**, 339 (1964), (*Chem. Abstr.*, **62**, 15994c (1965)).
165. O. Enriquez and V. Kubac, *Rev. Fac. Farm. Univ. Central Venezuela*, **3**, 249 (1962), (*Chem. Abstr.*, **61**, 10535f (1964)).
166. F. R. Fazzari and O. H. Riggleman, *J. Pharm. Sci.*, **58**, 1530 (1969).
167. G. Milch, K. Gyorbiro and B. Bittera, *Proc. Conf. Appl. Phys.-Chem. Methods Chem. Anal.*, Budapest **1**, 206 (1966), (*Chem. Abstr.*, **68**, 53286h (1968)).
168. A. Sugii, Y. Kabasawa and M. Hasegawa, *Nippon Daigaku Yakugaku Kenkyu Hokoku*, **7**, 1 (1965), (*Chem. Abstr.*, **65**, 15160f (1966)).
169. L. Vignoli, B. Cristau, F. Gouezo and C. Fabre, *Chim. Anal. (Paris)*, **45**, 499 (1963), (*Anal. Abstr.*, **11**, 5716 (1964)).
170. K. Wisser, W. Heimann and Ch. Fritsche, *Z. Anal. Chem.*, **230**, 189 (1967).
171. H. C. Chiang, *Hua Hsueh*, **1965**, 12, (*Chem. Abstr.*, **63**, 6789d (1965)).
172. K. Burger, *Talanta*, **10**, 573 (1963).
173. G. P. Tikhomirova, S. L. Belen'kaya, R. G. Madievskaya and O. A. Kurochkina, *Vopr. Pitaniya*, **24**, 32 (1965), (*Chem. Abstr.*, **62**, 12978d (1965)).
174. H. C. Chiang, *Hua Hsueh*, **1963**, 164, (*Chem. Abstr.*. **61**, 14472d (1964)).
175. L. Molle, G. J. Patriarche and A. A. Gerbaux, *J. Pharm. Belg.*, **20**, 263 (1965).
176. P. Gantés and J. P. Juhasz, *Annls. Pharm. Fr.*, **24**, 687 (1966), (*Anal. Abstr.*, **15**, 1008 (1968)).
177. A. Romano, J. Turczan and M. V. Polito, *J. Assoc. Offic. Agric. Chem.*, **48**, 1066 (1965).
178. P. Pflegel, *Pharm. Zentralhalle*, **106**, 509 (1967).
179. G. Bozsai-Bolda and L. Mosonyi-Sata, *Gyogyszereszet*, **1970**, 97, (*Chem. Abstr.*, **73**, 28955w (1970)).
180. L. Murea and F. V. Stoicescu, *Rev. Chim. (Bucharest)*, **16**, 231 (1965), (*Chem. Abstr.*, **63**, 8125d (1965)).
181. C. Russu, I. Cruceanu and M. Madgearu, *Rev. Chim. (Bucharest)*, **16**, 347 (1965), (*Chem. Abstr.*, **63**, 14639d (1965)).
182. B. J. Forman and L. T. Grady, *J. Pharm. Sci.*, **58**, 1262 (1969).
183. H. Sato and M. Shimamine, *Eisei Shikenjo Hokoku*, **83**, 59 (1965), (*Chem. Abstr.*, **65**, 19932e (1966)).
184. P. Vácha, P. Čuba, Vl. Preininger, L. Hruban and F. Šantavý, *Planta Med.*, **12**, 406 (1964), (*Chem. Abstr.*, **62**, 10825b (1965)).

185. A. Calderbank, *Residue Rev.*, **12**, 14 (1966), (*Chem. Abstr.*, **65**, 7895d (1966)).
186. G. Matta and E. S. Lopes, *Anais Azevedos*, **16**, 184 (1964), (*Chem. Abstr.*, **63**, 17800c (1965)).
187. T. M. Hopes, *J. Assoc. Offic. Agric. Chem.*, **48**, 585 (1965).
188. W. Kemula, D. Sybilska and K. Chlebieka, *Rocz. Chem.*, **39**, 1499 (1965), (*Anal. Abstr.*, **14**, 1002 (1967)).
189. J. M. Moore, *J. Pharm. Sci.*, **58**, 1117 (1969).
190. Yu. E. Orlov and N. P. Dzyuba, *Farmatsevt. Zh.* (*Kiev*), **20**, 36 (1965), (*Chem. Abstr.*, **64**, 7966h (1966)).
191. Yu. E. Orlov and N. P. Dzyuba, *Molekul. Khromatogr.*, *Akad. Nauk SSSR, Inst. Fiz. Khim.*, **1964**, 158, (*Chem. Abstr.*, **62**, 12132f (1965)).
192. J. Volke and H. Oelschläger, *Sci. Pharm.*, *Proc. 25th*, **1965**, 2, 105, Butterworths, London 1966.
193. N. M. Mazur and V. A. Devyatnin, *Med. Prom. SSSR*, **20**, 50 (1966), (*Chem. Abstr.*, **65**, 13455d (1966)).
194. J. Hakl, *Czech. Pat.*, 119, 354, 15 July 1966, (*Chem. Abstr.*, **66**, 111332s (1967)).
195. J. Hakl, *Chem. Listy*, **60**, (1966), (*Chem. Abstr.*, **64**, 12460f (1966)).
196. J. Hakl, *J. Electroanal. Chem.*, **11**, 31 (1966).
197. A. Kotarski, *Chem. Anal.* (*Warsaw*), **11**, 629 (1966), (*Chem. Abstr.*, **65**, 20774c (1966)).
198. M. V. Nikolaeva, L. N. Vertyulina, N. I. Malyugina and D. A. Vyakhirev, *Tr. Khim. Khim. Tekhnol.*, **1967**, 127, (*Chem. Abstr.*, **68**, 101640t (1968)).
199. G. F. Reynolds, *J. Polarogr. Soc.*, **12**, 27 (1966).
200. W. Kemula, *Rocz. Chem.*, **26**, 281 (1952), (*Chem. Abstr.*, **47**, 7344d (1953)).
201. W. Kemula, *Rocz. Chem.*, **29**, 653 (1955), (*Chem. Abstr.*, **50**, 3956h (1956)).
202. W. Kemula and Z. Stachurski, *Rocz. Chem.*, **30**, 1285 (1956), (*Chem. Abstr.*, **51**, 18479c (1957)).
203. E. K. Dobronravova and A. L. Markman, *Khim. Prir. Soedin.*, **2**, 333 (1966), (*Chem. Abstr.*, **66**, 79615y (1967)).
204. J. W. Sutherland, D. E. Wilson and J. G. Theivagt, *Anal. Chem.*, **43**, 206 R (1971) and earlier reviews.
205. J. Morvay and L. Kiss, *Acta Pharm. Hung.*, **39**, 198 (1969), (*Chem. Abstr.*, **72**, 359k (1970)).
206. D. R. Dzhalilov, N. G. Ermachenkova and M. I. Goryaev, *Aptechn. Delo*, **15**, 49 (1966), (*Chem. Abstr.*, **64**, 19318g (1966)).
207. Yu. E. Orlov and N. P. Dzyuba, *Aptechn. Delo*, **14**, 52 (1965), (*Chem. Abstr.*, **64**, 1903f (1966)).
208. A. Danek and J. Brandys, *Dissert. Pharm.*, **15**, 427 (1963), (*Chem. Abstr.*, **61**, 2910e (1964)).
209. P. Zuman and V. Horák, *Collect. Czechoslov. Chem. Commun.*, **27**, 187 (1962).
210. W. F. Smyth, *Proc. Anal. Chem. Conf.*, 3rd., **1970**, 2, 123; *Akad. Kiado, Budapest*, (*Chem. Abstr.*, **74**, 15750w (1971)).
211. S. K. Tiwari and T. D. Seth, *Vijnana Parishad Anusandhan Patrika*, **6**, 127 (1963), (*Chem. Abstr.*, **62**, 1518g (1965)).
212. W. Kahl and W. Pasek, *Rocz. Chem.*, **44**, 2425 (1970), (*Chem. Abstr.*, **74**, 82489f (1971)).
213. P. Pflegel, *Pharmazie*, 22, 643 (1967).
214. J. Gonzalez and J. I. Fernandez-Alonso, *An. Quim.*, **66**, 931 (1970), (*Chem. Abstr.*, **75**, 25478y (1971)).
215. N. A. Kudryavtseva, Z. V. Pushkareva and V. F. Gryazev, *Zh. Obshch. Khim.*, **35**, 14 (1965), (*Chem. Abstr.*, **62**, 13024b (1965)).

216. S. Usami, *Bunseki Kagaku*, **10**, 137 (1961), (*Chem. Abstr.*, **55**, 23187d (1961)).
217. H. Oelschläger, *Arch. Pharmaz.*, **296**, 396 (1963).
218. B. Z. Senkowski, M. S. Levin, J. R. Urbigkit and E. G. Wollish, *Anal. Chem.*, **36**, 1991 (1964).
219. H. Oelschläger and H. Hoffman, *Arch. Pharmaz.*, **300**, 817 (1967).
220. J. Volke, H. Oelschläger and G. T. Lim, *J. Electroanal. Chem.*, **25**, 307 (1970).
221. H. Oelschläger and H. P. Oehr, *Pharm. Acta Helv.*, **45**, 708 (1970).
222. G. Cimbura and R. C. Gupta, *J. Forensic Sci.*, **10**, 282 (1965).
223. J. Volke, L. Wasilewská and A. Ryvolová-Kejharová, *Pharmazie*, **26**, 399 (1971).
224. P. Pflegel and G. Wagner, *Pharmazie*, **22**, 147 (1967).
225. G. Dušinský and P. Antolík, *Česk. Farm.*, **15**, 139 (1966), (*Chem. Abstr.*, **65**, 3669f (1966)).
226. I. F. Jones, J. E. Page and C. T. Rhodes, *J. Pharm. Pharmacol.*, **20**, 45 S (1968).
227. G. Dušinský and P. Antolík, *Česk. Farm.*, **16**, 461 (1967), (*Chem. Abstr.*, **68**, 72285y (1968)).
228. N. Narasimhachari, G. R. Rao and K. S. V. Santhanam, *Current Sci. (India)*, **34**, 309 (1965), (*Chem. Abstr.*, **63**, 16756a (1965)).
229. D. Dumanović, J. Volke and R. Jovanović, *J. Assoc. Offic. Anal. Chem.*, **54**, 884 (1971).
230. V. Cieleszky and L. Sebessy, *Abhandl. Deutsch. Akad. Wiss. Berlin, Kl. Chem., Geol., Biol.*, **1964**, (1) 94.
231. I. M. Tereshin, *Lab. Delo*, **1969**, 44, (*Chem. Abstr.*, **70**, 90786h, (1969)).
232. R. Goodey, T. E. Couling and J. E. Hart, *J. Pharm. Pharmacol.*, **14**, 122 T (1962).
233. K. Kramarczyk and H. Berg, *Abhandl. Deutsch. Akad. Wiss. Berlin, Kl. Chem., Geol., Biol.*, **1964**, (1), 23.
234. L. J. Heuser, M. A. Dolliver and E. T. Stiller, *J. Am. Chem. Soc.*, **75**, 4013 (1953).
235. G. Milch, J. Hollós, K. Aczél and K. Bittera, *Pharm. Zentralhalle*, **104**, 564 (1965).
236. B. M. Jones, R. J. M. Ratcliffe and S. G. E. Stevens, *J. Pharm. Pharmacol. Suppl.*, **17**, 52 S (1965).
237. D. Dumanović, J. Volke and V. Vaigand, *J. Pharm. Pharmacol.*, **18**, 507 (1966).
238. M. Fedoronko and H. Berg, *Z. Physik. Chem.*, **220**, 120 (1962).
239. B. Nygard, J. Olofsson and M. Sandberg, *Acta Pharm. Suecica*, **3**, 343 (1966), (*Chem. Abstr.*, **66**, 49277x (1967)).
240. I. M. Roushdi, H. Abdine and W. S. Abdel Sayed, *J. Pharm. Sci. United Arab Rep.*, **4**, 171 (1963), (*Chem. Abstr.*, **64**, 3287d (1966)).
241. P. T. Cotrell and C. K. Mann, *J. Am. Chem. Soc.*, **95**, 3579 (1971).
242. L. Schlitt, M. Rink and M. v. Stackelberg, *J. Electroanal. Chem.*, **13**, 10 (1967).
243. J. Volke and V. Volková, *Collect. Czechoslov. Chem. Commun.*, **20**, 908 (1955).
244. J. W. Turczan, *J. Assoc. Offic. Anal. Chem.*, **50**, 652 (1967).
245. J. Brandys, *Dissert. Pharm.*, **18**, 319 (1966), (*Chem. Abstr.*, **66**, 22276j (1967)).
246 A. Kolusheva and N. Ninuo, *Farmatsiya (Sofia)*, **16**, 9 (1966), (*Chem. Abstr.*, **65**, 16794e (1966)).
247. E. Knobloch, *Physikalisch-chemische Vitaminbestimmungsmethoden*, Akademie Verlag, Berlin, 1963.
248. R. Strohecker and H. M. Hennig, *Vitaminbestimmungen*, Verlag Chemie, Weinheim, 1963.
249. I. E. Kruze, *Tartu Riikliku Ulikooli Toim*, **210**, 361 (1967), (*Chem. Abstr.*, **71**, 128791s (1969)).
250. M. E. Schertel and A. J. Sheppard, *J. Pharm. Sci.*, **60**, 1070 (1971).
251. E. Knobloch, *Collect. Czechoslov. Chem. Commun.*, **31**, 4503 (1966).
252. J. Liptak and E. Ulrich, *Proc. Anal. Chem. Conf.*, *3rd*, **1970**, 2, 157, Akad. Kiado, Budapest, (*Chem. Abstr.*, **74**, 45637g (1971)).

253. I. E. Kruze, *Farmatsiya* (*Moscow*), **18**, 59 (1969), (*Chem. Abstr.*, **71**, 116574k (1969)).
254. I. A. Solunina, V. A. Devyatnin and T. N. Kuznetsova, *Farmatsiya* (*Moscow*), **17**, 45 (1968), (*Chem. Abstr.*, **69**, 5249x (1968)).
255. J. C. Abbot, *Dissert. Abstr.*, **25**, 6185 (1965), (*Chem. Abstr.*, **63**, 8172f (1965)).
256. D. W. Imhoff, *Dissert. Abstr.*, **27 B**, 3811 (1967).
257. J. F. V. Serrano and A. M. Roque da Silva, *Rev. Port. Farm.*, **14**, 281 (1964), (*Chem. Abstr.*, **62**, 8937h (1965)).
258. V. A. Devyatnin and L. A. Kuznetsova, *Med. Prom. SSSR*, **18**, 58 (1964), (*Chem. Abstr.*, **61**, 2908f (1964)).
259. M. L. Girard, *Prod. Probl. Pharm.*, **19**, 47 (1964), (*Chem. Abstr.*, **60**, 14334b (1964)).
260. N. A. Bogoslovskii, V. G. Mairanovskii and E. N. Kuznetsova, *Khim. Farm. Zh.*, **1968**, 39, (*Chem. Abstr.*, **69**, 61566k (1968)).
261. M. Cospito, G. Raspi and L. Lucarini, *Anal. Chim. Acta*, **47**, 388 (1969).
262. S. L. Belen'kaya and G. P. Tikhomirova, *Novoe v Tekhnol. Pishch. Proizv., Sb.*, **1965**, 14, (*Chem. Abstr.*, **65**, 16931a (1966)).
263. H. Schmandke, *Intern. Z. Vitaminforsch*, **3**, 237 (1965), (*Chem. Abstr.*, **64**, 2403g (1966)).
264. G. J. Patriache and J. J. Lingane, *Anal. Chim. Acta*, **49**, 241 (1970).
265. R. F. Graner, *Circ. Farm.*, **24**, 213 (1966), (*Chem. Abstr.*, **67**, 25411f (1967)).
266. N. Shinriki and T. Nambara, *Yakugaku Zasshi*, **91**, 151 (1971), (*Chem. Abstr.*, **74**, 103062d (1971)).
267. K. Mnoucek, *Česk. Farm.*, **17**, 173 (1968), (*Chem. Abstr.*, **69**, 99471p (1968)).
268. M. Skóra-Zietek, *Farmacja Pol.*, **22**, 254 (1966), (*Anal. Abstr.*, **14**, 3547 (1967)).
269. R. Patzak and L. Neugebauer, *Monatsh. Chem.*, **83**, 776 (1952).
270. A. Danek, *Dissert. Pharm.*, **16**, 323 (1964), (*Chem. Abstr.*, **62**, 15993h (1965)).
271. A. Danek, *Acta Polon. Pharm.*, **19**, 345 (1962), (*Chem. Abstr.*, **61**, 1516c (1964)).
272. J. Stradinš, I. K. Tutane and G. Vanags, *Zh. Analit. Khim.*, **20**, 1239 (1965), (*Anal. Abstr.*, **14**, 4151 (1967)).
273. R. Benesch and R. E. Benesch, *J. Am. Chem. Soc.*, **73**, 3391 (1951).
274. J. Birner and J. R. Garnet, *J. Pharm. Sci.*, **53**, 1264 (1964).
275. J. Birner and J. R. Garnet, *J. Pharm. Sci.*, **53**, 1266 (1964).
276. T. M. Hopes, *J. Assoc. Offic. Anal. Chem.*, **49**, 840 (1966).
277. G. S. Porter, *J. Pharm. Pharmacol. Suppl.*, **20**, 43 S (1968).
278. G. Horn and P. Zuman, *Collect. Czechoslov. Chem. Commun.*, **25**, 3401 (1960).
279. J. Vachek, *Česk. Farm.*, **14**, 216 (1965), (*Anal. Abstr.*, **13**, 5161 (1966)).
280. G. Horn, *Abhandl. Deutsch. Akad. Wiss. Berlin, Kl. Chem., Geol., Biol.*, **1964**, (1), 37.
281. B. Jambor, I. P. Horvath and L. Institoris, *Acta Chim. Acad. Sci. Hung.*, **53**, 85 (1967), (*Chem. Abstr.*, **68**, 89896w (1968)).
282. E. Ochiai, *Aromatic Amine Oxides*, Elsevier Publishing Co., Amsterdam, 1967.
283. B. Kakác, K. Mnoućek, M. Semonský, V. Zikan and A. Cerný, *Pharmazie*, **20**, 320 (1965).
284. E. R. Garret and A. G. Cusimano, *J. Pharm. Sci.*, **55**, 703 (1966).
285. L. E. Kuchma and F. M. Shemyakin, *Farmatsiya* (*Moscow*), **19**, 59 (1970), (*Chem. Abstr.*, **73**, 69878s (1970)).
286. K. Kornfeld, M. Rink, G. van Riesenbeck and M. v. Stackelberg, *J. Electroanal. Chem.*, **6**, 54 (1963).
287. P. N. Smirnov, *USSR Pat.* 173 998, 6. August 1965, (*Chem. Abstr.*, **64**, P 4261c (1966)).
288. E. K. Dobronravova, K. A. Sabirov and T. T. Shakirov, *Khim. Prir. Soedin*, **6**, 79 (1970), (*Chem. Abstr.*, **73**, 38581k (1970)).

289. V. Parrák, E. Radejová and F. Machovićová, *Chem. Zvesti*, **18**, 369 (1964), (*Chem. Abstr.*, **61**, 9363e (1964)).
290. V. Parrák and E. Radejová, *Česk. Farm.*, **16**, 447 (1967), (*Chem. Abstr.*, **68**, 72272s (1968)).
291. H. W. Nürnberg and B. Kastening, *Polarographic and Voltammetric Techniques*, Chap. 8.1, *Methodicum Chimicum*, Houben-Weyl, Ed. F. Korte, G. Thieme Verl.; Stuttgart 1973, Academic Press, New York, 1974.
292. K. Seelert, K. Vetter and G. Schenk, *Arch. Pharmaz.*, **298**, 758 (1965).
293. M. Deželić, J. Grujić-Vasić and R. Popović, *Tehnika* (*Belgrade*), **19**, 2087a (1964), (*Chem. Abstr.*, **65**, 15163e (1966)).
294. F. Peter and G. Pályi, *Abhandl. Deutsch. Akad. Wiss. Berlin, Kl. Chem., Geol., Biol.*, **1964**, (1), 152.
295. Y. Asahi and K. Terada, *Takeda Kenkyusho Nempo*, **23**, 86 1964, (*Chem. Abstr.*, **63**, 16131f (1965)).
296. D. J. Cavanaugh, *J. Am. Chem. Soc.*, **81**, 2507 (1959).
297. P. Balatre, J. C. Gujot and M. Traisnel, *Ann. Pharm. France*, **24**, 425 (1966), (*Chem. Abstr.*, **66**, 5733n (1967)).
298. J. Volke, *Collect. Czechoslov. Chem. Commun.*, **33**, 3044 (1968).
299. G. J. Niemann and J. Dekker, *Neth. J. Plant. Pathol.*, **72**, 213 (1966), (*Chem. Abstr.*, **66**, 85018x (1967)).
300. C. G. Butler, *Kongr. Pharm. Wiss., Vortr. Originalmitt.*, **23**, Münster **1963**, 391, (*Chem. Abstr.*, **62**, 7586h (1965)).
301. H. Berg and K. H. Konig, *Analyt. Chim. Acta*, **18**, 140 (1958).
302. G. L. Booman, *Anal. Chem.*, **38**, 1141 (1966).
303. G. Lauer and R. A. Osteryoung, *Anal. Chem.*, **38**, 1137 (1966).
304. S. P. Perone, J. E. Harrar, F. B. Stephens and R. E. Anderson, *Anal. Chem.*, **40**, 899 (1968).
305. F. B. Stephens, F. Jakob, L. P. Rigdon, and J. E. Harrar, *Anal. Chem.*, **42**, 764 (1970).
306. E. R. Brown, D. E. Smith and D. D. De Ford, *Anal. Chem.*, **38**, 1130 (1966).
307. R. S. Nicholson and I. Shain, *Anal. Chem.*, **36**, 706 (1964).
308. F. Santavý, *Abhandl. Deutsch. Akad. Wiss. Berlin, Kl. Chem., Geol., Biol.*, **1964**, (1), 1.
309. G. P. Tikhomirova and S. L. Belen'kaya, *Zh. Analit. Khim.*, **20**, 727 (1965), (*Chem. Abstr.*, **63**, 3285g (1966)).
310. V. Parrák, E. Radejová and O. Mohelska, *Česk. Farm.*, **18**, 309 (1969), (*Chem. Abstr.*, **72**, 47385f (1970)).
311. A. M. Contri, *Farmaco, Ed. Prat.*, **1970**, 231, (*Chem. Abstr.*, **73**, 102102j (1970)).
312. A. Regosz and R. Kaliszan, *Farm. Pol.*, **26**, 1039 (1970), (*Chem. Abstr.*, **75**, 80313r (1971)).
313. V. B. Bezuglyi and Yu. I. Beilis, *Zh. Anal. Khim.*, **20**, 1000 (1965), (*Chem. Abstr.*, **64**, 5756e (1966)).

CHAPTER V

Analytical Aspects of Voltammetry in Non-aqueous Solvents and Melts

Janine Badoz-Lambling, *Laboratoire de Chimie Analytique Générale, associé au C.N.R.S., E.P.C.I., 10, rue Vauquelin, Paris (5 ème), France, and* Georges Cauquis, *Département de Recherche Fondamentale, Centre d'Etudes Nucléaires de Grenoble, BP 85, Grenoble, France*

I. INTRODUCTION

Voltammetry will be defined as covering all electrochemical methods of analysis dealing with the recording of the electrical intensity variations of a working electrode as a function of its applied potential (1). The simplest way in which to consider the technique or to interpret the results, is to look at the graphs, so-called current–potential curves, which represent the variation of the intensity during a slow potential sweep of either a rotating or a non-rotating microelectrode. Polarography is the name of this method when the electrode is a constantly renewed mercury drop (2).

These methods have seen new and considerable development due to the use of non-aqueous media, organic or inorganic molecular solvents and molten salts. Their properties being very different from those of water, allowed the observation of new phenomena or the study of species which could not exist in water, a universal but too reactive solvent (3–25).

How these 'analytical aspects' of voltammetry in non-aqueous media will be considered is a point which must be explained more precisely. 'Analytical aspects' include classical titrations and qualitative analysis as well as studies of reactions in solution. It is known that this is the general tendency for progress in analytical chemistry. Undoubtedly, the analytical chemist has first to understand the exact nature of the occurring phenomena before setting up the best conditions for new analytical methods. This is especially true for electrochemical techniques where reaction analysis and chemical analysis are merging.

In order to explain the electrode phenomena and to interpret the other electroanalytical methods, one should know the use and the theoretical basis of current–potential curves under a stationary diffusion state. The experimental conditions for recording current–potential curves and the role of the strong electrolyte, referred to as the supporting electrolyte, should also be known. The curves have a simple meaning only if this electrolyte is highly concentrated in the medium (1, 2).

Illustrative examples will often be taken from the authors' own research. It is not the authors' purpose to present an exhaustive bibliography of this important section of electrochemistry. Instead, a didactic introductory survey is presented with frequent references to the authors' limited personal experience in non-aqueous solvent electrochemistry. The authors hope that this approach will help the chemist beginning in this field.

Another feature is that the examples have been selected from organic and inorganic chemistry. In fact, electrochemistry in organic solvents and molten salts makes the difference between these two fields less: an organic solvent can be used to study the electrochemical properties of typical mineral species which cannot be observed in aqueous media. On the other hand some low melting point mineral solvents are useful for studying various organic substances.

For didactic reasons, this chapter will emphasize the developments related to the practical aspects of electrochemistry in non-aqueous media rather than physicochemical considerations of solutions or fused salt media which can be found in various publications and review articles (26–28).

Nevertheless, a minimum basic knowledge is needed to understand the new possibilities offered by the use of these media. The various solvent properties and their influence on electrode reactions will therefore be presented in the first part of this chapter. Phenomena related to the solvation of dissolved species and to the medium donor or acceptor properties towards various particles are considered first. They can modify the nature or the properties of species that participate in, or that are produced by, the charge transfer, causing the medium to have an indirect influence on the electrochemical reaction.

The medium can direct the charge transfer itself by influencing the electrode processes. The electrode reaction kinetics, the double layer structure and the adsorption phenomena are considerably influenced by the nature of the solvent (Section II-C). All these influences are modified by various physical factors such as temperature (Section II-D).

The most striking example of medium involvement is the electroactivity range appearing experimentally in the voltammetric curves. The electroactivity range is the potential zone where electrochemical reactions can take place without being disturbed by oxidation or reduction of the medium. In fact (see Section IV) this range is defined by the properties of the solvent and of the electrolytes dissolved for using the various electrochemical techniques, as well as by the nature of the electrode itself. As soon as the electroactivity ranges are defined for various media, the problem of media comparison arises; is it possible to select the best medium for studying a given electrochemical reaction known to take place at a given electrode potential value without oxidizing or reducing the medium at this potential? The comparison of electroactivity ranges and their respective positions in the potential scale is only one aspect of the comparison of the various media. All of the properties described earlier should be considered in selecting a medium (see Section III).

Various applications of electrochemistry in non-aqueous media to the analysis of *chemical* reactions in solution as well as to classical chemical analysis will be considered in Section V.

The electrochemical methods will then be described (Section VI) and a few auxiliary techniques will be briefly reviewed.

A table of the major properties of about 20 solvents (Section VII) will conclude this introduction of voltammetry in non-aqueous media.

II. INFLUENCE OF THE SOLVATION AND OF THE PROPERTIES OF THE MEDIUM

The influence of solvation and solvolysis phenomena on the chemical and electrochemical equilibria have to be analysed carefully in order to understand completely the analytical interest of non-aqueous media voltammetry. These phenomena are the basis of a large section of analytical chemistry. On the other hand, the effect of the same phenomena on the electrochemical reaction kinetics is very important, but unfortunately the present knowledge in this field of study is very empirical and fragmentary. Because of this, the highly desirable synthetical view of this domain is difficult to establish.

A. Solvation. Its Effect on the Ionization and Dissociation of Solutes

Solvation is defined as the interaction state when solvent molecules remain in contact with the dissolved ions or molecules. These solvent–solute interactions are of varying types: ion–dipole and dipole–dipole interactions, complex formation (especially charge-transfer complexes) and hydrogen bonding are mainly considered. Often, the exact solvation mechanism is unknown, but we must remember that the energy involved in these interactions is large enough to change considerably the solubility of the species and the nature of the bonds preexisting in the solute.

For a given A—B covalent compound, contact with the solvent first establishes the dissolution equilibrium:

$$AB_{solid} + Solvent \rightleftharpoons AB \; Solvent$$

but frequently, the dissolved molecules undergo an ionization into ion pairs which dissociate more or less completely into independent ions (dissociation). The resulting series of homogeneous equilibria should be considered.

$$AB_{solvated} \overset{K_i}{\rightleftharpoons} (A^+ . B^-)_{solvated} \overset{K_d}{\rightleftharpoons} A^+_{solvated} + B^-_{solvated}$$

K_i represents the ionization constant and K_d the dissociation constant. These constants are defined by the following equations:

$$K_i = \frac{|(A^+ . B^-)_{sol}|}{|AB_{sol}|} \qquad K_d = \frac{|A^+_{sol}| \, |B^-_{sol}|}{|(A^+ . B^-)_{sol}|}$$

The distinction between 'intimate' and 'solvent separated' ion pairs which is useful sometimes to explain reaction mechanisms, is unnecessary for our purposes.

Molecular iodine dissolved in acetic acid provides an example of a dissolution limited by an ionization equilibrium

$$I_2 + nAcOH \rightleftharpoons (I^+ . I^-)nAcOH$$

with the following constant:

$$K_i = \frac{|(I^+ . I^-)nAcOH|}{|I_{2_{sol}}|}$$

Its pK_i has been determined (29a):

$$pK_i = -\log K_i = 0.53$$

On the other hand, triphenylmethyl chloride dissolves in nitromethane according to the two following equilibria (the solvent molecules are not written):

$$(C_6H_5)_3CCl \overset{K_i}{\rightleftharpoons} (C_6H_5)_3C^+ . Cl^- \overset{K_d}{\rightleftharpoons} (C_6H_5)_3C^+ + Cl^-$$

The pK_i and pK_d values have been determined (30)

$$pK_i = -3.4 \quad \text{and} \quad pK_d = 8.0$$

In the case of species already ionized in the solid state such as NaCl the following equilibria should be considered:

$$Na^+Cl^- \text{ solid} + \text{Solvent} \rightleftharpoons (Na^+.Cl^-) \text{solvated}$$

$$(Na^+.Cl^-) \text{solvated} \rightleftharpoons Na^+ \text{ solvated} + Cl^- \text{ solvated}$$

The solubility of crystalline solids depends mainly on the solvent solvating properties and on the salt lattice energy. The dissociation constant K_d depends mainly on the solvent dielectric constant ε and on the ionic radii of the two solvated ions. The behaviour of ions and molecules in solvents is one of the most important problems in physical chemistry. The reader is referred to recent review articles (31).

The purpose of this chapter is to study the effect of solvent nature on the applications of voltammetry to analytical chemistry. The three following points should be emphasized:

1. The solubility of the species can be increased by selecting an appropriate solvent. Consequently, numerous water-insoluble substances which cannot be studied by electrochemical techniques are soluble in various solvents in which they exhibit well-defined current–potential curves. For example dibisphenylene ethylene is characterized in dimethylformamide (32) by two well-defined one-electron reduction waves. The same is true for other poly-nuclear hydrocarbons such as dibisphenylene butadiene, 1-phenyl-4-bi-phenylene butadiene, perylene, anthracene and tetracene (33). Other examples of the variation of species solubility with the nature of the solvent will be found in the section on applications to analytical chemistry.

2. Solvation is extremely important for obtaining the conducting medium necessary for voltammetry. The solvent acts on the electrolyte dissociation responsible for the medium conductivity in two different ways. First, by solvation of both electrolyte ions, the solvent determines the solvated anion and cation radii which are important factors in determining the salt dissocia-tion constant K_d.

Bjerrum first, then Denison and Ramsey and Gilkerson (34) suggested the following relationship between the dissociation constant K_d and the ionic radii r_1 and r_2 of the two ions*

$$pK_d = pK_0 + \frac{0.43 \, Ne^2}{RT} \times \frac{Z_1 Z_2}{r_1 + r_2} \times \frac{1}{\varepsilon}$$

* Other formulas for the pK_d were proposed since using the same and other parameters. The simple formula presented above is adequate for demonstration purposes.

In this formula Z_1 and Z_2 represent the ionic charges of both ions, N represents Avogadro's number, e is the electron charge and ε represents the solvent dielectric constant. pK_0 is a constant that depends on the salt considered. The pK_d decreases as the salt dissociation increases, i.e. when $r_1 + r_2$ increases. This occurs when the ions have larger solvation atmospheres. This formula also demonstrates the effect of the dielectric constant on dissociation. Dissociation increases with increasing ε.

It is important to note that ε alone cannot characterize the capacity of a given solvent to cause an ionized salt to dissociate, and to produce a conducting medium, since electrolyte solubility and solvated ion radii are also very important. For example tetrahydrofuran has considerable solvating properties with a dielectric constant of only seven, and provides a higher conducting media than dichloroethane, which is less solvating and which has a higher dielectric constant ($\varepsilon = 9$).

Nevertheless it is convenient to classify solvents into two main groups according to their dielectric constants. The first group is made up of 'dissociating' solvents with high dielectric constants ($\varepsilon > 30$) in which ion pairs are supposed to be completely dissociated. The best example is water in which ion pairs occur rarely. Methanol, ethanol, numerous amides, acetonitrile, nitromethane, etc., are also found in this group. The solvents of the second group are referred to as 'non-dissociating' solvents. These solvents cannot cause a dissociation of ion pairs because their dielectric constants are generally less than ten. Acetic acid, ethanol/benzene mixtures, etc., are examples of this group. A whole range of solvents are found between these two groups. Free ions and ion pairs are found in approximately equal proportions in this intermediate range. The limits between these groups are, of course, very indefinite because the same solvent can cause a solute to dissociate to a variable extent according to its nature or its concentration. Further information on this subject can be found in the book by Charlot and Tremillon (31). Weak current intensities are commonly used in voltammetry ($i < 20\,\mu A$). Both types of solvents can therefore be employed for recording voltammetric curves.* One should remember that in non-dissociating solvents, ohmic-drop corrections are more important and more difficult to evaluate.

Ionized molten salts are the third group of solvents. An additional electrolyte is unnecessary since the medium is generally conductive enough owing to the fused salt ionization. The solubility of species in these media depends on their solvation by the solvent ions. Thus, various metal chlorides are soluble in the fused LiCl/KCl eutectic because of metal ion solvation by the chloride ions. Highly concentrated strong electrolyte solutions in a molecular solvent, such as water or ammonia, can also be considered as solvents

---

* Conversely, other electrochemical techniques such as coulometry or controlled-potential electrolysis requiring high current intensities are often very delicate operations in non-dissociated media.

belonging to the third group. Ammonium nitrate ammoniate NH_4NO_3.
$1.3\,NH_3$ (Divers' liquid) is an example of this type of solvent. It is liquid at
room temperature and pressure and it forms an ionized medium which can
be used for voltammetric studies (35).

3. Solvation can modify solute electrochemical behaviour. Thus, for a
given redox couple, the solvation of the oxidized and reduced species varies
from one solvent to another. This produces the two following results: The
free energy ΔG_0 related to the Ox + $ne \rightleftharpoons$ Red reaction can be separated
into two parts, namely, the energy in relation to the electron exchange itself,
and the energy in relation to the fact that the solvation state of the oxidized
and reduced forms are different. Consequently, the total free energy of a redox
reaction varies with the solvent. The same is true for the corresponding
standard potential defined by $nFE_0 = -\Delta G_0$. This is expressed voltammetrically by a variation of the half-wave potentials of reversible systems
when the solvent is changed. Further consideration of this problem is found
in the section on reference electrodes. The solvation of electroactive species
has also an influence on electrochemical reaction kinetics. Thus, voltammetric
wave shifts and deformations are observed for the same redox couple when
passing from one solvent to another. This is further discussed in Section
II-C-1.

The supporting electrolyte ions can behave in the same way as the solvent
molecules with respect to solutes. The reduction of neutral molecules provides anions which form complexes of varying stabilities with the supporting electrolyte cations Li^+, Na^+, K^+ or R_4N^+. Peover observed the influence of lithium ions on the quinone reduction. This effect produces a
half-wave potential shift with respect to the potential found in quaternary
ammonium medium (36).

B. Influence of the Chemical Properties of the Solvent as a Particle Donor. Solvent Dissociation

So far, only the interactions between solvent molecules and solute molecules have been considered. However, solvents can affect the behaviour of
solute through the chemical and electrochemical properties of the ions or
molecules derived from solvent dissociation. It is well known that the chemical properties of aqueous solutions are greatly influenced by the H^+ and
OH^- ionic effects resulting from water dissociation. This phenomenon is
applicable to all solvents which can be dissociated. Thus protic solvents
such as HS (methanol, ethanol, liquid ammonia, etc.) are in this category and
the following equilibria apply:

$$HS \rightleftharpoons H^+ + S^- \quad \text{or} \quad 2HS \rightleftharpoons H_2S^+ + S^-$$

The second equilibrium expresses proton solvation while the first does not.

Similarly, antimony trichloride $SbCl_3$ or acetic anhydride undergo self-ionization.

$$SbCl_3 \rightleftharpoons SbCl_2^+ + Cl^-$$
$$(CH_3CO)_2O \rightleftharpoons CH_3CO^+ + CH_3COO^-$$

This ionization produces ionic species capable of modifying the solute properties, and even capable of reacting with them to produce new species. The same is true for ionized fused salts. The carbonate anion of fused sodium carbonate is partially dissociated:

$$CO_3^{2-} \rightleftharpoons CO_2 + O^{2-}$$

while the tetrahaloaluminates liberate aluminium trihalide.*

$$AlX_4^- \rightleftharpoons AlX_3 + X^-$$

Several examples will be presented now of ions or molecules produced by solvent dissociation with respect to voltammetric applications. These ions often participate in electrochemical oxidation and reduction reactions of the solvent itself. They contribute in defining the electroactivity range limits (Section IV). The examples mentioned here are considered only with respect to the electrochemical behaviour of the solute.

In organic chemistry, two or several one-electron steps occur very often during electrochemical oxidation or reduction of molecular species in solvents. Benzoquinone (Q) reduction in dimethylformamide (aprotic solvent) (38, 39) in neutral medium results successively in the radical anion $Q\cdot^-$ and the hydroquinone dianion Q^{2-} as follows:

$$Q + e \rightleftharpoons Q\cdot^-$$
$$Q\cdot^- + e \rightleftharpoons Q^{2-}$$

$Q\cdot^-$ and Q^{2-} are basic species and the following reactions can take place according to the medium acidity:

$$Q^{2-} + H^+ \rightleftharpoons HQ^-$$
$$HQ^- + H^+ \rightleftharpoons H_2Q$$

Acidification of the solution can transform the $Q\cdot^-$ radical into quinone and hydroquinone according to the following disproportionation:

$$2Q\cdot^- + 2H^+ \rightleftharpoons Q + H_2Q$$

It is necessary to add that quinone cannot be protonated by a strong acid in dimethylformamide (40).

---

* The molecules formed by such dissociations can be solvated by the solvent ions. $AlCl_3$ solvation by an $AlCl_4^-$ anion (tetrachloroaluminate) seems to produce an $Al_2Cl_7^-$ anionic species (37). However, when referring to aqueous solutions the H^+ ion is considered without solvating water molecules. In the same way, when considering reactions in fused $NaAlCl_4$ solvent, it is simpler to write the $AlCl_3$ molecule without writing its solvation.

All of these reactions can be followed by voltammetry. Figure 1 presents the voltammograms obtained in neutral (curve a) and acid media (curve b). The data are summarized in an $E_{1/2} = f(\text{pH})$ diagram (Figure 2). The electrochemical systems are fast or slow according to the medium. Often cathodic half-wave potentials ($E_{1/2}\text{cath}$) appear which are distinct from the

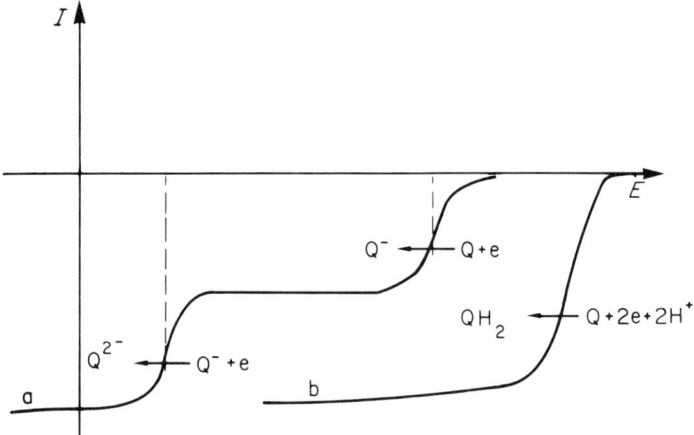

Fig. 1. Reduction of quinone in DMF. Curve a: neutral medium; Curve b: acid medium (see reference 40).

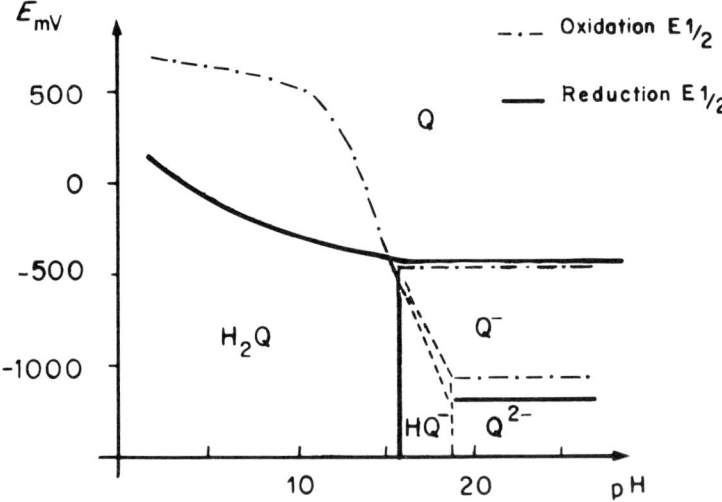

Fig. 2. Half-way potential/pH diagram: Q/Q^- and Q^-/Q^{2-} couples in D.M.F. (See reference 40).

corresponding anodic potentials $E_{1/2}\text{an}$. In aqueous solution the reduction proceeds through a two-electron step on a platinized platinum electrode;

this is the basis for the quinhydrone electrode process. Analogous phenomena are observed during aromatic amine oxidation or nitrogen heterocyclic derivatives such as phenothiazine and its derivatives (Section V-B-3). These two-step reaction types are symbolically represented by:

$$R_3\bar{N} - e \rightleftharpoons R_3\dot{N}^+$$

$$R_3\dot{N}^+ - e \rightleftharpoons R_3N^{2+}$$

When a primary amine (RNH_2) or a secondary amine (R_2NH) is considered, the acidic properties of the primary cation radical can be demonstrated.

$$R_2N\dot{H}^+ \rightleftharpoons H^+ + R_2N\cdot$$

The neutral radical thus formed can sometimes be oxidized.

$$R_2\dot{N} - e \rightleftharpoons R_2N^+$$

The R_3N^{2+} or R_2N^+ species are highly reactive with respect to some anions, especially towards the OH^- anion. Consequently, such species cannot exist in aqueous solutions but are fairly stable in solvents such as acetonitrile, nitromethane and dimethylformamide which are slightly dissociated. Under these conditions, no reaction takes place between nitrogenated cations formed at the electrode and the possibly-formed solvent anions. Potential/pH diagrams represent the areas in which these various species are predominant. Let us consider oxidation of tetramethylbenzidine (T). The molecules bears two amine groups with two basic functions and can undergo the following acid/base equilibria:

$$T + H^+ \rightleftharpoons HT^+$$

$$TH^+ + H^+ \rightleftharpoons H_2T^{2+}$$

Its oxidation in acetonitrile (41) takes place in two one-electron steps:

$$T - e \rightleftharpoons R$$

$$R - e \rightleftharpoons Q$$

which can be associated with proton exchanges.

Figure 3 shows the characteristic $i = f(E)$ curves for acetronitile solutions of T (curve a), R (curve b) and Q species, and Figure 4 presents a partial experimental potential/pH diagram in acetonitrile.

Until now, the behaviour of various cations in relatively inert solvents has been considered. However, in highly reactive solvents such as methanol the same nitrogen cations are capable of reacting with the solvent CH_3O^-

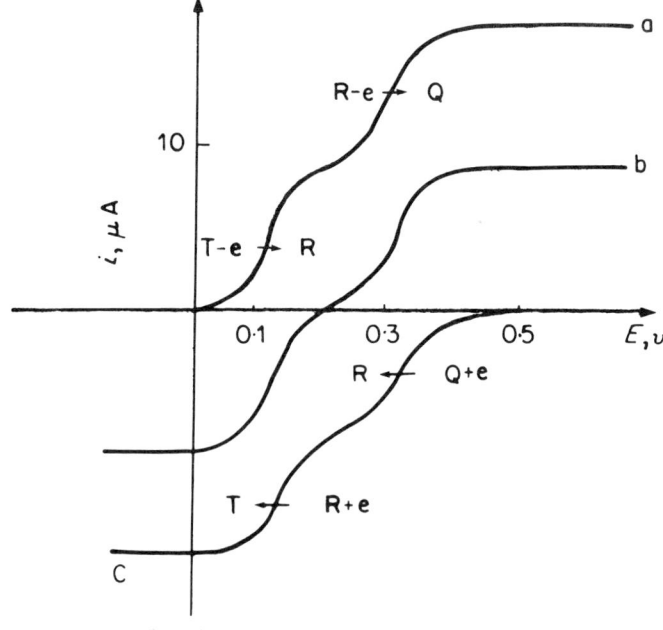

Fig. 3. Current–potential curves of tetramethylbenzidine in aceto-nitrile.
(See reference 41).

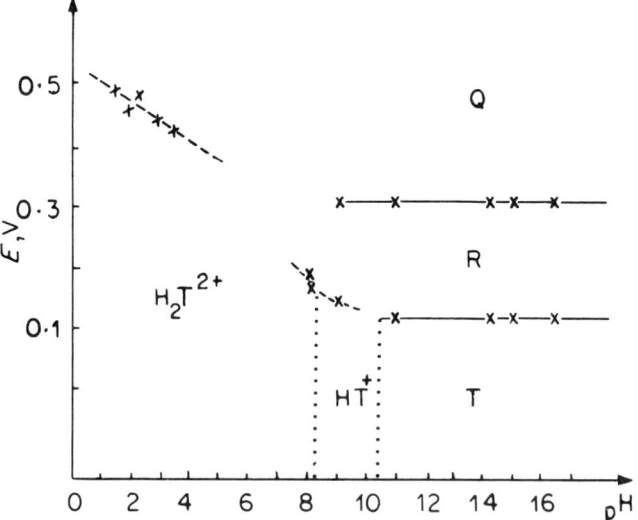

Fig. 4. Potential/pH diagram of tetramethylbenzidine in acetonitrile.
(See reference 41).

anion and imino ether formation is observed during electrochemical oxidation of tritertiobutylaniline in methanol

$$
\begin{array}{c}
NH \\
Y\text{---}\bigcirc\text{---}Y \\
Y \quad OCH_3
\end{array}
$$

where Y is $C(CH_3)_3$. This can be explained by the methylate anion attack on the carbonium ion produced by the anodic oxidation of the amine (42).

In real aprotic solvents such as fused antimony trichloride, analogous reactions are observed (43). Here the $SbCl_2^+$ cation is obtained from the dissociated solvent instead of the H^+ cation. The Cl^- and $SbCl_2^+$ ions are solvated by $SbCl_3$ molecules. The dissociation equilibrium of this solvent is

$$SbCl_3 \rightleftharpoons SbCl_2^+ + Cl^-$$

Under these conditions the oxidation of a metallic cation such as Fe^{2+} involves a simultaneous chloride exchange because this cation is a strong chloride acceptor. This is represented in the following reactions:

$$Fe^{2+} + Cl^- \rightleftharpoons FeCl^+$$
$$FeCl^+ + Cl^- \rightleftharpoons FeCl_2 \downarrow$$

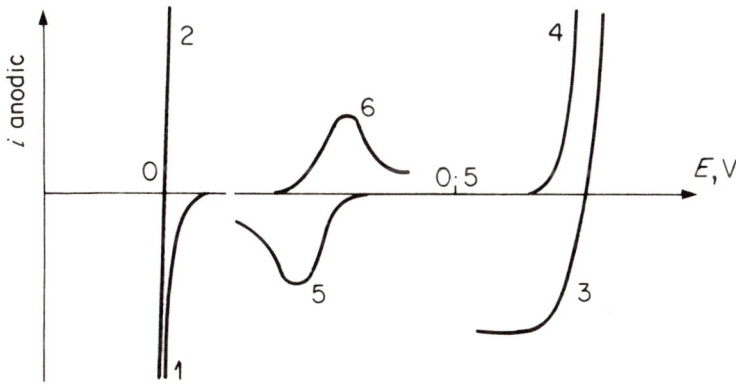

Fig. 5. Current-potential curves in fused $SbCl_3$ (99°C).

Curve I : $SbCl_3 + 3e \rightarrow Sb(s) + 3Cl^-$;

Curve II : $Sb(s) + 3Cl^- \rightarrow SbCl_3 + 3e$;

Curve III : $SbCl_5 + 2e \rightarrow SbCl_3 + 2Cl^-$;

Curve IV : $SbCl_3 + 2Cl^- \rightarrow SbCl_5 + 2e$;

Curve V : $FeCl_4^- + e \rightarrow FeCl_2(s) + 2Cl^-$;

Curve VI : $FeCl_2(s) + 2Cl^- \rightarrow FeCl_4^- + e$.

(See reference 43).

The ferric ion forms $FeCl_4^-$ in this medium. The following redox reactions are observed for the Fe(III)–Fe(II) couple according to the chloride ion availability in the medium:

$$FeCl_2 \downarrow - e + 2\,Cl^- \rightleftharpoons FeCl_4^-$$
$$FeCl^+ - e + Cl^- \rightleftharpoons FeCl_2^+$$

Figure 5 shows voltammetric curves of this couple and Figure 6 summarizes the data obtained in the form of a potential/pCl diagram. pCl^- is defined by the following relation:

$$pCl = -\log |Cl_{solvated}^-|$$

Here again, the Fe(II) and Fe(III) solute behaviour in fused $SbCl_3$ is influenced by the particle donor properties inherent in the solvent. This is a very general solvent property; only the nature of the exchanged particle can vary from one solvent to another.

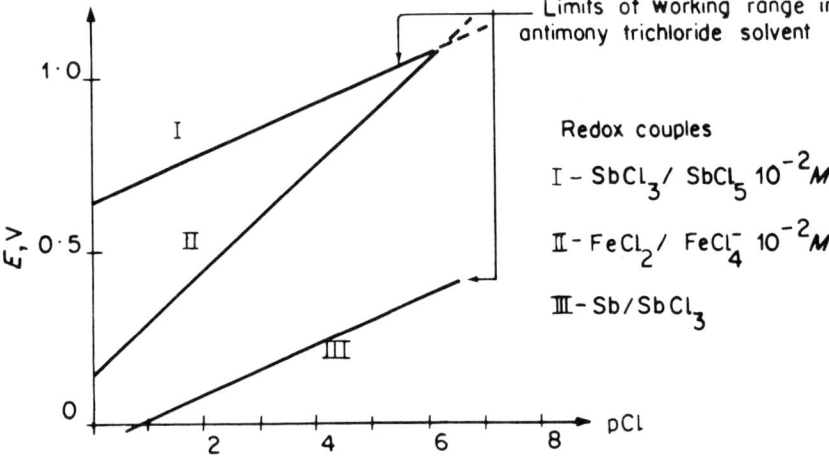

Fig. 6. E/pCl diagram for redox systems in antimony trichloride: Fe(II)/Fe(III), Sb(III)/Sb(V) and Sb(0)/Sb(III). (See reference 43.)

C. Influence of the Medium on the Global Kinetics of the Electrochemical Reaction

The global kinetics of an electrochemical reaction depends on the electron-transfer rate, and also on the preceding and following phenomena. Let us consider the general reaction process:

$$A + B \rightleftharpoons Ox \qquad \text{Chemical reaction}$$
$$Ox + e \rightleftharpoons Red \qquad \text{Electrochemical reaction}$$
$$Red \rightleftharpoons C + D \qquad \text{Chemical reaction}$$

The preceding and consecutive chemical reaction rates, as well as the electroactive species adsorption or desorption rates, have then to be taken into account.

The transfer of species from the bulk of the solution towards the electrode should also be considered. When this step is a rate-determining one, the system is said to be fast. The electrochemical reaction is reversible, and its rate does not affect the voltammetric curve position or shape. The solvent can modify the solute diffusion coefficient related to this transfer step. The diffusion coefficient reflects the medium viscosity as well as the variations in size of the moving species that result from solvation. But in fact the solvent influence on this transfer step is generally small when considering the electrochemical reaction global kinetics. Thus in the following section, the solvent influence on the kinetics of the electron transfer, of the coupled chemical reactions and of adsorption phenomena is only briefly considered. The role played by impurities is also mentioned.

1. Influence of the Medium on the Electron-transfer Rate

The determination of the standard rate constant $k°$ of the electron transfer is difficult. There are very few systematic studies of the medium influence on this constant. The work of Hoijtink (44) reviewed by Peover (45) in a general paper on the electrochemistry of aromatic hydrocarbons and their derivatives should be referred to. In contrast with water where charge transfer is generally a slow phenomenon ($k°$ is approximately 10^{-3} cm sec^{-1} for the quinone reduction in acid medium), it is not unusual to find $k°$ values in organic solvents which are too large to be measured by conventional electrochemical techniques. The $k°$ constant is greater than 4 cm sec^{-1} for the reduction of perylene to its anion radical in dimethylformamide at 25°C in the presence of tetrabutylammonium iodide. Such rates correspond to fast electrochemical systems, and under these conditions the charge-transfer process is never the rate-determining step of the overall kinetics.

These high transfer rate values probably result from the fact that the solvation states of the oxidized and the reduced forms are very close to one another. Consequently, no significant rearrangement of the solvation sphere occurs during the transfer. This remains true for numerous media with the result that the solvent nature probably has no important effect at this level. In the same way, it can be noticed that in water, proton reduction occurs with a large overvoltage on mercury, while it is a fast process on platinized platinum. Both phenomena can be found in numerous solvents.

Systematic studies which attempt to relate electron-transfer kinetics to the solvation of oxidized and reduced species in various types of solvents are highly desirable.

The supporting electrolyte which is the other main constituent of the media used in voltammetry, can also have an effect on the electrochemical kinetics. Thus, the second one-electron reduction wave of some quinones corresponding to the transformation of the radical anion $Q\cdot^-$ into the dianion Q^{2-} is slow or fast according to the nature of the supporting electrolyte. This phenomenon was explained by associations between the supporting electrolyte cations and the aromatic anions issued from the reduction.

2. Influence of the Medium on the Kinetics of Chemical Reactions Coupled with Electron Transfer

The existence of chemical reactions coupled with electron transfer can be demonstrated voltammetrically only if these reactions are fast enough to take place in the diffusion layer. If for a given electrochemical reaction, a solvent change produces considerable relative variations in chemical kinetics and mass-transfer kinetics, this change will modify the corresponding voltammetric curves. The advantage of using these curves for analytical purposes would then be considerably modified. Frequently a two-electron reduction of a molecule A occurs with the fixation of two protons. Theoretically, nine different species can be involved in this process. They are related to each other by a series of electron- and proton-exchange steps that can be illustrated by the following diagram:

$$
\begin{array}{ccccc}
A & \overset{+e}{\rightleftarrows} & A\cdot^- & \overset{+e}{\rightleftarrows} & A^{2-} \\[4pt]
\Big\downarrow\scriptstyle{+H^+} & & \Big\downarrow\scriptstyle{+H^+} & & \Big\downarrow\scriptstyle{+H^+} \\[4pt]
AH^+ & \overset{+e}{\rightleftarrows} & AH\cdot & \overset{+e}{\rightleftarrows} & AH^- \\[4pt]
\Big\downarrow\scriptstyle{+H^+} & & \Big\downarrow\scriptstyle{+H^+} & & \Big\downarrow\scriptstyle{+H^+} \\[4pt]
AH_2^{2+} & \overset{+e}{\rightleftarrows} & AH_2\cdot^+ & \overset{+e}{\rightleftarrows} & AH_2
\end{array}
$$

Thereoretically, six possible reaction pathways could occur for molecule A to pass to the completely reduced molecule AH_2. If this reduction takes place in a real aprotic medium, the voltammetric curve will show only two one-electron transfer steps corresponding to the transformation of A to $A\cdot^-$ and $A\cdot^-$ to A^{2-}. On the other hand, in a proton-donor solvent the reduction can lead directly to the AH_2 molecule; thus, at a sufficient rate of proton exchange, a two-electron voltammetric curve is recorded, according to:

$$A + 2e + 2SH \rightleftarrows AH_2 + 2S^-$$

where SH represents the solvent. Even in this case, there are several competitive reaction pathways, and changing the medium could modify their respective importance because the protonation kinetics of the various species involved in the general scheme are changed.

These different possibilities were studied for aromatic hydrocarbon (45) and quinone (46) reductions. The reader is referred to the original papers for a detailed discussion of this phenomenon.

If we call the mixture of the solvent and the supporting electrolyte the 'medium', it is important to note that simply changing the supporting electrolyte could change some coupled chemical reaction kinetics. Thus, substitution of a quaternary ammonium salt by an alkaline salt very often modifies the protonation kinetics resulting from the proton-donor properties of the always present residual water (36).

3. Influence of the Medium on Adsorption Phenomena

There is no need to demonstrate the important effect of these phenomena in electrochemical mechanisms. However, it is more difficult to understand the exact role of the solvent and still more difficult to predict the changes that a new solvent can cause in the adsorption process. The methods used for demonstrating these phenomena are the same in non-aqueous solvents and in aqueous media. The determination of electrocapillary curves and of double-layer capacities can be mentioned (47, 48).

Payne (49) used these methods to classify several anions in dimethylsulphoxide according to the relative strength of their adsorption at a mercury electrode. The following sequence was obtained:

$$I^- > Br^- > Cl^- > NO_3^- > ClO_4^- > PF_6^-$$

Similar determinations were undertaken in anhydrous formic acid. Lawrence and Persons (50) demonstrated a specific adsorption of alkaline cations increasing from lithium to cesium.

In the case of water/organic solvent mixtures, specific adsorption of the solvent can greatly modify the electrode surface and consequently, changing the solvent will disturb the electrochemical process. This is the case, for Zn^{2+} cation reduction on mercury (51).

Adsorption phenomena often make the use of voltammetric curves for analyses extremely delicate. It is one of the advantages of the use of various solvents to avoid these phenomena by the analyst carefully selecting the medium.

4. Influence of Medium Impurities and Especially of Residual Water

When an electroinactive impurity present in the medium is able to oxidize chemically an electrochemical reduction product, or inversely, to reduce the electrochemical oxidation product, one can observe a wave called a catalytic wave in polarography. This phenomenon is due to the regeneration of the electroactive solute and occurs naturally in non-aqueous media as well as in aqueous solutions.

The regeneration of the electroactive species can result from a more complex process than the mere oxidation or reduction of the electro-chemical reaction product. For example residual water is always present as an impurity in all non-aqueous media, and it can play an important role.

Let us consider the example of phenoxathine (P) oxidation (52) at a smooth platinum rotating disk electrode in acetonitrile. The voltammetric curve recorded in the medium made as anhydrous as possible (Figure 7, curve 1) exhibits two one-electron waves corresponding to the following reactions

$$P - e \rightleftharpoons P^{\cdot +}$$
$$P^{\cdot +} - e \rightleftharpoons P^{2+}$$

$$P = \text{(structure of phenoxathine)}$$

The equilibrium of the radical cation disproportionation

$$2 P^{\cdot +} \rightleftharpoons P + P^{2+}$$

is strongly shifted toward the left and the two one-electron waves are well separated. If the residual water concentration is much higher (for example, $4 \times 10^{-1} M$) a deformation is observed in the voltammetric curve (Figure 7, curve 2) which reflects phenoxathine regeneration in the vicinity of the electrode. This phenomenon can be explained by admitting the formation

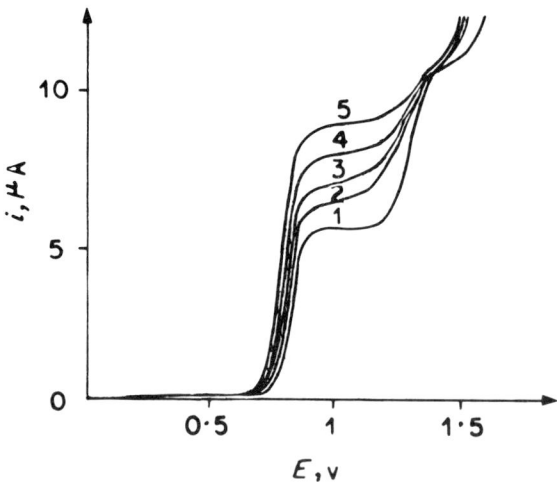

Fig. 7. Oxidation of phenoxathine $10^{-3} M$ in acetonitrile. Effect of water concentration. Curve 1: $10^{-3} M$. Curve 2: $4 \times 10^{-1} M$; Curve 3: $5.5 \times 10^{-1} M$; Curve 4: $8 \times 10^{-1} M$; Curve 5: $1.25 M$. (C. Barry, G. Cauquis and M. Maurey, *Bull. Soc. Chim. France,* **1966**, 2510.)

of the sulphoxide (actually isolated during controlled-potential electrolysis) by the reaction of the dication P^{2+} with water

$$P^{2+} + H_2O \rightarrow PO + 2H^+$$

This reaction shifts the equilibrium of the radical cation disproportionation toward the right-hand side thereby regenerating phenoxathine.* The height of the kinetic wave thus observed is a function of the rate of the slower sulphoxide formation reaction, i.e. it increases with the water concentration. This is the reason why, during studies in water-rich hydroorganic media direct two-electron oxidation is only observed (52).

This is a good example of non-aqueous solvent utility with respect to the electrochemistry of organic compounds. Intermediate reaction species are observed in most cases with non-aqueous solvents. However, reactions with residual water are not completely eliminated. It is important to point out that, if the electrochemically generated species react with the impurities slowly enough, the voltammetric curve does not show additional complications. Only the long time necessary for a controlled–potential electrolysis will allow the consequences of secondary reactions to be observed. For instance, the nature of the isolated species, or the coulometric yield, will not be in agreement with the voltammetric predictions. Let us again consider the oxidation of a 10^{-3} M 2,4,6-tritertiobutylaniline solution in acetonitrile on a smooth platinum electrode (42). The voltammetric curve, cyclic voltammetry and electron paramagnetic resonance confirm the formation of the $ArNH_2 \cdot^+$ radical cation of this primary amine at the electrode. This is true even for low concentrations of residual water between 0·002 to 0·005 M. Nevertheless, controlled-potential electrolysis at a potential corresponding to this one-electron transfer requires four faradays per mole of amine. The primary cation radical undergoes a degradation which leads finally to 2,6-ditertio-butyl iminobenzoquinone. This occurs because of the *residual water*, the effect of which cannot be observed on the voltammetric curve.

Sometimes the chemical analyst uses only voltammetric curves for titration or characterization purposes. Consequently it should be emphasized by these examples in non-aqueous media and in water, that these curves describe only partially the transformations observed during controlled-potential electrolysis.

---

* Some other possible mechanisms have been proposed (52). (Added to the proofs).

D. Influence of the Solvent Physical Properties

Two properties are especially important: the temperature range where the solvent is liquid and its viscosity.

The viscosity of the solvent especially concerns the solute diffusion coefficient and consequently the limiting diffusion current. This limiting current i_d is proportional to the square root of the diffusion coefficient D in the case of polarography and is also a function of D for the various types of solid electrodes. The following relation is found from the Stokes–Einstein and Ilkovic equations:

$$i_d . \eta^{1/2} = \text{constant}$$

This formula predicts a linear relationship between i_d and $\eta^{-1/2}$ for the same polarographic analysis performed in different solvents. This relationship suggests that the sensitivity of a polarographic analysis increases as the medium viscosity decreases. This is not always the case. The sensitivity of polarographic or voltammetric analyses is theoretically a function of the proportionality coefficient between the current and the concentration, but this sensitivity is limited for practical purposes by the residual current fluctuations which depend on the degree of purity of the solvent and the electrode pretreatment.

Most of the solvents used for voltammetry are liquid at room temperature. It would be incorrect to limit voltammetric analytical applications to room temperature only. Voltammetry in fused salts or in liquid ammonia is an expanding field. The temperature influence on the electrochemical behaviour of a solute is one of the means for determining the reaction mechanism. For example a kinetic wave can be distinguished from a diffusion wave, and protonation reactions with favourable kinetics can be demonstrated. Those solvents which are liquid in a large temperature range are very interesting. For example diallycyanamide has a useful temperature range extending from -70 to $+220°C$ (53).

III. THE REFERENCE ELECTRODE AND THE COMPARISON OF POTENTIALS DETERMINED IN VARIOUS SOLVENTS

Two aspects must be considered for defining a reference electrode suitable for non-aqueous media. A redox couple whose potential is stable and reproducible must be defined in each solvent. It serves as a reference potential. It is then necessary to establish the relations between the different reference potentials defined in the various solvents. This is related to specific solvation phenomena for each solvent and requires a definition of a reference electrode which is common to all the solvents.

A. Reference Electrodes in a Given Solvent

It is not intended to give a complete list of the various types of reference electrodes used in non-aqueous media. The common properties of satisfactory reference electrodes are presented in the tables at the end of the section which describe the electroactivity ranges of about twenty solvents. The nature of the reference electrode used in each case is specified. Practical indications are found in Section VI (technology) or can be found in specialized texts (54–58).

The most essential quality for reference electrodes in voltammetric applications is to exhibit a well-defined potential which is stable for long periods in aqueous or non-aqueous media. The reference electrode potential should be reproducible with a precision that varies with the experimental requirements and the type of problems that are to be resolved with voltammetry. The precision usually found in most non-aqueous media is generally between one and ten millivolts.

All redox couples with fast electrochemical kinetics at a given electrode are characterized by a definite equilibrium potential which varies very little when a weak current is applied to the electrode. This is the same as in aqueous solutions.

The current–potential curve for the electrode and the couple under consideration easily demonstrates this property. The following two redox couples can be considered as reference systems in a molten antimony trichloride medium:

$$\text{Sb} - 3\,e \rightleftharpoons \text{Sb(III)} \quad \text{and} \quad \text{Sb(III)} - 2\,e \rightleftharpoons \text{Sb(V)}$$

The corresponding current–potential curves are presented in Figure 8. Curve (a) represents the Sb(0)/Sb(III) couple which crosses the abscissa with a steep slope. This indicates a well-defined equilibrium potential. Curve (b) represents the Sb(III)/Sb(V) couple and illustrates a poorly defined equilibrium potential. The Sb/SbCl_3 couple is therefore the best suited reference electrode in this medium. A second example is provided by the Ag/Ag^+ couple which is fast in acetonitrile (Figure 9) and slow in liquid ammonia (Figure 10). It can only be used in acetonitrile media for this reason.

A well-defined potential is not the only requirement. The electrode potentials will remain stable for long periods only if the electroactive substances have constantly maintained concentrations. Despite its reversibility the Ag/Ag^+ electrode is unsatisfactory in dimethylformamide medium because the Ag^+ ion is progressively reduced by the solvent itself (59a). This can be avoided if the Ag/Ag^+ couple potential is lowered by precipitating the Ag^+ ion in the form of the chloride thereby forming an $\text{Ag/AgCl}{\downarrow}/\text{AgCl}_2{}^-$ electrode which can be satisfactorily used (59).

Some couples have well defined, stable potentials but they do not form satisfactory reference electrodes. These electrode potentials are not

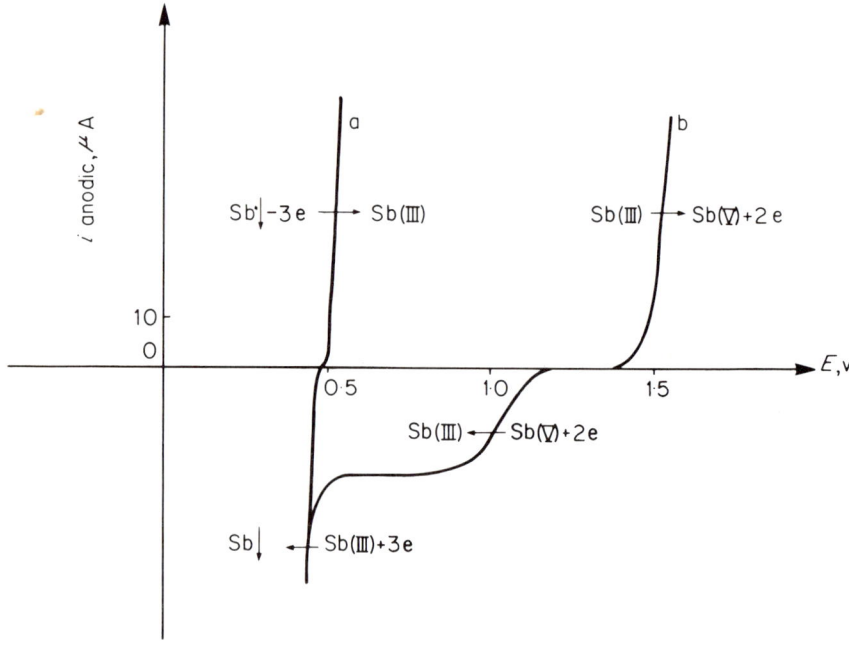

Fig. 8. Curve a : Current–potential curve at an antimony rotating disc electrode in fused $SbCl_3$ (99°C); Curve b: Current–potential curve at a rotating disc glassy carbon electrode in fused $SbCl_3$ (99°C): $AlCl_4^-$, $SbCl_2^+$, 0·1 M, Sb(V) 10^{-2} M. P. Texier, *Bull. Soc. Chim. France*, **1968**, 4716.

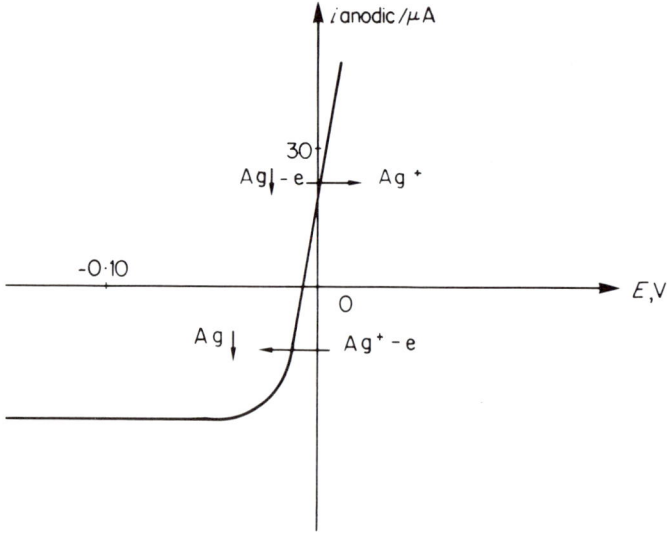

Fig. 9. Reduction curve of Ag^+ 10^{-3} M at a silver rotating disc electrode in acetonitrile. (J.-P. Billon, *J. Electroanal. Chem.*, **1**, 486 (1960).)

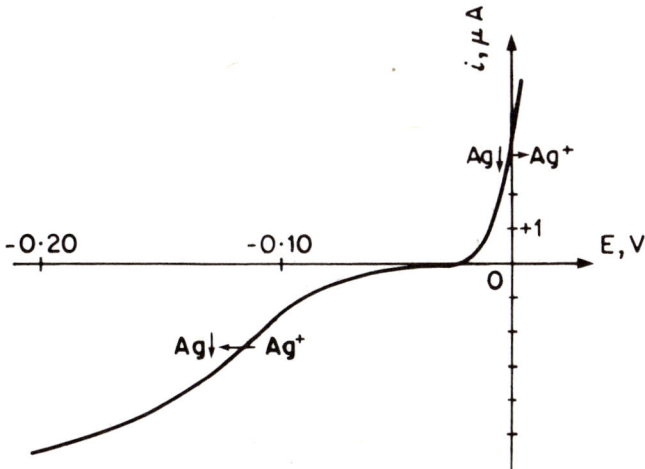

Fig. 10. Reduction curve of Ag^+ 10^{-3} M at a silver rotating disc electrode in liquid ammonia ($-60°C$). (M. Herlem, *Bull. Soc. Chim. France*, **1967**, 1687.)

reproducible from one preparation to another. This is observed especially when the couple ions are more strongly solvated by residual water or by another impurity than by the solvent. It is the case of the couple Ag/Ag^+ in nitromethane. Whatever the source it comes from, this solvent contains propionitrile as impurity, the concentration of which is between 10^{-4} M and 10^{-3} M. This nitrile complexes Ag^+ ions. It follows that the potential of the couple Ag/Ag^+ is a function of the ratio Ag^+/propionitrile, ratio which varies with the origin of the solvent.

A large selection of reference electrodes can be used in all new solvents because nearly all redox systems with stable oxidized and reduced forms and fast electrochemical kinetics are acceptable. All the electrodes thus obtained should be used only with a three electrode set with a very weak current passing through the reference electrode inserted in a high impedance circuit. Thus, its potential remains stable and very close to the zero current equilibrium potential.

Sometimes aqueous calomel electrodes are connected to non-aqueous media by a liquid junction. This should be avoided because of the risk of contaminating the anhydrous medium by diffusion of the water present in the reference electrode compartment.

B. Comparison of Potential Determinations in Various Solvents

This comparison is possible only if the relation between the potentials of a given couple in the various solvents is known. This relation reflects the effects of the solvation phenomena of each solvent. Several review articles were written on this topic (57, 61, 257). A simplified presentation illustrates more clearly the practical but limited interest of a single reference electrode common to all solvents.

Experimental potential scales for acetonitrile and nitromethane solvents are shown in Figure 11, the Ag/Ag$^+$ couple being arbitrarily selected as the potential origin. The relative strength of the oxidizing agents increases towards

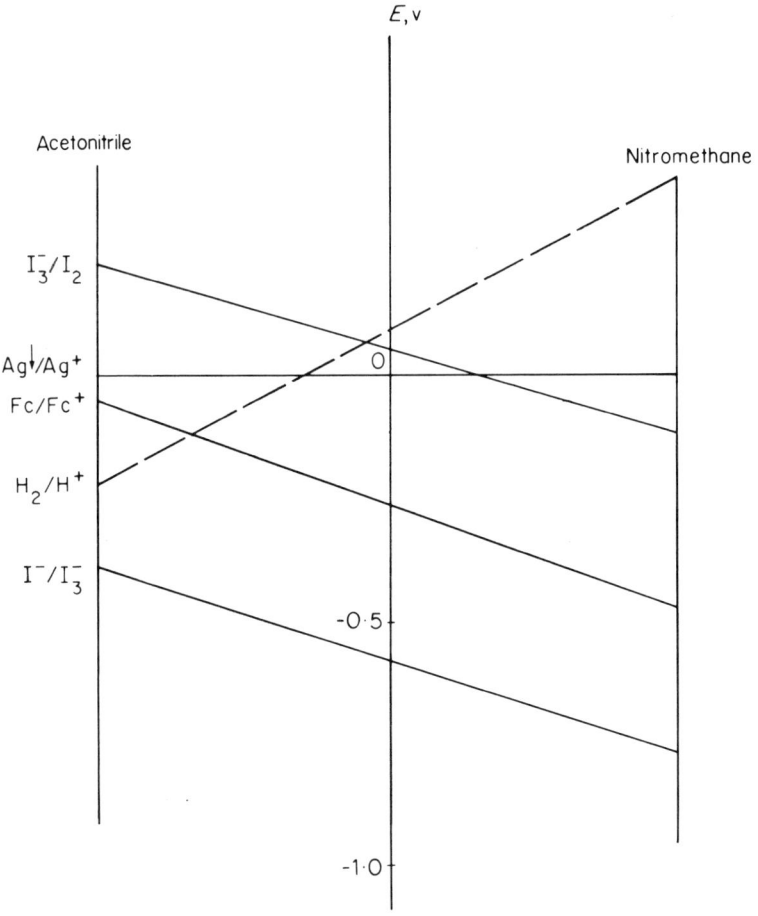

Fig. 11. Comparison of redox-potential scales in acetonitrile and in nitromethane— scale origin: potential of a 1 M solution of Ag$^+$ in each solvent.

the top of the scale and for reducing agents towards the bottom. Generally, the relative strength of oxidizing and reducing agents is not the same when passing from one solvent to another. In nitromethane the Ag^+ ions oxidize the triiodide I_3^- ion to iodine and the ferrocene Fc into the ferricinium Fc^+ ion. The opposite reactions are observed in acetonitrile. Further examination of these scales reveals two types of redox couples. Group I includes couples whose relative positions of redox potentials are maintained when passing from one solvent to another (I_2/I_3^-, I_3^-/I^-, Fc/Fc^+, etc., couples). The linear segments between the plotted points corresponding to these couples are parallel. For the couples of group II, the sequence of relative potentials varies (Ag/Ag^+, H_2/H^+, etc., couples). This empirical distinction is interesting because it applies to many media.

This distinction can be used to compare potential determinations in various solvents. The physical significance of the possibility of classifying redox couples into two groups will be explained later. The validity and the utility of electroactivity range comparisons in various solvents are also considered at the end of this section.

Let us consider Figure 12, illustrating the potential scales found in a series of solvents $S_1, S_2, \ldots$. These scales have been established by arbitrarily selecting the potential of a group I couple as the origin potential for each solvent. The ferrocene Fc/Fc^+ couple is a typical example. Consequently each group I couple is represented in the various solvents by points located on the same horizontal line. On the other hand, group II couple potentials vary from one solvent to another. The solvent oxidation and reduction potentials are also found on these scales. These potentials form the electroactivity range limits when neither the supporting electrolyte nor the electrode metal participate in defining these limits (see p. 364).

The selection of a reference couple among group I couples allows various semiquantitative predictions. The potential scales of various group I couples in different solvents can be predicted if its scale is already known in one of the solvents. Some species will be predicted to be unstable in some solvents because they oxidize or reduce these solvents. The Fc^+ ferricinium ion is unstable in solvent S_4 because it oxidizes it. In the same way rubidium will not be stable in solvent S_n because it reduces it, giving the Rb^+ cation. This type of phenomena occurs, for example, between water (solvent S_n) and rubidium metal. Conversely such predictions cannot be made for group II couples. By comparison (Figure 12) of the oxido-reduction potential of the Ag/Ag^+ couple in solvent S_1 and the oxidation potential of solvent S_4, it would seem that the Ag^+ cation could not exist in this easily oxidizable medium. This does not happen because the group II Ag/Ag^+ couple has a considerable potential decrease in solvent S_4. The Ag^+ ions are in fact extremely stable in this medium. This special behaviour of the Ag/Ag^+ couple is observed for example in water (solvent S_1) and liquid ammonia (solvent S_4). The relative potentials of Fc/Fc^+ and Ag/Ag^+ couples in various solvents are found in following Table I.

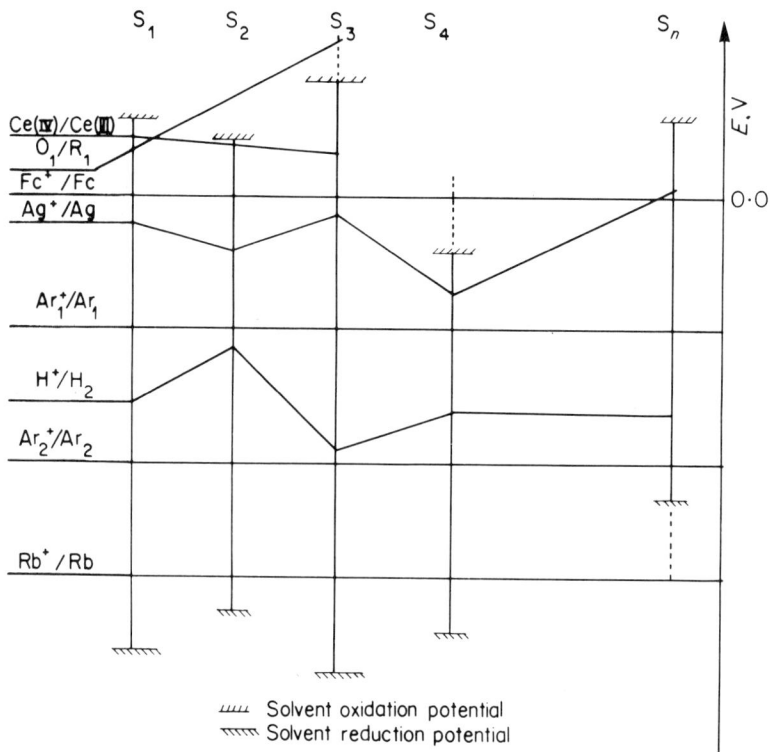

　　　ⅬⅬⅬⅬ Solvent oxidation potential
　　　ⅦⅦⅦ Solvent reduction potential

Fig. 12. Redox potential scales of different solvents $S_1, S_2, \ldots S_n$. Fc: ferrocene; Ar_i: aromatic hydrocarbons.

TABLE I

Relative Potentials of Fc/Fc$^+$ and Ag/Ag^{+a}
($E_{Fc^+} = 0$ in each Solvent)

Solvent	$E_{Ag^+}{}^0$ V	Solvent	$E_{Ag^+}{}^0$, V
H_2O	0·41	Sulpholan	0·38
CP	0·50	DMF	0·12 ⎱ 0·13 ⎰
MeCN	0·03	MeOH	0·35
DMSO	− 0·06	HMPT	−0·17 ⎱ −0·33 ⎰
$MeNO_2$	0·47	NMP	0·04
		DMAC	0·08

　^{a} See reference 62.

This illustrates how careful one must be when trying to draw conclusions based on the relative positions of electroactivity ranges for predicting the electrochemical behaviour of group II couple species.

We shall try to see, now, whether this classification of redox couples in two groups based, at first, on experimental findings, can be justified theoretically. It should be noted that this classification is approximate. Strictly speaking, the difference existing between the potentials of two group I couples is not exactly retained when passing from one solvent to another. There are, however, numerous couples whose behaviour is very similar to this ideal behaviour. Strehlow (57) and other authors (61, 61a) demonstrated that numerous organometallic complexes can be considered as belonging to group I couples. Recently, the potentials of couples from some polynuclear aromatic hydrocarbons and their radical cations or radical anions were shown (63) to satisfy the requirements of group I couples.

Let us consider, in the following lines, couples composed of organic molecules or neutral organometallic compounds $A_1, A_2, \ldots$ and the corresponding monovalent cations $A_1^+, A_2^+, \ldots$ associated with the same X^- anion. The $A_1, A_1X, A_2, A_2X, \ldots$ species are supposed to exist in the solid state and the salt crystals A_nX to have an ionic structure. The A_1/A_1X and A_2/A_2X couples are selected among the various possibilities so that the following redox reaction takes place because of the two redox potential values:

$$A_1 + A_2X \rightarrow A_1X + A_2$$

The energies are calculated when this reaction takes place between the solids (ΔG energies) and when this reaction takes place between the species dissolved in solvent S (ΔG_S energies) (see the diagram below). The relation existing between ΔG and ΔG_S reveals that the solvation energies of the various species in the solvent are involved in the reactions

$$A_1 + A_2X \xrightarrow{\Delta G} A_1X + A_2$$
$$\text{Solids} \qquad\qquad \text{Solids}$$

$$A_1 + A_2^+ + X^- \xrightarrow{\Delta G_S} A_1^+ + X^- + A_2$$
$$\text{In solution} \qquad\qquad \text{In solution}$$

The second transformation can be divided into three steps:

1. The A_1 and A_2X solids are dissolved in solvents. This first step is accompanied by the effect of A_1 and A_2X dissolution energies ($\Delta G_{sol}(A_1)$ and $\Delta G_{sol}(A_2X)$). This second energy can be considered as the sum of the solvation energies of the A_2^+ and X^- ions minus the crystal lattice energy ΔG_{crys} of A_2X.

$$\Delta G_{sol}(A_2X) = \Delta G_{sol}(A_2^+) + \Delta G_{sol}(X^-) - \Delta G_{crys}(A_2X)$$

2. Reaction in the solution between the A_1 and the A_2^+ ions producing the A_1^+ ion. The energy ΔG_S (defined above) is exchanged during this reaction.

3. The desolvation of the A_1^+, X^- and A_2 species is accompanied by the exchange of the energies $\Delta G_{sol}(A_1^+)$, $\Delta G_{sol}(X^-)$ and $\Delta G_{sol}(A_2)$ followed by A_1X crystal formation which liberates the crystal lattice energy $\Delta G_{crys}(A_1X)$. A comparison between the two transformations leads to:

$$\Delta G = \Delta G_{sol}(A_1) + \Delta G_{sol}(A_2^+) + \Delta G_{sol}(X^-) - \Delta G_{crys}(A_2X) + \Delta G_s$$

$$- \Delta G_{sol}(A_2) + \Delta G_{sol}(A_1^+) + \Delta G_{sol}(X^-) - \Delta G_{crys}(A_1X)$$

The $\Delta G_{sol}(X^-)$ terms cancel each other and the various other terms are regrouped according to whether or not they depend on the solvent. In the following equation giving the value of the energy of the reaction taking place in the solvent only the values of the terms contained in the first brackets do not depend on the solvent.

$$\Delta G_S = [\Delta G + \Delta G_{crys}(A_2X) - \Delta G_{crys}(A_1X)] + [\Delta G_{sol}(A_1^+) - \Delta G_{sol}(A_1)]$$

$$- [G_{sol}(A_2^+) - G_{sol}(A_2)]$$

On the contrary, the values of the terms contained in the last two brackets generally vary with the solvent.

However it is reasonable to think that the A_1/A_1^+, A_2/A_2^+ couples have nearly equal solvation energies for the neutral species A_1 and A_2 and for the corresponding ions. Hence, changing the solvent will only have a slight effect on the ΔG_S values. Such couples are classed in group I previously defined because ΔG_S is related to the ΔE difference between the redox potentials of the A_1/A_1^+ and A_2/A_2^+ couples. This is expressed by the equation $\Delta G_S = -F\Delta E$. Strehlow assumed that the $\Delta G_{sol}(A)$ and $\Delta G_{sol}(A^+)$ terms are about the same (57, 61) when the A molecule and, consequently, the cation A are sufficiently large and when the cation charge is delocalized. Under these conditions, the values in the last two brackets of the above equation remain small whatever the solvent is, and consequently, ΔG_S and the difference between the two redox potentials are also independent of the nature of the solvent.

This hypothesis is responsible for the extrathermodynamic explanation for the existence of two groups (I and II) of redox couples. Moreover, this hypothesis justifies the selection of couples, such as ferrocene/ferricinium, as common reference electrodes for various solvents. These permit the comparison between the potential scales determined in these solvents as well as between their electroactivity range limits. Other so-called extrathermodynamic assumptions have been proposed for the same purpose. One must be aware, however, that the use of common references for several solvents cannot entirely solve the problem of predicting all possible redox reactions in these solvents (independently of the additional difficulties inherent in reaction kinetics). In fact when changing the solvent it is impossible to predict the variations of redox potentials of couples with a large difference of solvation between the oxidized and the reduced forms.

IV. ELECTROACTIVITY RANGE

A. Definition

The electroactivity range of a medium is defined as the potential zone within which an electrochemical reaction can take place without being disturbed by massive oxidation or reduction of one of the medium components.

Experimentally, this zone is limited by the rapid increase of the current on current–potential curves (wall or barrier) in the absence of electroactive species (Figure 13). The range $\Delta E = E^2 - E^1$ thus observed is entirely relative and its measurement necessitates a definition of the current density value beyond which the residual current is thought to be important enough to conceal the electrochemical properties of the dissolved substances. For practical purposes, the limits ought to correspond to a few $\mu A \, cm^{-2}$ for a dropping mercury electrode and to about $10 \, \mu A \, cm^{-2}$ for a rotating platinum disk electrode.

Such a range can express in the first place the redox properties of the solvent itself. In this case two possibilities can be considered according to whether the solvent reduction or oxidation reactions limiting the zone are reversible or not (64).

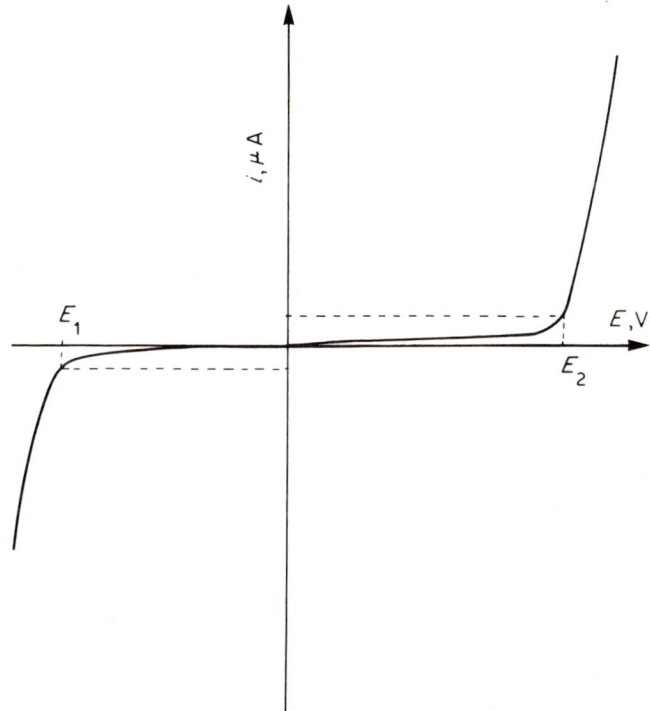

Fig. 13. The concept of electroactivity range.

In the first case it is said to be the *thermodynamic range* of the solvent. In the second case the kinetics of the solvent electrochemical reaction controls the potential and the range limits. The potentials depend on the nature of the considered electrode and the range thus obtained experimentally by plotting the voltammetric curve, can vary greatly. Thus, the electrodes corresponding to the fastest kinetics provide the smallest ranges.

Practically the range limits are rarely defined only by the solvent electrochemical properties. Frequently the metal of the electrode or the redox properties of the supporting electrolyte ions also participate. It is more constructive to speak of a *practical range* characteristic of the medium (solvent, supporting electrolyte, residual water and other impurities) and the electrode.

It must be noticed how the term 'electroactivity range' is not properly used. The medium/electrode set should be better described as the 'non-electroactivity range'.

The idea of the *electroactivity of a substance* should be associated with the range concept. It is said that a substance is electroactive in a given medium, at a given electrode when its voltammetric curve (oxidation or reduction) can be recorded within the characteristic range of such a defined whole set. A substance can appear non-electroactive in this set for thermodynamic or kinetic reasons. In the first case electroactivity does not occur because the normal oxidation (or reduction) potential of the species is clearly superior to the anodic limit (or clearly inferior to the cathodic limit) potential. In the second case electroactivity is absent when the electrochemical transformation kinetics is too slow to be revealed by plotting the voltammetric curve even though the normal potential is within the range.

It is important to note that some substances found as impurities in the medium can be included in this last-mentioned category. These substances which seem incapable of being oxidized or reduced because they do not deform the residual current, can participate in redox reactions with species produced at the electrode. Oxidation or reduction of these impurities which is not revealed at the electrode because of over-voltages, can occur in homogeneous phase and considerably change the controlled potential electrolysis data. Various authors thus explained the disappearance of some radical cations obtained during anodic oxidation of organic compounds in organic solvents by reduction caused by residual water. The presence of residual water and its reducing properties often do not appear on the electroactivity range and voltamperometric curves (65–67).

B. Determining Factors for Practical Electroactivity Range Limits and Residual Current Values

1. Influence of the Electrochemical Properties of the Solvent

A very interesting classical example of an electroactivity range limited by the solvent electrochemical properties is provided by water (68, 69). The kinetics of the reactions

$$H_2O - 2e \rightarrow 1/2 O_2 + 2 H^+$$
$$H_2O + e \rightarrow 1/2 H_2 + OH^-$$

which limit this range for most experimental conditions depend to a great extent on the chemical nature of the electrode. Thus on the cathodic side the range limit can perhaps have its thermodynamic value on a platinized platinum electrode whereas, on smooth platinum or mercury electrodes, the relatively slow reaction kinetics pushes this limit to more negative potentials (68). The anodic limit which also depends on the nature of the electrode and on its surface state in the case of platinum is always beyond the theoretical limit because of the slowness of the corresponding reaction (68).

In fused salts medium, when the indicating electrode is made of inert material such as vitreous carbon, one often finds an overlapping of the experimental range with the thermodynamic range. Molten sodium tetrachloroaluminate $Na^+ AlCl_4^-$ at 175°C has a range with this vitreous carbon electrode limited by the following fast reversible reactions (70):

$$AlCl_4^- - e \rightleftharpoons 1/2 Cl_2 + AlCl_3$$
$$AlCl_4^- + 3e \rightleftharpoons Al \downarrow + 4 Cl^-$$

Non-ionized antimony trichloride $SbCl_3$ becomes conductive at 99°C after addition of the electrolyte KCl. The reactions

$$SbCl_3 + 2 Cl^- - 2e \rightleftharpoons SbCl_5$$
$$SbCl_3 + 3e \rightleftharpoons Sb \downarrow + 3 Cl^-$$

responsible for the range limits on vitreous carbon (71) are fast and reversible. Consequently, the practical range and thermodynamic range are superimposed under these conditions.

The redox properties of various organic solvents under special conditions can also define their range limits; however, slow or even irreversible reactions generally take place. Electron spin resonance (ESR) and voltammetry demonstrated that the solvent radical anion formation plays an important role in limiting the range on the cathodic side at a bright platinum electrode as this is the case for nitrobenzene and $0.1 M$ quaternary ammonium solutions (72, 73). On the other hand, a controversy exists as to whether or not other solvents are involved in their range limitation. Despite the reports of Schmidt and Noack (74), the anodic barrier in $0.1 M$ perchlorate solutions in acetonitrile has not definitely been proved (75–77) to result from solvent oxidation according to the following overall processes:

$$2 CH_3CN \rightarrow NCCH_2CH_2CN + 2 H^+ + 2e$$

The cathodic limit of electroactivity range in water results from proton reduction. The same phenomenon can take place for sufficiently acidic solvents according to a reaction of the following type:

$$SH + e \rightarrow 1/2 H_2 + S^-$$

This has been demonstrated for solvents such as acetic and trifluoro-acetic acid (78), methanol (79), ammonia, etc., but they are generally slow reactions, at least with a smooth platinum or a glassy carbon electrode. However on a given electrode, the cathode barrier potential is related to the solvent acidity; the least acid solvents produce the largest cathodic ranges. Finally the radical anion formation reaction is responsible for the limit for very weakly acid solvents

$$SH + e \rightarrow SH \cdot^-$$

The potential of this radical formation is evidently not related to solvent acidity. Nitrobenzene was the first example. Benzonitrile has the same characteristics (80).

2. Influence of the Redox Properties of the Supporting Electrolyte and the Role of Residual Water

The use of an organic solvent requires the presence of a highly concentrated strong electrolyte in the medium (0·1 to 0·5 M generally). This is the reverse to fused salt requirements. The electrolyte ions often participate in the definition of the electroactivity range of the medium. The selection of sufficiently soluble salts with difficultly oxidizing cations (Li^+, Na^+, quaternary ammonium ions, etc.) and with difficultly reducing anions (Cl^-, ClO_4^-, BF_4^-, etc. ...) is relatively easy for most solvents. We shall discuss these cation reductions and anion oxidations which, as an approximation to the facts, behave as they do in aqueous media, except for modifications of potentials due to solvation energy variations.

The cathodic limit of the range is most frequently defined by the supporting electrolyte if it is an alkaline metal salt. This is true if the solvent is not too acid, in which case the proton discharge could take place prior to that of the metal cations. Thus, a bright platinum electrode is covered with lithium beyond $-2\cdot4$ volts with respect to the silver Ag/AgCl electrode in a 0·1 M lithium perchlorate solution in nitromethane as shown by the reoxidation peak which occurs in the return sweep of the current–potential curve (Figure 14, curve a′) (81).* Conversely, the cathodic limit of acetic acid with the same electrode was attributed to proton reduction (78). Amalgam formation takes place with a dropping mercury electrode. This promotes cation reduction and the range is shorter on the cathodic side. This is how the 0·1 M lithium perchlorate limit in sulpholane occurs at $-2\cdot6$ volts on mercury with reference to the Ag/Ag$^+$ system, whereas smooth platinum permits $-3\cdot7$ volts (82) to be reached. See tables in Section VII.

---

* The phenomenon of depositing and redissolving of very electropositive metals such as lithium in the most inert organic solvents permits the development of one of the two electrodes of high energy density generators (85, 86).

Experts in aqueous media polarography have considered for a long time that quaternary ammonium cations make it possible for more negative potentials to be reached on mercury electrodes than the alkaline cations. This observation is usually explained by the fact that the cathodic barrier is not defined by water reduction which exhibits a high over voltage on the mercury electrode and that the quaternary ammonium cations are more difficult to reduce than the alkaline cations, even though some of them can be reduced under special conditions (83). These observations are not applicable to all organic solvents and to all types of electrodes. Comparisons of cathodic limits described in paragraph VII are applicable to mercury electrodes in a great number of solvents. It is not true for bright platinum, gold and silver electrodes. This is shown in Figure 15 where a semiquantitative comparison of water (68), acetonitrile (75), nitromethane (81) and

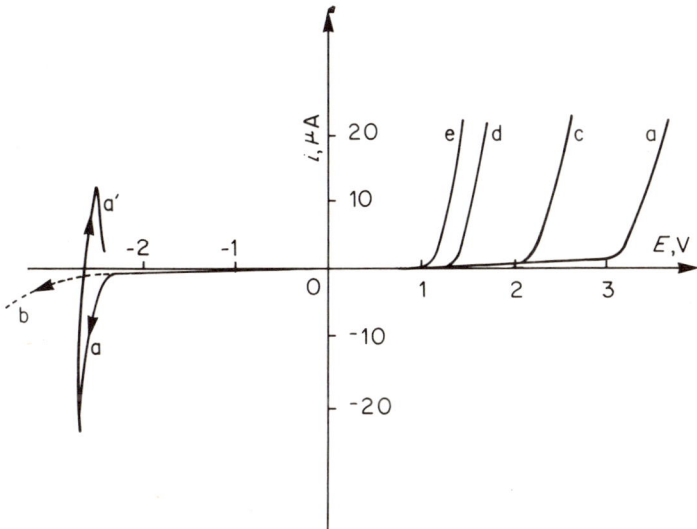

Fig. 14. The electroactivity range of nitromethane. Curve a : Smooth platinum electrode, background electrolyte $LiClO_4$ 0·1 M; Curve a′ : Reoxidation of the lithium layer; Curve b : As a but with water 0·2 M; Curve c : Smooth platinum electrode, background electrolyte $(C_2H_5)_4NNO_3$ 0·1 M; Curve d : DME, background electrolyte $LiClO_4$ 0·1 M; Curve e : Smooth silver electrode, background electrolyte $LiClO_4$ 0·1 M.

dimethyl sulphoxide (84) ranges on smooth platinum and mercury electrodes illustrate this fact very well. Each of the three organic solvents has a more restricted cathodic range with tetraethylammonium perchlorate than with lithium perchlorate on platinum electrodes. It was sometimes possible to demonstrate that quaternary ammonium cation reduction limited the range toward the cathodic side (84).

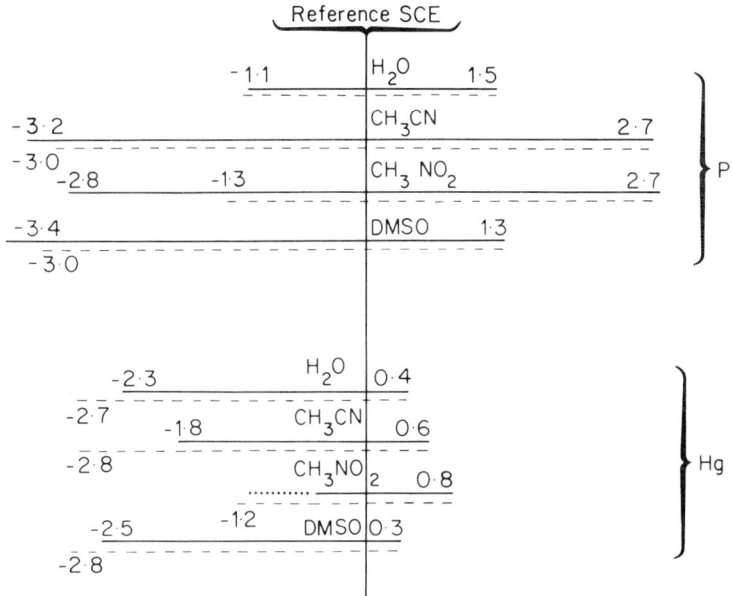

Fig. 15. Comparison of electroactivity ranges on platinum and mercury electrodes. The background electrolyte is LiClO₄ 0·1 M (——) or (C₂H₅)₄NClO₄ 0·1 M (– – –).

Practically in most solvents the cathode barrier position is very dependent on the residual water concentration. This type of comparison should be done with caution because this concentration is not precisely indicated in most publications. Dimethylsulphoxide is an illustrative example of the influence of water. There is considerable disagreement in the literature as to the extent of the electroactivity ranges of this solvent. Thus the range of a 0·1 M sodium perchlorate solution can vary at the same electrode (bright platinum) from 2·55 volts (87) to 4·25 volts (89). Moreover, the cathodic limit of this solution was demonstrated (84) to shift more than one volt towards positive potentials when the water concentration varies from 0·2 to 1 M. It was concluded from this result that the cathodic limit (84) cannot be attributed only to the metal cation reduction as it had been suggested earlier (88). The authors' feeling is that the problem of residual water participation in defining residual current values and cathodic limits in alkaline metal salt solutions is very complex* and is not completely solved. In the most anhydrous solvents, the presence of residual water and its reduction are observed by abnormalities appearing in cyclic voltammetry or chrono-

---

* This happens even when the metal resulting from the supporting electrolyte reduction is unable to react with the solvent. This is an additional complication which cannot be excluded for sodium salts in dimethylsulphoxide (84).

potentiometry recordings. These methods were used advantageously for studying the electrochemical behaviour of small concentrations of water in various solvents (89, 90). Higher and higher water concentrations are revealed simply by plotting the voltammetric curve. The cathodic barrier can be modified because massive reduction of this water becomes the predominant phenomenon towards the cathodic side or, more frequently because an insulating coating of alkaline hydroxide is deposited on the electrode at the beginning of water reduction. This phenomenon ought to explain the appearance of a maximum in the vicinity of the barrier potential observed in anhydrous media. It could also explain the apparent disappearance of all cathodic barriers observed in a solvent such as nitromethane (81) (Figure 14, curve b), dimethylsulphoxide (84), etc., as soon as the water concentration of a 0·1 M alkaline cation perchlorate solution approaches 0·2 M.

Water participation in defining the cathodic limit of solvents in the presence of alkaline cation salts is dominated by an association between these cations and residual water. This is true for sodium perchlorate solutions in hexamethylphosphorotriamide, where the simple shift of the cathodic barrier as a function of the water concentration demonstrates the presence of a $NaClO_4$, H_2O association (91). It is interesting to note that these well-defined stoichiometric associations can modify the chemical properties of the water contained in the medium. These associations were studied by other methods such as proton (92) or lithium (93) magnetic resonance. It is convenient thus to explain that some highly reactive (towards water) organic cation radicals are more stable in a solvent such as nitromethane when the supporting electrolyte is a lithium salt than when it is a quaternary ammonium salt (94).

The anodic limit of the electroactivity range of an organic medium seems to be frequently defined by the oxidation of the supporting electrolyte anion. This fact has been generally well established for easily oxidizable anions such as bromide or chloride ions. Sometimes, oxidation of a very usual ion which normally is not present in aqueous media is demonstrated in some organic solvents. Thus the anodic limit of 2·1 volts with reference to an Ag/AgCl system at platinum in nitromethane in the presence of 0·1 M tetraethylammonium nitrate (Figure 14, curve c) is due to the nitrate ion oxidation. The nitrate ion oxidation was studied in 10^{-3} M tetraethylammonium nitrate in nitromethane which contained tetraethylammonium perchlorate, a less oxidizable salt, as the supporting electrolyte. The results thus obtained are in good agreement with those obtained for the anodic limit of 0·1 M nitrate (95).

The literature is less explicit concerning an eventual participation of anions which are difficult to oxidize. The frequently used perchlorate anion is a typical example of this inaccuracy. Various authors have argued that in some solvents the anodic limit of various 0·1 M perchlorate solutions is found at the same potential on platinum regardless of the cation nature.

They concluded that the ClO_4^- anion oxidation is responsible for this common limit. This fact seemed to be confirmed when an ESR spectrum was recorded and was attributed to the $ClO_4\cdot$ radical during oxidations taking place in the spectrometer resonance cavity (95, 96). Then the very controversial question as to the existence of the $ClO_4\cdot$ radical (77) was again considered. The problem is further complicated by the fact that the ESR spectrum recorded could belong to ClO_2 (77) which is a paramagnetic species. All of this has been presented without positive proof. Hiraswa and coworkers (97) observed that the coupling constant, which is approximately 17 G in acetonitrile (96) and in nitromethane (95), is well inferior to the 50 and 72 G which were attributed to $ClO_4\cdot$ radicals trapped in various irradiated potassium perchlorate crystal sites (98). On the other hand, it approaches the constants measured in solutions of ClO_2 gas in various solvents. Other observations were unable to clarify the problem. Thus the same ESR spectrum was obtained during oxidations at controlled potentials well below the anodic limit normally found in perchlorate solutions in nitromethane. The same spectrum was also obtained by simply dissolving some species which are reduced far below this limit. When this happens, a relatively stable yellow coloration develops in the solution which could be attributed to ClO_2 (99).

As far as we know the problem of the eventual oxidation of ClO_4^- ions in organic solvents and of the chemical nature of the resulting species has not been resolved. Systematic studies of ESR spectra using the O_8^{17} isotope exhibiting a nuclear spin, should be undertaken. This example demonstrates the difficulties that could arise in characterizing the phenomena that are really responsible for the electroactivity range limits.

The salts containing some special anions make it possible for high anodic potentials to be reached whatever the nature of the phenomena might be. Thus the ClO_4^- ion seems less interesting than the tetrafluoroborate BF_4^- and hexafluorophosphate PF_6^- ions (100, 101). Their limits on platinum in acetonitrile are respectively at 2·99 and 3·02 volts with reference to a 10^{-2} M Ag/Ag^+ system. Limits higher than 2·48 volts are not reached with perchlorate ions under the same conditions. This increase of the range toward the positive potentials is very interesting.

3. Influence of the Electrode Nature. Insoluble Film Formation and Dissolution

The electrode can also contribute in defining the electroactivity range of the medium/electrode set in two different ways, according to whether the electrode is an easily oxidizable metal or a metal considered as 'inert'.

The massive oxidation of an oxidizable metal electrode and the formation of a soluble cation in the medium is frequently responsible for the anodic limit. Thus in 0·1 M nitromethane in lithium perchlorate, the silver and mercury electrodes fix the limits at 1·15 and 1·10 volts with respect to the Ag/AgCl reference electrode (Figure 14, curves d and e). The presence of a complexing agent such as the chloride ion can obviously considerably narrow these ranges (81). Massive electrode oxidation can explain the anodic limit in fused cryolite in the presence of alumina on carbon at 1000°C.* Metals generally considered as 'noble' can also be massively oxidized. Thus, a platinum anode dissolves in fused alkaline hydroxides (103) because of the reaction:

$$Pt + 2O^{2-} - 2e \rightarrow PtO_2^{2-}$$

The electrode metal is very rarely responsible for the cathodic limit of the medium. Nevertheless this seems to be the case for a lead electrode used in $0.1 M$ potassium bromide in liquid ammonia. The compound $K_4Pb(Pb)_8$ is produced by lead reduction (104).

The term 'inert electrode' applies to two different electrode types.

Electrodes the material of which has a standard oxidation potential high enough to exceed the anodic limit of most media currently used, are classed in the first group. The vitreous carbon electrode is in this category. It can nevertheless be oxidized in special media such as fused cryolite in the presence of oxide ions (105).

The formation of a conducting protective film caused by the beginning of the material oxidation renders some other electrodes inert. Thus, the mercury drop electrode, oxidized in chloride medium, is covered with calomel which is a salt conductive enough to permit the use of the protected drop as an indicating electrode (106). The conducting protective film can be an oxide as is the case for the tungsten electrode covered with a conducting hemi-pentoxide in some media (107) or for the classical smooth platinum electrode used in non-aqueous media in such a general way.

These noble metal oxide films generally do not participate in defining the anodic limit. However, their formation or their reduction considerably deforms the voltammetric and cyclic voltammetric recordings and thus render an interpretation difficult. The few studies that have been done until now about this oxide formation in non-aqueous media were generally not performed using classical electrochemical analytical methods. It is interesting to note the following qualitative results of this oxide formation during voltammetric and cyclic voltammetric analytical studies.

The general problem of anode oxide films (108) and especially platinum oxides (109, 110) were extensively studied in aqueous media. Our knowledge is still very limited concerning non-aqueous media where, frequently, the unavoidable residual water is responsible for film formation. Among many others, the works of Breiter systematically using cyclic voltammetry (111) in aqueous media demonstrated that the potential zone where the platinum is

---

* This phenomenon is well known in electrometallurgy of aluminium. See also (102).

uncovered (double layer zone) is relatively limited. This zone covers only about 0·4 volts for a precathodized electrode (Figure 16, curve a) in 1 N perchloric acid solution. The irreversibility of the oxide film reduction process delays the appearance of the corresponding peak during the cathodic sweep (Figure 16, curve b) and consequently, only a double layer region extending approximately 0·2 volts appears on the preanodized electrode. Proton reduction peaks and absorbed hydrogen reoxidation peaks are observed toward negative potentials. The symmetry observed between the anodic curve and the cathodic curve in this region illustrates the reversibility of the phenomena occurring there.

Popov and Geske reported as early as 1958 (112) that a smooth platinum electrode prepolarized at the anodic limit potential of 0·1 M LiClO$_4$ acetonitrile reveals a voltammetric cathodic peak located at $-0·7$ volts with

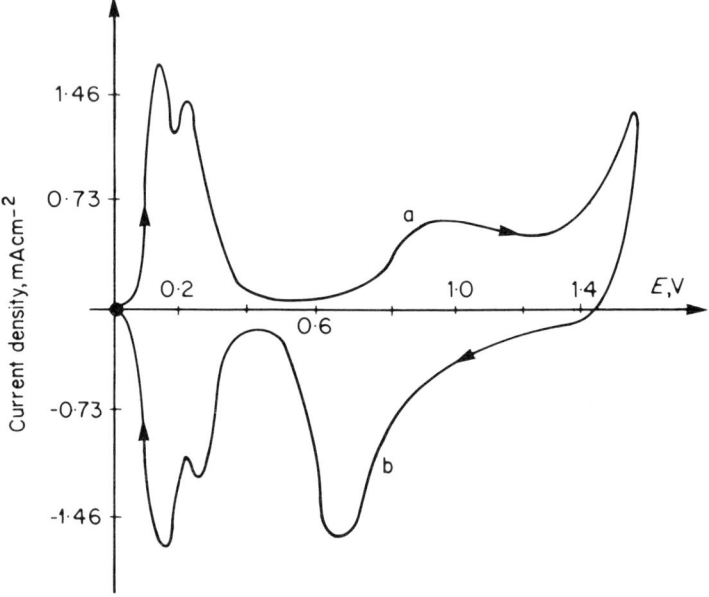

Fig. 16. Cyclic voltammetry on a platinum electrode in aqueous medium (see the text for explanations).

reference to a 0·2 M Ag/Ag$^+$ system; the intensity of this peak was an increasing function of the residual water concentration. This peak was attributed to the reduction of the platinum oxide formed at the high electrode potential. Recently, Conway and coworkers used various transitory methods to demonstrate that the appearance of platinum oxide in acetic (113) and trifluoroacetic (114) acids was related to the presence of residual water. The same is also true for propylene carbonate and acetonitrile (115).

At a scan rate of 0·250 volts min^{-1}, the current–potential curve characterizing the electro-activity range of a 0·1 M tetraethylammonium perchlorate
solution in nitromethane, about 0·005 M in residual water and deoxygenated
with argon, is an almost straight line between −0·9 and 2·8 volts with reference
to Ag/AgCl (Figure 17, curve a) (81). On the other hand, at 5 volts min^{-1},
the cyclic voltammetric curve recorded at the same stationary precathodized
electrode at −0·8 volts reveals a modification of the electrode surface (116).
Two rapidly increasing zones of residual currents are noted toward 1·8 volts
and then toward 2·4 volts (Figure 17, curve b). If the anodic sweep extends
beyond 2·4 volts, a reduction peak appears on the cathodic half-cycle at about
1·10 volts (Figure 17, peak b'). The same recording with or without stirring
of the solution reveals that the anodic phenomena observed at about 1·8
volts do not depend on the stirring while the cathodic peak b' disappears

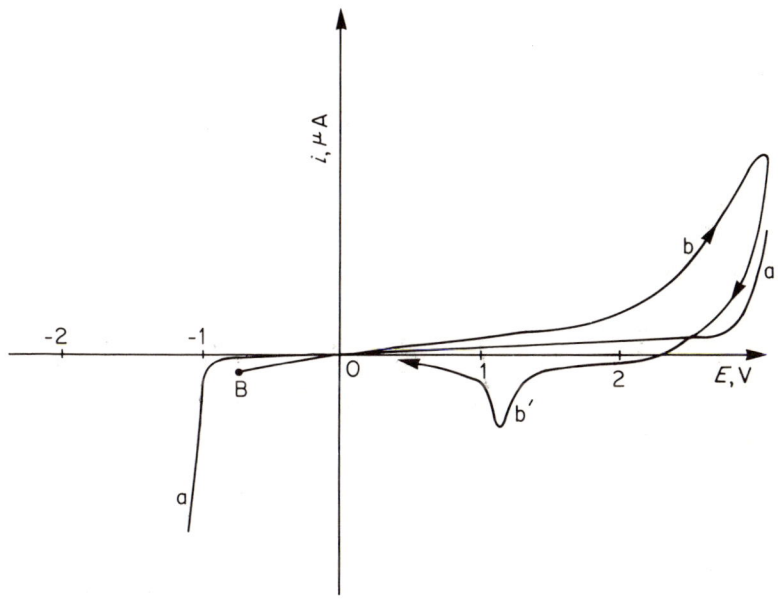

Fig. 17. Oxides formation and reduction on a smooth platinum electrode dipped in
nitromethane $(C_2H_5)_4NClO_4$ 0·1 M. Curve a: Electroactivity range on a rotating
electrode; Curve bb': Cyclic voltammetry on a stationary electrode.

upon stirring. The amplitudes of the anodic current beyond 1·8 volts and the
cathodic peak increase with increasing residual water concentration.

The phenomena observed on a preanodized electrode are different from
those described above. No cathodic peak is observed when the sweep is
running towards decreasing potentials from 1·8 to −0·6 volts (Figure 18,
curve a). A supplementary cathodic peak b located at −0·44 volts is observed

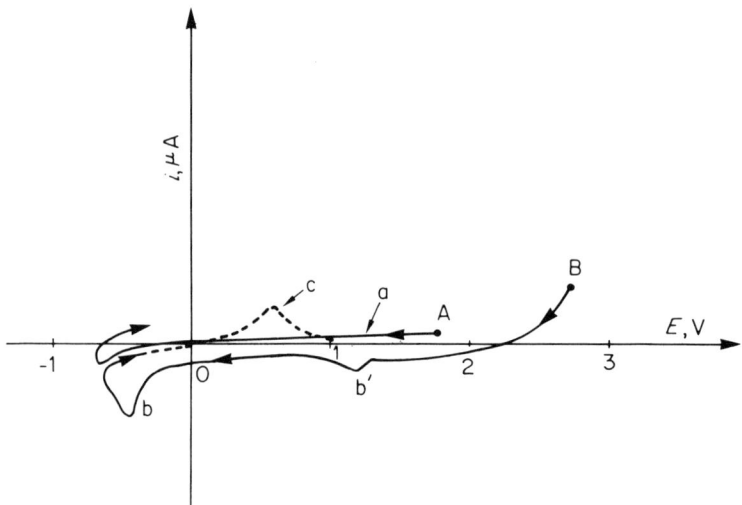

Fig. 18. Oxides formation and reduction on a smooth platinum electrode dipped in nitromethane $(C_2H_5)_4NClO_4$ 0.1 M (cyclic voltammetry). Curve a: After anodization at 1.8 V: Curve bb': After anodization at 2.6 V; Curve c: Oxidation peak of the adsorbed hydrogen.

if the sweep is starting at or beyond 2.2 volts. Unlike peak b', this peak is not eliminated but is slightly increased by stirring. Its amplitude increases with prolonged duration of the anodization time at 2.6 volts.

This collection of data and various other observations (116) suggest that in this medium and in water (109, 110), platinum can be covered with one or two oxides. The protons liberated during oxide formation are necessary for the reduction to take place at 1.10 volts. This is shown by the disappearance of the 1.10 volts peak upon stirring. The reduction observed at 0.44 volts is influenced by the water present in the medium as shown when the water concentration is increased. The cathodic peak found at -0.44 volts is very intense in a sufficiently hydrated medium (0.01 to 0.03 M). On the return anodic half-cycle this peak is followed by a 0.66 volts peak (Figure 18, curve c) attributed to the oxidation of absorbed hydrogen (116). It is tempting to suppose that the -0.44 volts peak is due to the residual water reduction. When using electrochemical techniques in non-aqueous media it is important to know that platinum oxides can considerably disturb the phenomena studied with noble metal electrodes. This has been proved for nitromethane as a solvent and should be remembered by chemists using electroanalytical techniques in non-aqueous media especially when using cyclic voltammetry.

Thus, during a sweep towards decreasing potentials, the oxides formed at the beginning of the sweep can form a platinized platinum surface when they are reduced.* An irreversible peak representing perchloric acid proton reduction at a polarized (at 1·8 volts) platinum electrode is found in 0·1 M nitromethane in quaternary ammonium perchlorate solution. However, an electrode polarized at 2·4 volts and that was initially covered with oxides which were reduced during the sweep towards the negative potentials permits the recording of both a cathodic and an anodic peak which form a quasi-reversible system for the same acid.

Inversely, some electrode processes are eliminated on 'reduced electrodes'. Popov and Geske related that the I_3^- reduction in acetonitrile apparently does not take place on a platinum electrode at a potential below 0·9 volts with reference to a 0·01 M Ag/Ag$^+$ electrode.

Modifications of the reversibility and even of the limiting current resulting from oxide formation or reduction are also observed during studies of organic compounds. Vermillon noticed a marked influence of the surface condition of a rotating platinum electrode on voltammetric curves for various phenol oxidations in acetonitrile (118).

The inverse phenomenon is sometimes observed, in that oxide formation or reduction can be modified by the presence of substances whose electrochemical properties were studied. Thus cyclic voltammetry reveals that platinum oxide formation is facilitated by the presence of the relatively strong basic NO_3^- ion in 0·001 M tetraethylammonium nitrate solution in 0·1 M tetraethylammonium perchlorate in nitromethane (116). This result is similar to the observation of pH influences on platinum oxide formation and reduction in aqueous media (119) and demonstrates that the anodic film behaviour is not strictly due to the medium/electrode set. Sometimes the oxide reduction and the reduction of an electroactive species present in the medium or produced during an earlier anodic sweep can be observed at the same potential giving a single cathodic peak. Although this question needs further studies one can assume that this happens under two different conditions. First, it could be a 'modified' oxide similar to that suggested by Conway and coworkers (120) which could explain that only one peak corresponds to the simultaneous reduction of oxygen and of the $CH_3COO\cdot$ radical (co-adsorbed on platinum) during the cathodic sweep of cyclic voltammetry of an aqueous potassium acetate solution. This type of modified oxide is found only if the substance present in the medium has a favourable composition and molecular geometry. Some indices seem to indicate that this could be true for nitric anhydride in a 0·1 M tetraethylammonium perchlorate solution in nitromethane (116).

---

* The exact electrode surface condition after anodization followed by cathodization is still a controversial subject. Some authors feel that the main effect of this treatment in aqueous media is to liberate the electrode surface from adsorbed impurities and not to produce a superficial layer of platinum having a special structure (117). A similar study has not been carried out for non-aqueous media.

A single cathodic peak can also be observed if the reduction of an electro-active species proceeding normally very slowly at an oxidized electrode, becomes rapid only after the oxide is reduced. This is found in cyclic voltammetry if the sweep is first extended until sufficiently anodic potentials are attained to induce oxide formation. The cathodic peak has a characteristic form which demonstrates the 'release' of the electrode.

These various phenomena complicate and even render uncertain comparative studies of electrochemical kinetics determined in different media.

Until now, no systematic study of noble metal oxides in fused salts has been undertaken. Actually such oxides dissolve in most of these media. Nevertheless oxide formations on gold electrodes were observed in the ternary eutectic KNO_3–$NaNO_3$–$LiNO_3$ fused at 140°C and containing traces of water (121).

C. Comparison of Electroactivity Ranges and Selection of a Medium for a Given Problem

When diagrams similar to those of Figure 15 or the tables of Section VII are available, it is tempting to compare electroactivity ranges in order to determine which medium allows for the highest potentials at a given electrode. Consequently, such a medium would allow a species to be oxidized, the oxidation of which is very difficult, in order to develop an electrochemical titration method, for example.

Such a comparison, however, is difficult and should be attempted only with caution.

The problem of comparing a reference electrode in different media should have been solved beforehand. Solutions of this problem have been proposed in Section III. The respective positions of the various ranges considered in an absolute potential scale were thus obtained. A diagram similar to that of Figure 15 is the result. Comparison of the various limits does not have any real significance unless the limits are very separated from each other. For example one can assume that acetonitrile and nitromethane permit oxidations on platinum electrodes that cannot be observed in water. However when compared with each other, the anodic limits of these two solvents at the same electrode are insignificant for many reasons. If a species the oxidation potential of which is very inferior to the potentials of these limits is oxidized, neither of the two solvents presents preferential advantages with respect to the extent of their ranges. On the contrary, if the oxidation of the species seems to take place close to these limits, the slight difference that exists between acetonitrile and nitromethane is hardly significant. It is not evident that this species will or can be oxidized at similar reaction rates. Comparisons of half-wave potentials are without interest under these conditions.

Reactions taking place when approaching the range limits can present unpredictable complications. Oxidation of the studied species can sometimes cause a massive catalytic oxidation of one of the two media. A phenomenon of this type, not revealed graphically by the voltammetric curves, seems to take place during iodine oxidation in acetic anhydride (122). Under these conditions an amperometric titration is still valid in the medium considered if the catalytic reaction is slow enough, whereas a coulometric titration is impossible.

Moreover one must remember that in some media various compounds can be oxidized or reduced at potentials the access of which seems to be impossible when considering the electroactivity range. A classic example is Kolbe's reaction which takes place most frequently in water or hydro-organic media on smooth platinum at potentials where oxidation of the solvent should be the predominant reaction. Preferential absorption of the carboxylate anions inhibit water oxidation preferentially to that of these anions (123). In some organic solvents the electrochemical process can be even simpler. The presence of an anodic barrier towards high potentials and the absence of, or only slight effect of adsorption phenomena permit the plotting of the carboxylate ion voltammetric curve. This is true for acetonitrile (124) and nitromethane (125).

This example pertaining to Kolbe's reaction illustrates the real interest in comparing electroactivity ranges and also their limits. Comparison of the water and acetonitrile ranges indicates the preference for acetonitrile as a medium for the study of slightly oxidizable species such as the carboxylate ions. But this comparison cannot predict that a preferential adsorption takes place which permits the oxidation of these ions in aqueous media.

V. APPLICATIONS TO ANALYTICAL CHEMISTRY

The expression 'analytical chemistry' used in its largest sense includes qualitative and quantitative chemical analysis as well as the study of solution reactions, separation methods and various electrochemical techniques, etc. Thus, chemical analysis is only one of the applications of the basic analytical studies.

The generalization of voltammetry had different repercussions on its analytical applications because of its use in non-aqueous solvents and molten salts. Four important aspects should be considered:

1. Voltammetry applications for analysing chemical solution reactions in non-aqueous media.

2. Use of voltammetry for developing other electroanalytical techniques; amperometry, potentiometry, coulometry, etc.

3. Identification of species in non-aqueous media by voltammetry (qualitative analysis).

4. Titrations of various substances in non-aqueous media by voltammetry (quantitative analysis).

Voltammetry has the same advantages and yields the same results in water or other solvents with respect to the first two above-mentioned applications. Thus, voltammetric determinations of thermodynamic constants, classically used by the chemical analyst, i.e. acid/base reaction pK, or complex formation pK, standard potentials and solubility products, are identical in non-aqueous solvents and in water. The same is true for the use of voltammetric curves in order to develop other analytical techniques. This is the reason why only some examples are described here for the two first analytical aspects.

A. Determination of Thermodynamic Constants in Non-aqueous Media by a Voltammetric Method

1a. Determination of pK values of weak acids in liquid ammonia and liquid ammoniates. Use for the hydrogen electrode.

Curves 1 and 2 of Figure 19 represent the reduction of strong acid (NH^{4+}) and weak acid solutions in sodium iodide ammoniate at a platinized platinum rotating disk electrode (126). These electrochemical systems are fast and consequently the difference between the two half-wave potentials is related in solvents, as in aqueous solution, to the weak acid pK. The constant K is defined by the following formula:

$$K = \frac{|H^+|_S|B|_S}{|HB^+|_S}$$

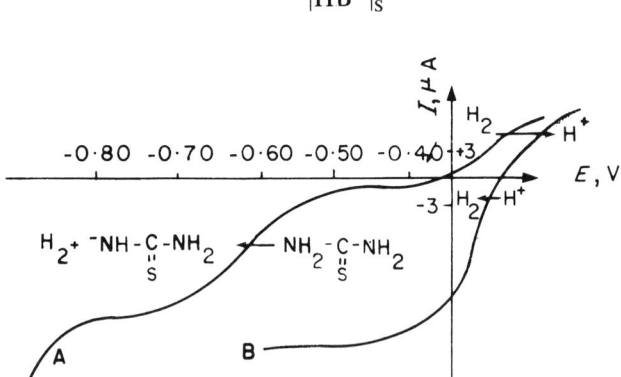

Fig. 19. Current–potential curves of the H_2/H^+ couple. NH_3, $-20°C$. (Hydrogen atmosphere.) Platinized platinum electrode. Curve A: weak acid; Curve B: strong acid. (M. Herlem, *Bull. Soc. Chim. France*, **1965**, 3329.)

where $|H^+|_S$ represents the solvated proton activity in this medium (NH_4^+ cation), $|B|_S$ and $|HB^+|_S$ the activities of the acid and the base in the same medium.

1b. The I_3^- complex dissociation pK and the standard potentials of the I_2/I_3^- and I_3^-/I^- redox couples were determined in acetonitrile using voltammetry at a rotating platinum electrode (127). The curve in Figure 20 shows the two oxidation waves of an iodide solution. Their respective heights (2:1) illustrate the presence of two successive oxidation reactions

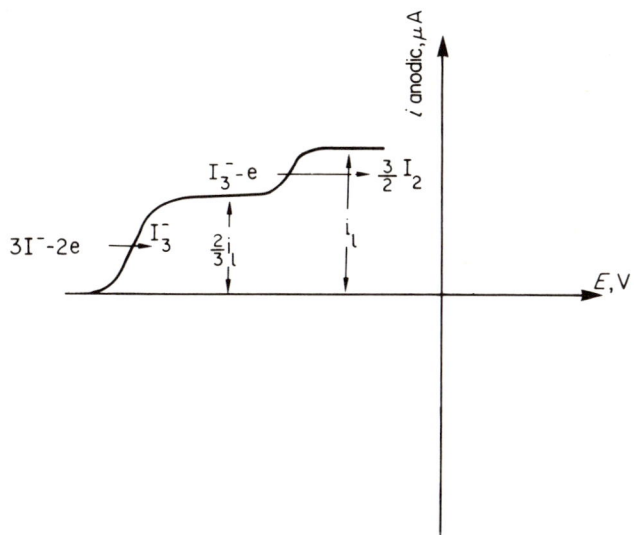

Fig. 20. Oxidation of an iodide solution in acetonitrile at a smooth platinum rotating disc electrode. (See reference 127).

$3 I^- - 2 e \rightarrow I_3^-$ and $2 I_3^- - 2 e \rightarrow 3 I_2$. The half-wave potential difference is simply related to the I_3^- complex dissociation pK defined by the following equation:

$$pK = -\log K = -\log \frac{|I_2| |I^-|}{|I_3^-|}$$

and thus the pK value has been determined to be 7.4 (128). A determination of this type cannot be carried out in water because the complex is too unstable (the pK is 3 (129)).

1c. The anodic oxidation of metals such as silver and mercury is facilitated by precipitation or complex formation. The presence of the resulting compounds is revealed by the appearance of anodic waves whose heights are proportional to the complexing ion concentration. Figure 21 illustrates silver dissolution in the presence of chloride ions in nitromethane (62, 130).

The following constant is obtained from these curves:

$$K = \frac{|Ag^+| \, |Cl^-|^2}{|AgCl_2^-|} = 10^{-19.5}$$

The solubility constant is found to be

$$S = Ag \downarrow / AgCl_2^- = 10^{-19.2}$$

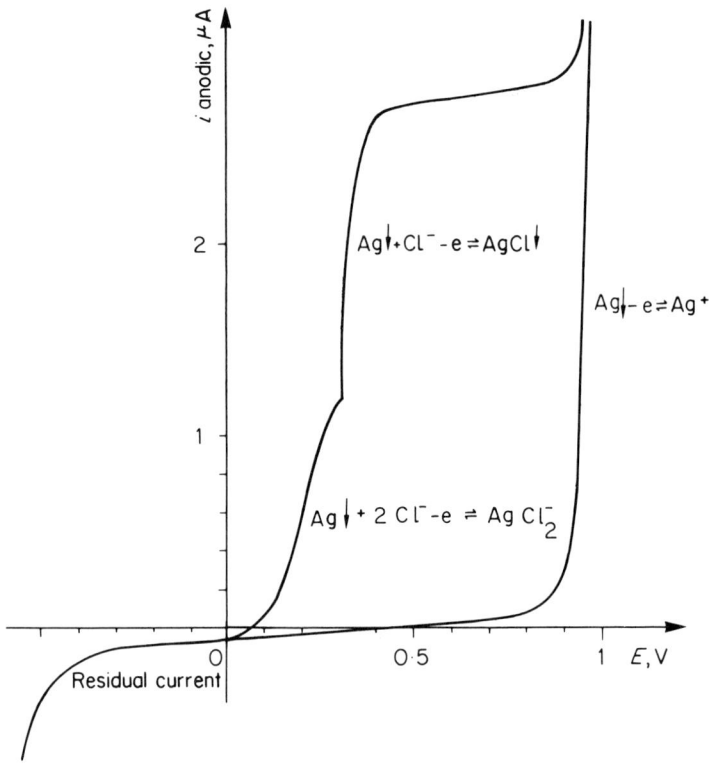

Fig. 21. Oxidation of a silver rotating disk electrode in a chloride solution in nitromethane. (J. Bardin, Thesis, Paris 1968.)

Stability constants for many silver and mercury complexes in various solvents were obtained in a similar way.

B. Use of Voltammograms in Electroanalytical Applications

As in aqueous solutions, the experimental conditions necessary for amperometric or potentiometric studies of the concentration of a given substance during a chemical reaction, are obtained from current–potential curves.

Two examples of voltammogram applications are presented. The first pertains to the H_2/H^+ couple voltammetric behaviour in liquid ammonia and its use for potentiometric titrations of very weak acids. The other considers voltammetric behaviour of phenothiazine in acetonitrile and its use for amperometric and potentiometric titrations of phenothiazine.

The $i = f(E)$ curves for the H_2/NH_4^+ (strong acid) and H_2/H_2O (weak acid) redox couples in liquid ammonia at $-60°C$ as observed at a suitably pretreated platinized platinum rotating disk electrode (130a) are shown in Figure 22. The hydrogen oxidation/proton reduction process is evidenced. The solvated proton reduction (a) is much easier than the reduction of protons from water (b). The corresponding electrochemical reactions are:

$$2 H^+ \text{ (solvated)} + 2 e \rightleftharpoons H_2 \qquad \text{curve a}$$
$$2 H_2O + 2 e \rightleftharpoons H_2 + 2 OH^- \qquad \text{curve b}$$

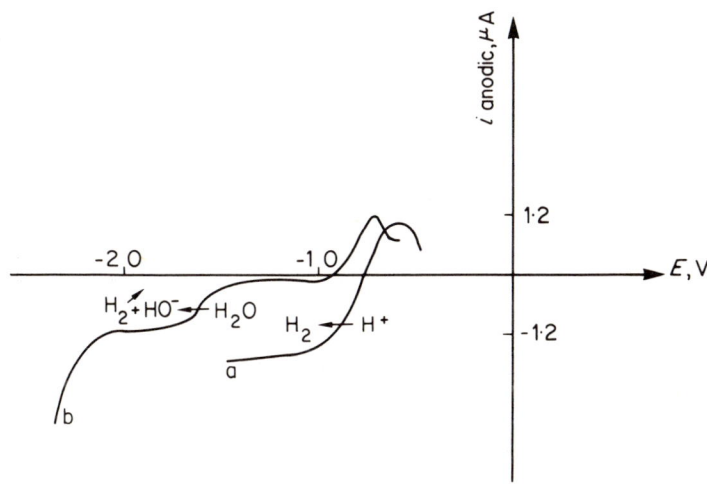

Fig. 22. Current–potential curves in liquid ammonia at $-60°C$, platinized platinum rotating disk electrode, hydrogen atmosphere. Curve a: NH_4^+ reduction; Curve b: H_2O reduction. (M. Herlem, *Bull. Soc. Chim. France*, **1967**, 1687.)

These reactions are fast enough for the current–potential curves to intersect the potential axis with a considerable slope. The potential at zero current follows Nernst's law. It is therefore possible to follow various acid titrations in liquid ammonia with potassium amide solution prepared coulometrically *in situ* (Figures 23 and 24). These titrations are carried out with zero current potentiometry. Water reacts quantitatively with the amide according to the following reaction:

$$H_2O + NH_2^- \rightleftharpoons NH_3 + OH^-$$

Its pK is 16. Aniline is a weaker acid and reacts incompletely with the amide ion: $pK = 28.8$ (131).

$$C_6H_5NH_2 + NH_2^- \rightleftharpoons C_6H_5NH^- + NH_3$$

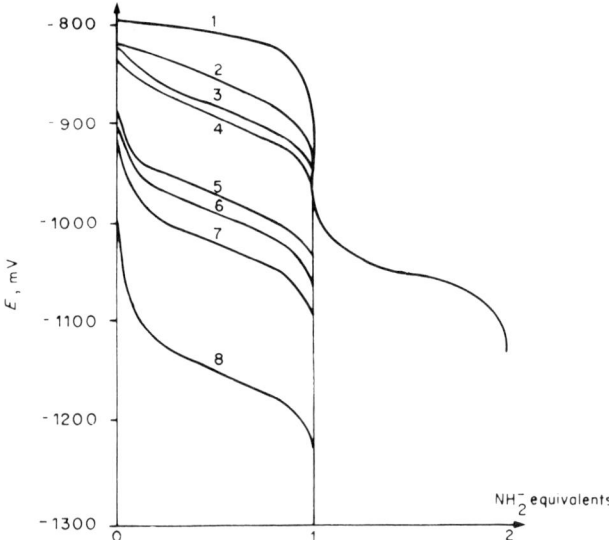

Fig. 23. Potentiometric titrations of acids in liquid ammonia ($-60°C$). Curve 1:
NH_4I, thiophenol, diphenyl thiourea thiocarbohydrazide; Curve 2: Phenyl thiourea
$pK = 3·8$; Curve 3: Thioacetamide $pK = 4·5$; Curve 4: Thiosemicarbazide $pK_1 =$
$4·7$, $pK_2 = 8·5$; Curve 5: Thiosinamine $pK = 6·6$; Curve 6; Diphenyl urea $pK =$
$7·0$; Curve 7: Diphenyl carbazide $pK = 7·7$; Curve 8: Diphenyl guanidine $pK =$
$10·8$. (M. Herlem and A. Thiebault, *Bull. Soc. Chim. France*, **1970**, 383.)

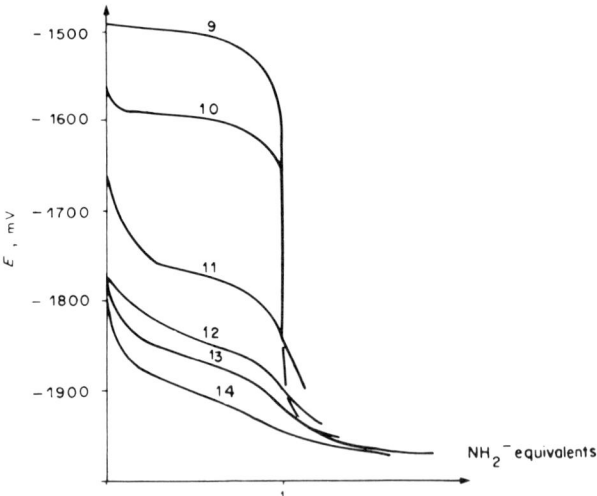

Fig. 24. Potentiometric titrations of acids in liquid ammonia ($-60°C$). Curve 9:
Water $pK = 18·9$; Curve 10: Acetone $pK' = 18·9$; Curve 11: Dichloro-2,5-aniline pK
$= 25·5$; Curve 12: p-Chloroaniline $pK = 27·4$; Curve 13: t-Butylalcohol $pK = 28$;
Curve 14: Aniline $pK = 28·8$. (M. Herlem and A. Thiebault, *Bull. Soc. Chim. France*,
1970, 383.)

Analogous studies of very weak acid titrations were possible in dimethyl sulphoxide (132) as illustrated by Figure 25.

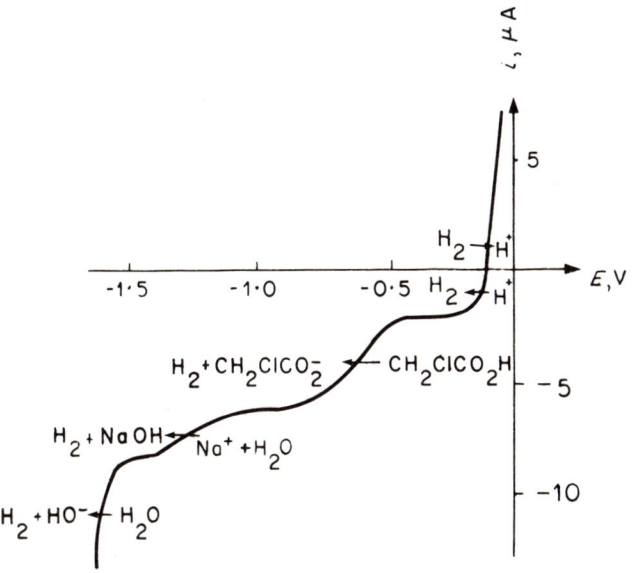

Fig. 25. Current–potential curve of a mixture of acids in DMSO at a rotating platinized platinum disk electrode. Hydrogen atmosphere. (J. Courtot-Coupez and M. Le Demezet, *Bull. Soc. Chim. France*, **1969**, 1033.)

The second example is dealing with amperometric and potentiometric titrations of phenothiazine in acetonitrile.

The phenothiazine (P) oxidation curve at a smooth platinum rotating disk electrode in acetonitrile reveals two one-electron steps (133) (Figure 26).

The phenazothionium cation (S) reduction (curve 3) and the oxidation and reduction curves of the radical (R) (curve 2) are also presented. The various possible types of curves for potentiometric or amperometric titrations of phenothiazine by an oxidant can easily be drawn from this figure. The oxidizing titration reagent used by the authors (134) is the phenazothionium

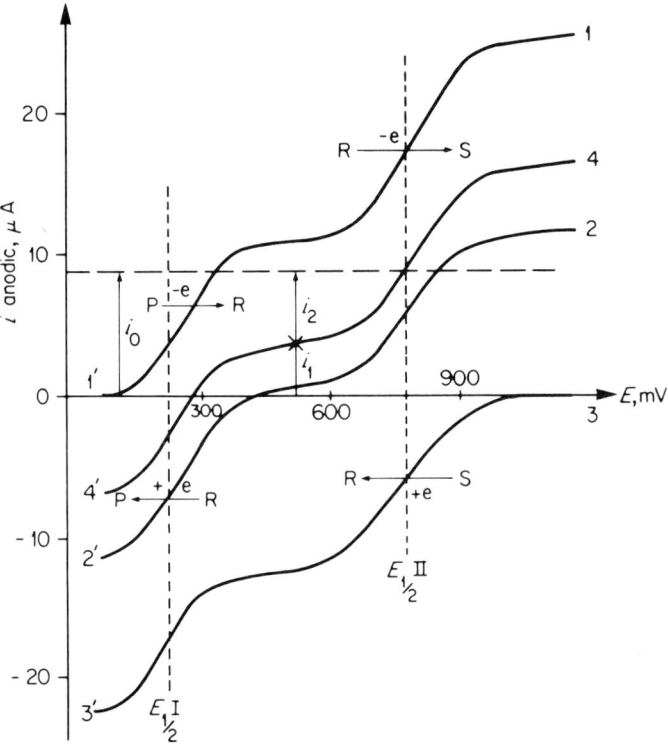

Fig. 26. Current–potential curves of phenothiane P (1), the radical R (2) and the phenazothionium ion S (3).

salt prepared *in situ* coulometrically. The chemical reaction of titration corresponds to phenothiazine (P) oxidation by the phenazothionium salt (S) to give the radical (R) according to the following equation:

$$P + S \rightarrow 2R$$

The phenothioazine solution to be titrated is therefore enriched in radical during the titration and the reaction end-point is characterized by the presence of only the radical in solution. When the end-point is passed the cation (S) appears in the solution. Zero current potentiometry produces a classical bilogarithmic titration curve $E = f$ (concentration of titrating reagent S):

$$E = E_0(R/P) + \frac{RT}{F} \ln \frac{C_R}{C_P}$$

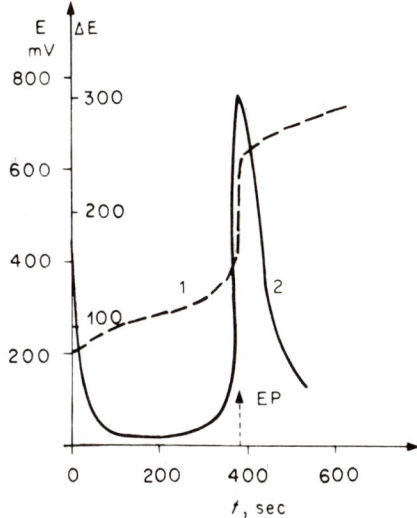

Fig. 27. Coulometric titration of phenothiazine. Potentiometric end-point determination
(EP). Curve 1: Zero current potentiometry:
$E = f$ (time); Curve 2: Fixed current potentiometry ($i = 30 \, \mu A$): $\Delta E = E_a - E_c = f$ (time).

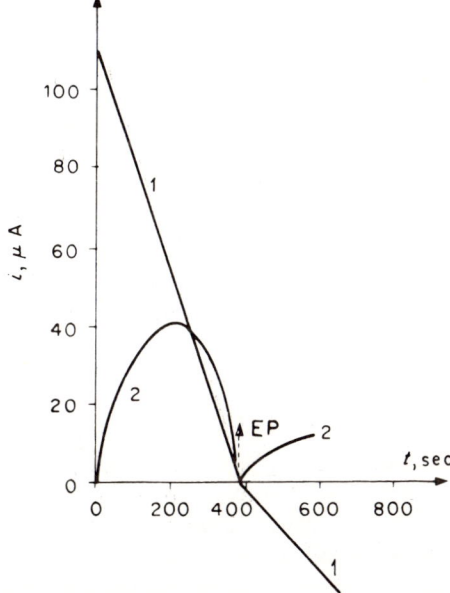

Fig. 28. Coulometric titration of phenothiazine. Amperometric end-point determination (EP). Curve 1: Indicative
anode $E_a = +0.05 \, V/Ag\downarrow/Ag^+ - 10^{-2} \, M$; Curve 2:
Two indicative electrodes $E_a - E_c = 100 \, mV$.

before the end point EP, and

$$E = E_0(R/S) + \frac{RT}{F} \ln \frac{C_S}{C_R}$$

after the end-point, where C_P, C_R and C_S are the P, R and S species concentrations in the solution. Curve 1 of Figure 27 is an example of this type of titration curve. Curve 2 presents the ΔE potential difference variations found between the two platinum electrodes with a 30 μA current during the same titration (potentiometry with two imposed current indicating electrodes). The same reaction can be followed amperometrically (Figure 28).

C. Voltammetric Identification of Species in Solvents

The main interest in using non-aqueous solvents is the possibility of generating species unstable in water, by new chemical or electrochemical reactions. The nature of these species should be specified and various physical methods should be associated with voltamperometry (see Section VI-D).

The I_3^- species, easily formed by mixing stoichiometric quantities of iodine and iodide, is evidenced in numerous solvents [Table II (135)] whereas under these conditions it is unstable in water.

TABLE II

Solvents S	CH_3NO_2	$C_3H_6CO_3$	$(CH_2)_4SO_2$	$HCON(CH_3)_2$	CH_3CN	$(CH_3)_2SO$	CH_3OH
$E(I_2/I_3^-)$	$+0.36_5$	$+0.39$	$+0.35$	$+0.245$	$+0.16$	$+0.16$	$+0.24$
$E(I_3^-/I^-)$	-0.27_5	-0.28	-0.31	-0.39	-0.31_5	-0.32	-0.13
$\beta(I_3^-)$	$10^{7.4}$	$10^{7.6}$	$10^{7.5}$	$10^{7.2}$	$10^{7.3}$	$10^{5.4}$	$10^{4.2}$

This species is easily characterized by voltagram 1a (Figure 20). The two step oxidation of iodide gives two well-separated waves. Their respective heights are directly related to the two reaction coefficients:

first step: $3 I^- \longrightarrow I_3^- + 2 e$

second step: $2 I_3^- \longrightarrow 3 I_2 + 2 e$

Some other examples of the applications of voltammetry in qualitative solution analysis will be given.

(a) Voltammetric NO_2^+ ion characterization in nitromethane. The addition of concentrated sulphuric acid to a nitric acid solution in nitromethane yields NO_2^+ ions which can be characterized (and analysed quantitatively) by plotting a reduction curve at a smooth platinum electrode (136).

(b) Characterization of the $O_2^{.-}$ ion in acetonitrile (137), dimethylformamide (138), fused alkaline hydroxides (139) or fused nitrate (140). The presence of this superoxide anion is demonstrated by an anodic wave, observed at a platinum electrode, the wave height being a function of the species concentration.

D. Voltammetric Titrations of Various Substances in Non-aqueous Solvents

It is not our aim to present a complete inventory of titrations in non-aqueous media. The reader is referred to various review articles (18–20, 141, 142), but we shall emphasize the main advantages of using non-aqueous media which seem to be the following:

1. The solubility of water insoluble species can be increased.
2. Species which do not react electrochemically in water can become electroactive.
3. Species which undergo hydrolysis in aqueous solutions can be stabilized.
4. Electrochemical reaction kinetics can be accelerated, their reaction mechanisms simplified, or complications resulting from adsorption phenomena can be eliminated.

1. Increasing Species Solubility

This advantage is obtained not only in anhydrous solvent media, but also in hydroorganic media. It has been successfully used for many years in potentiometric and polarographic titrations. Noteworthy examples are the good solubility of waxes (143) and kerosene (petroleum derivatives) (144) in methanol/benzene mixtures, that of fatty acids (145) and oils (146) in ethanol/benzene mixtures, nylon (147) in glycol/phenol mixtures, lubricants in ethanol/carbon tetrachloride mixtures (148) as well as some oxides in various fused salts or molten alkaline hydroxides (149, 150).

Takahashi's review article (151) illustrates the interest of these solvents in gas analyses because their solubility is greater than in water. Dimethylsulphoxide (152) has been used for various gas analyses.

Gas solubilities increase as the temperature decreases. The use of solvents or solvent mixtures with low melting points is advantageous. Low temperature oxygen analyses were carried out in methanol/water mixtures (153).

The good solubility of some species in various solvents can be used for extractions from aqueous media followed by direct voltammetric titrations in the extracting solvent. Several of these titrations were cited in a general review in 1960 (153). Aluminium and beryllium were titrated by acetylacetonate extractions (154). Similarly, silica was titrated after extracting a silico–molybdic acid complex with dichloroethane, followed by voltammetric reduction (154a).

2. Electroactivity of Species which are Usually not Electroactive in Water

This is one of the greatest advantages of using solvents other than water. The proton reduction limit for HS type solvents is shifted toward the cathodic side when these solvents are less acid than water. Very weak acid reduction waves including the water reduction wave can be evidenced on platinum electrodes (generally after platinizing). The reader is referred to the example of liquid ammonia and dimethylsulphoxide described above (see Section V-B-2). These same solvents permit voltamperometric titrations of substances not reducible in water. Dimethylsulphoxide permits titrations of aluminium and beryllium ions. Solvents which resist oxidation allow titrations of some species which would oxidize water.

3. Stabilization of Species which do not Exist in Aqueous Solution

This property is interesting because new species can be evidenced and it can be profitably used for chemical analysis.

The blue coloured I_2^+ species does not occur in aqueous solutions. It was described by Gillespie and coworkers (155). Adhami characterized this species by its current–potential curve in fluorosulphuric acid $HFSO_3$ (156). The medium is conductive enough without the addition of a supporting electrolyte so that a $0.05\ M\ I_2^+$ ion solution can be characterized by its oxidation wave according to the following formula:

$$I_2^+ - 5e \rightarrow 2I^{3+} \qquad E_{1/2} = +1.1\ V/\text{reference Au} \downarrow /Au(CN)_3$$

It can also be characterized by its reduction wave

$$I_2^+ + e \rightarrow I_2 \qquad E_{1/2} = -0.05\ V/\text{reference Au} \downarrow /Au(CN)_3$$

The $O_2 \cdot^-$ superoxide ion which is unstable in aqueous solution, was characterized in a NaOH–KOH eutectic at 227°C by a well-defined reduction wave at a platinum electrode which corresponds to the following reaction: reaction:

$$O_2^- + e \rightarrow O_2^{2-} \qquad E_{1/2} = 1.75\ V/\text{reference Na/Na}^+ (157)$$

This anion is also characterized in dimethylformamide, acetonitrile, dimethylsulphoxide, acetone and pyridine by cyclic voltammetry. An oxidation peak reflecting the generation of this anion was obtained by oxygen reduction according to the following reaction:

$$O_2 + e \rightarrow O_2 \cdot^-$$

(see Peover and White (158)).

Some species too basic to exist in water are stable in liquid ammonia. The NH_2^- ion is an example of this type. This anion has been used as a titrating reagent for a large number of very weak acids (see Section V-2a). Similarly, acids entirely ionized and dissociated in water such as hydrochloric acid $(H_3O^+ + Cl^-)$ can exist in trifluoroacetic acid in the molecular state (159). Highly reactive reducing agents unstable in water such as metallic sodium or potassium are stable in liquid ammonia. Similarly, substances such as NO_2^+ ions which oxidize water are stable in nitromethane (136). These substances are characterized in these respective media by their oxidation and reduction curves.

4. Increasing Rate of Electrochemical Reaction Kinetics

When the electrochemical kinetics are slow, the current–potential curves are often badly defined. This is especially true when there are two successive waves which are difficult to separate. Changing the species solvation can cause the appearance of a new electrochemical reduction mechanism involving fast kinetic steps.

Thus, the reduction of rare earths in different aprotic solvents such as acetonitrile (160), acetone (161), dimethylsulphoxide (146) or dimethylformamide (162) produce well-defined waves. Figure 29 permits a comparison

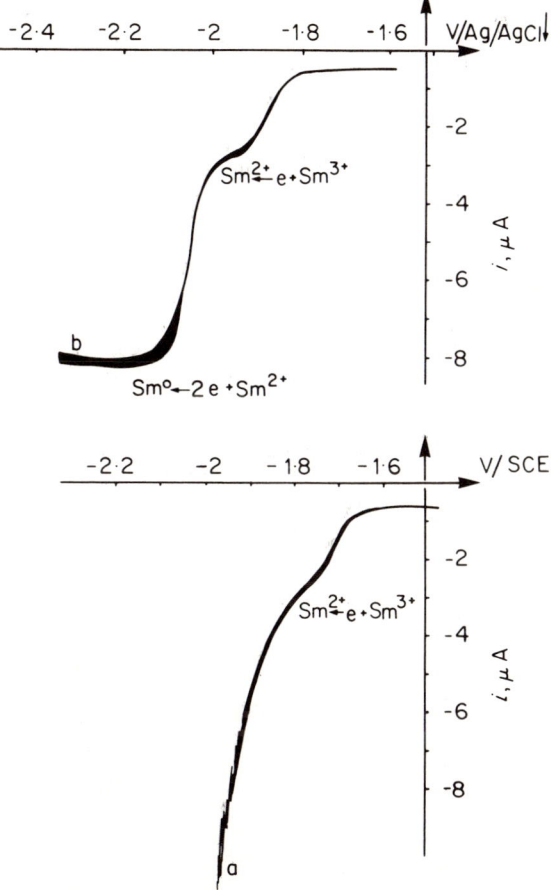

Fig. 29. Reduction of Sm^{3+} in water (curve a) and in dimethylformamide (curve b). (A. Leveque, Thesis, Paris 1970.)

of samarium reduction polarograms in water and dimethylformamide (162). Magnesium, beryllium and aluminium also have well-defined waves in the solvents mentioned above.

VI. TECHNIQUES

A. Recording of Voltamperometric Curves and Problem of Cell Resistance

The main difficulty encountered in applying electrochemical analytical methods in non-aqueous media is due to the ohmic drop across the cell. This phenomenon is particularly important in weak dielectric constant solvents. It causes considerable difficulties (163) in data interpretation for modern methods such as cyclic voltammetry, superimposed alternating current polarography, etc. Under unfavourable conditions, a too high cell resistance can affect also the simple plot of a voltamperometric curve.

A three-electrode set-up must be used in all cases. This arrangement does not completely compensate the ohmic drop between the reference electrode *extremity* and the indicating electrode and, naturally, the reference electrode must not be placed between the indicating and the auxiliary electrodes. In spite of selecting the most favourable position, it may be necessary to consider this ohmic drop.

Firstly, the non-compensated ohmic drop value can be determined. Various methods were proposed, but Booman and Holbrook's (164) method seems to be used most frequently (165). When the ohmic drop is known, the potential applied to the indicating electrode can then be corrected. However, this technique can be used only for ordinary voltamperometry because the ohmic drop is present more insidiously in the other methods. Thus, in the methods with a rapid potential sweep, the true value of the sweep speed is affected with an error $R\dfrac{di}{dt}$ which varies throughout the tracing. Two attitudes can be considered in this event. The ohmic drop participation can be included in the theory of the method, but this may be difficult (163, 166). However, many authors compensate this ohmic drop experimentally. But many controversies have arisen on the various methods and the reader is asked to consult the original articles (163). Controlled-potential electrolysis (167, 168) reveals that the ohmic drop has a considerable influence on the true value of the potential of the working electrode.

B. Reference Electrodes

The requirements for reference electrodes were presented in Section III. We shall not discuss this question again in more detail especially since each solvent has special problems with respect to this subject. The references cited in Section VII will permit the reader to find the methods for preparing reference systems used in various solvents (169).

The following process can be used to determine which reference electrode should be used in a new solvent. A redox system generally furnishes a satisfactory reference when based on known solvent properties or on analogies between satisfactory systems used for solvents with similar structures. The reversibility of the system should be studied in the new solvent by tracing its voltamperometric curves. To perform this, a temporary reference should be used, even one for aqueous media can be used if it is sufficiently separated from the non-aqueous medium.

Several examples of the new reference should be prepared to make sure of its reproducibility. The measurement of the half-wave potential of a reversible system *versus* these electrodes thereby reveals the stability of the selected reference with respect to time. It is advisable to prove that slight modifications of the residual water content do not modify the potential.

It would be desirable for each author to measure and publish the approximate potential difference between his new reference and a standard reference electrode which is easy to reproduce, such as the ferrocene–ferricinium ion couple or even a calomel electrode in aqueous media. Systematic publication of these data would greatly facilitate data comparison in the various solvents despite the inaccuracies of these results (see Section III).

It is often interesting to prepare the reference electrodes by an electrochemical method. The Ag/Ag^+ system, for example, can be obtained in most solvents by simple anodic oxidation of a pure silver wire. If this oxidation is carried out with an amperostat for a given period of time and if the reference electrode compartment has a reference level line which facilitates measuring the same volume of the solvent-indifferent electrolyte mixture every time, one thus has the best method for a silver wire reference immersed in an Ag^+ ion solution with reproducible results. This method avoids introducing supplementary residual water by dissolving silver salts.

In spite of the opinion of some authors, the present authors prefer not to use aqueous media electrodes whenever possible. Aqueous media references always risk contaminating the solvent which has been carefully dehydrated beforehand. Moreover, an unknown junction potential which is not very reproducible is present between the reference and the anhydrous medium. Our experience demonstrates that reference systems can be defined as easily in non-aqueous media as in water.

The bridge connecting the reference compartment to the cell should contain an appropriate electrolyte to reduce the junction potentials (170). Besides, an intermediary compartment separating the reference electrode from the solution by sintered glass disks will avoid the reciprocal contamination of the reference and the solution (171).

C. Indicating Electrodes

Organic solvents do not present new fundamental difficulties with respect to the use of indicating electrodes which normally function in aqueous media.

Thus, the dropping mercury electrode used for polarography generally gives satisfactory results (168). Sometimes, residual current anomalies are observed which seem to depend on the nature of the indifferent electrolyte (170) and which are not found in aqueous media.

The different solid electrolyte types are widely used in organic solvents. An excellent description of these electrodes and their practical application has been written by Adams (172). Many details are given and the problem of obtaining reproducible surfaces is discussed. The problem of oxide formation on noble metal electrodes in organic solvents was considered above (Section IV). The main difficulty found with these solvents is that the various waxes or insulating varnishes used in electrode construction are soluble in most of them. This difficulty seems to be overcome since a wax impregnated graphite electrode which functions in acetonitrile (173) and a compressed carbon electrode whose binder is insoluble in acetonitrile, nitromethane and propylene carbonate have been described (174).

Electrode insulation can be a difficult problem to solve for melted salts used at relatively high temperatures. Teflon is satisfactory for temperatures lower than 200–230° C. Pyrex can be used above this range if the bath does not attack it.

Special insulating materials have been proposed for temperatures beyond the melting point of Pyrex. For corrosive baths, the electrode is only slightly immersed. Under these conditions, however, the active surface of the electrode is badly defined.

D. Various Auxiliary Methods

1. Electron Spin Resonance (ESR)

The ESR spectroscopy has been very helpful in elucidating electrode reactions especially in the electrochemistry of organic compounds in organic solvents since Austen, Given, Ingram and Peover (175) first used them to demonstrate that controlled-potential reduction of aromatic hydrocarbons and quinones can produce radical species.

We will not go into details on this subject because this is covered in another chapter of this volume (176). Various books have been written on the ESR principles and its technological aspects (177, 178) and reviews devoted to its chemical applications (179–181) and electrochemical applications (182–184) also exist.

There are two types of experimental devices for observing by ESR the radicals generated electrochemically. One method uses an electrode within the spectrometer resonant cavity and radicals are immediately generated at this place (internal generation). In the second type, the radicals are produced outside of the compartment and then transferred to it by an adequate circulation of the electrolysed solution (external generation). Maki and Geske (185) used an internal generation in their experiments which were the first satisfactory associated application of ESR and electrochemistry. This technique has the advantage of detecting very unstable radicals but it presents various experimental difficulties related to the introduction of a metallic electrode in the resonant cavity. Moreover, it is often very difficult to obtain perfect deoxygenation of the solution when using this method. This perfect deoxygenation is necessary to avoid an increase of the width of the original line of the spectrum which can be encountered in the presence of oxygen due to its paramagnetic properties, as well as to avoid the possible decomposition of the radicals formed at the electrode by oxygen. External generation does not exhibit these drawbacks but its application offers more difficulties in the observation of short-lived radicals. Various devices were conceived which tried to apply the advantages of both techniques.

The association of ESR and electrochemical methods resulted in three different types of work. Firstly, ESR is used to demonstrate the transitory radical species formation in an electrochemical process. Obtaining a well-defined spectrum and its analyses are generally sufficient for characterizing this intermediate species. This is true even though ESR special analyses do not always present a single solution. One should remember the two following points: the radical thus identified in a complex process is not always the electrode reaction main species, moreover, the fact that no ESR spectrum is observed does not eliminate the possibility that a radical species occurs during the reaction. Experimental results and quantitative studies (183) demonstrated the low sensibility of ESR detection of radicals of electrochemical origin. Some of them can thus be detected by rapid cyclic voltammetry whereas they cannot be observed with ESR methods.

One can find an example of this application of ESR to elucidate an electrochemical mechanism in reference 186.

In a second type of publication, the authors use electrochemical methods for preparing free radicals to study their chemical properties. Thus, a chemical reaction mechanism can be studied by electrochemical formation of radicals thought to participate in this reaction (187).

In a third type of publication, the simultaneous use of electrochemistry and ESR methods is developed for quantum chemistry studies. Simple electrochemical radical preparation is used for comparing spectral interpretations with quantum chemistry calculations carried out on the same radical (188, 189). Electrochemistry has the advantage, in this field, of allowing the generation of radicals in various solvents and in the presence of various ions. Thus the effect of the medium interactions with the radical on the ESR spectrum can be studied (190).

2. *In Situ Visible and Ultraviolet Spectroscopy*

Very often, it is necessary to follow the change in composition of the solution during electrolysis by visible and ultraviolet spectroscopy to reveal complex processes taking place at the electrode or in solution. Periodic sampling of the solution and placing the sample in the spectrometer cell is generally unsatisfactory, especially when working in organic solvents. Solvent evaporation, absorption of atmospheric water, the possible effect of oxygen, etc., are phenomena which accompany these sample transfers and make it impossible to obtain a series of reliable spectra. For example isosbestic points are difficult to observe even when the reaction is simply a compound transformation into another one without intermediate accumulation or complex reactions and it is practically impossible to evaluate the relation between the number of coulombs which passed through the solution and the optical densities.

An adapted device installed in a commercial spectrophotometer proved satisfactory in detecting very reactive compounds. Thus, it is possible to show that phenylhydrazine oxidation (191) in neutral non-buffered aceto-nitrile produces phenyldiazene which is highly reactive with respect to oxygen:

$$3\,C_6H_5NHNH_2 \rightarrow 2\,e + 2\,C_6H_5NHNH_3{}^+ + C_6H_5N{=}NH$$

Some rapid scanning spectrophotometers (192) with similar devices can detect relatively unstable intermediates in the solution.

Kuwana and coworkers developed an interesting transparent electrode method (193). The surface of a glass or quartz slide is covered with a doped tin oxide (194), a noble metal film (195) or a very fine metallic grid (196). The use of this device permits observation of the appearance of more or less unstable compounds in the solution but its most successful application is its use in internal reflection spectrophotometry. The concentration of compounds in the immediate vicinity of the electrode can thus be determined (197). One can foresee the eventual possibilities in using this technique for studying reaction intermediates and their absorption state at the electrode.

Spectrophotometric studies of compounds appearing during electrolysis performed in salt melts encounter serious experimental difficulties.

3. Separation and Identification of Reaction Products

The separation and identification of electrolysis products can contribute greatly to understanding the corresponding reaction processes especially in organic electrochemistry. However, the overall electrolysis phenomenon is not necessarily simply related to primary electrode processes revealed by the various electrochemical analytical methods described earlier. Moreover, the separation of a compound may transform it in a way that the pure final product obtained may have only a very remote relationship with the electrode processes. The analytical chemist should seek advice from organic chemists for this part of his work.

Two illustrative examples (186, 187) describe the importance that the identification of reaction products has for the understanding of electrochemical processes studied beforehand by ESR and electrochemical methods.

Organic perchlorates are so frequently used as indifferent electrolytes that it is necessary to call attention to the dangers encountered in the isolation of a perchlorate resulting from the electrolysis. A perchlorate salt can be even more dangerous if its structure and consequently the stabilizing or non-stabilizing influence of the organic cation are unknown. It is preferable to carry out controlled-potential electrolysis followed by product separation with a fluoborate background electrolyte. Quaternary ammonium fluoborates are preferentially selected because they are sufficiently soluble in most solvents.

E. Solvent Preparation and Control

1. Purity Tests

The impurities found in electrochemical media can be classified into two categories according to whether they are electroactive or only reactive with the compounds whose electrochemical properties are studied or with their transformation products. Consequently, purity tests cannot be limited to the recording of residual currents. Various other analytical tests must be added to this first test to detect impurities which are able to modify the electrochemical and chemical behaviour of the compound being studied.

Residual water is the predominant impurity constantly present in all organic solvents and to a lesser degree in the melted salts. It is rare that the presence of water affects the residual current curve. Nevertheless, one must remember that classical treatment (action of dehydrating agents, distillation in sealed apparatus, etc.) produce solvents which contain about 10^{-3} mole of water per liter (about 20 p.p.m.). Generally, the addition of an indifferent electrolyte and manipulation of the medium thus obtained considerably increase this concentration which can reach from 5×10^{-3} to $10^{-2} M$ according to the hygroscopicity of the solvent and the degree of dehydration of the indifferent electrolytes. Actually, non-aqueous electrochemistry takes place, during the use of electrochemical analytical methods, in a medium containing between 10 and 100 times more water than the compound studied. This must always be remembered. Consequently, residual water titration is the second test that should be done on the medium. The Karl Fischer method is very frequently used for organic solvents with indifferent electrolytes added. Various types of automatic titration apparatus have been described (198). Special water analysis techniques can sometimes be used for melted salts because of the O^{2-} ion donor properties. For example the free water is easily titrated in melted antimony trichloride by following the addition of aluminium trichloride by conductometry. The reaction which is as follows

$$H_2O + AlCl_3 \rightarrow AlOCl + 2\,HCl$$

produces species like AlOCl and HCl which are not dissociated in the medium. The conductivity of the solution remains weak until the end-point is reached and beyond this end-point the reaction

$$SbCl_3 + AlCl_3 \rightarrow AlCl_4^- + SbCl_2^+$$

increases the medium conductivity (198a).

Additional tests can also be used to identify and analyse the other non-electroactive impurities. Various modern analytical methods are used for this purpose. In a logical work, each bath of electrolyte solution should be classified and analysed with the various tests and the results should be kept until all experimentation using the same medium is finished.

2. Purification Methods

Each solvent is a special case. The reader should consult the original work and Mann's article (199) which covers the main features of the literature prior to 1967. Some books on solvents and their various applications (200, 201) also provide some information. The choice of the commercial origin of a solvent is not without importance. Thus, we observed that problems presented by the purification of sulpholane differs largely with the various suppliers.

Generally, the solvent and the indifferent electrolyte should be dehydrated independently in order to have maximum dehydration in the final medium obtained by dissolving the electrolyte in the solvent. The medium can be stored on dehydrating columns. Molecular sieves with acetonitrile loaded with various perchlorates (171) are satisfactory even though some cation exchange takes place between the molecular sieves and the solution for some electrolytes. Some solvents undergo transformation when in contact with molecular sieves. Thus, nitromethane becomes a yellow colour probably due to the appearance of its aci form. Drying the electrolyte solution with an anhydrous calcium sulphate column proved satisfactory (170). A continuous distillation–dehydrating process for electrolyte solutions in a Soxhlet apparatus containing a dehydrating agent was described (202, 169).*

The use of strongly dehydrated media (water concentration about $10^{-4}\,M$ or less) requires the use of special installations and operating techniques (203, 204).

Critical evaluation of the various purification processes described in the literature for a given solvent and the choice of the best process are difficult problems. Coetzee (205) carried out critical studies of various acetonitrile purification processes. Mann's review, already cited, describes purification methods for various indifferent electrolytes.

Purification of melted salt media or melted salt mixtures are always special cases. Generally, each of the constituents are purified independently by recrystallization, zone fusion, sublimation, etc.

3. Security

Solvent toxicity can become an important problem when a research worker must work several months or even years with the same electrochemical medium. Mann's review touches this subject. Most often, the effects of prolonged inhalation of the vapours of a new solvent systematically studied in electrochemistry are completely unknown. Very strict precautions should be taken.

On the other hand, the general use of perchlorates as indifferent electrolytes is not satisfactory owing to the explosion hazards resulting from the drying of these salts by heat and from the possible generation of unknown perchlorates during controlled-potential electrolysis (Section VI-D).

VII. TABLES OF THE ELECTROACTIVITY RANGES OF VARIOUS SOLVENTS AND MELTS

Approximately twenty solvents tabled here are examples selected for different temperature ranges and chemical behaviour, just to show how wide the notion of solvent may be. Solvent mixtures which represent a very important field of applications, have been left out. A systematic compilation of electroactivity ranges would be very useful and should be made. We inform the reader that the paper of C. K. Mann also contains the electroactivity ranges of about forty solvents with their physical constants (199).

---

* Moreover a recent article by Butler (169) contains several practical indications on the preparation of reference electrodes in apronic organic solvents and on the purification and analysis of these solvents. See also a recent review in (259).

ORGANIC SOLVENTS

Solvent formula	Dielectric constant	Temperature, °C	Density	Water content, M	Electrolyte	Electrode	Reference electrode	E Fc/Fc$^+$, volts	Total range, volts	Anodic limit, volts	Cathodic limit, volts	Ref.
Acetamide $CH_3-CO-NH_2$	59·2	100										206
		98			Me_4NCl 1 M Neutral medium	Pt (platinized)	Ag/AgCl				$-1·0$	207
					$AcNH_3{}^+Cl^-$ Acid medium	Pt (platinized)	Ag/AgCl NMe_4Cl				$-0·8$	207
								$(E_{0_{Fe}} - E_{0_M}) =$ $0·641 \pm 0·08$				207
Acetic acid CH_3-COOH	6·2	25										208
				10^{-2}	$NaClO_4$ 0·5 M $+ HClO_4$ Acid medium	Hg	Hg/HgOAc NaOAc		0·8	$+0·3$	$-0·5$	209
				10^{-2}	$NaClO_4$ 0·5 M $+$ NaOAc Basic medium	Hg	Hg/HgOAc NaOAc		1·6	$+0·1$	$-1·5$	209
				10^{-2}	$NaClO_4$ 0·5 M Neutral medium	Hg	Hg/HgOAc NaOAc		1·7	$+0·2$	$-1·5$	209
Acetonitrile CH_3-CN	36·2	25										210
					$LiClO_4$ 0·2 M	Pt (bright)	SCE_{aq}	$+0·34$				211
								$E_{1/4} \mid SCE_{aq} =$ 0·300 Chronopot.				212
					$NaClO_4$ 10^{-1} M	Pt (platinized)	Ag/Ag$^+$ 10^{-2}		4·5	$+3$	$-1·5$	213
					$HClO_4$	Pt (platinized)	Ag/Ag$^+$ 10^{-2}		3·0	$\sim +3$	~ 0	
				0 per cent	$LiClO_4$ 10^{-1} M	Pt (bright)	Ag/AgNO$_3$ 10^{-2}		5·5	$+2·2$	$-3·3$	214

Organic Solvents—continued

Solvent formula	Dielectric constant	Temperature, °C	Density	Water content, M	Electrolyte	Electrode	Reference electrode	E Fc/Fc$^+$ volts	Total range, volts	Anodic limit, volts	Cathodic limit, volts	Ref.
Acetonitrile $CH_3—CN$ (continued)				10^{-2}	$NaClO_4$ 10^{-1} M	Pt (bright)	Ag/AgClO$_4$ 10^{-2} M		4.5	+2.5	−2	215
				10^{-2}	Bu_4NClO_4 10^{-1} M	Pt (bright)	Ag/AgClO$_4$ 10^{-2} M		5.4	+2.4	−3	215
				5×10^{-2}	$LiClO_4$ 10^{-1} M	Pt (bright)	Ag/AgClO$_4$ 10^{-2} M		5.5	+2.5	−3	215
				10^{-2}	$LiClO_4$ 10^{-1} M + LiH	Pt (bright)	Ag/AgClO$_4$ 10^{-2} M		5.5	+2.5	−3.5	215
1,2-Dichloroethane	10.2	25	1.24	7×10^{-2}	$(C_7H_{14})_4NClO_4$ 10^{-1}	Pt (polished)		0.00				
$CH_2Cl—CH_2Cl$					Neutral[b]	Pt (polished)			3.9[a]	1.4[a]	−2.5[a]	
					Acid[b]	Pt (polished)			3.5[a]	1.4[a]	−2.1[a]	
					Neutral	C (vitreous)			4.0[a]	1.3[a]	−2.7[a]	216
					Acid	C (vitreous)			3.4[a]	1.3[a]	−2.1[a]	
					Neutral	DME			3.0[a]	0.4[a]	−2.6[a]	
					Acid	DME			2.5[a]	0.4[a]	−2.1[a]	
					Acid + Cl$^-$	DME			1.5[a]	−0.7[a]	−2.2[a]	
					Acid + Br$^-$	DME			1.3[a]	−0.9[a]	−2.2[a]	

[a] All potentials are given versus $E_{1/2}$ Fc/Fc$^+$ on platinum electrode in neutral medium.
[b] Neutral medium: unbuffered. Acid medium: electrolyte + (isooctyl)$_3$NHClO$_4$, 4·10^{-2} M.

Organic Solvents—continued

Solvent formula	Dielectric constant	Temperature, °C	Density	Water content, M	Electrolyte	Electrode	Reference electrode	E Fc/Fc⁺, volts	Total range, volts	Anodic limit, volts	Cathodic limit, volts	Ref.
N,N-Dimethylformamide HCON(CH₃)₂	36.71 (E11)	25	0.9445_4^{25}	5×10^{-2}	NEt₄ClO₄ 10^{-1} M NEt₄ClO₄ 10^{-1} M	Pt (polish) Hg	Hg/Hg₂²⁺ Ag/AgCl_Sₜ KCl(St) KClO₄ 0.8 M		4.95	+1.3 +0.4	−3.65	217
				5×10^{-2}	{ NEt₄ClO₄ 10^{-1} M monochloracetic	Hg	KClO₄ 0.8 M		1.45	−0.1	−1.55	
				5×10^{-2}	{ NEt₄ClO₄ 10^{-1} M OHNEt₄ 10^{-2} M	Hg	KClO₄ 0.8 M			−0.55		
				5×10^{-2} 5×10^{-2}	LiClO₄ 10^{-1} M NET₄ClO₄ 10^{-1} M { benzoic acid pHS⁺ = 9.5	Pt (polish)	KClO₄ 0.8 M		1.90	+0.35	−1.55 −0.84	
				10^{-2}	LiClO₄ 10^{-1} M	Pt (polish)	KClO₄ 10^{-1} M		{ 2.7 2.3	+1.7	−1.0 −0.6	
				5×10^{-2}	NEt₄ClO₄ 10^{-1} M { NEt₄ClO₄ 10^{-1} M HClO₄ 10^{-1} M	Pt (platinized) Pt (platinized)	KClO₄ 0.8 M Ag/AgCl_Sₖ KCl(St) KClO₄ 0.8 M	0.456	3.7	1.7	−2 −0.3	
				5×10^{-2}	{ NEt₄ClO₄ 10^{-1} M pHS = 6.7 dichloracetic acid { NEt₄ClO₄ 10^{-1} M 2,6-dinitrophenol pHS⁺ = 6.5	Pt (platinized) Pt (platinized)	Ag/AgCl_Sₖ KCl(St) KClO₄ 0.8 M Ag/AgCl_Sₖ KCl(St) KClO₄ 0.8 M	0.456	1.55	+1.0 −0.33	−0.55	217

Organic Solvents—*continued*

Solvent formula	Dielectric constant	Temperature, °C	Density	Water content, M	Electrolyte	Electrode	Reference electrode	E Fc/Fc⁺ volts	Total range, volts	Anodic limit, volts	Cathodic limit, volts	Ref.
					NE_4ClO_4 10^{-1} M 6-chlorobenzoic acid pHS⁺ = 10.2	Pt (platinized)	Ag/AgClₛₜ KCl(St) KClO₄ 0.8 M		1.76	+0.94	−0.82	
					NE_4ClO_4 10^{-1} M phenol/phenate	Pt (platinized)	Ag/AgClₛₜ KCl(St) KClO₄ 0.8 M			+0.1		218
					NE_4ClO_4 10^{-1} M acetic acid NE_4ClO_4 $OHNE_4$ 10^{-3} in methanol	Pt (platinized)	Ag/AgClₛₜ KCl(St) KClO₄ 0.8 M			+0.17 −1		
					$HClO_4$	Pt	Hg/Hg_x^{2+} 10^{-2} M	−0.01				218
				5×10^{-1}	$LiClO_4$	Hg	Hg/Hg_x^{2+} 10^{-2} M		2.7	0.0	−2.7	219
					H_2SO_4				1.55	0.0	−1.55	
					HNO_3				1.55	0.0	−1.55	
					HCl, HBr, HI				0.55	−1.0	−1.55	
				5×10^{-2}	$HClO_4$	Pt	Hg/Hg_x^{2+} 10^{-2} M		2.05	+1.3	−0.75	220
					$LiClO_4$				5.1	+1.3	−3.8	
					KBF_4					+1.3		
					H_2SO_4				1.95	+1.2	−0.75	
					HNO_3				1.95	+1.2	−0.75	
					LiCl				3.8	+0.4	−3.8	
					NaBr					+0.15		
					KI					−0.35		

Organic Solvents —continued

Solvent formula	Dielectric constant	Temperature, °C	Density	Water content. M	Electrolyte	Electrode	Reference electrode	E Fc/Fc+. volts	Total range, volts	Anodic limit, volts	Cathodic limit, volts	Ref.
Dimethyl sulphoxide CH_3-S-CH_3 ‖ O					LiClO4	Pt	Ag/AgCl2$^-$ 10^{-3} M LiClO4, I(3)	+802 mV/ref	4.8	1.7	−3.1	221
					KClO4	Pt		+732/H2H+	4.4	1.7	−2.7	
					NaClO4	Pt		+0.802/ref	4.2	1.7	−2.5	
					KNO3	Pt		+0.732/H2H	4.4	1.7	−2.7	
					KBF4	Pt			4.4	1.7	−2.7	
				2 × 10^{-2}	K2S2O8				4.4	1.7	−2.7	
					LiCl	Pt			4.2	1.15	−3.05	
		25			Me4NCl	Pt			3.9	1.25	−2.8	
					Et4NClO4				4.4	1.7	−2.7	
					Bu4NBr	Pt			3.9	1.1	−2.8	
					LiClO4	Ag			3.7	0.65	−3.05	
					KClO4	Ag			3.35	0.65	−2.70	
					NaClO4				3.10	0.65	−2.45	
					KNO3				3.35	0.65	−2.70	
					LiCl				3.10	0.05	−3.05	
					Me4NCl	Ag			2.65	0.05	−2.60	
		25		2 × 10^{-2}	Et4NClO4				3.35	0.65	−2.70	
					Bu4NB4	Ag			2.85	0.15	−2.70	
				2 × 10^{-2}	LiClO4	Hg	Ag/AgCl2$^-$ 10^{-3} M LiCl 0.1 M		2.95	0.75	−2.20	
					KClO4	Hg			2.60	0.75	−1.85	
		25			NaClO4	Hg			2.65	0.75	−1.90	
					Me4NCl	Hg			2.35	0.05	−2.30	
					Bu4NBr	Hg			2.50	0.00	−2.10	
	46.6	25	1.100	10^{-3}	Et4NCl	Hg			3.15	0.75	−2.40	222 223

Organic Solvents—continued

Solvent formula	Dielectric constant	Temperature, °C	Density	Water content, M	Electrolyte	Electrode	Reference electrode	E Fc/Fc+, volts	Total range, volts	Anodic limit, volts	Cathodic limit, volts	Ref.
Ethylammonium chloride		127				C^c (vitreous)	Fc^d/Fc^+	0.00		+0.8		
$C_2H_5NH_3{}^+Cl^-$						Pt^c				+0.5		224
						Hg				−0.5		
Ethylpyridinium bromide		127					Io/I^-					225
												225
						Au			0.9	−1.0	−0.1	225
						Pt			1.3	−1.0	+0.3	225
						Graphite			1.3	−1.0	+0.3	225
						Hg			0.4	−1.0	−0.6	225
N-Methylacetamide	178.9	30										226
CH_3—CO—NH—CH_3		40			Et_4NBr 0.12 M	Hg	Hg pool		1.8	0	−1.8	227
		40			$NaNO_3$ 10^{-1} M	Hg	Ag/AgCl	+0.52		+0.2 (with Cl−)		228

(Ethylpyridinium bromide structure: pyridinium ring with N^+—Et, Br^-)

^{c} Rotation speed = 500 t/m.
^{d} Potential origin = $E_0(Fc/Fc^+)$. Reference electrode (Bi/Bi(III)) potential = -0.656 ± 0.005 V/Fc, Fc+.

Organic Solvents —continued

Solvent formula	Dielectric constant	Temperature, °C	Density	Water content, M	Electrolyte	Electrode	Reference electrode	E Fc/Fc+, volts	Total range, volts	Anodic limit, volts	Cathodic limit, volts	Ref.
N-Methyl 2-pyrrolidone	32.0	25										229
	32.9	20										230
	31.5	30										230
	32.2	25										231
		20	1.0327									232
		25	1.027									233
		25	1.0286									234
		Room		2.5×10^{-1}	HClO$_4$	Pt	Hg/Hg$_2{}^{2+}$ 10^{-2} M	+0.04				
		Room			LiClO$_4$		Hg/Hg$_2{}^{2+}$ 10^{-2} M in HClO$_4$ 0.1N medium		5.2	+1.3	-3.9	
					Ba(NO$_3$)$_2$				4.4	+1.25	-3.15	
					N(CH$_3$)$_4$I	Pt (Bright)					-3.0	257
					KClO$_4$				3.65	+1.3	-2.35	
					KNO$_3$				3.6	+1.25	-2.35	
					KBF$_4$						-2.35	
					NaBr				2.5	+0.3	-2.2	
					NaI				2.4	+0.2	-2.2	
					Na$_2$S$_2$O$_8$				3.5	+1.3	-2.2	
					NH$_4$ClO$_4$				2.4	+1.3	-1.1	
					(NH$_4$)$_2$SO$_4$				2.4	+1.3	-1.1	
					HClO$_4$				2.05	+1.3	-0.75	
					KBF$_4$(sat.)					+1.35		
					LiCl					+0.6		
				5×10^{-1}	N(CH$_3$)$_4$ClO$_4$	Hg			3.1	0.0	-3.1	
					LiClO$_4$						-2.6	257
					NaBr				1.65	-0.8	-2.45	
					KI				1.55	-0.9	-2.45	

Organic Solvents—continued

Solvent formula	Dielectric constant	Temperature °C	Density	Water content, M	Electrolyte	Electrode	Reference electrode	E Fc/Fc+, volts	Total range, volts	Anodic limit, volts	Cathodic limit, volts	Ref.
		Room		5×10^{-1}	NH_4NO_3	Hg	Hg/hg_x^{2+} 10^{-2} M in $HClO_4$ 0.1 N medium		2.25	−0.05	−2.3	
					$HClO_4$				1.85	0	−1.85	
					K_2SO_4					0		
					KBF_4					+0.05		
					$LiCl$					−0.7		
				2.5×10^{-1}	$LiClO_4$	Ag	Hg/Hg_x^{2+} 10^{-2} M in $HClO_4$ 0.1 N medium		3.75	−0.05	−3.8	
					$KClO_4$						−2.5	
					$NaBr$				1.9	−0.7	−2.6	
					$HClO_4$				1.3	−0.05	−1.25	
					KBF_4					−0.1		
					NH_4NO_3					−0.1		
					K_2SO_4					−0.1		
					$LiCl$					−0.8		
					KI					−0.7		
				$<2.5 \times 10^{-1}$	$LiClO_4$	Pt	Hg/Hg_x^{2+} 10^{-2} M $HClO_4$ 0.1 N medium		5.1	+1.2	−3.9	
					$KClO_4$	platinized					−2.3	
					$NaBr$				2.1	+0.1	−2.2	
					$HClO_4$					+0.4	−0.8	257
					$LiCl$					−0.25		
					KI							
Nitromethane CH_3NO_2	38	25	1.131	5×10^{-3} to 10^{-2}	$LiClO_4$ 0.1 N	Pt (bright)	Ag/AgCl, Me_4NCl sat. in $MeNO_2$	0.705	5.4	3.0	−2.4	235
					Et_4NClO_4 0.1 M	Pt (bright)		0.675	3.95	2.95	−1.0	235
					Me_4NBF_4 0.1 M	Pt (bright)			3.95	2.95	−1.0	236
					Bu_4NPF_6 0.1 M	Pt (bright)			4.15	3.15	−1.0	236
					Et_4NNO_3 0.1 M	Pt (bright)			3.1	2.1	−1.0	237
					Et_4NClO_4 0.1 M	Hg			2.0	1.15	−0.95	237
					$LiClO_4$ 0.1 M	Ag			3.50	1.10	−2.4	235

Organic Solvents—*continued*

Solvent formula	Dielectric constant	Temperature, °C	Density	Water content, M	Electrolyte	Electrode	Reference electrode	E Fc/Fc+, volts	Electroactivity range			Ref.
									Total range, volts	Anodic limit, volts	Cathodic limit, volts	
Propylene carbonate	69.0 (23°C)	20–25	1.21	10^{-3}	LiClO$_4$	Pt	Ag/Ag+ 10^{-2} M	−0.390	5.9	+2.2	−3.7	
						Pt^c			5.7	+2.0	−3.7	
						Ag			3.6	0.0	−3.6	
						C vitreous			5.6	+2.0	−3.6	
						Hg			2.7	+0.1	−2.6	
$CH_3-CH-CH_2$					KPF$_6$	Pt			6.3	+2.7	−3.6	
						Pt^c					−3.5	
						Ag			3.8	0	−3.8	
						C vitreous			5.5	+1.7	−3.8	
						Hg			2.6	+0.1	−2.5	238
					NaBØ$_4$	Pt			4.0	0.0	−4.0	
						Pt^c			3.6	−0.1	−3.7	
						Ag			3.8	0	−3.8	
						C vitreous			2.5	−0.4	−2.9	
						Hg			3.4	0.1	−3.3	
					Et$_4$NCl	Pt			2.9	0	−2.9	
						Pt^c				−1.0		
						Ag			3.4	0.0	−3.4	
						C vitreous			2.4	−1.2	−3.6	
						Hg			4.0	−0.0	−4.0	
					Bu$_4$NBr	Pt^c			3.1	−0.3	−3.4	
						Ag			2.4	−0.9	−3.3	
						C vitreous			3.5	−0.2	−3.7	
						Hg			2.7	−1.2	−3.9	

Solvent formula	Dielectric constant	Temperature, °C	Density	Water content, M	Electrolyte	Electrode	Reference electrode	E Fc/Fc$^+$, volts	Electroactivity range			Ref.
									Total range, volts	Anodic limit, volts	Cathodic limit, volts	
Sulpholane	43·3	25	0·262	5×10^{-4}			Ag/Ag$^+$ (10^{-2} M) LiClO$_4$ 0·1 M	−0·26				239
					LiClO$_4$	Pt			6	+2·3	−3·7	
					NaClO$_4$	Pt			6·35	+2·3	−4·05	
					Et$_4$NClO$_4$	Pt			5·4	+2·3	−3·1	
					NH$_4$PF$_6$	Pt			5·5	+2·5	−3·0	240
					LiClO$_4$	Hg			2·78	+0·18	−2·6	
					NaClO$_4$	Hg			2·73	+0·18	−2·55	
					Et$_4$ClO$_4$	Hg			3·68	+0·18	−3·5	
Tetrahydrofurane (THF)	7·4	25			LiClO$_4$ 0·3 M	Pt (bright)	Ag/Ag$^+$		5·5	+1·5	−4·0	241
					LiClO$_4$ 0·3 M	Ag			4·5	+0·5	−4·0	242
					LiClO$_4$ 0·3 M	Hg			3·0	+0·5	−2·5	
					Bu$_4$NI	Pt or Au	Ag/Ag$^+$		3·2	−0·8	−4·0	
					Bu$_4$NClO$_4$	Pt or Au			5·5	+1·5	−4·0	
					Bu$_4$NF	Pt or Au			5·5	+1·5	−4·0	
					LiClO$_4$	Pt or Au			5·1	+1·5	−3·6	
					NaClO$_4$	Pt or Au			5·1	+1·5	−3·6	
				5×10^{-4} M	Bu$_4$NI	Hg and Au–Hg			2·8	−1·2	−4·0	243
					Bu$_4$NI	Ag			3·2	−0·8	−4·0	
					Bu$_4$NClO$_4$	Mg and Au–Hg			4·1	+0·1	−4·0	
					Bu$_4$NClO$_4$	Ag			4·0	0	−4·0	
					LiClO$_4$	Hg and Au–Hg			1·7	+0·1	−1·6	
					LiClO$_4$	Ag			3·6	0	−3·6	
					NaClO$_4$	Hg and Au–Hg			1·5	−0·1	−1·6	
					NaClO$_4$	Ag			3·6	0	−3·6	

Organic Solvents—continued

Solvent formula	Dielectric constant	Temperature, °C	Density	Water content, M	Electrolyte	Electrode	Reference electrode	E Fc/Fc$^+$, volts	Total range, volts	Anodic limit, volts	Cathodic limit, volts	Ref.
Trifluoroacetic acid CF$_3$COOH	8·22	30 25	1·53	10^{-2}	(C$_2$H$_5$)$_4$NClO$_4$ 10^{-1} M	Pt (bright)	Ag↓/AgClO$_4$ (C$_2$H$_5$)$_4$ NClO$_4$ 10^{-1} M		2·8	+2·1	−0·7	
						Pt (platinized)			2·2	+1·7	−0·5	
						Au			3	+2·1	−0·9	
						Ag			1·3	+0·1	−1·2	
						Hg			1·4	+0·1	−1·3	244
					HClO$_4$ 1·2 M	Pt (bright)	Ag↓/AgClO$_4$ (C$_2$H$_5$)$_4$ NClO$_4$ 10^{-1} M		1·9	+2·2	+0·3	
						Pt (platinized)			0·9	+1·8	+0·9	
						Au			1·6	+1·9	+0·3	
				Anhydrous solvent (anhydride traces)	CF$_3$COONa M	Pt (bright)	Ag↓/AgClO$_4$ (C$_2$H$_5$)$_4$ NClO$_4$ 10^{-1} M		3·4	+2·3	−1·1	
						Ag			1·1	+0·1	−1·0	

INORGANIC SOLVENTS

Solvent formula	Dielectric constant	Temperature, °C	Density	Water content, M	Electrolyte	Electrode	Reference electrode	E Fc/Fc$^+$, volts	Total range, volts	Anodic limit, volts	Cathodic limit, volts	Ref.
Ammonia NH_3	25	−60	0.72	5×10^{-3}	KI	Pt (bright)	Ag/Ag$^+$ $10^{-2}\,M$		2.8	+0.5	−2.3	
Ammoniate $NH_4NO_3.3NH_3$		+20	1.18		NH_4NO_3	Pt (bright)	Ag/Ag$^+$ $10^{-2}\,M$		1.3	+0.60	−0.70	245 246
					NH_4NO_3	Hg			1.15	−0.15	−1.3	247
Ammoniate $NaI.3,3NH_3$		+20	1.60	5×10^{-2}	NaI	Pt (bright)	Ag/Ag$^+$ $10^{-2}\,M$		1.9	+0.6	−1.3	
					NaI	Hg			1	0	−1.0	
Ammonia	25	−60	0.72		KBr		Ag/Ag$^+$		3	+0.7	−2.3	
Antimony trichloride $SbCl_3$	30.4	99	2.62	2×10^{-2}	$K^+ Cl^-$ $SbCl_2^+ AlCl_4^-$ $K^+ AlCl_4^-$	C (vitreous)	Sb↓/SbCl$_3$ KCl$_{sat}$		0.700 1.000 1.5	+0.7 +1.5 +1.5	0.0 +0.5 0.0	248
Molten LiCl/KCl eutectic (LiCl 59 moles) (KCl 41 moles per cent)						Pt	Pt/Pt^{2+}		2.8	+0.2	−2.2	249
						Au	Pt/Pt^{2+}		2.5	0	−2.6	250
						Pt	Pt/Pt^{2+}		2.5	0	−2.5	
						W	Pt/Pt^{2+}			+0.4	−2.5	
						Graphite	Pt/Pt^{2+}					
Molten NaOH/KOH eutectic NaOH = 53.1 per cent (weight) KOH = 46.9 per cent (weight)		248 260 310 230	1.88 1.865 1.84 1.89 ±0.01									
		227		3 M Acid medium		Pt			1.5^a			251
				0 neutral and basic medium		Pt			2.1^a			

^{a} Current density = 15 mA/cm^2.

Inorganic Solvents—continued

Solvent formula	Dielectric constant	Temperature, °C	Density	Water content, M	Electrolyte	Electrode	Reference electrode	E Fc/Fc⁺, volts	Total range, volts	Anodic limit, volts	Cathodic limit, volts	Ref.
Sodium tetrabromo-aluminate NaAlBr₄	2·9	210					Al/Al(III) (Al↓/NaBr)					
					Basic medium	Cᵃ (vitreous)	NaAlBr₄		1·7	1·7	0	252
					Basic medium	Ptᵃ			1·6	1·6	0	
Sodium tetrachloro-aluminate NaAlCl₄		175 ± 2					Al/Al(III) (Al↓, NaCl, NaAlCl₄)					
					Basic medium (excess NaCl)	Cᵇ (vitreous)			2·3	+2·3	0	253
						Wᵇ			2·5	+2·5	0	
						Hg			1·1	+1·1	0	
						Auᵃ			1·8	+1·8	0	
						Ptᵃ			1·8	+1·8	0	

ᵃ Rotation speed = 300 t/mn.
ᵇ Rotation speed = 600 t/mn.

Inorganic Solvents—*continued*

Solvent formula	Dielectric constant	Temperature, °C	Density	Water content, M	Electrolyte	Electrode	Reference electrode	E Fc/Fc$^+$, volts	Total range, volts	Anodic limit, volts	Cathodic limit, volts	Ref.
Water H_2O					HCl 10^{-1} M	Pt (platinized)	SCE				-0.25	254
					NaCl M							
					NaOH 10^{-1} M							
					NaCl M						-0.9	
					pH0	Pt	SCE	+0.75	2.25	1.45	-0.8	255
					pH2				2.22	1.42	-0.8	
					pH5				2.20	1.20	-1	
					Neutral unbuffered				2.38	1.38	-1	
					pH12				1.75	0.75	-1	
					pH14				1.59	0.59	-1	
					pH0	PbO$_2$	SCE		0.22	1.55	1.33	
					pH2				0.30	1.45	1.15	
					pH5				0.57	1.23	0.66	
					Neutral unbuffered				0.97	1.45	0.48	
					pH14				0.48	0.58	0.10	
					pH2	Tl$_2$O$_3$	SCE		0.35	1.27	0.92	
					pH5				0.28	0.90	0.62	
					Neutral unbuffered				0.78	1.13	0.35	
					pH12				0.47	0.66	0.19	
					pH14				0.56	0.54	-0.02	
Fluorosulphuric acid FSO_3H	62	20	1.726	No water	NaFSO$_3$	Pt	Au/Au(CN)$_3$	0.3	2.5*	+1.8*	+0.7*	256
						Graphite			2.5*	+1.9*	-0.6*	
						Ni			3.2*	+2.2*	-1*	

ACKNOWLEDGMENTS

The authors wish to take the opportunity to thank Mrs. Collumeau for tabulating the electroactivity ranges and Dr. Demange and Mr. R. Combes who checked the translation of the technical items of this paper.

REFERENCES

1. G. Charlot, J. Badoz-Lambling and B. Tremillon, *Electrochemical Reactions*, Elsevier, Amsterdam, 1962; R. W. Murry and C. N. Reilley, *Electroanalytical Principles*, John Wiley and Sons, New York, 1963.
2. J. Heyrovsky and J. Kuta, *Principles of Polarography*, Academic Press, New York, 1966.
3. G. Charlot, J. Badoz-Lambling and B. Tremillon, *Les réactions électrochimiques. Méthodes électrochimiques d'analyse*, Masson and Cie, Paris, 1959, p. 341. *Electrochemical Reactions*, Elsevier, Amsterdam, 1962, p. 337.
4. R. Takamashi, *Talanta*, **12**, 1211 (1965).
5. S. Wawzonek, *Talanta*, **12**, 1229 (1965).
6. H. A. Laitinen, *Talanta*, **12**, 1237 (1965).
7. S. G. Mairanovskii, 'The effect of the composition of aqueous organic solvents on the polarographic behaviour of organic compounds,' *Polarography 1964*, G. J. Hills, Ed., Vol. 2, Macmillan, London, 1966, p. 719.
8. M. E. Peover and J. D. Davies, 'Complex formation in the polarography of Aromatic Compounds in Organic Solvents,' *Polarography 1964*, G. J. Hills, Ed., Vol. 2, Macmillan, London, 1966, p. 1003.
9. T. A. Gough and M. E. Peover, 'Voltammetric oxidation and reduction of aromatic compounds in acetonitrile,' *Polarography 1964*, G. J. Hills, Ed., Vol. 2, Macmillan, London, 1966, p. 1017.
10. K. Schwabe, 'Polarographie in nicht wässrigen Lösungen,' *Progress in Polarography*, Vol. 1, p. 333, P. Zuman and I. M. Kolthoff, Eds., Interscience Publishers, New York, 1962, p. 333.
11. G. J. Hills, 'Reference electrodes in non-aqueous solutions,' *in Reference Electrodes. Theory and Practice*, Academic Press, New York, 1961, p. 433.
12. R. W. Laity, 'Electrodes in fused salts systems,' *in Reference Electrodes. Theory and Practice*, Academic Press, New York, 1961, p. 524.
13. G. J. Janz and R. D. Reeves, 'Molten-salt electroyltes. Transport properties,' *Advances in Electrochemistry and Electrochemical Engineering*, P. Delahay and C. W. Tobias, Eds., Interscience Publishers, New York, Vol. 5, 1967, p. 137.
14. A. Brenner, 'Electrolysis of non-aqueous systems,' *Advances in Electrochemistry and Electrochemical Engineering*, P. Delahay and C. W. Tobias, Eds., Interscience Publishers, New York, 1967, p. 205; A. D. Graves, G. J. Hills and D. Inman, *Advances in Electrochemistry and Electrochemical Engineering*, Vol. 4, P. Delahay and C. W. Tobias, Eds., Interscience Publishers, New York, 1965, p. 117.
15. H. Bloom and J. O'M. Bockris, *in Modern Aspects of Electrochemistry*, J. O'M. Bockris and B. E. Conway, Eds., Vol. 2, Butterworths, London, 1959.
16. Yu K. Delimarskii and B. F. Markov, *Electrochemistry of Fused Salts*, translated by A. Peiperl and R. Wood, Sigma Press, New York, 1961.
17. C. K. Mann, 'Non-aqueous solvents for electrochemical use,' *in Electroanalytical Chemistry*, A. J. Bard, Ed., Vol. 3, Marcel Dekker, New York, 1965, p. 57.
18. D. N. Hume, *Anal. Chem.*, **38**, 261 R (1966); *Anal. Chem.*, **40**, 174 R (1968).

19. C. J. Pietrzyk, *Anal. Chem.*, **38**, 278 R (1966); *Anal. Chem.*, **40**, 194 R (1968).
20. G. A. Halow and D. H. Morman, *Anal. Chem.*, **38**, 485 R (1966); **40**, 418 R (1968).
21. H. A. Laitinen and R. A. Osteryoung, 'Electrochemistry in molten salts,' *in Fused Salts*, B. R. Sundheim, Ed., McGraw-Hill Book Co., New York, 1964, p. 255.
22. B. R. Sundheim, 'Transport properties of liquid electrolyte,' *in Fused Salts*, B. R. Sundheim, Ed., McGraw-Hill Book Co., New York, 1964, p. 165.
23. J. Lumsden, *Thermodynamics of Molten Salt Mixtures*, Academic Press, London, 1966.
24. L. F. Audrieth and J. Kleinberg, *Non-aqueous Solvents*, John Wiley and Sons, New York, 1953.
25. V. Gutmann and G. Schöber, *Angew. Chem.*, **70**, 98 (1958).
26. J. F. Coetze and C. D. Ritchie, Eds., *Solute-solvent Interactions*, Marcel Dekker, New York, 1969.
27. J. J. Lagowski, Ed., *The Chemistry of Non-aqueous Solvents*, Academic Press, New York, Vol. 1, 1966, and Vol. II, 1967.
28. G. Charlot and B. Tremillon, *Les réactions chimiques dans les solvants et les sels fondus*, Gauthier-Villars, Paris, 1963; *Chemical Reactions in Solvents and Melts*, Pergamon Press, Oxford, 1969.
29a. G. Charlot and B. Tremillon, *Les réactions chimiques dans les solvants et les sels fondus*, Gauthier-Villars, Paris, 1963, p. 223.
29b. R. E. Buckles and J. F. Mills, *J. Am. Chem. Soc.*, **75**, 552 (1953).
30a. R. E. Buckles and J. F. Mills, *J. Am. Chem. Soc.*, **75**, 364 (1953).
30b. A. Bentley, A. G. Evans and J. Halpern, *Trans. Faraday Soc.*, **47**, 711 (1951); A. Bentley and A. G. Evans, *J. Chem. Soc.*, **1952**, 3468.
31. G. Charlot and B. Tremillon, *Les Réactions Chimiques dans les Solvants et les Sels Fondus*, Gauthier-Villars, Paris, 1963; *Chemical Reactions in Solvents and Melts*, Pergamon Press, London, 1969; J. J. Lagowski, Ed., *The Chemistry of Non-aqueous Solvents*, Vol. II, Academic Press, New York, 1967; A. J. Parker, *Chem. Rev.*, **69**, 1 (1969); C. Agami, *Bull. Soc. Chim.*, 4031 (1967); 2033 (1968); C. Agami and M. Caillot, *Bull. Soc. Chim.*, 1990 (1969).
32. A. C. Aten, C. Butler and A. J. Hoijtink, *Trans. Faraday Soc.*, **55**, 324 (1959).
33. A. C. Aten and G. J. Hoijtink, *Advances in Polarography*, Vol. 2, p. 777, I. S. Longmuir, Ed., Pergamon Press, Oxford, 1961.
34. Authors cited by G. Charlot and B. Tremillon in (29a).
35. J. Badoz-Lambling and M. Herlem, *Bull. Soc. Chim. France*, **1964**, 90.
36. M. E. Peover and J. D. Davies, *J. Electroanal. Chem.*, **6**, 46 (1963).
37. B. Tremillon and G. Letisse, *J. Electroanal. Chem.*, **17**, 371 (1968).
38. S. Wawzonck, R. Berkey, E. W. Blaha and M. E. Runner, *J. Electrochem. Soc.*, **103**, 456 (1956).
39. P. H. Given and M. E. Peover, *J. Chem. Soc.*, **1960**, 385.
40. J. Badoz-Lambling and H. Demange Guerin, *Anal. Letters*, **2**, 123 (1969).
41. J. P. Saget and V. Plichon, *Bull. Soc. Chim. France*, **1969**, 1395.
42. G. Cauquis, G. Fauvelot and J. Rigaudy, *Bull. Soc. Chim. France*, **1968**, 4928.
43. J. Badoz-Lambling, D. Bauer and Ph. Texier, *Anal. Letters*, **2**, 411 (1969).
44. A. C. Aten, J. Dieleman and G. J. Hoijtink, *Disc. Faraday Soc.*, **1960–1961**, 182.
45. M. E. Peover *in Electroanalytical Chemistry*, A. J. Bard, Ed., Vol. 2, Marcel Dekker, New York, 1967, p. 1.
46. K, J, Vetter, *Electrochemical Kinetics*, Academic Press, New York, 1967.
47. A. N. Frumkin, *Z. Phys. Chem.*, **103**, 43 (1923).
48. R. Payne, 'The electrical double layer in non-aqueous solutions,' *Advances in Electrochemistry and Electrochemical Engineering*, P. Delahay and C. W. Tobias, Eds., Vol. 7, Interscience Publishers, New York, 1970.

49. R. Payne, *J. Am. Chem. Soc.*, **89**, 489 (1967).
50. J. Lawrence and R. Parsons, *Trans. Faraday Soc.*, **64**, 1656 (1968).
51. W. Jaenicke and P. H. Schweitzer, *Z. Phys. Chem.*, NF, **52**, 104 (1967).
52. C. Barry, G. Cauquis and M. Maurey, *Bull. Soc. Chim. France*, **1966**, 2510.
53. J. J. Aubry, 'Propriétés chimiques et électrochimiques du diallyl-cyanamide,' Thèse de spécialité, Paris, 1969.
54. H. Schneider and H. Strehlow, *J. Electroanal. Chem.*, **12**, 530 (1966).
55. A. F. Alabyshev, M. F. Lantratov and Morachevskii, *Reference Electrodes for Fused Salts*, Sigma Press. Translator, A. Peiperl, 1900.
56. See also (11), (12) and (17).
57. H. Strehlow, 'Electrode potentials in non-aqueous solvents,' *in The Chemistry of Non-aqueous Solvents*, J. J. Lagowski, Ed., Vol. 1, Academic Press, New York, 1966, p. 129.
58. J. N. Butler, see (169).
59. C. P. Kumar and D. A. Pantony, *Polarography 1964*, G. J. Hills, Ed., Macmillan, London, 1966, p. 1061.
59a. G. Demange-Guerin, unpublished results.
60. J. C. Bardin, *Analusis*, **1**, 140 (1972).
61. O. Popovych and A. J. Dill, *Anal. Chem.*, **41**, 456 (1969).
61a. B. Case, N. S. Hugh, P. Parson and M. E. Peover, *J. Electroanal. Chem.*, **10**, 360 (1965).
62. J. C. Bardin, *J. Electroanal. Chem.*, **28**, 157 (1970).
63. D. Bauer, J. P. Beck and P.·Texier, *C.R. Acad. Sci. Paris*, **265**, 1335 (1968).
64. G. Charlot, J. Badoz-Lambling and B. Tremillon, *Electrochemical Reactions*, Elsevier, Amsterdam, 1962.
65. R. F. Dapo and C. R. Mann, *Anal. Chem.*, **35**, 677 (1963).
66. C. D. Russel, *Anal. Chem.*, **35**, 1291 (1963).
67. G. Cauquis and M. Genies, *Bull. Soc. Chim. France*, **1967**, 3220.
68. G. Charlot, J. Badoz-Lambling and B. Tremillon, *Electrochemical Reactions*, Elsevier, Amsterdam, 1962, p. 73.
69. K. J. Vetter, *Electrochemical Kinetics*, Academic Press, New York, 1967.
70. B. Tremillon and G. Letisse, *J. Electroanal. Chem.*, **17**, 371 (1968).
71. P. Texier, *Bull. Soc. Chim. France*, **1968**, 4716.
72. J. E. Dubois and A. Marie de Ficquelmont, *J. Chim. Phys.*, **64**, 904 (1967).
73. G. Briere, G. Cauquis, B. Rose and D. Serve, *J. Chim. Phys.*, **66**, 44 (1969).
74. J. H. Schmidt and J. Noack, *Z. Anorg. Allgem. Chem.*, **296**, 262 (1958).
75. J. P. Billon, *J. Electroanal. Chem.*, **1**, 486 (1960).
76. C. K. Mann, *in Electroanalytical Chemistry*, A. J. Bard, Ed., Vol. 3, Marcel Dekker, New York, 1969, p. 57.
77. C. D. Russel, *Anal. Chem.*, **35**, 1291 (1963).
78. J. Bessiere and G. Petit, *J. Electroanal. Chem.*, **25**, 317 (1970).
79. G. Cauquis, G. Fauvelot and J. Rigaudy, unpublished data.
80. P. H. Rieger, I. Bernal, W. H. Reinmuth and G. F. Fraenkel, *J. Am. Chem. Soc.*, **85**, 683 (1963).
81. G. Cauquis and D. Serve, *Bull. Soc. Chim. France*, **1966**, 302.
82. J. Desbarres, P. Pichet and R. L. Benoit, *Electrochim. Acta*, **13**, 1899 (1968).
83. J. S. Mayell and A. J. Bard, *J. Am. Chem. Soc.*, **85**, 421 (1963).
84. J. Courtot-Coupez and M. Le Démézet, *Bull. Soc. Chim. France*, **1967**, 4744.
85. R. Jasinski, *High-energy Batteries*, Plenum Press, New York, 1967.
86. J. F. Laurent, *Rev. Géné. Electr.*, **76**, 1397, 1406 (1967).
87. T. B. Reddy, Thesis, University of Minnesota, U.S.A., 1960.

88. M. C. Giordano, J. C. Bazan and A. J. Arvia, *Electrochim. Acta*, **11**, 741 (1966).
89. B. Burrows and S. Kirkland, *J. Electrochem. Soc.*, **115**, 1164 (1968).
90. A. N. Déy, *J. Electrochem. Soc.*, **114**, 823 (1967).
91. J. E. Dubois, P. C. Lacaze and A. Marie de Ficquelmont, *C.R. Acad. Sci.*, **262**, 249 (1966).
92. R. A. Craig and R. E. Richards, *Trans. Faraday Soc.*, **59**, 1972 (1963).
93. G. A. Maciel, *J. Inorg. Chem.*, **5**, 554 (1966).
94. G. Cauquis and coll., unpublished data.
95. G. Cauquis and D. Serve, *C.R. Acad. Sci.*, **262**, 1516 (1966).
96. A. H. Maki and D. H. Geske, *J. Chem. Phys.*, **30**, 1356 (1959).
97. R. Hirasawa, T. Mukaibo, H. Hajegawa, N. Odan and T. Maruyama, *J. Phys. Chem.*, **72**, 2541 (1968).
98. J. R. Morton, *J. Chem. Phys.*, **45**, 1800 (1966).
99. G. Cauquis and D. Serve, *J. Electroanal. Chem.*, **27**, App. 3 (1970).
100. M. Fleischmann and D. Pletcher, *Tetrahedron Letters*, **1968**, 6255.
101. T. Osa, A. Yildiz and T. Kuwana, *J. Am. Chem. Soc.*, **91**, 3994 (1969).
102. A. J. Arvia, W. E. Triaca and H. A. Videla, *Electrochem. Acta*, **15**, 9 (1970).
103. J. Goret and B. Tremillon, *Electrochim. Acta*, **12**, 1065 (1967).
104. J. J. Minet, Thesis, Paris, 1970; A. Kerovanton, Thesis, Paris, 1972; E. Zintl and A. Harder, *Z. Physik. Chem.*, **154A**, 47 (1931).
105. J. Badoz-Lambling and J.-P. Sajet, *C.R. Acad. Sci.*, Ser. C., 250 (1970).
106. T. Kuwana and R. N. Adams, *Anal. Chim. Acta*, **20**, 51, 60 (1959).
107. J. Jordan and L. R. Jiminez, unpublished data, see R. N. Adams, *Electrochemistry at Solid Electrodes*, Marcel Dekker, New York, 1969, p. 28.
108. L. Young, *Anodic Oxide Films*, Academic Press, New York, 1961.
109. S. Gilman, *in Electroanalytical Chemistry*, A. J. Bard, Ed., Vol. 2, Marcel Dekker, New York, 1967, p. 111.
110. P. Delahay, *Double Layer and Electrode Kinetics*, John Wiley and Sons, New York, 1965, Chap. 10.
111. M. Breiter, *Electrochim. Acta*, **8**, 925 (1963).
112. A. I. Popov and D. H. Geske, *J. Am. Chem. Soc.*, **80**, 1340 (1958).
113. B. E. Conway and A. K. Vijh, *Z. Anal. Chem.*, **224**, 149 (1967).
114. B. E. Conway and M. Dzieciuch, *Can. J. Chem.*, **41**, 38 (1963).
115. B. E. Conway, N. Marincic, D. Gilroy and E. Rudd, *J. Electrochem. Soc.*, **113**, 1144 (1966).
116. G. Cauquis and D. Serve, unpublished data.
117. S. D. James, *J. Electrochem. Soc.*, **114**, 1113 (1969).
118. F. J. Vermillon, Thesis, Lawrence College, Appleton, Wisconsin, 1963; F. J. Vermillon and I. A. Pearl, *J. Electrochem. Soc.*, **111**, 1392 (1964).
119. W. Bold and N. Breiter, *Electrochim. Acta*, **5**, 145 (1961); S. Gilman *in Electroanalytical Chemistry*, A. J. Bard, Ed., Vol. 2, Marcel Dekker, New York, 1967, p. 140.
120. A. K. Vijh and B. E. Conway, *Z. Anal. Chem.*, **224**, 160 (1967).
121. R. Pineaux, *C.R. Acad. Sci.*, Ser. C, **268**, 788 (1969).
122. V. Plichon, J. Badoz-Lambling and G. Charlot, *Bull. Soc. Chim. France*, 287 (1964).
123. For a general review, see A. K. Vijh and B. E. Conway, *Chem. Rev.*, **67**, 623 (1967).
124. D. H. Geske, *J. Electroanal. Chem.*, **1**, 502 (1960).
125. G. Cauquis and D. Serve, *Bull. Soc. Chim. France*, **1966**, 310.
126. M. Herlem, *Bull. Soc. Chim. France*, **1965**, 3329.

127. A. I. Popov and D. H. Geske, *J. Am. Chem. Soc.*, **80**, 1340 (1958).
128. J. Desbarres, *Bull. Soc. Chim. France*, **1961**, 502.
129. L. I. Katzin and E. Gebert, *J. Am. Chem. Soc.*, **76**, 2049 (1954).
130. G. Cauquis and D. Serve, *Bull. Soc. Chim. France*, **1966**, 302.
130a. M. Herlem, *Bull. Soc. Chim. France*, **1967**, 1687.
131. M. Herlem and A. Thiebault, *Bull. Soc. Chim. France*, **1970**, 383.
132. J. Courtot-Coupez and M. Le Demezet, *Bull. Soc. Chim. France*, **1969**, 1033.
133. J. P. Billon, *Bull. Soc. Chim. France*, **1961**, 1923.
134. J. Badoz-Lambling and Stojkovic, *Bull. Soc. Chim. France*, **1963**, 2709.
135. R. L. Benoit, *Inorg. Nucl. Chem. Letters*, **4**, 723 (1968).
136. G. Cauquis and D. Serve, *C.R. Acad. Sci. Paris*, **262**, 1516 (1966).
137. D. L. Maricle and W. G. Hodgson, *Anal. Chem.*, **37**, 1562 (1965).
138. M. E. Peover and B. S. White, *Electrochim. Acta*, **11**, 1061 (1966).
139. J. Goret and B. Tremillon, *Bull. Soc. Chim. France*, **1966**, 67.
140. J. Jordan, *J. Electroanal. Chem.*, **29**, 127 (1971).
141. S. Wawzonek, *Talanta*, **12**, 1229 (1965).
142. R. Takahashi, *Talanta*, **12**, 1211 (1965).
143. J. Radell and E. T. Donahue, *Anal. Chem.*, **26**, 590 (1954).
144. S. Sergicuko, P. Galich, N. A. Izmaïlov and L. Spivak, *Zhur. Anal. Khim.*, **10**, 315 (1955); **11**, 73x (1956).
145. ASTM, *Proc.*, **25**, 1, 282 (1925).
146. L. J. David, *Peint. pigm. vernis*, **33**, 629 (1957).
147. S. R. Palit and U. N. Singh, *J. Indian Chem. Soc.*, **33**, 507 (1956).
148. C. C. Washbrook, *Analyt.*, **78**, 254 (1953).
149. G. Delarue, Thesis, Paris, 1960.
150. J. Goret and B. Tremillon, *Bull. Soc. Chim. France*, **1966**, 2872.
151. R. Takahashi, *Talanta*, **12**, 1211 (1965).
152. H. Dehn, V. Gutmann, H. Kirch and G. Schöber, *Monatsh. Chem.*, **93**, 1348 (1962).
153. Z. P. Zagorski and M. Cyrankowska, *Advances in Polarography*, I. S. Longmuir, Ed., Vol. 2, Pergamon Press, London, 1960, p. 584.
154. H. Dehn, V. Gutmann, H. Kirch and G. Schöber, *Monatsh. Chem.*, Sec. V, **93**, 453, 877 (1962).
154a. R. Kollar, V. Plichon and J. Saulnier, *J. Electroanal. Chem.*, **27**, 233 (1970).
155. R. J. Gillespie and J. B. Milne, *Inorg. Chem.*, **5**, 1236 (1966); **7**, 1577 (1968).
156. G. Adhami, Thèse de spécialité, Paris, November 1969.
157. J. Goret and B. Tremillon, *Bull. Soc. Chim. France*, **1966**, 67.
158. M. E. Peover and B. S. White, *Electrochim. Acta*, **11**, 1061 (1966).
159. J. Bessiere, *Bull. Soc. Chim. France*, **1969**, 3356.
160. I. M. Kolthoff and J. F. Coetzee, *J. Am. Chem. Soc.*, **79**, 1852 (1957).
161. J. F. Coetzee and W. S. Siao, *Inorg. Chem.*, **2**, 14 (1963).
162. A. Leveque, Thèse de spécialité, Paris, 1970.
163. E. R. Brown and coll., *Anal. Chem.*, **38**, 1119 (1966).
164. G. L. Booman and W. B. Holbrook, *Anal. Chem.*, **35**, 1793 (1963); **37**, 795 (1965).
165. See, for example, M. S. Shuman and I. Shain, *Anal. Chem.*, **41**, 1818 (1969).
166. See, for example, R. S. Nicholson, *Anal. Chem.*, **37**, 667 (1965).
167. J. E. Harrar and I. Shain, *Anal. Chem.*, **38**, 1148 (1966).
168. D. Peltier and C. Moinet, *Bull. Soc. Chim. France*, **1968**, 2657; **1969**, 690.
169. J. N. Butler, in *Advances in Electrochemistry and Chemical Engineering*, P. Delahay and C. W. Tobias, Eds., Vol. 7, Interscience Publishers, New York, 1970, p. 77.
170. G. Cauquis and D. Serve, *Bull. Soc. Chim. France*, **1966**, 302.
171. J.-P. Billon, *J. Electroanal. Chem.*, **1**, 486 (1960).

172. R. N. Adams, *Electrochemistry at Solid Electrodes*, Marcel Dekker, New York, 1969.
173. G. A. Ward, *Talanta*, **10**, 261 (1963).
174. L. S. Marcoux, K. B. Prater, B. G. Prater and R. N. Adams, *Anal. Chem.*, **37**, 1446 (1965).
175. D. E. Austen, P. H. Given, D. J. E. Ingram and M. E. Peover, *Nature*, **182**, 1784 (1958).
176. B. Kastening, *Electroanalytical Chemistry*, Nurnberg, Ed., John Wiley and Sons, New York, 1974, p. 421.
177. C. P. Poole, Jr., *Electron Spin Resonance*, John Wiley and Sons, New York, 1967.
178. R. S. Alger, *Electron Paramagnetic Resonance*, John Wiley and Sons, New York, (1968).
179. A. Carrington, *Quart. Rev.*, **17**, 67 (1963).
180. A. D. Forrester, J. M. Hay and R. H. Thomson, *Organic Chemistry of Stable Free Radicals*, Academic Press, London, 1968.
181. R. O. C. Norman and B. C. Gilbert, *Advances in Physical Organic Chemistry*, V. Gold, Ed., Vol. 5, Academic Press, London, 1967.
182. G. Cauquis, *Bull. Soc. Chim. France*, **1968**, 1618.
183. R. N. Adams, *J. Electroanal. Chem.*, **8**, 151 (1964).
184. K. Möbius, *Z. Naturforsch*, **20a**, 1093, 1102 (1965).
185. D. H. Geske and A. H. Maki, *J. Am. Chem. Soc.*, **82**, 2671 (1959).
186. G. Cauquis, G. Fauvelot and J. Rigaudy, *Bull. Soc. Chim. France*, **1968**, 4928.
187. G. Cauquis, J. Badoz-Lambling and J.-P. Billon, *Bull. Soc. Chim. France*, **1965**, 1433.
188. See, for example, L. O. Wheeler, K. S. V. Santhanam and A. J. Bard, *J. Phys. Chem.*, **71**, 2223 (1967).
189. See, for example, G. Cauquis, M. Genies, H. Lemaire, A. Rassat and J.-P. Ravet, *J. Chem. Phys.*, **47**, 4642 (1967).
190. See, for example, L. H. Piette, P. Ludwig and R. N. Adams, *J. Am. Chem. Soc.*, **84**, 4212 (1962).
191. G. Cauquis and M. Genies, *Tetrahedron Letters*, **1968**, 3537.
192. G. Gruver, T. Osa and T. Kuwana, in the Proceedings of Symposium on 'The Synthetic and Mechanistic Aspects of Electroorganic Chemistry,' U.S. Army Research Office, Durham, North Carolina, U.S.A., October 14–16, 1968.
193. T. Kuwana, R. K. Darlington and D. W. Leedy, *Anal. Chem.*, **36**, 2023 (1964).
194. J. W. Strojek and T. Kuwana, *J. Electroanal. Chem.*, **16**, 471 (1968).
195. A. Yildie, P. T. Kissinger and C. N. Reilley, *Anal. Chem.*, **40**, 1018 (1968).
196. R. Murray, W. Heineman and G. W. O'Dom, *Anal. Chem.*, **39**, 1666 (1967).
197. W. N. Ansen, R. A. Osteryoung and T. Kuwana, *J. Am. Chem. Soc.*, **88**, 1062 (1966).
198. J. Bizot, *Bull. Soc. Chim. France*, **1967**, 151.
198a. P. Texier and J. Desbarres, *C.R. Acad. Sci.*, **255**, 602 (1968).
199. C. K. Mann, *in Electroanalytical Chemistry*, A. J. Bard, Ed., Vol. 3, Marcel Dekker, New York, 1969, p. 57.
200. J. A. Riddick and E. E. Toops, *Organic Solvents, Vol. VII of Technique of Organic Chemistry*, A. Weissberger, Ed., Interscience Publishers, New York, 1955.
201. See, for example, J. F. Coetzee and C. D. Ritchie, Eds., *Solute–Solvent Interactions*, Marcel Dekker, New York, 1969, p. 607.
202. D. S. Rulison, P. Arthur and K. D. Berlin, *Anal. Chem.*, **40**, 1015, 1250 (1968).
203. K. S. V. Santhanam and A. J. Bard, *J. Am. Chem. Soc.*, **88**, 2669 (1966).
204. P. Champion and J. Royon, *C.R. Acad. Sci.*, **261**, 4744 (1965).

205. J. F. Coetzee, G. P. Cunningham, D. K. McGuire and G. R. Padmanabhan, *Anal. Chem.*, **34**, 1139 (1962).
206. J. Jander and G. Winkler, *J. Inorg. Nucl. Chem.*, **9**, 24 (1959).
207. S. Guiot, Thèse Doctorat Sciences Physiques, Paris, 1968.
208. J. M. Tedder, *J. Chem. Soc.*, **1954**, 2646.
209. G. Durand, Diplôme d'Etudes Supérieures, Paris, 1963; G. Durand and B. Tremillon, *Bull. Soc. Chim. France*, **1963**, 2855.
210. A. M. Brown and R. M. Fuoss, *J. Phys. Chem.*, **64**, 1341 (1960).
211. H. M. Koepp, H. Wendt and H. Strehlow, *Z. Elektrochem.*, **64**, 483 (1960).
212. T. Kuwana, D. E. Bublitz and H. Hoh, *J. Am. Chem. Soc.*, **82**, 5811 (1960).
213. J. Vedel and B. Tremillon, *J. Electroanal. Chem.*, **1**, 241 (1960).
214. A. J. Popov and D. H. Geske, *J. Am. Chem. Soc.*, **80**, 1340 (1958).
215. J. P. Billon, *J. Electroanal. Chem.*, **1**, 486 (1960).
216. R. Kollar, V. Plichon and J. Saulnier, *Bull. Soc. Chim. France*, **1969**, 2193.
217. G. Demange-Guerin, unpublished results.
218. M. Breant, M. Bazouin, Cl. Buisson, M. Dupin and J. M. Rebattu, *Bull. Soc. Chim. France*, **1968**, 5065.
219. Nguyen van Kiet, Thèse, Lyons, 1967.
220. Ch. Sinicki, Thèse en préparation, Lyons.
221. J. Courtot-Coupez and M. Le Demezet, *Bull. Soc. Chim. France*, **1967**, 4744.
222. P. G. Sears, G. R. Terter and L. R. Dawson, *J. Phys. Chem.*, **60**, 1433 (1956).
223. Bulletin technique SNPA.
224. G. Picard, Thèse Doctorat 3ème cycle, Spécialité Chimie Analytique, Paris, 1968.
225. G. Vedel and B. Tremillon, *Bull. Soc. Chim. France*, **1966**, 220.
226. G. Charlot and B. Tremillon, *Les Réactions Chimiques Dans les Solvants et les Sels Fondus*, Masson, Paris, 1963, p. 360.
227. D. E. Sellers and G. W. Leonard, *Anal. Chem.*, **33**, 334 (1961).
228. B. Agurto-Cid and M. Machtinger, *Bull. Soc. Chim. France*, **1965**, 1915.
229. P. G. Sears, *J. Chem. Eng. Dela.*, **11**, 406 (1966).
230. P. Olavi and I. Virtanen, *Suomen Kemistilehti B*, **40**, 313 (1967).
231. R. Reynaud, *C.R. Acad. Sci.*, **266**, 489 (1968).
232. V. A. Granzhan, S. V. Semenenko and O. G. Kirillova, *Zh. Prikl. Kim.*, **39**, 1399 (1966).
233. E. Fisher, *J. Chem. Soc.*, **1955**, 1382.
234. P. Olavi and I. Virtanen, *Suomen Kemistilehti B*, **40**, 313 (1967).
235. G. Cauquis and D. Serve, *Bull. Soc. Chim. France*, **1966**, 302.
236. G. Cauquis and D. Serve, unpublished results.
237. G. Cauquis and D. Serve, *Compt. Rend. Acad. Sci., Ser. C.*, **2621**, 1516 (1966).
238. J. Courtot-Coupez and L'Her, *Bull. Soc. Chim. France*, **1970**, 1631.
239. R. L. Benoit, M. Guay and J. Desbarres, *Canad. J. Chem.*, **46**, 1261 (1968).
240. J. Desbarres, R. Pichet and R. L. Benoit, *Electrochim. Acta*, **13**, 1899 (1968).
241. F. E. Critchfield, J. A. Gibbon and J. C. Hall, *J. Am. Chem. Soc.*, **75**, 6044 (1953).
242. J. Badoz-Lambling and M. Sato, *Acta Chim. Acad. Sci. Hung.*, **32**, 191 (1962).
243. J. Perichon and R. Buvet, *Bull. Soc. Chim. France*, **1968**, 1279.
244. J. Bessiere and G. Petit, *J. Electroanal. Chem.*, **34**, 489 (1972).
245. M. Herlem, Thesis, Paris, 1966, and *Bull. Soc. Chim. France*, **1965**, 221.
246. H. Smith, *Organic Reactions in Liquid Ammonia*, Interscience Publishers, New York, 1963.
247. J. Jander, *Anorganische und allgemeine Chemie in flüssigem Ammoniak*, Interscience Publishers, New York, 1966.
248. D. Bauer and P. Texier, *Compt. Rend. Acad. Sci., Ser. C*, **266**, 602 (1968).

249. H. A. Laitinen, W. S. Ferguson and R. A. Osteryoung, *J. Electrochem. Soc.*, **104**, 516 (1957).
250. G. Delarue, Thèse Docteur-Ingénieur, Paris, 1960.
251. J. Goret and B. Tremillon, *Electrochim. Acta*, **12**, 1065 (1967).
252. M. Arnac, Thèse Doctorat 3ème cycle, Spécialité Chimie Analytique, Paris, 1968.
253. B. Tremillon and G. Letisse, *J. Electroanal. Chem.*, **17**, 371 (1968).
254. J. Courtot-Coupez, Thèse Doctorat es-Sciences Physiques, Paris, 1960.
255. M. Machtinger-Convers, Thèse Doctorat es-Sciences Physiques, Paris, 1961.
256. G. Adahmi, Thèse de doctorat, Paris, 1969; M. Convers, *Bull. Soc. Chim. France*, **1959**, 792.
257. E. Grunwald and J. Berkowitz, *J. Am. Chem. Soc.*, **73**, 4939 (1951); I. M. Kolthoff, J. J. Lingane and W. D. Larson, *J. Am. Chem. Soc.*, **60**, 2512 (1938); R. G. Bates, *Determination of pH*, John Wiley and Sons, New York, 1964; A. J. Parker, *J. Chem. Soc.*, **1966**, 220; N. A. Izmaylov, *Zh. Fiz. Khim.*, **34**, 2414 (1960); P. Popovych, *Anal. Chem.* **68**, 558 (1966).
258. T. C. Waddington, *Non-aqueous Solvent Systems*, Academic Press, New York, 1965; G. Jander, H. Spandau and C. C. Addison, *Chemistry in Non-aqueous Ionizing Solvents*, Vol. I, Part 1 (1966); Vol. II, Part 2 (1963); Vol. III, Part 1, Part 2 (1968), Pergamon Press, London; D. Bauer, Ph. Texier, A. Collumeau, V. Plichon, M. Breant and G. Demange-Guerin, *Bull. Soc. Chim. France*, **1**, 4313 (1968); **2**, 4315 (1968); **3**, 4317 (1968); **4**, 5087 (1968); **5**, 3369 (1969); **6**, 2935 (1969); D. S. Reid and C. A. Vincent, *J. Electroanal. Chem.*, **18**, 427–465 (1968); J. N. Butler, *J. Electroanal. Chem.*, **14**, 89 (1967).
259. D. Bauer and M. Breant, 'Solute behavior in solvents and melts, a study by use of transport activity coefficients', in *Electroanalytical Chemistry*, A. J. Bard, Ed., M. Dekker, New York, 1972.

CHAPTER VI

Joint Application of Electrochemical and ESR Techniques *

B. Kastening, *Forschungsabteilung Angewandte Elektrochemie,*
Zentralinstitut für Analytische Chemie, Kernforschungsanlage Jülich, Germany

I. INTRODUCTION

It is well established that molecules (or ions) with an odd number of electrons, viz. free radicals (ion radicals), play an important role as intermediates of chemical reactions, and may be stable constituents of solutions or even stable solids. In the course of the renaissance of organic electrochemistry during the last three decades, it became evident that radicals are also involved in most organic reduction and oxidation reactions at electrodes. Apart from the quinone/semiquinone/hydroquinone system, well established by the work of Michaelis (1932–1938), some polarographic investigations may be mentioned here in which radicals were claimed as intermediates of electrode processes, viz. for the reduction of aromatic hydrocarbons (1), heterocyclics (2), carbonyl compounds (3–5), organic halogen compounds (6, 7) and aromatic nitro compounds (8). While most of these investigations were done in essentially aqueous media, subsequent progress in the elucidation of radical mechanisms was due—next to the improvement of experimental techniques—to the application of organic, especially aprotic, solvents

* To Professor L. Holleck on his 70th birthday.

(9) in which radicals are in general more stable than in aqueous solutions. The most frequently applied methods are controlled-potential electrolysis (10) by which radicals can be generated as intermediate or stable products and, in some cases, even isolated as solid material, as well as voltammetric techniques (11) like polarography; this is illustrated by reviews on mechanisms (12, 13) and free radicals (14) in organic polarography.

Due to the uncompensated spin of the odd electron, radicals are paramagnetic. As a specific tool for the investigation of paramagnetic material, the electron spin resonance (ESR) or electron paramagnetic resonance (EPR) technique has been developed twenty-five years ago. While the study of paramagnetic solids was predominant at the beginning, growing application has been made since then to free radicals in solution as well. The principles of the method have been presented within the framework of spectroscopy (15–17) as well as in more or less comprehensive monographs (18–22); smaller books were devoted to the experimental techniques (23, 24) and the basic theory (25). The application to chemistry has been dealt with in a number of books (26–29) and review articles (30–33). Special problems in the chemistry of radicals as studied by ESR have been reviewed, e.g. conformation and structure (34), ion pairing (35), some structural problems reflected in ESR linewidths (36) and the application to heterogeneous catalysis (37, 38). ESR spectra and/or their location in the literature have been compiled in tables (39–42). More recent work may be gathered from the current reviews (43–48). No extensive application seems to have been made in connection with electrochemical work of more sophisticated techniques like ENDOR (electron nuclear double resonance); for a survey of this technique cf. reference 49.

Because of the considerable significance of radicals in electrochemistry and the specific abilities of the ESR technique, the logical idea was created 1958/1959 to combine both methods (cf. Section II), thus mutually serving for progress in both fields. Apart from some early reviews in Japanese (50, 51), the matter has been summarized in 1964 (52). Since the literature has increased considerably, and the advantage of joint application of ESR and electrochemical techniques has been stressed repeatedly (53–59), the need seems to exist for a more comprehensive review; an extract of the material presented in this chapter has recently been published by the author (60).

Electrochemistry may serve as an elegant tool for the generation of free radicals in solution and their investigation by ESR; this kind of generation frequently exhibits considerable advantages compared with other methods, viz. homogeneous chemical reactions, reduction by alkali metals, photolysis and radiolysis, etc. By the control of electrode potentials, reactions can in general be restricted to a definite charge-transfer process and the generation of a particular radical. A large variety of applicable solvents and the possibility of applying different temperatures allows the proper conditions for stability

to be selected and a more or less pronounced interaction with the environment to be effectuated at will. The unavoidable presence of an additional
supporting electrolyte in general does not mean an interference; ion pairing—
if not just under investigation—can be avoided by the application of tetraalkylammonium salts the cations of which practically do not exhibit ion
pairing, in contrast with the alkali metal ions formed during the generation
of anion radicals by means of these metals.

On the other hand, ESR techniques offer the possibility of detecting and
identifying radicals involved in electrochemical reactions, thus elucidating
unknown mechanisms or confirming those derived from merely electrochemical and/or other additional techniques. Even mechanisms involving
more than one radical species have been investigated in several cases.
Moreover, details of ESR spectra will give profound information about the
structural state of the radicals and may serve for a better understanding of
their reactivity. Finally, both the detailed characteristics of ESR spectra
(e.g. hyperfine coupling constants and g factors) and reduction/oxidation
potentials are closely related to quantum chemical parameters, e.g. molecular-
orbital energies; hence, immediate correlations exist and have been
established between electrochemical and ESR data (61–65).

Besides organic radicals which will be the essential subject of this review,
inorganic paramagnetic ions have also been studied by ESR and electrochemical techniques, e.g. the role of the ligand (complexing agents like
ethylenediamine) in anodic dissolution and electrodeposition of metals,
making use of the paramagnetism of the metal ions (66).

This review is essentially devoted to investigations of radicals generated
by electrolysis. Several studies, however, have been carried out with
simultaneous application of ESR and electrochemical techniques, in which
radicals are immediately applied (67) or generated by other methods (68, 69);
further work of this type is referred to in the course of this review.

Furthermore, it should be kept in mind that non-electrochemical
techniques may actually involve heterogeneous charge-transfer processes
which immediately resemble those of electrochemical reactions. Thus, the
frequently applied generation of anion radicals by the action of alkali
metals (or other active metals, like zinc) actually represents an oxidation
and a reduction taking place at the same interface,

$$Met \rightarrow Met^+ + e^- \tag{1}$$

$$M + e^- \rightarrow M^- \tag{2}$$

where Met is the metal atom and M an organic molecule, the sum of reactions
(1) and (2) representing the overall process. The same is true for organic
anion radicals which are generated in contact with tetraalkylammonium
(70) and -phosphonium (71) amalgams and have been studied by ESR.

Similarly, the heterogeneous oxidation of organic molecules at lead and silver oxides (72–74) constitutes the anodic part of a corresponding overall reaction. Radical structures and/or paramagnetic states may also be involved in the surface structure of metallic (75) as well as of semiconductor electrodes like germanium (76) or phthalocyanine (77). A single ESR line observed upon anodic oxidation of hydrogen ($H_2 \rightarrow H^+$) at platinized carbon electrodes was attributed to free electrons in a nearly empty conductance band of an extrinsic semiconductor (78). Finally, charge transfer with the formation of adsorbed radicals—sporadically observed with ESR upon electrochemical generation (79)—plays a role in heterogeneous catalysis and has been studied by ESR methods (37, 38, 80, 81); investigations with silica, alumina and zinc oxide (82–87), titanium dioxide (88), and group VIII transition metal oxides (89) may be mentioned in this connection.

II. EXPERIMENTAL TECHNIQUES

The electrochemical techniques employed in combination with ESR investigations are, in general, polarography and other voltammetric methods making use of different types of electrodes and electrode materials (preferably mercury for reduction and platinum for oxidation reactions) on the one hand, and controlled-potential electrolysis on the other. These methods are thought to be well known or dealt with in other articles of this volume.

For more comprehensive information about the techniques of ESR measurements, the reader is referred to monographs and review articles mentioned in the preceding section. The method is based upon the magnetic momentum of an unpaired electron spin occupying two different orientations within an external magnetic field (with quantum numbers $+\frac{1}{2}$ and $-\frac{1}{2}$, respectively), differing by a small amount of energy and, hence, showing a small difference of population. When an energy corresponding to this difference is employed through an electromagnetic wave (in general a microwave of frequency 9000 MHz for external fields of about 3000 gauss), then the population of the two quantum states is shifted somewhat towards an equal distribution. Since, however, an energy exchange takes place by a relaxation mechanism preferably with the environment ('spin–lattice interaction'), a steady absorption of microwave energy takes place. This absorption is observed as a more or less sharp line, if the external magnetic field is slowly shifted at a constant microwave frequency. (In practice, the first derivative is generally recorded by making use of a small modulation of the magnetic field at medium frequencies, e.g. 10^5 Hz.)

Those nuclei in the respective molecule exhibiting a magnetic momentum (e.g. ^1H, ^{14}N, ^{13}C, ^{15}N, ^2H, ^{19}F, ^{17}O, etc.) will produce internal magnetic fields. If there is a certain probability of finding the odd electron at such a nucleus ('spin density' at this nucleus), it experiences a magnetic interaction the extent of which depends on the nuclear quantum state. Instead of a single absorption line, therefore, a variety of lines will occur (hyperfine splitting, HFS), one line for each combination of quantum states of all interacting nuclei. The intensities of these lines will differ from one another, if there are two or more nuclei of equivalent structural position, due to different statistical weights of combinations. Thus, one proton gives rise to two lines of equal intensities (for the two quantum states $\leftarrow$ and $\rightarrow$), two equivalent protons produce three lines with intensities, $1:2:1$ (for quantum states $\leftarrow\,\leftarrow$; $\rightleftarrows$ or $\leftrightarrows$; $\rightarrow\,\rightarrow$). The distance of these lines (HFS coupling constant: a_H for protons, a_N for a nitrogen nucleus, etc.) directly corresponds to the spin density at this nucleus and the bonding situation of the respective atom. As an example, the ESR spectrum (first derivative of the absorption intensity versus magnetic field strength) of the nitrobenzene anion radical in aqueous solution is shown in Figure 1. The way in which the fifty-four observed lines result from the interaction of the nuclear quantum states of N and H atoms is schematically shown above the observed spectrum.

The width of the lines is in the first place related to the relaxation mechanism mentioned above. (Spin–spin interaction, constituting another source of linewidth, is not of particular importance for electrochemical investigations, as long as radical concentrations are sufficiently small.) While the linewidths of a particular spectrum would superficially be expected to be (and in most cases, approximately are) constant, we shall see in a later section that pronounced variations of linewidths may occur, reflecting peculiar rapid changes of intramolecular structures.

All types of relaxation, though necessary for the resonance phenomenon, may prevent hyperfine structure from being resolved if—as a consequence of small lifetimes of spin states—the lines are broader than the distance between them. Thus, cation radicals of pentaphenylpyrroles (90) of the type

exhibit only one comparatively broad ESR line* (somewhat deformed at the centre indicating unresolved HFS) if R is H, CH_3 or OCH_3. In these cases, the unpaired electron is delocalized throughout the ring systems, and a large number of weakly interacting protons produces a corresponding number of lines as close to one another as to prevent resolution. (Precipitation of blue-black crystals of the cation radical perchlorate upon electrochemical oxidation inside the microwave cavity gives rise to a narrow single line

---

* Similarly, a broad single line was observed with porphyrin compounds (91).

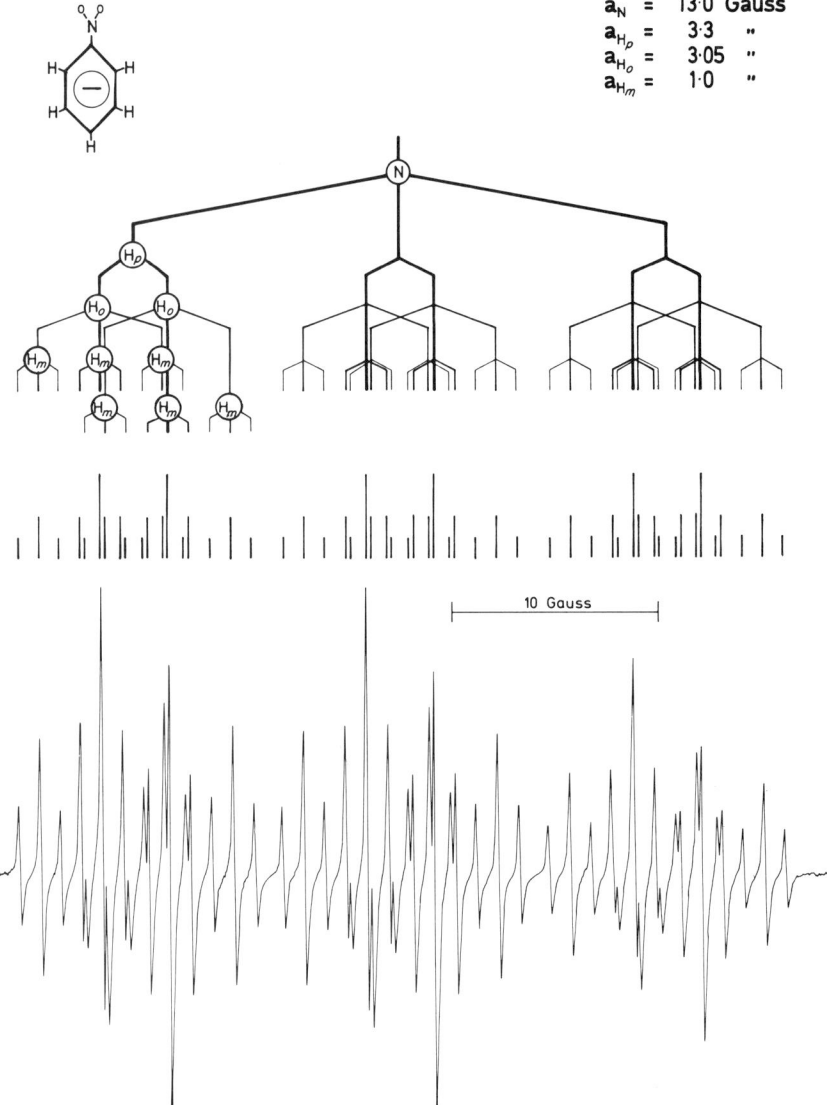

Fig. 1. ESR spectrum of nitrobenzene anion radicals in aqueous solution. Above: Schematic representation of the generation of the various lines by the combination of nuclear spin states.

interfering with the broader line of dissolved species.) If on the other hand, $R = N (CH_3)_2$, in the pyrrole derivative shown above, then the unpaired electron is localized in the group and a HFS spectrum similar to that of *para*-substituted dimethylaniline cation radicals is observed; simultaneously, half-wave potentials are considerably less positive than those for the former type of compounds, indicating the greater ease of releasing an electron from the *p*-dimethylaminophenyl system than from the pyrrole system.

The resonance phenomenon is in general observed by transmitting a microwave through rectangular waveguides to a rectangular resonant cavity, though cylindrical resonant cavities have also been employed in electrochemical investigations (92). The cavity in which standing electromagnetic waves are produced, is placed in the external field. The paramagnetic sample under investigation has to be placed within a small range of maximum magnetic field of these standing waves; this is of particular importance for the investigation of electrochemically generated radicals.

In the first investigation in this field (93) samples were withdrawn from the electrolysis cell into tubes, which were immediately chilled in liquid nitrogen and later on, at low temperatures, placed in the microwave cavity. (Thawing of such frozen samples has also been employed for an estimate of half-lives of the corresponding radicals (94).) This 'external' method has subsequently been improved in different ways in order to reduce the time taken to transfer the radicals from the electrolysis cell to the resonant cavity and to allow for investigations to be made even at room temperature. Some examples will be given below.

Instead, an elegant method has been introduced soon afterwards in which the electrolytic generation of radicals takes place within the resonant cavity proper (92, 95). This 'internal' or 'intramuros' technique has been employed since then in most investigations. A typical assembly is shown in Figure 2 (64). Mercury has been used for the working electrode in this case which may, however, be replaced by platinum wires or foils, etc. The electrode interface is placed at the point of maximum sensitivity, viz. at the intersection of both waveguide and sample tube axes. The sensitivity decreases from this centre along the sample tube axis. Moreover, extending the sample cell in the direction of the waveguide axis reduces the 'quality' of the resonant cavity and hence the sensitivity, especially if solvents of high dielectric constants, such as water, are employed. Cylindrical tubes of small diameter or flat cells (extending perpendicular to the drawing plane only) have been used, preferably made from quartz glass.*

From the geometric distribution of sensitivity it is clear, that the ESR signal intensity is not a simple function of the number of radicals or their concentration within the resonant cavity, but depends on the particular geometry of the cell and its proper adjustment. Moreover, the electrolytic

---

* A useful do-it-yourself sample cell made from heat-shrinkable transparent Teflon tubing, however, has been described and applied for internal electrochemical generation of radicals (96).

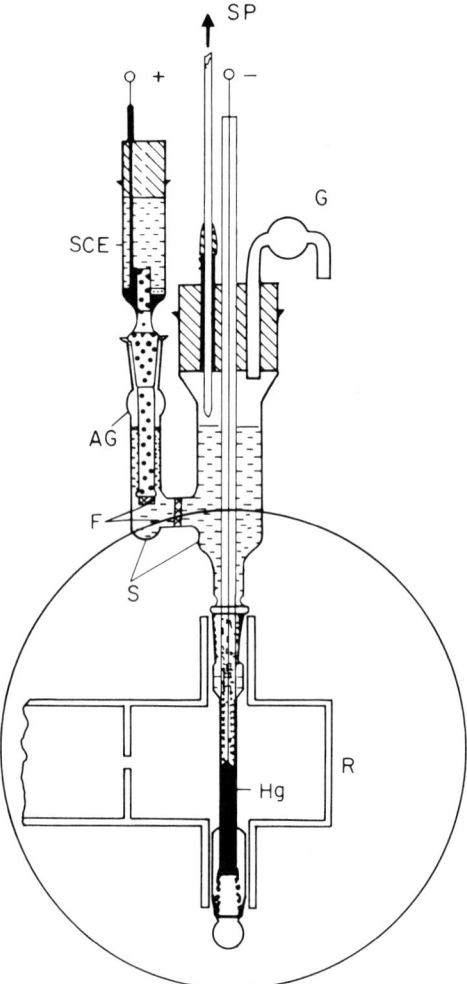

Fig. 2. Electrolytic cell for internal generation of
radicals; according to reference 64. R: microwave
cavity. F: glass frit. AG: agar gel salt bridge. SCE:
saturated calomel electrode. G: gas outlet. S: sample
solution. SP: sample preparing equipment.

generation produces an inhomogeneous distribution of concentrations
following, in the absence of convection, the laws of diffusion (97). During
prolonged electrolysis, however, natural convection may give rise to a
steady state, serving for steady conditions during the time necessary for
recording the ESR spectrum. The method, though excellent for the detection
and identification of radicals, is not quite as advantageous for an estimation

of absolute radical concentrations nor for the observation of decay kinetics after current interruption, since the ESR signal will decrease also due to diffusional homogenization. *In situ* methods for the determination of absolute amounts of electrochemically generated radicals, however, have also been described (98).

For the determination of absolute concentrations of radicals, a forced flow of the electrolyte through the electrolysis cell is preferable serving for more definite steady-state conditions. This method is even applicable, if the radicals decay by a chemical reaction sufficiently rapid to reduce the signal intensity during maintained generation (99). Such an assembly is shown in Figure 3; a critical estimation of its abilities has been given (99),

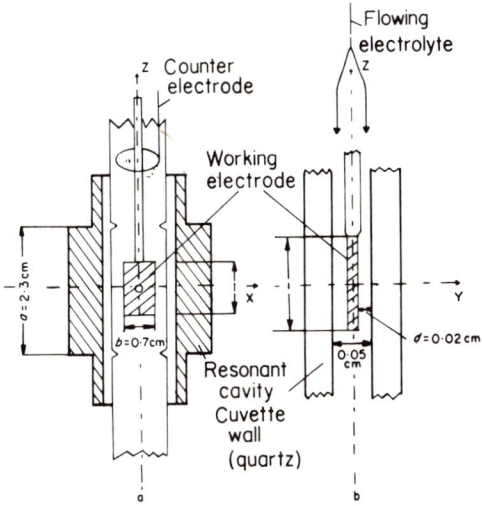

Fig. 3. Electrolytic cell for internal generation with flowing electrolyte; according to reference 99.

although the facilities seem to be somewhat overestimated for practical application.

Assemblies with internal generation of radicals have been described in several works (100–102). A particular difficulty arises from the inevitably small dimensions of the electrode compartment, whereas large electrode areas and large currents would be desirable in order to improve the signal-to-noise ratio and the resolution of ESR spectra. Because of the particular geometry of the sample cells and the limited solubilities of supporting electrolytes (resulting in poor conductivities), a rather high voltage drop across the cell may occur (103) deteriorating potential control or causing inhomogeneous potential distribution across the working electrode. Special designs have, therefore, been proposed with a separate reference electrode (instead of using the counter electrode as a reference as shown in Figure 2) situated next to the working electrode (104, 105); an example is given in Figure 4. Reference and counter electrodes have also been placed at opposite sides of the working electrode (106, 107); this arrangement was also applied with the method of continuously flowing electrolyte (108) illustrated in Figure 3.

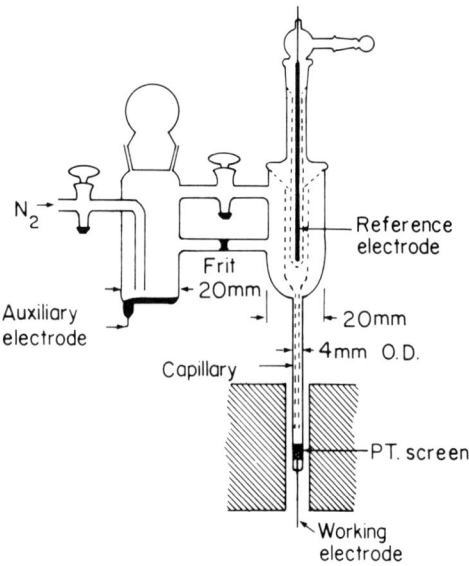

Fig. 4. Electrolytic cell for internal generation, with Luggin capillary of the reference electrode placed near the working electrode; according to reference 105.

A special cell with a surrounding Dewar is shown in Figure 5, allowing for investigations in liquid ammonia (109) as well as in other solvents at low temperatures (110).

Two more recent attempts to avoid the inhomogeneous current distribution across the working electrode, arising from the voltage drop along the sample cell axis, should be mentioned here. In the first case (111) the working and counter electrodes were simultaneously placed inside the sample cell as shown in Figure 6. With this construction, a voltage drop across the electrode can only develop perpendicular to the sample cell axis and will be—at comparable electrode areas and currents— considerably smaller than with the counter electrode situated outside or at the end of the sample cell. Special care is required with respect to the products of the counter electrode process. Because the counter electrode is situated within the sensitive part of the ESR cavity, these products must not be radicals. Moreover, any convective motion of the solution should be prevented in order to avoid reactions of the radicals under investigation with the counter electrode products (unless the study of such reactions is of special interest).

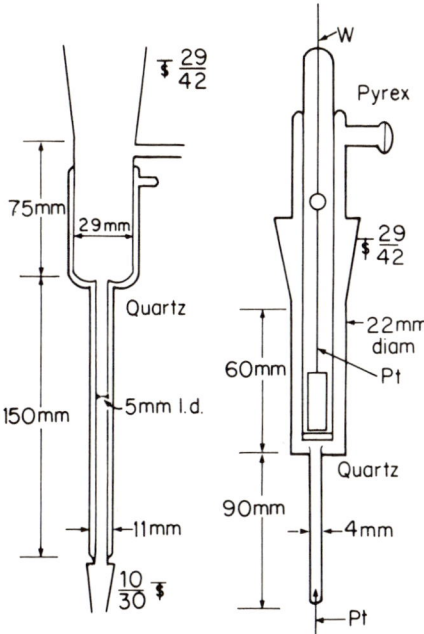

Fig. 5. Cell and Dewar for internal generation in liquid ammonia; according to reference 109.

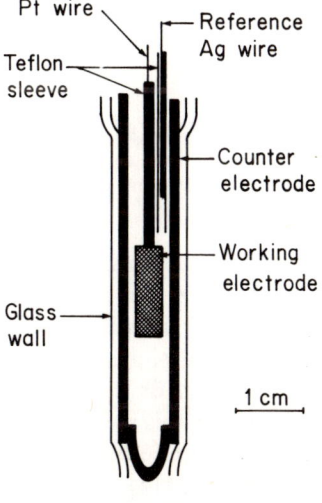

Flat cell

Fig. 6. Electrolytic cell for internal generation with both working and counter electrodes placed within the sample cell; according to reference 111.

The capabilities of another attempt at avoiding voltage drops are still under investigation (112). While in this case the counter electrode is again placed somewhere outside the sample cell, the working electrode has been divided into several sections insulated from one another, each having its own current supply as shown in Figure 7. By proper choice of the individual currents, the potentials of the various parts of the working electrode can be adjusted to a uniform value.

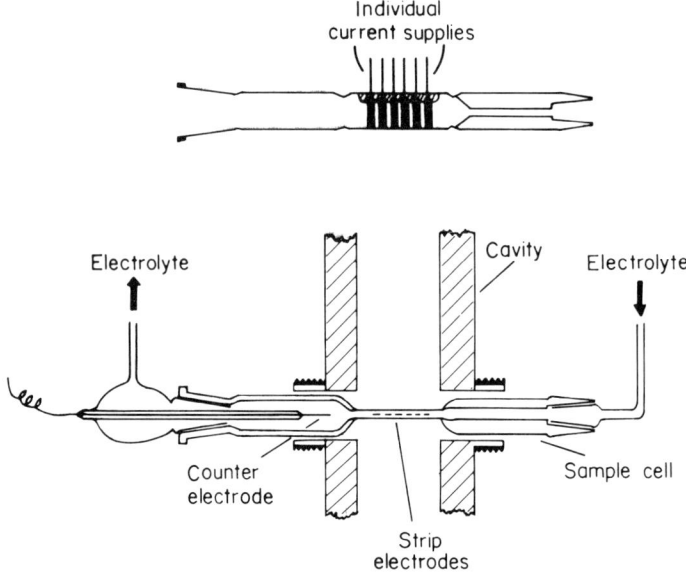

Fig. 7. Electrolytic cell for internal generation with a set of working electrodes having individual current supplies (for compensation of the Ohmic drop across the cell); according to reference 112.

If the radicals tend to move in a downward direction, it was found useful to place the working electrode at the top of the electrolysis cell, just above the sensitive part of the cavity (113); in order to profit at the same time from the advantages of mercury electrodes, amalgamated silver electrodes were applied instead of hanging mercury drops which proved not to be stable enough. Strictly speaking, however, this is an external method of generation.

Instead of waveguide cavities which have been used in general, the application of a helix has been proposed (114). The helix which is placed in the external magnetic field and matched to the ESR spectrometer by a coaxial line, may serve as the working electrode proper, because the magnetic component of the microwave exhibits a maximum value next to the helix surface. A suitable arrangement is shown in Figure 8.

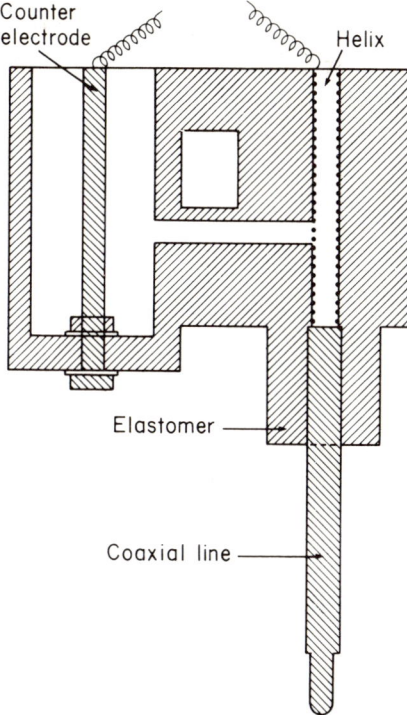

Fig. 8. Electrolytic cell for internal generation
at the surface of a helix applied instead of a
microwave cavity; according to reference 114.

The method of internal generation dealt with so far, offers the advantages of a rather uncomplicated experimental technique readily allowing for the observation of ESR spectra with comparatively small volumes of solution. For short-lived radicals, pulse electrolysis along with rapid scanning of the magnetic field or observation of signal intensities at fixed field strength have been applied, occasionally improved by sampling techniques (computers of average transients), for entire spectra or rapid decay kinetics (108, 111, 115, 116).

As has been mentioned above, however, some difficulties are encountered: the determination of radical concentrations as well as of decay kinetics may deteriorate due to the inhomogeneous distribution of concentrations. Moreover, electrolysis is not complete within the sensitive part of the sample; hence, rapid electron exchange with the parent molecule

$$M + M^- \rightleftharpoons M^- + M \tag{3}$$

may constitute another relaxation mechanism decreasing the lifetime of a particular radical and increasing the linewidth with the consequence of worse resolution (cf. Section IX). Furthermore, the progress of electrolysis is prevented from being visually observed, which may be desirable (colour of the solution, etc.).

These disadvantages are overcome by the method of external generation, although this method may be restricted somewhat more with respect to short-lived radicals. The solution has to be transferred from the electrolysis cell to the resonant cavity allowing for a lower limit of about 0·1 second for the lifetime, whereas this limit may be about 0·01 second or somewhat lower (115–117) for the internal method (cf. even more optimistic estimates (99)).

The above-mentioned technique of withdrawing samples during or after complete electrolysis from the electrolysis cell (93) has been improved in later work as shown in Figure 9 (118). Here, after electrolysis, the solution

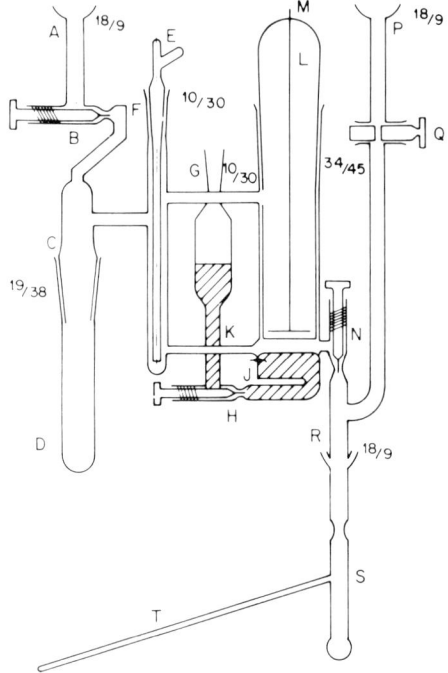

Fig. 9. Vacuum electrolytic cell for external generation. A: connection to manifold. B: Teflon needle valve. C: mixing chamber. D: solvent collection tube. E: connection to reference electrode. G: mercury storage bulb. H: mercury admission valve. J: connection to cathode. K: cathode compartment. L: anode compartment. M: connection to anode. N: cathode compartment exit valve. P: connection to manifold. Q: stopcock. R: sample tube connection point. S: sample tube. T: ESR side arm. See text for details; according to reference 118.

is tapped off via the valve N into a previously evacuated sample tube S, frozen in liquid nitrogen and, after sealing off the tube with a flame at the constriction and warming, tipped into the sidearm T for ESR measurements. Similar equipment has been described by other authors (119).

Instead of discontinuous withdrawing of samples, several methods have been developed for continuous flow of the solution through the microwave cavity, either with circulation systems or with irreversible passage from the electrolysis cell through the ESR sample tube.

A circulation system (120, 121) has been applied almost as early as the first application of internal techniques. An improved assembly allowing for both internal and external generation is shown in Figure 10 (54); circulation of the solution is effectuated by bubbling nitrogen through the sidearm. A similar technique is shown in Figure 11 (122). An assembly in which circulation is effectuated by the hydrostatic pressure produced by a large

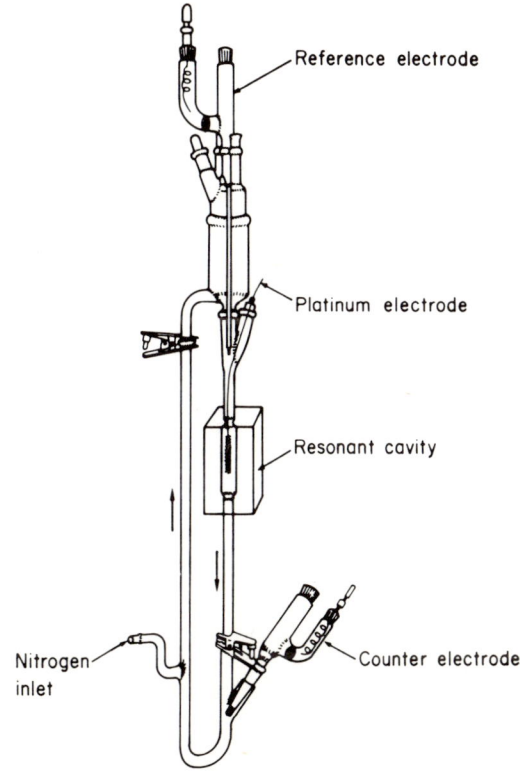

Fig. 10. Electrolytic cell for external and internal generation.
Circulation of the solution is effectuated by bubbling nitrogen
into the left-hand side arm; according to reference 54.

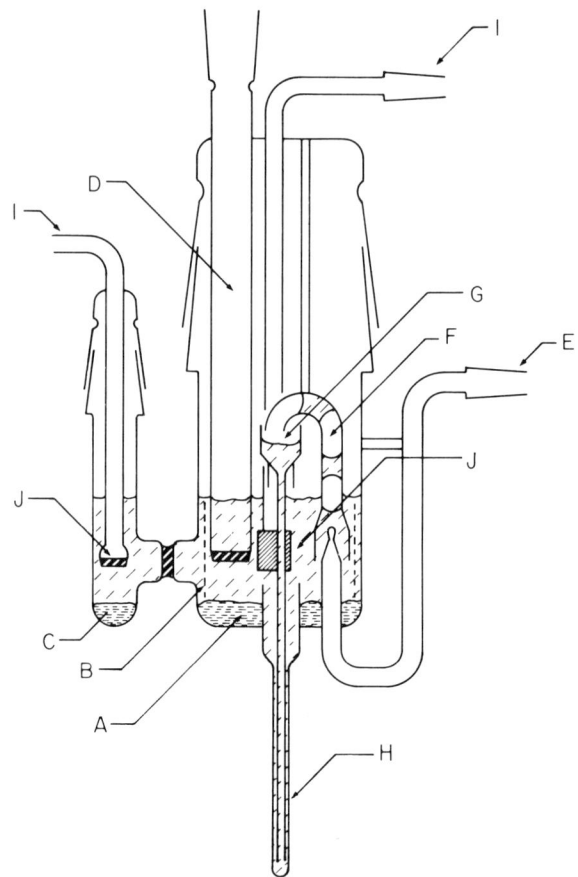

Fig. 11. Electrolytic cell for external generation with circulating solution. A: Hg pool electrode. B: Pt gauze working electrode. C: counter electrode. D: chamber for reference electrode. E: admission of nitrogen so that bubbles will pump solution through F filling the reservoir G. H: outer tube through which solution circulates under hydrostatic pressure, extending into the microwave cavity. I and J: admission of nitrogen to exclude air from the system. According to reference 122.

and rapidly rotating working electrode—first applied to the observation of light absorption of radicals (123)—has also been applied to ESR measurements (124). This technique was shown to serve for a sufficient enlargement of ESR signals which were found too weak during internal generation for complete analysis (125).

Continuous flow of the solution from a reservoir successively through the electrolysis cell and the microwave cavity has also been effectuated by the hydrostatic pressure of the solution column (126). In such arrangements,

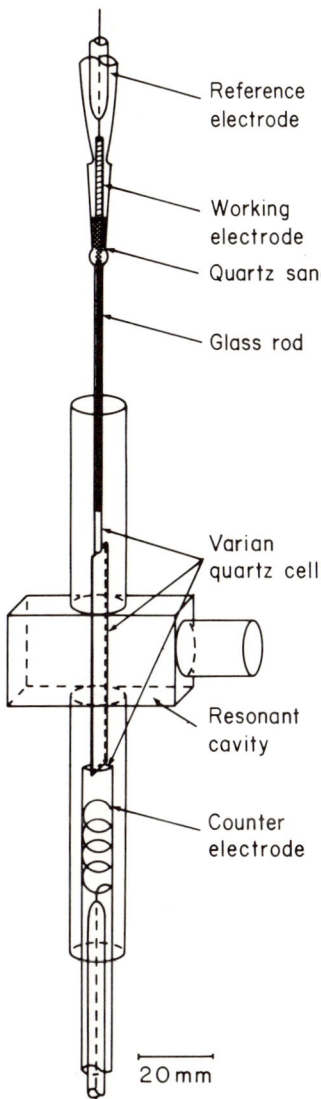

Reference electrode

Working electrode

Quartz sand

Glass rod

Varian quartz cell

Resonant cavity

Counter electrode

20 mm

Fig. 12. Electrolytic cell for external generation. Under hydrostatic pressure from a reservoir, the solution flows—after homogenization by the layer of quartz sand and the glass rod—through the microwave cavity; according to reference 129.

complete electrolysis can be approximated by proper adaptation of flow rates. Similar equipment (127, 128) as well as circulation systems mentioned above (123, 124) have been applied to the investigation of decay kinetics. An assembly with the solution flowing under hydrostatic pressure is shown in Figure 12 (129, 130). A layer of quartz sand and a glass rod with milled-in threads homogenizes the solution before it enters the resonant cavity; the counter electrode is placed beneath the cavity for facilitating the potential control of the working electrode.

For the investigation of rapid reactions of radicals, the electrolysis cell mentioned above (123) has been combined with a flow apparatus as shown in Figure 13 (131). After electrolytic generation under comparatively stable

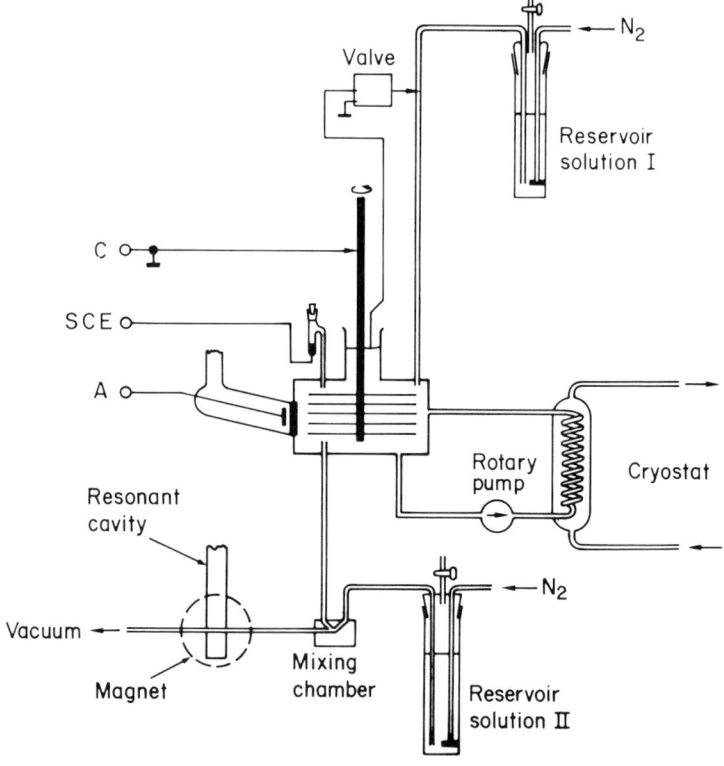

Fig. 13. Electrolytic cell for external generation allowing for admixing a reagent solution for the study of fast radical reactions; according to reference 131. Similar equipment was used with circulating solution, the rapidly rotating large-area working electrode serving as a circulation pump. C: working electrode. A: counter electrode. SCE: reference electrode.

conditions (lifetime some seconds or more), the solution of radicals is admixed in a mixing chamber with a proper reagent solution, the mixed solution flowing through a capillary tube inserted in the microwave cavity. Variation of the distant mixing chamber/cavity during continuous flow allows for the observation of rapid reactions that result in a decrease of the lifetime to 0·01 second, whereas the stopped flow technique is applied for slower reactions. Flow is effectuated by sucking off with vacuum. Replacement of the solution in the electrolysis cell is controlled by an automatic valve, while an additional circulation system connected to a cooling unit serves for maintaining constant temperatures. Variation of concentrations as well as of ESR sensitivity along the tube axis has been taken into consideration by theoretical calculation for laminar as well as turbulent flow (132). Work is in progress with an improved version making use of a pressure pump to increase flow rates and allow for the investigation of more rapid reactions (133). The technique has proved to be useful also for the observation of ESR spectra in aqueous–organic solvents of radicals which cannot directly be generated in aqueous media; in this case, the radicals are generated in the aprotic organic solvent, and water is added in the mixing chamber (60). A similar technique has been applied (134) in which the solution of an electrochemically generated radical M_1^- and a solution of a molecule M_2 were mixed resulting in the fast electron exchange reaction

$$M_1^- + M_2 \rightarrow M_1 + M_2^- \qquad (4)$$

and allowing for the study of the decay kinetics of radical M_2^-.

The presence of oxygen in general interferes with organic electrochemical studies; it may be even more harmful in ESR investigations. On the one hand, many radicals react with oxygen, on the other hand the paramagnetic oxygen molecule opens an additional path of relaxation resulting in increased linewidths and a worse resolution of ESR spectra. While exhaustive reduction during external generation frequently also removes traces of oxygen, this is a more delicate problem for the internal method; deaeration with commercially available prepurified gases is necessary, though sometimes additional purification in the laboratory may be advantageous (135).

It has been mentioned above that a variety of solvents—extending from water and alcohols to aprotic solvents like acetonitrile, dimethylformamide, dimethylsulphoxide and even tetrahydrofuran—have been applied. The geometry of sample cells is more critical with water, while an enhanced tendency of ion pairing has to be taken into consideration with solvents of low dielectric constants. The application of propylene carbonate has also been recommended (136); the high dielectric constant of this solvent serves for proper solvation of ion radicals and better resolution of HFS, while the high viscosity may avoid homogenization by convection during internal generation, thus facilitating an exhaustive electrolysis in the sensitive region of the resonant cavity and preventing exchange broadening by the mechanism of equation (3). On the other hand, the effect of increased viscosity on molecular motions may be reflected in HFS structures.

A particular aspect (137) for the electrolytic reduction in aprotic solvents giving rise, at separated potentials, to the anion radical and dianion, respectively, should be mentioned here since it is frequently not well understood. (The same is true for an oxidation giving rise to cation radicals and dications, respectively.) If subsequent irreversible changes (e.g. dimerization of radicals and/or chemical reactions of the dianion due to proton abstraction from the solvent) are absent or sufficiently slow, then a homogeneous equilibrium exists

$$M + M^{2-} \rightleftharpoons 2 M^{-} \tag{5}$$

and—corresponding to the potential difference of both electron-transfer steps—the formation constant of the radical is rather large. Consequently, the dianion cannot exist at perceptible concentrations as long as the parent molecule is present. Even if the dianion is generated at the electrode at more negative potentials, it will react with the parent molecules in the diffusion layer to give the anion radical. Radicals will even be detected, therefore, if the potential is raised beyond that of the second electron transfer. Such behaviour has been observed during the oxidation of 9-arylaminoanthracenes, although there is a chemical reaction by which the dication is irreversibly removed at moderate rates (138). With phthalocyanine systems it was shown (139), that the ESR signal decreases upon reduction at potentials of the second wave, but increases again after current interruption because the reaction of equation (5) takes place. During exhaustive electrolysis with external techniques at such potentials, anion radicals will be the product until the parent molecules have almost been consumed. Hence, maintenance of electrode potentials is not so crucial under these conditions. The situation is, of course, quite different if rapid irreversible reactions of the dianion take place.

ESR has occasionally been combined with other spectroscopic techniques in the investigation of electrochemically generated radicals. Thus, electronic absorption spectra developing parallel with ESR spectra could be attributed to the corresponding radicals during electrochemical studies in the violene series (140) and with anthraquinone (141).

III. CORRELATIONS OF EXPERIMENTAL DATA WITH THEORETICAL CALCULATIONS

In this review, though devoted to the joint application of electrochemical and ESR techniques, no distinction has been made in the selection of papers dealt with, as to whether the particular work considered the electrochemical or ESR aspects or both. Much work has been done to evaluate electrochemical mechanisms, and the ESR technique merely served for the confirmation that radicals are involved and/or for their identification. On the other hand, the electrochemical generation of radicals has rather frequently been applied for the purpose of ESR investigations without particular consideration of the electrochemical aspects; this applied to numerous works in which HFS coupling constants have been correlated with calculations of spin densities from molecular orbital (MO) theory.

While more detailed information about the applicability of theoretical methods can be achieved from ESR data, electrochemical measurements—especially those of reversible reduction–oxidation potentials—nevertheless are also correlated to MO calculations. Several authors have, therefore, simultaneously compared ESR and electrochemical data with MO theory; based on such theories, even immediate correlations of electrochemical potentials and ESR data have been realized, cf. below.

The most valuable information comes from the ESR hyperfine coupling constants. While the number and relative intensities as well as the relative positions of HFS lines are connected with the molecular framework (e.g. the number of equivalent H and N atoms), the absolute values of coupling constants reflect the electron distribution in the paramagnetic molecule.

Without going into details, the correlation of HFS coupling constants with electron distributions should be mentioned in a few words. Consider the two ion radicals derived from the benzene molecule, $C_6H_6^+$ and $C_6H_6^-$, with five and seven electrons, respectively, in the π system. One sixth of the unpaired electron spread over the ring can be attributed to each of the carbon atoms: the 'spin density' at these atoms is $\frac{1}{6}$. (Adopting the practice in the literature, the term 'density' is used here, although strictly speaking it should be replaced by 'population', because density would have the dimension volume^{-1}, whereas the quantity considered here means an integration of spin densities of orbitals centred at the respective atom.) While the π-electronic wave functions would reveal a node in the plane of the molecule, thus excluding an interaction with the nuclear spins, the unpaired electron serves for a certain polarization of the σ bond C—H, thus effecting a certain spin density at the proton. Hence, each proton gives rise to a hyperfine splitting, according to McConnell's equation (142)

$$a_{H_i} = Q_{CH}\rho_i \qquad (6)$$

where a_{H_i} is the coupling constant, ρ_i the spin density at carbon atom 'i' and Q_{CH} is the σ–π parameter amounting to about -23 gauss. (The negative sign results from a detailed consideration of the polarization mechanism; since only absolute values are reflected in the HFS spectrum, the sign is irrelevant unless a more detailed analysis—for instance of linewidth variations within a particular spectrum—is carried out.) The relatively low value of this constant, as compared with a splitting of about 510 gauss observed with free hydrogen atoms, is due to the restricted interaction by means of the polarization mechanism.

Whereas the six carbon atoms (and hence, the six protons) of the simple benzene radicals are all equivalent and spin densities ρ_i can be predicted without calculation, this is not the case with other hydrocarbons, e.g. naphthalene or substituted benzenes, in which spin densities differ according to the position, though sets of two or more equivalent protons may occur according to the molecular symmetry. With the help of equation (6), spin densities calculated from MO theories—the Hückel MO method and McLachlan's modification have been applied in most cases—can therefore be compared with experimental HFS data. (It should be noted that even negative spin densities may occur; this is the reason for more significant deviations of the overall splitting—the distance between the outermost high- and low-field lines—from the value of about 23 gauss.)

Apart from ring protons those in alkyl groups attached to the ring will also effect a splitting. Thus, a parameter Q_{CCH_3}, inserted in equation (6) instead of Q_{CH}, would correlate the corresponding coupling constant of methyl protons of toluene radicals with the spin density at the carbon atom to which the methyl group is attached.

Next to protons, the nitrogen nucleus ^{14}N is most important in organic molecules. Since the nuclear spin quantum number is 1, a single nitrogen atom gives rise to three equidistant lines of equal intensities, corresponding to quantum states $-1, 0$ and $+1$. Relations like equation (6) are more complex in this case and differ for different bonding situations. This is due to the fact that not only a polarization of bonding σ electrons, as is the case with protons, must be taken into account, but also an analogous polarization of non-bonding electrons including a lone pair of electrons at the nitrogen atom. Coupling constants are, therefore, dependent on the spin densities at the nitrogen atom and at the atoms to which it is bonded. For a nitrile, $-C\equiv N$, the coupling constant is related by

$$a_N = Q_N\rho_N + Q_{CN}\rho_C \tag{7}$$

to the spin densities ρ_N and ρ_C at both the nitrogen and carbon atoms of the CN group. The corresponding relation for a nitrogen atom within a heterocyclic system

would be

$$a_N = Q_N\rho_N + Q_{CN}(\rho_{C_1} + \rho_{C_2}) \tag{8}$$

ρ_{C_1} and ρ_{C_2} being the spin densities at the two carbon atoms attached to the nitrogen atom. The parameters Q_N and Q_{CN}, however, differ from those in equation (7), Q_{CN} in equation (8) in general being suggested to be small as compared with Q_N thus leading to an approximate relation

$$a_N \simeq Q_N \rho_N \qquad (9)$$

Finally for nitro compounds, the nitrogen coupling constant depends upon three kinds of spin densities

$$a_N = Q_N \rho_N + Q_{CN} \rho_C + 2 Q_{ON} \rho_O \qquad (10)$$

(the factor 2 of the last term indicates the equivalency of both oxygen atoms); again Q_{CN} is suggested to be sufficiently small to be neglected.

Whereas spin densities are calculated from MO theories, there are no well-founded quantum chemical relations for an evaluation of the Q parameters. These parameters, therefore, have frequently been determined by calculating spin densities and measuring HFS coupling constants for a variety of analogous radicals (e.g. a series of substituted nitrobenzenes), and by evaluating those Q values which give the best fit of experimental and theoretical data.

Such a procedure has often been carried out with electrochemically generated radicals. Thus, from measurements with numerous substituted nitrobenzene anion radicals (143), it has been concluded that

$$Q_N = \pm(99 \cdot 0 \pm 10 \cdot 2) \text{ gauss}$$

$$Q_{ON} = \pm(35 \cdot 8 \pm 5 \cdot 9) \text{ gauss}$$

in equation (9), whereas the term with Q_{CN} is sufficiently small to be neglected. (The signs of Q_N and Q_{ON} were not known from these investigations, but they must be opposite.) The value of Q_{CH} in equation (6) was found to depend upon the bond angle at the respective carbon atom (144); for azulenes these bond angles differ from the value of 120° valid for benzenoid aromatics, and from one another. Hence, different values of Q_{CH} apply to the different groups of protons, whereas Q_{CCH_3} for methyl substituents does not depend upon the position. Qualitative agreement with theoretical calculations has been established.

In an investigation of the anion radical derived from 1,6-methano{10}-annulene

rather low coupling constants a_H have been observed for the ring protons except those of the methylene group (145). This is attributed to the non-planarity of the π system. Since the coupling constants a_{13C}, however, are very similar to those observed with naphthalene—suggesting a similar spin distribution among the perimeter C atoms—the σ–π polarization seems to be quite different from planar aromatics suggesting the inapplicability of McConnell's equation (5) in such cases.

A value of $Q_{CF} = -47.5$ gauss for splitting due to a fluorine nucleus, analogous to equation (6), has been estimated from a comparison of a_F in p-fluoronitrobenzene with $a_{H(para)}$ in nitrobenzene anion radicals (63). Further work on fluorine coupling constants has been done with ketones (146, 147) and triphenylmethyl derivatives (148). The parameter Q_{OC} for oxygen coupling constants has been determined with the help of ^{17}O enriched quinones (149).

As has been pointed out above, the parameter Q_{CN} in equation (8) is of minor importance for heterocyclic nitrogen splitting. Consequently, there is some disagreement in the literature as to the appropriate value. Thus, values of $Q_N = +27.3$ gauss and $Q_{CN} = -1.7$ gauss (150) as well as $Q_N = +19.3$ and $Q_{CN} = +9.1$ gauss (101) have been suggested by different authors; further values given for Q_{CN} were -2 ± 2 (151) and $0 \leq Q_{CN} \leq 2.5$ gauss (152). While all of these values were obtained with anion radicals, $+2.62$ gauss was found with dihydropyrazine cation radicals (153); the difference of Q_{CN} for anion and cation radicals, respectively, may be attributed to the different charges (152). For further determinations of Q parameters of heterocyclics cf. reference 154. Moreover, the nitrogen coupling constant is affected if the lone pair of electrons is replaced by an N—H bond (151).

From splittings due to ^{13}C atoms, corresponding parameters have also been evaluated for carbon atoms in heterocyclic ring systems (151). Further investigations were related to the π bonding C—Si and C—Ge, silicon being found to be more electropositive than germanium in agreement with polarographic half-wave potentials (155).

Some investigations were related to the 'pairing theorem' predicting identical spin densities at corresponding positions in positive and negative ions derived from the same alternant hydrocarbon. To account for different proton splitting constants, a relation

$$a_{H_i} = -\{Q_{CH}(O) + K \cdot \varepsilon_i\}\rho_i = -\{Q_{CH}(O) \cdot \rho_i \pm K \cdot \rho_i^2\} \tag{11}$$

has been proposed (156) instead of equation (6), where ε_i is the excess charge at the carbon atom, $Q_{CH}(O)$ the parameter for zero charge and K a constant; the positive sign within the last brackets is for the cation and vice versa. In some cases, both positive and negative ion radicals have been generated electrochemically; this applies to the investigation of 9,10-diphenylanthracene (102). Further studies were done with the positive ion generated in sulphuric acid, including correlations of ^{13}C splittings with spin densities and corresponding relations for a_C (118). The pairing theorem, valid for alternant hydrocarbons, cannot be applied to other types of compounds.

Thus, quite different ESR spectra have been observed with negative and positive ions of nitrosobenzene, not accountable for by relations analogous to equation (11). The anion (131, 135, 157, 158) shows a nitrogen splitting of 8 to 11 gauss, according to the solvent, and—at sufficient resolution (157, 158)—proton splittings for five non-equivalent protons (cf. Section V), whereas the cation radical (159) reveals a nitrogen splitting constant of 37 gauss and essentially only one proton splitting due to a single (*meta*) proton. This has been explained (159) by the fact that the anion is a π radical, whereas the cation is a σ radical with localization of the odd electron at the nitroso group.

The radical obtained during oxidation of *N*-(*p*-nitrosophenyl-) *N*-phenyl-hydroxylamine (I) was shown to be the π-radical (II) and not the σ radical (III) from the nitrogen coupling constants ($a_{N_1} = 9\cdot14$ and $a_{N_2} = 1\cdot36$ gauss), because one considerably larger coupling constant (27 to 55 gauss) would be expected for the σ radical with the odd electron essentially localized at the NO group (160).

(I)   (II)   (III)

Studies of thiophenes and other compounds containing sulphur atoms were concerned with the question whether the conjugation of sulphur atoms with ring systems is due to p or d electrons. While MO calculations based on p and d models, respectively, were both found to be compatible with experimental results of nitrothiophenes (161), only the p model was in agreement with observed splittings in dibenzothiophenes (162). Other authors, however, believe that HFS interpretations are not a sufficiently conclusive criterion for this question (163).

Some further investigations with electrochemically generated radicals or including electrochemical studies of radicals, in which correlations of HFS with MO considerations have been carried out, will be mentioned here without more details being given. These studies were concerned with hydrocarbons like polyphenyls (164), chrysene (125), 1,8-bisdehydro-14-annulene (165), non-alternant hydrocarbons (166), butadiene (109), amino-ethylenes (167), heterocyclic compounds (168–170c), nitro compounds (171–174), ketones (104), unsaturated ketones (175), benzocyclobutadienoquinone (176), dibenzopentaleno (4,8)-quinone (177), furaquinones (178), dipheno-quinones (179), naphthoquinones (180), bicyclic hydrazines (181), vitamin A aldehyde and related Schiff bases (182) and organometal complexes with de-localized electrons (139, 183, 184). (For further work on metal–organic complexes see references (185–188).)

Polarographic half-wave potentials have often been correlated by semi-empirical and theoretical considerations to the electronic structure of the respective molecules. Extensive application has been made of the Hammett–Taft relation to account for the effect of substituents on half-wave potentials. Such correlations were established also in work with simultaneous ESR investigations, e.g. for substituted anthraquinones (189), azomethine compounds (190), p'-substituted p-nitrobiphenyls (191) and for the formation of different aromatic anion radicals (192) in which the pronounced effect of *para* substituents was explained by the increased statistical weight of quinoid structures.

Since, however, experimental data of ESR investigations are frequently correlated with MO calculations, the simultaneous correlation of such calculations with electrochemical data is near at hand, once the radicals are produced by electrochemical methods. Correlations of electrochemical reduction–oxidation potentials (especially polarographic half-wave potentials) with Hückel MO calculations and refined methods have been developed independent from ESR work, cf. the literature given in references 64, 65, 193–195. These correlations are based upon the fact that an immediate relation must exist between the electrochemical potentials, $E_{1/2}$—representing energies if given in eV—at which an electron is transferred to ($R \longrightarrow R^-$) or withdrawn from a molecule ($R \longrightarrow R^+$), and the π-electron energies, E_π, of the lowest vacant (LVMO) and highest occupied molecular orbitals (HOMO), respectively:

$$E_{1/2} = -E_\pi + \text{constant} \tag{12}$$

LVMO and HOMO energies, however, are given in quantum chemical calculations by an expression

$$E_\pi = \alpha + \beta x \tag{13}$$

in which α and β are Coulomb and bond integrals, respectively, given in energy units, and x is a dimensionless number characterizing the energy coefficient in units of β. Combining equations (12) and (13), we have

$$E_{1/2} = A - Bx \tag{14}$$

Plots of $E_{1/2}$ versus the coefficient x, calculated from MO theories, therefore reveal straight lines in a fairly good approximation, the slopes B of which are a measure of the integral β.

The validity of such a relationship has been proved for the reduction of various aromatic systems (196), of aromatic nitriles (197) and nitro compounds (143), imides such as phthalimide (198), azomethines (190), methyl-substituted sulphones (199), dibenzofuran (162) and benzoxadiazole (163) as well as the

S and Se isologs of the two latter compounds, and the oxidation of methoxy-benzenes and a corresponding S isolog (200). Similarly, LVMO energies have been calculated from $E_{1/2}$ as well as from ESR data for substituted biphenyls (including germyl and silyl compounds), and satisfactory agreement was established (155). Further studies were carried out on anion radicals of trimethylsilyl- and trimethylgermyl-substituted polycyclics including MO calculations and the determination of heteroatom parameters (201). Again, the participation of p or d electrons of sulphur atoms has been discussed in terms of a comparison of changes of $E_{1/2}$ and of LVMO energies, respectively, for nitrothiophenes; this was in favour of the d-electron model (161).

For substituted nitrobenzenes (63), an immediate correlation between electrochemical and ESR data was given

$$a_N^2 = mE_{1/2} + b \qquad (15)$$

where m and b are constants. While this is only an empirical relation without theoretical background, fitting the data better than a linear relationship, a theoretical relation has been derived for the same type of compounds,

$$(a_{H(ortho)}/a_{H(para)})^{1/2} = (x^2 - 2)/2 \qquad (16)$$

which describes the simultaneous effect of electron-withdrawing substituents on the LVMO energy x and on the spin densities at $ortho$ and $para$ protons, reflected in the corresponding HFS constants $a_{H(ortho)}$ and $a_{H(para)}$, respectively. Combining this relation with equation (14), we have

$$E_{1/2} = -B\sqrt{2}\left[1 - \sqrt{\frac{a_{H(ortho)}}{a_{H(para)}}}\right]^{1/2} + A \qquad (17)$$

Experimental data were shown (61) to fit this equation.

A linear relation between half-wave potentials and nitrogen coupling constants a_N has been observed (62) for amino-substituted semiquinones (IV),

(IV)

$E_{1/2}$ being shifted to more negative values as a_N increases (-0.2 V gauss^{-1}). This was explained in terms of a_N being a measure of the ability of the nitrogen to donate electrons to the ring by resonance which in turn makes the reduction more difficult. Simultaneously, a relationship was established between the coupling constant a_H of the α proton (indicated in IV) and the heterogeneous electron-transfer rate constant the logarithm of which increases with increasing a_H values. This implies a correlation between the spin distribution in the radical and the energy of the electrochemical transition state and may be attributed to an increased stabilization of the radical or to a lowering of the solvation energy of the transition state with increasing delocalization of the charge.

Linear relations between half-wave potentials and coupling constants a_N and a_H have also been observed for various substituted nitrofurans (202); even the plot of $\Delta E_{1/2}$, the difference of half-wave potentials in water and dimethylformamide, respectively, versus Δa_N and Δa_H, the corresponding differences of coupling constants in these solvents, were found to yield straight lines.

Another correlation exists between electrochemical potentials and the g factor. The meaning of this factor should be mentioned here with some simplifications. A free electron with the magnetic momentum β_0 (Bohr magneton) due to its spin will, in an external magnetic field H, occupy one of two quantum states exhibiting an extra energy of $+\beta_0 H$ and $-\beta_0 H$, respectively. Hence, the energy difference of these states is $2\beta_0 H$. This energy corresponds to the energy $h\nu$ of the microwave giving rise to the resonance phenomenon. Due to a relativistic correction, the factor 2 has to be replaced by 2·0023, or more generally

$$h\nu = g\beta_0 H \tag{18}$$

with $g = 2\cdot0023$ for the free electron. The unpaired electron in a free radical has a g value which deviates somewhat from that of the free electron, due to a small spin–orbit interaction. Thus, the significant difference of g values observed with diaminosemidiones and dialkoxysemidiones has been attributed to the difference of spin–orbit coupling constants affected by N and O atoms, respectively (203). The difference Δg is linearly related to MO energy parameters under certain restricting conditions (204). Considering equation (14), we can also expect, therefore, a linear correlation of Δg and $E_{1/2}$. Such a relation, viz.

$$\Delta g = (27\cdot8 \pm 0\cdot5)10^{-5} - (5\cdot6 \pm 0\cdot2)10^{-5}E_{1/2} \tag{19}$$

was shown to hold for a variety of aromatic hydrocarbons (64, 65), $E_{1/2}$ being given in volts (versus SCE).

IV. APPLICATION OF ISOTOPES

Fortunately, the ^{12}C nucleus exhibits a zero magnetic momentum, otherwise most ESR spectra would be too complex to be analysed. (The same is true, to some extent, for the ^{16}O nucleus.) On the other hand, this means a certain lack of information, since no spin density can be determined from the experimental data for a carbon atom to which no proton is attached. Various applications have, therefore, been made of ^{13}C the magnetic quantum number of which is $\frac{1}{2}$ like that of protons. In most cases, use has been made of the natural abundant isotopic concentration (1·1 per cent), though 50 per cent enriched ^{13}C samples of anthracene have been applied for a study supporting the pairing theorem of positive and negative ion radicals mentioned above (118). The effect of alternating linewidths (cf. Section VI) has been used in this case for the assignment to the appropriate C atoms (205). At natural abundance, a fraction of 1·1 per cent of the total number of radicals contains, at a particular position of C atoms, a ^{13}C atom in this position giving rise to a doublet accompanying each main absorption line if there is a finite spin density at this C atom (one line on either side at equal distances, each of them having 1/200 of the intensity of the main line; if there are two C atoms of identical structural position, the intensities are doubled, and so on).

These satellites can, of course, only be observed if they are not masked by another main line. Observation of ^{13}C splitting at natural abundance is, therefore, mostly restricted to radicals the spectra of which are not too involved by ^1H and ^{14}N splittings. As an example, the anion radical of bis (carbodyicyano)cyclobutanedione

$$
\begin{array}{c}
\mathrm{CN} \\[-2pt]
\quad\diagdown \\
\qquad \mathrm{C}=\mathrm{C}-\mathrm{C}=\mathrm{O} \\
\quad\diagup \qquad | \qquad | \\
\mathrm{CN} \\[4pt]
\mathrm{CN} \qquad\quad | \qquad | \\
\quad\diagdown \\
\qquad \mathrm{C}=\mathrm{C}-\mathrm{C}=\mathrm{O} \\
\quad\diagup \\
\mathrm{CN}
\end{array}
$$

exhibits only nine main lines resulting from coupling with four equivalent nitrogen nuclei. At sufficiently high amplification, two sets of satellites are observed corresponding, respectively, to ^{13}C atoms in the CN groups and at the positions attached to the CN groups (assignment by theoretical considerations) (206). For the tetracyanoquinodimethane anion radical,

$$
\left[
\begin{array}{c}
\mathrm{NC}\diagdown \\
\qquad \mathrm{C}{=}\!\!\left\langle\!\!\bigcirc\!\!\right\rangle\!\!{=}\mathrm{C} \\
\mathrm{NC}\diagup
\end{array}
\begin{array}{c}
\diagup\mathrm{CN} \\[10pt]
\diagdown\mathrm{CN}
\end{array}
\right]^{\bar{\ }}
$$

though showing 45 main lines, all four ^{13}C splitting constants (from two doublets and two quartets of equivalent C atoms) could be determined and attributed to the corresponding positions considering: (a) the relative intensities, and (b) MO calculations (207). Further observations of ^{13}C satellites—partly with correlation to MO calculations and corresponding assignment of C atom positions—have been reported for ion radicals of cyclooctatetraene (208), phthalic anhydride (209), pyromellitic dianhydride and imide anhydride (198), p-benzoquinone (210, 211), aromatic nitriles (for C atoms to which the cyano group is attached) (197), tetracyanoethylene (212) and N-heterocyclics (151).

In several anion radicals derived from aromatic nitro compounds, the nitro group is twisted out of the plane of the ring system due to steric effects of substituents. Consequently, coupling with the π electrons of the ring is poor, and hyperfine splitting—except for the nitrogen nucleus—is absent. The spectrum is then very simple and the observation of ^{13}C splittings is facilitated. This was observed with nitrodurene and 2,6-di-t-butylnitro-benzene (92), pentamethylnitrobenzene, diaminonitromesitylene and amino-nitrodurene (61), 2,4,6-tri-t-butyl- and tetra-isopropylnitrobenzene (213) and with pentachloronitrobenzene (214). Similarly, 9-nitrotriptycene—the nitro group not being bonded directly to the rings—resembles the behaviour of aliphatic nitro compounds with the spin being essentially confined to the nitro group; two groups of ^{13}C satellites were observed with an intensity ratio of 1:3, corresponding to the C atom in the 9-position and the three C atoms attached to it, respectively (215).

Assignment to the appropriate C atoms is often based upon the intensities, partly by comparison with the additional ^{15}N satellites. The latter nucleus exhibits a spin of $1/2$ and gives rise to two lines instead of the three lines from ^{14}N. Distinction between ^{15}N and ^{13}C satellites is facilitated by the fact that the positions of ^{15}N lines can be calculated from the coupling constants of ^{14}N and the ratio of nuclear g factors of both N nuclei ($g_{^{14}N} = 0.4036$, $g_{^{15}N} = -0.5661$; hence $|a_{^{15}N}/a_{^{14}N}| = 1.4$). The concentration of ^{15}N in natural abundance is 0.365 per cent. Each of the two lines, therefore, gives rise to an intensity of 0.55 per cent of that of the three main lines due to ^{14}N. Comparable intensities of ^{13}C and ^{15}N satellites, therefore, indicate splitting by a single C atom.

Deuterium (nuclear spin quantum number 1) gives rise to three lines instead of two from protons. The nuclear g value, however, is much smaller, and $a_D : a_H = 0.1535$. Splitting due to ^{2}H is, therefore, more likely to be unresolved. Due to the small natural concentration (0.016 per cent), only deuterium-enriched samples allow, in general, for the observation of the appropriate splitting. Samples deuterated at definite positions, however, may serve as a valuable means if the assignment of proton coupling is uncertain. The ordinary azobenzene anion radical exhibits more than 180 lines and an analysis is almost impossible, whereas the perdeuterated com-pound gives only five lines (from the two equivalent N atoms, ^{2}H splitting being unresolved) and allows for the determination of a_N (216). Similarly, an analysis for the ordinary tetramethylbenzidine cation radicals (in which N splitting is not resolved) is rather difficult because of the large number of lines, whereas the compound deuterated in both rings shows only the 13 lines due to the 12 equivalent protons of the methyl groups (217). Assignment of the two splitting constants for *p*-phenylenediamine cation radicals became possible by the application of the 2,3,5,6- and N,N,N',N'-tetradeutero compounds, respectively (218). The cation radical of tetrakis (dimethyl-amino)ethylene shows approximately 300 of the possible total of 1521 lines. An assignment of coupling constants is facilitated by application of the perdeuterated product from the spectrum of which the a_N value for the four equivalent nitrogen atoms can easily be obtained. This allowed for the evaluation of two a_H coupling constants, each for twelve equivalent protons, for the non-deuterated compound. The existence of two different kinds of protons is due to steric hindrance of rotation about the $C—N(CH_3)_2$ bond (219). Further application of deuterated samples in order to facilitate assignment of coupling constants has been made with radicals derived from azulene and trimethylazulene (144), phthalaldehyde and terephthal-aldehyde (220), N,N-dialkylthiobenzamides (221), non-cyclic enones (222), 4,4'-dinitrothiobenzophenone (223), triphenylphosphine and -phosphine-oxide (224) and polyazines (151).

In diaminodurene and N,N'-dihydropyrazine cation radicals, deuteration at the nitrogen atoms caused a shift in both the nitrogen and ring proton coupling constants, the a_D values simultaneously differing significantly from what would be expected as compared with the a_H values in the non-deuterated compounds. A vibrational mechanism has been suggested as a possible source of these effects (225).

Assignment of proton coupling constants in anion radicals generated by electrolysis of naphthoquinones and naphthazarines in dimethylformamide was facilitated by the addition of some D_2O. Moreover, an exchange of protons attached to C atoms has been proved in this way (226). This exchange is most likely due to an enolization reaction, viz.

$$(20)$$

Thus, the multiline spectrum of anion radicals derived from 2,7-dihydroxy-naphthazarine showing splitting by all protons, changes to a three line spectrum with intensities 1:2:1 immediately after addition of some D_2O, due to the exchange of all hydroxyl protons. After some time, the intensities of both outer lines decrease in favour of the centre line, since even the two ring protons are slowly exchanged by the mechanism shown above. (Deuteron splitting is not resolved.)

Some work (149, 227) has been done with ^{17}O-enriched quinones in organic solvents as well as in water, in order to determine coupling constants a_O, spin densities at O atoms and the σ–π parameters Q_{OC} and Q_{CO} according to the relation

$$a_O = Q_{OC} \cdot \rho_O + Q_{CO} \cdot \rho_C \qquad (21)$$

While the nuclear magnetic momentum of ^{16}O is zero, that of ^{17}O is finite with a spin quantum number of $\frac{5}{2}$. Natural abundance (0·037 per cent), however, is too poor for observation of oxygen splitting.

An unusual line-shape has been found with 2-chloro-5-nitropyridine anion radicals, due to the splitting by the two chlorine isotopes both of which have spin quantum number $\frac{3}{2}$ but different nuclear g values (152).

HFS due to titanium isotopes has been observed with organometallic compounds (183). Two of the five natural isotopes (not including the predominant component ^{48}Ti) show non-zero magnetic momenta.

V. CONFORMATION OF RADICALS

Much work has been devoted to the conformation of radicals, because ESR can provide valuable information in this respect. An important source of such information is the fact that ESR can distinguish between equivalent and non-equivalent protons, because non-equivalent protons will only accidentally exhibit equal splitting constants. Moreover, detailed information can arise from comparison with theoretical calculations.

The anion radicals of cyclooctatetraene were shown to have planar conformation, since the ESR spectrum consisted of nine lines with relative intensities corresponding to eight equivalent protons; from electrochemical measurements it was concluded that even the transition state for the first electron transfer approaches the planar conformation, since the charge-transfer coefficient α is considerably smaller than 0·5 (208). Similarly, the ESR spectrum of sym-dibenzcyclooctatetraene anion radicals indicates considerable spin delocalization about the central ring and the two benzene rings; this is also apparent from the relative ease of polarographic reduction indicating strong interaction of the two central double bonds with the aromatic rings (228). On the contrary, the anion radicals of cycloheptatriene (electrolytically generated in liquid ammonia) show an ESR spectrum indicative of a boat conformation like that of the parent molecule (229). Further conformational investigations concerned 1,3-dienes (230).

Hindered rotation of substituent groups at aromatic ring systems has been observed in numerous cases due to steric and/or electronic interaction. Thus, for p-nitrobenzaldehyde anion radicals, it has been concluded from the inequivalency of the two protons *ortho* to the nitro group, that the —CHO group cannot freely rotate (63). Six individual proton coupling constants observed for benzaldehyde as well as for acetophenone anion radicals show that the same is true for these compounds. (Only the methyl protons of acetophenone are equivalent (231).) Hindered rotation has also been found for terephthalaldehyde for which the *trans* rotamer is the more stable one, and for phthalaldehyde anion radicals; in the latter case only the 'meso' form exists from three possible isomers, because it is stabilized by an internal hydrogen bond (220). Benzoic and isonicotinic acid (232) as well as nitrosobenzene anion radicals (154, 157, 158) show ESR spectra indicating the corresponding groups to be locked in positions planar with the ring. Even the nitro group of nitrobenzene anion radicals is locked to the ring plane; however, due to the symmetry of this group, *ortho* and *meta* protons constitute two sets of equivalent protons. Rotation about the N—C bond of alkyl groups in *N*-alkylphthalimide anion radicals was discussed in terms of the α-proton coupling constants; whereas a methyl group can freely rotate, this is not the case with ethyl and larger groups due to steric hindrance and is indicated by smaller coupling constants (209).

The rotation frequency of the COOR group of anion radicals, derived from benzoic and phthalic acid esters, around the bond with the aromatic ring has been determined from an analysis of HFS to be $1.2 \times 10^7 \, \text{sec}^{-1}$ (233). For further studies on carboxylic esters and anhydrides cf. reference 234.

Steric effects are also encountered in the HFS due to the phosphorus atom in *p*-nitrophenylphosphoric esters (235). Hyperfine interaction of methyl protons in methoxy substituted naphthosemiquinones is, in part, a function of the spin density at the ether oxygen, a favoured—but not fixed—conformational structure serving for π-electronic overlap (236). Coupling with methoxy protons is, however, in general by far weaker than with methyl protons under similar conditions. Thus, for 4,4'-dimethoxy- and 4,4'-dimethylbenzophenone anion radicals, the respective constants have been determined to be $a_{H(OCH_3)} = 0.27$ and $a_{H(CH_3)} = 3.64$, respectively (237). The theory of coupling constants for protons at C atoms attached to a π-electronic system has been developed and applied to nitroalkanes (238); rotation about the plane of the π-system is hindered, a barrier separating two conformational structures. The occurrence of two different proton coupling constants, each for twelve equivalent protons, for tetrakis(dimethyl-amino)ethylene cation radicals (219) has already been mentioned above.

Simultaneous occurrence of two conformational isomers has been observed several times by ESR spectra composed from contributions of two kinds of species. From the relative intensities of the respective HFS lines, the equilibrium constants have been determined, assuming equal linewidths. Such behaviour was observed with anion radicals of terephthaldehyde (concentrations of *cis*:*trans* = 1·4; from the linewidth of approximately 0·12 gauss a

lifetime of more than 10^{-5} seconds has been concluded) (239), p-diacetyl-
benzene ($trans:cis = 1.4$) m-cyanoacetophenone (240), and 3-acetylpyridine
($cis:trans = 2.1$; the spectrum is composed of more than 200 lines all of
which have been assigned) (152). Electrolysis of dialkylmaleates and
-fumarates first produced a spectrum A which during five to ten minutes
changed to a spectrum B, both of which were composed from contributions
of two kinds of species. Spectrum A was attributed to cis and $trans$ forms of
the corresponding anion radicals ($cis:trans = 2.0$), whereas spectrum B is
due to the two dealkylated species ($cis:trans = 2.4$) (241). Two conformational isomers were also observed on reduction of tetraisopropylnitrobenzene
due to steric hindrance, with the nitro group being twisted out of the ring
plane at two different angles, reflected in nitrogen coupling constants
$a_N = 23.6$ and 22.0 gauss, respectively. The free enthalpy of interconversion
was determined to be -0.5 kcal/mole at $25°C$ in acetonitrile (213). The
trimethylene chain in the anion radical of the thiathiophthene derivative

exhibits two conformations which are transformed into one another at a
rate of $10^7 \, sec^{-1}$ at room temperature, this rate being sufficiently low at
$-40°C$ to show an ESR spectrum attributable to two separate conformers
(242).

In a comparison of the interaction between the two aromatic rings in
paracyclophanes (V) on the one hand and diphenylethane (VI) on the other,

(V)   (VI)

(including compounds with napthalene and anthracene instead of the
benzene rings), both polarographic half-wave potentials and ESR spectra
of the corresponding anion radicals indicated that a strong spatial overlap
is present in V, while electron exchange between the rings in VI is considerably
slower due to a negligible direct overlap in this case (243).

Twist angles have frequently been determined from HFS data and MO
theoretical calculations. The proton coupling constants of a phenyl group

as a substituent of an aromatic hydrocarbon, for instance, would increase with increasing coplanarity with the hydrocarbon ring system, due to π-electronic interaction. Thus, twist angles of 68° have been determined for the phenyl rings of 9-phenyl- and 9,10-diphenylanthracene anion radicals (119, 244), whereas for 5,6,11,12-tetraphenylnaphthacene (rubrene) the angles were 60° for the cation and 65° for anion radicals (245). An even larger twist angle of about 75° for both the anion and cation radicals as well as for the neutral form of 9,10-dinaphthylanthracene has been attributed to the predominance of steric effects (246). Only one ring system of biazulenyl anion radicals gives rise to HFS coupling, because the lack of π-electronic overlap allows only for a slow spin exchange between the two halves of the molecule (247). The twist angle of the two rings of biphenyl anion radicals has been calculated, accounting for correlations of the corresponding g factor with MO calculations, to be 38°, whereas the anion radical of phenanthrene has planar conformation, with bond angles deviating from 120° (248). The ester groups in o-phthalic acid and o,o'-biphenyldicarboxylic acid diester anion radicals were found to be twisted due to steric hindrance by 35 to 45 and 35 to 60°, respectively, out of the ring planes (249). The carbamoyl group in N,N-dialkylthiobenzamide anion radicals is twisted out of the plane of the aromatic ring, the spin being localized in the aryl ring (221, 250). The ESR spectrum of the cation radical of 9-amino-10-phenylanthracene has been obtained by electrolytic oxidation (251); the coupling constants were discussed in terms of conformational structure with the amino group in the plane of the ring system, whereas the phenyl group is twisted by 55° out of this plane (252). Twist angles between two aromatic ring systems have also been investigated for nitrophenyl-benzenes, -naphthalenes and -anthracenes (253).

Attention has been paid repeatedly to nitro compounds the coupling constants a_N of which reflect the degree of spin localization. Apart from changes due to the solvent, the value of this constant for anion radicals changes from about 10 gauss for nitrobenzene (with considerable delocalization of the spin into the π-electron system of the ring) to about 25 gauss for nitro aliphatics, where the spin is essentially localized at the nitro group. In nitro aromatics with more or less bulky substituents in the *ortho* position, the nitro group is twisted to some extent out of the plane of the benzene ring, thus weakening the π-electron interaction; hence, the a_N value increases. This was shown for 2,6-dimethylnitrobenzene ($a_N = 17.8$) and nitrodurene ($a_N = 20.4$) (254) in acetonitrile, and, with similar values, in dimethylformamide (143), for other methylsubstituted nitrobenzenes (255), 2,6-dihalonitrobenzenes (dichloro in dimethylformamide $a_N = 14.5$; in a mixture of dimethylformamide and water: dichloro 20.8, dibromo 21.4 gauss) (256), pentachloronitrobenzene (20.5 gauss in methanol) (214), and for *ortho-t*-butyl-nitrobenzene (12.73 in dimethylformamide, 14.9 in acetonitrile), whereas other alkyl substituents including isopropyl were ineffective, presumably because the bulky group is oriented for minimum interaction with the nitro group (257). It was shown that twisting of the nitro group out of the plane of the benzene ring simultaneously increases the rate of halide ion elimination upon reduction of halo nitrobenzenes (258). In the 2,4-dinitrobenzoic acid anion radical, the nitro group *ortho* to the carboxyl group is twisted out of the ring plane and does not interact, the ESR spectrum resembling that of *p*-nitrobenzoic acid anion radical; moreover, no reduction of the nitro group in the *ortho* position takes place (259). Twist angles in several methyl and *t*-butyl substituted nitro- and dinitrobenzenes have been correlated with coupling constants as well as with shifts of half-wave potentials (92).

Two rapidly exchanging conformations of the dinitrodurene anion radical, with the first nitro group being twisted out of the plane and the second being planar with the ring and vice versa, are responsible for an alternation of linewidths in the corresponding ESR spectrum (cf. Section VI) (260). Steric conditions favouring hydrogen bonding are responsible for hyperfine coupling with hydroxyl protons in *o*-nitrophenol anion radicals, while no such coupling is observed with the *meta* and *para* isomers (259, 261). Upon reduction of 1-nitro-2,4,6-triphenylbenzene a 'normal' anion radical (A) is first observed which is, however, shortlived ($\tau_{1/2} \sim 12$ sec); within a few minutes the ESR spectrum of a new radical (B) is superimposed to the first and increases at the expense of radical A. The small nitrogen coupling constant of radical B ($a_N = 6\cdot74$ as compared with $a_N = 12\cdot56$ for radical A), has been attributed to an interaction of the nitro group with the *ortho* phenyl group

either by hydrogen bonding or by orbital overlap between the *p* orbitals of the oxygens and the π systems of the two rings (262).

VI. LINEWIDTH EFFECTS

According to simple theoretical considerations, the different lines from which an ESR spectrum is composed would show amplitudes the ratio of which can be expressed by integers, reflecting the statistical weight (degree of degeneration) of nuclear spin combinations. While such behaviour is observed in numerous cases with good approximation, some factors are responsible for deviations from simple amplitude ratios. The physical background of such alterations of amplitudes is better interpreted in terms of changes of linewidths within a particular spectrum. If the population of a particular spin state is strictly proportional to its statistical weight, the area under the absorption curve proper is independent of the shape and width of the line. If the shape of the lines does not change, then the amplitude S (the height from peak to peak) for the generally recorded first derivative is correlated with the linewidth ΔH (difference of field strength between positive and negative peaks of that line) according to

$$S\Delta H^2 = \text{constant} \tag{22}$$

The relation would be more involved, if there is also a change of line shapes; this makes, however, in general only a small contribution (263).

There are two most effective sources of line-width alterations (and, consequently, of amplitude ratios differing from simple values) within a particular spectrum:

1. Anisotropic g tensors and intramolecular electron–nuclear magnetic dipole interactions associated with molecular tumbling within the surroundings of solvent molecules. This effect may be described by an equation relating the linewidth ΔH (or the reciprocal appropriate relaxation time) with the quantum number m of a particular nucleus

$$\Delta H = A + Bm + Cm^2 \tag{23}$$

Hence, a set of lines, differing only by the value of m, shows a continuous variation of linewidths with increasing field strength, superimposed by a broadening of the lines towards the wings.

2. Fluctuations in the isotropic hyperfine splittings induced by intramolecular motions or interactions between the radical and the solvent or non-radical solutes. This effect may cause dramatic deviations of the HFS pattern as compared with that expected from the simple theory. Groups of lines may be much broadened as to show lack of resolution of the components or to be concealed completely from observation. In simple cases, broad and narrow lines alternate, from which the term 'alternating linewidths' has been derived.

For a review of theory and application of effects [1] and [2] cf. reference (36).

The effect described under [1] has repeatedly been observed also with electrochemically generated radicals and especially dealt with for mono-(264, 265) and dinitrobenzenes (266). The amplitudes, for instance, in the spectrum of nitrobenzene anion radicals differ for the three groups of lines: the central and low field groups are approximately of equal intensity, whereas the high field group lines are considerably smaller, reflecting a combined action of the linear and quadratic terms in equation (23), m representing the quantum number of the nitrogen nucleus.

While the theory of effect [1] is rather formalistic and not easily associated with an illustrative model (due to the so far restricted knowledge of molecular motions in liquids), effect [2] is always attributable to a definite molecular process. One of these processes has already been mentioned in the preceding section, referring to the two conformational isomers of nitrodurene anion radical with either nitro group A or B planar with the ring (260, 267). We shall discuss this example at some length, because it may stand for the entire group of processes effecting the phenomenon of alternating linewidths. Since the coupling constant a_N differs for planar and non-planar conformation, we have two different values, a_1 and a_2. If the lifetime of a conformational structure were long, then—neglecting the HFS which results from the protons—we would expect a nine-line spectrum, all lines showing equal intensities (from two non-equivalent nitrogen nuclei), whereas two equivalent N would give a five-line spectrum with amplitudes $1:2:3:2:1$. What is observed, however, is a spectrum composed of five groups with approximate ratios of amplitudes (in dimethylformamide at room temperature) $1:0.67:1.45:0.65:0.73$. This has been attributed to a rapid exchange of conformations of nitro groups A and B. If both nuclei have the same quantum numbers (either -1, -1, or 0, 0 or $+1$, $+1$), then there is no change of the resulting splitting constant ($-a_1-a_2$, or 0, or $+a_1+a_2$) during the conformation jump. If they differ, however, there is a change during the jump of the resulting splitting constant amounting to either $|a_1 - a_2|$ or $|a_1 + a_2|$ associated with a mean splitting constant of $\pm\frac{1}{2}(a_1 + a_2)$ or 0, respectively. Hence, we have to expect five groups of lines: three groups with large amplitude (small line widths and good resolution) at positions $-(a_1 + a_2)$, 0 and $+(a_1 + a_2)$ with respect to the centre, each of statistical weight 1 (equal amplitudes), and three groups (one of which is superimposed to the centre group mentioned before) with small amplitudes (large line widths due to change of the resulting splitting constant with time, and bad resolution) at positions $-\frac{1}{2}(a_1 + a_2)$, 0 and $+\frac{1}{2}(a_1 + a_2)$ with respect to the centre, each of statistical weight two, in accord with the observed spectrum. Since the rate of conformational jump changes with temperature, the linewidths, amplitudes and resolution of the badly resolved groups are temperature-dependent; the nine-group spectrum for very low rates of jump, however, has not been observed even at the lowest temperatures.

Another example for processes causing alternating linewidths is that of m-dinitrobenzene (267–271), where one nitro group is interacting with solvent molecules, whereas the other interacts by ion-pairing with a cation; again these different interactions cause different coupling constants and rapid exchange causes alternating linewidths as explained above. The same mechanism applies to 1,8-dinitronaphthalene anion radicals if generated by alkali metal reduction in dimethoxyethane, whereas the spectrum from electrolytic reduction in dimethylformamide is not perturbed (272). A 'four jump model' has been developed for the discussion of linewidth effects observed for dinitro compounds in solvent mixtures (water/dimethylformamide), where two types of nitro groups and two types of solvent molecules interact (273). A 'three jump mechanism' was suggested for 3-trifluoromethyl-4-nitrophenol (274):

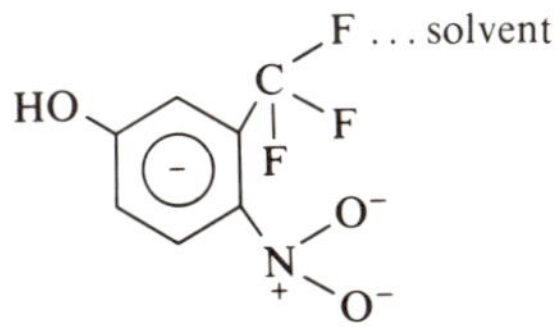

One of the fluorine atoms (nuclear spin quantum number $\frac{1}{2}$; resulting quantum numbers for three equivalent fluorine atoms: $-\frac{3}{2}$, $-\frac{1}{2}$, $+\frac{1}{2}$, $+\frac{3}{2}$) is interacting with the solvent, while the other two interact with the nitro group; due to the rotational jump of the CF_3 group, the lines corresponding to fluorine quantum numbers $\pm\frac{1}{2}$ are broadened. Since the interaction depends on the solvent, the alternating linewidth effect becomes perceptible to a different degree in different solvents. Thus, the broadened lines mentioned above are still observable in acetonitrile, rather weak in dimethylsulphoxide, and absent in dimethylformamide. In addition to the change of linewidths, there is also a significant 'dynamic frequency shift', by which the individual lines are shifted off the positions expected from simple HFS theory (268).

Analyses of linewidth effects have been applied for the determination of the signs of coupling constants and Q parameters (227, 266, 275), of spin densities not obtainable due to the lack of Q parameters (276), and for the assignment of coupling constants, because linewidth variations depend upon spin densities (205).

VII. ION-PAIRING AND SOLVATION

It has already been mentioned in the preceding section, that solvent molecules and/or metal cations (by ion-pairing with anion radicals) may affect the ESR spectra by interaction with the radical species. The linewidth variations mentioned above, however, represent only one of several effects resulting from such interactions.

Interactions due to ion-pairing have been reviewed (35) and the essential effects can be summarized as follows:

1. Since alkali metal nuclei possess a nuclear magnetic momentum (spin quantum number $\frac{3}{2}$ for potassium, sodium and lithium-7), the ESR spectrum of ion pairs may (but must not) show additional HFS with the appropriate coupling constant a_{Me}.

2. The interaction with the cation gives rise to a shift of spin densities within the radical species and a corresponding change of hyperfine splitting constants.

3. Change of the molecular symmetry by rapid fluctuation of cations between different positions of negative charge gives rise to the linewidth variations as described above; this affects also the HFS due to the cation. Additional effects may result from exchange and dissociation reactions.

Since reductions with alkali metals are in most cases performed in solvents of low dielectric constants (and/or low solvating properties), e.g. tetrahydrofuran, the extent of ion-pairing under such conditions is large, and additional HFS by the metal nucleus is observed in numerous cases. The method of electrochemical generation of radicals has, therefore, the advantage of observing the free anion radical, by application of either more suitable solvents or of tetraalkylammonium salts as supporting electrolytes with low tendency for ion-pairing. Large differences of the observed ESR spectra from alkali metal and electrochemical reductions have often been reported. Some examples are given here.

The free anion radical of acenaphthylene was observed by electrochemical reduction, while the ion pair occurs in alkali metal reduction; while in the former case no significant change of HFS is observed upon variation of the solvents, such changes and additional splitting due to a_{Me} are shown when reduction is performed with metals (277). HFS from K or Na was observed during reduction of tetraazanaphthalene with these metals in dimethoxyethane, this splitting being absent when electrochemical reduction is performed in dimethylformamide (278). Similar behaviour was observed with triphenylene, where prolonged reduction with alkali metals gives rise to the ESR active dianion triplet state associated with metal ions (279). Upon reduction of m-dinitrobenzene with sodium or potassium in dimethoxyethane, an ESR spectrum is observed showing HFS from two non-equivalent nitrogen nuclei (with a_N values of 9 to 10 and 0·2 to 0·3 gauss, respectively), whereas reduction with caesium metal gives rise to a spectrum which resembles that obtained upon electrolytic reduction and is interpreted in terms of the line width variation effect described in the preceding section, due to rapid motion of the cation between the two nitro groups, with a resulting coupling constant of 4·66 gauss, attributable to $\frac{1}{2}(a_1 + a_2)$. On cooling the caesium salt solution, however, the caesium ion motion becomes slow, and two unequivalent nitrogen coupling constants are again observed (280). Further information was obtained by a detailed study of the temperature dependence (281). Significant differences of the ESR pattern obtained for phenanthrene upon electrochemical and potassium reductions, respectively, were attributed to a rather small change of one of the proton coupling constants and a somewhat different linewidth (282). Differences of ESR spectra from electrochemical and alkali metal reductions were also observed with terephthalic acid diesters and attributed to ion-pairing in the latter case (249). Sometimes the involved spectra obtained during alkali metal reduction could only be interpreted after an electrochemical generation of the free ion radicals was performed, e.g. with benzophenone ketyls (283). Further investigations in which the results from reduction by alkali metals and by electrolysis have been compared were devoted to anion radicals of methoxyazulene (284), triphenylphosphine and tripenylphosphineoxide (224) as well as orbitally degenerate hydrocarbon free radicals (285).

As was shown by the effect of alternating linewidths, ion-pairing may also affect the ESR spectra of electrolytically generated ion radicals. Whereas the two sites of the cation are equivalent in the case of m-dinitrobenzene, this is not so with 2,6-dimethyl-p-benzosemiquinone anions. Here the cation is preferably situated at the oxygen in 4-position; line-broadening, however, indicates the contribution of the isomer with the cation at oxygen-1 and a rapid exchange of its position (286). The ESR spectra of trinitrobenzene are rather sensitive to the cation of supporting electrolytes; complexes of long lifetimes form in the presence of divalent cations (Ca^{2+}, Sr^{2+}, Ba^{2+}) with a large splitting by one nitrogen nucleus. Occasionally, spectra from both the free anion and the ion pair are superimposed in these cases. From the significantly low g value in the case of the barium complex it was concluded, that the complex is very tight and some leakage of spin into the vacant $6p$ orbital of the barium ion takes place (287). The nitrogen coupling constant a_N of p-chloronitrobenzene anion radicals, generated electrolytically in dimethylformamide, shows considerable dependence on the kind and concentration of the supporting electrolyte cation; a_N increases with increasing concentration and in the series K < Na < Li. (Increasing line-broadening for the lines corresponding to $m_N = \pm 1$ is also observed.) Simultaneously, the second waves of voltammetric curves are shifted to more positive potentials. The effects are due to ion-pairing and are absent, if NR_4^+ is used as supporting electrolyte cation (288). Another ion-pairing effect is observed with benzil in the presence of lithium ions. Whereas in the absence of these ions, reduction proceeds through the intermediate radical stage as indicated by two separated one-electron waves in cyclic voltammetry and the observed ESR spectra, the presence of Li^+ gives rise to a single reversible two-electron reduction wave and the absence of an ESR spectrum, due to rapid association of the dianion with this cation, giving, presumably, $\phi-C(OLi)=C(OLi)=\phi$ (289).

As a result of ion-pairing, the four methylene protons of benzocyclobutene anion radicals (VII) form two sets of equivalent protons at temperatures below $-50°C$ in the presence of alkali metal ions, whereas these four protons

(VII)   (VIII)

become equivalent above $-10°C$. In the intermediate temperature range, an alternating linewidth phenomenon (cf. Section VI) is observed. This was attributed to an oscillation of the metal cation between positions 'above' and 'below' the aromatic ring (VIII), the interconversion frequency being in the range of the ESR time scale (i.e. about 10^6 sec^{-1}, cf. Section IX) within the temperature range mentioned. This interpretation was confirmed by the equivalency of the four protons even at $-90°C$ upon electrolytic reduction with NR_4^+ ions as supporting electrolyte (290).

Similar to the interaction of metal ions with anion radicals, halide ions have also been observed to interact as a result of ion-pairing with metal porphyrine cation radicals, which were shown to exhibit the corresponding halogen HFS components (291).

Beside the rapidly changing ion pair conformation mentioned above, di-nitrobenzene anion radicals also form short-lived complexes with solvent molecules at one of the nitro groups, resulting in a time-dependent increase of the respective a_N value and a corresponding linewidth variation, as has been shown by investigating the temperature dependence (266, 269).

Interactions with solvent molecules have frequently been observed by variations of coupling constants. Such variations were shown to arise from the formation of localized complexes between the solvent and polar substituents or heteroatoms in the radical. The magnitude of the change of proton coupling constants will, therefore, be small as compared with the variations of the coupling constants for the nuclei of many-electron atoms, because the coupling constants of the latter are very critical functions of the spin density (210). Thus, significant changes have been found for the coupling constant of the C-1 atom (from ^{13}C satellites) of semiquinones in dimethylsulphoxide on addition of water, whereas the proton coupling constants do not show marked variations; also the oxygen coupling constant (from ^{17}O enriched samples) varies considerably with the water content of an organic solvent (227). Similarly, there is a marked difference of a_N values for nitrobenzene (143) as well as dinitro- and dicyanonaphthalene anion radicals (292) in dimethylformamide and acetonitrile, respectively, whereas the proton coupling constants are only slightly affected. These variations were attributed to the influence of complex formation on the electronegativity of the oxygen atoms and could be accounted for in MO calculations by a variation of the oxygen Coulomb integral parameter. A redistribution of spin densities within the nitro group has also been considered to be more significant than a withdrawal of the spin from the ring during investigations of trinitrobenzene anion radicals (287). The value of a_N of nitro aromatics is even more significantly changed, if water is applied as a solvent (106, 293), and upon addition of water to dimethylformamide solutions, a_N increases rapidly along with a simultaneous decrease of the wavelength of optical absorption (294, 295). Since there is again no simultaneous decrease of proton coupling constants, the effect cannot be attributed to a twisting of the nitro group out of the plane of the aromatic ring, but is most probably due to hydrogen bonding, the effect of alcohols being smaller than that of water due to the weaker hydrogen bonding. Since there is no such change of a_N with nitroaliphatic anion radicals in different solvents, a change of geometry could not be excluded, regarding the free rotation of the nitro group about the C—N bond in nitro aliphatics as compared with nitro aromatics (294). The ESR spectrum of nitrobenzene anion radicals in liquid $LiNO_3, NH_3$ resembles that in water rather than in aprotic solvents, suggesting a similar solvation, though the stability is considerably enhanced as compared with water (296). Quantitative formulations based on a_N values have been given for the equilibrium of anion radical/water and anion/radical dimethylformamide complexes for nitrophenol anion radicals in mixtures of these solvents (297). At low temperatures, superimposed spectra of both complexes were observed due to the low rate of exchange of solvent molecules (297, 298). Linewidth alternations observed with anion radicals of dinitrobenzenes and -naphthalenes have been attributed to an asymmetric solvation (271), cf. Section VI. In some cases, there is a considerable increase of resolution upon addition of water to organic solvents, as has been shown with fluorenone anion radicals in dimethylformamide (104). The ESR spectra of semiquinone anion radicals (with

$X = H, Cl, Br, I)$ in dimethylsulphoxide exhibit a positive or negative shift, respectively, of the g factor (see Section III) according to the nature of X upon the addition of water. This shift results from hydration via hydrogen bonding at the oxygen atoms with a corresponding decrease of spin density at the O atoms in favour of the ring and the X atoms; these results have been accounted for (299) by the appropriate theoretical considerations (210).

VIII. ELUCIDATION OF MECHANISMS

Under suitable experimental conditions (low concentrations of radicals and parent molecules, say $\leq 10^{-3}$ M, absence of oxygen, etc.), ESR spectra of organic radicals show in most cases fairly narrow lines and good resolution of HFS. Then it is in general possible to obtain rather detailed information with respect to the molecular structure of the radical, especially about the number of protons and nitrogen atoms (as far as they interact with the electronic spin) and about how many of them are structurally equivalent. Such an analysis may be rather tedious and require some experience if the number of such atoms is large and if there are several sets of equivalent atoms, since the amplitude of the individual lines will then differ largely and small lines will tend to be masked by large ones or even be missing due to restricted sensitivity. However, together with chemical considerations about what type of radicals may be expected, there is a good chance of recognizing the species, beyond the mere statement that a radical is present at all. This possibility can turn out to be rather helpful in the elucidation of mechanisms of electrochemical and subsequent chemical reactions. A variety of examples will be presented here.

One question of some interest which can be answered by ESR investigations, concerns the participation of protons in electrochemical reduction and oxidation (cf. also reference 300). Thus, it was shown that the reduction of phenazine in an aprotic solvent yields the unprotonized anion radical, whereas the oxidation of dihydrophenazine as well as the reduction of phenazinium dications give rise to the phenazylium cation with one proton at each of the nitrogen atoms (103). Upon oxidation of 9-arylamino-anthracene, a stable cation radical was obtained, if the molecule was substituted with phenyl in the 10-position; from the ESR spectrum it was shown that the radical retained the proton at the nitrogen atom (301). During reduction of 9,9'-hydrazoacridine in acetonitrile, coulometry as well as the determination of liberated acid protons showed, that the uncharged radical was generated:

$$MH_2 \rightarrow MH + H^+ + e^- \tag{24}$$

This was in accord with the ESR activity of the isolated brown-black crystals consisting of the uncharged radicals. On the contrary, the anion radical was formed upon oxidation in alkaline methanol (302):

$$MH_2 \rightarrow M^- + 2H^+ + e^- \tag{25}$$

On oxidation of tri-t-butylaniline in nitromethane or acetonitrile, the cation radical $R-NH_2^+$ was formed, whereas in acetonitrile in the presence of a base (diphenylguanidine), the uncharged radical $R-NH$ was observed (303, 304). Oxidation of p-phenylenediamine in acidic aqueous solution, though showing a two-electron oxidation wave, resulted in the cation radical, presumably resulting from the reactions (107, 305) shown in reaction (26).

$$
\underset{\underset{NH_2}{\overset{\overset{+}{N}H_3}{\bigcirc}}}{} \;\rightleftharpoons\; \underset{\underset{NH}{\overset{\overset{+}{N}H_2}{\bigcirc}}}{} +2\,e^- + 2\,H^+ \quad \underset{\underset{NH_2}{\overset{\overset{+}{N}H_3}{\bigcirc}}}{} + \underset{\underset{NH}{\overset{NH_2}{\bigcirc}}}{} \;\rightleftharpoons\; 2\,\underset{\underset{NH_2}{\overset{NH_2}{\bigodot}}}{} \tag{26}
$$

No radical was observed on oxidation of the *ortho* and *meta* isomers (nor from the *para* compound if the *ortho* and/or *meta* isomers were present; a chemical reaction between the isomers giving rise to an oxidizable product was suggested to take place. Oxidation of di-*p*-anisylamine in nitromethane resulted in the corresponding cation radical, $R_2—NH^+$, bearing an easily exchangeable proton as shown by the modifications observed on addition of heavy water; this result contributed to the understanding of the reaction mechanism upon dissolution of nitroxide radicals in acidic media (306). Phthalimide exhibits a reversible one-electron wave in dimethylformamide, and the ESR spectrum of the corresponding anion radical was observed (198). In this work (cf. also reference 307), an ESR spectrum reported earlier (308) which was composed of contributions from more than one radical, was interpreted in terms of proton rearrangement within the anion radical (with the participation of an unreduced depolarizer molecule as a proton donor):

$$
\underset{CO}{\overset{CO}{\bigcirc\!\!>\!\!NH}} \;\xrightarrow{+e^-}\; \underset{CO}{\overset{CO}{\bigcirc\!\!\ominus\!\!>\!\!NH}} \;\xrightarrow{+H^+}\; \underset{\underset{OH}{C}}{\overset{CO}{\bigcirc\!\!\ominus\!\!>\!\!NH}} \;\xrightarrow{-H^+}\; \underset{\underset{OH}{C}}{\overset{CO}{\bigcirc\!\!\ominus\!\!>\!\!N}} \tag{27}
$$

For further polarographic and ESR investigations on aromatic carbonic and sulphonic amides cf. reference 309.

The intermediate attachment of a proton during successive reduction and oxidation of tetraphenylallene, $\phi_2C{=}C{=}C\phi_2$, according to the sequence

$$
M \xrightarrow{+e^-} M^- \xrightarrow{+H^+} MH \xrightarrow{+e^-} MH^- \tag{28}
$$

$$
MH^- \xrightarrow{-e^-} MH \tag{29}
$$

was established by the corresponding proton HFS (310–312). Similarly, diphenyltetrahydrocinnoline is oxidized reversibly at sufficient positive potentials to the corresponding cinnolinium cation which then constitutes a reversible redox-couple with the corresponding free radical (313):

$$
\underset{\underset{\phi}{\overset{|}{\underset{CH}{\overset{|}{CH_2}}}}}{\overset{\overset{\phi}{\overset{|}{N}}}{\bigcirc\!\!<\!\!{}^{N}_{}NH}} \;\xrightarrow[-2H^+]{-2e}\; \underset{\underset{\phi}{\overset{|}{CH}}}{\overset{\overset{\phi}{\overset{|}{N}}}{\bigcirc\!\!<\!\!{}^{N}_{CH}}} \;\xrightarrow[-H^+]{-2e^-}\; \underset{\phi}{\overset{\overset{\phi}{\overset{|}{N}}}{\bigcirc\!\!<\!\!{}^{+N}_{}}} \;\underset{-e^+}{\overset{+e^-}{\rightleftharpoons}}\; \underset{\phi}{\overset{\overset{\phi}{\overset{|}{N}}}{\bigcirc\!\!<\!\!{}^{N^\cdot}_{}}}
$$

$$\tag{30}$$

Neither during oxidation of triphenylhydrazine in the presence of pyridine as a proton acceptor

$$\phi_2N-NH\phi \xrightarrow{-e^+} (\phi_2N-NH\phi)^+ \xrightarrow{-H^+} (\phi_2NN\phi)\cdot \xrightarrow{-e^-} (\phi_2N=N\phi)^+ \qquad (31)$$

nor during subsequent reoxidation of the triphenyldiazenium cation can the intermediate radical species be observed; however, if sodium carbonate is employed as an (insoluble) proton acceptor, then reoxidation proceeds with the transfer of one electron and the ESR spectrum observed is attributable to the uncharged radical (314). The primary oxidation product, undergoing subsequent chemical reactions, of tetraphenylpyrrole was shown to be the cation radical with the proton, attached to the nitrogen atom, being preserved; the ESR spectrum differed from that of the corresponding un-charged radical (315). Semiquinone anion radicals and neutral semiquinones derived from quinones of the triptycene series could be distinguished according to the proton HFS of the corresponding OH group (316).

While the reversible transfer of one electron is in general associated with the formation of a radical and an irreversible transfer of one electron indicates subsequent stabilization reactions, e.g. dimerization, this is not an entirely conclusive proof. ESR investigations may, therefore, serve for distinguishing between radicals and dimers. Thus, ESR spectra were obtained on reduction of xanthone and anthrone, but not from 2,6-dimethylpyrone-4, and it was concluded that the dimeric pinacolate is formed in the latter case (317). (From the ESR spectrum observed upon the reduction of anthrone it has been concluded that, instead of the anthrone anion radical, the anthra-semiquinone is generated through some subsequent chemical reaction (318).) Similarly, the reduction of compounds $R-CO-PO(R')(R'')$ showed reversible one-electron reduction and an ESR signal (including HFS from protons of the phenyl ring as well as from the P nucleus) if R was aromatic, whereas irreversible reduction and no ESR signal was observed if R was methyl (319). Dimerization was also suggested to account for the weakness of the single-line ESR signal upon reduction of hexafluoracetone with alkali metals, while electrolytic reduction resulted in a strong 7-line spectrum due to the anion radical (320); cf. also reference 321. ESR lines from a second radical species were superimposed on those of the ordinary cation radical during oxidation of 1,4-dimethoxybenzene; this was supposed to be due to dimerization (possibly a dimeric radical) (200). The oxidation of aromatic amines was shown to result in stable cation radicals only, if the *para* position was blocked by substitution (occupation of one *para* position is sufficient in triphenylamine), otherwise dimerization at the *para* position took place to give the benzidine compound (322–324) as had already been observed with *N,N*-dimethylaniline (325) and triphenylmethane dyes (326). (For triphenylmethane derivatives cf. also references 327 and 328.) In some cases, the radical derived from the corresponding benzidine derivative was observed by ESR confirming the proposed mechanism. Solutions of tetracyanoquinodimethane anion radicals in aprotic solvents show one-electron oxidation and reduction waves and exhibit a strong ESR spectrum composed of more than forty lines, whereas the dimer is prevailingly present in aqueous solutions showing only a weak ESR response which, however, can be increased by heating due to the dissociation of the dimer (329). Stable radicals have been observed upon the reduction of some immonium cations, $>\overset{+}{N}=C<$, in acetonitrile, while dimerization proceeded with others depending on the aromaticity and the volume of the groups attached to the C atom (330).

Both 5,10-diaryldihydrophenazines (IX) and tetraarylhydrazines (X) result

(IX)   (X)

from oxidation of diarylamines, $\phi—NH—\phi$; ESR spectra of the cation radicals of both types of dimers have been observed. The cation radical of X was shown to be transformed under certain conditions to that of IX (331). In the course of extensive investigations (170, 332) of systems schematically represented as

$$-X-(\overset{|}{C}=\overset{|}{C})_n-X- \;\overset{-e}{\rightleftharpoons}\; -\dot{X}-(\overset{|}{C}=\overset{|}{C})_n-\overset{+}{X}- \;\overset{-e^-}{\rightleftharpoons}\; -\overset{+}{X}=(\overset{|}{C}-\overset{|}{C})_n=\overset{+}{X}- \qquad (32)$$

where X is a heteroatom (O, S, NR) and $-(C=C)_n-$ is some π-electron transmitting system, electrochemical investigations of the two redox couples as well as ESR investigations of the intermediate cation radicals were included; the cation radical of 2,2′,6,6′-tetraphenyl-4,4′-dipyrylene, was shown

to undergo reversible dimerization in methylene chloride on varying the concentration between 10^{-6} and $2 \times 10^{-4} M$, whereas no indication of dimerization was observed in acetonitrile even at $2 \times 10^{-4} M$ (170b).

Polarographic reduction of diphenylacetylene in dimethylformamide proceeded in a two-electron reduction wave to give diphenylethylene; only a very weak signal was observed at room temperature, whereas the observation of a strong ESR signal (which was identified to be that of the diphenylacetylene anion radical) at very low temperatures showed, that the anion radical was an intermediate in the reduction process (318).

An extensive reversible redox system is established by the indophenoxyl radical (R), resulting from lead dioxide oxidation of the corresponding phenol, and its reduction products, viz. the phenolate anion (R$^-$), the dianion radical (R^{2-}) and the trianion (R^{3-}), respectively:

$$(33)$$

$$(R) \qquad (R^-) \qquad (R^{2-}) \qquad (R^{3-})$$

The reversibility was confirmed by polarographic measurements, while the ESR spectra of both radical species, R and R^{2-}, could be observed. The placement of the dots indicates the different spin distributions which are reflected in the corresponding coupling constants ($a_N(R) = 2 \cdot 25$; $a_N(R^{2-}) = 7 \cdot 3$ gauss) (333).

The presence of water and/or other impurities in aprotic solvents may considerably change the mechanism of electrochemical reductions or oxidations, particularly because of the proton-donating property of such impurities. Favouring of subsequent reactions may decrease the stability of the intermediate radicals and prevent their observation by ESR. Prepurification of the solvent will not always be effective. This is of particular importance if internal generation is applied and the solution is flowing through the sample cell at flow rates sufficient for effecting a stationary concentration distribution. Thus, the SO_2^- anion radical could only be observed in acetonitrile or dimethylformamide at very low flow rates (334) insufficient for a homogeneous distribution but allowing for the generation of more radicals than are necessary for removing the impurity. High flow rates, however, could be applied if the solution, containing the compound under investigation, was preelectrolysed in an external cell for some time prior to being introduced into the sample cell (335). Examples of the interference of water are given in the following.

Oxidation of phenothiazine in acetonitrile was shown to give the cation radical during the first polarographic wave. Water present in the solvent reacts with the cation radical to give the corresponding sulphoxide; a method of preparing the sulphoxide, otherwise difficult to obtain free from the sulphone, was based upon these investigations (336). Whereas the benzopyrene cation radical dimerizes too rapidly in acetonitrile for observing its ESR spectrum, traces of water open another pathway resulting in 6-hydroxybenzopyrene radicals which were observed with ESR (337). The presence of water resulted on the other hand in the instability of the diphenylenedioxide cation radical, which in the absence of water could even be isolated as black-brown crystals consisting of the cation radical perchlorate (338). (*o*-Tolidine (339), perylene (340) and 1,2-dimethoxy-4-propenylbenzene (341) are further examples for compounds the anodic oxidation of which resulted in the production of paramagnetic solids, viz. the corresponding cation radical perchlorates, exhibiting single-line ESR-spectra. The latter applies also to the precipitate formed on one-electron reduction of 1-methyl-4-cyanoquinolinium cations (342).)

p-Benzoquinone was shown to be the product of hydrolysis of the corresponding quinoneimine generated by oxidation of *p*-aminophenol, by observation of the p-benzosemiquinone ESR spectrum (343). Similarly, the corresponding semiquinone was observed upon the oxidation of *p*-hydroxydiphenylamine due to hydrolysis of the intermediate oxime (344). For the electrolytic reduction and ESR investigation of quinoneimines cf. reference 345.

The mechanism of reduction of *p*-nitrosodimethylaniline (with rate-determining release of water from the hydroxylamine stage to give the imine which is reduced to *N,N*-dimethyl-*p*-phenylenediamine) was confirmed by complete reduction to the amine and subsequent oxidation of the latter to give the corresponding cation radical which was observed with ESR (346).

The two-electron wave observed during polarographic reduction of 2-phenylindandione-1,3 in alkaline aqueous solutions was suggested to give a biradical:

$$(34)$$

Since no details of the observed ESR signal were given (347), one may suspect, as an alternative, that the ESR activity is due to the monoradical, which is in equilibrium with the parent molecule and the two-electron product, the latter having more probably a quinoid structure. Biradicals were also suggested to form during the second one-electron polarographic wave of methylene-linked polybenzoquinones; that the transfer of a second electron took place into a ring system other than that which accepted the first electron, was concluded from the values of half-wave potentials. The ESR spectra observed under the conditions of the formation of these biradicals were highly complex—in contrast with the spectrum of the mononegative radical—and did not allow appropriate interpretation (348). On the other hand, ESR spectra of diquinone anion radicals with immediately linked ring systems, indicated delocalization of the electron throughout both rings, in contrast with what had been observed upon chemical reduction of the same diquinone; formation of secondary radicals has, therefore, been suggested to take place in the latter case (349).

Azopyridine-dioxide in dimethylformamide shows two one-electron waves the first of which corresponds to the formation of the anion radical. Protonation causes elimination of the oxygen atoms and formation of azopyridine which again shows two one-electron waves with the corresponding anion radical as the intermediate product; both radical species contribute to the observed ESR spectrum (350). Even three types of ESR spectra have been observed upon reduction at three different potentials of pyrazine-di-N-oxide corresponding to the anion radical of the dioxide, blue monoxide and pyrazine, respectively; equilibrium constants of complexes resulting from hydrogen bonding have been evaluated from HFS data for pyridine and pyrazine-N-oxides (351). No ESR signal is observed at the first reduction wave of benzonitrile N-oxide, $\phi-C\equiv N \rightarrow O$, while the spectrum observed during reduction at the second wave corresponds to the benzonitrile anion radical $(\phi-C\equiv N)^-$ (352).

Elimination of alkyl groups has been observed by ESR during the reduction of dialkylmaleates and -fumarates. While the spectrum of the corresponding anion radical is observed immediately after reduction, elimination of one alkyl group is indicated after 5 to 10 minutes by missing of the corresponding proton HFS; both types of radicals are present in two conformational isomers, cf. Section V (241). The second pair of radicals was suggested to result from dimerization following elimination, and electron transfer to the dimer, while the structure of a third type of radicals observed on reduction at more negative potentials could not unambiguously be elucidated (353). Oxidation of dihyrdodimethylphenazine proceeds in two one-electron waves in acetonitrile, and the cation radical was observed upon oxidation at potentials of the first wave; if oxidation at potentials of the second wave is carried out in the presence of nucleophilic solvents or additives, including water, a methyl group is eliminated, since subsequent reduction gives the monomethylphenazinium cation radical as indicated by ESR (354). Both electrochemical and ESR investigations of tetracyanoethylene and 1,1,2,2-tetracyanocyclopropane showed, that the reduction products of both compounds in dimethylformamide are identical, viz. the tetracyanoethylene anion radical; elimination of the methylene group, therefore, must take place in the latter case (212). The ESR spectrum observed upon reduction of phenyldimethylphosphine, $\phi P(CH_3)_2$, and resulting from the corresponding anion radical was shown to decay rapidly; an ESR spectrum observed previously upon alkali metal reduction of this compound was, therefore, suggested to result from secondary reactions, e.g. elimination of one or both methyl groups (355).

Elimination of alkyl groups was shown to take place during the reduction of aryl alkyl sulphones:

$$\phi-SO_2-CH_3 \xrightarrow[H^+]{2e^-} \phi-SO_2^- + CH_4 \tag{35}$$

Even the presence of proton donors, however, is not sufficient to serve for protonation of the sulphinite anion, because no further reduction takes place in spite of the fact that the corresponding sulphinic acid, $\phi-SO_2H$, exhibits a one-electronic wave at even more positive potentials with the simultaneous occurrence of an ESR spectrum which may be attributed to the corresponding anion radical $\phi-SO_2H^-$ (356).

Elimination reactions have also been observed upon reduction of aromatic nitriles in dimethylformamide. Thus, phthalonitrile anion radicals occur on reduction of this compound at potentials of the first polarographic wave, whereas the ESR spectrum turns into that of benzonitrile anion radicals if the potential is raised to that of the second wave (126):

$$\tag{36}$$

Reduction of p-aminobenzonitrile immediately proceeds with the elimination of NH_2^- and dimerization to give 4,4'-dicyanobiphenyl which is reversibly reducible to its anion radical, the latter being observed with ESR (126):

$$\tag{37}$$

In some cases ESR investigations have given evidence for chemical reactions taking place with traces of oxygen present in the solution. Thus, naphthacene (tetracene) showing the ESR spectrum of the corresponding anion radical R^- on reduction in acetonitrile at -0.8 volts, exhibited a new spectrum on reduction at -2.5 volts which had first been attributed to the trinegative ion R^{3-} (357); actually, however, this spectrum arises from the naphthacenequinone anion radical (358, 359) due to a reaction with oxygen (provided the quinone was not originally present in small quantities). Similarly, dibiphenyleneethylene in dimethylformamide shows—according to the applied potential—the ESR spectra of either its anion radical or that of fluorenone (166). ESR spectra of the anion radical as well as of the corresponding *ortho*-semiquinone system were obtained on reduction of 9,10-dihydrophenanthrene (360). The ESR spectrum observed on reduction of fluorene (and derivatives substituted in the 9-position) was that of the fluorenone anion radical (360); cf. also reference 361 for the latter radical. Similarly, anthraquinone anion radicals result from the reduction of anthrone (362). Upon reduction of *p*-nitrobenzylchloride in aprotic solvents the anion radicals of *p*-nitrobenzaldehyde, *p*-nitrotoluene, 4,4′-dinitrobibenzyl (short-lived) and the dinegative *p*-nitrobenzoate radical were observed, resulting presumably in reactions involving oxygen and proceeding via the primary radical $O_2N-\phi-CH_2\cdot$ (363). Oxygen proper gives rise to a reversible O_2/O_2^- couple in aprotic solvents; the ESR signal of the superoxide ion O_2^- has been observed in pyridine at low temperatures (364).

The elimination of halide ions upon reduction of different types of compounds is a frequently observed phenomenon. Thus, the benzophenone anion radical has been observed upon reduction of *p*-chloro, *p*-bromo and *m*-bromo benzophenones, while the halogen atom was retained on reduction of other substituted benzophenones (365). The concerted elimination of two chloride ions was suggested to take place during reduction of substituted furaquinones (366).

Beyond the elimination of single atoms, fragmentation of molecules upon reduction or oxidation is frequently observed due to a corresponding weakening of bonds. Thus, the ESR spectrum observed on reduction of dibenzoyl methide ions, $\phi-CO-CH-CO-\phi$, originally ascribed to the corresponding dianion radical (367), were shown to result from 1-phenyl-1,2-propanedione anion radicals, $(\phi-CO-CO-CH_3)^-$, while the dibenzoyl methide ion reduction in dimethylsulphoxide solution yielded acetophenone anion radicals (368).

Subsequent chemical reactions have frequently been observed on reduction of nitro compounds. Elimination of halide ions from halonitrobenzenes takes place, as was easily seen from the fact that the ESR spectrum of the nitrobenzene anion radical appeared on reduction of iodo- and *o*-bromonitrobenzenes, whereas a composite spectrum from the anion radicals of both nitrobenzene and the parent compound was observed with *m*-bromonitrobenzene (369). With iodonitrobenzenes, elimination of halide and formation of the anion radical was shown to take place at different potentials, no ESR spectrum being observed on reduction at the first wave:

(38)

In the presence of cyanide ions, reduction of p-iodonitrobenzene resulted in the anion radical of p-nitrobenzonitrile, suggesting the mechanism (370) shown in reaction (39). No elimination was observed with chloro (cf. also

$$I-\langle\bigcirc\rangle-NO_2 \underset{+e^-}{\overset{}{\rightleftarrows}} I-\langle\bigcirc\rangle-NO_2 \longrightarrow \cdot\langle\bigcirc\rangle-NO_2 + I^-$$

(39)

$$\langle\ominus\rangle-NO_2 \rightleftharpoons \langle\bigcirc\rangle-NO_2 \quad NC-\langle\ominus\rangle-NO_2$$

reference 371) and fluoro compounds. The same is true for 2,6-dichloro- and 2,6-dibromonitrobenzenes; instead the dihalo anion radicals were obtained in a 1:1 dimethylformamide/water mixture showing rather large values of a_N due to twisting of the nitro group out of the ring plane because of steric hindrance (cf. Section V) (256). With o-chloronitrobenzene in acetonitrile, the corresponding anion radical was observed on reduction at -1.3 volts, which changed, however, to that of nitrobenzene in the course of 45 minutes, whereas at -1.5 volts the latter species was immediately obtained; for p-bromonitrobenzene it was shown that elimination depends upon the solvent and potential (172, 372). Elimination was also observed with 2-chloro-5-nitropyridine (and similarly with the 2-chloro-3-nitro isomer); the anion radical of the parent compound was observed on reduction at -0.8 volts, whereas that of 3-nitropyridine occurred at -1.45 volts (152). A three-line spectrum was observed with pentachloronitrobenzene in methanol containing sodium methoxide, the large value of a_N indicating steric hindrance of the nitro group; after prolonged electrolysis as well as on reduction of a solution which had been allowed to stand for some days (or was refluxed for one hour) prior to reduction, the spectrum was composed of contributions from two (or even three) three-line spectra, which was attributed to a dehalogenation reaction (presumably an exchange by CH_3O-) taking place even in the parent solution without reduction (214). On reduction of pentafluoronitrobenzene in aprotic solvents, no ESR signal was observed at potentials of the first reduction wave, while a spectrum consisting of three 1:2:1 triplets obtained at considerably more negative potentials might be assigned to coupling with one nitrogen ($a_N = 18$ gauss) and two equivalent fluorine nuclei ($a_F = 1.6$ gauss); whether this spectrum is due to the expected anion radical, however, was questionable (373). Reduction of o-trifluoromethylnitrobenzene at -1 volt gives rise to the anion radical. At -2 volts, however, the ESR spectrum is superimposed by contributions from another radical species which, upon cooling, is the only radical species present; the nature of this species obviously showing no nitrogen coupling is unexplained so far (274).

Decarboxylation was observed with trinitrobenzoic acid showing the ESR spectrum of the trinitrobenzene anion radical, whereas at more negative potentials the dinegative trinitrobenzoate radical was formed. On the contrary, trinitrotoluene exhibited the expected spectrum on reduction at -0.8 volts, whereas at more negative potentials the observed spectrum indicated elimination of a proton to give, presumably (287):

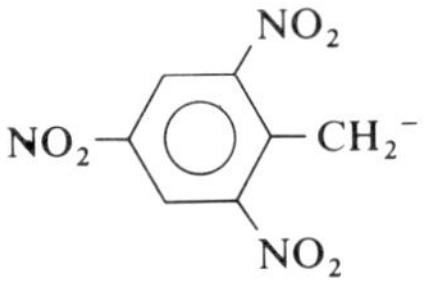

Different spectra have been obtained at different potentials with polynitro-durenes and -mesitylenes; it was shown that at less negative potentials the corresponding anion radicals were found, while at more negative potentials one (or two) of the nitro groups were reduced to amino groups and the ESR spectra were due to the aminonitro radicals (61), in accord with what was observed but not attributed previously (255).

Peculiar behaviour was observed on the reduction of o-nitrobenzoic acid in dimethylformamide; whereas reduction at potentials of the first wave resulted in the ordinary anion radical, the spectrum observed at more negative potentials indicated the presence of a very strongly coupling proton, which was attributed to further reduction beyond the nitroso stage to give (259):

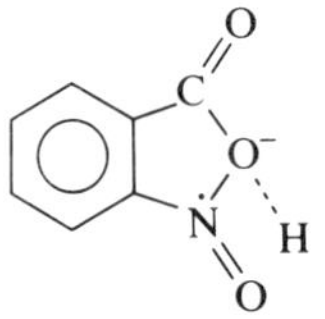

While the ordinary anion radical was observed during electrochemical reduction of o-nitrobenzaldehyde, a spectrum with an HFS rather similar to that just mentioned for the reduction of o-nitrobenzoic acid at more negative potentials was obtained during the photochemically induced rearrangement of o-nitrobenzaldehyde to give o-nitrosobenzoic acid; the structure shown above (and besides one with $-\dot{N}H$ instead of $-\dot{N}(O)H$) has been attributed to the corresponding radical which was suggested not to be an intermediate of the rearrangement but of a side reaction, according to the low radical yield (374).

Elimination of nitrite ions has been observed with aliphatic nitro com-pounds. Thus, reduction of t-nitrobutane first gives rise to the three-line spectrum of the corresponding radical anion ($a_N \simeq 26.5$ gauss) which changes within seconds to a new spectrum on interrupting the electrolysis (or upon prolonged reduction), also consisting of three lines ($a_N \simeq 15.5$ gauss).

$$R-NO_2 \xrightarrow{+e^-} R-NO_2^- \longrightarrow R\cdot + NO_2^- \tag{40}$$

$$R-NO_2^- + 2R\cdot \rightarrow R_2NO\cdot + R-O^- \tag{41}$$

The latter spectrum was attributed to the formation of di-t-butylnitroxide, according to reactions (40) and (41). Di-t-butylnitroxide can be isolated upon reduction of t-nitrobutane as a red liquid showing an ESR spectrum with a linewidth of 8.5 gauss; its solution in acetonitrile exhibits one-electron polarographic reduction and oxidation waves (375). A similar sequence of reactions like that of t-nitrobutane takes place with t-nitrocumene; the corresponding anion radical, however, was too short-lived (about 1 millisecond) for an ESR observation and only the nitroxide was observed (376). Nitrite elimination was also indicated by ESR spectra during reduction of 1,1-dinitro- and 1,1,1-trinitroethane. Whereas the spectrum on reduction of the former compound consists of two components one of which is due to the dianion $CH_3-C(NO_2)_2^{2-}$, the second component—which is identical with that observed on reduction of the trinitro compound—could not clearly be attributed, although there is no doubt about the elimination of NO_2^- (377). Reduction accompanied by C—C, C—Cl and C—NO_2 bond cleavages was observed with several polynitro compounds; the spectrum obtained on reduction of hexanitroethane (a triplet with $a_N = 40$ gauss) is evidently due to some N-oxide, probably $\cdot NO_3^{2-}$, which is also formed during reduction of silver and mercurous nitrates (378). No elimination reaction was observed on reduction of the stable nitroform anion, $(NO_2)_3C^-$, giving $(NO_2)_3C^{2-}$ in alkaline solution, in contrast with the elimination of NO_2^- upon chemical reduction (379). The ordinary anion radicals instead of an elimination reaction have also been observed during the reduction of gem-dinitro alkanes in aprotic solvents (380). For electrolytically generated anion and dianion radicals of aliphatic nitro compounds cf. also reference 381.

Large differences exist between nitrogen coupling constants a_N observed on electrolytic reduction of aliphatic nitro compounds (or chemical reduction in the presence of primary alcohols) on the one hand (23 to 28 gauss), and chemical reduction in the presence of isopropylalcohol (15 to 16 gauss) on the other; this is obviously due to the formation of nitroxide species in the latter case, which probably have the structure (382):

$$\begin{array}{c} R-N-O\cdot \\ | \\ (CH_3)_2C-OH \end{array}$$

Differences between electrochemical and chemical electron transfer reactions were also observed with 4,4'-dinitrobenzophenone in dimethylformamide. Whereas electrolytic reduction generated the anion radical (exhibiting a coupling with two equivalent nitrogen nuclei), chemical treatment with tetrabutylammonium hydroxide resulted in a spectrum which indicated coupling with only one N (probably due to the formation of peroxide type radicals resulting in the elimination of the conjugation between the two rings); after some time the spectrum changed to that of the anion radical (383).

Anion radicals from aliphatic nitro compounds as well as from nitrobenzenes can be generated in aqueous alkaline solutions (106, 293), in accord with the previously observed one-electron polarographic wave (8), whereas at low pH values pH < 6 (293), as well as in aprotic solvents upon addition of proton donors (384), no ESR signals were observed. The lack of observation of the radical species is due to the rapid dismutation reaction induced by the formation of the uncharged radical $R-\overset{\cdot}{N}O_2H$, the kinetics of which has been investigated with the help of the ESR method and served for elucidating the dismutation mechanism (117, 124, 129–132, 385), cf. Section IX. The fact that no ESR spectrum has been observed with nitrobenzoic acid is obviously due to the proton-donating property of the carboxylic group (293); since these protons can, however, only serve for the complete reduction of a fraction of the molecules, the dinegative radical should be observable at more negative potentials.

The lack of HFS due to coupling with ring protons on reduction of the furan derivative

has been suggested to result from some chemical reaction preceding the electroreduction and resulting in the reduction of the olefinic double bond; coupling with N and one proton was only observed (386).

Different hyperfine structures from anion radicals of triphenylhydrazine substituted with the nitro group at different positions, rendered valuable support to the analysis of electrolytic reduction products of dinitrotriphenylhydrazine in aprotic solvents (387).

Finally, some electrochemical and ESR investigations of inorganic systems may briefly be mentioned here. Two different species $B_{10}H_{14}^-$ with different stabilities seem to be formed on reduction of $B_{10}H_{14}$; the difference is possibly caused by a rearrangement of terminating and bonding hydrogen atoms. Stabilization proceeds by disproportionation to give $B_{10}H_{13}^-$ and $B_{10}H_{15}^-$; the ESR signal is broad and does not show hyperfine splitting (388).

Two ESR lines were observed upon the electrolytic reduction of sulphur dioxide in dimethylformamide; whereas for one line the deviation Δg of its g factor from that of the free electron is in agreement with MO calculations of SO_2^-, the second line has been attributed to the species $[(SO_2)_2SO_2]^-$; the dissociation constant of the latter complex ion has been determined (389).

Electrochemical measurements have been employed to study the predominant paramagnetic species present in the $Ti(IV)-H_2O_2$ system frequently used for the generation of free radicals in flow systems; the results support the formulation of this species as $TiOO\cdot{}^{3+}$ (390).

Oxidation of perchlorate as a supporting electrolyte anion was observed in acetonitrile, a four-line spectrum with a coupling constant of about 15 gauss indicating the coupling with a chlorine nucleus; it was first ascribed to ClO_4 radicals (95):

$$ClO_4^- \rightarrow ClO_4 + e^- \qquad (42)$$

On account of the fact, however, that the g value as well as the coupling constant resemble the corresponding values for ClO_2 it was suggested that the spectrum is due to chlorine dioxide (391), cf. also reference 116. Further support for ClO_2 as the observed radical was derived from a more detailed study (392); ClO_2 may be an immediate decomposition product of the unstable ClO_4 radical

$$ClO_4 \rightarrow ClO_2 + O_2 \qquad (43)$$

or result from reactions via Cl_2O_7. The radical ClO_4 has also been discussed to be involved in the anodic oxidation of hydrocarbons (393). The same phenomenon (with $a_{Cl} = 17.5$ gauss) has also been observed in nitromethane as a solvent, whereas no indication for a radical formation was present with nitrate instead of perchlorate as the supporting electrolyte anion, due to an immediate two-electron oxidation to give NO_2^+ and O_2 (394).

On applying a high voltage field across the pure solvent nitrobenzene with the cathode inside the microwave cavity, a well-developed single ESR line has been observed due to the formation of the anion radical; based on this observation, a general mechanism of conductivity was suggested for solvents capable of forming anion or cation radicals, resulting in a high mobility of charge due to the electron exchange with parent molecules, and resembling the mechanism of the proton mobility in water (395).

IX. KINETICS OF RADICAL REACTIONS

There are two types of applications of ESR investigations for studying the kinetics and free-radical reactions. The first makes use of the fact that the lifetimes of spin states are reflected in linewidths. The other mode of application is a more classical one, ESR signals being merely used to follow the change of concentrations of radicals.

As to the response of linewidths to relaxation processes, we must realize that there are several factors, quite dissimilar in nature, which can contribute to linewidths (including, for instance, the inhomogeneity of the magnetic field or magnetic saturation at increased microwave power); critical consideration is, therefore, required if linewidth effects are discussed. Spin–spin relaxation and exchange processes (becoming an essential source of linewidth and shape at increased radical concentrations or in the presence of other paramagnetic species like oxygen) and spin–lattice relaxation effectuated by spin–orbit coupling mechanisms are the main contributions to linewidths. The comparatively low degree of spin–orbit coupling in organic free radicals in solution accompanied by relatively long lifetimes of the electronic spin states is a fundamental prerequisite for well-resolved HFS. Furthermore, good resolution of HFS requires that there is no rapid change of the pattern of nuclear spins with which the odd electron is associated and which is the origin of HFS: a definite resonance condition arises from a particular combination of nuclear spin states and gives rise to a particular HFS line. Now, if this pattern of nuclear spins changes rapidly, the uncertainty of the resonance condition will increase and the lines will broaden. In solutions of elevated radical concentrations, such changes of the nuclear spin pattern can be effected by rapid exchange of electronic spin states between two radicals of different spin orientation; with increasing concentrations the HFS lines broaden and finally collapse to give a single broad line. At still higher concentrations, this single line becomes narrower, due to the averaging of the nuclear interactions. Thus, electrolytically generated beryllium ethyl radicals BeC_2H_5 (stabilized by complexation with one pyridine molecule), which can be isolated as a black substance, show one narrow line in concentrated benzene solutions, whereas a HFS pattern with about 72 resolved lines is observed if a diluted solution in pyridine is prepared (396). A second source for a change of the nuclear spin pattern has already been dealt with in Section VI and was associated with the term 'alternating linewidths'; it was due to rapid changes of conformational structures (cf. also Section V) or solvation states resulting in corresponding changes of the interactions of electronic and nuclear spins.

Another origin of changes of nuclear spin patterns is a rapid exhange of electrons between the radicals and diamagnetic species of the same molecular type; an anion radical may, for instance, undergo charge-transfer reactions with its parent molecule or with the corresponding dianion, respectively:

$$M^- + M \rightleftarrows M + M^- \tag{44}$$

$$M^- + M^{2-} \rightleftarrows M^{2-} + M^- \tag{45}$$

Because the combination of nuclear spins will in general differ for the two species participating in the exchange reaction, the resonance condition will, in general, be different for the radical before and after charge transfer. With the usually applied X band spectrometers (microwave frequency about 9000 MHz, magnetic field about 3300 gauss), a linewidth of, say, 50 milligauss corresponds to a relaxation time of about 10^{-6} seconds. If the lifetime of M^- is decreased by the mechanisms shown above to a value of this order of magnitude, line-broadening may, therefore, become appreciable provided the other sources of linewidth allow for a 'natural' width of about 50 milligauss.

A quantitative relation between the linewidth and the rate constants of reactions (44) or (45) can be derived from a comparison of the characteristic time constants. The lifetime of M^- due to these reactions, τ_{ex}, depends upon the concentration of the interacting diamagnetic species, e.g.

$$(\tau_{ex})^{-1} = k_2[M] \tag{46}$$

where k_2 is the second order rate constant. On the other hand, the time constant, $\tau_{\Delta H}$, associated with a linewidth ΔH (field distance between the peaks of the derivative spectrum; at X band conditions) is

$$(\tau_{\Delta H})^{-1} = 1\cdot54 \times 10^7 . \Delta H \tag{47}$$

(τ in seconds, ΔH in gauss). Comparing the time constants, we have

$$\Delta H = 0\cdot65 \times 10^{-7} . k_2 . [M] \tag{48}$$

(k_2 in $M^{-1} s^{-1}$; [M] in M). On plotting the linewidth versus the concentration of M (or M^{2-}, respectively), the rate constant k_2 can be derived from the slope of the experimental line.

This correlation has been applied to the determination of rate constants for electron exchange reactions of aromatic nitriles (126, 397), anthracene, quinones and nitrobenzenes (398, 399), p-dinitrobenzene (400), duroquinone (401), as well as quinones of vitamin E and K type (402) and binuclear metallic complexes (403). (A difficulty which is encountered if the method of internal generation is applied, arises from some uncertainty as to the concentration of M which, due to the electrode process, is depleted to some extent within the diffusion layer where the concentration of M^- exhibits its maximum value; external generation may be preferable for this reason.)

There are some interesting features which have been observed during such investigations. Thus, the rate constant k_2 of the nitrobenzene system was shown to decrease by about two orders of magnitude upon addition of 10 per cent water to dimethylformamide; this was attributed to the hindrance of the exchange reaction by the strong solvation of the nitro group with water molecules (398). A correlation was established between the nitrogen coupling constant a_N and $\log k_2$ for different nitrobenzenes, reflecting the effect of solvation on both spin densities (cf. Section VII) and rates of exchange; the large value of k_2 for p-dinitrobenzene corresponds to the delocalization of the electron, thus allowing the exchange reaction to escape the decelerating barrier of solvation at the nitro groups (399). A comparison of the rates of homogeneous electron exchange and heterogeneous electron transfer at electrodes, which is of considerable interest in the theory of these elementary steps, was made for some aromatic hydrocarbon reductions; the homogeneous rate constants were evaluated from ESR linewidth variations (404).

The exchange broadening of the HFS lines at further increase of the concentration of the parent compound finally results in a total collapse of HFS, and a single broad line is observed instead. Upon further increase of concentrations, this line is subject to the 'exchange narrowing' effect; this narrowing arises from the fact that, at sufficiently high exchange rates, the electronic spin experiences only an average interaction of the different nuclear spin state combinations. The effect has been observed, for instance, with the benzonitrile/anion radical system (397), and has been analysed quantitatively for the p-dinitrobenzene/anion radical system for the evaluation of exchange rates, the results being in accordance with what was calculated from the broadening effect at low concentrations (400).

The effect of hyperfine line-broadening due to intramolecular conformational or solvation/ion-pairing phenomena and resulting in the broadening only of certain groups of the spectral lines ('alternating linewidths') has been dealt with in Section VI. Since this effect is also correlated with the rate of structural exchange, such rates have been determined from the observation of the line broadening, especially when observed under various conditions (e.g. temperatures); reference is made to the literature cited in that section.

Perturbation of HFS caused by intramolecular electron exchange has been observed with anion radicals of the type

$$\left[O_2N-\hspace{-2pt}\left\langle\bigcirc\right\rangle\hspace{-2pt}-X-\hspace{-2pt}\left\langle\bigcirc\right\rangle\hspace{-2pt}-NO_2 \right]$$

($X = O$, S, CH_2, CH_2CH_2). If a rapid exchange between the left- and right-hand systems (with $X = CH_2$ or with dinitrobiphenyl) takes place, coupling with two equivalent nitrogen nuclei is observed, while slow exchange (for $X = CH_2CH_2$ and, in acetonitrile, also $X = O$ or S) leaves only coupling with one N (and coupling with protons from one ring system). At intermediate exchange rates (some $10^6 \, s^{-1}$), perturbation of HFS due to the different nuclear spin orientation in both ring systems is observed. In the same work, radical formation constants (relative to the precursor and the dinegative ion) have also been calculated from the shape of the polarographic waves on the supposition of overall reversibility (122). For this type of compounds cf. also reference 405.

Kinetic investigations of other types of radical reactions are based upon the proportionality of ESR signal amplitudes and radical concentrations. This proportionality is, in general, observed at low concentrations of radicals (where spin–spin interactions causing variations of lineshapes and linewidths do not interfere) and low, or at least constant, concentrations of species with which electron exchange may occur, cf. equations (44) and (45). The latter condition is frequently difficult to establish with the method of internal generation of radicals. Moreover, difficulties are encountered when determining reaction rates by this method due to an inhomogeneous distribution of concentrations in the vicinity of the electrode and to the perturbation of concentrations upon interrupting the generating current (97) as well as, at prolonged observation times, convective motions (107). Steady-state distribution can be obtained by rapid flow of the solution (112, 334, 335). Instead, an external generation with subsequent transfer of homogeneous solutions of radicals by circulating or other flow systems has frequently been applied to such investigations. In general, decay kinetics of radicals are observed, either by interrupting the electrolysis current if either the internal method or circulating systems are applied, or by stopping the flow of the solution through the microwave cavity. Time constants ranging from some tenths of a second to several hours have been determined in such ways.

Rate constants and reaction orders have been determined for anion radicals of polycyclic and heterocyclic compounds (113). Anthrasemiquinone exhibited a half lifetime of about 75 minutes (406). While the acetophenone anion radical decayed with a time constant of 20 minutes in dimethyl-formamide, the benzaldehyde anion radical was considerably more unstable; cyclic voltammetry revealed an irreversible reaction at 0·3 c.p.s., but reversible conditions at 30 c.p.s., thus indicating a time constant of some seconds (231). The rate of oxygen exchange between p-benzosemiquinone and water in acetonitrile (with 0·9 M water) has been investigated applying ^{17}O enriched benzoquinone and making use of the ^{17}O HFS decay after current inter-ruption; the rate constant was shown to be 0·192 hours^{-1}. The ^{17}O multiplets appeared again upon repeated electrolysis of the quinone solution, indicating that the exchange with the quinone molecule proceeds at a considerably slower rate (227). Decay kinetics have been observed of free-radical inter-mediates in the oxidation of 1-phenyl-3-pyrazolidone (407). The decay of radicals derived from camphorquinone was studied in aqueous alkaline solution and at lower pH values; from a simultaneous observation of polarographic waves due to the radical and to the parent molecule, respec-tively, it was concluded that the decay of the former was not associated with an increase of the concentration of the latter, thus indicating that the decay reaction did not proceed by dismutation (124). The stability of cation radicals of polycyclic hydrocarbons was shown to depend upon whether or not the molecules were substituted at positions of high electron density (408). Annelation of the quinone part of the molecules of various triptycene type quinones was found to decrease the stability of the corresponding semi-quinone and simultaneously render the polarographic reduction more difficult (409). The decay of radicals derived from the azo-/hydrazo-acridine system proceeds faster in dimethylsulphoxide solution than with the isolated solid radicals (302). Rates of the nucleophilic substitution reaction taking place between organic halides and the SO_2^- anion radical have been determined from the decrease, in the presence of benzyl bromide, of the ESR signal of SO_2^- generated by the internal method at rapid flow of the solution (335). The anodic dissolution and cathodic deposition of copper in aqueous solutions of complexing agents (ethanolamine, polyethylene-polyamine) has been followed by observing the copper HFS (410). Kinetics of electron transfer from electrochemically generated metallic complex ions to acceptor molecules have been determined by observing the increase of ESR signals due to the anion radicals of the acceptor molecules (403). Anion radicals of anthracene and benzophenone in dimethylformamide exhibited increased instability upon addition of water, whereas no decay of anthraquinone anion radicals was observed up to 50 per cent of water; the decay kinetics were compared with the simultaneous increase of the first polarographic waves due to reduction beyond the first electron transfer (128) resembling the behaviour of nitrobenzene anion radicals (123).

Decay kinetics of nitroaromatic anion radicals have been investigated by several authors. In aprotic solvents, these species are frequently very stable. Some substituted nitrobenzenes in dimethylformamide exhibited slow first order decay (264). The half lifetimes of both conformers (cf. Section V) from tetra-isopropylnitrobenzene in acetonitrile were 14 minutes (213), while those of some 4-nitrophenyl phosphoric esters (pesticides) were between 3 and 19 minutes in this solvent (411). Nitrofuran anion radicals in aqueous solution exhibited lifetimes of one second or even less (100, 173), whereas the stability was considerably increased with increasing content of ethanol in the solvent (412). ESR investigations of decay kinetics of anion radicals derived from nitrobenzene and substituted nitrobenzenes were helpful in elucidating the mechanism of stabilization. This reaction, proceeding by dismutation as suggested (413) and confirmed (123) previously, is rather slow in aqueous alkaline solution, whereas at lower pH values the rate increases (123, 124). While the alkaline reaction is induced by direct electron transfer between two anion radicals,

$$2\,R-NO_2^- \rightarrow R-NO_2 + R-NO_2^{2-} \tag{49}$$

presumably followed by

$$R-NO_2^{2-} + H_2O \rightarrow R-NO + 2\,OH^- \tag{50}$$

and immediate reaction of nitrosobenzene with anion radicals to give phenylhydroxylamine and nitrobenzene, the reaction at lower pH values is preceded by proton transfer

$$R-NO_2^- + H^+ \rightleftharpoons R-NO_2H \tag{51}$$

$$R-NO_2^- + R-NO_2H \rightarrow R-NO_2 + R-NO + OH^- \tag{52}$$

Bimolecular rate constants of reactions (49) and (52) were about $1\ M^{-1}\,s^{-1}$ and $1 \cdot 7 \times 10^8\ M^{-1}\,s^{-1}$, respectively (123, 414). Rates at different pH values have been determined with different techniques. At low pH values, down to pH 6 where the first half lifetime was about 0·01 seconds, a special technique has been applied including a flow system according to Hartridge and Roughton (131, 414), cf. Section II. Second-order kinetics as observed in these investigations have also been shown to govern decay reactions of nitrophenols in alkaline solutions (127) and of nitrofuran derivatives, the anion radicals of which exhibit considerable lifetimes even in slightly acidic solutions (415). From measurements with internal generation of radicals it was shown, that nitrobenzene anion radicals may undergo stabilization also by heterogeneous reactions at the electrode (385). Furthermore, the homogeneous reaction was sensitive to the concentrations of alkali metal cations (129, 385) as well as to the content of alcohol in the solvent (129, 416), similar to the influence of ethanol on the stability of nitrofuran anion radicals mentioned above (412). The effect of cations has been attributed (129) to the electrostatic favouring of the encounter of two negatively charged species according to equation (49). Traces of oxygen were found to catalyse the dismutation reaction (130). Furthermore, reactions of nitrobenzene anion radicals with reducible species, e.g.

$$R-NO_2^- + S_2O_8^{2-} \rightarrow R-NO_2 + SO_4^- + SO_4^{2-} \tag{53}$$

$$R-NO_2^- + SO_4^- \rightarrow R-NO_2 + SO_4^{2-} \tag{54}$$

where the system $R-NO_2/R-NO_2^-$ serves as a catalyst for the reduction of these species, were also followed by ESR investigations including flow techniques mentioned above (131, 414). The $S_2O_8^{2-}$ ion gives rise also to the formation of nitrosobenzene anion radicals through oxidation of phenyl-hydroxylamine which is a product of the dismutation of nitrobenzene anion radicals (131); the nitrosobenzene anion radical, cf. also Section III, was found to be extremely sensitive to traces of oxygen (135).

REFERENCES

1. H. A. Laitinen and S. Wawzonek, *J. Amer. Chem. Soc.*, **64**, 1765 (1942).
2. P. C. Tompkins and C. L. A. Schmidt, *Univ. Calif. Pub. Physiol.*, **8**, 237, 247 (1944), (cited according to I. M. Kolthoff and J. J. Lingane, *Polarography*, Interscience Publishers, New York, 1952, p. 814).
3. M. Ashworth, *Collect. Czech. Chem. Commun.*, **13**, 229 (1948).
4. L. Holleck and H. Marsen, *Z. Elektrochem.*, **57**, 301, 944 (1953).
5. R. Pasternak, *Helv. Chim. Acta*, **31**, 753 (1948).
6. P. J. Elving and C. S. Tang, *J. Amer. Chem. Soc.*, **72**, 3244 (1950).
7. M. v. Stackelberg and W. Stracke, *Z. Elektrochem.*, **53**, 118 (1949).
8. L. Holleck and H. J. Exner, *Proceed. I. Internat. Congr. Polarography*, Prague 1951, Vol. I., p. 97; *Z. Elektrochem.*, **56**, 46 (1952); L. Holleck; *Z. Naturforsch. A*, **7**, 282 (1952).
9. G. Le Guillanton, *Bull. Soc. Chim. France*, **1963**, 2359.
10. H. Lund, *Österr. Chem. Ztg.*, **68**, 152 (1967); P. Zuman, *J. Polarogr. Soc. (London)*, **13**, 53 (1967).
11. H. W. Nürnberg and B. Kastening, 'Polarographic and voltammetric techniques', in *Methodicum Chimicum*, F. Korte, Ed., Georg Thieme Verlag/Academic Press, Stuttgart/New York, 1973.
12. B. Kastening, *Chem.-Ing. Techn.*, **44**, 199 (1972).

13. C. L. Perrin, *Progress in Physical Organic Chemistry*, Vol. 3, S. G. Cohen, A. Streitwieser and R. W. Taft, Eds., Interscience Publishers, New York, 1965, p. 165.
14. B. Kastening, *Progress in Polarography*, Vol. 3, P. Zuman, L. Meites and I. M. Kolthoff, Eds., John Wiley and Sons, New York, 1972, p. 195.
15. A. Carrington and A. D. McLachlan, *Introduction to Magnetic Resonance*, Harper and Row, New York, 1967.
16. D. J. E. Ingram, *Spectroscopy at Radio and Microwave Frequencies*, Butterworths, London, 1967.
17. S. Walker and H. Straw, *Spectroscopy*, Vol. I, Chapman and Hall, London, 1961, p. 154.
18. R. S. Alger, *Electron Paramagnetic Resonance*, Interscience Publishers, New York, 1968.
19. S. A. Altschuler and B. M. Kosyrew, *Electron Paramagnetic Resonance*, Academic Press, New York, 1964.
20. H. M. Assenheim, *Introduction to Electron Spin Resonance*, Hilger and Watts, London, 1966.
21. P. B. Ayscough, *Electron Spin Resonance in Chemistry*, Methuen and Co., London, 1967.
22. C. P. Poole, *Electron Spin Resonance*, Interscience Publishers, New York, 1967.
23. F. Schneider and M. Plato, *Elektronenspinresonanz*, Verlag Karl Thiemig, München, 1971.
24. T. L. Squires, *Introduction to Microwave Spectroscopy*, George Newnes Ltd., London, 1963.
25. H. G. Hecht, *Magnetic Resonance Spectroscopy*, John Wiley and Sons, New York, 1967.
26. P. W. Atkins and M. C. R. Symons, *The Structure of Inorganic Radicals*, Elsevier, Amsterdam, 1967.
27. L. A. Bljumenfeld, W. W. Wojewodski and A. G. Semjonow, *Die Anwendung der Paramagnetischen Elektronenresonanz in der Chemie*, Akadem. Verlagsges., Frankfurt (Main), 1966.
28. F. Gerson, *Hochauflösende ESR-Spektroskopie*, Verlag, Chemie, Weinheim, 1967.
29. G. Schoffa, *Elektronenspinresonanz in der Biologie*, Verlag G. Braun, Karlsruhe, 1964.
30. B. D. Flockhart and R. C. Pink, *Talanta*, **12**, 529 (1965).
31. R. O. C. Norman and B. C. Gilbert, *Advances in Physical Organic Chemistry*, Vol. 5, V. Gold, Ed., Academic Press, London, 1967, p. 53.
32. M. Soutif, *J. Chim. Phys.*, **61**, 1549 (1964).
33. M. C. R. Symons, *Advances in Physical Organic Chemistry*, Vol. 1, V. Gold, Ed., Academic Press, New York, 1963, p. 283.
34. D. H. Geske, *Progress in Physical Organic Chemistry*, A. Streitwieser and R. W. Taft, Eds., Vol. 4, Interscience Publishers, New York, 1967, p. 125.
35. M. C. R. Symons, *J. Phys. Chem.*, **71**, 172 (1967).
36. G. K. Fraenkel, *J. Phys. Chem.*, **71**, 139 (1967).
37. L. L. van Reijen, *Ber. Bunsenges. Phys. Chem.*, **75**, 1046 (1971).
38. V. V. Voevodskii, Proceed. Intern. Congr. Catalysis, Amsterdam, 1964, p. 88, (pub. 1965).
39. B. H. J. Bielski and J. M. Gebicki, *Atlas of Electron Spin Resonance Spectra*, Academic Press, New York, 1967.
40. K. W. Bowers, *Advances in Magnetic Resonance*, J. S. Waugh, Ed., Vol. I, Academic Press, New York, 1965, p. 317.

41. H. Fischer, *Magnetic Properties of Free Radicals, in: Landolt–Börnstein, Numerical Data and Functional Relationships in Science and Technology*, New Series, K. H. Hellwege, Ed., Group II, Vol. 1, Springer-Verlag, Berlin, 1965.

42. H. M. Hershenson, *Nuclear Magnetic Resonance and Electron Spin Resonance Spectra*, Index for 1958–1963, Academic Press, New York, 1965.

43. A. Carrington and G. R. Luckhurst, *Ann. Rev. Phys. Chem.*, **19**, 31 (1968).

44. D. H. Eargle, *Anal. Chem.*, **38**, 371 R (1966); **40**, 303 R (1968)

45. E. G. Janzen, *Anal. Chem.*, **44**, 113 R (1972).

46. M. T. Jones and W. D. Phillips, *Ann. Rev. Phys. Chem.*, **17**, 323 (1966).

47. A. H. Maki, *Ann. Rev. Phys. Chem.*, **18**, 9 (1967).

48. M. C. R. Symons, *Ann. Rev. Phys. Chem.*, **20**, 219 (1969).

49. A. L. Kwiram, *Ann. Rev. Phys. Chem.*, **22**, 133 (1971).

50. Y. Asahi, *Rev. Polarography* (*Kyoto*), **10**, 135 (1962).

51. T. Kitagawa, *Rev. Polarography* (*Kyoto*), **12**, 11 (1964).

52. R. N. Adams, *J. Electroanal. Chem.*, **8**, 151 (1964).

53. G. Cauquis, *Bull. Soc. Chim. France*, **1966**, 459; *Revue des Techniciens du Pétrole*, **177**, 67 (1966).

54. G. Cauquis, *Bull. Soc. Chim. France*, **1968**, 1618.

55. P. J. Elving, *Pure Appl. Chem.*, **15**, 297 (1967).

56. M. E. Peover, *Electroanalytical Chemistry*, Vol. 2, A. J. Bard, Ed., Marcel Dekker, New York, 1967, p. 1.

57. K. J. Vetter, *Chem. Ing. Tech.*, **35**, 343 (1963).

58. S. Wawzonek, *Science*, **155**, 39 (1967).

59. P. Zuman, *Progress in Physical Organic Chemistry*, Vol. 5, A. Streitwieser and R. W. Taft, Eds., Interscience Publishers, New York, 1967, p. 81.

60. B. Kastening, *Chem.-Ing. Techn.*, **42**, 190 (1970).

61. R. D. Allendoerfer and P. H. Rieger, *J. Amer. Chem. Soc.*, **88**, 3711 (1966).

62. J. L. Huntington and D. G. Davis, *J. Electrochem. Soc.*, **118**, 57 (1971).

63. A. H. Maki and D. H. Geske, *J. Amer. Chem. Soc.*, **83**, 1852 (1961).

64. K. Möbius, *Z. Naturforsch. A*, **20**, 1093 (1965).

65. K. Möbius, *Z. Naturforsch. A*, **20**, 1102 (1965).

66. N. V. Gudin, M. S. Shapnik, A. V. Il'yasov and N. S. Gerif'yanov, *Nekotorye Vopr. Teorii i Praktiki Ispol'z v Gal'vanotekhn. Neyadovit. Elektrolitov, Kazan*, **1964**, 103, (*Chem. Abstr.*, **64**, 1618 d (1966)).

67. D. A. Hall and P. J. Elving, *Electrochim. Acta*, **12**, 1363 (1967).

68. H. Berg and K. Weller, Sixth International Symposium on Free Radicals, University of Cambridge, 1963 (paper G).

69. A. D. Broadbent and H. Zollinger, *Helv. Chim. Acta*, **47**, 2140 (1964).

70. J. Myatt and P. F. Todd, *Chem. Commun.*, **1967**, 1033.

71. W. R. T. Cottrell and R. A. N. Morris, *Chem. Commun.*, **1968**, 409.

72. W. G. B. Huysmans and W. A. Waters, *J. Chem. Soc. B*, **1967**, 1163.

73. V. D. Pokhodenko and V. A. Khizhnyi, *Teor. Eksp. Khim.*, **2**, 700 (1966), (*Chem. Abstr.*, **66**, 90050 (1967)).

74. F. G. Valitova and Yu. M. Ryzhmanov, *Dokl. Akad. Nauk SSSR*, **170**, 1124 (1966).

75. M. Bonnemay and C. Lamy, *C. R. Acad. Sci.* (*Paris*) *Ser. C*. **265**, 695 (1967).

76. H. Gerischer, A. Mauerer and W. Mindt, *Surface Sci.*, **4**, 431 (1966).

77. M. Bonnemay, C. Lamy and M. Savy, *C. R. Acad. Sci.* (*Paris*), *Ser. C*, **266**, 879 (1968).

78. C. Lamy and P. Malaterre, *Surface Sci.*, **22**, 325 (1970).

79. Yu. V. Vodzinskii, A. A. Vasil'eva, I. A. Korshunov and G. A. Abakumov, *Electrokhimiya*, **7**, 24 (1971).

80. V. B. Kazanskii, *Probl. Kinet. Katal.*, *Akad. Nauk SSSR*, **12**, 36 (1968), (*Chem. Abstr.*, **69**, 54586 (1968)).
81. G. M. Schwab, *Chem. Ing. Tech.*, **39**, 1191 (1967).
82. H. Arai, Y. Saito and Y. Yoneda, *Bull. Chem. Soc. Japan*, **40**, 312 (1967).
83. B. D. Flockhart, I. R. Leith and R. C. Pink, *J. Catal.*, **9**, 45 (1967); *Chem. Commun.*, **1966**, 885.
84. K. Hirota, K. Kuwata and Y. Akagi, *Bull. Chem. Soc. Japan*, **38**, 2209 (1965).
85. G. M. Muha, *J. Phys. Chem.*, **71**, 633, 640 (1967).
86. Y. Ono and T. Keii, *J. Phys. Chem.*, **72**, 2851 (1968).
87. G. M. Zhabrova, V. I. Vladimirova, B. M. Kadenatsi, V. B. Kazanski and G. B. Pariiskii, *Zh. Fiz. Khim.*, **41**, 1898 (1967).
88. D. N. Misra, *Nature*, **214**, 1108 (1967).
89. I. T. Ernst, J. L. Garnett and W. A. Sollich-Baumgartner, *Austral. J. Chem.*, **19**, 529 (1966); I. T. Ernst, B. Fisher, J. L. Garnett and W. A. Sollich-Baumgartner, *Austral. J. Chem.*, **19**, 877 (1966).
90. G. Cauquis and M. Geniès, *Bull. Soc. Chim. France*, **1967**, 3220.
91. G. Peychal-Heiling and G. S. Wilson, *Anal. Chem.*, **43**, 545, 550 (1971).
92. D. H. Geske and A. H. Maki, *J. Amer. Chem. Soc.*, **82**, 2671 (1960); D. H. Geske, J. L. Ragle, M. A. Bambenek and A. L. Balch, *J. Amer. Chem. Soc.*, **86**, 987 (1964).
93. D. E. G. Austen, P. H. Given, D. I. E. Ingram and M. E. Peover, *Nature*, **182**, 1784 (1958).
94. A. D. Korsun and L. N. Nekrasov, *Elektrokhimiya*, **5**, 212 (1969).
95. A. H. Maki and D. H. Geske, *J. Chem. Phys.*, **30**, 1356 (1959).
96. J. F. Ambrose, D. Dillard, A. K. Carpenter and R. F. Nelson, *Anal. Chem.*, **42**, 814 (1970).
97. I. B. Goldberg, A. T. Bard and S. W. Feldberg, *J. Phys. Chem.*, **76**, 2550 (1972).
98. J. K. Dohrmann and F. Gallusser, *Ber. Bunsenges. phys. Chem.*, **75**, 432 (1971).
99. J. K. Dohrmann and K. J. Vetter, *J. Electroanal. Chem.*, **20**, 23 (1969).
100. R. Gavars, J. Stradins and S. Hillers, *Zavodsk. Lab.*, **31**, 41 (1965), (*Chem. Abstr.*, **62**, 9959 h (1965)).
101. J. C. M. Henning, *J. Chem. Phys.*, **44**, 2139 (1966).
102. R. E. Sioda and W. S. Koski, *J. Amer. Chem. Soc.*, **87**, 5573 (1965).
103. K. H. Hausser, A. Häbich and V. Franzen, *Z. Naturforsch. A*, **16**, 836 (1961).
104. C. Corvaja, P. L. Nordio, M. V. Pavan and G. Rigatti, *Ric. Sci. Rend. Sez. A*, **4**, 297 (1964).
105. M. T. Jones, E. A. LaLancette and R. E. Benson, *J. Chem. Phys.*, **41**, 401 (1964).
106. L. H. Piette, P. Ludwig and R. N. Adams, *J. Amer. Chem. Soc.*, **83**, 3909 (1961).
107. L. H. Piette, P. Ludwig and R. N. Adams, *Anal. Chem.*, **34**, 916 (1962).
108. R. Koopmann and H. Gerischer, *Ber. Bunsenges. Phys. Chem.*, **70**, 118 (1966).
109. D. H. Levy and R. J. Myers, *J. Chem. Phys.*, **41**, 1062 (1964).
110. W. M. Tolles and D. W. Moore, *J. Chem. Phys.*, **46**, 2102 (1967).
111. I. B. Goldberg and A. J. Bard, *J. Phys. Chem.*, **75**, 3281 (1971).
112. B. Kastening, J. Divišek and B. Gostiša-Mihelčić, 23rd ISE Meeting, Stockholm, 1972 cf. also refs. 334, 335.
113. R. Pointeau, J. Favede and P. Delhaes, *J. Chim. Phys.*, **61**, 1129 (1964).
114. P. Boyer and J. Dericbourg, *C. R. Acad. Sci. (Paris) Ser. C*, **265**, 429 (1967).
115. R. Hirasawa, T. Mukaibo, H. Hasegawa, Y. Kanda and T. Maruyama, *Rev. Sci. Instr.*, **39**, 935 (1968).
116. R. Hirasawa, T. Mukaibo, H. Hasegawa, N. Odan and T. Maruyama, *J. Phys. Chem.*, **72**, 2541 (1968).
117. R. Koopman, *Ber. Bunsenges. Phys. Chem.*, **72**, 32 (1968).
118. J. R. Bolton and G. K. Fraenkel, *J. Chem. Phys.*, **40**, 3307 (1964).

119. L. O. Wheeler, K. S. V. Santhanam and A. J. Bard, *J. Phys. Chem,*, **71**, 2223 (1967).
120. J. P. Billon, G. Cauquis, J. Combrisson and A. M. Li, *Bull. Soc. Chim. France,* **1960**, 2062.
121. J. P. Billon, G. Cauquis and J. Combrisson, *J. Chim. Phys.*, **61**, 374 (1964).
122. J. E. Harriman and A. H. Maki, *J. Chem. Phys.*, **39**, 778 (1963).
123. B. Kastening, *Electrochim. Acta*, **9**, 241 (1964).
124. B. Kastening, *Collection Czech. Chem. Commun.*, **30**, 4033 (1965).
125. M. Plato, *Z. Naturforsch. A*, **22**, 119 (1967).
126. P. H. Rieger, I. Bernal, W. H. Reinmuth and G. K. Fraenkel, *J. Amer. Chem. Soc.*, **85**, 683 (1963).
127. C. Corvaja, G. Farnia and E. Vianello, *Electrochim. Acta*, **11**, 919 (1966).
128. K. Umemoto, *Bull. Chem. Soc. Japan*, **40**, 1058 (1967).
129. D. Kolb, W. Wirths and H. Gerischer, *Ber. Bunsenges. Phys. Chem.*, **73**, 148 (1969).
130. D. Kolb and R. Koopmann, *Ber. Bunsenges. Phys. Chem.*, **73**, 284 (1969).
131. B. Kastening, *Z. Anal. Chem.*, **224**, 196 (1967).
132. B. Kastening, *Ber. Bunsenges. Phys. Chem.*, **72**, 20 (1968).
133. B. Kastening and B. Gostiša-Mihelčić, to be published.
134. C. P. Andrieux and J. M. Savéant, *J. Electroanal. Chem.*, **28**, 446 (1970; J. M. Fritsch, H. Weingarten and J. D. Wilson, *J. Amer. Chem. Soc.*, **92**, 4038 (1970).
135. K. Möbius, *Z. Angew. Physik*, **17**, 534 (1964).
136. R. F. Nelson and R. N. Adams, *J. Electroanal. Chem.*, **13**, 184 (1967).
137. B. Kastening, 20th CITCE-Meeting, Strasbourg, Sept. 1969.
138. J. P. Billon, G. Cauquis, J. Raison and Y. Thibaud, *Bull. Soc. Chim. France,* **1967**, 199.
139. L. D. Rollmann and R. T. Iwamoto, *J. Amer. Chem. Soc.*, **90**, 1455 (1968).
140. S. Hünig, D. Scheutzow, P. Čársky and R. Zahradnik, *J. Phys. Chem.*, **75**, 335 (1971).
141. R. E. Sioda and W. Kemula, *J. Electroanal. Chem.*, **31**, 113 (1971).
142. H. M. McConnell, *J. Chem. Phys.*, **24**, 632 (1956).
143. P. H. Rieger and G. K. Fraenkel, *J. Chem. Phys.*, **39**, 609 (1963).
144. J. Bernal, P. H. Rieger and G. K. Fraenkel, *J. Chem. Phys.*, **37**, 1489 (1962).
145. F. Gerson, K. Müllen and E. Vogel, *Helv. Chim. Acta*, **54**, 2731 (1971).
146. W. R. Knolle and J. R. Bolton, *J. Amer. Chem. Soc.*, **93**, 3337 (1971).
147. S. V. Kulkarni and C. Trapp, *J. Amer. Chem. Soc.*, **92**, 4809 (1970).
148. S. V. Kulkarni and C. Trapp, *J. Amer. Chem. Soc.*, **92**, 4801 (1970).
149. M. Broze, Z. Luz and B. L. Silver, *J. Chem. Phys.*, **46**, 4891 (1967).
150. C. L. Talcott and R. J. Myers, *Mol. Phys.*, **12**, 549 (1967).
151. E. W. Stone and A. H. Maki, *J. Chem. Phys.*, **39**, 1635 (1963).
152. P. T. Cottrell and P. H. Rieger, *Mol. Phys.*, **12**, 149 (1967).
153. B. L. Barton and G. K. Fraenkel, *J. Chem. Phys.*, **41**, 1455 (1964).
154. C. L. Talcott, U.S. At. Energy Comm., UCRL-17743 (1967), (*Chem. Abstr.*, **69**, 48134 (1968)).
155. M. D. Curtis and A. L. Allred, *J. Amer. Chem. Soc.*, **87**, 2554 (1965).
156. J. P. Colpa and J. R. Bolton, *Mol. Phys.*, **6**, 273 (1963).
157. E. J. Geels, R. Konaka and G. A. Russell, *Chem. Commun.*, **1965**, 13.
158. D. H. Levy and R. J. Myers, *J. Chem. Phys.*, **42**, 3731 (1965).
159. G. Cauquis, M. Genies, H. Lemaire, A. Rassat and J. P. Ravet, *J. Chem. Phys.*, **47**, 4642 (1967).
160. G. Cauquis, A. Rassat, J. P. Ravet and D. Serve, *Tetrahedron Letters*, **1971**, 971.
161. E. A. C. Lucken, *J. Chem. Soc. A*, **1966**, 991.
162. R. Gerdil and E. A. C. Lucken, *J. Amer. Chem. Soc.*, **88**, 733 (1966).

163. N. M. Atherton, J. N. Ockwell and R. Dietz, *J. Chem. Soc. A*, **1967**, 771.
164. A. L. Allred and L. W. Bush, *J. Phys. Chem.*, **72**, 2238 (1968).
165. N. M. Atherton, R. Mason and R. J. Wratten, *Mol. Phys.*, **11**, 525 (1966).
166. K. Möbius and M. Plato, *Z. Naturforsch. A*, **22**, 929 (1967).
167. B. C. Gilbert, R. H. Schlossel and W. M. Gulick, *J. Amer. Chem. Soc.*, **92**, 2974 (1970).
168. A. I. Brodskii, L. L. Gordienko and Yu. A. Kruglyak, *Teor. Eksp. Khim.*, **3**, 98 (1967), (*Chem. Abstr.*, **67**, 69337 (1967)).
169. D. H. Geske and G. R. Padmanabhan, *J. Amer. Chem. Soc.*, **87**, 1651 (1965).
170. S. Hünig, *Pure Appl. Chem.*, **15**, 109 (1967); cf. Theses, University Würzburg: (a) D. Scheutzow (1966), (b) G. Ruider (1967), (c) G. Kiesslich (1968).
171. P. H. H. Fischer and C. A. McDowell, *Mol. Phys.*, **8**, 357 (1964).
172. T. Fujinaga, Y. Deguchi and K. Umemoto, *Bull. Chem. Soc. Japan*, **37**, 822 (1964).
173. R. Gavars, J. Stradins and S. Hillers, *Dokl. Akad. Nauk SSSR*, **157**, 1424 (1964).
174. A. H. Maki and D. H. Geske, *J. Chem. Phys.*, **33**, 825 (1960).
175. G. A. Russell and G. R. Stevenson, *J. Amer. Chem. Soc.*, **93**, 2432 (1971).
176. D. H. Geske and A. L. Balch, *J. Phys. Chem.*, **68**, 3423 (1964).
177. P. S. Kinson and B. M. Trost, *J. Amer. Chem. Soc.*, **93**, 3823 (1971).
178. S. F. Nelsen and E. D. Seppanen, *J. Amer. Chem. Soc.*, **92**, 6212 (1970).
179. J. Petránek, J. Pilař and O. Ryba, *Collect. Czechoslov. Chem. Commun.*, **35**, 2571 (1970).
180. R. D. Rieke and W. E. Rich, *J. Amer. Chem. Soc.*, **92**, 7349 (1970).
181. S. F. Nelsen and P. J. Hintz, *J. Amer. Chem. Soc.*, **92**, 6215 (1970).
182. C. M. Lang, J. Harbour and A. V. Guzzo, *J. Phys. Chem.*, **75**, 2861 (1971).
183. R. E. Dessy, R. B. King and M. Waldrop, *J. Amer. Chem. Soc.*, **88**, 5112 (1966); R. E. Dessy, P. M. Weissman and R. L. Pohl, *J. Amer. Chem. Soc.*, **88**, 5117; R. E. Dessy, R. L. Pohl and R. B. King, *J. Amer. Chem. Soc.*, **88**, 5121; R. E. Dessy and P. M. Weissman, *J. Amer. Chem. Soc.*, **88**, 5124, 5129.
184. R. H. Felton and H. Linschitz, *J. Amer. Chem. Soc.*, **88**, 1113 (1966).
185. R. E. Dessy and L. Wieczorek, *J. Amer. Chem. Soc.*, **91**, 4963 (1969); R. E. Dessy, M. Kleiner and S. C. Cohen, *J. Amer. Chem. Soc.*, **91**, 6800; R. E. Dessy, J. C. Charkoudian, T. B. Abeles and A. L. Rheingold, *J. Amer. Chem. Soc.*, **92**, 3947 (1970).
186. J. Fajer, D. C. Borg, A. Forman, D. Dolphin and R. H. Felton, *J. Amer. Chem. Soc.*, **92**, 3451 (1970).
187. J. A. McCleverty and B. Ratcliff, *J. Chem. Soc. A*, **1970**, 1627.
188. A. Wolberg and J. Manassen, *J. Amer. Chem. Soc.*, **92**, 2982 (1970); *Inorg. Chem.*, **9**, 2365 (1970).
189. A. I. Brodskii and L. L. Gordienko, *Teor. i Eksperim. Khim., Akad. Nauk Ukr. SSSR*, **1**, 452 (1965), (*Chem. Abstr.*, **64**, 3054 f (1966)).
190. J. M. W. Scott and W. H. Jura, *Canad. J. Chem.*, **45**, 2375 (1967).
191. L. S. Degtyarev, N. Garnyuk, A. M. Golubenkova and A. I. Brodskii, *Dokl. Akad. Nauk SSSR*, **157**, 1406 (1964).
192. A. I. Brodskii, L. L. Gordienko and L. S. Degtyarev, *Zh. Vses. Khim. Obshchestva im. D. I. Mendeleeva*, **11**, 196 (1966), (*Chem. Abstr.*, **65**, 8730 d (1966)).
193. J. Koutecký, *Electrochim. Acta*, **13**, 1079 (1968).
194. A. Streitwieser, *Molecular Orbital Theory for Organic Chemists*, John Wiley and Sons, New York, 1961, p. 173.
195. R. Zahradník and C. Párkányi, *Talanta*, **12**, 1289 (1965).
196. A. I. Brodskii, L. L. Gordienko and L. S. Degtyarev, *Electrochim. Acta*, **13**, 1095 (1968).

197. P. H. Rieger and G. K. Fraenkel, *J. Chem. Phys.*, **37**, 2795 (1962).
198. R. E. Sioda and W. S. Koski, *J. Amer. Chem. Soc.*, **89**, 475 (1967).
199. R. Gerdil and E. A. C. Lucken, *Mol. Phys.*, **9**, 529 (1965).
200. A. Zweig, W. G. Hodgson and W. H. Jura, *J. Amer. Chem. Soc.*, **86**, 4124 (1964);
     A. Zweig and W. G. Hodgson, *Proc. Chem. Soc.*, **1964**, 417.
201. A. L. Allred and L. W. Bush, *J. Amer. Chem. Soc.*, **90**, 3352 (1968).
202. R. A. Gavars, W. K. Grin, G. O. Reikhman and J. P. Stradins, *Teor. Exp. Khim.*,
     6, 685 (1970).
203. J. Voss, *Tetrahedron*, **27**, 3753 (1971); **28**, 2627 (1972).
204. B. G. Segal, M. Kaplan and G. K. Fraenkel, *J. Chem. Phys.*, **43**, 4191 (1965).
205. J. R. Bolton and G. K. Fraenkel, *J. Chem. Phys.*, **41**, 944 (1964).
206. B. Lunelli, C. Corvaja and G. Farnia, *Trans. Faraday Soc.*, **67**, 1951 (1971).
207. P. H. H. Fischer and C. A. McDowell, *J. Amer. Chem. Soc.*, **85**, 2694 (1963).
208. R. D. Allendoerfer and P. H. Rieger, *J. Amer. Chem. Soc.*, **87**, 2336 (1965).
209. M. Hirayama *Bull. Chem. Soc. Japan*, **40**, 1557 (1967).
210. J. Gendell, J. H. Freed and G. K. Fraenkel, *J. Chem. Phys.*, **37**, 2832 (1962).
211. E. W. Stone and A. H. Maki, *J. Chem. Phys.*, **36**, 1944 (1962).
212. P. H. Rieger, I. Bernal and G. K. Fraenkel, *J. Amer. Chem. Soc.*, **83**, 3918 (1961).
213. T. M. McKinney and D. H. Geske, *J. Chem. Phys.*, **44**, 2277 (1966).
214. B. Kastening and S. Vavřička, *J. Electroanal. Chem.*, **29**, 195 (1971).
215. P. H. Heller and D. H. Geske, *J. Org. Chem.*, **31**, 4249 (1966).
216. G. H. Aylward, J. L. Garnett and J. H. Sharp, *Anal. Chem.*, **39**, 457 (1967).
217. Z. Galus and R. N. Adams, *J. Chem. Phys.*, **36**, 2814 (1962).
218. M. T. Melchior and A. H. Maki, *J. Chem. Phys.*, **34**, 471 (1961).
219. K. Kuwata and D. H. Geske, *J. Amer. Chem. Soc.*, **86**, 2101 (1964).
220. E. W. Stone and A. H. Maki, *J. Chem. Phys.*, **38**, 1999 (1963).
221. J. Voss and W. Walter, *Liebigs Ann. Chem.*, **734**, 1 (1970).
222. J. Harbour and A. V. Guzzo, *Mol. Phys.*, **23**, 151 (1972).
223. L. Lunazzi, G. Maccagnani, G. Mazzanti and G. Placucci, *J. Chem. Soc. B*, **1971**,
     162.
224. A. V. Il'yasov, Yu. M. Kargin, Ya. A. Levin, I. D. Morozova, B. V. Mel'nikov,
     A. A. Vafina, N. N. Sotnikova and V. S. Galeev, *Izv. Akad. Nauk SSSR, Ser.
     Khim.*, **20**, 770 (1971).
225. M. R. Das and G. K. Fraenkel, *J. Chem. Phys.*, **42**, 792 (1965).
226. L. H. Piette, M. Okamura, G. P. Rabold, R. T. Ogata, R. E. Moore and P. J.
     Scheuer, *J. Phys. Chem.*, **71**, 29 (1967).
227. W. M. Gulick and D. H. Geske, *J. Amer. Chem. Soc.*, **88**, 4119 (1966).
228. T. J. Katz, M. Yoshida and L. C. Siew, *J. Amer. Chem. Soc.*, **87**, 4516 (1965).
229. D. H. Levy and R. J. Myers, *J. Chem. Phys.*, **43**, 3063 (1965).
230. D. H. Levy and R. J. Myers, *J. Chem. Phys.*, **44**, 4177 (1966).
231. N. Steinberger and G. K. Fraenkel, *J. Chem. Phys.*, **40**, 723 (1964).
232. M. Hirayama, *Bull. Chem. Soc. Japan*, **40**, 1822 (1967).
233. A. V. Il'yasov, Yu. M. Kargin, Ya. A. Levin, I. D. Morozova and N. N. Sotnikova,
     Izv. Akad. Nauk SSSR, Ser. Khim., **1968**, 1030; *Dokl. Akad. Nauk SSSR*, **179**, 1141
     (1968).
234. A. V. Il'yasov, Yu. M. Kargin, Ya. A. Levin and V. Kh. Ivanova, *Izv. Akad. Nauk
     SSSR, Ser. Khim.*, **1966**, 583; A. V. Il'yasov, Yu. M. Kargin, Ya. A. Levin, I. D.
     Morozova, N. N. Sotnikova, V. Kh. Ivanova and N. I. Bessolitsyna, *Izv. Akad.
     Nauk SSSR, Ser. Khim.*, **1968**, 740, (*Chem. Abstr.*, **65**, 6723c (1966); **69**, 24004
     (1968).
235. W. M. Gulick, *J. Amer. Chem. Soc.*, **94**, 29 (1972).

236. G. P. Rabold, R. T. Ogata, M. Okamura, L. H. Piette, R. E. Moore and P. J. Scheuer, *J. Chem. Phys.*, **46**, 1161 (1967).
237. P. L. Nordio, G. Giacometti and P. Favero, *Ric. Sci. Rend. Ser. A*, **3**, 107 (1963).
238. E. W. Stone and A. H. Maki, *J. Chem. Phys.*, **37**, 1326 (1962); **38**, 1254 (1963).
239. A. H. Maki, *J. Chem. Phys.*, **35**, 761 (1961).
240. P. H. Rieger and G. K. Fraenkel, *J. Chem. Phys.*, **37**, 2811 (1962).
241. S. F. Nelsen, *Tetrahedron Letters*, **1967**, 3795.
242. F. Gerson, R. Gleiter, J. Heinzer and H. Behringer, *Angew. Chem. Internat. Ed.* **9**, 306 (1970).
243. D. J. Williams, J. M. Pearson and M. Levy, *J. Amer. Chem. Soc.*, **93**, 5483 (1971).
244. L. O. Wheeler, K. S. V. Santhanam and A. J. Bard, *J. Phys. Chem.*, **70**, 404 (1966).
245. L. O. Wheeler and A. J. Bard, *J. Phys. Chem.*, **71**, 4513 (1967).
246. L. S. Marcoux, A. Lomax and A. J. Bard, *J. Amer. Chem. Soc.*, **92**, 243 (1970).
247. Y. Ikegami and S. Seto, *Bull. Chem. Soc. Japan*, **43**, 2409 (1970).
248. K. Möbius, *Z. Naturforsch. A*, **20**, 1117 (1965).
249. M. Hirayama, *Bull. Chem. Soc. Japan*, **40**, 2234, 2530 (1967).
250. J. Voss and W. Walter, *Liebigs Ann. Chem.*, **743**, 177 (1970).
251. G. Cauquis, J. Badoz-Lambling and J. P. Billon, *Bull. Soc. Chim. France*, **1965**, 1433.
252. F. Tonnard and S. Odiot, *J. Chim. Phys.*, **63**, 227 (1966); F. Tonnard, *C. R. Acad. Sci. (Paris)*, **260**, 2793 (1965).
253. G. R. Underwood, D. Jurkowitz and S. C. Dickerman, *J. Phys. Chem.*, **74**, 544 (1970).
254. D. H. Geske and J. L. Ragle, *J. Amer. Chem. Soc.*, **83**, 3532 (1961).
255. I. Bernal and G. K. Fraenkel, *J. Amer. Chem. Soc.*, **86**, 1671 (1964).
256. T. Kitagawa and R. Nakashima, *Rev. Polarography (Kyoto)*, **13**, 115 (1966).
257. T. M. McKinney and D. H. Geske, *J. Amer. Chem. Soc.*, **89**, 2806 (1967).
258. W. C. Danen, T. T. Kensler, J. G. Lawless, M. F. Marcus and M. D. Hawley, *J. Phys. Chem.*, **73**, 4389 (1969).
259. P. L. Nordio, M. V. Pavan and C. Corvaja, *Trans. Faraday Soc.*, **60**, 1985 (1964).
260. J. H. Freed and G. K. Fraenkel, *J. Chem. Phys.*, **37**, 1156 (1962).
261. K. Umemoto, Y. Deguchi and T. Fujinaga, *Bull. Chem. Soc. Japan*, **36**, 1539 (1963).
262. M. J. Feighan and M. T. Jones, *J. Amer. Chem. Soc.*, **92**, 6756 (1970).
263. J. Gendell, J. H. Freed and G. K. Fraenkel, *J. Chem. Phys.*, **41**, 949 (1964).
264. R. D. Allendoerfer and P. H. Rieger, *J. Chem. Phys.*, **46**, 3266 (1967).
265. F. Millet and J. E. Harriman, *J. Chem. Phys.*, **44**, 1945 (1966).
266. J. H. Freed and G. K. Fraenkel, *J. Chem. Phys.*, **40**, 1815 (1964).
267. J. H. Freed and G. K. Fraenkel, *J. Chem. Phys.*, **41**, 699 (1964).
268. R. J. Faber and G. K. Fraenkel, *J. Chem. Phys.*, **47**, 2462 (1967).
269. J. H. Freed, P. H. Rieger and G. K. Fraenkel, *J. Chem. Phys.*, **37**, 1881 (1962).
270. C. J. W. Gutch and W. A. Waters, *Chem. Commun.*, **1966**, 39.
271. C. J. W. Gutch, W. A. Waters and M. C. R. Symons, *J. Chem. Soc. B*, **1970**, 1261.
272. F. Gerson and R. N. Adams, *Helv. Chim. Acta*, **48**, 1539 (1965).
273. R. D. Allendoerfer and P. H. Rieger, *J. Chem. Phys.*, **46**, 3410 (1967).
274. J. W. Rogers and W. H. Watson, *J. Phys. Chem.*, **72**, 68 (1968).
275. B. L. Barton and G. K. Fraenkel, *J. Chem. Phys.*, **41**, 695 (1964).
276. M. Kaplan, J. R. Bolton and G. K. Fraenkel, *J. Chem. Phys.*, **42**, 955 (1965).
277. M. Iwaizumi and T. Isobe, *Bull. Chem. Soc. Japan*, **37**, 1651 (1964).
278. F. Gerson and W. L. F. Armarego, *Helv. Chim. Acta*, **48**, 112 (1965).

279. H. van Willigen, J. A. M. van Broekhoven and E. de Boer, *Mol. Phys.*, **12**, 533 (1967).
280. Chi-Yuan Ling and J. Gendell, *J. Chem. Phys.*, **46**, 400 (1967).
281. Chi-Yuan Ling and J. Gendell, *J. Chem. Phys.*, **47**, 3475 (1967).
282. S. H. Glarum and L. C. Synder, *J. Chem. Phys.*, **36**, 2989 (1962).
283. P. B. Ayscough and R. Wilson, *Proc. Chem. Soc.*, **1962**, 229; P. B. Ayscough, F. P. Sargent and R. Wilson, *J. Chem. Soc.*, **1963**, 5418.
284. Y. Ikegami and S. Seto, *Bull. Chem. Soc. Japan*, **44**, 1905 (1971).
285. M. R. Das, S. B. Wagner and J. H. Freed, *J. Chem. Phys.*, **52**, 5404 (1970).
286. T. A. Claxton, J. Oakes and M. C. R. Symons, *Trans. Faraday Soc.*, **64**, 596 (1968); T. A. Claxton and J. Oakes, *Trans. Faraday Soc.*, **64**, 607 (1968).
287. S. H. Glarum and J. H. Marshall, *J. Chem. Phys.*, **41**, 2182 (1964).
288. T. Kitagawa, T. Layloff and R. N. Adams, *Anal. Chem.*, **36**, 925 (1964).
289. R. H. Philp, T. Layloff and R. N. Adams, *J. Electrochem. Soc.*, **111**, 1189 (1964).
290. R. D. Rieke, S. E. Bales, P. M. Hudnall and C. F. Meares, *J. Amer. Chem. Soc.*, **93**, 697 (1971).
291. A. Forman, D. C. Borg, R. H. Felton and J. Fajer, *J. Amer. Chem. Soc.*, **93**, 2790 (1971).
292. E. Brunner, R. Mücke and F. Dörr, *Z. Phys. Chem. (Frankfurt)*, **50**, 30 (1966).
293. L. H. Piette, P. Ludwig and R. N. Adams, *J. Amer. Chem. Soc.*, **84**, 4212 (1962).
294. P. Ludwig, T. Layloff and R. N. Adams, *J. Amer. Chem. Soc.*, **86**, 4568 (1964).
295. J. Q. Chambers and R. N. Adams, *Mol. Phys.*, **9**, 413 (1965); J. Q. Chambers, T. Layloff and R. N. Adams, *J. Phys. Chem.*, **68**, 661 (1964).
296. T. Kuwana and R. K. Darlington, NOLC Report 697 (1967).
297. C. Corvaja and G. Giacometti, *Ric. Sci. Rend. Sez. A*, **8**, 1038 (1965).
298. C. Corvaja and G. Giacometti, *J. Amer. Chem. Soc.*, **86**, 2736 (1964).
299. T. Yonezawa, T. Kawamura, M. Ushio and Y. Nakao, *Bull. Chem. Soc. Japan*, **43**, 1022 (1970).
300. T. Fujinaga, K. Izutsu, K. Umemoto, T. Arai and K. Takaoka, *Nippon Kagaku Zasshi*, **89**, 105 (1968); (*Chem. Abstr.* **68**, 110833 (1968)).
301. G. Cauquis, J. P. Billon, J. Raison and Y. Thibaud, *C. R. Acad. Sci. (Paris)*, **257**, 2128 (1963).
302. G. Cauquis and G. Fauvelot, *Polarography 1964*, G. J. Hills, Ed., Macmillan, London, 1966, p. 847; *Bull. Soc. Chim. France*, **1964**, 2014.
303. G. Cauquis and M. Geniès, *C. R. Acad. Sci. (Paris)*, Ser. C, **265**, 1340 (1967).
304. G. Cauquis, G. Fauvelot and J. Rigaudy, *Bull. Soc. Chim. France*, **1968**, 4928.
305. H. Y. Lee and R. N. Adams, *Anal. Chem.*, **34**, 1587 (1962).
306. G. Cauquis and D. Serve, *Tetrahedron Letters*, **1970**, 17.
307. G. Farnia, A. Romanin, G. Capobianco and F. Torzo, *J. Electroanal. Chem.*, **33**, 31 (1971).
308. P. H. Rieger, Thesis, Columbia University, New York 1961 (cited according to Reference 198).
309. L. Horner and R. J. Singer, *Tetrahedron Letters*, **1969**, 1545.
310. R. Dietz and B. E. Larcombe, *J. Chem. Soc. Ser. B*, **1970**, 1369.
311. R. Dietz, M. E. Peover and R. Wilson, *J. Chem. Soc. B*, **1968**, 75.
312. A. E. J. Forno, *Chem. Ind.*, **1968**, 1728.
313. G. Cauquis and M. Geniès, *Tetrahedron Letters*, **1970**, 3403.
314. G. Cauquis and M. Geniès, *Tetrahedron Letters*, **1971**, 4677.
315. M. Libert and G. Caullet, *Bull. Soc. Chim. France*, **1971**, 1947.
316. A. I. Brodskii, L. L. Gordienko, A. G. Chuklantseva, A. A. Balandin, R. Yu. Alieva, E. I. Klabunovski and L. V. Antik, *Zh. Strukt. Khim.*, **11**, 604 (1970).

317. V. E. Sahini, L. Ciurea and E. Volanschi, *Rev. Roum. Chim.*, **12**, 355 (1967).
318. R. E. Sioda, D. O. Cowan and W. S. Koski, *J. Amer. Chem. Soc.*, **89**, 230 (1967).
319. G. A. Savicheva, M. B. Gazizov, A. V. Il'yasov and A. I. Razumov, *Zh. Obshch. Khim.*, **37**, 2785 (1967), (*Chem. Abstr.*, **68**, 110808 (1968)).
320. E. G. Janzen and J. L. Gerlock, *J. Phys. Chem.*, **71**, 4577 (1967).
321. V. V. Bukhtiyarov and N. N. Bubnov, *Teor. Eksp. Khim.*, **4**, 413 (1968), (*Chem. Abstr.*, **69**, 63461 (1968)).
322. B. M. Latta and R. W. Taft, *J. Amer. Chem. Soc.*, **89**, 5172 (1967).
323. M. Melicharek and R. F. Nelson, *J. Electroanal. Chem.*, **26**, 201 (1970).
324. W. H. Bruning, R. F. Nelson, L. S. Marcoux and R. N. Adams, *J. Phys. Chem.*, **71**, 3055 (1967); E. T. Seo, R. F. Nelson, J. M. Fritsch, L. S. Marcoux, D. W. Leedy and R. N. Adams, *J. Amer. Chem. Soc.*, **88**, 3498 (1966).
325. T. Mizoguchi and R. N. Adams, *J. Amer. Chem. Soc.*, **84**, 2058 (1962).
326. Z. Galus and R. N. Adams, *J. Amer. Chem. Soc.*, **86**, 1666 (1964).
327. I. Nemcová and I. Nemec, *J. Electroanal. Chem.*, **30**, 506 (1971).
328. P. H. Plesch, A. Stasko and D. Robson, *J. Chem. Soc.*, *B*, **1971**, 1634.
329. L. R. Melby, R. J. Harder, W. R. Hertler, W. Mahler, R. E. Benson and W. E. Mochel, *J. Amer. Chem. Soc.*, **84**, 3374 (1962).
330. C. P. Andrieux and J. M. Savéant, *J. Electroanal. Chem.*, **26**, 223 (1970).
331. G. Cauquis, H. Delhomme and D. Serve, *Tetrahedron Letters*, **1971**, 4649.
332. S. Hünig, F. Linhart and D. Scheutzow, *Angew. Chem. Internat. Ed.*, **10**, 275 (1971).
333. A. N. Prokof'ev, S. P. Solodovnikov, G. A. Nikiforov and V. V. Ershov, *Izv. Akad. Nauk SSSR, Ser. Khim.*, **1971**, 324.
334. B. Kastening, J. Divišek, B. Gostiša-Mihelčić and H. G. Müller, *Z. Phys. Chem. (Frankfurt)*, **87**, 125 (1973).
335. B. Kastening, B. Gostiša-Mihelčić and J. Divišek, *Faraday Discussions*, **56**, (1973), 341.
336. C. Barry, G. Cauquis and M. Maurey, *Bull. Soc. Chim. France*, **1966**, 2510.
337. L. Jeftić and R. N. Adams, *J. Amer. Chem. Soc.*, **92**, 1332 (1970).
338. G. Cauquis and M. Maurey, *C. R. Acad. Sci. (Paris), Ser. C*, **266**, 1021 (1968).
339. H. N. Blount and T. Kuwana, *J. Amer. Chem. Soc.*, **92**, 5773 (1970).
340. T. C. Chiang, A. H. Reddoch and D. F. Williams, *J. Chem. Phys.*, **54**, 2051 (1971).
341. M. Sainsbury, *J. Chem. Soc. C*, **1971**, 2888.
342. S. Kato, J. Nakaya and E. Imoto, *Bull. Chem. Soc. Japan*, **44**, 1928 (1971).
343. D. Hawley and R. N. Adams, *J. Electroanal. Chem.*, **10**, 376 (1965).
344. D. W. Leedy and R. N. Adams, *J. Amer. Chem. Soc.*, **92**, 1646 (1970).
345. B. I. Shapiro and V. M. Kazakova, *Zh. Strukt. Khim.*, **9**, 306 (1968), (*Chem. Abstr.*, **69**, 56163 (1968)).
346. D. W. Leedy and R. N. Adams, *J. Electroanal. Chem.*, **14**, 119 (1967).
347. J. Stradins, *Electrochim. Acta*, **9**, 711 (1964).
348. A. S. Lindsey, M. E. Peover and N. G. Savill, *J. Chem. Soc.*, **1962**, 4558.
349. E. W. Stone and A. H. Maki, *J. Chem. Phys.*, **41**, 284 (1964).
350. J. L. Sadler and A. J. Bard, *J. Electrochem. Soc.*, **115**, 343 (1968): *J. Amer. Chem. Soc.*, **90**, 1979 (1968).
351. T. Kubota, Y. Oishi, K. Nishikida and H. Miyazaki, *Bull. Chem. Soc. Japan*, **43**, 1622 (1970).
352. H. Miyazaki, K. Nishikida and T. Kubota, *Bull. Chem. Soc. Japan*, **44**, 277 (1971).
353. A. V. Il'yasov, Yu. M. Kargin, N. N. Sotnikova, V. Z. Kondranina, B. V. Mel'nikov and A. A. Vafina, *Izv. Akad. Nauk SSSR, Ser. Khim.*, **20**, 932 (1971).

354. R. F. Nelson, D. W. Leedy, E. T. Seo and R. N. Adams, *Z. Anal. Chem.*, **224**, 184 (1967).
355. F. Gerson, G. Plattner and H. Bock, *Helv. Chim. Acta*, **53**, 1629 (1970).
356. J. Simonet and G. Jeminet, *Bull. Soc. Chim. France*, **1971**, 2754.
357. K. Möbius and M. Plato, *Z. Naturforsch. A*, **19**, 1240 (1964).
358. K. Möbius and M. Plato, *J. Phys. Chem.*, **72**, 1830 (1968).
359. E. T. Seo, J. M. Fritsch and R. F. Nelson, *J. Phys. Chem.*, **72**, 1829 (1968).
360. R. Dehl and G. K. Fraenkel, *J. Chem. Phys.*, **39**, 1793 (1963).
361. P. B. Ayscough and R. Wilson, *J. Chem. Soc.*, **1963**, 5412.
362. B. J. Tabner and J. R. Zdysiewicz, *J. Chem. Soc. B*, **1971**, 1659.
363. P. Peterson, A. K. Carpenter and R. F. Nelson, *J. Electroanal. Chem.*, **27**, 1 (1970).
364. M. E. Peover and B. S. White, *Electrochim. Acta*, **11**, 1061 (1966).
365. L. Nadjo and J. M. Saveant, *J. Electroanal. Chem.*, **30**, 41 (1971).
366. S. F. Nelsen, E. F. Travecedo and E. D. Seppanen, *J. Amer. Chem. Soc.*, **93**, 2913 (1971).
367. N. L. Bauld and M. S. Brown, *J. Amer. Chem. Soc.*, **89**, 5413, 5417 (1967).
368. R. C. Buchta and D. H. Evans, *J. Org. Chem.*, **35**, 2844 (1970).
369. T. Kitagawa, T. P. Layloff and R. N. Adams, *Anal. Chem.*, **35**, 1086 (1963).
370. D. E. Bartak, W. C. Danen and M. D. Hawley, *J. Org. Chem.*, **35**, 1206 (1970).
371. N. N. Vylegzhanina, A. V. Il'yasov and Yu. P. Kitaev, *Zh. Strukt. Khim.*, **6**, 153 (1965), (*Chem. Abstr.*, **63**, 1378 c (1965)).
372. T. Fujinaga, K. Umemoto and T. Arai, *Denki Kagaku*, **34**, 135 (1966), (*Chem. Abstr.*, **66**, 81868 (1967)).
373. J. K. Brown and W. G. Williams, *Trans. Faraday Soc.*, **64**, 298 (1968).
374. A. J. Tench and P. Coppens, *J. Phys. Chem.*, **67**, 1378 (1963).
375. A. K. Hoffmann and A. T. Henderson, *J. Amer. Chem. Soc.*, **83**, 4671 (1961); A. K. Hoffmann, W. G. Hodgson and W. H. Jura, *J. Amer. Chem. Soc.*, **83**, 4675 (1961).
376. A. K. Hoffmann, W. G. Hodgson, D. L. Maricle and W. H. Jura, *J. Amer. Chem. Soc.*, **85**, 631 (1964).
377. M. Masui and H. Sayo, *Chem. Pharm. Bull.* (*Tokyo*), **14**, 306 (1966).
378. A. I. Prokof'ev, and S. P. Solodovnikov, *Izv. Akad. Nauk SSSR, Ser. Khim.*, **1967**, 428, (*Chem. Abstr.*, **67**, 38179 (1967)).
379. C. Lagercrantz, *Acta Chem. Scand.*, **18**, 1384 (1964).
380. B. I. Shapiro, V. M. Kazakova, Ya. K. Syrkin and L. V. Okhlobystina, *Dokl. Akad. Nauk SSSR*, **173**, 618 (1967).
381. V. V. Bukhtiyarov and N. N. Bubnov, *Teor. Eksp. Khim.*, **4**, 267 (1968), (*Chem. Abstr.*, **69**, 56147 (1968)).
382. W. E. Griffiths, G. F. Longster, J. Myatt and P. F. Todd, *J. Chem. Soc. B*, **1967**, 533.
383. B. I. Shapiro, V. M. Kazakova and Ya. K. Syrkin, *Dokl. Akad. Nauk SSSR*, **171**, 156 (1966).
384. S. H. Cadle, P. R. Tice and J. Q. Chambers, *J. Phys. Chem.*, **71**, 3517 (1967).
385. R. Koopmann and H. Gerischer, *Ber. Bunsenges. Phys. Chem.*, **70**, 127 (1966).
386. G. O. Reikhman, J. P. Stradins, R. A. Gavars and S. A. Hillers, *Zh. Obshch. Khim.*, **41**, 906 (1970).
387. F. G. Valitova, A. V. Il'yasov, N. N. Sotnikova and S. Yu. Bailgil'dina, *Zh. Strukt. Khim.*, **6**, 777 (1965), (*Chem. Abstr.*, **64**, 6093 f (1966)).
388. E. B. Rupp, D. E. Smith and D. F. Shriver, *J. Amer. Chem. Soc.*, **89**, 5562 (1967).
389. K. P. Dinse and K. Moebius, *Z. Naturforsch. A*, **23**, 695 (1968).
390. H. B. Brooks and F. Sicilio, *J. Phys. Chem.*, **74**, 4565 (1970).

391. C. D. Russel, *Anal. Chem.*, **35**, 1291 (1963).
392. G. Cauquis and D. Serve, *J. Electroanal. Chem.*, **27**, App. 3 (1970).
393. K. Koyama, T. Susuki and S. Tsutsumi, *Tetrahedron*, **23**, 2665 (1967).
394. G. Cauquis and D. Serve, *C. R. Acad. Sci.* (*Paris*), *Ser. C*, **262**, 1516 (1966).
395. G. Brière, G. Cauquis, B. Rose and P. Servoz-Gavin, *C. R. Acad. Sci.* (*Paris*), *Ser. C*, **265**, 503 (1967).
396. W. Strohmeier and G. Popp, *Z. Naturforsch. B*, **22**, 891 (1967).
397. P. Ludwig and R. N. Adams, *J. Chem. Phys.*, **37**, 828 (1962).
398. T. Layloff, T. Miller, R. N. Adams, H. Fäh, A. Horsfield and W. Proctor, *Nature*, **205**, 382 (1965).
399. P. A. Malachesky, T. A. Miller, T. Layloff and R. N. Adams, IAEA Symposium on Biochem. Reactions, Brookhaven, June 1965.
400. T. A. Miller, R. N. Adams and P. M. Richards, *J. Chem. Phys.*, **44**, 4022 (1966).
401. M. P. Eastman, G. V. Bruno and J. H. Freed, *J. Chem. Phys.*, **52**, 2511 (1970).
402. J. M. Fritsch, Sh. V. Tatwawadi and R. N. Adams, *J. Phys. Chem.*, **71**, 338 (1967).
403. R. E. Dessy and R. L. Pohl, *J. Amer. Chem. Soc.*, **90**, 1995, 2005 (1968); R. E. Dessy, R. Kornmann, C. Smith and R. Haytor, *J. Amer. Chem. Soc.*, **90**, 2001 (1968).
404. A. E. J. Forno, M. E. Peover and R. Wilson, *Trans. Faraday Soc.*, **66**, 1322 (1970).
405. B. I. Shapiro, V. M. Kazakova and Ya. K. Syrkin, *Zh. Strukt. Khim.*, **6**, 540 (1965), (*Chem. Abstr.*, **64**, 196 e (1966)).
406. H. Rinkel and W. Windsch, *Z. Phys. Chem.* (*Leipzig*), **227**, 281 (1964).
407. A. Castellan, F. Masetti, U. Mazzucato and E. Vianello, *J. Phot. Sci.*, **14**, 164 (1966).
408. J. Phelps, K. S. V. Santhanam and A. J. Bard, *J. Amer. Chem. Soc.*, **89**, 1752 (1967).
409. E. I. Klabunovskii, R. Yu. Mamedzade-Alieva and A. A. Balandin, *Zh. Fiz. Khim.*, **42**, 1177 (1968).
410. G. S. Vozdvizhenskii, N. V. Gudin, M. S. Shapnik, N. S. Garif'yanov, and A. V. Il'yasov, *Zh. Fiz. Khim.*, **38**, 1682 (1964) **39**, 64 (1965).
411. G. Lassmann, W. Damerau, K. Lohs, N. Klimes, and Z. Baldjeva, *Z. Chem.*, **10**, 297 (1970).
412. J. Stradins, R. Gavars, G. Reichmanis and S. Hillers, *Abh. Deut. Akad. Wiss. Berlin, Kl. Med.*, **1966**, 601.
413. B. Kastening and L. Holleck, *Z. Elektrochem.*, **63**, 166 (1959).
414. B. Kastening and S. Vavřička, *Ber. Bunsenges. Phys. Chem.*, **72**, 27 (1968).
415. J. P. Stradins, R. A. Gavars, W. K. Grin and S. A. Hillers, *Teor. Exp. Khim.*, **4**, 774 (1968).
416. D. Kolb, *Ber. Bunsenges. Phys. Chem.*, **73**, 980 (1969).

CHAPTER VII

pH Measurements and their Applications

K. Schwabe, *Meinsberg Research Institute of the Technical University, Dresden, G.D.R.*

I. INTRODUCTION

The application of the pH measurement has been extended more and more from the pure chemical laboratory to the biological–medical field and is further used for process control and plant tests in numerous kinds of industry. Mostly the pH value only serves as a characterizing number, and its exact physical-chemical significance is not important. Due to new applications in pH measurement, new electrode systems and new measuring devices have been developed. Requirements for accuracy and reproducibility have been raised but measurements under extreme conditions brought in new error sources.

Since Mattock (1) has described, in detail, pH measurement in the laboratory and has summarized all available electrodes and instrumentation in *Advances in Analytical Chemistry and Instrumentation, Vol. 2*, the present publication has the aim of indicating applications in biology and medicine as well as in process control including the main questions and problems which occur when pH measurement is employed. Although it is true that in medical chemistry and in process control the pH value serves only for characterizing a definite state it seems necessary to cast light on its correlation with the hydrogen ion concentration or activity. In cases where notable differences appear between the analytical concentration of strong acids and the resulting hydrogen ion activity from the measured pH value the reasons for these differences have to be discussed. This phenomenon was observed in concentrated aqueous electrolytic solution and partly in organic solvents too. Each result can serve to give more intensive information, when its physical significance is exactly known. Therefore even such results will be mentioned, which can help to solve the problem of how far the conventional pH value can serve as a criterion for hydrogen ion activity.

II. THEORETICAL BASIS OF THE ELECTROCHEMICAL pH MEASUREMENT

The conventional pH value, measured by cells with transference, is considered to be a numerical value without physical importance where the test method determines the value. This interpretation is allowed since the cell voltage, used for calculating the pH value, includes diffusion potentials with unknown values, which change with ionic strength and the ion type of the test solution.

liquid junction paths.

If, therefore, the activity of H^+ is considered to be a measure of the acidity of a non-aqueous solution, principally, this cannot be indicated by means of the conventional pH value and no thermodynamically exact method, of course, can serve for determination of individual ion activity.

For all attempts to determine the individual ion activity of hydrogen ions a hypothesis is required which cannot be thermodynamically-based. These attempts would be valid as methods for determination of the thermodynamically defined activity only if it could be shown that they provide identical values, but this, of course, is not possible. But when, independently of each other, different measuring systems, used to determine the pH value, yield the same result, one is entitled to attribute a certain physical significance to this value. In critical cases where remarkable discrepancies appear between $-\log c_{H^+}$ and the conventional pH value, especially in concentrated solution of neutral salts, it is suitable to measure with cells without transference and with convenient electrochemical methods as well as with spectrophotometrical methods (2).

Table I demonstrates the important characteristics of measurement and methods for acidity determination, where acidity is the value proportional to the molality of hydrogen ions. While pwH and paH are referred to a fictitious standard state where the e.m.f. can be obtained only by extrapolation, the conventional pH unit relates to the buffer solution of known pH.

TABLE I

Values	Methods (cells for measurement)
1. $paH = -\log a_{H^+} = -\log m_{H^+} f_{H^+}$	Cannot be measured exactly but in Debye–Hückel range identical with ptH
2. $pcH = pSH = -\log m_{H^+}$ (Sörensen pH)	Cannot be measured directly, at ionic strength $J \to 0 : pSH = paH$
3. $ptH = -\log m_{H^+} f_{\pm}$	$(Pt)H_2/HA, MeA(s)/Me$; $$ptH = \frac{E^0 - E}{2F_N}; \ F_N = \frac{2 \cdot 303 RT}{F}$$
$m_{\pm} = m_{H^+}$	$paH = 2ptH + \log m_{H^+} f_{A^-}$
4. $pwH = -\log(m_{H^+} f_{H^+} f_{A^-})$	$(Pt)H_2/HA, MeA(s)/Me$; $$pwH = \frac{E^0 - E}{F_N} + \log m_{H^-}$$ $paH = pwH + \log f_{A^-}$ $E^0 =$ standard e.m.f. of the cell for $m_{HA} = f_{HA} = 1$; HA = acid with the anion A^-
5. $H_0 = -\log m_{H^+} \dfrac{f_H \cdot f_B}{f_{BH^+}} = pk + \log \dfrac{m_B}{m_{BH^+}}$	Measured spectrophotometrically from dissociation equilibrium of the indicating acid: $$BH^+ \underset{f_B}{\overset{K}{\rightleftharpoons}} B + H^+ ; pK = -\log K,$$ $$paH = H_0 + \log \frac{f_B}{f_{BH^+}}$$
6. $pH_r = -\log K_r (K_r =$ velocity constant$)$ $pH_r = paH - \log K_r^0$	From the velocity constant of a reaction catalysed by H^+, $K_r (K_r^0 =$ velocity constant for $a_{H^+} = 1)$
7. Conventional pH value $pH = pH_s - \dfrac{(E - E_s)}{F_N}$ $pH_s =$ known pH value of a buffer solution. $E_s =$ e.m.f. of the cell including diffusion potential if a buffer solution is used, but E, if the test solution is used.	$(Pt)H_2/HA/_{Ed}/KCl(sat)Hg_2Cl_2(s)/Hg$. The hydrogen electrode can be replaced by any electrode, indicating H^+ and the calomel electrode can be replaced by any other reference electrode.

Different pH scales have been listed according to the respective buffer solution (1). The main difficulties in using a collective internationally recognized pH scale, which is very desirable, are that sometimes (e.g. British standards) the diffusion potentials are enclosed in the cell voltage, but other standard buffers are measured by means of cells without transference (e.g. the U.S. standards). Bates and Guggenheim proposed using two standard buffer solutions to measure their pH values in cells without transference which are near the pH_x of the test solution ($pH_1 > pH_x > pH_2$) and determining pH_x in cells with transference according to

$$\frac{pH_x - pH_1}{pH_2 - pH_1} = \frac{E_x - E_1}{E_2 - E_1}$$

but this is complicated and in this case it is presumed that the pH_x is approximately known and that the diffusion potential is constant between pH_1 and pH_2. But this assumption is not satisfied when the ionic strengths of the three solutions do not coincide.

Even when the reference pH value could be determined with extreme accuracy (even then the general mean error is about 0·002) and when $pH_s =$ adequate paH_s, the value of the test solution, obtained by comparison, represents only a numerical value, which is not a sufficiently accurate measure for a_{H^+}, since generally in the test solution the diffusion potential would differ from that of the buffer solution and can be obtained only by measurement with a cell without transference and by extrapolation of the ionic strength to zero (i.e. conventional pH units are numerical values, which can be compared reasonably accurately only with solutions which do not differ in other properties, especially in ionic strength (e.g. blood or other equivalent biological substrates)).

Whether carried out by redox electrodes or by membrane electrodes, the electrochemical pH measurement depends on the equilibrium of the electrochemical potentials between two phases ($\tilde{\mu}_I$ and $\tilde{\mu}_{II}$). From the difference between the inner electrical potentials, the Nernst equation results as well as the relationship between cell voltage and pH unit (3). With redox electrodes, especially with the hydrogen electrode, the equilibrium adjustment $\tilde{\mu}_I = \tilde{\mu}_{II}$ does not take place immediately and can be retarded significantly by easily adsorbable molecules, especially by macromolecular substances and colloids, as well as by high concentrations of electrolytic solution. Using glass electrodes, the equilibrium adjustment can be restrained considerably in the same manner. Then accuracy and correctness is required to be very high, regardless of the fact that the pH value is used only for comparison, the equilibrium adjustment must be as exact as possible, otherwise the results do not serve as a comparison and are influenced by the duration of the measurement. Since the cell voltage equalizes asymptotically to the value of the equilibrium, in unfavourable circumstances, even after hours, small deviations from the final value can occur. Since such creeping adjustments exist also when cells without transference are used together with hydrogen electrodes or glass electrodes, the reason for this cannot be the diffusion potential, which varies with time (4). Since spectrophotometric measurements do not show these retardations,* it is advisable, as far as possible, to use this method for control in those cases where retarding adjustment of the equilibrium cell voltage might be present (4).

Table I demonstrates the values of pwH and ptH as well as their correlation with the paH units which are not directly available. Since pwH and ptH values are obtained from cells without transference, these values can be considered to be thermodynamically correct and at decreasing ionic strength, they approach paH. If exact data of activity are required or if very high diffusion potentials are suspected, when the pH value is measured in the conventional way, it is advisable to determine these values also. For normal acidity measurements, these values are scarcely of any importance, but they are very useful for standardizing buffer solutions (6). In the following sections the electrodes for indicating and for reference purposes in conventional pH measurements are discussed only as far as new systems and new application methods have occurred since the survey given by Mattock (1).

III. MEASURING ELECTRODES

A. Redox Electrodes

In principle, each redox equilibrium is appropriate, if hydrogen ions take part, to determine the pH value.

$$A + nH^+ + ne^- \rightleftharpoons AH_n$$

$$\varepsilon = \varepsilon^0 + \frac{F_N}{n} \log \frac{a_{AH_n}}{a_A \cdot a_{H^+}^n}; \qquad F_N = \frac{2.303RT}{F} \qquad (1)$$

Except the already known redox electrodes (hydrogen, quinhydrone, metal/metal oxide) new systems have scarcely been suggested, probably since the pH measurement by means of these systems is disturbed by other redox processes; on the other hand, the glass electrode which is independent of redox systems is available. Moreover only a few redox processes can adjust quickly and reversibly.

1. The Hydrogen Electrode

In spite of many inconveniences (production of extremely pure hydrogen, consideration of the atmospheric pressure p, of the vapour pressure of the test solution p_d and of the hydrostatic over-pressure p_h; frequently the platinizing of the electrode must be regenerated, etc.) *the hydrogen electrode* has

---

* About test methods see [5a]. Concerning the Hammett function as a measure for acidity see [5b].

not lost its importance for the standardization of buffer solutions or for checking pH units, which have been obtained by means of other electrodes since the hydrogen electrode depends on a simple reversible equilibrium:

$$H_2 \rightleftharpoons 2H^+ + 2e^-$$

$$\varepsilon_H = \varepsilon_H^0 + F_N \log \frac{a_{H^+}}{(a_{H_2})^{1/2}} \tag{2}$$

Horiuti (7) and coworkers have studied in detail the reaction mechanism of the platinum hydrogen electrode and, considering the determination of the isotope exchange between hydrogen and deuterium, of the pH influence and the values of the activation enthalpy; they stated that the mechanism

$$H^+ + e \rightleftharpoons H_{ad}; \qquad H_{ad} + H^+ + e^- \rightleftharpoons H_2(g)$$

or

$$2H^+ + e^- \rightleftharpoons H_{2ad}^+; \qquad H_{2ad}^+ + e^- \rightleftharpoons H_2(g)$$

is rate-determining under equilibrium conditions in acid solution. Since convention stated that ε_H^0 is set to zero for all temperatures and since a_{H_2} is replaced by fugacity and since for $p_{H_2} \cong 1$ the fugacity coefficient is very close to unity, for H_2 at atmospheric pressure, the equation between ε_H and paH is simplified to be:

$$\varepsilon_H = -F_N \cdot pH$$

If very exact measurements are required, e.g. for determination of pwH and paH with cells without transference, due to Hills and Ives (8) the semi-empirical correction must be used for the measured potential ε_m for the vapour pressure p_d of the solvent and for the hydrostatic over-pressure p_h

$$\varepsilon_H = \varepsilon_m + F_N \log \frac{p - p_d + 0.42 \cdot p_h}{760} \tag{3}$$

p_h results from the over-pressure in mm water column p_n due to $p_h = p_n/\rho_{Hg}$; ρ_{Hg} = density of mercury.

Concerning p_d, chiefly, the correction cannot be neglected, especially at high temperature and when very volatile solvents are used. The equilibrium adjustment depends on platinizing and in aqueous organic solution and especially in non-aqueous organic solution as well as at high ionic strengths, especially when the salt concentration is very high ($J > 2$), the process proceeds very slowly. In such media, the hydrogen electrode is evidently very sensitive to polarization; therefore it is advisable to use the compensation method together with a high-ohmic amplifier (9). Moreover, especially in organic solvents a better potential adjustment is received, when using only slight (gree) platinizing, and when this platinizing is regenerated after each measurement (10). Polished platinum in the form of a cone on the bottom of the cell, together with fresh platinum mud (8) is very suitable when the electrode tends towards quick fatigue or contamination phenomena.

When platinized sheets are employed, it is very important to clean the sheet before platinizing. Palladium electrodes, covered with palladium black, mostly, furnish potentials which differ somewhat from those of the platinum electrode but recent studies have shown (13) that the phase-state of the palladium is very important. If cells without transference are used, the reproducibility of the voltage with hydrogen electrodes, of course, depends on the reference electrode too. Only if this electrode is prepared very carefully and is completely cleaned, an agreement within $10\,\mu V$ might be reached; comparing different hydrogen electrodes in one and the same aqueous solution, the potential differences are rarely smaller (1).

Bianchi (11) and coworkers have suggested the use of the hydrogen electrode in the form of a porous graphite disk electrode, which is activated by platinum black and needs no permanent hydrogen stream. This electrode shows a bias of 0·03 mV and behaves constantly for several days up to 0·01 mV. Figure 1 shows this electrode which needs only a very small liquid volume.

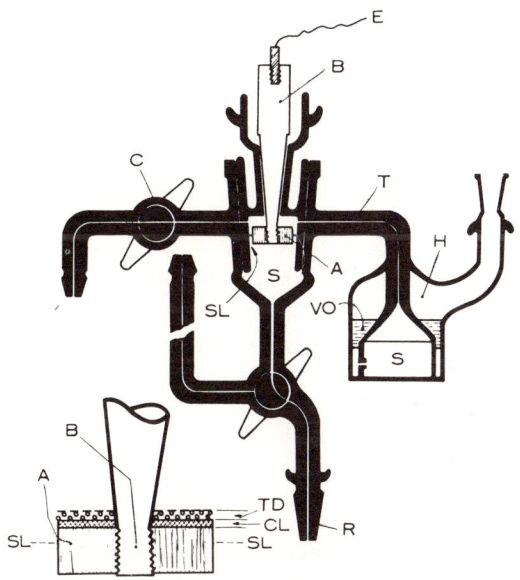

Structure of the hydrogen electrode: A, porous graphite disk; B, screw stem of impermeabilized graphite; C, gas inlet stopcock; CL, catalyst layer; E, electrical connection; H, hydraulic seal; R, to second half-cell; S, solution; SL, solution level; T, tube connecting to hydraulic seal; TD, Teflon drop network; VO, vacuum oil layer on solution.

Fig. 1. Structure of the hydrogen electrode.

The platinum black is not covered with liquid but is wetted by capillary action. A teflon screen protects the platinum black and prevents the formation of a liquid layer on the platinum black.

To avoid the passing of hydrogen, which can produce a pH shift by expelling volatile acids or bases Nylen (12) has already used a palladium electrode which is charged before measurement with H_2. Recently Vasile and Enke (13) have studied this electrode in detail. In the coexistence range of the α and β phases of palladium the potential is independent of the palladium composition and independent of the pH value, the potential changes linearly with temperature, $0.65 \text{ mV degree}^{-1}$.

A hydrogen electrode from Raney-platinum, which receives the hydrogen by cathodic evolution with a current density of 2 mA cm^{-2}, is described by Binder and coworkers (13a). The specific surface of the Raney-platinum is so large that the true current density is smaller than $1 \mu\text{A cm}^{-2}$ and therefore the polarization of the electrode is negligible. This electrode seems useful for comparison of polarization effects in electrode kinetics.

2. Quinhydrone Electrode

As well as the hydrogen electrode, the quinhydrone electrode is very sensitive to other redox systems as contaminations such as H_2S, As_2O_3, CO, CN^- and colloidal dispersions (1, 6, 14). But the quinhydrone electrode cannot be used in the alkaline range (pH $\geq$ 8) (12). This electrode has been carefully studied by Schwing (15) in connection with Ag/AgCl(s) and $Hg/HgCl_2$(s) electrodes in $3 M$ NaCl. By measurements of the Au-quinhydrone electrode versus the hydrogen electrode in hydrochloric acid, the standard potential was stated to be $+(699.74 \pm 0.03 \text{ mV abs.})$. This result was in very exact agreement with those noted in former investigations. Consequently, the salt effect is independent of pH (1 to 8.25 pH). Gabbard's (16) results, that the error increases considerably when pH > 5.5 was stated to be caused by the acid behaviour of the quinhydrone. The equation for correcting the salt error ΔpH already reported by Hovorka and Dearing (17), see also (17a),

$$\Delta\text{pH} = K \cdot m_{\text{salt}} \tag{4}$$

where the proportion coefficient K, varying with the salt species, can amount from -0.04 up to $+0.03$, is extended. But this enlargement is of little help since the correction factors are known only partly. It is best to use other electrodes which do not show this error—firstly caused by a different shift of solubility of quinone or hydroquinone with the concentration of salt—when measurements should be made in concentrated salt solutions. Generally, for pH measurements and for potentiometric titration in the chemical laboratory and chiefly for industrial pH measurement the quinhydrone electrode, as well as the hydrogen electrode, are of less importance today.

Probably, this point of view will not change if, as suggested by Fliss and Worobjew (18), instead of gold or platinum the cheaper titanium were used. A successful application of the quinhydrone electrode for pH-measurement in hydrogen fluoride solutions was recently described by Warren (16a). He could determine the association constant:

$$K_d \frac{[H_2F_2]}{[HF]^2} = 2\cdot7 \pm 0\cdot3$$

and found a good correspondence between pH, measured with the quinhydrone electrode and the glass electrode above pH $2\cdot8$ in pure HF ($\sim 3 \times 10^{-3}$ Mol 1^{-1}) and above pH $4\cdot9$ in $1\,M$ NaF.

3. Metal–Metal Oxide Electrodes

(a) *The antimony electrode.* On account of its mechanical stability antimony is sometimes used for industrial pH measurement. Also the bismuth electrode, occasionally, is employed especially in the alkaline range. But today generally, the glass electrode is preferred, because it is less responsive to redox systems and it is excellently reversible in the chemical laboratory as well as for process control and industrial tests. For soil tests recently Sovova (19) has recommended an electrode pressed at $1200\,\mathrm{kp\,cm^{-2}}$ out of antimony powder and antimony (III) oxide. Compared with cast antimony electrodes, they show a higher speed in potential adjustment and when containing 5–10 per cent Sb_2O_3, the maximum deviation was $\pm1\cdot5\,\mathrm{mV}$, while in the case of cast electrodes the deviation was $\pm8\cdot5\,\mathrm{V}$. Comparing tests of different soil samples has shown partly extreme differences from those results, if the glass electrode or colorimetric measurement were applied (e.g. pH (glass) $5\cdot42$, pH (pressed antimony electrode) $4\cdot78$).

Cuta and Havelka (20) have used cast electrodes of very pure antimony at temperatures above 80°C in the air supply and have noted very fast potential adjustment simultaneously with sufficient measuring accuracy. When the air supply was replaced by N_2, this type of electrode was completely useless. The mechanism of the antimony electrode has been studied by the same authors. In principle, the metal electrode follows the equilibrium:

$$m\,Me + n\,H_2O \rightleftharpoons Me_mO_n + 2n\,H^+ + 2n\,e^-$$

The reversible adjustment depends on formation and contact of both solid phases and can be influenced considerably by anion adsorption and contaminations.

(b) *The bismuth electrode.* This electrode has been used for potentiometric chromatography and for potentiometric titration in the form of microelectrodes molten in glass capillaries (21). Using this electrode in the range from pH 3 to 12 at 20°C the potential at $\varepsilon = 0\cdot045$ to $0\cdot055$ pH versus the $0.1\ M$ KCl, Hg_2Cl_2/Hg electrode was measured. Recently, bismuth microelectrodes have been used for pH measurements at the surface of ion exchange membranes where hydroxides of rare earth metals were electrolytically sedimented, to be used as β sources (22). Abdalla (23) carefully studied the pH behaviour of bismuth electrodes in dependence of electrolyte composition.

(c) *Other metal electrodes.* Other metal electrodes are, for example, made from molybdenum, tungsten, columbium, tantalum, germanium or arsenic and Raney-nickel do not indicate any advantage compared with antimony electrodes or bismuth electrodes (24). Their potential adjustment is smaller than the Nernst factor (59 mV at 25°C) and mostly linearity occurs only in a small pH range. Therefore they are used only occasionally for potentiometric pH titration. It must be noted that generally all redox electrodes are not convenient for pH measurement in the presence of strong oxidizing or reducing agents. Therefore, the glass electrode generally has been preferred for use in the chemical laboratory and for industrial pH measurement.

B. Membrane Electrodes

If between two phases (l and f) an equilibrium of one type of ions (J) is adjusted it follows from the condition of equilibrium that

$$_J\tilde{\mu}_l = {}_J\tilde{\mu}_f$$

$$_J\tilde{\mu}_l = {}_J\tilde{\mu}_1^0 + RT\ln\ {}_Ja_1 + nF\ {}_J\varphi_1 = {}_J\mu_f^0 + RT\ln\ {}_Ja_f + nF\ {}_J\varphi_f$$

for the equilibrium potential

$$_J\varphi_f - {}_J\varphi_1 = \varepsilon_J = ({}_J\mu_1^0 - {}_J\mu_f^0)(nF)^{-1} + \frac{RT}{nF}\ln\frac{{}_Ja_1}{{}_Ja_f} \tag{5}$$

ε_J at the phase boundary:

$$\varepsilon_J = \varepsilon_J^0 + \frac{RT}{nF}\ln\frac{{}_Ja_1}{{}_Ja_f} \tag{6}$$

where n indicates the charge number of the ion, and ε_J^0 is the standard potential, and all other symbols have the usual meaning. Assuming that the same reference state is used for both phases, i.e. ${}_J\mu_f^0 = {}_J\mu_1^0$, ε_J^0 is omitted, and the equation for the phase boundary potential is:

$$\varepsilon_J = \frac{RT}{nF}\ln\frac{{}_Ja_1}{{}_Ja_f} \tag{7}$$

From this equation the activity of the ion in one phase can only be calculated if the activity is known in the other one. Here, the equilibrium potential would represent a diffusion potential between two liquids of different activities and it would be omitted if the different concentrations are equalized by diffusion. Combining two phase boundaries, with an identical (unknown) value in the first one (e.g. f) and with an unknown ($_ja_1^x$) together with the known ($_ja_1$) in the second stage (e.g. l) from (6) a potential difference $\Delta\varepsilon_J$ is obtained and now the unknown value can be calculated

$$\Delta\varepsilon_J = \frac{RT}{nF} \ln \frac{_ja_1^x}{_ja_1} \tag{8}$$

As can be seen, ε_J^0 is omitted too, when $_{jl}\mu_f^0 \neq {}_{jl}\mu_1^0$ and when the difference is constant. When such a difference of membrane potentials should be used for activity determination of a certain ion, its activity in one phase must be proportional to that in the other one (independent of their values) and the potential must be influenced only by this ion. The constancy of the activity proportion a_1/a_f is due to partition equilibrium, where an activity difference in the equilibrium must be balanced on account of different states (μ^0 values)

$$\left(\frac{a_1}{a_f} = e^{(\mu_f^0 - \mu_1^0)/RT}\right)$$

Often, an attempt has been made to use membranes made out of different materials, especially of organic high polymers for the measurement of hydrogen activity or pH respectively, but up to now, the glass electrode, proposed firstly by Cramer (25), is doubtless the most suitable one.

1. The Glass Electrode

This is the generally used electrode for pH measurement in the range between 0 and 14 pH units. It can be used as well in the presence of strong oxidizing (e.g. chromic acid) or reducing agents (e.g. hydrazine) as under extreme conditions of temperature or pressure. In this pH range the change of potential difference of the glass electrode follows correctly the Nernst equation when the concentration of alkaline ions is small and the glass composition is adequate. Therefore, if on one side of a membrane from the electrode glass a buffer solution of known pH is present, pH_B, and on the other side the test solution exists with pH_x, due to (8) follows:

$$\Delta\varepsilon = F_N(pH_B - pH_x)$$

$$pH_x = pH_B - \frac{\Delta\varepsilon}{F_N} \tag{9}$$

Chiefly at high alkaline concentration, an error may occur which depends upon the glass composition. This error is explained by the ion-exchange theory of the glass electrode. This theory will be considered only informatively, if necessary, for the understanding of the application of the glass electrode for pH measurement.*

(a) *Theory of the glass electrode.* As already shown, in pure partition equilibrium of hydrogen ions between the glass phase and the aqueous solution, the ion activities must be different in both phases, if the chemical standard potential is different.

The potential constancy of glass electrodes over nearly any interval in aqueous solution, demonstrates that no diffusion potentials exist but that phase boundary potentials are present. The glass membrane adsorbs some water on the surface and forms a so-called swell film, but the hydrogen ion in glass is by no means the same as in aqueous solution since in glass, they are fixed to located anions of silicic acid. Glass membranes are not permeable to hydrogen ions, so they do not form diffusion potentials; this was shown by tritium marking (27). Recently by Doremus (28) on account of different mobilities of Na^+ and H^+ in glass, diffusion potentials in glass membranes are assumed. But experiments made to test the permeability of swollen glass membranes for H^+ contradict this assumption as the constancy of potentials does as well and the fact that the adjustment rate is extremely high. But the demand that only the hydrogen electrode may take part in formation of the phase boundary potential is not completely fulfilled by even glass electrodes. The cations in glass are bound electrostatically to the anions of silicic acid, forming the lattice and they can be replaced by other cations. If, for example, in the glass phase the sodium ions are marked with Na^+, of a sodium potassium silicate, and the hydrogen ions with H^+, and in solution the ions are Na_1^+ and H_1^+, according to the assumption of Nikolski (29), the ion exchange equilibrium between the two phases is adjusted

$$Na_1^+ + H^+ \rightleftharpoons H_1^+ + Na^+$$

and the equilibrium constant is:

$$K = \frac{[H_1^+][Na^+]}{[H^+][Na_1^+]} \tag{10}$$

Here $[H_1^+][Na^+]$, etc., should be the activities of the ions in the respective phases.

From (6) as an equilibrium condition must follow the phase boundary potential

$$\varepsilon = \varepsilon_{H^+}^0 + F_N \log \frac{[H_1^+]}{[H^+]} = \varepsilon_{Na^+}^0 + F_N \log \frac{[Na_1^+]}{[Na^+]} \tag{11}$$

* For the theory of the glass electrode see reference 26a and references in 26b.
* With respect to the transient effect of electrical behaviour of glass membranes not in accord with the diffusion potential theory see newer work 28b.

Using (10), $[H_i^+]/[H^+]$ can be written:

$$\frac{[H_i^+]}{[H^+]} = \frac{[H_i^+ + K \cdot [Na_i^+]}{[H^+] + [Na^+]}$$ (12)

When it is assumed that the total of all exchangeable cations in glass is constant, caused by anions which are fixed in the network and its activity is equal the number of ions (that means the activity coefficient is unity) then

$$\varepsilon = \varepsilon^{0'} + F_N \log (a_{H^+} + K a_{Na^+})$$ (13)

results* for the potential in the phase boundary; where

$$\varepsilon^{0'} = \varepsilon_{H^+}^0 - F_N \log [H^+ + Na^+]$$

It is easy to calculate the constant K from the potential change with a_{Na^+} at constant pH and it is dependent on the glass composition. Using pH glasses the value of K is about 10^{-11}. In the presence of hydrogen ion activity of 10^{-7}, a sodium ion activity of unity changes the term by 10^{-2} per cent but this small influence cannot be measured. If on one side of the glass membrane the solution is completely free of sodium with $a_{H^+} = 10^{-7}$ and on the other side the solution has $a_{H^+} = 10^{-7}$ but with $a_{Na^+} = 1$

$$\Delta\varepsilon = F_N \log \frac{a_{H^+} + K a_{Na^+}}{a_{H^+}}$$ (14)

should be zero.

Indeed this value would not be found, because Na^+ changes the activity of H^+ and since most of the glass membranes show 'asymmetry potentials' (see page 515). Furthermore the assumption that the activity coefficients of H^+ and Na^+ in the glass phase are one and that they are independent of interaction of ions, can be considered only for a first approximation and using metal ion responsive electrodes, useless values are obtained. Presenting an extended theory, Nikolski (30) assumed differential linkage of the hydrogen ions to the anions of the glass and was able to explain more correctly the transfer from the H^+ function to the Na^+ function, especially for glass containing Al_2O_3 and B_2O_3.

Equations (13) and (14) give sufficient explanation for the 'alkali error' of pH glass electrodes demonstrating that it is a function of the hydrogen activity and sodium ion activity as well. Therefore it is necessary to indicate this error for a fixed pH value and for a given concentration of sodium ions. As can easily be seen this can be calculated by

$$\Delta\varepsilon_F = F_N \log \left(1 + K \frac{a_{Na^+}}{a_{H^+}}\right)$$ (14a)

and is smaller, for smaller K. When in some glasses, Na^+ is replaced by Li^+,

_____
* a_{H^+} and a_{Na^+} are the activities of H_i^+ and Na_i^+ respectively in the solution.

Cs^+ or Rb^+, especially when two different alkali ions are used simultaneously, compared with the sodium ions, which mostly are used in aqueous
solution, the exchange constant can be made so small ($< 10^{-14}$), that even
at $pH = 14$ in $1\,M\,Na^+$ solution at room temperature the alkali error does
not exceed $10\,mV$, that is about $0 \cdot 2\,pH$ units. Other cations show an error
which is considerably smaller and especially the quaternary cations, which
evidently are not exchangeable, at pH 14 do not show any deviation from the
Nernst equation (9). On the other hand, the exchange constant increases
considerably with increasing temperature, so that already at pH 10 to 11, in
concentrated Na^+ solution the alkali error is too large and is not tolerable
for correct pH measurement. Especially at high temperature, in alkaline
solution the glass electrode potential is shifted temporarily on account of the
change in glass composition by attack of the solution.

The fact that the glass membrane forms on its surface a swell film by
absorption of water does not furnish identical chemical standard potentials
in glass and in solution, as already mentioned, but at high acid concentration,
by changing the water activity in the solution or by penetration of acid into
the swell film, an acid error can be caused (see page 510). The penetration of
acid ion pairs could be proved by radioactive marking in hydrochloric acid
(31). It increases with the fugacity of the acid, is largest when hydrochloric
acid is used and therefore the acid error is the largest too. If in the swell film,
which is in contact with the strong acid, the hydrogen ion activity (a_A) is
larger than in the swell film in contact with the comparing buffer solution
(a_B), according to equations (6) and (8) respectively the acid error must occur

$$\Delta\varepsilon_S = F_N \log \frac{a_B}{a_A} < 0 \tag{15}$$

When using non-volatile acids, such as sulphuric or phosphoric acid, this
phenomenon is not important. Dole (32) has shown that the acid error
measured at very high concentration of such acids is caused by a decrease
of water activity in the solution phase. But regarding this, he proposed the
glass electrode potential to be the diffusion potential with the transference
number of H^+ equal to unity in the glass phase and the acid error results to be

$$\Delta\varepsilon_S = F_N \log a_{H_2O} \tag{16}$$

i.e. a negative value is obtained. This relation can be obtained, also utilizing
the theory of phase boundary potentials, when assuming the corresponding
correlation between the differences of the chemical standard potentials. In
every case, the acid error, measured in non-volatile acids, coincides sufficiently with equation (16) (31). The influence of the water activity in the test
solution on the potentials of the glass electrode is very important for pH
measurement by means of glass electrodes in organic solutions or in mixtures

of them with water, the reduced water activity furnishing a supplementary potential difference according to equation (16). But in this case also the penetration of the corresponding solvent into the swell film must be considered. Correctly stated, the pH in such solvents can be measured only by comparison by means of the glass electrode with buffer solutions in the same solvent. But difficulties occur, for the pH value of the buffer solution must be exactly known (see page 551). Recently Booksay and coworkers (32a) tried to interpret the acid error in glass electrodes, applying the principle of the additivity of the activities in glass but no explanation about the specific differences of acid anions was given. In a more recent work Csàkvári and coworkers (32b) have dissolved the surface films of the glass gradually in hydrofluoric acid solutions and determined the content of Na^+ (resp. Li^+) in different depths of the surface by measuring the content in the hydrofluoric acid solution. After thirteen days' treatment in water they found a diminished content (50 to 90 per cent) of Na^+ in the first layers compared with the compact glass. The loss of Na^+ increases with the temperature of water treatment and reaches to larger depths (up to 2.5×10^{-3} mm). By treatment with ethanol the loss is restricted to small depths. Lithium glass gives by treatment with water a much smaller loss of Li^+ than sodium glass of Na^+. The authors believe, that this loss of cations is responsible for the errors incurred with the glass electrodes.

This change in the content of cations on the surface of glass electrodes by leaching was also investigated by Bach and Baucke (32c) by sputtering equipment and determining the ions (Li^+) by luminescence. After leaching with water solutions of different pH and Li^+ concentration they obtained a concentration profile, which indicates a loss of Li^+ decreasing with pH and with Li^+ concentration in the leaching solution. The results are interpreted by the assumption of an ion exchange equilibrium. On the other hand, Tadros and Lyklema (32d), measuring the surface charge of Na^+ and K^+ sensitive glasses in dependence of pH found no connection with the content of Li^+, Na^+, K^+, Cs^+ and $(C_2H_5)_4N^+$ in the solution. From pH 4 to 10 the surface charge changes independently of the cations, from $\sim +150$ to $-350 \ \mu C \ cm^{-2}$ in the same manner; this is not easy to understand by the ion exchange theory. The authors assume from their results that there is no relation between the response of the glass electrode (from the investigated glasses) toward a given cation and 'the affinity of that cation for the glass'.

2. Range of Measurement of the Glass Electrode (Acid Error and Alkali Error)

As already stated in previous chapters, the error shown with glass electrodes depends on hydrogen activity as well as on cations and anions present in the solution and on water activity as well. Generally no pH range can be indicated within which the glass electrode operates correctly; moreover, the total composition of the solution must be considered as well as the temperature.

(a) *Acid error.* At very high concentration of strong acid in aqueous solution, using the glass electrode, the measured pH is higher than that measured by a hydrogen electrode. (This acid error is smaller in non-volatile acids than in volatile ones and here the acid error increases with time at a higher rate than in non-volatile acid. This behaviour of the acid error corresponds, as already shown, with an explanation of the penetration of volatile acid into the swell film and the change of the water activity by concentrated non-volatile acids such as H_2SO_4 and H_3PO_4.) The acid error of the glass electrode can be correctly checked by a cell without transference

$$\text{Hg/Hg}_2(\text{Ac})_2(\text{s}), \text{HAc}, \text{NaAc/Glass/Acid X/H}_2(\text{Pt}) \quad (a) \quad U_1$$

when comparing its e.m.f. with the cell

$$\text{Hg/Hg}_2(\text{Ac})_2(\text{s}), \text{HAc}, \text{NaAc/H}_2(\text{Pt}) \quad (b) \quad U_2$$

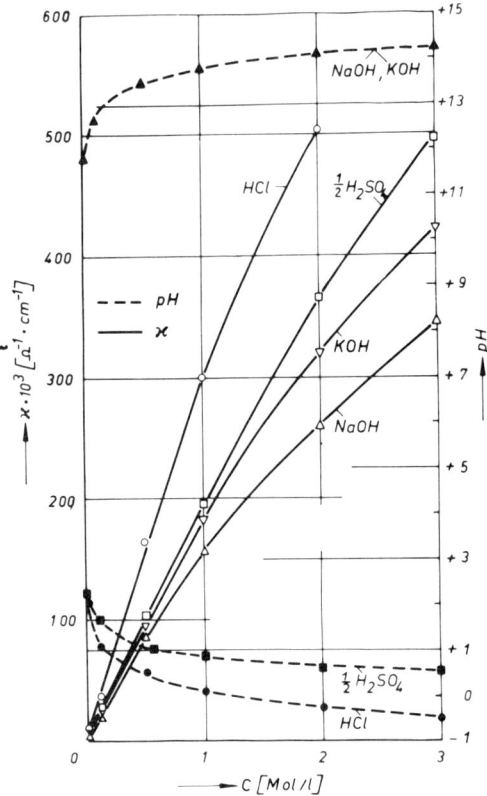

Fig. 2. Changes of the specific conductance κ and of the pH value with the concentration of the strong acid and base at 25°C.

At equal concentrations of acetic acid and sodium acetate the difference between the cell voltages U_1 and U_2 equal to $\Delta\varepsilon_s$ indicates the acid error (33). The tension of cell (a) U_1 is the sum of the potentials of the Hg/Hg_2Ac_2-electrode $\varepsilon_{Hg} = \varepsilon^0 - F_N \log a_{Ac^-}$, the glass electrode $\varepsilon_{ge} = F_N \log a_{H^+} - F_N \log a_{XH^+}$ and the hydrogen electrode $\varepsilon_H = F_N \log a_{XH^+}$, that means that U_1 is independent of a_{XH^+} (in the acid) and equal to U_2, the tension of cell (b), if there is no acid error.

In principle, the acid error of the glass electrode becomes negligible not before the concentration range is reached where concentration test by pH measurement is considerably more inexact (caused by the logarithmic relation between cell voltage and activity) than other analytical methods as conductivity measurement or other methods (34). Figure 2 demonstrates the change of specific conductivity κ and of the pH in strong acids and strong bases with concentration. Due to the more sensitive conductivity shift, the

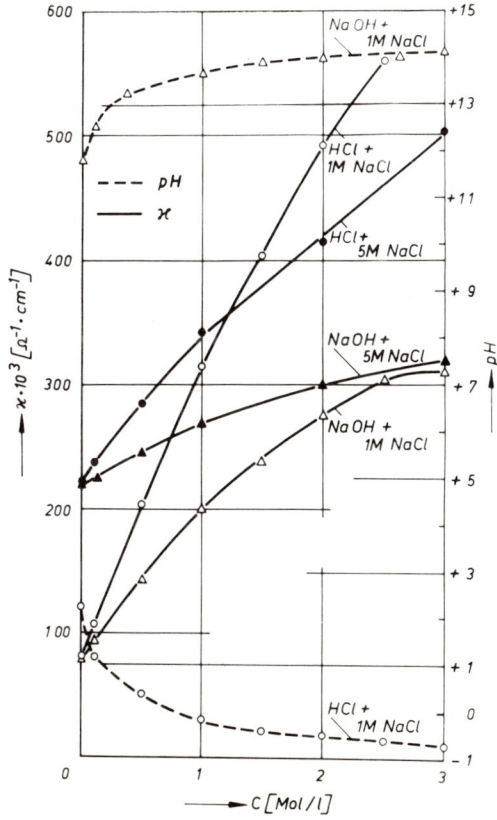

Fig. 3. Changes of the specific conductance κ and of the pH value with the concentration of HCl and NaOH in 1 M NaCl and in 5 M NaCl at 25°C.

concentration of acids and bases can be determined more correctly than by pH measurement. This is true even in the presence of very high concentrations of neutral salts, as can be seen in Figure 3. Here, the pH shift in 5 M NaCl (not drawn) is even smaller.

Even in hydrochloric acid, where the acid error is the largest one, the error is noticeable only after the concentration exceeds 1 mole l^{-1} especially if the measurement is carried out very quickly. Figure 4 shows various acid

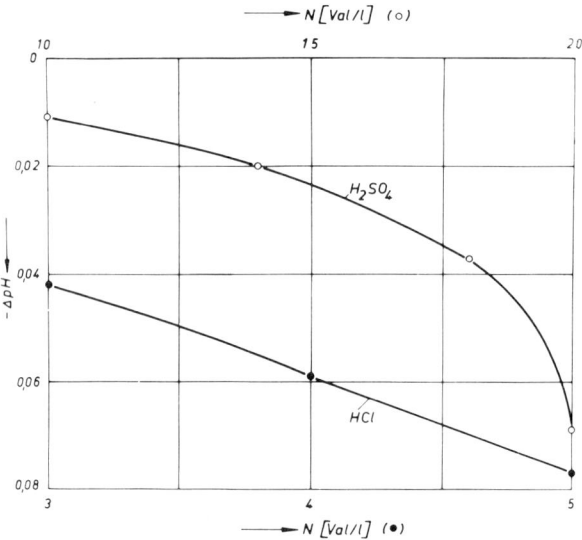

Fig. 4. Acid error ΔpH of HCl and H_2SO_4 in dependence of the acid concentration at 25°C. The measurement was carried out 15 min. after immersing the electrode.

errors (ΔpH) of hydrochloric acid and sulphuric acid as a function of acid concentration. The cell voltage was measured about 15 minutes after immersing the glass electrode into the acid solution. As electrode glass, the MacInnes glass was applied (72 per cent SiO_2, 22 per cent NaO and 6 per cent CaO). Figure 5 demonstrates that the acid error increases extremely quickly with time after the immersion of the electrode, especially in the presence of hydrochloric acid and can be indicated only for a certain time limit. Therefore, the pH measurement in such concentrated acid solution is rather useless. In addition, the acid error grows very quickly with increasing temperature. Figure 6 shows the relative change of rate of the acid error with the temperature. From the behaviour of the glass electrode one sees that it is impossible to perform suitable pH measurements in aqueous solution of strong acid for concentrations more than 1 M by means of this electrode. Utilizing less soluble glasses, the error can be reduced, but at least in

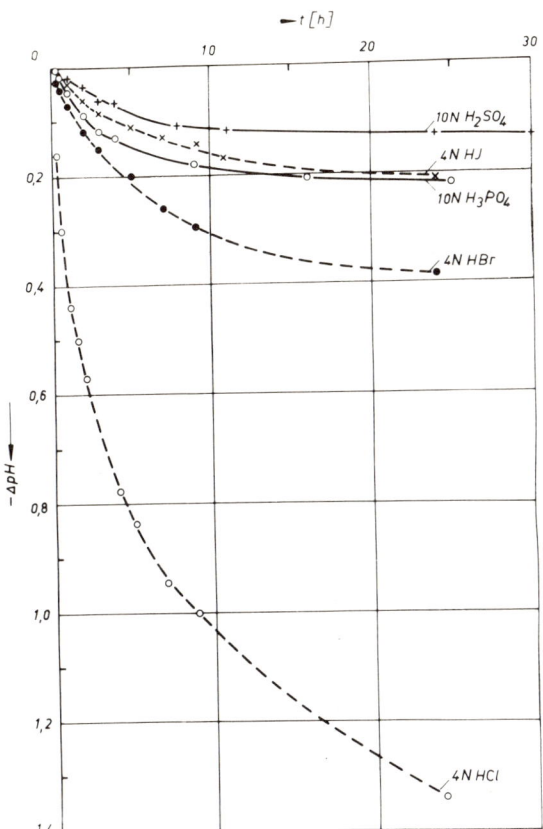

Fig. 5. Increase of the acid error of the glass electrode in
$4\,N$ HCl, HBr, HI and in $10\,N$ H_2SO_4 and in $10\,N$ H_3PO_4
in course of time, $\vartheta = 25°C$.

hydrochloric acid at a concentration above $1\,M$, especially at high temperature, no reasonable results are obtainable using the glass electrode. It
is moreover impossible to check the concentration of the acid by pH measurement, since at high ionic strength the activity coefficient of the hydrogen ions,
generally, is considerably higher by 1 (sometimes 2) orders of magnitude,
therefore $a_{H^+} \gg m_{H^+} = m_{acid}$.

(b) *Alkali error*. According to the Nikolski equation, the alkali error,
especially that caused by sodium ions, is noticeable above pH 9 if the exchange constant $K = 10^{-11}$ (see page 506). If the measurement is carried out
at room temperature, utilizing special glass material which is available
today, the alkali error can be neglected up to pH 13. Figure 7 shows the alkali
error of an electrode, suitable for pH range 0–14 at room temperature as a
function of the pH at different concentration of sodium ions. Using other

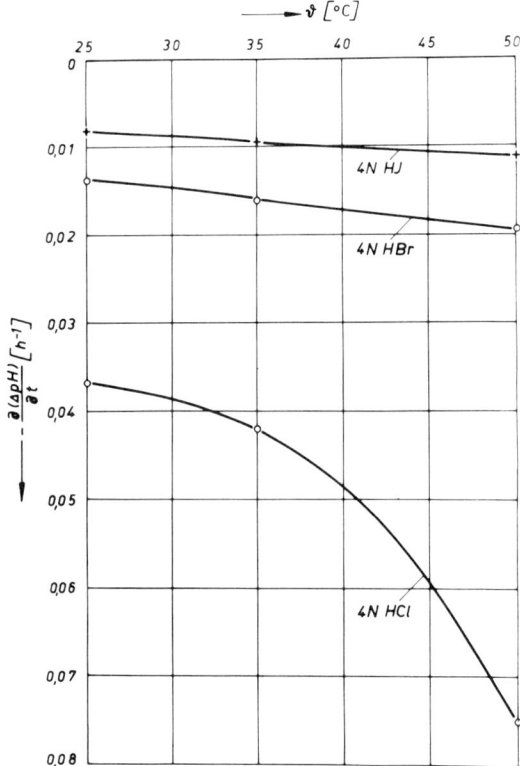

Fig. 6. Temporal changes of the acid error $-[\partial(\Delta\text{pH})/\partial t]$ of 4 N HCl, HBr and HI at $t = 10\,\text{h}$ as a function of the temperature.

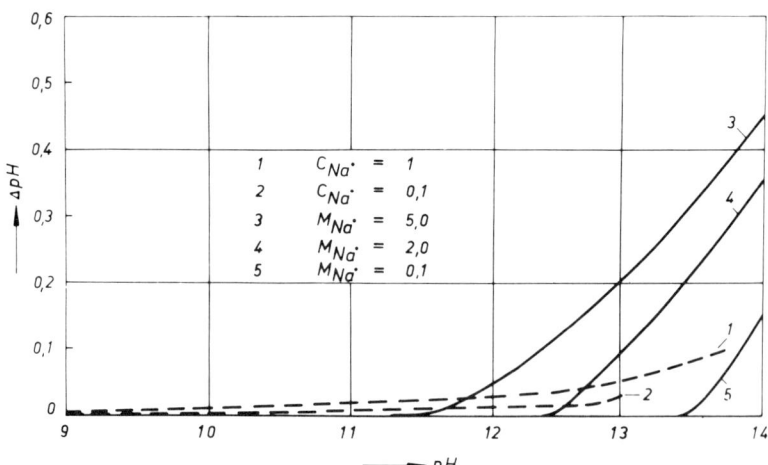

Fig. 7. Comparison of the alkali error—ΔpH of two types of commercial Li–Cs electrodes of special glass material as a function of the pH value at various sodium concentrations, $\vartheta = 25°C$.

bases or cations the error is even smaller. For the alkaline range, i.e. the range where the concentration test by pH measurement is reasonable, glass electrodes are available for measurements at room temperature with negligible alkaline error. (At pH 14 the concentration test by pH measurement is very inaccurate for the same reason as it is in concentrated acid solution.) But in strong alkaline solution (pH > 11) it must be considered, that perhaps the electrode may demonstrate temporary potential drift, due to the corrosion influence of the alkaline solution on the glass membrane. If there is no real pH drift, e.g. caused by carbon dioxide of the air, this potential drift may be produced by the electronic amplifier. Furthermore the 'asymmetry potential' of the electrode can change with time (see page 515). When continuous industrial pH measurement in strong alkaline solution is required, it is necessary in every case, to check within fixed intervals, whether the registered results correspond with the real ones by laboratory tests. This is most important, if the measurements are carried out at high temperature, since the alkaline error increases with temperature as Figure 8 demonstrates.

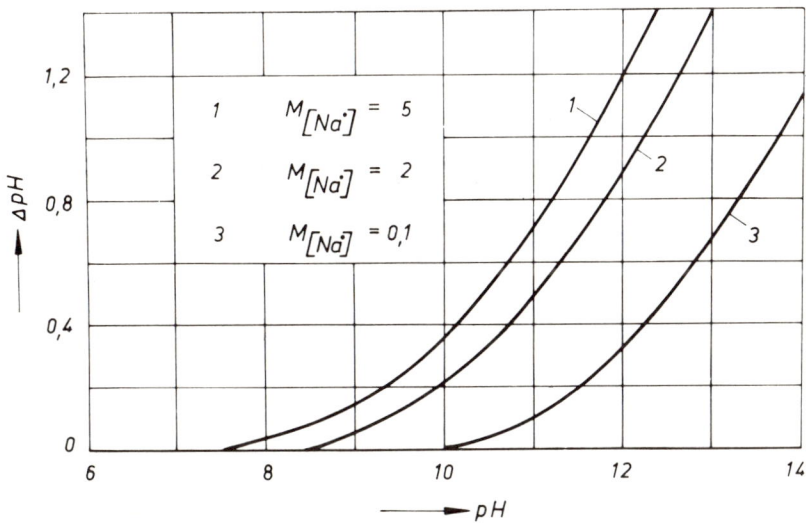

Fig. 8. Alkali error of a special Li–Cs electrode as a function of the pH value in solutions of various sodium concentrations at 80°C.

It is not advisable to use continuous measurements in strong alkaline solution at high ionic strength with respect to sodium salts. Furthermore, the pH value of the buffer solution at this ionic strength and at fixed temperature must be known, if precise pH measurement is required (see page 527).

(c) *The asymmetry potential of the glass electrode.* In principle, pH measurement using the glass electrode represents a comparison measurement with solutions of known pH value on the other side of the glass membrane. If the pH value of the test solution is exactly the same as that of the solution to be compared, the potential difference on the glass membrane should be zero. If the solutions on both sides of the membrane are identical, using the same reference electrodes, e.g. saturated calomel electrodes (see page 517) nevertheless, generally, a potential difference occurs, i.e. the asymmetry potential ε_a and equation (9) must be transformed to

$$pH_x = pH_B - \frac{(\Delta\varepsilon - \varepsilon_a)}{F_N} \tag{17}$$

The different formation of the swell film is considered to be the cause of the asymmetry potential. When a new electrode is filled and immersed in the same medium, e.g. water or aqueous solution, the formation of the swell film which is necessary for formation of the phase boundary potential, needs more time the less alkali is present in the electrode glass (35) (some days or even several weeks). Electrode glasses which are mostly used today and which are suitable for the pH range up to pH 13 or 14, are relatively poor in alkali and they are less soluble. Since, usually, electrodes are used which have already been filled with buffer solution the inner swell film possibly has formed some months ago, but the external one is produced only on immersion into water. Therefore, using so-called 'ready electrodes', i.e. electrodes which have already been prepared for measurement, it is still important to put them into water for a long time before measurement. Figure 9 demonstrates the

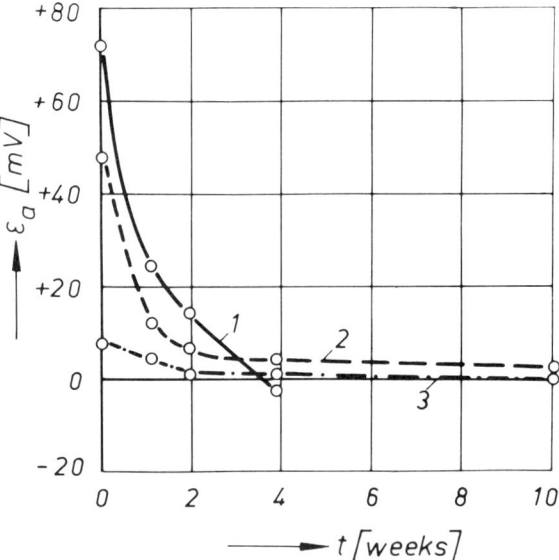

Fig. 9. Shift of the asymmetry potential E_a in course of time soaking electrodes containing various amounts of Na_2O and BaO measured in 0·1 M HCl at room temperature. Electrode 1: 30 per cent Na_2O–5 per cent BaO–65 per cent SiO_2; Electrode 2: 17, 5 per cent Na_2O–5 per cent BaO–77·5 per cent SiO_2; Electrode 3: 20 per cent Na_2O–6 per cent BaO–74 per cent SiO_2.

dependence of drift rate of the asymmetry potential on glass composition, if the electrode is immersed, for example, into 0·1 M hydrochloric acid. Formation of a stationary asymmetry potential can be accelerated when high temperatures are used. On the other hand, the asymmetry potential is changed at high temperature (36) and when a continuous test is made and if permanent operation lasts for some weeks at 95°C, values of more than 50 mV can be reached since the external swell film is continuously decomposed but the inner one and the buffer on the inner side of the glass membrane are in equilibrium, as was observed by Geisler (29b). As far as possible crushproof electrodes are used with thick glass walls since nowadays it is no problem to operate at resistance ranges of 1000 megohm and more. Petrovsky and Cuta (29a) have verified, that the asymmetry potential and the period necessary to establish a stationary value, increases proportionally with the thickness of the membrane. When thick-walled electrodes are used, the temporal change of the asymmetry potential during measurement must be considered. For correct measurements both reference electrodes must be exactly the same or the potentials must be exactly known. When diffusion potentials occur (see Section IV), the potentials of the reference electrodes are known at best to 1 mV of accuracy.

If very accurate measurements are required, using glass electrodes, especially with cells without transference, it is necessary to apply thin-walled electrodes (resistance of 50 M ohm) since they show temporally constant asymmetry potentials ≤ 2 mV, even when the alkaline content of the glass is small. Thin-walled electrodes are not crush-resistant, therefore it is advisable not to fill the electrodes before using them.

(d) *Reference electrodes for glass electrodes.* The phase boundary potentials occurring on both sides at the glass membrane need two reference electrodes, an internal and external one, and mostly secondary types of electrodes are used. In general, it is advisable to use the same electrode type for the internal and for the external side. Using this symmetrical reference system no additional potential difference occurs except some possible diffusion potentials when both reference electrodes show identical potentials. If measurements are made at very high or very low temperatures, the temperature coefficient of the reference electrodes is ensured to be without influence. For this purpose the cell

$$Ag/AgCl(s)KCl(1\ M)/_{\varepsilon_{da}}/pH_x/glass/pH_B/_{\varepsilon_{di}}/KCl(1\ M)AgCl(s)/Ag \quad (c)$$

would be a suitable one. Both diffusion potentials ε_{da} and ε_{di} differ the less and cancel the more completely the less the buffer solution of pH_B and the test solution of pH_x differ one from the other in ionic strength and in ionic species. Mostly Ag/AgCl electrodes are used as reference electrodes in commercial, closed glass electrodes. Adding KCl (e.g. 0·1 M) to the buffer solution the Ag/AgCl electrode can be immersed directly into buffer solution and the diffusion potential is avoided. Of course, now, ε_{da} is not eliminated and this diffusion potential can be avoided only by adding the same molality of chloride ions to the test solution, but generally this would be impossible. For very precise measurements of activity coefficients by means of glass electrodes this type of electrode is preferred (see page 527). Since the cell voltage of the glass electrode is standardized versus the buffer solution of known pH the diffusion potential ε_{da} involved in the measured cell voltage is not important. But since this is pH-dependent, it can simulate a gradient $\Delta\varepsilon/\Delta pH$, which is different from that of the Nernst equation. Furthermore even at identical cell voltage in the test solution and in the buffer solution as well, differences $pH_B - pH_x$ can be present. If those differences are caused by different ionic composition, a difference in diffusion potentials $\varepsilon_{da} - \varepsilon_{di}$ occurs, which compensates the potential difference, due to differences between the pH values. The conventional pH value, measured with the glass electrode represents only a numerical value but is no precise index of hydrogen ion activity, regardless of the fact that glass electrodes are true electrodes.

In the presence of asymmetrical grounding (e.g. inner side $Ag/AgCl$; external side Hg/Hg_2Cl_2 electrode) the voltage difference between both electrodes is not essential, since it is involved in the function $\varepsilon = f(pH)$, when standardizing the buffer solution. In this case, however, the standardization must be done at the corresponding temperature of measurement and the pH values of the standard buffer solution must be known for this temperature range. Generally, it is impossible to transform the pH value, measured at temperature T to that of the operating temperature T_a since the reference value ε_H^0 (standard potential of the hydrogen electrode) is assumed to be zero for any temperature and therefore for each temperature there is a different potential scale and pH scale as well. Using asymmetric glass cells, this transformation will be very inaccurate since temperature coefficients of both reference electrodes will be different; even their signs can be different.

In principle, nearly all reference electrodes, suitable for potentiometric measurements and summarized by Ives and Janz (37) can be used for reference electrodes in glass cells. The Hg/Hg_2Cl_2 (s) electrode with different concentrations of KCl is frequently used and is excellently reproducible but it causes many difficulties since a platinum wire must be used, which can produce its own potentials (14), furthermore even at a few degrees over room temperature ($>$ about 35°C), irreversible potential drift is caused by disproportion (37). Another disadvantage is that this electrode is not suitable for continuous current load ($> 10^{-6}\,A\,cm^{-2}$). The large influence of the hydrogen ion concentration in the measured solution on the liquid junction potential in connection with a saturated calomel electrode was shown by Shatkay (37a). He estimated, that this potential has a value of 4·7 mV at a H^+ concentration of $10^{-4}\,M$ diminishes to 3 mV at $10^{-2}\,M$ and then increases strongly to 8 till 9 mV at $10^{-1}\,M$.

Lietzke and Stoughton (38) checked that the $Hg/Hg_2SO_4(s)$ electrode was reversible in sulphuric acid solution at temperatures more than 100°C.

Every and Banks (39) recommended a series of reference electrodes for use in sulphuric acid solution. Very careful measurements have been made by Hills and coworkers (40) in the temperature range from 0 to 45°C using the Hg/Hg_2Br_2 (s) electrode. The mercury/mercury (I) acetate electrode (Hg/Hg_2-$(Ac)_2$ (s) electrode) is favourable, because it can be applied without diffusion potential in acetate buffer solution which is very often used (33), and also in organic solvents (41). The standard potential was fixed by Covington and coworkers (42) at 25°C to be 511·2 (abs.) mV. Schwabe and Glöckner (33) noted the potential in HAc/NaAc buffer solution (according to Michaelis) to be 851·3 (abs.) mV at 25°C; this furnished the value of 510·8 (abs.) mV for the standard potential. Using the cell

$$(Pt)H_2/HAc(m), Hg_2Ac(s)/Hg \quad (d)$$

Larson (43) has determined the temperature coefficient of the Hg/Hg_2Ac_2 (s) electrode in the temperature range between 5 and 37·5°C at different concentrations of acetic acid and by extrapolation, he stated the temperature coefficient of the standard electrode ($m_\pm = f_\pm = 1$)($\partial\varepsilon_H/dT = 0$). The potential can be calculated from the equation

$$\varepsilon = a + b\vartheta + c\vartheta^2 \quad (18)$$

where ϑ is the temperature in °C. Table II gives the constants (a, b, c) for different concentrations of HAc and for the standard potential ε^0. At least in this temperature range, the $Hg/Hg_2Ac_2(s)$ electrode is suitable for very precise measurements as well as the calomel electrode. The standard potential at 25°C, 511.36 (abs.) mV, resulting from (18) is in very good agreement with the value found by Covington.

TABLE II

Constants Used for Calculating the Potentials of the Mercury/Mercury (I) Acetate Electrode at Various Temperatures (ϑ°C) and Various Concentrations of Acetic Acid (m = mol kg^{-1}) due to $\varepsilon = a + b\vartheta + c\vartheta^2$

m(mol kg^{-1})	a	10^4b	10^6c
0·4057	0.80488	5·259	−3·209
0·5217	0·79872	5·361	−3·912
0·7736	0·78955	4·781	−3·354
1·039	0·78281	4·295	−2·846
ε^0	0·52441	−3·754	−5·859

With the second type of silver electrodes, the difficulties which occur when liquid metal is used are avoided but, on the other hand, these electrodes hardly ever can be produced to be strictly reversible. Formerly, many of the most famous authors in electrochemistry (44) noticed, that the standard potential is reproducible only to ±0·2 mV by means of the most often used Ag/AgCl electrode and recommended standardization together with a

0·01 M HCl solution since the medium activity coefficient of this solution is known very accurately ($f_\pm = 0.904\,\mathrm{kg\,mol^{-1}}$ at 25°C). Meanwhile two of the authors (45) advised, in detail, how to produce the Ag/AgCl as well as the Ag/AgBr and Ag/AgI electrodes and reported correct values of standard potentials. If very precise measurements are required, by means of cells without transference in a suitable temperature range, the mercury electrode should be given preference, since according to our own experiences, the potentials of some silver electrodes show high deviation, regardless of whether the silver electrode is prepared as carefully as the hydrogen electrode. Table III gives the standard potentials of three silver halide electrodes which were reported by the U.S. National Bureau of Standards (46). Lietzke and coworkers (47) have used Ag/AgCl and Ag/AgBr electrodes in the temperature range from 50°C and more up to 275°C. Kortüm and Häusermann (48) have studied Ag/AgI electrodes at temperatures up to 200°C. In Table IV, some standard potentials are listed at temperatures up to 200°C. All these values have been reached by the use of cells without transference of the type

$$(Pt)H_2/HX,\ AgX(s)/Ag \quad (e)$$

where X is a halide ion.

TABLE III

Standard Potentials of the Silver/Silver Halide Electrodes from 0° up to 50°C

°C	Ag/AgCl(abs)mV	Ag/AgBr(abs)mV	Ag/AgI(abs)mV
0	236·55	81·28	−146·37
5	234·13	79·61	−147·19
10	231·42	77·73	−148·22
15	228·57	75·72	−149·42
20	225·57	73·49	−150·81
25	222·34	71·06	−152·44
30	219·04	68·56	−154·05
35	215·65	65·85	−155·90
40	212·08	63·10	−157·88
45	208·35	60·12	−159·98
50	204·49	57·04	−162·19

TABLE IV

Standard Potentials of the Silver/Silver Halide Electrodes at High Temperature

°C	Ag/AgCl(abs)mV(47)	Ag/AgBr(abs)mV(47)	Ag/AgI(abs)mV(48)
25	+222·4	+71·2	−152·0
100	+160·0	+18	−192
150	+103·0	−32	−238
200	+ 35·0	−94	−307

These experiments show at the same time, that the hydrogen electrode is suitable at very high temperatures, if correct values for hydrogen partial pressure are applied (47, 48). For calculating the potential of the silver halide electrode in solutions of various halide concentrations the standard potentials can be used if the concentration and the activity coefficient are known and no masking of the silver ion occurs. But generally, these assumptions are not satisfied, when immersing the silver halide electrode directly into buffer solution, furthermore a formation of silver salt can take place, which is less soluble than halide itself. Khairy and Mahgoub (49) have studied the potentials of silver halide electrodes in various buffer solutions at different pH values and at different halide concentrations. They noticed a considerable influence on the pH value and temporal drift was observed too. But no exact potential values can be taken from the indicated diagrams. Generally, these electrodes are used in pure halide solution of known concentration and only the diffusion potential between this solution and the test solution or the buffer solution can exist but here the precise value of the potential of the reference electrode is also impossible to obtain. If measurements are required at temperatures up to 200°C, the Ag/AgAc electrode is useful as a reference electrode for a buffer solution in glass electrodes. This electrode can be immersed directly into the buffer solution consisting of acetic acid and acetate (50). Utilizing cells of the type (c) and replacing water by other solvents, especially organic solvents, neither pH $\equiv -\log a_{H^+}$ (reference state $m_{H^+} = f_{H^+} = 1$ in water) nor the potential Ag/AgX nor of any other electrode referred to ε_H^0 in water can be obtained from the measured cell voltage, since both electrode potentials are dependent on the solvent. Therefore, none of the numerous measurements (51) in various solvents, which have been carried out with these cells during the last years, are able to give precise data about paH drift caused by replacing the solvent, but cells without transference give information on the change of the thermodynamical values as $a_\pm$ of the acid concerned, and in general, of the corresponding electrolyte at transition from water to the specific solvent (52). But the problem is not completely solved concerning the reason for the change of the standard free enthalpy, represented by the so-called primary medium effect (53) at transition from one solvent to the other. The assumption, originated by Born, that the change of the dielectric constants of the solvent should be inversely proportional to ΔG^0, has not been verified. Recently (54) the change of the water activity in mixed solvents was assumed to be the essential reason for the shift of ε^0, but the question remains unsolved how ε^0 is determined in completely water-free solution.

If a comparison is required between the pH values in organic solvents and those in aqueous solution, it is advisable to use the conventional pH measurement with the hydrogen electrode or with a glass electrode which is swollen in water, i.e. to measure the potential in the solvent concerned versus the aqueous reference electrode of known potential. The phase boundary potentials are not higher than those between two aqueous solutions of different ionic strength. The measurement accuracy, concerning the pH value corresponds to that of the conventional pH measurement in aqueous solution (see Section V). Utilizing glass electrodes, it must be considered that the swell film, necessary for potential formation, absorbs organic solvent too (55) and therefore the phase boundary potential is changed and consequently, the pH value referred to the aqueous solution is not measured. This can be avoided by studying the potential shift with time and extrapolation to time zero. Since the conventional pH value is considered to be only a comparison value, of course the stationary pH values measured with the glass electrode versus the aqueous reference electrode in the medium concerned can be used only as comparison values. Therefore, the glass electrode must be stored in the same solvent, but when used for comparison in aqueous solvent between the measurements (as short as possible) it must be kept in water.

Besides silver reference electrodes, solid amalgam electrodes have been proposed recently. Fricke (56) suggested the use of a thallium amalgam electrode consisting of solid thallium chloride and potassium chloride solution which was found to be suitable up to 135°C. Solid cadmium electrodes of the secondary type can be produced from some kind of insoluble cadmium salt and organic acid used as basic material, as they are used for buffer solution. By this method the diffusion potential on the side of the buffer solution can be eliminated (57). Table V shows the potentials of some cadmium amalgam electrodes of the secondary type. In connection with the chapter about

TABLE V

Temperature Functions of the e.m.f. of Following Cells:
Cd(Hg)/Cd-citrate (s), citrate buffer//sat. calomel electrode
Cd(Hg)/Cd-succinate (s), succinate buffer//sat. calomel electrode
Cd(Hg)/Cd-phthalate (s), phthalate buffer//sat. calomel electrode
(the amalgam contains 12·5 per cent Cd)

Temperature, °C	ε (citrate), mV	ε (succinate), mV	ε (phthalate), mV
25	−722·3	−682·8	−691·6
35	−728·6	−679·7	—
40	—	—	−691·3
45	−725·1	−677·3	—
55	—	—	−690·7
70	—	—·	−687·4
85	—	—	−681·4

Composition of the buffer solutions:
Citrate buffer: 60 ml of 0·1 M disodium hydrogen citrate (21·008 g citric acid + 200 ml NaOH, diluted to 1 l) and 40 ml 0·1 N NaOH.
Succinate buffer: 0·025 M on sodium hydrogen succinate and disodium succinate
Phthalate buffer: 0·05 M potassium hydrogen phthalate

reference electrodes, the excellent monograph by Ives and Janz (37) should be referred to where nearly all electrodes are listed which have been proposed up to 1960, together with the range of functionality, temperature coefficients, etc.

By measuring the tension of the cell

$$(Pt)H_2/pH//KCl(s), TlCl(s)/HgTl(40 \text{ wt per cent})$$

from 5 to 95°C Baucke (56a) has determined the temperature coefficient of the thallium amalgam reference electrode (thalamidelectrode). Including the junction (ε_j) potential he gives an equation:

$$\varepsilon^{0\prime} + \varepsilon_j = a_0 + a_1(\vartheta - 25) + a_2(\vartheta - 25)^2 + a_3(\vartheta - 25)^3$$

(ϑ = temperature in degrees centigrade). The values of the constants a_0 still depend on the pH of the measured solution (for instance $a_0 = -0.5734$ V in 0.1 M HCl, -0.5766 V in phosphate buffer, 0.5795 V in borax buffer and $a_1 = -8.297 \times 10^{-4}$, -7.818×10^{-4}, 7.906×10^{-4} V°C^{-1} in the same buffer sequence).

A new silver reference electrode was proposed by Afsar Naqu and Mathur (56b): Ag/Ag tartrate (s), tartaric acid (0.1 to 0.01 M). Its standard potential was determined to be $\varepsilon^0 = +0.5610$ V (StHE). In the temperature range of 5 to 35°C the temperature gradient is -1.34 mV C°$^{-1}$.

IV. DIFFUSION POTENTIALS

As mentioned before, between test solution and reference electrode in conventional pH measurements a diffusion potential always appears, which prevents exact determination of the paH value. Considering the glass electrode with the inner reference electrode to be a unit, the external electrode can be considered to be the reference electrode. Indeed, the diffusion potential between the buffer solution and the internal reference electrode can be avoided if the electrode is immersed directly into the buffer solution which contains the defined concentration of that ion, which is potential-determining for the reference electrode (see above). In contrast to cells of the type (c) using cells of the type:

$$\text{Ag/AgCl(s), KCl//pH}_x/\text{glass/pH}_B\text{KA(m)AgA(s)/Ag} \quad (f)$$

only one diffusion potential occurs and no compensation at all takes place. The experimental determination of the diffusion potential is impossible, since the cell voltage presents always the sum of the potentials of two electrodes and of one or more diffusion potentials. The exact theoretical calculation is impossible since the single ion activity of those ions which are present in the transition phase must be known and this result cannot be attained by experiments. On the boundary of two solutions of the same electrolyte solution but of different concentrations, the diffusion potential disappears when the mobilities of both ions are identical (58) but this simple case will rarely appear in practical pH measurement. On the other hand, the mobility of all cations and anions is smaller than that of the ions of water, which are to be determined by pH measuring.

Picknett (59) has experimentally determined the approximate diffusion potentials in cells of the type (f), using solutions with correctly defined pH values. He used the saturated calomel electrode as an external reference electrode. Equimolar mixtures of HAc and NaAc of various concentrations and potassium hydrogen phthalate of various concentrations served for solutions with exactly defined pH values respectively. Also, the diffusion potential according to the Planck–Henderson equation was calculated approximately, where the single ion activities were drawn from an expanded Debye–Hückel equation for individual activity coefficients. Table VI shows

TABLE VI

Change of Cell e.m.f. with Concentration

| | | Electrometer reading | | |
| | | Experiment, mV | Theory, mV | Difference, mV |
Solution strength				
Equimolar acetic acid and sodium acetate	1	-1.0	-6.8	$+5.8$
	10^{-1}	-3.6	-5.0	$+1.4$
	10^{-2}	0	0	0
	10^{-3}	$+3.2$	$+3.4$	-0.2
	10^{-4}	$+10.6$	$+11.0$	-0.4
	10^{-5}	$+36.1$	$+35.7$	$+0.4$
Potassium hydrogen phthalate	10^{-1}	-7.1	$+10.6$	$+3.5$
	5×10^{-2}	-4.9	-7.0	$+2.1$
	10^{-2}	0	0	0
	10^{-3}	$+13.1$	$+13.3$	-0.2
	10^{-4}	$+38.9$	$+39.4$	-0.5
	10^{-5}	$+75.0$	$+74.6$	$+0.4$

the agreement between calculated and experimentally attained results, if for 0·01 M buffer solution the diffusion potential is defined to be zero. In Table VII the approximately calculated values for the diffusion potential between saturated KCl solution and various electrolyte solutions are listed. It is evident that the diffusion potentials increase with growing dilution, and that they are not zero, even between KCl solutions of different concentrations neither between $10^{-2}\,M$ HAc/NaAc nor in $10^{-2}\,M$ potassium hydrogen phthalate solution, as assumed in Table VI. Evidently it is impossible to eliminate completely the diffusion potential by insertion of 'salt bridges' with concentrated potassium chloride solution between measuring electrode and reference electrode.

TABLE VII

Diffusion Potentials between Various Solutions and Saturated KCl at 25°C

Molarity	Equimolar Na acetate + acetic acid, mV	Potassium hydrogen phthalate, mV	Sodium acetate, mV	HCl, mV	KOH, mV	KCl, mV
10^{-2}	3·20	3·49	3·23	2·85	1·92	2·78
10^{-3}	4·15	4·06	4·21	3·97	3·22	3·93
10^{-4}	5·00	4·87	5·27	4·77	4·48	5·02
10^{-5}	5·80	5·78	6·29	5·69	5·75	6·10
10^{-6}	6·72	6·71	7·23	6·70	6·88	7·07
10^{-7}	—	—	—	—	—	7·58

Even when the composition of the solution, where the diffusion potential occurs at the interface, is exactly known, the diffusion potentials calculated according to Henderson (60) do not correspond with those obtained from cell voltage and from single electrode potentials (61). Bianchi (62) and co-workers have tested whether the activity coefficients of the diffusing single ions, necessary for calculation due to Henderson can be obtained from the Guggenheim equation (63). According to that, the activity coefficient f_C can be calculated conventionally for the salt CA with the medium activity coefficient $f_{\pm CA}$ according to

$$n_C \ln f_C = n_A \ln f_A = \frac{n_C + n_A}{2} \ln f_{\pm CA} \tag{19}$$

where n_C and n_A are the number of the A and C ions, respectively. Using the cell

$$(Pt)H_2/HCl(m)//HBr(m)/H_2(Pt) \quad (g)$$

the differences between measured and calculated value for $m_{HCl} = m_{HBr} = 4 \; mol \; kg^{-1}$ are about 5 mV.

If the diffusion potentials are constant, no activities can be calculated from the measured cell voltage, but the cell voltage and consequently, the resulting conventional pH value, calculated according to (9) can be used as a comparison number. The time constancy of the diffusion potentials depends considerably on the interface between the two liquids. Besides stoppers made from porous material, such as cellulose, magnesium oxide, asbestos and Vycor-glass (64) which can be impregnated with concentrated potassium chloride, recently ground caps have been used (Figure 10). If they are not greased, the resistance hardly exceeds 1000 ohm. The contact between the liquids can be easily cleaned and if on the inner side of the reference electrode a small over-pressure is present, no test solution is able to penetrate, and the

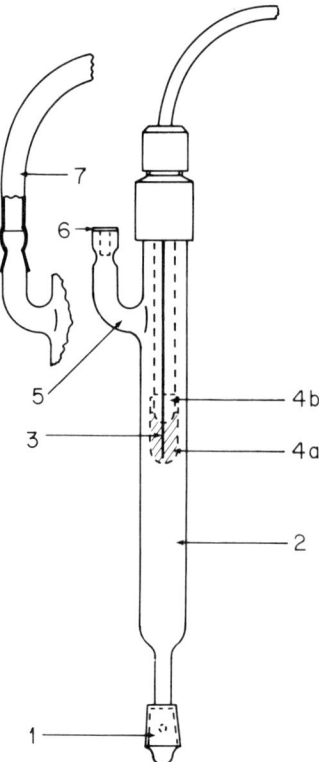

Fig. 10. Calomel electrode with
polished tube connection. 1 polished
plug; 2 KCl-solution; 3 Pt-wire; 4
calomel electrode 4a Hg, 4b Hg_2Cl_2;
5 filling port; 6 rubber plug; 7 connection for pressure compensation.

potential of the reference electrode is not changed. Such ground caps on
reference electrodes are very suitable for pH measurement in industry (65).

But such contact between two liquids of different composition and concentration does not represent well-defined phase boundaries with constant
potential difference between both solutions. In order to study diffusion
potentials in principle, as far as possible, the two solutions have to be able
to pass from different sides, so that their composition remains constant. A
suitable apparatus is described by Mattock (1) but for practical measurements in industry it cannot be used. The chief difficulty of the problem, i.e.
the influence of the diffusion potential on the measured cell voltage, is not
eliminated by the above mentioned design.

V. PRECISION MEASUREMENTS

Since diffusion potentials are unknown and not exactly determinable, the acidities of the solution must be measured in cells without transference of the type (c) and they give medium activities ($\sqrt{a_{H^+} a_{A^-}}$). For this purpose the hydrogen electrode serves excellently and supplies reproducible potentials accurate to 5 μV as is already well known. Reference electrodes are available for these cells with adequate accuracy of measurement and exact reproducibility; this was tested long ago (see page 517, Section III-B-2-d). Experiments performed with cells without transference during the last years have shown that the glass electrode follows the Nernst equation as well as the hydrogen electrode when acid activity is changed. According to former experiments, the glass cell (h) was believed to change the potential linearly with the pH value but the gradient $(d\varepsilon/dpH)_T$ should be different from the Nernst coefficient; mostly the gradient should be less. First, Leonhard and Sibbald (66) have pointed out that at least in the pH range 4–9 the glass electrode follows the Nernst equation. Zielen (67) calculated by extrapolation the standard potential of the Ag/AgCl electrode from the difference of the cell voltages of the cell

$$(Pt)H_2(1 \text{ atm})/HCl(m)/glass \quad (h^1)$$

and the cell

$$Ag/AgCl(s)/HCl(m)/glass \quad (h^2)$$

at different concentrations of hydrochloric acid ($m = 0.001$ up to 4 mol kg^{-1}) and at 25°C. He found the value 0.22235 (abs.) V, which is in perfect agreement with 0.22234 V, the value which had been found by Bates and Bower (68) with the cell

$$(Pt)H_2(1 \text{ atm})/HCl, AgCl(s)/Ag \quad (i)$$

From these results together with the one obtained from the cell voltage of the cell (h), which at least in the concentration range up to 0.5 M is independent of hydrochloric acid concentration, it is evident that the glass electrode shows an exact hydrogen function according the Nernst equation. Consequently, it was possible to apply the glass cell of the type (h) for determination of the medium activity coefficient of acids and for obtaining the pwH values. All experiments indicate perfect agreement with those values obtained by other methods, if the glass electrode was utilized within the limit given by its acid error or alkaline error, respectively (69). Even at high ionic strength, caused by addition of neutral salts, the pwH values, measured with cells of the type (i), must be in exact accord with those of the cell of the type (h) (70). It is advisable to use thin-walled glass electrodes, since their asymmetry potential is very low and constant. Prue (69) applied electrodes with the resistance of 0.5 megohm and was able to measure by compensation without amplification. Also at high ionic strength, the potential adjustments in glass electrodes take place very rapidly (within some minutes) in acidic or slightly alkaline ranges. But changing very rapidly from strong alkaline range to extremely lower pH values, the slower adjustment time must be considered, especially when unbuffered solutions of low pH values are used (24). Utilizing the hydrogen electrode, however, in concentrated electrolytic solutions a constant potential is attained only after some hours. It is profitable to immerse the electrode some hours before measurement in the test solution in a hydrogen atmosphere. Furthermore, it is advisable to bring into the compensation circuit a high ohmic amplifier since the sensitivity to current load is very high (86).

paH values can be obtained only in the Debye–Hückel range, when using cells without transference. At low ionic strength and in buffers of small capacity, the glass electrode does not work correctly because alkali ions diffuse from the glass. It is advisable to renew the test solution as far as necessary (71). Accuracy of measurement does not depend on reproducibility and on correct functioning of the electrode but on the precision, by which the pH value of the buffer is fixed and on diffusion potential, if conventional pH measurements are carried out. This definition is valid for the use of glass electrodes as well as of hydrogen electrodes. Since the conventional pH value is defined as a comparison number (see Table II) the pH value must be standardized with buffers of exactly known pH value. But since the pH values of these buffer solutions are not a correct measure of a_{H^+} and are used only as a comparison number, various pH scales are obtained for various standard buffer solutions. On account of the dependence of the diffusion potentials on the type of the salt bridge, which is included in the cell voltage, the measurement cell ought to be specified for standardizing the pH value and for tabulation of new pH scales. This system is used, for example, in British Scales. The problem of pH scales and buffer solutions for standardization is treated in detail in a paper by Mattock (1) and in books published by Bates (6), Schwabe (24) and others and the present paper refers to these. But one remark seems to be important: If a hydrogen, quinhydrone or glass electrode (for very accurate measurement, metal electrodes are useless) is combined with a reference electrode and possibly with a salt bridge and if for one run the cell voltage ε_B of this combination is measured in a buffer solution with pH$_B$ and at an identical temperature measurement is made in the test solution with pH$_x$ leading to a cell voltage ε_x, the pH$_x$ calculated from

$$pH_x = pH_B + \frac{\varepsilon_x - \varepsilon_B}{F_N} \qquad (17a)$$

is considered to be the more an accurate comparison number with respect to pH$_B$, the less the difference of the hydrogen ion activity and of the ionic strength between buffer solution and test solution is. This difference can be greatly reduced when the pH value and ionic strength of the test solution are nearly known. But even under these conditions it would be rather difficult to keep the diffusion potential constant to less than 0·5–1·0 mV, i.e. this comparison is unstable within pH 0·01 to 0·02. When considerable differences in pH units occur, the voltage error can reach values of an order of magnitude or more. Only by comparison of solutions of constant composition with small differences in hydrogen ion activity with the same buffer solution of similar pH value, as carried out in pH measurements of blood (see page 558) the comparison in the accuracy range pH 0·002 to 0·003 is possible, if extreme care is taken (see reference 73).

VI. BUFFER SOLUTIONS

Naturally, all buffer solutions which are to be used must be produced with pH values of adequate reproducibility. Recently, Bates (72) has rechecked all buffer solutions in the temperature range 0 to 95°C which have been used for tabulating the pH scale of NBS(U.S.A.) and has defined the pH values up to the third decimal place (Table VIII). These buffer solutions are of relatively small buffer capacity and are highly responsive to contamination (alkali from glass, NH_3, CO_2); consequently, these values are reproducible only if they are produced with extreme care and are used only together with those cells used by Bates (i.e. primary standards). The alkali ions from electrode glasses can cause significant errors, when used in solutions with small buffer capacity, which are not negligible in very accurate measurements. Covington and coworkers (73) have pointed out that the deviation between glass electrodes and hydrogen electrodes in buffer solutions of 0·025 equimolar Na_2HPO_4–KH_2PO_4 reaches 0·003 to 0·055 pH units and consequently, they refuse to apply this buffer as a primary standard for glass electrodes, when high accuracy is required. At increasing dilution, the error rises considerably and is dependent on electrode type (e.g. using a special type of electrode, the error in the Na_2HPO_4–KH_2PO_4 buffer solution was tested to be 0·01 to 0·022 pH units). Since this buffer solution is applied for the standardization of glass electrodes for pH measurements in blood, where the required accuracy ought to be 0·003 pH, electrodes with an alkaline error as small as possible must be utilized. On the other hand, according to Covington (73) an alkaline error up to 0·02 pH versus hydrogen electrode can occur in buffer solutions of high ionic strength, possibly caused by alkaline error of the glass electrode, since in solutions below pH 4, no deviation was observed by this author.

In general, for the purpose of conventional pH measurement, buffer solutions would be profitable with more intensive buffer capacity but less reproducibility. For the pH range between 1 and 12, numerous buffer

TABLE VIII

pH Values of National Bureau of Standards Buffers[a,b]

Temp., °C	0.05 M potassium tetro-xalate 1.	Saturated (25°C) potassium hydrogen tartrate 2.	0.05 M potassium hydrogen phosphate 3.	Potassium phosphate 0.025 M phosphate 0.025 M 4.	Dihydrogen 0.008695 M hydrogen 0.03043 M 5.	0.01 M borax 6.	Saturated (25°C) calcium hydroxide 7.
0	1.666	—	4.003	6.984	7.534	9.484	13.423
5	1.668	—	3.999	6.951	7.500	9.395	13.207
10	1.670	—	3.998	6.923	7.472	9.332	13.003
15	1.672	—	3.999	6.900	7.448	9.276	12.810
20	1.675	—	4.002	6.881	7.429	9.225	12.627
25	1.679	3.557	4.008	6.865	7.413	9.180	12.454
30	1.683	3.552	4.015	6.853	7.400	9.139	12.289
35	1.688	3.549	4.024	6.844	7.389	9.102	12.133
38	1.691	3.548	4.030	6.840	7.384	9.081	12.043
40	1.694	3.547	4.035	6.838	7.380	9.068	11.984
45	1.700	3.547	4.047	6.834	7.373	9.038	11.841
50	1.707	3.549	4.060	6.833	7.367	9.011	11.705
55	1.715	3.554	4.075	6.834	—	8.985	11.574
60	1.723	3.560	4.091	6.836	—	8.962	11.449
70	1.743	3.580	4.126	6.845	—	8.921	—
80	1.766	3.609	4.164	6.859	—	8.885	—
90	1.792	3.650	4.205	6.877	—	8.850	—
95	1.806	3.674	4.227	6.886	—	8.833	—

[a] In the buffer solution preparations, it is essential to use high purity materials (NBS samples for the most accurate work), and to employ freshly distilled or deionized water of specific conductivity not greater than 5 microsiemens. Solutions having a pH of 6 or above should be stored in plastic containers, and preferably with a soda-lime trap to prevent ingress of atmospheric carbon dioxide. They can normally be kept for 2–3 weeks, or slightly longer in a refrigerator.

[b] These values are derived from Bates and are consistent with the older NBS values which are defined to ± 0.01 pH unit.

1. 0.05 M potassium tetroxalate. Weigh out 12.61 g. $KH_3(C_2O_4).2H_2O$ and dissolve in water to one liter of solution at 25°C. The solution has excellent buffer properties, but has a relatively high dilution value, i.e., relatively subject to pH change on dilution.

2. Saturated (25°C) potassium hydrogen tartrate. An excess of the salt is shaken with water, and it can be stored in this way. Before use it should be filtered or decanted at a temperature between 22 and 28°C.

3. 0.05 M potassium hydrogen phthalate. Although this is not usually essential, the crystals may be dried at 110°C for an hour, and cooled in a desiccator. 10.12 g $C_6H_4(CO_2H)(CO_2K)$ are dissolved in water and the solution made up to one liter at 25°C.

4. 0.025 M disodium hydrogen phosphate. 0.025 M potassium dihydrogen phosphate. The anhydrous salts are best used, and each should be dried for two hours at 120°C, and cooled in a desiccator, since they are slightly deliquescent. Higher drying temperature should be avoided to obviate formation of condensed phosphates. Dissolve 3.53 g Na_2HPO_4 and 3.39 g KH_2PO_4 in water to give one liter of solution at 25°C.

5. 0.008695 M potassium dihydrogen phosphate 0.03043 M disodium hydrogen phosphate. Prepare as in (4) and dissolve 1.179 g KH_2PO_4 and 4.30 g Na_2HPO_4 in water to give one litre of solution at 25°C.

6. 0.01 M sodium tetraborate decahydrate. Dissolve 3.8 g $Na_2B_4O_7.10H_2O$ in water to give a litre of solution. This borax solution is particularly susceptible to pH change from carbon dioxide absorption, and should be correspondingly protected.

7. Saturated (25°C) calcium hydroxide. Pure calcium hydroxide can be shaken with water, or $CaCO_3$ ignited at 1,000°C for an hour to give CaO, which is then carefully slaked, the suspension boiled, cooled, filtered, dried in an oven, and crushed. The solution should be decanted at 25°C before use.

solutions are available, i.e. solutions of weak acids or bases mixed with their salts, where the pH value is known to sufficient accuracy and where ionic strength and buffer capacity are relatively large (24). Moreover, for use in an extended pH range, buffer solutions with constant ionic strength have been developed (6, 24) which are important, e.g. for studying the kinetics of electrodes especially for use in polarography. Buffer solutions produced by Thiel and coworkers (74) are useful since buffer solutions with pH units as integers are obtained but the buffer capacity is weak, especially in medium pH ranges. Utilizing higher concentrations, the initial solutions of those buffer solutions, served for mixtures with higher buffer capacity and their pH values have been established (75). In Table IX the most important mixtures are listed together with the appropriate pH values.

Measurements in biological systems require buffer solutions, especially in the neutral range, where amines and the corresponding hydrochlorides are suitable. Bates (76) recommended the Tris-buffer (0·1 M tris (hydroxymethyl)-aminomethane + hydrochloric acid) for the range pH 7–9 and Good (77) and coworkers suggested 12 buffer solutions of amine acids or of amines (pH 6·15–8·35) to be suitable for biological investigations, where secondary

TABLE IX

Buffer Solutions According to Thiel, Schulz and Coch at High Buffer Capacity

	ml of basic solution					Influence of dilution	Temperature coefficient	$10^2 \cdot p_w$ mol
							10^2 pH/°C	
pH	I	II	III	IV	V[a]	$pH_{1/2}$	(15–55°C)	1 pH
1·5	71·0	29·0				+0·20	+0·0$_5$	23·0
2·0	45·0	55·0				+0·16	+0·1$_5$	14·0
2·5	21·4	78·6				+0·09	+0·2	9·0
3·0	2·0	98·0				+0·06	−0·1	9·5
3·5		90·0	10·0			+0·02	−0·4	10·0
4·0		74·8	25·2			+0·03	−0·4$_5$	13·5
4·5		55·0	45·0			+0·05	−0·4$_5$	16·0
5·0		34·4	65·6			+0·05	−0·4$_5$	16·5
5·5		14·0	86·0			+0·08	−0·5	16·0
6·0			93·6	6·4		+0·10	−0·6$_5$	14·5
6·5			70·6	29·4		+0·11	−0·7	11·5
7·0			51·8	48·2		+0·10	−0·7	8·0
7·5			39·2	60·8		+0·10	−0·8	6·0
8·0			30·2	69·8		+0·09	−0·9	6·0
8·5			20·8	79·2		+0·04	−1·0$_5$	5·5
9·0			8·8	91·2		−0·03	−1·1	8·0
9·5				91·6	8·4	−0·06	−1·2$_5$	9·5
10·0				65·5	34·5	−0·03	−1·30	10·5
10·5				33·6	66·4	−0·01	−1·3$_5$	7·5
11·0				10·5	89·5	−0·02	−1·7	5·0
11·5				0	100	−0·13	−2·0$_5$	3·0

[a] Composition of basic solutions I–V are given in Table IXa.

TABLE IXa

Basic solution	Consists of	Advice for preparation; Dilute to 1 l	pH
I	0·2 M Oxalic acid	25·2134 g $(COOH)_2 . 2H_2O$	1·09
	0·2 M Boric acid	12·3666 g H_3BO_3	
II	0·2 M Potassium dihydrogen phosphate	27·2178 g KH_2PO_4	
	0·2 M Succinic acid	23·6180 g $(CH_2COOH)_2$	
	0·2 M Boric acid	12·3666 g H_3BO_3	
	0·2 M Potassium sulphate	34·8532 g K_2SO_4	3·05
III	0·2 M Prim. potassium succinate	23·6180 g $(CH_2COOH)_2$	
	0·2 M Dipotassium hydrogen phosphate	27·2178 g KH_2PO_4	
	0·2 M Boric acid	12·3666 g H_3BO_3	
		200 ml 2 n KOH	5·86
IV	0·2 M Potassium dihydrogen phosphate	27·6180 g KH_2PO_4	
	0·05 M Potassium tetraborate $(K_2B_4O_7 . 8H_2O)$	150 ml 2 n KOH	
		12·3666 g H_3BO_3	
	0·05 M Potassium sulphate	8·7133 g K_2SO_4	9·27
V	0·2 M Potassium dihydrogen borate	12·3666 g H_3BO_3	
		100 ml 2 n KOH	
	0·2 M Potassium carbonate	27·6426 g K_2CO_3	11·48

biological effects have been tested. According to this author Hepes viz. Table X is biologically more inert than Tris. This table shows these buffer materials as well as pK values and the temperature coefficients of pH. The substance 'Hepes' which can act both as an acid and a base, is utilized in 0·1 molal solution and the pH at 20°C is fixed to be 7·55. For pH measurements in blood, in general, $NaHPO_4$–KH_2PO_4 is used, approximately in equimolal solution. At the normal temperature of measurement (38°C) the pH value is established as 6·840. For every exact standardization a second solution is required, which consists of 0·008695 molal KH_2PO_4 and 0·03043 molal Na_2HPO_4 and furnishes the pH 7·384 at 38°C. If the error is required not to exceed 0·0036 pH, the weight of the buffer substance must be determined extremely precisely, as shown by experiments of Neumann and Geisler (78) (variations from accurate weight of 10 mg kg^{-1} cause an error in pH units of about 0·004 at pH 7·384). Furthermore, the temperature coefficients of these solutions are large, therefore standardization of the glass electrode, as far as possible, should be performed at the temperature of measurement. Since the hydrogen ion activity is normalized on the standard hydrogen electrode and its potential is assumed to be zero for all temperatures, in principle the pH values cannot be converted from the temperature of measurement to other temperatures. Therefore buffer solutions must be available with known

TABLE X

Structure	Proposed name	pK_a at 20°C	$\Delta pK_a/°C$	Saturated solution at 0°C (M)
O⟨⟩N⁺HCH$_2$CH$_2$SO$_3^-$	MES	6·15	−0·011	0·65
H$_2$NCOCH$_2$N⁺⟨CH$_2$COO⁻ / H \ CH$_2$COONa⟩	ADA	6·6	−0·011	—
NaO$_3$SCH$_2$CH$_2$N⟨⟩N⁺HCH$_2$CH$_2$CO$_3^-$	PIPES	6·8	−0·0085	—
H$_2$NCOCH$_2$NH$_2$CH$_2$CH$_2$SO$_3^-$	ACES	6·9	−0·020	0·22
(CH$_3$)$_3$≡N⁺—CH$_2$CH$_2$NH$_2$Cl⁻	Cholamine chloride	7·1	−0·027	4·2[a]
(HOCH$_2$CH$_2$)$_2$=N⁺HCH$_2$CH$_2$CO$_3^-$	BES	7·15	−0·016	3·2
(HOCH$_2$)$_3$=N⁺HCH$_2$CH$_2$SO$_3^-$	TES	7·5	−0·020	2·6
HOCH$_2$CH$_2$N⁺⟨⟩NCH$_2$CH$_2$SO$_3^-$ / H	HEPES	7·55	−0·014	2·25
H$_2$NCOCH$_2$NH$_2$CH$_2$COO⁻	Acetamido-glycine	7·7	—	Very large
(HOCH$_2$)$_3$≡C⁺NH$_2$CH$_2$COO⁻	Tricine	8·15	−0·021	0·8
H$_2$NCOCH$_2$NH$_2$	Glycin-amide	8·2	−0·029	4·6[a]
(HOCH$_2$)$_3$≡CNH$_2$	Tris	8·3	−0·031	2·4
(HOCH$_2$CH$_2$)$_2$=NHCH$_2$COO⁻	Bicine	8·35	−0·018	1·1
H$_3$N⁺CH$_2$CONHCH$_2$COO⁻	Glycyl-glycine	8·4	−0·028	1·1

[a] As the hydrochloride.

pH values, measured at different temperatures. Petersen (79), using comparison with primary NBS standard buffer solutions, has determined the pH values for many so-called technical buffer solutions up to 95°C and recorded them in a plot (see Figure 11). Buffer solutions, which are most often used in practice, are measured within a small temperature range (24) and are not referred to as a primary standard, viz. also Table XI.

Kriukov and coworkers (80) have given information about the pH values of some equimolar mixtures of weak acids together with their salts in the temperature range up to 150°C (Table XII). If special measurements require operation at temperatures more than 150°C, the pH values, determined by Lietzke (81) and coworkers in the temperature range up to 275°C in hydrochloric acid of various concentrations can be applied. (The standard potential of the Ag/AgCl electrode in this temperature range is given by the same author.) Table XII gives information about a series of buffer solutions, measured with the cell (h) at temperatures up to 200°C.

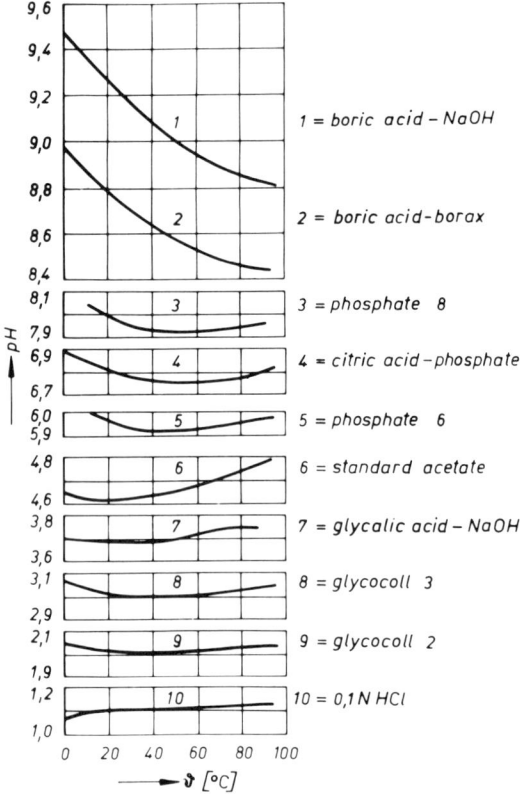

Fig. 11. pH values of various buffer solutions, used in
industry as a function of temperature.

TABLE XI
pH values of Various Solutions from 0 to 60°C

Temperature °C	0·05 M potassium diphthalate	0·025 M KH_2PO_4 0·025 M Na_2HPO_4	0·01 M sodium tetraborate
0	4·01	6·9	9·46
5	4·01	6·95	9·38
10	4·00	6·92	9·33
15	4·00	6·90	9·27
20	4·00	6·88	9·22
25	4·01	6·86	9·18
30	4·01	6·85	9·14
35	4·02	6·84	9·10
40	4·03	6·84	9·07
45	4·04	6·83	9·04
50	4·06	6·83	9·01
55	4·08	6·83	9·89
60	4·10	6·84	8·96

TABLE XII

pH Values of Various Buffer Solutions from 25 up to 200°C

°C	26	40	60	80	100	120	140	160	180	200
0·01 M HCl	2·04	2·05	2·05	2·05	2·05	2·05	2·06	2·06	2·06	2·07
KH—tartrate Saturated at 25°C	3·56	3·54	3·56	3·61	3·68	3·77	3·89	4·04	4·22	—
0·1 N CH$_3$COOH ⎱^{a} 0·1 N CH$_3$COONa ⎰	4·62	—	4·65	4·67	4·70	4·79	4·88	4·97	—	—
0·1 N CH$_3$COOH ⎱^{a} 0·01 N CH$_3$COONa ⎰	4·72	4·73	4·76	4·78	4·82	4·91	4·99	5·08	5·19	5·35
0·01 M Na$_2$B$_4$O$_7$	9·18	9·07	8·96	8·89	8·82	8·76	8·69	8·63	8·59	8·56

^{a} Inner buffer of the glass electrode.

VII. pH MEASUREMENTS IN CONCENTRATED ELECTROLYTIC SOLUTIONS

As is evident, the glass electrode shows correct functionality within vast ranges and the pwH or ptH values measured in concentrated salt solutions with the hydrogen electrode corresponds entirely with those measured under equivalent conditions with the glass electrode. Therefore, the conventional pH values measured with this electrode are considered to be a precise measure of acidity. Long before it was known (82) that additions of salt (e.g. alkaline halides) change considerably the conventional pH value, especially in acid solutions, mostly in a decreasing direction. Since added salts also change the diffusion potential and since the hydrogen electrode was assumed to show a salt error itself, the often enormous pH drop was not with certainty considered to correspond to the calculated increase of hydrogen ion acitivity, but no satisfactory theoretical interpretation was available. The diffusion potentials of the cell

$$(Pt)H_2/HA(m), \; NaB(m')/_{\varepsilon_d}/KCl(sat.)Hg_2Cl_2(s)/Hg \quad (k)$$

or

$$Ag/AgCl(s)pH_B/glass/HA(m), \; NaB(m')/_{\varepsilon_d}/KCl(sat.)Hg_2Cl_2(s)/Hg \quad (l)$$

(HA is the acid with the anion A, NaB is the salt with the anion B) surely are not greater but smaller (if m′ > m) than in the same cell without added neutral salt, since the differences in ionic strength and mobility of ions diminish on both sides of the phase boundary, when neutral salt is added. On the other hand, it is absolutely improbable that the glass electrode and hydrogen electrode should both possess the same salt error, since their potentials refer to entirely different electrode processes. Therefore, it seems very unlikely that the potential of the platinized hydrogen electrode, which is surrounded by H$_2$, is changed by adsorbed anions or ion pairs of the salt

in the same manner as the potential of the glass electrode, where the surface
is much less adsorbent. Furthermore, quite identical pH drift and corresponding changes of the Hammett function have been determined by spectrophotometric methods, where an adsorption effect is impossible (83). Finally,
a comparison between pH and pwH or ptH values in adequate cells has
shown that these values change completely in parallel, when the salt concen-
tration exceeds 1 mol kg^{-1}, i.e. the increase of a_{H^+} and $m_{H^+} \cdot f_{H^+} f_{Cl^-}$ must be
identical with growing salt concentration. Usually a linear drop in the
conventional pH value and the pwH value is recorded, regardless of whether
the anions of the salt and those of the acid are identical or not. Figure 12,

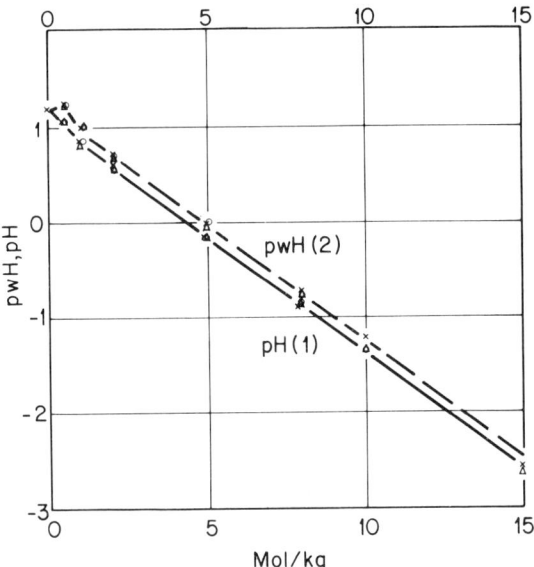

Fig. 12. Shifts of the pH value and the pwH value in 0·1 M
HBr, when LiBr is added. The cells used are : (1) △ Pt(H$_2$)/0·1
M HBr, LiBr//KCl$_{sat}$, Hg$_2$Cl$_2$/Hg; (1) ⊙ XII/25/glass elec-
trode, × GB/glass electrode; (2) △ Pt(H$_2$)/0·1 M HBr, LiBr,
Hg$_2$Br$_2$/Hg, Hg/Hg$_2$Cl$_2$, KCl$_{sat}$//0·1 M HCl/glass/0·1 M
HBr, LiBr, Hg$_2$Br$_2$/Mg; ⊙ XII/25 glass electrode; × GB/
glass electrode.

for example, shows the decrease of the pH and the pwH value in 0·1 M HBr
caused by LiBr, measured with various cells, utilizing the hydrogen electrode
and the glass electrode as well. Obviously, the hydrogen electrode and the
glass electrode furnish identical pH and pwH values and they decrease in the
same way from 1 to about −2·6. In 15 molal LiBr solution, which simultaneously is 0·1 molal in HBr, the hydrogen ion activity is about 4000 times
more than in pure 0·1 molal HBr solution. In Figure 13 the decrease of pH and

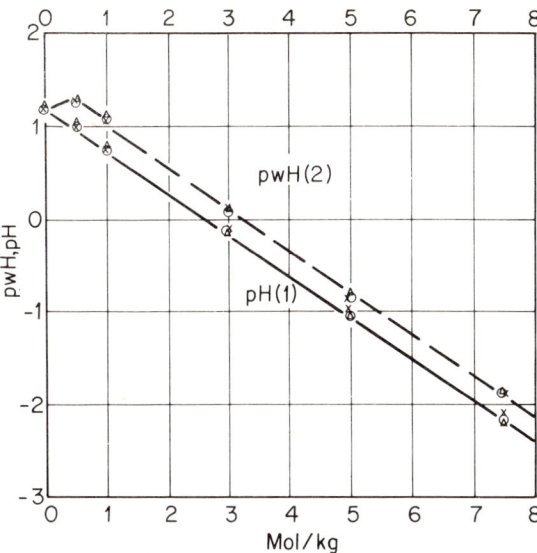

Fig. 13. pH and pwH shift in 0·1 M HCl on addition of Sr(ClO$_4$)$_2$. The cells used are: (1) $\triangle$ Pt(H$_2$)/0·1 M HCl, Sr(ClO$_4$)$_2$//NaCl$_{sat}$, Hg$_2$Cl$_2$/Hg; (1) Hg/Hg$_2$Cl$_2$, KCl$_{sat}$//0·1 M HCl/glass/0·1 M HCl, Sr(ClO$_4$)$_2$, AgCl/Ag; $\odot$ XII/25 —glass electrode; $\times$ GB—glass electrode; (2) $\triangle$ Pt(H$_2$)/0·1 M HCl, Sr(ClO$_4$)$_2$, AgCl/Ag, Hg$_2$Cl$_2$, KCl$_{sat}$//0·1 M HCl/glass/0·1 M HCl, Sr(ClO$_4$)$_2$, AgCl/Ag; $\odot$ XII/25/glass electrode; $\times$ GB/glass electrode.

pwH is shown, which is measured with the hydrogen electrode and various glass electrodes in 0·1 M HCl when Sr(ClO$_4$)$_2$ has been added. The linear and parallel drop is identical to that which occurs when salt is added to the acid, both containing the same ion. The slope dpH/dm_{salt} or $dpwH/dm_{salt}$ is independent of the concentration of the acid, as can be seen in Figure 14, where UO$_2$(ClO$_4$)$_2$ is added to various 0·1 and 1 molal acids. Referring to this and other results (2, 4) it is evident, that the concentration of the hydrogen ion does not change but the activity coefficient does. Further conclusions may be drawn from the parallel behaviour of pH and pwH and the correlation between paH and pwH (see Table I):

$$p a H = p w H + \log f_{A^-} \tag{19}$$

The activity coefficient of the anion f_{A^-} does not change within wide ranges of salt addition, when neglecting the diffusion potential, and the conventional pH value is equalized approximately to the paH value. Moreover, only slight variations in values are obtained which correspond relatively well

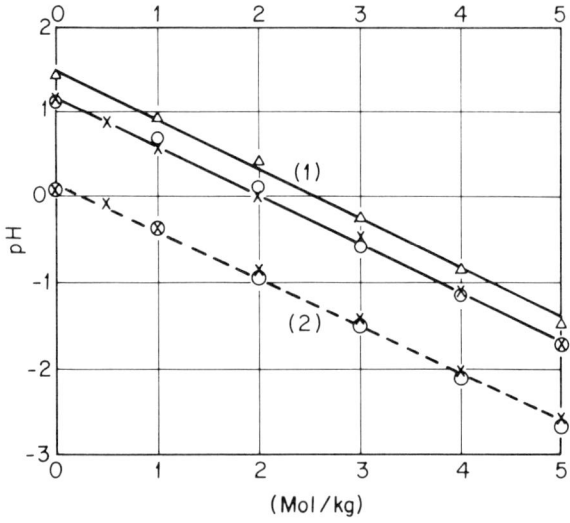

Fig. 14. Dependence of the conventional pH value on addition
of $UO_2(ClO_4)_2$ to various acids of different concentration.

0·1 M HCl	
0·1 M HClO$_4$	(1) 0·1 M acid
0·1 M HNO$_3$	(2) 1 M acid
0·1 M H$_2$SO$_4$	

TABLE XIII

Activity coefficients Determined with Cells of the Type
Me/MeA(s), HA, salt//KCl$_{sat}$, Hg$_2$Cl$_2$/Hg
Compared with the Calculated Value of the Difference between pH and pwH

Cell: Hg/Hg$_2$Cl$_2$, KCl$_{sat}$//0·1 M HCl, CaCl$_2$, AgCl/Ag
Buffer solution: 0·1 M HCl + M CaCl$_2$

M CaCl$_2$	0·5	2	3	5	7
pH	0·92	0·33	−0·12	−1·06	−1·68
pwH	1·34	0·71	0·18	−0·75	−1·40
f_{H^+}	1·20	4·18	13·2	115	479
f_{Cl^-} from pH–pwH	0·38	0·42	0·50	0·49	0·53
f_{Cl^-} from the cell	0·39	0·45	0·54	0·56	0·59

Cell: Hg/Hg$_2$Cl$_2$, KCl$_{sat}$//0·1 M HBr, Hg$_2$Br$_2$/Hg
Buffer solution: 0·1 M HBr + M NaBr

M NaBr	0·5	1	3	5	6	9
pH	1·02	0·89	0·42	0·05	−0·10	−0·80
pwH	1·05	1·00	0·53	0·09	−0·02	−0·74
f_{H^+}	0·955	1·01	3·16	7·42	12·6	56·2
f_{Br^-} from pH–pwH	0·93	0·78	0·78	0·87	0·83	0·87
f_{Br^-} from the cell	0·84	0·79	0·66	0·79	0·91	0·93

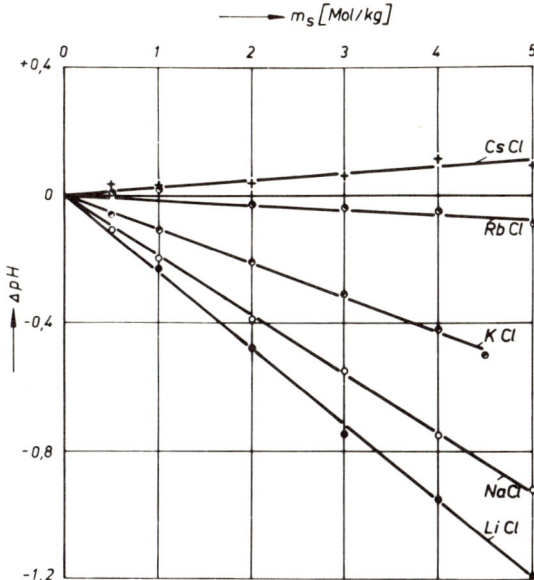

Fig. 15. pH shift of 0·01 M HCl caused by addition of LiCl, NaCl, KCl, RbCl and CsCl.

with those received from equation (19), when $(m' \gg m)\, f_{Cl^-}$ or f_{Br^-} is calculated with cells of the type

$$\text{Ag/AgCl(s)NaCl(m')HCl(m)//KCl(sat.)Hg}_2\text{Cl}_2\text{(s)/Hg} \quad \text{(m)}$$

neglecting the diffusion potential (see Table XIII). In Table XIII the activity coefficients received from pH $= -\log m_{H^+}\, f_{H^+}$ are also listed. As can be seen, they increase in several magnitudes but f_{Cl^-} or f_{Br^-} does not change by more than 50 per cent. The slope $d\text{pH}/dm_{\text{salt}}$ rises with growing cation radius of the salt and can become positive when the cations are sufficiently large. When alkali salts are added to 0·01 M HCl, the largest decrease of pH is seen in LiCl, while RbCl has the smallest and CsCl shows a small increase of the conventional pH value (84) (Figure 15). When tetraalkyl salts are added to acids, the large quaternary cations cause considerable pH increase, but this has not a linear dependence (Figure 16). Furthermore, in alkaline solutions, using the glass electrode and the hydrogen electrode as well, when neutral salts are added, a pH change is measured, but this does not depend linearly on salt concentration (2, 4). The reason for this change in acidity, which sometimes is very large, cannot be treated in detail and we refer to other reports (e.g. reference (4)). Of course this cannot be interpreted only by hydration of the dissolved salt and the resulting increase of the concentration, as sometimes assumed (85), since the f_{H^+} or $f_{\pm}$ of the acid changes much more than the medium activity coefficient of this salt at the same concentration.

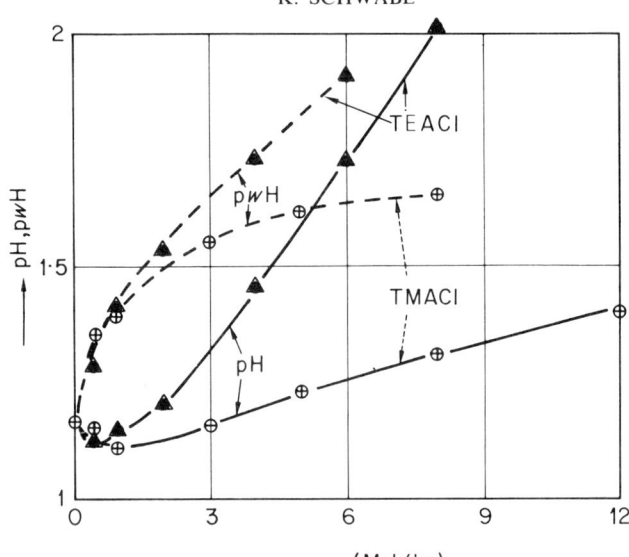

Fig. 16. Comparison of pH and pwH shifts of $0.1\ M$ HCl caused by addition of TEACl and TMACl. The cells used are:

(TEACl) Pt(H$_2$)/$0.1\ M$ HCl, TEACl, AgCl/Ag (pwH)

Pt(H$_2$)/$0.1\ M$ HCl, TEACl//KCl$_{sat}$, Mg$_2$Cl$_2$/Hg (pH)

Hg/Hg$_2$Cl$_2$, KCl$_{sat}$//$0.1\ M$ HCl/glass/$0.1\ M$ HCl, TEACl, AgCl/Ag (pwH)

Hg/Hg$_2$Cl$_2$, KCl$_{sat}$//$0.1\ M$ HCl/glass/$0.1\ M$ HCl, TEACl//KCl$_{sat}$, Hg$_2$Cl$_2$/Hg (pH)

(TMACl) Pt(H$_2$)/$0.1\ M$ HCl, TMACl//KCl$_{sat}$, Hg$_2$/Hg (pH)

Pt(H$_2$)/$0.1\ M$ HCl, TMACl, AgCl/Ag (pwH)

Hg/Hg$_2$Cl$_2$, KCl$_{sat}$//$0.1\ M$ HCl/glass/$0.1\ M$ HCl, TMACl, AgCl/Ag (pwH)

Hg/Hg$_2$Cl$_2$, KCl$_{sat}$//$0.1\ M$ HCl/glass/$0.1\ M$ HCl, TMACl//KCl$_{sat}$, Hg$_2$Cl$_2$/Hg (pH)

The hydration shells of the salt cations which are considered to be very polarized, possess positive poles, which are directed outwards, and experience intensive repulsion by hydrogen ions, since, on account of their high exchange rate with the H atoms of the water, the protons are dispersed all over the water and consequently, a positive free excess enthalpy is obtained, which will be larger, the smaller the cation is and therefore the larger the polarization and volume of the hydration shell is. Concentrated solutions of neutral salts, compared with dilute solutions, are considered to be a different solvent, since the water structure is disturbed and most of the H$_2$O molecules are required for hydration. The standard state of the acid or basic test solution in this electrolyte at constant ionic strength can be defined by a standard e.m.f.. From the difference between this value and that of the standard e.m.f. of a fictitious solution where $m_{HA} = 1$, $f_{\pm} = 1$ and $J = 0$, a primary medium effect can be calculated (86).

As is evident, the often enormous acidity changes due to addition of so-called neutral salts is of practical importance for equilibria as well as for kinetics in electrolyte solutions. Very frequently, the acidity can be changed without modifying the hydrogen concentration by addition of salt. At constant H^+ concentration the concentration of the salt can be defined by the linear relationship between the pH value and the concentration of the salt, for $m_s > 1$

$$pH = pH_0 - K \cdot m_s \tag{21}$$

pH_0 = pH value without salt, $K = dpH/dm_{salt}$ = factor of proportionality.

VIII. pH MEASUREMENTS AT HIGH TEMPERATURES AND AT HIGH PRESSURES

In Section VI, buffer solutions have been treated for standardization of pH values at high temperatures and in Section III glass electrodes and reference electrodes were described for identical conditions. When 'isotherm' cells are used, i.e. the working electrode and the reference electrode operate at the same temperature, the difference between the cell voltages due to (17) must increase linearly with the temperature

$$\Delta\varepsilon = (\varepsilon_x - \varepsilon_B) = KT(pH_x - pH_B); \qquad K = \frac{2 \cdot 303R}{F} \tag{17b}$$

If the measurement is carried out with a symmetric glass cell and if the pH value of the buffer inside the glass electrode is independent of temperature, then the cell voltage is independent of the temperature and becomes zero if the pH value of the inner solution is identical to that of the test solution or to the outer buffer solution and the asymmetry potential is zero. But in practice, mostly, asymmetric glass cells are applied. When the same assumption is valid and further, the total voltage, which includes possible asymmetry potentials and diffusion potentials, changes linearly with temperature, then one has a pH_{is} value, where the cell voltage is independent of temperature, i.e. that the straight lines of pH versus voltage at all temperatures cross at one point with the coordinates pH_{is} and ε_{is} (87). The pH value at a certain temperature can be calculated from the voltage at the 'isotherm intersection' and the respective pH_{is} value from the cell voltage ε_T due to:

$$pH_T = pH_{is} + \frac{\varepsilon_{is} - \varepsilon_T}{F_N} \tag{22}$$

In this way, the cell voltage $\Delta\varepsilon$ referred to the isotherm intersection is proportional to the gradient $d\Delta\varepsilon/dT = 2 \cdot 303R/F = 0 \cdot 1983 \text{ mV deg}^{-1}$.

Under this presumption an automatic temperature-compensation in plant control, where temperatures of the test solution often cannot be kept constant is possible, e.g. using a resistance thermometer, the increase of cell voltage can be compensated by the growing resistance of the thermometer due to rising temperature. Of course the temperature compensation must be adapted to the corresponding cell and is valid only for this one. But the presumptions made above are not always valid. The pH value of the inner buffer is temperature dependent and the several potential differences including the asymmetry potentials and the diffusion potentials are not linearly dependent upon temperature, they may even change sign (88). For this reason no exact isotherm intersection with precisely defined pH_{is} and ε_{is} occurs but a more or less large intersection range is observed. Therefore, this method serves only for correcting very small temperature differences ($\sim 10°$) and if very accurate measurements at high temperatures are required, symmetric isotherm cells should be used and the pH dependence of the inner buffer must be considered. Of course in this case, equation (17) can be applied together with the Nernst factor ($2\cdot303RT/F$) valid for this temperature. It is even more complicated to use non-isothermal cells for measurement, when the reference electrode (or the external reference electrode of the glass electrode) is kept at room temperature and the temperature of the measuring electrode is very high. The constant temperature difference between both electrodes causes temporarily variable thermo-diffusion voltages up to $0\cdot6$ mV degree^{-1} and the stationary value E_{st} is reached after a relatively long time. The time constant ϑ in equation (23), where E_0 signifies an initial value

$$E_t = E_{st} + \frac{8}{\pi^2}[E_0 - E_{st}]\,e^{-t/\vartheta} \qquad (23)$$

increases with the square of the length of the diffusion path and is inversely proportional to the diffusion coefficient (89); but under normal conditions the calculation is nearly impossible. Therefore it is not probable that at large temperature intervals an isotherm intersection would occur, as is sometimes assumed in order to use the $Hg/Hg_2Cl_2(s)$ electrode as a reference electrode (87c). Since reference electrodes are available, e.g. Ag/AgCl electrodes and Ag/CH_3CO_2Ag electrodes, which are reversible up to high temperatures (see Tables III and IV) non-isothermal cells should be avoided.

pH Measurements at high temperatures and also at low temperatures and pressures can be used, to determine the dissociation constant as a function of temperature and, consequently, for the determination of the free standard enthalpy, and the heat and entropy of dissociation. Thermodynamically correct values are obtained using cells without transference (81), but only ptH or pwH values are obtained. Non-thermodynamic additional assumptions are necessary to obtain paH values. As mentioned above, the diffusion potentials cannot be eliminated completely using cells with transference by means of salt bridges, but they can be reduced enough to reach a sufficient precision of the dissociation constants as a function of temperature. Utilizing glass cells for this purpose the alkaline loss of electrode material creates some disturbance, which can cause considerable pH change when weak acids or small volumes of electrolyte solution are used. Nevertheless, the pH value in aqueous solution of boric acid and its dissociation constant has been measured at temperatures up to 200°C by means of glass cells and has been compared with the values obtained from conductivity measurements for dissociation constants (90). For this purposes the symmetric glass cell

$$Ag/AgAc(s), NaAc(0.1\ M)HAc(0.1\ M)//KNO_3(3.8\ M)//$$

$$//H_3BO_3/glass/NaAc(0.1\ M)HAc(0.1\ M)AgAc(s)/Ag \quad (n)$$

$$(Ac = acetate)$$

has been used. Estimating the diffusion potential, the value of about 2 mV was found for normal temperatures, at very high temperature they are believed to be even smaller, on account of the slight difference in ionic mobility. Le Peintre (91) has demonstrated that under these conditions useful values could be obtained by means of salt bridges, compared with the pH values which Lietzke (81) obtained from cells without transference, measurements were made of HCl up to 200°C with the cell

$$(Pt)H_2/HCl(0.01\ M)//KCl//HCl(0.1\ M)/H_2(Pt) \quad (o)$$

The results agreed within a maximum deviation of 0.05 pH units. Utilizing this cell the pH values of various buffer solutions up to 200°C for standardizing the glass electrodes have been determined* (Table XII). In this way the temperature dependence of the asymmetry potential and of the pH value of the inner buffer could be eliminated. The pH change caused by CO_2 was eliminated when the measurement was carried out in completely deionized water and in nitrogen atmosphere. Although the electrode glass used is very temperature-resistant (29b), the alkali loss affected some difficulties, which were studied in various concentrations of boric acid using lithium glass; Figure 17 shows the slope with the concentration of boric acid and the time at 160°C, in $\mu gLi_2O\ cm^{-2}$. With rising temperature the alkali loss increases nearly exponentially (90). This allows estimation of the change of pH values, produced by alkali loss, when the hydrogen ion concentration is calculated from the alkali loss at the time t and from the dissociation constant K_c, which is assumed to be known. This was possible, since K_a

---

* The initially used molal concentrations (c) have been converted to M (mol/kg^{-1}) on account of the pH scale which refers to molality. This could be carried out at low ionic strength with $c \cong m\rho_{H_2O}$ (ρ_{H_2O} = density of water). Below 100°C the difference is negligible, at 140°C, 0.03 pH units and at 200°C, 0.06 are calculated, if $paH_{(m)} = paH_{(c)} + \log \rho_{H_2O}$ is valid.

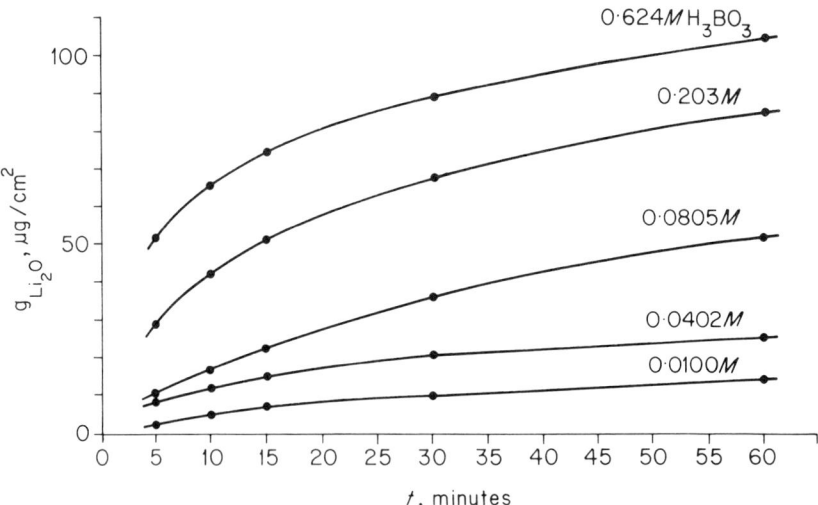

Fig. 17. Alkali loss from lithium glass as a function of the time and of the concentration of
the boric acid ($\vartheta = 160°C$).

and K_c were obtained from conductivity measurements. In order to obtain
K_a directly from pH measurements, these have been extrapolated to the time
zero or to the initial time, where no alkali loss could cause an appreciable
pH drift. Since the graphic extrapolation and the pH value, obtained from
K_c correspond up to 0·02 pH, both methods are considered to be useful.
The pH values, obtained from the extrapolation method have been used for
calculating the pK_a values for boric acid between 25 and 200°C at various
concentrations. As Table XIV shows, the agreement with values obtained
from measurements of conductivity is sufficient. Obviously, the differences
are not caused by the alkali loss of the glass, since they do not rise significantly
at high temperatures. The strong concentration dependence of $pK_a =
-\log K_a$ is caused by formation of polyboric acid at high concentration (92)*.
At high temperatures a depolymerization occurs and therefore pK_a increases.
The measurements were carried out in an autoclave, where the glass electrodes
were resistant to the pressure of 25 atm and they were fastened in a stuffing
box (Figure 18). This design has been successfully used also for other measurements, when high pressure was required (93). The reference electrode is a
simplified type of that reference electrode, formerly suggested for use at
high pressure with pressure compensation (94) and was useful for many
measurements (93), e.g. for pH measurements in sugar solution at more than

---

* For a complete discussion of the polymerization of boric acid in connection with the new
pK_a values see reference 90.

TABLE XIV

pK_a Values Obtained from Conductivity Measurements (con) and from pH Measurements (pH). M (mol/kg sol.)

°C M	0.0100	0.0201	0.0301	0.0402	0.0502	0.0603	0.0805
25 con		9.41				9.25	9.16
pH		9.22				9.19	9.17
60 con	9.42	9.26				9.06	
pH	9.46	9.56				8.90	
80 con	9.21		9.06	9.19	9.12	9.07	8.99
pH	9.54		9.20	9.22	9.12	9.09	9.93
100 con	9.31	9.07				8.92	
pH	9.20	9.20				9.13	
120 con	9.16	9.14		9.03		8.94	8.73
pH	9.06	9.32		9.22		9.00	8.96
140 con		9.39				9.02	
pH		9.37				0.14	
160 con	9.30			9.17		9.09	9.07
pH	9.50			9.30		9.17	9.00
180 con		9.22					
pH		9.69					
200 con						9.19	
pH						9.45	

°C M	0.1008	0.152	0.203	0.306	0.411	0.517	0.624	0.788
25 con			8.76	8.52	8.22	7.97	7.76	7.49
pH			8.73	8.33	7.95	7.71	7.51	7.22
60 con	8.96		8.78		8.33		7.94	
pH	8.86		8.49		8.08		7.91	
80 con	8.97	8.93		8.55	8.40		8.02	7.80
pH	8.92	8.94		8.53	8.13		7.78	7.41
100 con	8.95				8.37		7.97	7.75
pH	9.06				8.57		8.08	7.77
120 con	8.95	8.96	8.67		8.44		8.13	7.92
pH	9.07	8.67	8.83		8.31		7.97	7.82
140 con	9.00		8.78		8.47		8.23	7.96
pH	9.07		8.67		8.75		8.34	7.91
160 con	9.09		8.85		8.59		8.34	8.03
pH	8.86		8.67		8.54		8.35	8.13
180 con	9.29		8.86		8.67			8.13
pH	9.45		8.75		9.00			8.13
200 con	9.24		9.04				8.54	8.14
pH	9.13		8.90				8.36	8.30

100°C (Figure 19) where the current supply was made by a spark plug.

pH measurements by means of glass electrodes are of practical importance for use in the cellulose industry at high temperature, where in the production of sulphite cellulose the decomposition process of wood pulp can be controlled by pH change, and consequently interesting deductions have been made about the mechanism of lignin solution. In addition to the isothermic cells

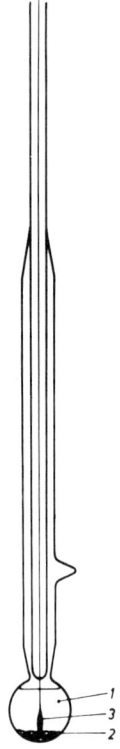

Fig. 18. Glass electrodes resistant versus over-pressure: 1 standard acetate, saturated with CH_3COOAg; 2 CH_3COOAg, solid; 3 Pt-wire, silvered.

described already in connection with reference electrodes with pressure compensation (93) other electrode designs have been applied. Ingruber (95) suggested a water-cooled calomel electrode, where nitrogen over-pressure has to prevent the penetration of the test solution but this design was unsuitable since often the circuit was interrupted by bubbles of water at the capillary end of the junction (96). For this type of measurement another cell is proposed by Wirz (97) however, it operates isothermally. But generally, the reference electrode with pressure compensation recently has been preferred and also glass electrodes have been furnished with pressure compensation for use at high pressure. Le Peintre (91) applied an electrode, completely filled with buffer solution and silicone oil, where a U-tube fastened on one side contained mercury for pressure compensation (Figure 20). A mechanical stabilization carried out by metal coating (98) or by filling with lead amalgam (99) obviously was not favourable on account of the very retarded and uncertain potential adjustment (100) of this electrode type, as has been known for a long time. Undoubtedly, the direct pH measurement in reaction vessels of the chemical industry at high temperature and pressure is generally

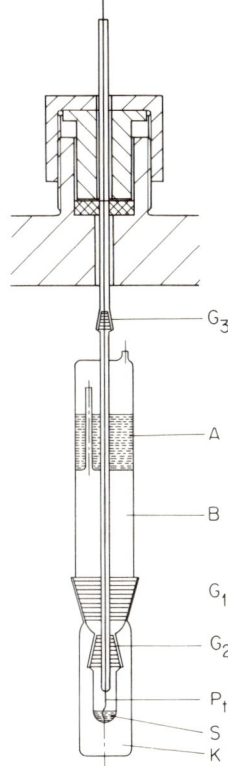

Fig. 19. Silver/silver acetate reference electrode for measuring at high pressure: A Siliconoil; B Standard acetate solution saturated with CH_3COOAg; S Solid silver acetate; K KNO_3 solution, saturated at 25°C; G_1 Ground glass ring diaphragm for connecting KNO_3 with the test solution; G_2 Ground glass ring diaphragm for connecting KNO_3 with silver acetate solution; G_3 Grinding connection Pt-wire silvered.

very important for process control but the durability of the glass electrodes at high temperature, especially in strong acid or alkaline solution is very small (93). This disadvantage can be corrected using thick-walled electrodes and the life-time of the electrodes can be increased up to several days and even weeks (35b).

Distèche (101) has been occupied with pH measurements at very high pressure up to 1000 kp cm^{-2} at normal temperatures with respect to conditions at great sea depths. Here too, pressure compensation is applied to the complete cell (Figure 21) and the electrodes are separated from the test solution by silicon oil. The total design with the equipment for protecting the electrodes, etc., is shown in Figure 22. Utilizing this device, the pressure dependence of the dissociation constants of weak acids or the activity coefficients were studied. One has

$$\left(\frac{\partial \ln K_a}{\partial p}\right)_{T,m} = -\frac{\Delta \overline{V}^0}{RT} \tag{24}$$

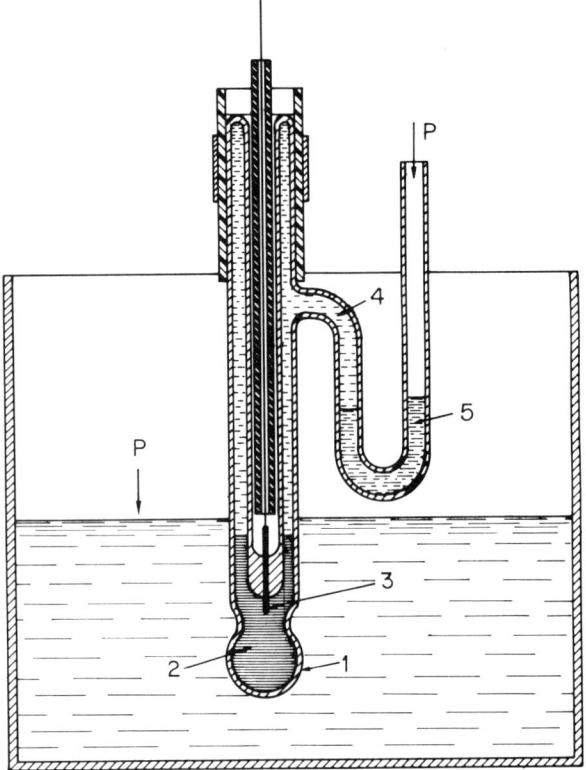

Fig. 20. Pressure compensation, due to Le Peintre: 1 glass membrane; 2 HCl–KCl-solution; 3 silver/silver chloride electrode; 4 silicon oil; 5 mercury.

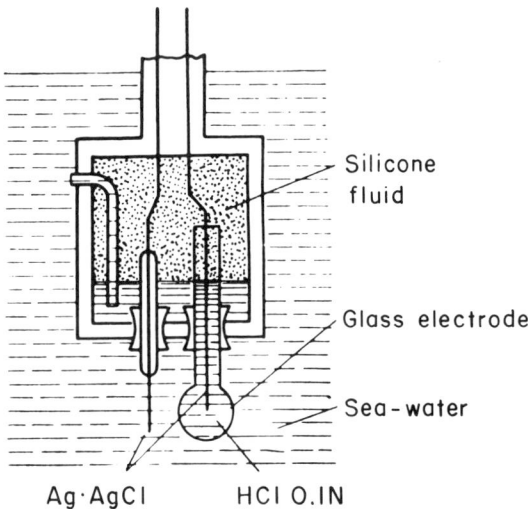

Fig. 21. Cell resisting versus over pressure, due to Distèche.

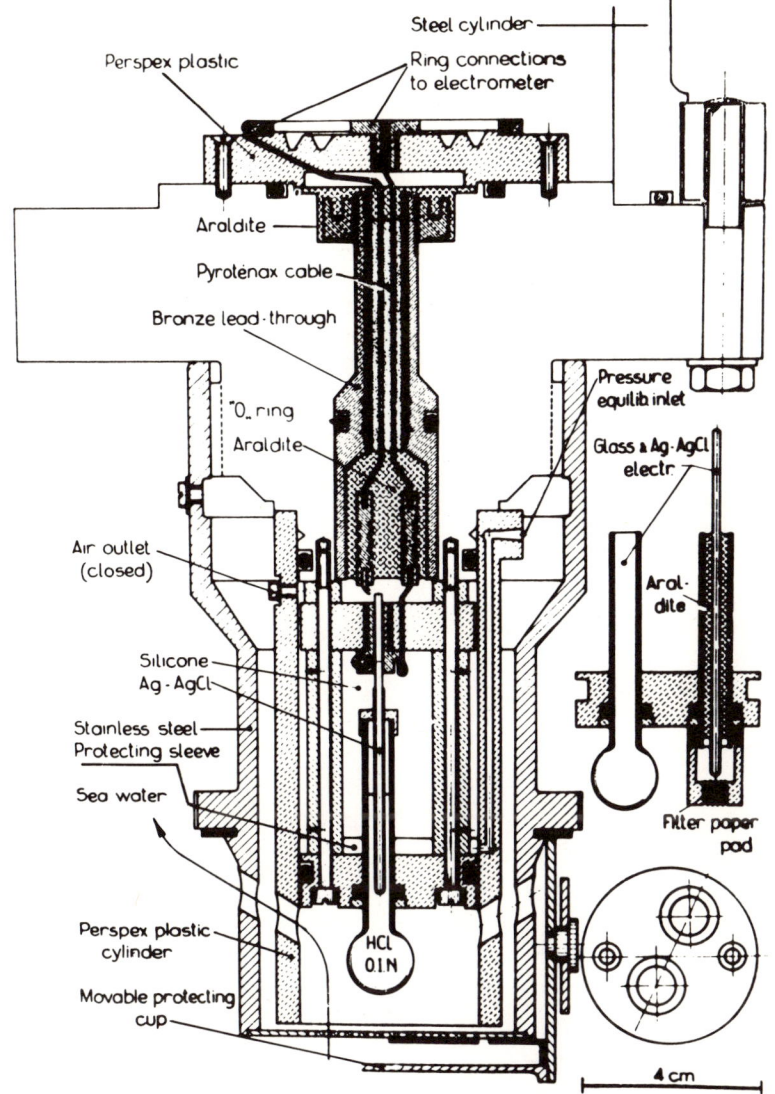

Fig. 22. Glass electrode assembly for deep sea investigation from Distèche and Dubuisson.

or

$$\left(\frac{\partial \ln f_i}{\partial p}\right)_{T,m} = \frac{\overline{V}_i - \overline{V}_i^0}{RT} \tag{25}$$

where $\Delta \overline{V}^0$ is the difference of the partial mol volumes at the standard state of ions and the acid molecules, $\overline{V}_i$ is the partial mol volume of the ion type i, and $\overline{V}_i^0$ is the volume at standard state. In connection with conditions in sea water, the influence of pressure on the dissociation of carbonic

acid and the effect of salt and boric acid (102) have been studied. The $\Delta \bar{V}^0$
values calculated from cell voltage change at transition from 1 atm to 1000
bar (after extrapolation of $\Delta \varepsilon$ to the ionic strength 0) are in sufficient agreement with those values obtained from measurements of density and conducti-
vity, concerning the investigated equilibria. On increasing the pressure
from 1 atm to 1000 bar the p_aH drift in sea water (Cl^- content is about
2 per cent) is practically pH independent within the studied pH range
(5–9.6). These results are in very good agreement with those obtained by
Buch and Grippenberg (103) from the pressure dependence of the dissociation
constants of carbonic acid but they differ remarkably from values which
recently Park (104) has obtained by direct measurement in the north
eastern Pacific, at a sea depth of about 6000 m, by the method suggested by
Strickland and Parsons (105). Figure 23 shows a maximum near to the
surface, then the values drop rapidly till a minimum at about 7.6 pH is
attained, now a slow rise is noted up to the depth of about 3500 m and
afterwards again the pH decreases slowly. Investigations carried out by
Distèche and Dubuisson (106) in the Mediterranean down to the depth of

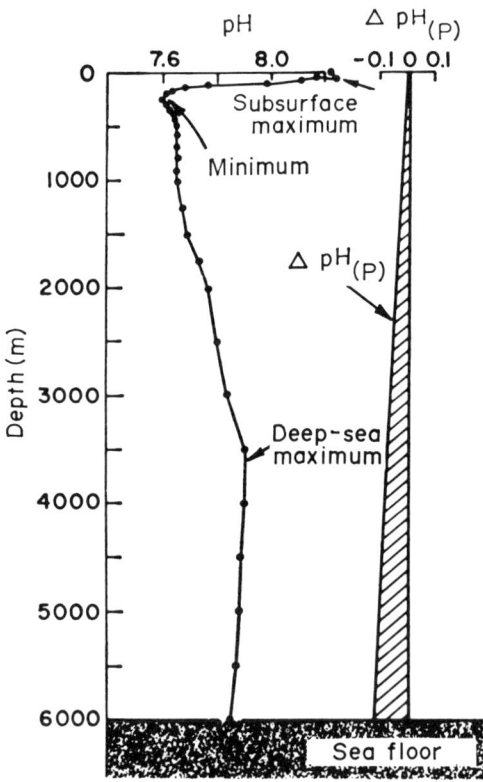

Fig. 23. Vertical profiles of pH at 53°46′ N and 158°36′ W
on 7 July 1966 from K. Park.

2350 m have observed analogous results utilizing the device shown in Figure 22. Recently Ben-Yaakov and Kaplan (107) have designed a cell which, up to the pressure at 150 kp cm^{-2}, furnishes a reproducibility of 0·02 pH units and this was used for measurements down to the depth of 270 m. The Ag/AgCl electrode used contains KCl with a layer of silicon oil. KCl is pressed by a plastic syringe through a ceramic diaphragm and silicon oil is used for compensating the pressure inside the glass electrode.

pH measurements at low temperatures do not meet special difficulties, since the hydrogen electrode, even at 0°C, adjusts the potential within 10–15 minutes and the increasing resistance of the glass electrode does not disturb measurement, because today suitable pH amplifiers are available. Krishnar and coworkers (108a)* have demonstrated that even at -6°C, in frozen water the hydrogen electrode behaves reversibly. The authors determined the dependence of the potential difference of two hydrogen electrodes on pressure in a frozen 10^{-4} M NaCl solution. The potential differences, calculated from the pressure difference between the two Pt electrodes, which are surrounded by platinum powder, agree with experimental results.

IX. pH MEASUREMENTS IN NON-AQUEOUS SOLVENTS

According to the definition of the pH value measurements can be carried out only in media containing hydrogen ions. Moreover, for the reference state of hydrogen ion activity solutions are taken, with the concentration 1 mol kg^{-1}(mol l^{-1}) and $f_{H^+} = 1$, i.e. this solution behaves like infinitely diluted aqueous solutions. Therefore, the hydrogen ions must be hydrated, if their activity in every medium should be compared with that of the reference state. The potential measurement in completely non-aqueous solvent using any electrode which indicates hydrogen ions in aqueous solution, cannot furnish a pH value suitable for comparison with the aqueous solution. Therefore, to each medium belongs its own pH scale with a special reference state if the 'pH value' should serve as an indication of the hydrogen ion activity. This is valid too, when the medium is not completely anhydrous. Of course, it is impossible to determine the single ion activity, and only the mean activity can be obtained. This has been carried out, frequently, in aqueous and non-aqueous organic solvents by means of cells without transference of the type (i) (51). If conventional pH measurements are carried out, i.e. as well as in aqueous solution, standard buffers can be used, of which the pH values agree as far as possible with paH. Bates and coworkers (109) have produced standard buffer solutions for mixtures of 50 per cent methanol and 50 per cent water. The authors determined the paH$_x$ (paH referred to this solvent) measuring the pwH value and adding log f_{Cl^-} which

---

* Concerning the tension between ice and an electrolytic solution see (108b) but compare with (108c).

was calculated from the Debye–Hückel equation (26)

$$-\log f_{\text{Cl}^-} = \frac{AJ^{1/2}}{1 + BaJ^{1/2}} \tag{26}$$

J = ionic strength,

$$A = \frac{1 \cdot 8246 \cdot 10^6}{(\varepsilon \cdot T)^{1/2}} \rho_0^{1/2}, \qquad B = \frac{50 \cdot 29 \cdot 10^8}{(\varepsilon \cdot T)^{1/2}} \rho_0^{1/2},$$

ε = dielectric constant, ρ_0 = density of the solvent, a = ion size parameter (for Cl^- 4·5 Å).

Table XV shows values obtained for 50 per cent methanol and various buffer solutions containing NaCl at different temperature. For D_2O such

TABLE XV

Values of pa_H ($p_s(a_\text{H} f_\text{Cl})$ + $\log f_{s\,\text{Cl}}$) in 50 weight per cent Methanol

I	Buffer:	$KH_2PO_4\,(M)$ + $Na_2HPO_4\,(M)$ + $NaCl\,(M)$;			$J = 5\,M$		
	10	15	20	25	30	35	40°C
0·01	8·277	8·259	8·243	8·232	8·225	8·221	8·221
0·02	8·204	8·184	8·168	8·157	8·148	8·144	8·142
0·03	8·151	8·131	8·115	8·102	8·093	8·088	8·086
0·04	8·109	8·088	8·071	8·058	8·049	8·043	8·040
0·05	8·072	8·051	8·034	8·021	8·011	8·004	8·001
0·06	8·040	8·019	8·002	7·988	7·978	7·971	7·967
0·07	8·011	7·990	7·973	7·959	7·948	7·940	7·936
0·08	7·984	7·963	7·946	7·932	7·921	7·913	7·908
0·09	7·960	7·939	7·921	7·907	7·896	7·887	7·882
0·10	7·937	7·916	7·898	7·884	7·872	7·863	7·858

Buffer: $HAc\,(M)$ + $NaAc\,(M)$ + $NaCl\,(M)$; $J = 2M$

I	10	15	20	25	30	35	40°C
0·01	5·608	5·599	5·593	5·591	5·592	5·597	5·604
0·02	5·586	5·577	5·571	5·568	5·569	5·573	5·580
0·03	5·571	5·561	5·555	5·552	5·553	5·556	5·563
0·04	5·560	5·549	5·543	5·540	5·540	5·543	5·550
0·05	5·550	5·539	5·532	5·529	5·529	5·533	5·540
0·06	5·542	5·531	5·524	5·520	5·520	5·524	5·530
0·07	5·535	5·523	5·516	5·512	5·512	5·515	5·522
0·08	5·529	5·517	5·509	5·505	5·505	5·508	5·515
0·09	5·523	5·511	5·503	5·499	5·499	5·502	5·508
0·10	5·518	5·506	5·498	5·493	5·493	5·496	5·502

Buffer: $NaHSuc$ (sodium hydrogen succinate) (M) + $NaCl\,(M)$; $J = 2M$

I	10	15	20	25	30	35	40°C
0·01	5·902	5·884	5·870	5·860	5·854	5·850	5·850
0·02	5·863	5·844	5·829	5·818	5·811	5·806	5·806
0·03	5·832	5·812	5·797	5·785	5·777	5·772	5·770
0·04	5·806	5·786	5·770	5·757	5·748	5·743	5·741
0·05	5·784	5·764	5·747	5·734	5·725	5·719	5·716
0·06	5·766	5·745	5·728	5·714	5·705	5·698	5·696
0·07	5·750	5·729	5·712	5·698	5·688	5·680	5·676
0·08	5·738	5·716	5·698	5·684	5·675	5·668	5·666
0·09	5·728	5·705	5·688	5·674	5·664	5·658	5·656
0·10	5·720	5·697	5·680	5·666	5·656	5·650	5·648

standard solutions are also necessary; they have been produced and their paD values, i.e. $-\log a_{D^+}$ (D^+ = deuterium ion), have been determined by Bates (110a). The pK values of acetic acid and phosphoric acid in D_2O are quite different from those in water (acetic acid: H_2O, p$K = 4.756$, D_2O, p$K = 5.313$ both at 25°C). Recently Covington and coworkers (110b) have studied the behaviour of the glass electrode in deuterium oxide together with the relation between the standardized paD scale and the operational pH in heavy water. From direct and indirect comparative studies of commercial glass electrodes with the deuterium gas electrode at 25°C in buffered solution of pD from 1 to 13 it is confirmed, that the glass electrode is suitable for measurements in heavy water as well as in normal water. The experiments have been carried out by means of four different cells furnishing identical results.

Using the pH value as a comparison number for practical purposes or in potentiometric titration in non-aqueous solution, the relation between the measured number and the activity of the hydrogen ions, when water is used as the solvent, is not important. But for all measurements the same cell must be used and the comparison must refer to the same solvent. Moreover, there is no difference whether a reference electrode with aqueous electrolyte (e.g. SCE) or with non-aqueous electrolyte is used (50), since the phase boundary potentials between test solution and electrolyte of the reference solution do not exceed that which is present at the contact between two aqueous solutions of the same ionic concentrations (111). For physical-chemical investigations, defining that effect, caused by medium, solvent (51) etc., generally the hydrogen electrode has been used. In organic solvents a slower potential adjustment is observed than in water. Due to Hills and Ives (8), grey platinized electrodes are more suitable than black platinized ones, when quick and reversible potential adjustment is required. Recently, the glass electrode has been used more and more for this purpose, when high reproducibility is desired. Comparison with the hydrogen electrode in cells without transference of the type (h) or (i) showed sufficient agreement when the content of the organic solvent was 80 per cent and less (112). It is advisable to keep the electrode constantly in the same medium, if reproducible pH measurements are required, since the external swell film changes with changing medium and creeping shift of the cell voltage occurs. If, for example, an electrode stored in water is immersed in methanol containing 2.7 per cent water, the cell voltage changes markedly with time. The value, extrapolated to the time zero, yields the pH value in sufficient agreement with the ptH value, referred to the aqueous solution (Figure 24). By penetration of the methanol, the H^+ activity of the swell film is changed and, consequently, the cell voltage changes too. But when an entirely dry (water-free) electrode is brought into carbon tetrachloride, saturated with T_2O (0.02 per cent), the glass electrode is charged very rapidly, with tritized water

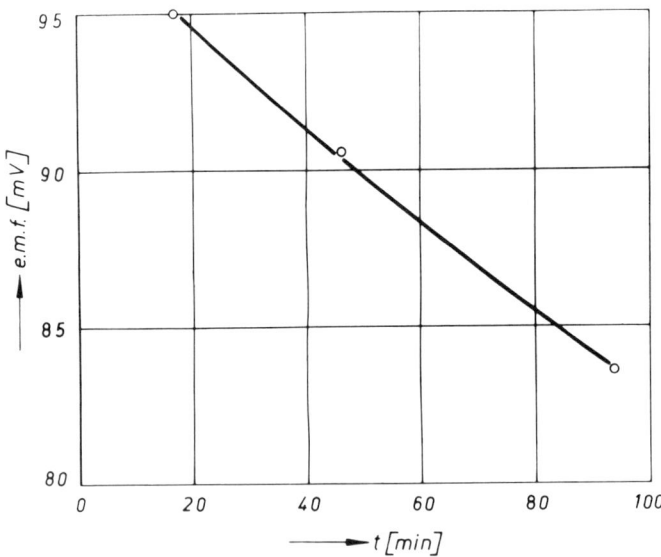

Fig. 24. Time function of the e.m.f. of a glass electrode in 97·3 per cent
methanol.

and about $2 \mu g \, cm^{-2}$ are adsorbed (113a).* It must be remarked, that the
type of the electrolyte used is of considerable importance for reproducibility
of glass electrode potentials in organic solutions. Glass electrodes show larger
alkaline error in organic solvent salts at high concentrations of alkali metals
($>0·1 M$) than in water. Such high concentrations are only reached in
mixtures of water with organic solvents. This should be considered even in
the neutral range, when standardizing the electrode (55). Kolthoff and
Chantooni, Jr. (114) did not succeed in obtaining reproducible glass electrode
potentials in non-aqueous acetonitrile with $HClO_4$, $N(C_2H_5)_4ClO_4$ and in
mixtures of H_2SO_4 and $N(C_2H_5)_4HSO_4$. But suitable results have been
attained in 2,5-dichlorobenzene sulphonic and methane sulphonic acid
together with the $N(C_2H_5)_4$ salts in acetonitrile. These solutions could be
used for buffer mixtures when standardizing the glass cell, where $Ag/0·01 M$
$AgNO_3$ (in acetonitrile) was used as reference electrode. The pK_{HA} values
of picric acid (11·0), obtained from potentiometric measurements in the
same medium agree very well with spectrophotometric results (11·1).
Kolthoff and coworkers (115) and recently, Wegmann, Escarfail and Simon
(116) studied in detail the behaviour of glass electrodes in acetic acid, since in
this medium the glass electrode, frequently, is used for titration of weak bases
with perchloric acid (117). When only the determination of the concentration

* Concerning the uptake and diffusion of T^+ at high temperatures (800–1300°C) in glass
see (113a).

of acids or bases is required, by potentiometric titration in organic medium, metal electrodes can be used as well (118).

Johannson and Norberg (118a) studied the dynamic response on pH change of the glass electrode in organic solvents (isopropanol, methylethyl ketone). The time constant was found to be of the order of 1000 times slower than in water. In organic solvents the logarithm of the time constant was proportional to pH up to neutrality, then the time constant was practically independent of a_{H^+}. It seems not improbable, in the opinion of the authors of reference (118a), that this time constant is a function of the rate of solvent exchange between the solution and the 'swell-layer' of the glass.

Sørensen and Lundgaard (119) have demonstrated that glass electrodes can be used in diluted solutions of hydrofluoric acid at least for potentiometric titrations in aqueous solutions. In $0.1\,M$ HCl, containing HF, the potential dependence of HF concentration occurs with the Nernst factor of about 42 mV at 25°C, when the HF concentration changes up to 10 times. In a salt melt, of course, no pH values can be measured with the glass electrode. But we should refer to the possibility of using the glass electrode in analytical and physical-chemical investigations of such melts (120).

X. pH MEASUREMENT IN BIOLOGY AND IN MEDICINE

In biological and medical laboratories, recently, pH measurements have been used more and more, and nearly always the glass electrode is applied, since the redox systems which are always present in biological substrates disturb measurements performed by other electrodes. Spectrophotometric methods for this purpose are not suitable, since the substrates themselves are coloured, moreover automation of these methods is more difficult than in the case of potentiometric methods, and special requirements are necessary, when pH measurement is applied. For this purpose, one problem must often be solved—the volume of the liquid is mostly very small and for in vivo measurements special precautions must be taken, concerning screening and isolation. On the other hand, cells for measurements are required which can be sterilized at high temperature. Moreover, for medical purposes the desired reproducibility of the results is extremely high. These problems required the development of special forms of glass electrodes and cells for measurements.

For extremely small volumes, microelectrodes have been produced, where the membrane for measuring is as small as possible, e.g. it is molten at the tip of a glass capillary with no response on pH, or the test solution is taken into a thin capillary out of the electrode glass material. This design demands that the swell film on the outer side of the capillary is surrounded by the buffer solution of known pH value. Both forms of the microelectrode meet some difficulties concerning the reference electrode which is immersed into the test solution. Firstly, the connection with the test solution must be as small as possible, but on the other hand solution from the reference electrode or from the salt bridge should not penetrate into the test solution, otherwise an undesired pH shift can be caused on account of the small volume and a possibly low buffer capacity. In the first design it is favourable, when the glass electrode, prepared in the form of a small plane membrane or a tip, is brought into a small vessel, which at the bottom is furnished with the contact supply for the external reference electrode (Figure 25). Using this

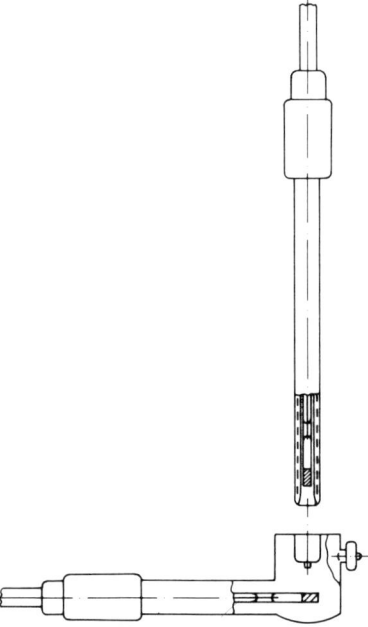

Fig. 25. Cell for micro measurements, consisting of a screened plane glass electrode and a reference electrode in form of a cup.

design, a volume of the test-solution of 0·05 ml is quite sufficient. Utilizing a plane membrane electrode as a cell, in the form of a combined electrode where the reference electrode is inside the shaft, connected with the test solution by a small porous pin, the complete design can be thermostated and the test solution can flow through a small test space of 0·1 ml or less (Figure 26). Generally, the glass electrode can be used for in vivo measurements too, if the capillary is incorporated into the circuit of the organ liquid.

Capillary electrodes are available in various forms. Figure 27 shows a relatively simple form, which is a kind of combined electrode and has a suitable capillary diameter; the necessary volume is not more than 20 μl. Of course, this type is suitable for thermostating. But it must be borne

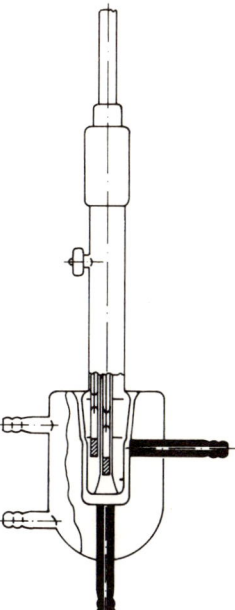

Fig. 26. Combined electrode with NS in a flow chamber, suitable for thermostating.

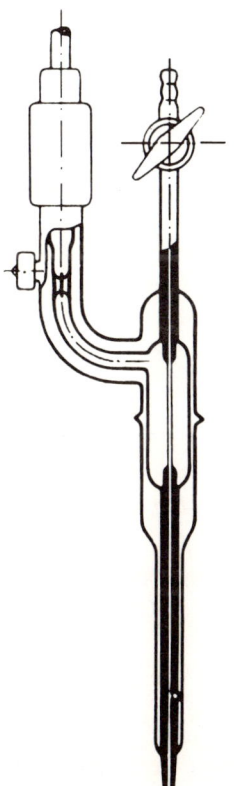

Fig. 27. Capillary combined electrode.

in mind that these electrodes have a very large ratio between electrode surface and the volume of the test solution (~ 1000). Therefore, the pH shift, caused by alkali loss of the glass material in unbuffered solution must be considered. This type of capillary glass electrode has been successfully used for pH measurements of blood for control of acid–base contents, especially in cases of illness, surgical operation, etc. (121), since they allow the blood to be taken anaerobically. Astrup applied a cell, especially designed for pH measurement of blood (122) (Figure 28). This type of electrode is used to check the CO_2 content of blood (121). For this purpose, the pH value of the blood is determined and then the blood is brought into equilibrium with gaseous mixtures of CO_2 and O_2 at various partial pressures of CO_2 and the pH value

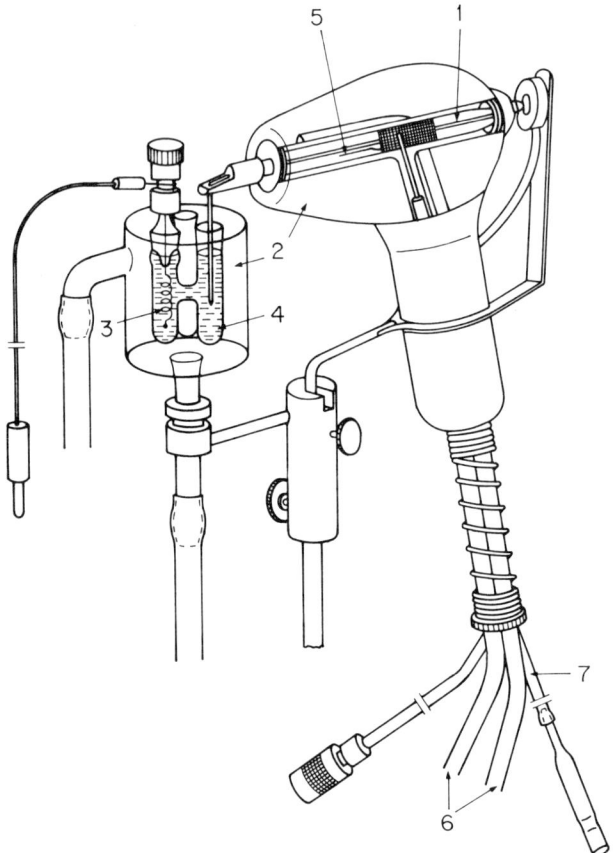

Fig. 28. Microcapillary cell: 1 capillary glass electrode; 2 chamber for thermostating; 3 inner reference electrode; 4 electrolyte bridge; 5 screened outer reference electrode; 6 tubing for thermostating water; 7 tubing for suction of blood.

is determined once more. From the straight-line plot of pH versus log p_{CO_2}, the CO_2 pressure can be determined that corresponds to the initial pH value of the blood. Figure 29 shows another electrode for pH measurement in blood, which can be applied in a simple way and which serves for other biological investigations too. The volume of the cell is $20\,\mu l$ and it can be thermostated. Special methods for the determination of bicarbonate in plasma and for pH measurement in blood have been suggested by Freier and co-workers (123)* and the results have been compared with those obtained from other methods or by calculation.

* About the problems concerning tapped blood and the errors which occur there, especially at pH measurements of fetal blood caused by CO_2 loss see (123b).

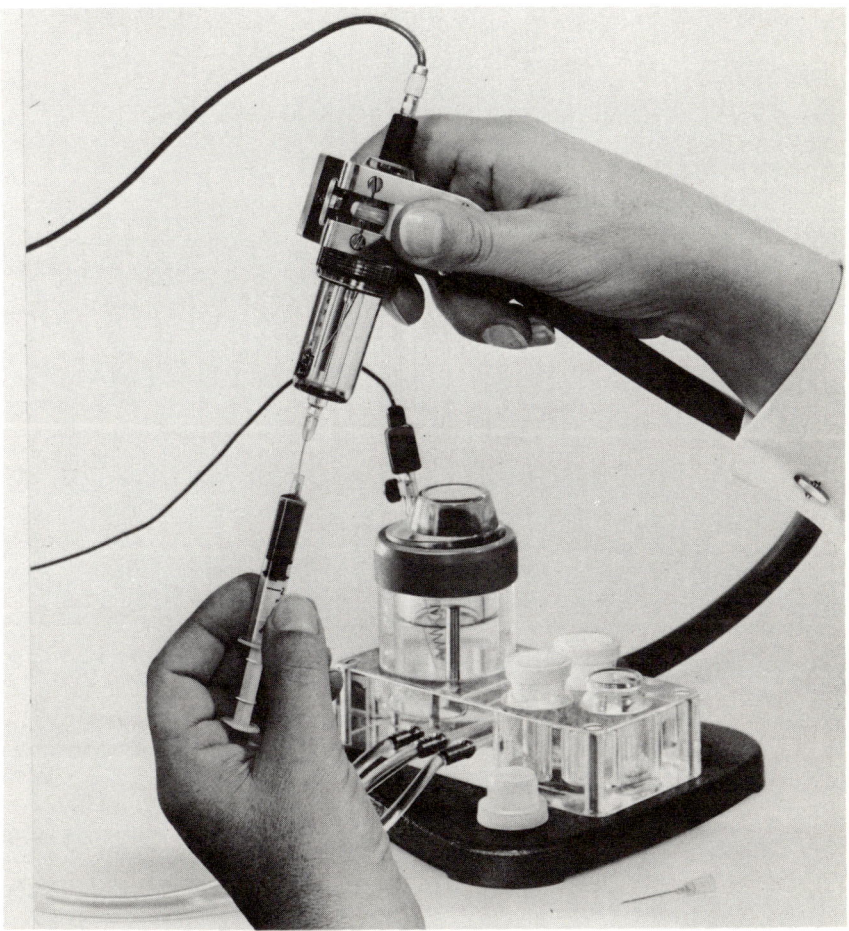

Fig. 29. Microcapillary cell for blood test.

Frequently for biological-medical investigations, the applied instruments must be sterile. When sterilization is performed at a temperature of 120°, closed glass electrodes can be used without difficulty, since they are resistant against the resultant over-pressure (inside). Glass material is available which does not lose measuring capability when treated at very high temperatures (124). But the reference electrodes $Hg/Hg_2Cl_2(s)$ are unsuitable for this purpose, since they show irreversible potential shift (see page 518). Mostly

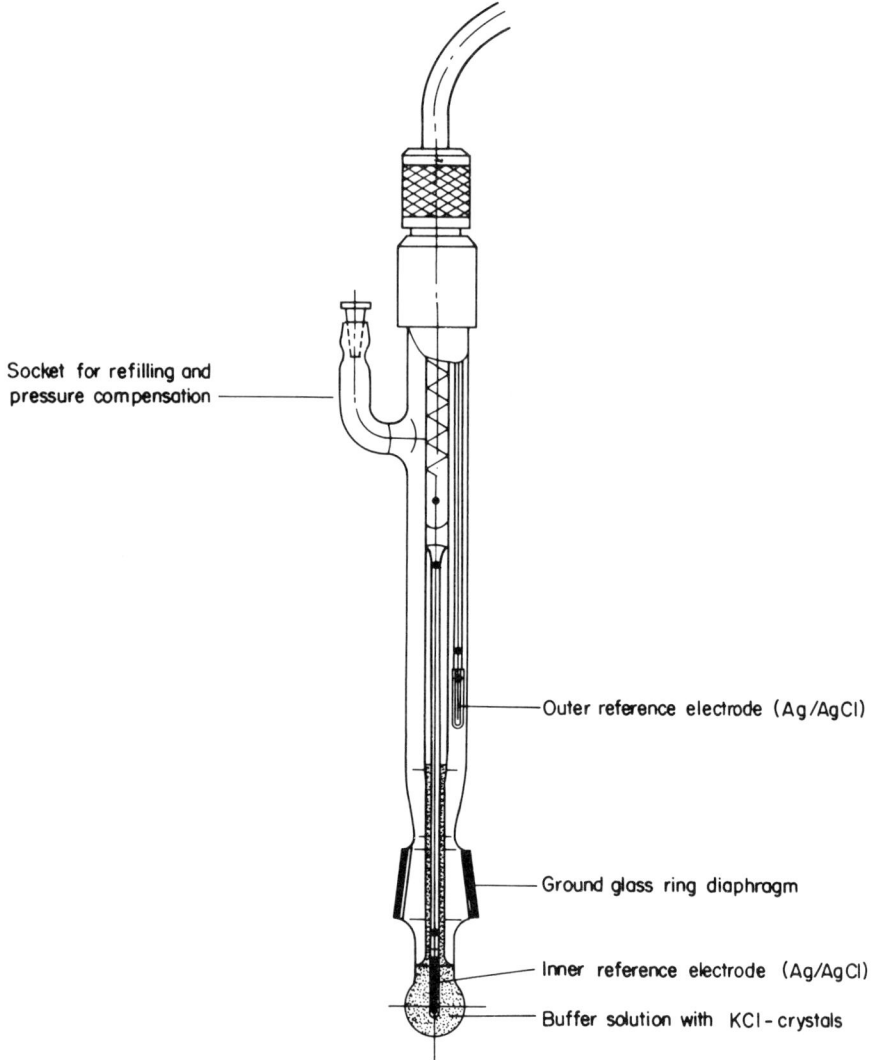

Socket for refilling and
pressure compensation

Outer reference electrode (Ag/AgCl)

Ground glass ring diaphragm

Inner reference electrode (Ag/AgCl)

Buffer solution with KCl-crystals

Fig. 30. Combined electrode, with Ag/AgCl electrodes, usable for heat sterilization.

Ag/AgCl electrodes are used and $Ag/AgCH_3CO_2$ electrodes have been shown to be successful. Ingold (125) suggested an Ag/AgCl electrode, especially for this purpose, which does not show any hysteresis after being cooled down to room temperature. Figure 30 shows a type of combined electrode which is resistant against sterilization at high temperature.

Sometimes the reference electrode, on immersion into the test solution, provokes some disturbance in biological substrates, especially in colloid substrates for they strongly influence the ionic mobility since diffusion potentials can occur or the electrolyte solution penetrates into the test solution which produces some coagulation. Therefore it was suggested to apply two glass electrodes (126) with different slopes, i.e., to replace the external reference electrode with a glass electrode of smaller slope. Generally, a potential difference is produced, which is proportional to the difference within the electrode gradient and in any case this is smaller than with an electrode of the secondary type as external reference electrode. This grants less accuracy in measurements, moreover it is very difficult to prepare electrodes which need only a small amount of test solution. It is also necessary to know the concentration of other cations, especially Na^+ in the test solution or to keep them at a constant level. Very careful investigations (127) have shown, that this type of electrode with small pH response is always very responsive versus cations and the desirable constancy is attained only at constant pH, if the concentration in alkali ions does not change.

Recently, a 'micro glass electrode for use in vivo' has been described suitable for insertion into the organism of biological objects, where the tip for measurement is only 10–20 μ long and 1 μ thick (128). Occasionally, the preparation and application are described in great detail. A calomel electrode serves as an external reference electrode; it is incorporated into the sidewall of a microburette filled with coloured KCl solution with a tip also inserted into the organism (Figure 31). The reproducibility is shown to be ± 0.01 pH in buffer solutions and 0.03 pH when in vivo measurements are carried out. This precision can be attained only if no electrolyte solution creeps under the polystyrene paint, but general experience shows that this is hardly avoidable, when the electrode is used continuously for a long time. In this case, a mean value is measured between the pH at the top of the electrode which is immersed in the test solution and the pH value under the polystyrene paint, which is progressively changed, due to the alkali loss of the glass material. Scheid and Kunze (129) reported their experiences, obtained in experiments with pH measurements in tumors of living animals, using microglass electrodes. In biological substrates, frequently, it will be of general importance to keep the pH value at a constant level, e.g. in order to study the influence of other parameters on biological processes. Soltero and Lee (130) have described a quite simple design, where the glass electrode

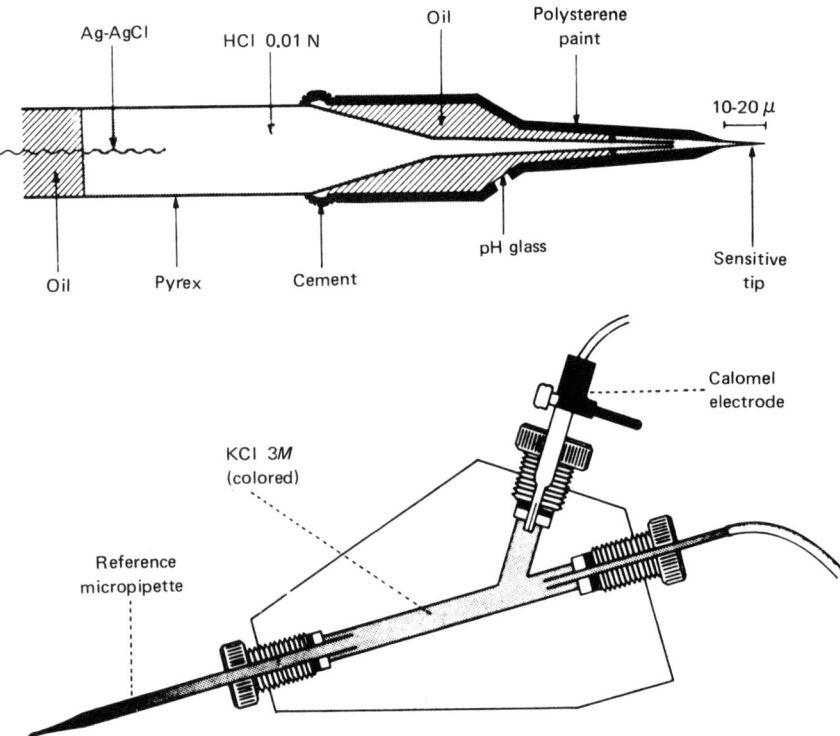

Fig. 31. pH glass microelectrode and reference electrode for in vivo application.

with the reference electrode controls the supply of air enriched by CO_2 (9·2 or 9·5 per cent CO_2) since the pH value should be kept constant between 6 and 9 $\pm$ 0·1 pH units, in algae cultures. Utilizing this apparatus a maximum rate of growth was found to be at pH 7.

In order to keep a constant pH value under laboratory conditions, i.e. inside small volumes, various designs of a so-called pH-stat are described. In principle, they do not differ from that, which we call mechanic control-plant for pH values (131) which is outside the scope of this paper. In the pH-stat the addition of acids or bases, necessary for maintaining the required value, is realized by a syringe-drive mechanism. Cotmar and Smith (132) demonstrate an electromechanical pH-stat as follows: The glass electrode cell is connected with an electronic pH-meter-amplifier, where the output from another amplifier controls the motor of the syringe-drive mechanism, so that the difference of the voltage between the glass cell and the required pH value, is always kept at zero. Various reagents for neutralization are used for the individual pH ranges which should be controlled. The control precision is $\pm$0·05 pH. Malmstadt and Piepmeier (133) have described in detail a pH-stat with digital indication. A constancy of 0·002 pH is attained and the device simply consists of units, which are available in every modern laboratory. The authors have developed a microburette with automatic filling and exhausting under photoelectrical control. The pH indication is obtained by a combination of glass and calomel electrodes. The pH value was kept at a constant level by automatic control and its influence on various enzyme reactions was studied. This device is suitable, too, when an automatic acid–base titration is carried out.

XI. SOME SPECIAL pH MEASUREMENTS IN TECHNOLOGY

Since hydrogen ion activity has attained significant importance not only for the chemical industry but others too, the electrometric pH measurement is applied in most parts of industrial plants and consequently special electrodes are used. Usage of glass electrodes meets difficulties, since high temperature and pressure must be taken into account and sometimes special precautions are necessary.

In less buffered aqueous solutions, as in water for boiler feed which is completely desalted by ion exchanger, the alkali loss from glass material must be considered. Therefore, electrode glass is used with small alkali content and with high electric resistance and moreover the test water must pass the electrode. Industrial control in water circuits for boiler feed furnishes other difficulties, on account of the small conductivity of such water ($\varkappa = 10^{-7}\,\mathrm{ohm^{-1}\,cm^{-1}}$ (134)). Immersed into water with high contaminations of oil and other liquids insoluble in water, the electrode is rapidly covered with a film of these substances and incorrect pH indications are obtained, or at least the indication is retarded. As far as possible, the glass electrode must be cleaned with organic solvents to dissolve the film after each measurement and must be soaked in water. It is advisable to check the functionality of the glass electrode in buffer solutions of known pH value. For continuous plant control in very dirty waste water, various types of electrodes are used which are continuously cleaned by mechanical means (135). Figure 32 shows an electrode, which is cleaned by rotating brushes of plastic material. The very slowly rotating brushes periodically change their rotation direction and the cleaning effect is strengthened but the stability of the electrode, which rotates slowly too, is not much shortened. When salt layers are built up on the electrode, due to over-saturation of the liquid, the functionality of the electrode may also be reduced. In the sugar-industry, the calcium carbonate, formed during the saturation process, adheres to the electrode and special precautions are necessary. At high flow velocity of the sugar solution which flows against the electrode in a tangential direction the separation can be reduced but not avoided completely. It is not advisable to use brushes, since $CaCO_3$ or other salts adhere to the brushes. It is favourable to balance the supersaturation by the addition of

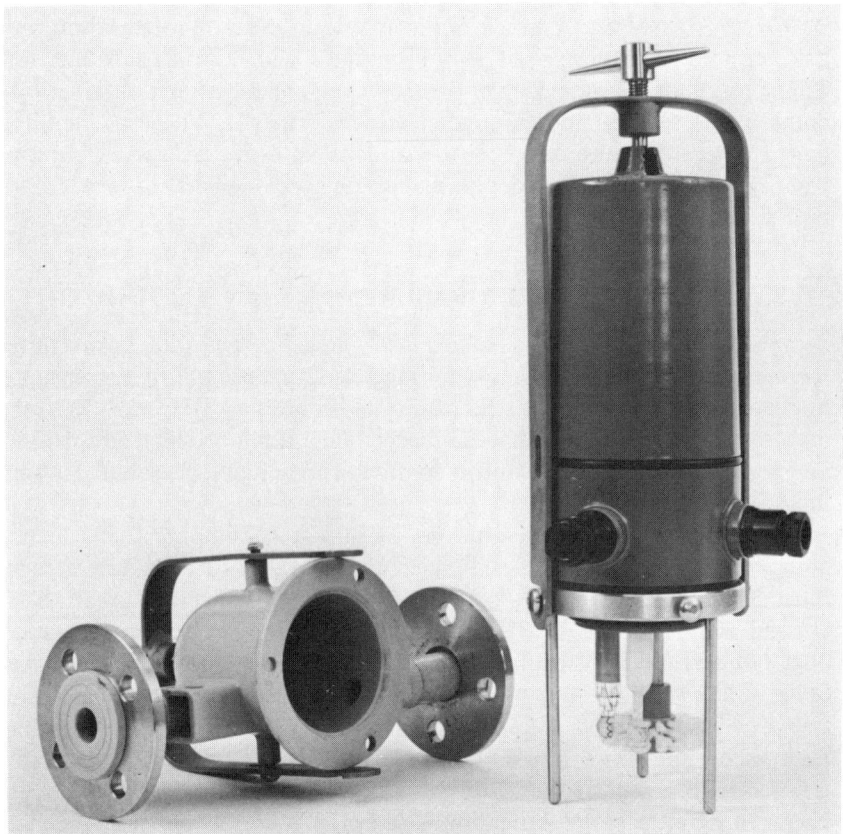

Fig. 32. Electrode with self-cleaning by brushes.

nucleating substances to the part which serves for measurement, in the form of small crystals of the salt concerned; in the above case crystals of $CaCO_3$ are used. In this way, the formation of salt layers on the electrode is avoided completely.*

If solid particles, suspended in the waste water, are present, mechanical cleaning is quite sufficient but care must be taken that no interruption between the flow of the liquid and the reference electrode occurs; for this purpose bridges with ground covers are used. Pallmann (137) noticed the 'dispersion effect' which can provoke considerable differences of pH values between the colloidal dispersion and its dispersion medium (up to 1 pH unit). Therefore, it is necessary to use either pure dispersion reagent when suspended particles are separated, or the measurement must be carried out at the same

* The Forschungsinstitut Meinsberg/Sa has applied this method already in sugar refineries with great success (136).

dispersion state, i.e. the measured pH unit depends on the depth of immersion of the electrodes. Lehmann (138) and coworkers have carried out numerous experiments in suspensions of felspar and studied the influence of the dispersion agent and the concentration of felspar but on account of the marked dependence of the dispersion effect on the charge of the particles, these results can scarcely be applied to other suspensions.

Ultrasonics have been used for cleaning the electrode, especially when fibrous material adheres to the electrode (139).

When measurements are carried out in carbonated water, special precautions must be taken, since the CO_2 content must remain unchanged during measurement. Krjukov (140) has described a special cell and a glass electrode for this purpose. For measurements in waste water, containing hydrofluoric acid, antimony electrodes have been suggested, treated with oxides, which furnish correct values to ± 0.15 pH units even at higher concentration of HF but no redox systems, which disturb the measurement (141), may be present.

In many areas of industries, especially in the food industry, it is useful to apply pH measurements, e.g. in pasty and semisolid material, but some water must be present in order to compare the pH value with that in aqueous solutions. Various kinds of milk products, e.g. in curds, cheese and butter, pH measurements were used for quality testing in order to guarantee the time the food keeps its quality (142). In meat and its products (sausages, ham, etc.) sometimes the pH value was used as a criterion of quality. For this purpose combined electrodes are applied in the form of syringe electrodes (so that they can be pricked into the material) (Figure 33).

Fig. 33. Needle-shaped combined electrode.

If the pH value has to be measured on films, foils, in textures and on paper, glass electrodes have been developed, furnished with plane membranes and plane reference electrodes, where plane and porous glass sheets or other diaphragm furnish the connection between test object and reference electrode (Figure 34). Palenius and coworkers (143) have demonstrated, that the pH value, resulting from measurement of paper, obtained by various extractions in cold or hot water, may differ markedly from that value, obtained by direct measurements (up to 1 pH unit). It is suggested that this result is valid when other material is tested (foils, etc.). Since the results obtained from one and the same material show large variation, statistic tests have been carried out which show identical standard deviations using corresponding methods, i.e. the same guarantee limit was attained. Electrodes as shown in Figure 34 have also been used successfully for pH measurement of skin.

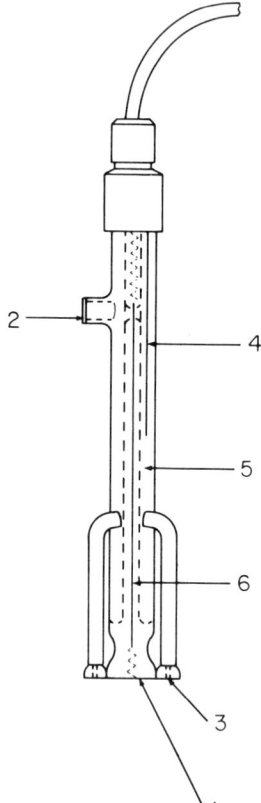

Fig. 34. Combined electrode for pH measurement on wet surface: 1 pH-sensitive membrane; 2 potassium chloride filling port; 3 porous ceramic plug; 4 reference electrode; 5 KCl-solution; 6 internal buffer solution.

XII. DEVICES FOR MEASUREMENTS IN CONTINUOUS PLANT CONTROL

Generally, only glass electrodes are used in apparatus for plant control. Of course, the cells for measurement must be treated as for laboratory purposes. Before use, the electrodes must be standardized with buffer solutions

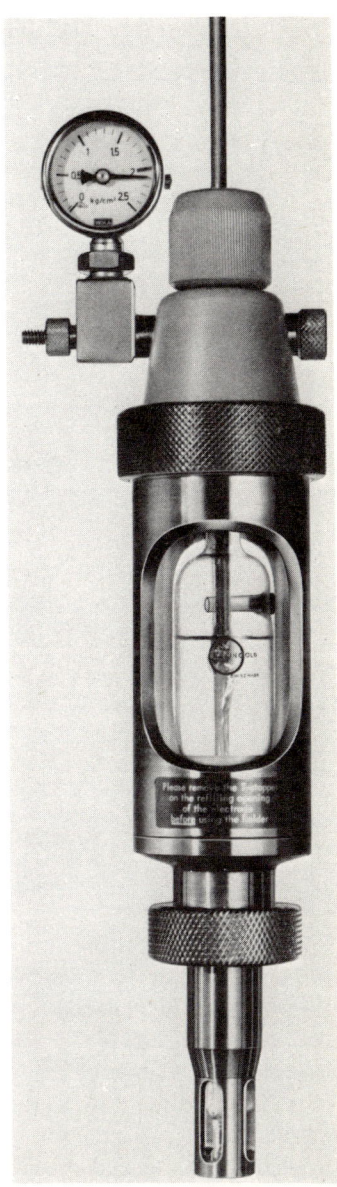

Fig. 35. Immersion electrode to insert into pipe lines.

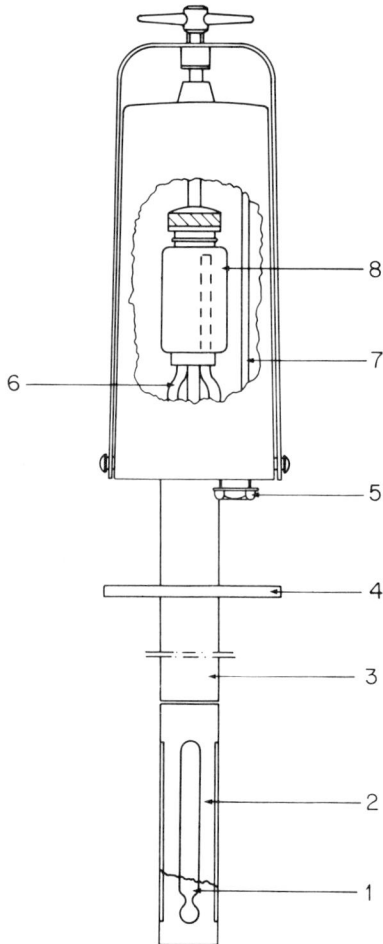

Fig. 36. Immersion electrode for open cells: 1 combined glass electrode; 2 screening; 3 shaft; 4 protection sheet; 5 screw joint; 6 tubing for pressure compensation; 7 coaxial cable; 8 KCl supply.

and the electric recorder must be adjusted in such a way, that the measured pH unit is in correct agreement with that value shown by the recorder. For this purpose, two special types of electrodes are used. One is the immersed one, where the cell for measurement is immersed directly into the test solution either in a vessel or the inside of a tube, or in an open passage. The other is a unit used for flow testing, where a partial current of the test material passes through a small vessel or tube which is furnished with the cell for measurement. Special precaution must be taken that no liquid may pene-

Fig. 37. Flow electrode box of polypropylene.

trate in an undesired direction. Immersion electrodes are used in various forms, according to the construction of the vessel. Figure 35 shows a design of an immersion electrode, which is directly incorporated into a pipe line by a flange and where the level of the solution for the reference electrode can be observed through a small window. To avoid the test solution, penetrating into the reference electrode, which possibly can be caused by over-pressure, the solution of the reference electrode (e.g. KCl) is kept under slight over-pressure too, which can be controlled by a manometer. Usually, the glass material of the electrode is protected against crushing by a metal screen, since sharply shaped particles in the fluid circuit frequently crush the electrodes. Figure 36 shows an immersion electrode for use in open vessels with variable filling level made of plastics. For this purpose, sometimes the cells for measurement are incorporated in float chambers, e.g. for use in rivers, so that its position is automatically coordinated to the fluid level (144).

Electrodes for the flow test are also used in different performances, and precautions are taken against corrosion effects of the test solution. In Figure 37 an equipment for the flow test is shown also; here the electrolyte solution of the reference electrode is automatically delivered from a reservoir.

Generally, it must be remarked that pH measurements in connection with the recording of the pH values indicated by glass cells, a continuous control is unavoidable which compares the recorded pH with pH values, obtained by laboratory tests with another assembly, standardized with buffer solutions. When large differences occur the cell for measurement in the flow-test apparatus must be restandardized, and if necessary it must be replaced. Süss (145) has supplied a list of various devices for industrial pH measurement together with glass electrodes and has pointed out their importance as a part of the electric circuit, which must have a realiable ground connection.

The influence of magnetic fields on the pH of water, which could be important specially for industrial measurements, was investigated by Joshi and Kamal (146); they measured a permanent pH change up to $+0.62$ by strong magnetic fields. But Quickenden and coworkers (147) repeating these measurements with all necessary care and could not find any pH change up to magnetic fields of 24,000 G. They show, that such a pH change by a magnetically induced change in the ionization constant of water (146) is thermodynamically extremely improbable.

XIII. INSTRUMENTS FOR MEASUREMENT

The development of instruments for pH measurement, especially with glass electrodes, has recently been brought about by the need to use as far as possible high ohmic electrodes (800–1000 megohm) without drift of the zero point and at a constant measuring precision. Moreover, instruments are desired which are furnished with battery feed and which can be used everywhere, in order to be suitable for every situation even in the sick-bed. Finally, in a medical laboratory a very high measuring sensitivity at correct constancy of zero point is necessary for pH measurement in blood.

All these demands are answered by the development of transistor amplifiers. These instruments, when furnished with a suitable circuit, possess an input resistance of 10^{12}–10^{14} ohms, especially when field effect transistors are used, i.e. the resistances are higher by three magnitudes than glass cells with 1000 megohms. As is known, the input resistance should be higher by two orders of magnitudes compared with the cell resistance. Transistor amplifiers need less space than tube amplifiers, at less current consumption, moreover batteries with small capacity and consequently with less weight can be used for current supply. When strong negative feedback is applied, completely satisfactory constancy of zero can be obtained and no difficulty occurs when a measuring sensitivity of 0.001 pH is required. More details are not within the range of this paper. In Table XVI the characteristic values of various modern pH-meter amplifiers are compared. Some of these instru-

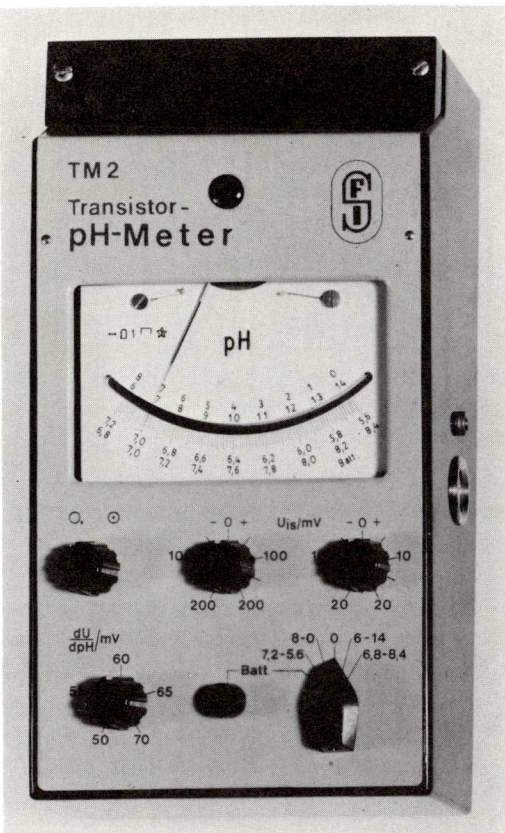

Fig. 38. Transistor pH meter.

ments are already equipped with digital indication of the pH value. Figure 38 shows the design of a completely transistorized pH-meter amplifier with current supply from batteries, which is able to measure pH values in the range between 0 and 14 pH units. With the extended scale, the pH in the interval of 1·6 can be determined accurately to 0·05 pH. Its dimensions are identical with that of an amperometer or a voltmeter (215 × 116 × 80 mm).

In order to make pH changes corresponding to a given, but variable programme in different solutions Carpeni and coworkers (148) have developed equipments consisting of an electrochemical programmer with a set of syringes, a pH meter and a recorder. The application for reactions with a slow rate, for instance titration of Na_2WO_4 with $HClO_4$, is described by many examples.

K. SCHWABE

TABLE XVIa
Table of Various Pocket pH Meters with Battery Supply

Instrument produced in	Type of amplifier	Range of pH measurement, pH	Range of mV measurement, mV	Precision of measurement, pH	Input resistance ohm	Length of the scale, mm	Scale division, pH	How long the battery can be used, h	Power supply	Supply for recorder	Dimensions, mm	Weight, kg
TM 2 GDR	M T	8–8 6–14 5·6–7·2 6·8–8·4	±350 ±70	0·02	5×10^{10}	75	0·1 0·02	100	–	–	215 116 80	1·4
MV 12 GDR	D R	2–12	–	0·1	10^{11}	70	0·1	200	–	–	196 106 75	1·2
Portatest 901 GFR	M T	0–14	±1400	0·1	2×10^{12}	57	0·2	2400	–	+	230 100 45	0·7
Portamess 902 GFR	M T	0–14 12 divisions pH = 1·4	12 divisions 140 mV	0·02	2×10^{12}	57	0·05	2400	–	+	230 100 45	0·7
Nr. 4391301 GFR	D R	1–13	0–700	0·1	10^{11}	100	0·2		–	–	206 224 105	2·6
pH 54 GFR	D R	2–12	–	0·1	10^{12}	57	0·1	250	–	–	161 87 85	1·5
Nr. 1817000 GFR	D R	0–12·5	–	0·1	10^{11}	68	0·2		+	–	170 100 70	1·0

TABLE XVIa—*continued*

Table of Various Pocket pH Meters with Battery Supply

Instrument produced in	Type of amplifier	Range of pH measurement, pH	Range of mV measurement, mV	Precision of measurement, pH	Input resistance ohm	Length of the scale, mm	Scale division, pH	How long the battery can be used, h	Power supply	Supply for recorder	Dimensions, mm	Weight, kg
GBA Austria	D R	1–13	–	0·1		85	0·2		–	–	97 114 42	0·5
55 B Switzerland	D T	0–14	±1400	0·1	10^{12}	110	0·1	5000	–	+	240 145 98	1·6
E 280 A Switzerland	D R	0–14	±1000	0·1	2×10^{11}	110	0·1	100	–	–	195 170 120	
30 c England	D T R	0–8 6–14	0–800 600–1400	0·1	10^{11}	75	0·1	120	–	–	83 140 184	1·3
Nr. 34460 Netherlands	D R	2–12	–	0·1		58	0·2		–	–	149 76 57	
40 A Netherlands	D R	1–13	–	0·1	10^{12}	76	0·2		–	–	112 225 82	
PIM France	D T	0–7 7–14	–	0·03	10^{11}	85		500	–	–	160 110 80	0·9

TABLE XVIa—*continued*

Table of Various Pocket pH Meters with Battery Supply

Instrument produced in	Type of amplifier	Range of pH measurement, pH	Range of mV measurement, mV	Precision of measurement, pH	Input resistance ohm	Length of the scale, mm	Scale division, pH	How long the battery can be used, h	Power supply	Supply for recorder	Dimensions, mm	Weight, kg
PP 26 France	D R	0–14	±500	0·03	10^{13}	110	0·1	100	–	–	232 132 80	2·3
PHM 24 Denmark	D R	0–8 6–14	–	0·05	10^{12}	100	0·1	400	–	–	210 150 100	2·4
PHM 29 Denmark	M T	0–10 4–14	+700 −300 −700 +300	0·05	10^{11}	130	0·1	800	+	–	230 150 120	2·4
OP-105 Hungary	D R	0–12	–	0·1	10^{10}	120	0·1	50	–	–	200 110 70	1·4
Pocket pH Meter U.S.A.	D R	2–12	–	0·1		60	0·2	400	–	–	150 75 50	0·9
37 A U.S.A.	M T	0–14	±1500	0·005		–	0·005		+	–	190 300 130	3·7

TABLE XVIa—*continued*
Table of Various Pocket pH Meters with Battery Supply

Instrument produced in	Type of amplifier	Range of pH measurement, pH	Range of mV measurement, mV	Precision of measurement, pH	Input resistance ohm	Length of the scale, mm	Scale division, pH	How long the battery can be used, h	Power supply	Supply for recorder	Dimensions, mm	Weight, kg
210 U.S.A.		0–14	±700 ±1400	0·05	10^{11}	180	0·1	25	+	+	280 280 100	4·5
175 U.S.A.	T	0–14	—	0·07	10^{13}	180	0·1	2000	—	—		
PB U.S.A.	D R T	0–14	—	0·05	10^{14}	140	0·1	800	+	+	300 300 100	
401 U.S.A.	M T	0–14	±70 ±700	0·02	10^{13}	170	0·1	1000	—	+	146 200 100	2·0
TM4 GDR	M T	0–14 3·5–10·5	+1400 −1400 +140 −140	0·05	10^{12}	130	0·1 0·05	500	+	+	191 164 89	1·6
TM5 GDR	M T	0–14 5·6–8·4	+140 −140	0·02	10^{12}	130	0·1 0·02	500	+	+	191 164 89	1·6

TABLE XVIa—*continued*

Table of Various Pocket pH Meters with Battery Supply

Instrument produced in	Type of amplifier	Range of pH measurement, pH	Range of mV measurement, mV	Precision of measurement, pH	Input resistance ohm	Length of the scale, mm	Scale division, pH	How long the battery can be used, h	Power supply	Supply for recorder	Dimensions, mm	Weight, kg
TM6 GDR	M T	0–14	±500	0·1	10^{12}	130	0·1	500	+	+	191 164 89	1·6
MV 81 GDR	M T	0–14 3·5–10·5	±700 ±350	0·05	10^{12}	80	0·2 0·1	300	+	+	215 150 95	1·6
901 GFR	M T	0–14	±1400	0·1	10^{12}	57	0·2	5200	–	+	230 100 45	0·7
902 GFR	M T	0–14 12 div. pH = 1·4	12 div. 140 mV	0·02	10^{12}	57	0·05	4500	–	+	230 100 45	0·7
pH 56 GFR	D T	0–14 4·5–9·5	±1000 ±500	0·02	10^{12}	75	0·2 0·05	180	–	+	150 90 60	0·5
55 B Switzerland	D T	0–14	±1400	0·1	10^{12}	110	0·1	5000	–	+	240 145 98	1·6
610 England	D T	0–14 1·4 pH	±700 ±70	0·01	10^{12}	140	0·1 0·01	800	–	+	310 120 210	2·0

TABLE XVIa—*continued*

Table of Various Pocket pH Meters with Battery Supply

Instrument produced in	Type of amplifier	Range of pH measurement, pH	Range of mV measurement, mV	Precision of measurement, pH	Input resistance ohm	Length of the scale, mm	Scale division, pH	How long the battery can be used, h	Power supply	Supply for recorder	Dimensions, mm	Weight, kg
PHM 29 Denmark	M T	0–10 4–14	+700 −300 −700 +300	0·05	10^{13}	130	0·1	800	+	−	230 150 120	2·4
OP-106 Hungary	D T	1–13	—	0·1	10^{11}	100	0·1		−	−	120 135 45	1·0
H5 USA	D T	0–14 2·8pH	±1400 ±140	0·01	10^{13}	148	0·1 0·02	50	+	+	265 152 55	2·3
401 USA	M T	0–14	±700 ±70	0·02	10^{13}	170	0·1	1000	−	+	146 200 100	2·0

Abbreviations: D = direct coupled amplifier   T = transistor connection
M = modulation amplifier   + = with supply
R = tube connection   − = without supply

TABLE XVIb
Table of Various Digital pH Meters

Instrument produced in	Range of pH measurement, pH	Range of mV measurement, mV	Precision of measurement, pH	Input resistance ohm	Digital output	Analogous output	Dimensions, mm	Weight, kg
DIGI 610 GFR	0–14	±1999	0.01	10^{12}	+	+	380 220 140	5·0
PW 9408 Netherlands	0–14	±1999	0.01	10^{12}	+	+	444 310 146	9·5
PHM 52 Denmark	0–14	±1400 ±7000	0.002	10^{12}	+	+	340 230 190	5·6
OP-206 Hungary	0–14	±1400	0.02	10^{12}	–	–	270 210 170	4·0
DAT 2002 Italy	0–14	±1500	0.01	10^{15}	+	+	400 350 150	8·0
Model 205 U.S.A.	0–14	±1400	0.007	10^{14}	–	–	330 216 152	
Model DR U.S.A.	0–14	1400	0.002	10^{13}	–	–	314 317 140	9·5

TABLE XVIb—*continued*
Table of Various Digital pH Meters

Instrument produced in	Range of pH measurement, pH	Range of mV measurement, mV	Precision of measurement, pH	Input resistance ohm	Digital output	Analogous output	Dimensions, mm	Weight, kg
Model 801 U.S.A.	0–14	±1000	0.002	10^{14}	+	+	390 305 157	7.3
MV 87 GDR	0–14 6–8	±1999 ±199.9	0.01 0.002	10^{13}	+	+	105 350 280	6.0
641 GFR	0–14	±1999.9	0.002	10^{12}	+	+	240 115 205	1.3
Digi 510 GFR	0–14	±1999	0.01	10^{12}	+	+	300 170 260	4.3
5105 Switzerland	0–14	±1500	0.01	10^{12}	+	+	350 260 150	
E 500 Switzerland	0–14	±2000 ±1000	0.01	10^{13}	+	+	300 210 240	4.6

TABLE XVIb—*continued*
Table of Various Digital pH Meters

Instrument produced in	Range of pH measurement, pH	Range of mV measurement, mV	Precision of measurement, pH	Input resistance ohm	Digital output	Analogous output	Dimensions, mm	Weight, kg
PHM 52 Denmark	0–14	±1400 ±7000	0·002	10^{12}	+	+	340 230 190	5·6
DAT 2002 Italy	0–14	±1400	0·01	10^{15}	+	+	400 350 150	8·0
7421 USA	0–14	±1400	0·002	10^{13}	+	+	300 240 200	5·8
801 USA	0–14	±999·9	0·002	10^{14}	+	+	390 305 157	7·3

REFERENCES

1. G. Mattock, 'Laboratory pH Measurements' *in Advances in Analytical Chemistry and Instrumentation*, C. N. Reilley, Ed., Vol. 2, p. 35–122, John Wiley, New York, 1963.
2. K. Schwabe, *Electrochim. Acta*, **12**, 67 (1967).
3. G. Kortüm, *Lehrbuch der Elektrochemie*, 4th ed. Verlag Chemie, Weinheim, 1966; E. Lange, *Thermodynamische Elektrochemie*, Hüthig Verlag, Heidelberg, 1962; K. Schwabe, *pH-Messtechnik*, Steinkopf Verlag, Dresden, 1963.
4. K. Schwabe, *Österr. Chemiker Ztg.*, **65**, 339 (1964); A. Ferse, Dissertation, T. U. Dresden, Institut für Elektrochemie u. physikalische Chemie, 1964.
5. (a) G. Kortüm, *Kolometrie, Photometrie u. Spektrometrie*, Springer Verlag, Göttingen, 1955; (b) L. P. Hammett, *Physical Organic Chemistry*, McGraw-Hill, New York, 1940; E. A. Braude, *J. Chem. Soc. (London)*, **1948**, 1971, 1976; P. Salomaa, *Acta Chem. Scand.*, **11**, 127 (1957); D. Rosenthal and J. S. Dwyer, *Canad. J. Chem.*, **41**, 89 (1963); *Anal. Chem.*, **35**, 161–166 (1963).
6. R. G. Bates, *Determination of pH*, John Wiley and Sons, New York, 1963.
7. J. Horiuti, A. Matsuda, M. Enyo and H. Kita, The First Australian Congress on Electrochemistry, 1963; A. Matsuda and J. Horiuti, *J. Res. Inst. Catalysis (Sapporo)*, **6**, 231 (1958); M. Enyo, M. Hashi and H. Kita, *J. Res. Inst. Catalysis (Sapporo)*, **10**, 153 (1962).
8. G. J. Hills and D. J. G. Ives, *J. Chem. Soc. (London)*, **1951**, 305, 311.
9. A. Ferse and K. Schwabe, *Z. Physik. Chem. (Leipzig)*, **230**, 20 (1965).
10. K. Schwabe and S. Ziegenbalg, *Z. Elektrochem.*, **62**, 172 (1958).
11. G. Bianchi, A. Barosi, G. Faita and T. Mussini, *J. Electrochem. Soc.*, **112**, 921 (1965).
12. P. Nylen, *Svensk. Kem. Tidskr.*, **48**, 76 (1936).
13. M. J. Vasile and C. G. Enke, *J. Electrochem. Soc.*, **112**, 865 (1965).
13a. H. Binder, A. Köhling and A. Sandstede, *Chem. Ing. Technik*, **43**, 1084 (1971).
14. K. Schwabe, *pH-Messtechnik*, Steinkopff Verlag, Dresden, 1963.
15. J. P. Schwing, *J. Chim. Phys.*, **61**, 491–507 (1964).
16. J. L. Gabbard, *J. Amer. Chem. Soc.*, **69**, 533 (1947).
16a. L. O. Warren, *Anal. Chim. Acta*, **53**, 199 (1957); K. Schwabe, *pH Messung*, Steinkopff, Dresden, 1963, p. 75.
17. F. Hovorka and W. C. Dearing, *J. Amer. Chem. Soc.*, **57**, 446 (1935).
17a. B. D. Struck and O. Schneider, *J. Electroanal. Chem.*, **36**, 41 (1972).
18. J. de Fliss and J. M. Worobjew, *Betriebslaboratorium USSR (Russian print)*, **29**, 538 (1963).
19. A. Sovová, *Collect. Czech. Chem. Commun.*, **28**, 739 (1963).
20. F. Čuta and S. Havelka, *Collect. Czech. Chem. Commun.*, **28**, 3005 (1963).
21. Z. Dolinski and B. Waligova, *Prace Chemiczne*, **5**, 87 (1959).
22. J. Römer, Dissertation T. U. Dresden, Institut für Elektrochemie und physikalische Chemie, 1967, J. Römer and K. Schwabe, *Electrochim. Acta*, 1s, 885 (1970).
23. M. R. Abdalla, Thesis Cairo University (1965).
24. More detailed information, see: K. Schwabe, *pH-Messtechnik*, Steinkopff Verlag, Dresden, 1963, p. 82–99.
25. M. Cremer, *Z. Biol.*, **47**, 562 (1906); F. Haber and E. Klemensiewicz, *Z. Physik. Chem.*, **67**, 385 (1909).
26a. K. Schwabe and H. D. Suschke, *Angew. Chem.*, **76**, 39 (1964).
26b. G. A. Rechnitz, *Z. Analyt. Chem.*, **214**, 252 (1965); R. A. Durst, *J. Chem. Educat.*, **44**, 175 (1967).
27. K. Schwabe and H. Dahms, *Mber. dtsch. Akad. Wiss. Berlin*, **I**, 279 (1959).

28a. R. M. Doremus, Gen. El. Report No. 67 C. 296 (1967).

28b. R. B. Buck and J. Krull, *J. Electroanalyt. Chem.*, **18**, 363 (1968).

29. B. P. Nikolski, *J. Physic. Chem.* (*Russian print*), **10**, 495 (1937).

30. B. P. Nikolski and M. M. Schulz, *Westnik. Lenisgradskogo Univ.*, **4**, 74 (1963); *J. Physic. Chem.* (*Russian print*), **36**, 1327 (1962); Z. Bocksay and B. Czakvari, *Ann. Univ. Sci. Budapestinensis. Sect. Chim.*, **10**, 125 (1968); B. S. Smoljakov, V. D. Perkovec and P. A. Krjukov, *Nachr. sibir. Abt. Akad. Wiss. UdSSR* (*Russian print*), **4**, 2, 3–6 (1967).

31. K. Schwabe, N. Dahms, Q. Nguyen and G. Hoffmann, *Z. Elektrochem., Ber. Bunsenges. Physik. Chem.*, **66**, 304 (1962).

32. M. Dole, *J. Amer. Chem. Soc.*, **54**, 2120 (1932).

32a. Z. Bocksay and B. Czavari, *Ann. Univ. Sci. Budapestinensis. Sect. Chim.*, **10**, 125 (1968).

32b. B. Csakian, Z. Boksay and G. Bouquer, *Anal. Chim. Acta*, **56**, 279 (1971).

32c. H. Bach and F. G. K. Baucke, *Electrochim. Acta*, **16**, 1311 (1971).

32d. H. F. Tadros and I. Lyklema, *J. Electroanal. Chem.*, **22**, 96 (1969).

33. K. Schwabe and G. Glöckner, *Z. Elektrochem.*, **59**, 504 (1955); Z. Bocksay, G. Bouquet, B. Czakvari and T. Garai, *Ann. Univ. Sci. Budapestinensis Sect. Chim.*, **10**, 119 (1968).

34. See, e.g., K. Schwabe, *Dechema-Monogr.*, Vol. 31, No (426–50) p. 151 (1959); A. S. Benevolskij, A. I. Parfenov, M. M. Schulz and V. P. Juchnovskij, *Westnik Leningradskogo Univ. Ser. Physik. Chem.*, **10**, 2, 111–117 (1966).

35a. G. T. Petrovsky and F. Čuta, *Collect. Czech. Chem. Commun.*, **27**, 25 (1962).

35b. H. Geisler, *Chem. Techn.*, **20**, 233 (1968).

36. About the temperature coefficient of electrode potentials, see A. J. De Bethune, T. S. Licht and N. Svendeman, *J. Electrochem. Soc.*, **106**, 616 (1959).

37. D. J. G. Ives and G. J. Janz, *Reference Electrodes, Theory and Practice*, Academic Press, New York, 1961.

37a. A. Shathag, *Anal. Biochem.*, **30**, 287 (1909).

38. M. H. Lietzke and R. W. Stoughton, *J. Amer. Chem. Soc.*, **75**, 5226 (1953).

39. R. L. Every and W. P. Banks, *Electrochem. Technol.*, **4**, 275 (1966).

40. S. R. Gupta, G. J. Hills and D. J. G. Ives, *Trans. Faraday Soc.*, **59**, 1886 (1963).

41. K. Schwabe, *Naturwiss*, **44**, 350 (1957).

42. A. K. Covington, P. K. Talukdar and H. R. Thiask, *Trans. Faraday Soc.*, **60**, 412 (1964).

43. W. D. Larson, *J. Phys. Chem.*, **67**, 937 (1963).

44. R. G. Bates, E. A. Guggenheim, H. S. Narned, I. J. G. Ives, G. J. Janz, C. B. Monk, R. A. Robinson, R. N. Stokes and W. F. K. Wynne-Jones, *J. Chem. Phys.*, **25**, 361 (1956).

45. R. G. Bates *in Determination of pH*, (p. 6) see reference 6, a. *I. J. G. Ives* in 'Reference Electrodes' (p. 37) see reference 37.

46. R. G. Bates and V. E. Bower, *J. Res. Nat. Bur. Standards*, **53**, 283 (1954); H. B. Hetzer, R. A. Robinson and R. G. Bates, *J. Physic. Chem.*, **66**, 1423 (1962); R. G. Bates, NSB Technical Note, **271**, 33 (1965).

47. R. S. Greeley, W. T. Smith, R. W. Stoughton and M. H. Lietzke, *J. Physic. Chem.*, **64**, 642 (1960); M. B. Towns, R. S. Greenley and M. H. Lietzke, *J. Physic. Chem.*, **64**, 1861 (1960); see also *J. Physic. Chem.*, **67**, 2573 (1963); **69**, 2395 (1965); **70**, 756 (1966); **72**, 257 (1968).

48. G. Kortüm and W. Häusermann, *Ber. Bunsen Ges.*, **69**, 594 (1965).

49. E. H. K. Lairz and A. E. Mahgouh, *J. Electroanalyt. Chem.*, **8**, 482 (1964).

50. K. Schwabe, *pH-Messungen unter extremen Bedingungen*, Verlag Chemie,

Weinheim, 1960; M. Böttcher, Dissertation T. U. Dresden. Institut f. Elektrochemie u. physikalische Chemie, 1967.

51. See, e.g., R. G. Bates *in* A. K. Covington and P. Jones, *Hydrogen-Bonded Solvent Systems*, Taylor and Francis, London, 1968, p. 49; L. R. Dawson, K. H. Kim and H. C. Eckstrom, *J. Phys. Chem.*, **70**, 775 (1966), and former work of L. R. Dawson, C. L. de Ligny and A. A. Wieneke, *Rec. Trav. Chim. Pays-Bas.*, **79**, 268 (1960); D. Feakins and C. M. French, *J. Chem. Soc.* (*London*), **1957**, 2581, see further references noted there, K. Schwabe, *Österr. Chemiker Ztg.*, **65**, 339 (1964); K. Schwabe and R. Müller, *Z. Elektrochem. Ber. Bunsenges. Physik. Chem.*, in press; M. Paabo, R. G. Bates and R. A. Robinson, *Anal. Chem.*, **37**, 462 (1965); W. Marple, *J. Chemic. Engin. Data*, **12**, 437 (1967); D. Feakins, K. G. Lawrence and R. P. T. Tomkins, *J. Chem. Soc.* (*London*), **1967**, 753; L. R. Dawson, K. H. Kim and H. C. Eckstrom, *J. Phys. Chem.*, **70**, 775 (1966); K. Schwabe and E. Ferse, *Z. Elektrochem., Ber. Bunsenges. Physik. Chem.*, **70**, 849 (1966); K. W. Morcom and B. L. Muiju, *Nature*, **217**, No 5133, 1048 (1968); H. P. Bennetto, D. Feakins and D. J. Turner, *J. Chem. Soc.* (*A*), **1966**, 1211.
52. H. S. Harned and B. B. Owen, *The Physical Chemistry of Electrolytic Solutions*, 3rd, ed., Reinhold Publishing Corporation, New York, 1958.
53. D. Feakins and C. M. French, *J. Chem. Soc.* (*London*), **1957**, 2581; W. Graichen, Dissertation T. U. Dresden, Institut f. Elektrochemie u. physikalische Chemie, 1963; E. S. Amis, *J. Electroanalyt. Chem.*, **8**, 413 (1964); see also J. Juillard, Thèse Univ. Clermont, Faculté des Sciences. 'Etude de la dissociation des acides en solvants organiques et hydroorganiques.' 1968.
54. A. S. Quist and W. L. Marshall, *J. Phys. Chem.*, **72**, 1536 (1968).
55. Dipl. Arbeit. R. Behrens, T. U. Dresden, Institut f. Elektrochemie u. physikalische Chemie, 1962.
56. H. K. Fricke, *Beitr. Angew. Glasforschung* (*Schott Stuttgart*), **1959**, 175; *Zucker.* **14**, 162, 188 (1961).
56a. F. G. K. Baucke, *J. Electroanal. Chem.*, **33**, 135 (1971).
56b. S. M. Afsar Naqu and P. B. Mathur, *Electrochim. Acta*, **13**, 1369 (1968).
57. Dipl. Arbeit M. Seidler, T. U. Dresden, Institut f. Elektrochemie u. physikalische Chemie, 1967; K. Schwabe and M. Seidler, *Chem. Techn.*, in press.
58. G. Kortüm, *Lehrbuch der Elektrochemie*, 4th ed., Verlag Chemie, Berlin, 1966, p. 275f.
59. R. O. Picknett, *Trans. Faraday Soc.*, **64**, 1059 (1968).
60. See, e.g., J. J. Hermans, *Rec. Trav. Chim. Pays-Bas*, **57**, 1373 (1968); G. Kortüm, *Rec. Trav. Chim. Pays-Bas*, **58**.
61. P. A. Rock, *Electrochim. Acta*, **12**, 1531 (1967).
62. G. Bianchi, G. Faita, R. Galli and T. Mussini, *Electrochim. Acta*, **12**, 439 (1967).
63. E. A. Guggenheim, *J. Phys. Chem.*, **34**, 1758 (1930); Z. Szabo, *Z. Physik. Chem. A*, **174**, 22, 33 (1935); **A176**, 125 (1936); **A181**, 169 (1938).
64. W. N. Carson, J. E. Michelson and K. Koyama, *Anal. Chem.*, **27**, 472 (1955).
65. G. Brunisholz, *Anal. Chim. Acta* (*Amsterdam*), **10**, 470 (1954); K. Schwabe, *pH-Messtechnik*, Steinkopff Verlag, Dresden, 1963.
66. 16th National Meeting, American Association of Clinical Chemists, Boston, Mass. 1964.
67. A. J. Zielen, *J. Phys. Chem.*, **67**, 1474 (1963).
68. R. G. Bates and V. E. Bower, *J. Res. Nat. Bur. Stds.*, **53**, 283 (1964).
69. A. K. Covington and J. E. Prue, *J. Chem. Soc.*, **1955**, 3696; A. K. Covington, *J. Chem. Soc.*, **1060**, 4441; K. Schwabe, *Österr. Chemiker Ztg.*, **65**, 339 (1964); T. S. Light and K. S. Fletcher, *Anal. Chem.*, **39**, 70 (1967).

70. K. Schwabe, *Electrochim. Acta*, **12**, 67 (1967); Dipl. Arbeiten Almut and Andrej Stasko, 1964 and G. Steyer, 1966, T. U. Dresden, Institut f. Elektrochemie u. physikalische Chemie.

71. K. Schwabe, *Z. Elektrochem.*, **42**, 147 (1936); **43**, 874 (1937).

72. R. G. Bates, *J. Res. Nat. Bur. Stds.*, **66A**, 179 (1962).

73. W. M. Beck, A. E. Bottom and A. K. Covington, *Anal. Chem.*, **40**, 501 (1968).

74. *Z. Elektrochem. Angew. Physik. Chem.*, **40**, 150 (1934).

75. W. Wilke, Dipl. Arbeit, T. U. Dresden, Institut f. Elektrochemie u. physikalische Chemie, 1967.

76. R. G. Bates, *Ann. N.Y. Acad. Sci.*, **92**, 341 (1961); see also reference 6, p. 120.

77. N. E. Good, G. D. Winget, W. Winter, Th. W. Connolly, S. Izawa and R. M. M. Singh, *Biochem.*, **5**, 467 (1966).

78. H. Neumann and H. Geisler, *Deut. Gesundheitswes.*, **23**, 2414 (1968).

79. O. Petersen, *Chemie-Ing. Techn.*, **40**, 76 (1968).

80. P. A. Kriukov, W. D. Perhoves, I. J. Starestina and B. J. Smolsikov, *Nachr. Sibir. Abt. Akad. Wiss. UdSSR* (*Russian print*), **7**, 30 (1966).

81. M. H. Lietzke and J. V. Vaughen, *J. Amer. Chem. Soc.*, **77**, 876 (1955); R. S. Greeley, W. T. Smith, Jr., R. W. Stoughton and M. H. Lietzke, *J. Phys. Chem.*, **64**, 652 (1960); see also reference 14, Table 19.

82. W. D. Treadwell and L. Weiss, *Helv. Chim. Acta*, **3**, 433 (1920); E. Schreiner, *Z. Anorg. Allgem. Chem.*, **116**, 102 (1921); F. Reiff, *Z. Anorg. Allgem. Chem.*, **208**, 321 (1932).

83. See, e.g., F. E. Critchfield and J. B. Johnson, *Anal. Chem.*, **31**, 570 (1959).

84. G. Beinroth, K. Schwabe and H. D. Suschke, *Z. Physik. Chem.* (*Leipzig*), **235**, 133 (1967).

85. R. H. Stokes and R. A. Robinson, *J. Phys. Chem.*, **61**, 1132 (1967); *Trans. Faraday Soc.*, **53**, 305 (1957); E. Glueckauf, *Trans. Faraday Soc.*, **51**, 1235 (1955); **53**, 305 (1957).

86. K. Schwabe and E. Ferse, *Ber. Bunsenges. Physik. Chem.*, **69**, 383 (1965); Dipl. Arbeit J. J. Heinrich, T. U. Dresden, Institut f. Elektrochemie u. physikalische Chemie, 1964.

87a. H. Engelhardt, *Dechema Monograph*, **25**, 600 (1955).

87b. D. Wegemann and W. Simon, *Helv. Chim. Acta*, **47**, 1181 (1964).

87c. J. T. Clerc, Z. Štefanec and W. Simon, *Helv. Chim. Acta*, **48**, 54 (1965); G. D. T. *Fachz. Lab.*, **10**, 1033 (1966).

88. K. Schwabe, *Elektrochemische pH-Messungen unter extremen Bedingungen*, Verlag Chemie, Weinheim, 1960, p. 60.

89. See, e.g., L. N. Agar and W. G. Breck, *Trans. Faraday Soc.*, **53**, 167, 179, (1957); J. Jackson, *J. Polarogr. Soc.*, **12**, 111 (1966).

90. Manfred Böttger, Dissertation, T. U. Dresden, Institut für Elektrochemie u. physikalische Chemie, 1967.

91. M. Le Peintre, *Bull. Soc. Franc. Électriciens*, **8**, 584 (1960).

92. P. J. Antikainen, *Ann. Acad. Sci. Fennicae. Ser. A*, **II.56**, 7 (1954); *Suomen Kemistilehti*, **B 30**, 74 (1957); J. O. Edwards, *J. Amer. Chem. Soc.*, **75**, 6151 (1953).

93. G. Fiehn, *Zellstoff Papier*, **10**, 449 (1961); **11**, 84, 455 (1962); **13**, 136 (1962); *Chem. Techn.*, **16**, 369 (1964); Dissertation, T. U. Dresden, Institut f. Elektrochemie u. physikalische Chemie, 1963.

94. K. Schwabe, *Chemie Ing. Techn.*, **36**, 228 (1958).

95. O. V. Ingruber, *Pulp Paper*, **58**, 131 (1957).

96. K. A. E. Blackmore and A. E. Markham, *Tappi*, **41**, 138 A (1958).

97. W. W. Wirz, *Dechema Monograph*, **43**, 145 (1962).

98. A. J. Shukow, *J. Phys. Chem.* (*UdSSR*) (*Russian print*), **37**, 235 (1963).
99. R. Fournie, P. LeClerc and Saint-James, *Silicates Ind.*, **27**, 33 (1962).
100. K. Schwabe, *Z. Elektrochem.Angew. Physik. Chem.*, **41**, 681 (1935).
101. A. Distèche, *J. Electrochem. Soc.*, **109**, 1084 (1962).
102. A. Distèche and S. Distèche, *J. Electrochem. Soc.*, **114**, 330 (1967).
103. K. Buch and S. Grippenberg, *J. Cons. Permanent Int. Explorat. Mer.*, **7**, 233 (1932).
104. K. Park, *Science*, **154**, 1540 (1966); see also C. Culbertson, O. R. Kester and R. M. Pytkovitz, *Science*, **157**, 59 (1967).
105. A. Manual of Seawater Analysis in *Bull. Fisheries Res.*, *Board Canada*, No. 125, Ottawa 1965, pp. 31–33.
106. A. Distèche and B. Dubuisson, *Bull. Inst. Oceanogr. Monaco*, **57**, 1174 (1960); **64**, 1320 (1964).
107. J. R. Kaplan, *Rev. Sci. Instr.*, **39**, 1133 (1968).
108a. P. N. Krishnan, L. Young and R. E. Salomon, *J. Phys. Chem.*, **70**, 1595 (1966).
108b. F. Heinmetz, *Trans. Faraday Soc.*, **58**, 788 (1962).
108c. B. K. Jindal and W. A. Tiller, *Surface Science*, **9**, 137 (1968).
109. M. Paabo, R. A. Robinson and R. G. Bates, *J. Amer. Chem. Soc.*, **87**, 415 (1965); R. G. Bates, M. Paabo and R. A. Robinson, *J. Phys. Chem.*, **67**, 1833 (1963).
110a. R. G. Bates, Technical Note NBS, **271**, 25 (1965).
110b. A. K. Covington, M. Paabo, R. A. Robinson and R. G. Bates, *Anal. Chem.*, **40**, 700–706 (1968).
111. K. Schwabe and H. Geisler, *Electrochim. Acta*, **12**, 147 (1967); M. Alfenaar and C. L. de Ligny, *Electrochim. Acta*, **13**, 662 (1968); K. Schwabe, H. Geisler and H. Steinhauer, *Electrochim. Acta*, **13**, 663 (1968).
112. K. Schwabe and K. Wankmüller, *Ber. Bunsenges*, **69**, 528 (1965).
113a. Unpublished experiments in the Institute of Electrochemistry T. U. Dresden, 1965.
113b. J. Burn, T. Drury and J. P. Roberts, *Silicates Ind.*, **30**, 403 (1965).
114. I. M. Kolthoff and M. K. Chantooni Jr., *J. Amer. Chem. Soc.*, **87**, 4428 (1965).
115. I. M. Kolthoff and S. Bruckenstein, *J. Amer. Chem. Soc.*, **78**, 1, 10, 2974 (1956); **79**, 1, 5915 (1957).
116. W. Simon, *Helv. Chim. Acta*, **45**, 826 (1962).
117. J. D. K. Roe, *Anal. Chem.*, **38**, 461 R (1966).
118. G. Jander, H. Spandau and C. C. Addison, *Chemistry in Nonaqueous Ionizing Solvents*, Vol. 4, Interscience Publishers, New York, 1963.
118a. G. Johannson and K. Norberg, *J. Electroanal. Chem.*, **18**, 239 (1968).
119. E. Sørensen and P. Lundgaard, *J. Electroanal. Chem.*, **9**, 128 (1965).
120. See the numerous papers of K. H. Stern and coworkers, e.g. K. H. Stern and E. Meador, *J. Res. Nat. Bur. Stds*, A **69**, 553 (1965); *J. Electrochem. Soc.*, **111**, 208 (1965); **114**, 1257 (1967) and former work R. H. Doremus, *J. Phys. Chem.*, **72**, 2877 (1968).
121. See, e.g., H. J. Dulce, *Der Internist*, **5**, 51 (1964) or the excellent book *Acid Base Physiology in Medicine* from R. W. Winters, K. Engel and R. B. Dell, this is very suitable for self-instruction, because programmed by R. P. Berkson, (1967), to get from Radiometer A/S Emdrüpvej 72, DK-2400 Copenhagen, N$\bar{V}$, Denmark, or the London Company, 811 Sharon Drive, Westlake, Ohio.
122. M. C. Sanz, *Clin. Chem.* (*New York*), **3**, 406 (1957); P. Astrup, *J. Clin. Invest.*, **8**, 33 (1956).
123a. R. F. Freier, K. J. Clayson and E. S. Benson, *Clin. Chim. Acta* (*Amsterdam*), **9**, 348 (1964).

123b. H. Bellée, R. Franke and F. Stösslein, *Deut. Gesundheitswes*, **22**, 353 (1968).

124. A. Fletcher, W. Ingold and A. Baerfuss, *Chemie Ing. Techn.*, **36**, 1000 (1964).

125. W. Ingold, *Dechema Monograph*, **43**, 153 (1961).

126. D. W. Lübbers, *Nature*, **49**, 493 (1962).

127. K. Schwabe and H. D. Süschke, unpublished results.

128. R. N. Khuri, *Rev. Sci. Instr.*, **39**, 730 (1962); see also P. C. Caldwell, *J. Physiol. (London)*, **126**, 169 (1954); N. W. Carter, F. C. Rector, D. S. Campion and D. W. Seldin, *J. Clin. Invest.*, **46**, 920 (1967).

129. P. Scheid and P. Kunze, *Unio Int. Cancrum Acta (Louvain)*, **18**, 256 (1962).

130. R. V. Soltero and G. E. Lee, *Environmental Science Technol.*, **1**, 503 (1967).

131. See, e.g., J. Hengstenberg, B. Sturm and D. Winkler, *Messen u. Regeln in der chem. Technik*, Springer Verlag, Berlin, 1961, p. 639; R. Schröder, *Glas Instrumenten Techn.*, **6**, 377 (1962).

132. C. W. Cotmar and D. M. Smith, *J. Sci. Instr.*, **47**, 561 (1964).

133. T. V. Malmstadt and E. H. Piepmeier, *Anal. Chem.*, **37**, 34 (1965).

134. R. Süss, *Wasser, Luft, Betr.*, **6**, 285 (1962).

135. W. Hirsch, *Chem. Techn.*, **14**, 36 (1962).

136. K. Schwabe, unpublished results.

137. H. Pallmann, *Kolloid-Beih.*, **30**, 334 (1930). About the theory of 'dispersion effect' see, e.g., J. Th. G. Overbeck, *Progr. Physics Biophysic. Chem.*, **6**, 57 (1956) and M. Kahlweit, *Z. Physik. Chem. (Frankfurt/M.)*, **15**, 196 (1958).

138. H. Lehmann and W. Lorenz, TIZ-Zbl., **85**, 325, 349 (1961); **86**, 79 (1962).

139. F. Church, *Instrument. Pract.*, **1965**, 1091.

140. P. P. Krjukov, *Hydrochem. Mater.* (*Russian print*), **33**, 155–159 (1960).

141. F. Oehme, Kommunalwirtsch. **1964**, vol. 9.

142. K. W. Roennefahrt, *Fette, Seifen, Anstrichmittel*, **60**, 284 (1964).

143. Papper och Trä, **44**, 85 (1962).

144. Forschungsinstitut Meinsberg, unpublished.

145. ATM 332-24.

146. K. M. Joshi and P. U. Kamal, *J. Indian Chem. Soc.*, **43**, 620 (1966).

147. T. I. Quickenden, D. M. Belts, B. Cole and M. Noble, *J. Phys. Chem.*, **75**, 2830 (1971).

148. A. Carpen, A. Double, G. Ferroni, C. Rovike, P. Asensi and G. Nao Wyol, *Electrochim. Acta*, **14**, 587 (1969).

Index

Advances in Analytical Chemistry and Instrumentation

CUMULATIVE INDEX, VOLUMES 1–11

Author Index

Subject Index